CFD Applications in Energy Engineering Research and Simulation—Volume I

CFD Applications in Energy Engineering Research and Simulation—Volume I

Editor

Alfredo Iranzo

MDPI

Basel • Beijing • Wuhan • Barcelona • Belgrade • Novi Sad • Cluj • Manchester

Editor
Alfredo Iranzo
University of Sevilla
Sevilla
Spain

Editorial Office
MDPI AG
Grosspeteranlage 5
4052 Basel, Switzerland

This is a reprint of articles from the Special Issue published online in the open access journal *Processes* (ISSN 2227-9717) (available at: https://www.mdpi.com/journal/processes/special_issues/CFD_Energy_Engineering).

For citation purposes, cite each article independently as indicated on the article page online and as indicated below:

Lastname, A.A.; Lastname, B.B. Article Title. *Journal Name* **Year**, *Volume Number*, Page Range.

Volume I
ISBN 978-3-7258-2315-4 (Hbk)
ISBN 978-3-7258-2316-1 (PDF)
doi.org/10.3390/books978-3-7258-2316-1

Set
ISBN 978-3-7258-2221-8 (Hbk)
ISBN 978-3-7258-2222-5 (PDF)

Contents

About the Editor ix

Alfredo Iranzo
Editorial for Special Issue on Computational Fluid Dynamics (CFD) Applications in Energy Engineering Research and Simulation
Reprinted from: *Processes* **2024**, *12*, 1744, doi:10.3390/pr12081744 1

Shenming Ren, Yuan Zheng, Cong Yuan, Bin Liu, Emmanuel Fernandez-Rodriguez and Yuquan Zhang
The Influence of Hydraulic Characteristics on Structural Performance in a Pump-Turbine under No-Load Conditions
Reprinted from: *Processes* **2023**, *11*, 3422, doi:10.3390/pr11123422 4

Petrică Cană, Razvan George Ripeanu, Iulian Pătîrnac, Alin Diniță and Maria Tănase
Investigating the Impact of Operating Conditions on Relief Pressure Valve Flow through CFD and Statistical Analysis
Reprinted from: *Processes* **2023**, *11*, 3396, doi:10.3390/pr11123396 23

Hongchang Ding, Fei Ge, Kai Wang and Fanyun Lin
Influence of Blade Trailing-Edge Filing on the Transient Characteristics of the Centrifugal Pump during Startup
Reprinted from: *Processes* **2023**, *11*, 2420, doi:10.3390/pr11082420 53

Yiyang Wei, Binglin Li, Xiaomei Xu, Minxiang Wei and Chengdong Wang
Design of Electric Supercharger Compressor and Its Performance Optimization
Reprinted from: *Processes* **2023**, *11*, 2132, doi:10.3390/pr11072132 72

Zhiliang Wang, Yu Lei, Zhenhua Wu, Jian Wu, Manlai Zhang and Ruiquan Liao
Structure Size Optimization and Internal Flow Field Analysis of a New Jet Pump Based on the Taguchi Method and Numerical Simulation
Reprinted from: *Processes* **2023**, *11*, 341, doi:10.3390/pr11020341 90

Tao Bai, Qingzhen Yang and Jian Liu
Numerical Study of the Purge Flow's Effect on the Loss Mechanism of the Blocking and Shear Effects
Reprinted from: *Processes* **2023**, *11*, 50, doi:10.3390/pr11010050 114

Yuxuan Deng, Xiaodong Wang, Jing Xu, Yanna Li, Yanli Zhang and Chunyan Kuang
Gas–Liquid Interaction Characteristics in a Multiphase Pump under Different Working Conditions
Reprinted from: *Processes* **2022**, *10*, 1977, doi:10.3390/pr10101977 130

Hyun-Su Kang, Sung-Yeon Kim and Youn-Jea Kim
Wet Compression Study for an Aero-Thermodynamic Performance Analysis of a Centrifugal Compressor at Design and Off-Design Points
Reprinted from: *Processes* **2022**, *10*, 936, doi:10.3390/pr10050936 146

Yue Chen, Wei Zhang, Dehong Fang, Mingkang Sun, Jian Liu, Daoyuan Song and Xiaoping Zhang
Vortex Suppression and Flow Pattern Analysis of a Hydrofoil with Parallel Grooves
Reprinted from: *Processes* **2021**, *9*, 816, doi:10.3390/pr9050816 167

Thodsaphon Jansaengsuk, Mongkol Kaewbumrung, Wutthikrai Busayaporn and Jatuporn Thongsri
A Proper Shape of the Trailing Edge Modification to Solve a Housing Damage Problem in a Gas Turbine Power Plant
Reprinted from: *Processes* **2021**, *9*, 705, doi:10.3390/pr9040705 . . . . **184**

Seung Il Baek and Joon Ahn
Large Eddy Simulation of Film Cooling Involving Compound Angle Holes: Comparative Study of LES and RANS
Reprinted from: *Processes* **2021**, *9*, 198, doi:10.3390/pr9020198 . . . . **201**

Fei Zhao, Fanyu Kong, Xiaohui Duan, Huiyuan Wu and Jun Wang
Numerical Investigation on the Transient Flow of a Boiler Circulating Pump Based on the Shear Stress Transport Turbulence Model
Reprinted from: *Processes* **2020**, *8*, 1279, doi:10.3390/pr8101279 . . . . **221**

Zhehong Li, Xinxue Ye and Yikun Wei
Investigation on Vortex Characteristics of a Multi-Blade Centrifugal Fan near Volute Outlet Region
Reprinted from: *Processes* **2020**, *8*, 1240, doi:10.3390/pr8101240 . . . . **238**

Hongqiang Chai, Guolai Yang, Guoguo Wu, Guixiang Bai and Wenqi Li
Research on Flow Characteristics of Straight Line Conjugate Internal Meshing Gear Pump
Reprinted from: *Processes* **2020**, *8*, 269, doi:10.3390/pr8030269 . . . . **256**

Xu Meng, Zhigang Zuo, Michihiro Nishi and Shuhong Liu
A Numerical Study on the Flow Mechanism of Performance Improvement of a Wide-Angle Diffuser by Inserting a Short Splitter Vane
Reprinted from: *Processes* **2020**, *8*, 143, doi:10.3390/pr8020143 . . . . **284**

Weidong Cao and Jiayu Mao
Study of the Affinity Law of Energy and Cavitation Characteristics in Emergency Drainage Pumps at Different Rotating Speeds
Reprinted from: *Processes* **2019**, *7*, 932, doi:10.3390/pr7120932 . . . . **298**

Rong Guo, Rennian Li, Renhui Zhang and Wei Han
Numerical Study of the Unsteady Flow Characteristics of a Jet Centrifugal Pump under Multiple Conditions
Reprinted from: *Processes* **2019**, *7*, 786, doi:10.3390/pr7110786 . . . . **315**

Jin-Hyuk Kim, Sang-Bum Ma, Sung Kim, Young-Seok Choi and Kwang-Yong Kim
Design and Verification of a Single-Channel Pump Model based on a Hybrid Optimization Technique
Reprinted from: *Processes* **2019**, *7*, 747, doi:10.3390/pr7100747 . . . . **334**

Zikang Li, Hongchang Ding, Xiao Shen and Yongming Jiang
Performance Optimization of High Specific Speed Centrifugal Pump Based on Orthogonal Experiment Design Method
Reprinted from: *Processes* **2019**, *7*, 728, doi:10.3390/pr7100728 . . . . **355**

Yuxin Lun, Xinxue Ye, Limin Lin, Cunlie Ying and Yikun Wei
Unsteady Characteristics of Forward Multi-Wing Centrifugal Fan at Low Flow Rate
Reprinted from: *Processes* **2019**, *7*, 691, doi:10.3390/pr7100691 . . . . **374**

Yikun Wei, Cunlie Ying, Jun Xu, Wenbin Cao, Zhengdao Wang and Zuchao Zhu
Effects of Single-arc Blade Profile Length on the Performance of a Forward Multiblade Fan
Reprinted from: *Processes* **2019**, *7*, 629, doi:10.3390/pr7090629 . . . . **408**

Bowen Li, Xiaojun Li, Xiaoqi Jia, Feng Chen and Hua Fang
The Role of Blade Sinusoidal Tubercle Trailing Edge in a Centrifugal Pump with Low Specific Speed
Reprinted from: *Processes* **2019**, *7*, 625, doi:10.3390/pr7090625 . . . . **432**

Si Huang, Yifeng Wei, Chenguang Guo and Wenming Kang
Numerical Simulation and Performance Prediction of Centrifugal Pump's Full Flow Field Based on OpenFOAM
Reprinted from: *Processes* **2019**, *7*, 605, doi:10.3390/pr7090605 . . . . **448**

Mohd Fadzli Hamid, Mohamad Yusof Idroas, Mazlan Mohamed, Shukriwani Sa'ad, Teoh Yew Heng, Sharzali Che Mat, et al.
Numerical Investigation of the Effect of Incorporated Guide Vane Length with SCC Piston for High-Viscosity Fuel Applications
Reprinted from: *Processes* **2020**, *8*, 1328, doi:10.3390/pr8111328 . . . . **459**

Nam-Hun Kim and Kwang-Yong Kim
Effects of Bending of Fluidic Oscillators on Aerodynamic Performance of an Airfoil with a Flap
Reprinted from: *Processes* **2021**, *9*, 1429, doi:10.3390/pr9081429 . . . . **477**

About the Editor

Alfredo Iranzo

Alfredo Iranzo is a Chemical Engineer from the University of Zaragoza (2000), MSc in Thermal Energy Systems from the University of Seville (2010), and PhD from the University of Seville (2011). He worked at ANSYS Germany as Customer Support for CFD software. His research activity focuses on Computational Fluid Dynamics applications in energy and chemical engineering, hydrogen technologies, green hydrogen production, and, in particular, PEM fuel cell technology, including CFD (computational fluid dynamics) models and simulations and experimental research.

He has published 70 articles in JCR journals with high international impact (h index = 25). He has participated in several public and private funded projects in CFD and hydrogen technologies. He is currently CSO (Chief Scientific Officer) of the Innovation Hub of the University of Seville, within the European University Ulysseus, in Sustainable Energy, Transport, and Mobility for Smart Cities. He is member of the Editorial Board of the journal *Engineering Applications of Computational Fluid Mechanics* and Assistant Editor of the *International Journal of Hydrogen Energy*.

processes

MDPI

Editorial

Editorial for Special Issue on Computational Fluid Dynamics (CFD) Applications in Energy Engineering Research and Simulation

Alfredo Iranzo

Thermal Engineering Group, Energy Engineering Department, School of Engineering, University of Sevilla, Camino de los Descubrimientos s/n, 41092 Sevilla, Spain; airanzo@us.es; Tel.: +34-954487471

1. Introduction

Computational fluid dynamics (CFD) has been firmly established as a fundamental discipline to advancing research on energy engineering. Since the early work by D. Brian Spalding [1–4], B.E. Launder [1,4–6], S. V. Patankar [2,4,7], and many other contributors to the establishment of computational fluid dynamics as a discipline to solve real engineering problems, there has been significant progress during recent decades. Numerical methods and physical models have been developed, and computer power, graphics, and 3D CAD design have also developed impressively. The major progress achieved during the last two decades both on software modeling capabilities and hardware computing power have resulted in the considerable and wide spread of CFD interest among scientist and engineers. Numerical modeling and simulation developments are increasingly contributing to the current state of the art in many energy engineering aspects, such as power generation, combustion, wind energy, concentrated solar power, hydro power, gas and steam turbines, fuel cells, and many others.

Today, CFD is a very well-established discipline and tool in both industry and academia. Commercial vendors have developed software with highly relevant capabilities, able to cover a wide range of applications. Graphical user interfaces are much improved, enabling the expansion of the business, and allowing more users to become involved in CFD. Nevertheless, CFD remains a very specific and highly demanding tool in terms of knowledge requirements for ensuring the quality and reliability of the simulation results (it is not a plug and play tool). A very high level of knowledge on discretization schemes, numerical methods, fluid mechanics, heat transfer, and chemical reactions is needed for the successful and valuable use of any CFD software.

With over 1000 journal publications on CFD yearly, it is obvious that industry and academia are highly involved in research activities both for the development and application of CFD. This Special Issue on "CFD Applications in Energy Engineering Research and Simulation" is aimed at covering some of the most recent advances in the applications of CFD in the field of energy engineering.

Citation: Iranzo, A. Editorial for Special Issue on Computational Fluid Dynamics (CFD) Applications in Energy Engineering Research and Simulation. *Processes* **2024**, *12*, 1744. https://doi.org/10.3390/pr12081744

Received: 3 August 2024
Accepted: 14 August 2024
Published: 20 August 2024

2. Review of Contributions

This Special Issue contains a wide range of topics within the contributions (Figure 1), with a particular focus on rotating machinery (pumps, turbines, and compressors). Oil and gas applications are also well covered, together with novel technologies such as electrolysers and redox flow batteries. The overview of the main topics covered is depicted in Figure 1, but nevertheless, additional topics are present in this Special Issue as well, such as hydrogen pre-cooling for fast filling, pressure valve modeling, CFD numerics and codes, CFD-DEM coupling, fiber spinning processing, shell and tube heat exchangers, indoor air quality, steam soil disinfection, desalination, waterjet propulsion, and solar radiation heat transfer for spacecraft applications. A perspective on "CFD Applications in Energy Engineering Research and Simulation" is also included in this Special Issue.

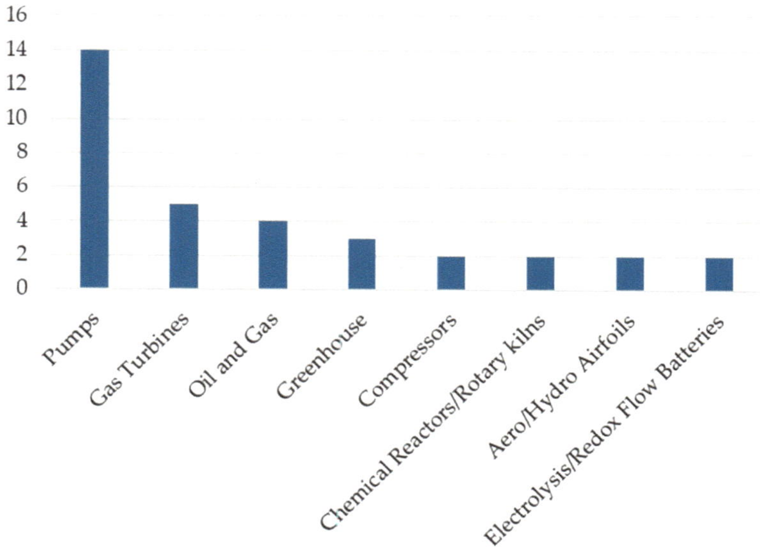

Figure 1. Overview of the main topics covered by this Special Issue.

In terms of geographical contributions, the high relevance of China (Figure 2) is noteworthy, along with a significant number of contributions from the USA, Korea, Mexico, and Spain.

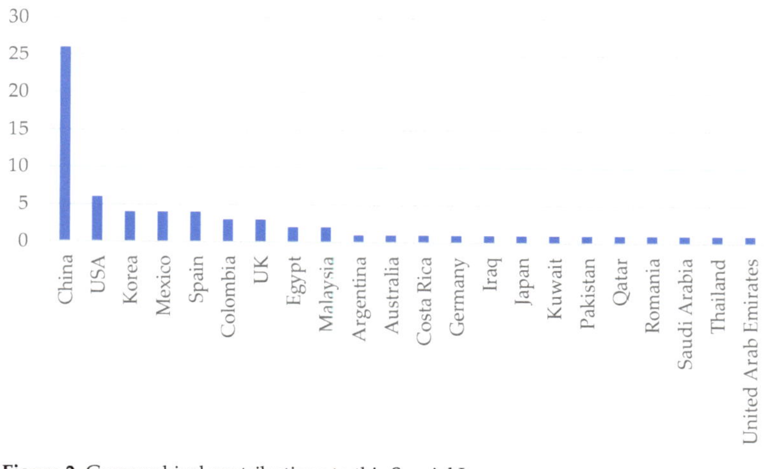

Figure 2. Geographical contributions to this Special Issue.

Overall, this Special Issue contains contributions from 22 countries, involving a wide representation of CFD activities in different countries and sectors.

It is worth mentioning that some of the contributions have high potential for subsequent research activities, which represents a success, as the contents of this Special Issue have fostered further research and progress in the field. In this sense, it is worth mentioning the contributions to the field of pumps by Zikang Li et al. [8], Bowen Li et al. [9], and Hongqiang Chai et al. [10], centrifugal fans (Zhehong Li et al. [11]), shell and tube heat exchangers (Xuejun Qian et al. [12]), alkaline water electrolysis (Jesús Rodríguez and Ernesto Amores [13]), redox flow batteries (García-Salaberri et al. [14]), indoor environmental quality (Payam Nejat et al. [15]), energy performance in greenhouses (Aguilar-Rodriguez et al. [16]), and waterjet propulsion systems (Chuan Wang et al. [17]).

It is expected that these research articles, as well as all the others published in this Special Issue, will significantly contribute to the further research of CFD in different energy engineering applications.

Conflicts of Interest: The author declares no conflicts of interest.

References

1. Launder, B.E.; Spalding, D.B. The numerical computation of turbulent flows. *Comput. Methods Appl. Mech. Eng.* **1974**, *3*, 269–289. [CrossRef]
2. Patankar, S.V.; Spalding, D.B. A calculation procedure for heat, mass and momentum transfer in three-dimensional parabolic flows. *Int. J. Heat Mass Transf.* **1972**, *15*, 1787–1806. [CrossRef]
3. Spalding, D.B. A novel finite difference formulation for differential expressions involving both first and second derivatives. *Int. J. Numer. Methods Eng.* **1972**, *4*, 551–559. [CrossRef]
4. Artemov, V.; Beale, S.B.; de Vahl Davis, G.; Escudier, M.P.; Fueyo, N.; Launder, B.E.; Leonardi, E.; Malin, M.R.; Minkowycz, W.J.; Patankar, S.V.; et al. A tribute to D.B. Spalding and his contributions in science and engineering. *Int. J. Heat Mass Transf.* **2009**, *52*, 3884–3905. [CrossRef]
5. Launder, B.E.; Reece, G.J.; Rodi, W. Progress in the development of a Reynolds-stress turbulence closure. *J. Fluid Mech.* **1975**, *68*, 537–566. [CrossRef]
6. Launder, B.E.; Sharma, B.I. Application of the energy-dissipation model of turbulence to the calculation of flow near a spinning disc. *Lett. Heat Mass Transf.* **1974**, *1*, 131–137. [CrossRef]
7. Patankar, S.V. *Numerical Heat Transfer and Fluid Flow*, 1st ed.; Hemisphere Publishing Corporation, McGraw-Hill Book Company: Washington, DC, USA; New York, NY, USA; London, UK, 1980; pp. 1–197.
8. Li, Z.; Ding, H.; Shen, X.; Jiang, Y. Performance Optimization of High Specific Speed Centrifugal Pump Based on Orthogonal Experiment Design Method. *Processes* **2019**, *7*, 728. [CrossRef]
9. Li, B.; Li, X.; Jia, X.; Chen, F.; Fang, H. The Role of Blade Sinusoidal Tubercle Trailing Edge in a Centrifugal Pump with Low Specific Speed. *Processes* **2019**, *7*, 625. [CrossRef]
10. Chai, H.; Yang, G.; Wu, G.; Bai, G.; Li, W. Research on Flow Characteristics of Straight Line Conjugate Internal Meshing Gear Pump. *Processes* **2020**, *8*, 269. [CrossRef]
11. Li, Z.; Ye, X.; Wei, Y. Investigation on Vortex Characteristics of a Multi-Blade Centrifugal Fan near Volute Outlet Region. *Processes* **2020**, *8*, 1240. [CrossRef]
12. Qian, X.; Lee, S.W.; Yang, Y. Heat Transfer Coefficient Estimation and Performance Evaluation of Shell and Tube Heat Exchanger Using Flue Gas. *Processes* **2021**, *9*, 939. [CrossRef]
13. Rodríguez, J.; Amores, E. CFD Modeling and Experimental Validation of an Alkaline Water Electrolysis Cell for Hydrogen Production. *Processes* **2020**, *8*, 1634. [CrossRef]
14. García-Salaberri, P.A.; Gokoglan, T.C.; Ibáñez, S.E.; Agar, E.; Vera, M. Modeling the Effect of Channel Tapering on the Pressure Drop and Flow Distribution Characteristics of Interdigitated Flow Fields in Redox Flow Batteries. *Processes* **2020**, *8*, 775. [CrossRef]
15. Nejat, P.; Hussen, H.M.; Fadli, F.; Chaudhry, H.N.; Calautit, J.; Jomehzadeh, F. Indoor Environmental Quality (IEQ) Analysis of a Two-Sided Windcatcher Integrated with Anti-Short-Circuit Device for Low Wind Conditions. *Processes* **2020**, *8*, 840. [CrossRef]
16. Aguilar-Rodriguez, C.E.; Flores-Velazquez, J.; Ojeda-Bustamante, W.; Rojano, F.; Iñiguez-Covarrubias, M. Valuation of the Energy Performance of a Greenhouse with an Electric Heater Using Numerical Simulations. *Processes* **2020**, *8*, 600. [CrossRef]
17. Wang, C.; He, X.; Cheng, L.; Luo, C.; Xu, J.; Chen, K.; Jiao, W. Numerical Simulation on Hydraulic Characteristics of Nozzle in Waterjet Propulsion System. *Processes* **2019**, *7*, 915. [CrossRef]

processes MDPI

Article

The Influence of Hydraulic Characteristics on Structural Performance in a Pump-Turbine under No-Load Conditions

Shenming Ren [1], Yuan Zheng [1], Cong Yuan [2], Bin Liu [3], Emmanuel Fernandez-Rodriguez [4] and Yuquan Zhang [1,*]

1 College of Energy and Electrical Engineering, Hohai University, Nanjing 211100, China; rensm1999@hhu.edu.cn (S.R.); zhengyuan@hhu.edu.cn (Y.Z.)
2 Huaishu New River Management Office of Jiangsu Province, Huaian 223001, China
3 Luoyun Water Conservancy Project Management Divison in Jiangsu Province, Suqian 223800, China
4 Technological Institute of Merida, Technological Avenue, Merida 97118, Mexico
* Correspondence: zhangyq@hhu.edu.cn

Abstract: The operating state of a pump-turbine unit under no-load conditions is directly related to its safe and stable operation. In order to probe into the influence of hydraulic characteristics on structural performance, a pump-turbine assembled in China is selected for research by using CFD (computational fluid dynamics) and unidirectional FSI (fluid–structure interaction) methods. The vortex distribution and the law of pressure pulsation propagation are analyzed to capture the peculiar flow phenomena. The results show that the vortex distribution in the runner channel appears initially at the suction side of the blades but then propagates toward the pressure side with GVO. This produces rotating stall frequencies ($0.7f_n$) and a combination of the RSI, asymmetry of the water ring in vaneless space, and high-amplitude pressure pulsations in the downstream channel close to the runner inlet and elbow section of the draft tube. This, in turn, is associated with the structural stress of the runner and guide vane. The stress level of the guide vane becomes alleviated under no-load conditions with large GVO, but the stress distribution of the runner is no longer symmetrical, which aligns with the vortex evolution in the runner passage. The stress concentration that develops further along the blade root increases the structural failure, which is also captured and verified as a crack in the prototype runner. The phenomena suggest that the RPT should avoid operating under no-load conditions with large GVO as far as possible. Therefore, in the design or optimization of the pump-turbine unit, the structures of the guide vanes and runner could be treated as a whole to investigate the resulting internal flow and structure characteristics.

Keywords: pump-turbine; no-load condition; rotating stall; pressure pulsation; structural stress

Citation: Ren, S.; Zheng, Y.; Yuan, C.; Liu, B.; Fernandez-Rodriguez, E.; Zhang, Y. The Influence of Hydraulic Characteristics on Structural Performance in a Pump-Turbine under No-Load Conditions. *Processes* **2023**, *11*, 3422. https://doi.org/10.3390/pr11123422

Academic Editor: Krzysztof Rogowski

Received: 8 November 2023
Revised: 6 December 2023
Accepted: 11 December 2023
Published: 13 December 2023

1. Introduction

With the increasing proportion of renewable energy in electric power systems, the set goals mainly include the improvement of energy conversion [1–3] and the development of pumped storage units to ensure the power generation quality and safety of the reversible operating condition [4]. Nevertheless, this demands greater operating intensity in the reversible pump-turbines (RPTs), leading to failure if not properly contemplated. Therefore, the characteristic of the pumped storage technology determines the complexity of the operating conditions in the unit, such as off-design conditions [5,6] and extreme working conditions. Particularly, during the no-load condition, after the full-load rejection in generating mode, the RPT can suffer flow instability under the influence of "S" characteristics [7–9]; this is manifested as a high rotation speed and large flow fluctuation. If the slope of the point discharge relative to the rotation speed change under no-load conditions [10] is positive (i.e., the full characteristic curve is represented as the "S" region), the same rotation speed could correspond to multiple different flows. In other words, the internal flow rate of the unit changes, resulting in rotor–stator interaction (RSI) [11] and a severe pressure pulsation and rotating stall [12].

With regard to the above problems, Tanaka [13] proposed a radial vibration of a high head pump-turbine through theoretical and experimental tests. The results indicated that the inlet edges and the areas near the inlet, which were about 0.5 pitch chord length of the runner, concentrated the stress. Hu et al. [14] probed into the transition process of amplitude-frequency characteristics of a pump-turbine with splitter blades (4 + 4) under rated head and high discharge conditions, which are unlike equal-blade pump-turbines. As the flow rate decreased, the dominant frequency of pressure pulsation in the vaneless zone changed from $4f_n$ to $8f_n$, corresponding to half and whole numbers of runner blades, respectively. Han et al. [15] investigated the influence of the distance between the rotor and stator on the RSI strength of mixed-flow centrifugal pumps. They demonstrated that the increasing rotor–stator distance could reduce the pressure change amplitude without varying the hydraulic performance of centrifugal pumps significantly. Some scholars [12,16,17] utilized numerical calculation and model experiments in research for rotating stalls in RPTs and concluded that the characteristic frequency of a rotating stall is approximately 50~70% of the runner rotational frequency. Moreover, the high-speed water retaining ring [18] in the vaneless zone is gradually blurred with the broadening of the GVO, and the blockage of the flow channel increases. This is manifested as the symmetry of the vortex structure distribution inside the runner worsens [5,19] and the pressure pulsation intensity augments.

For the enhancement of operating instability and improvement of the "S" characteristic of pump-turbine, Fu et al. [20] offered some corresponding measures using 1D + 3D coupled simulation, such as decreasing the runaway rotational speed, increasing the minimum head of pump-turbine, and increasing the height of guide vanes, etc. Some runners with equal-length blades were designed and modified with splitter blades by Hu [14,21], and the runner with splitter blades could markedly improve the S-shaped characteristic but decrease the hydraulic efficiency under design conditions. As the "S" characteristic is closely related to the geometric parameters of the leading edge of the runner blade, Li et al. [22,23] optimized the blade's high-pressure side to reduce the pressure pulsation and suggested that the runner with the large negative blade lean behaves better in terms of operational stability.

As stated above, most of the currently published papers concentrate on the analysis of inner flow characteristics and pressure pulsation under operating conditions in the "S" region and propose improvement measures [24,25] correspondingly. The research object of structural performance is usually aimed at a single component, thus limiting knowledge on the effect of internal flow on the structural response of combined components (e.g., runner and guide vane) under no-load conditions.

Therefore, we model a prototype RPT assembled in China to investigate its internal flow characteristics, pressure pulsation under no-load conditions, and influence of hydraulic characteristics on the structural performance of the guide vane and runner. This paper is divided into five sections. Section 2 deals with the numerical calculations of working conditions using different GVO. Then, the internal flow characteristics and the law of pressure pulsation propagation of pump-turbine are discussed in Sections 3 and 4, respectively. The influence on the structure of guide vanes and runner is analyzed in Section 5. This work could provide a reference for the design and optimization of pump-turbine structures and for the improvement of operational stability as well.

2. Numerical Model and Settings

2.1. Illustration of Numerical Model

The Chinese prototype pump-turbine is constituted by a spiral casing, tandem cascade (stay vanes and guide vanes), runner, draft tube, and the extended pipes set at the inlet of the spiral casing and the outlet of the draft tube, respectively. This is shown in Figure 1, with the main parameters being shown in Table 1. After weighing the computational accuracy against computing resource consumption, all of the above hydraulic components were divided with a structured mesh [15,20,26], and the grids at the near-interface layer, profiles of vanes, and blades were refined by O-blocks and Y-blocks and converted into an unstructured mesh for the following calculation. The total number of meshes was finally determined to be 7.3×10^6 by virtue of a mesh independence test (shown in Section 2.3).

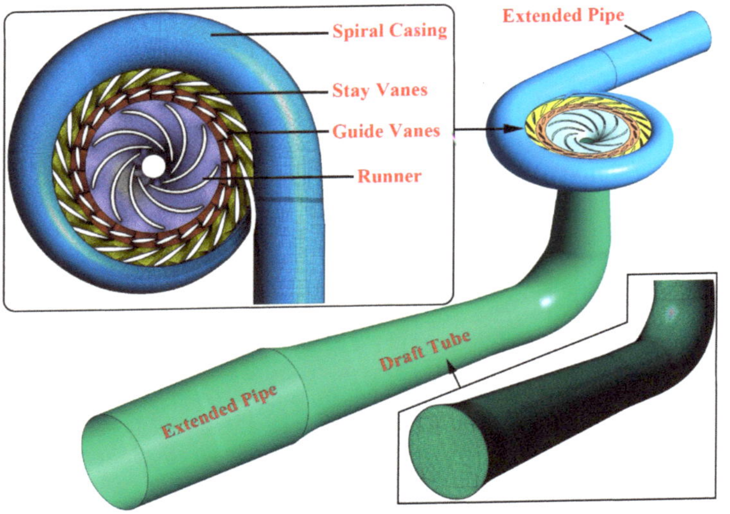

Figure 1. The 3D model and mesh of a pump-turbine.

Table 1. Basic parameters of a pump-turbine.

P_r (MW)	n_r (rpm)	H_r (m)	D (m)	Z_r	Z_g	Z_s
306	428.6	430	2.238	9	20	20

2.2. Turbulence Model and Boundary Conditions

In light of the complexity of internal flow under no-load conditions, the selection of a turbulence model is considered for its influence on the calculation accuracy [8,9] so as to capture the generation and evolution of the vortex structure in the flow passage. Although the dimensionless distance y^+ between the first layer nodes of mesh and the wall is required to be close enough, the SST $k-\omega$ model has been widely put to use for numerous numerical simulations of hydraulic machinery, like centrifugal pump [27], pump as turbine (PAT) [28], floating tidal stream turbine [29], etc. Since $k-\omega$ and $k-\varepsilon$ models are deployed in the SST model near and away from the wall, respectively, the transport of shear stress in the inverse pressure gradient is adequately taken into consideration without over-predicting the turbulent viscosity. Thus, the momentum equation is closed by this model for the solution. The k (Equation (1)) and ω equations (Equation (2)) can be depicted as follows:

$$\frac{\partial \rho k}{\partial t}+\frac{\partial \rho k u_j}{\partial x_j}=\frac{\partial}{\partial x_j}\left[(\mu+\sigma_k\mu_t)\frac{\partial k}{\partial x_j}\right]+P_k-D_k \tag{1}$$

$$\frac{\partial \rho \omega}{\partial t}+\frac{\partial \rho \omega u_j}{\partial x_j}=\frac{\partial}{\partial x_j}\left[(\mu+\sigma_\omega\mu_t)\frac{\partial \omega}{\partial x_j}\right]+P_\omega-D_\omega+S_\omega \tag{2}$$

where ρ denotes the fluid density, k denotes the turbulent kinetic energy, (T.K.E), ω is the ratio dissipation rate of T.K.E, u_j is the average turbulent velocity, μ is the coefficient of viscosity, P_k and D_k are the generating term and dissipative term of turbulence, respectively,

S_ω is the cross-diffusion term, and σ is a coefficient where the subscript k and ω represent the k equation and ω equation, respectively.

Based on the actual operation of the pumped storage unit, the boundary condition's setup is as follows: the mass flow rate is specified at the upstream inlet of the spiral casing, and the downstream outlet of the draft tube is set with the static pressure. The flow surface of each component is assumed as a no-slip wall formulation, and the dimensionless distance y^+ is controlled in the range of 15 to 200. Moreover, data transmission between stationary parts is accessed by general grid interfaces (GGIs). Considering that the circumferential speed distribution in the vicinity of the runner domain will be uneven due to the action of blades, the frozen rotor model takes into account the relative positions of both sides of the rotor and stator interface, allowing the flow parameters to vary greatly in the circumferential direction. This is adopted for interfaces of a runner with guide vanes upstream and with draft tubes downstream, but it is replaced with a transient rotor–stator in unsteady computation settings. The total simulation time corresponds to the period of 10 runner rotations, and the runner revolves about 2° per time step. In order to obtain the law of pressure pulsation change during the simulation process, several pressure monitoring points are located in the entire flow channels (Figure 2). A total of six monitoring points from SC1 to SC6 are arranged in the spiral casing so as to capture the upstream transmission capacity of pressure pulsation from the downstream hydraulic components, like guide vanes, vaneless space, and runner. The GR monitoring points are located uniformly at the vaneless region between the guide vanes and the runner, whose number is consistent with the number of guide vanes. In contrast to the arrangement of a single point or several symmetrical points, the points GR1~GR20 can simultaneously observe the pressure pulsation changes in both adjacent and symmetrical positions. Based on the above considerations, the RN points in the runner and the DT points in the draft tube are also assigned in this manner. Due to the difference in the flow pattern of the runner inlet and outlet, there are five monitoring points set in order from the high- to low-pressure sides in each single-blade channel, which are named RNx-1 to RNx-5, respectively, where x corresponds to the blade number. In the draft tube, six cross-sections are chosen from the conical to the diverging section. Each cross-section contains a center monitoring DTM point for capturing the vortex rope frequency and 12 DT points for the rotational flow near the wall of the draft tube.

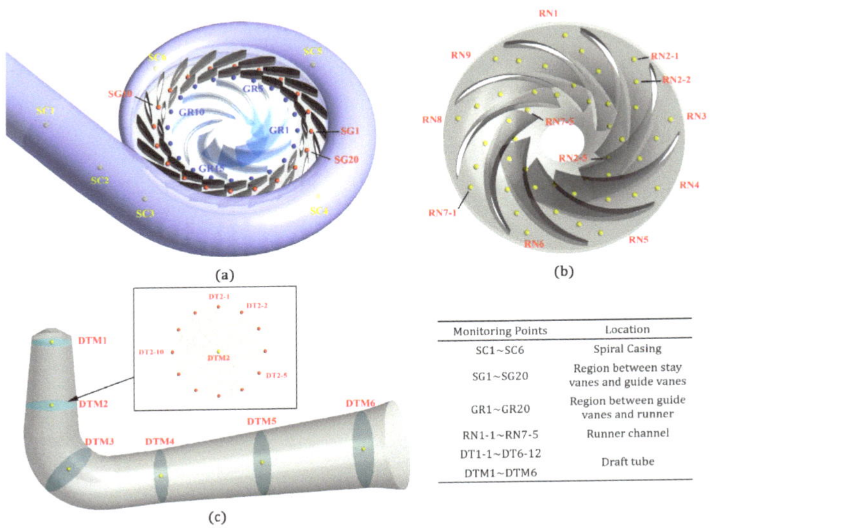

Monitoring Points	Location
SC1~SC6	Spiral Casing
SG1~SG20	Region between stay vanes and guide vanes
GR1~GR20	Region between guide vanes and runner
RN1-1~RN7-5	Runner channel
DT1-1~DT6-12	Draft tube
DTM1~DTM6	

Figure 2. Location of pressure monitoring points in spiral casing and double-row cascade (**a**), runner (**b**), and draft tube (**c**).

2.3. Verification of Numerical Simulation

In this section, mesh sensitivity tests and a comparison of numerical and experimental data are discussed in detail so as to ensure the reliability of computation results. Table 2 presents the selected operating points in this paper, and the BEP stands for the mesh sensitivity test, where four no-load condition points (OP1~OP4) are opted to verify and analyze the results. To facilitate the performance trend, two parameters are defined as follows:

$$n_{11} = nD/\sqrt{H} \tag{3}$$

$$Q_{11} = Q/(D^2H^{1/2}) \tag{4}$$

where n is the rotation speed of a pump-turbine, rpm; D is the runner outlet diameter in turbine mode, m; H is the operating head of a pump-turbine, m; Q is the volume discharge, l/s.

Table 2. The parameters of selected operating points.

Operating Points	GVO/°	n_{11}/(r/min)	Q_{11}/(l/s)
OP1	4	40.92	51.18
OP2	10	47.13	100.1
OP3	14	50.17	125.43
OP4	22	55.66	222.63
BEP	28	44.15	727.47

Figure 3 shows the efficiency of the pump-turbine at BEP using five sets of different mesh sizes; the element number of each component can be found in Table 3. It is known that the discretization of the physical model is refined better with the increase of grid density, providing more information on flow characteristics but also demanding large computing resources and time. In this study, the calculated efficiency at BEP approaches the experimental value (91.43%), and the difference is minimized with grid numbers of 9.1×10^6 (Figure 3C). However, to simulate all the transient processes in the in-house computer of eight independent processing cores would require 18 days. Therefore, the eventual scheme with 7.3×10^6 elements (Figure 3A) is used after weighing the pros and cons; the deviation is within 0.25%, which is small enough to meet the accuracy requirements.

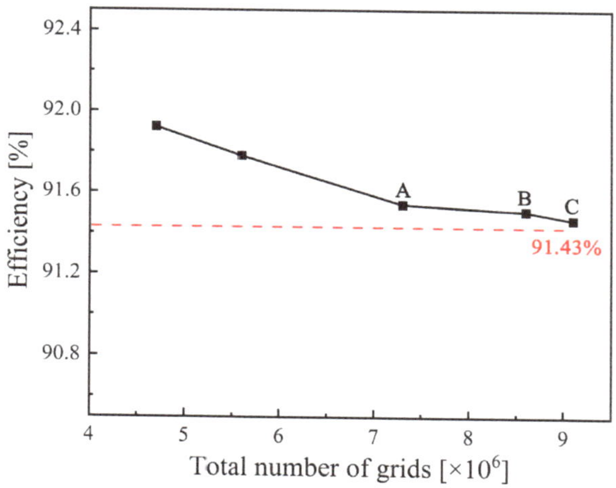

Figure 3. Mesh sensitivity test based on efficiency: A 7.3×10^6, B 8.6×10^6, C 9.1×10^6.

Table 3. The element number of each component under five schemes.

Scheme	1	2	3	4	5
Spiral casing	0.7	0.8	1.0	1.2	1.3
Stay vanes	0.9	1.0	1.2	1.4	1.5
Guide vanes	1.0	1.2	1.5	1.6	1.7
Runner	1.3	1.6	2.0	2.3	2.5
Draft tube	0.6	0.7	1.0	1.2	1.2
Extended pipes	0.2	0.3	0.6	0.9	0.9
Total ($\times 10^6$)	4.7	5.6	7.3	8.6	9.1

In Figure 4, a slight deviation between the numerical and experimental data can clearly be found. Under the no-load condition with small GVO, the numerical simulation results are in good agreement with the corresponding experimental data, whose relative error reaches 1.68%. However, when the GVO is larger, the channel and stall vortices in the runner passage vary dramatically, which gives rise to the obvious fluctuation in the external characteristics. Despite the calculation error increasing to 3.73%, it is still enough to guarantee the reliability of numerical simulation. In summary, the relative error of all operating points in this study is kept within 4%, and the general change trend of numerical simulation is consistent with that of the experimental test.

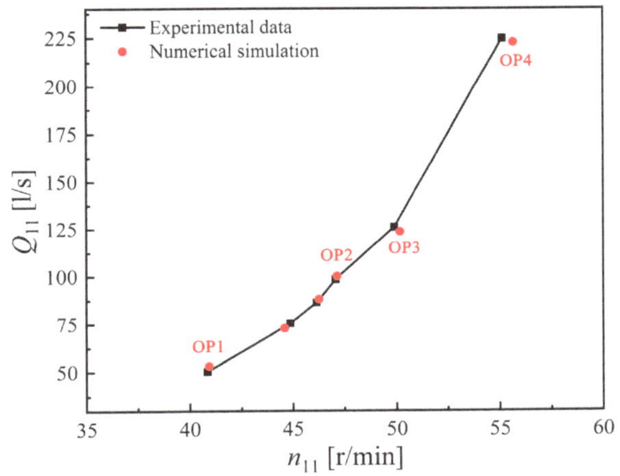

Figure 4. Comparison of experimental data and numerical simulation.

3. Analysis of Internal Flow Characteristics

The flow state of the vaneless zone between guide vanes and runner is bound up with the main sources of pressure fluctuation, such as rotor–stator interaction, rotating stall, and so on. Figure 5 demonstrates the flow pattern change of guide vanes and runner under four no-load conditions in a streamlined manner. Under the no-load conditions with lower GVO, the circumferential velocity of the water flow is larger because of the restraint and guidance of guide vanes. Hence, the water ring with a certain thickness is mainly gathered in the vaneless zone, resulting in the blockage of the flow channel. Owing to the large incidence angle of water flow at the runner inlet, the flow gradually separates and vortices are generated on the blade suction side, although the flow state of the overall channel still maintains the circular symmetry well. With the increase of the GVO, the thickness of the water ring decreases; meanwhile, vortices augment in runner channel transit from the suction side (Figure 5a,b) to the pressure side (Figure 5c,e) of the blades. Under the impression of ever-changing vortices, the contrast of the flow patterns in the single channel of each blade becomes more apparent, which destroys the symmetrical distribution of the vortex structure inside the runner. When the GVO is 22°, increasing the flow velocity (Figure 5g,h) leads to more intense vortex evolution. Hence, compared with low GVOs, the unit stability under no-load conditions with large GVOs is more serious.

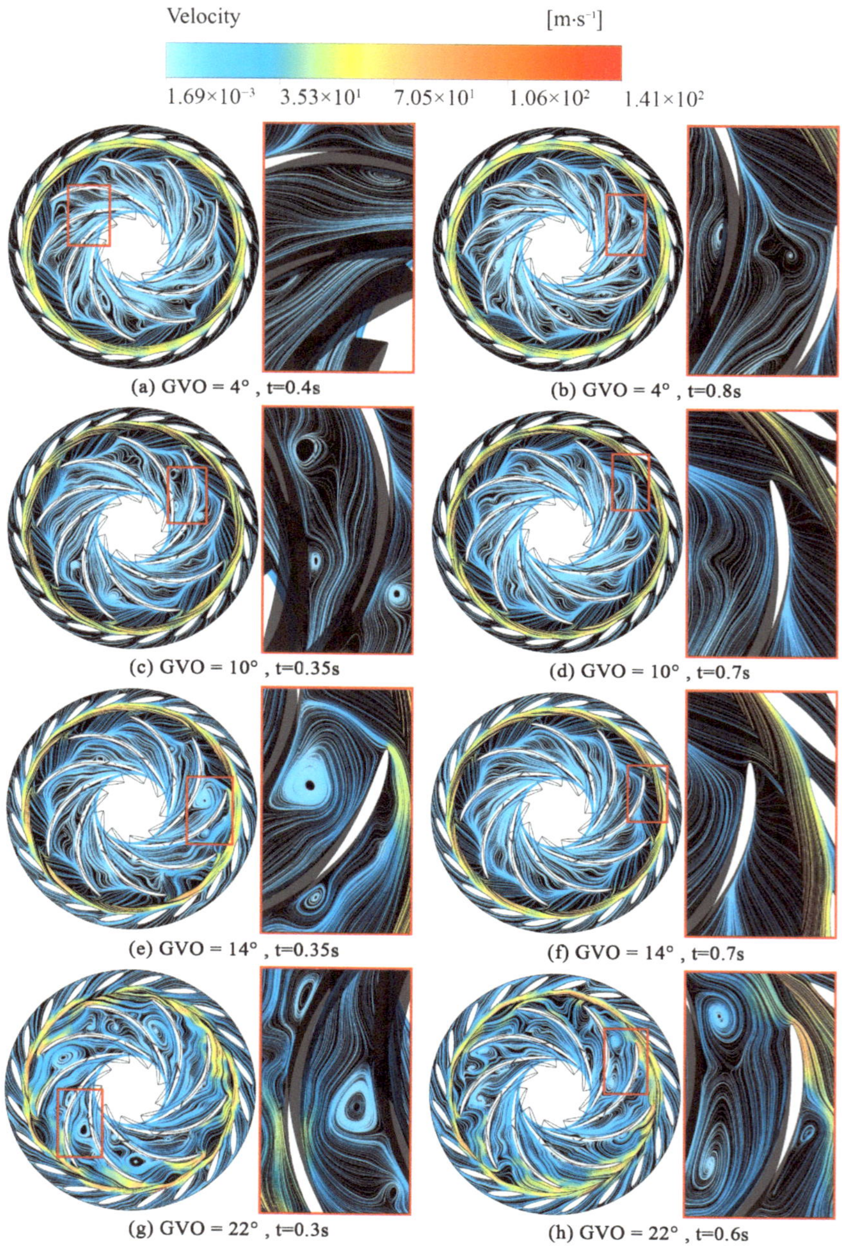

Figure 5. The distribution of streamlines in guide vanes and runner.

For observing the distribution of the vortex structure in guide vanes and runner, the third-generation vortex identification method, Ω method [30,31], is used due to the ability to capture both strong and weak vortices and the applicability in reversible pump-turbines. Figure 6 shows this. The flow at the inlet of the runner is separated by the blades, and a series of channel vortices twist, shear, break, and spread in the downstream narrow runner passage. In contrast to the symmetrical vortex structure inside the runner under small GVO (Figure 6a), the number of vortices increases significantly after the GVO is larger, and the vortices tend to spread to the vaneless zone and guide vanes passage. In addition, the evolution of the vortex structure is no longer synchronous in the flow channels between different blades. As a result, vortices gather at the runner inlet of some flow channels, which is not conducive to the balanced operation of the runner.

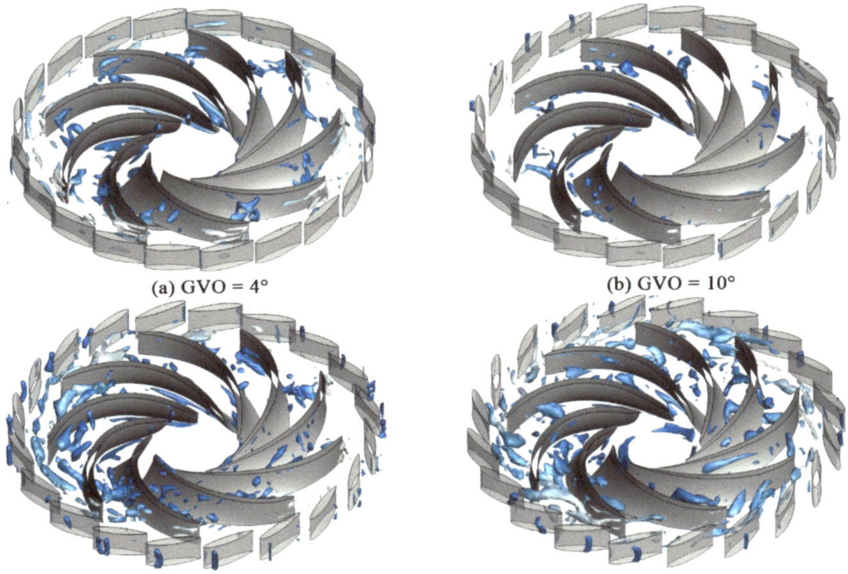

Figure 6. The distribution of a vortex structure in guide vanes and runner.

In order to analyze the pressure distribution on the runner blade surface, the cross line between the blade surface and the middle turbo surface (Span = 0.5) is taken as the blade load line, as shown in Figure 7. Then, the pressure distribution on the pressure and suction side of the blade is obtained along the streamwise direction.

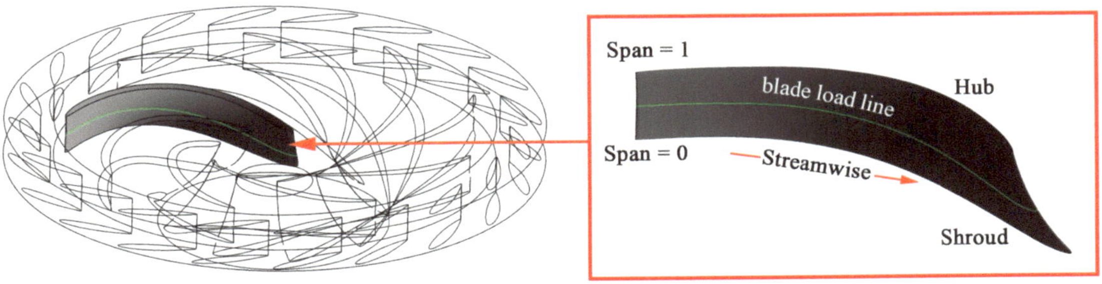

Figure 7. Schematic diagram of a blade load line.

The pressure distribution of a blade load line under no-load conditions with different GVO is depicted in Figure 8a. In general, the pressure shows a decreasing trend along the streamwise, while the pressure difference between the pressure side and suction side of the blade is larger under the effect of the increasing GVO. At the inlet of the runner, the pressure fluctuates but to a different extent. According to the previous analysis in this paper, the formation of a high-speed water ring in a vaneless zone impedes the flow capacity of the runner, and flow separation occurs at the pressure side of the blade's leading edge (see Figure 6). Therefore, the reverse pressure distribution forms at the runner inlet, and the reverse pressure turning point gradually moves downstream when the GVO broadens. On account of the serious disturbance in runner channels caused by eddies, the blade leading edge even displays a negative pressure distribution, and the severe pressure difference change tends to concentrate the stresses. Furthermore, Figure 8b presents the pressure distribution at different times at the OP4. It can be seen that the blade pressure at the runner inlet changes irregularly compared to the other three conditions, and the relative length of the blade affected by pressure fluctuation also extends from 0.1 to 0.22.

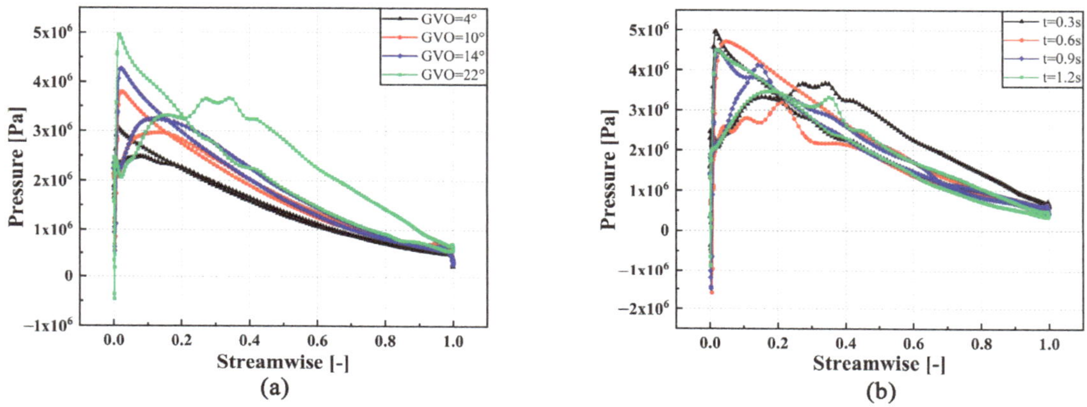

Figure 8. The pressure distribution on the blade surface. (**a**) Under different GVO; (**b**) At different times of OP4.

4. Analysis of Pressure Pulsation

The fluctuation signal of pressure change with time can be obtained by setting up pressure monitoring points in the flow passage of each component (Figure 2). For facilitating observation and analysis, the pressure pulsation coefficient C_p is introduced in this paper to nondimensionalize the time domain signal; then, the amplitude-frequency characteristic of the pressure pulsation can be obtained by the Fast Fourier Transform (FFT). Equation (5) provides the dimensionless conversion formula of C_p:

$$C_p = \frac{p - \overline{p}}{\rho g H} \times 100\% \quad (5)$$

where p is the instantaneous pressure value, P_a; $\overline{p}$ is the time-averaged pressure value, P_a; ρ denotes the fluid density, kg/m^3; g is the gravity acceleration, 9.8 m/s^2; and H is the head of the operating point, m.

4.1. Pressure Pulsation in the Vaneless Zone

Twenty monitoring points GR1~GR20 are arranged counterclockwise along the circumference in the vaneless zone, which can be seen in Figure 2a. Figure 9 shows the time domain diagrams of the pressure pulsation in the vaneless zone at OP1~OP4. As stated, under the small GVO conditions, the pressure pulsation in the vaneless zone shows a clear periodicity and propagates clockwise at a certain speed in the circular direction. This circumferential propagation rate is preliminarily estimated using the number of monitoring points of peak or valley propagation of the pressure pulsation per unit time as the offset angular velocity θ. It is found that the increase of GVO promotes the circumferential propagation of pressure pulsation in the vaneless zone. Moreover, as shown in Figure 5g,h, the water ring is no longer distinct with different thicknesses in the circumferential direction, and the corresponding pressure pulsation components in the vaneless zone are more complex (Figure 9d) than those in other small GVO conditions.

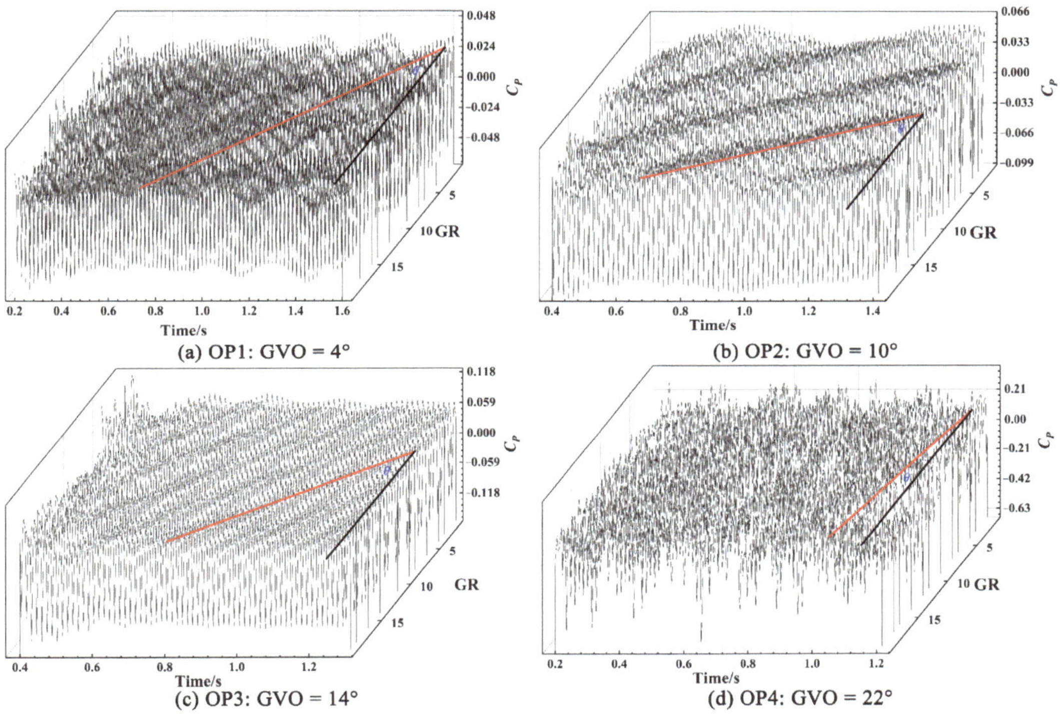

Figure 9. The time domain diagrams of pressure in the vaneless zone (GR).

Therefore, the frequency domain signals of adjacent distributed monitoring points GR1 and GR2 and symmetrical points GR12, GR16, and GR20 are selected for further analysis of pressure pulsation components. In agreement with OP1 and OP3, the pressure pulsation frequency of each monitoring point in the vaneless zone under OP2 (Figure 10a) is mainly composed of the blade-passing frequency and its harmonic components, i.e., $9f_n$ and $18f_n$; thus, the pressure pulsation is principally affected by RSI. Meanwhile, due to the poor flow pattern between the trailing edge of the guide vanes and the inlet of the runner blade under the interference of the water ring, the water impacts the blade, hence generating low-frequency components. However, when GVO is 22°, in addition to the typical RSI frequency, various low-frequency signals represented by $0.7f_n$ begin to strengthen (Figure 10b), and the rotating stall effect in the vaneless region intensifies. In other words, the appearance of the rotating stall leads to the flow separation at the leading edge of the runner. Subsequently, the channel vortices form and develop into the runner passage.

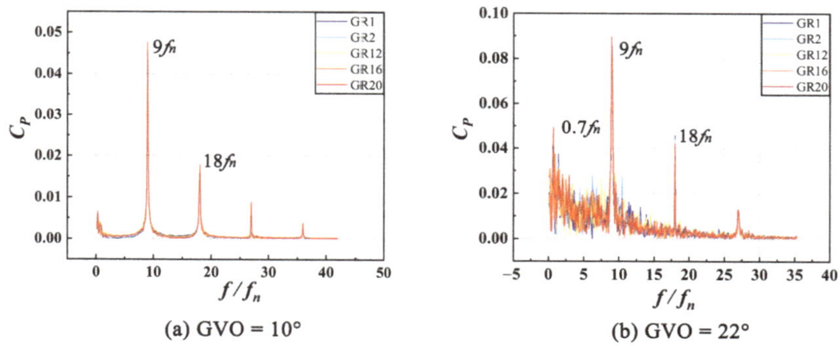

Figure 10. The frequency domain diagrams of pressure in the vaneless zone (GR).

4.2. Pressure Pulsation in the Runner Channel

Figure 2b shows the layout of pressure monitoring points in the runner channel. After comparing the data of pressure pulsation in the time and frequency domain under different conditions, it was found that although there were differences in the time sequence of pressure pulsation in the single runner channel, the signals in the frequency domain were basically the same. Thus NO. 4 blade is chosen for further analysis of the pressure pulsation in the frequency domain, as shown in Figure 11. Obviously, the high-frequency signal $20f_n$ exists in the runner passage at OP1~OP4, which is also the characteristic frequency caused by the RSI, namely, the guide vane passing frequency. With the increase of the GVO, the amplitude of its C_p increases from 0.004 in GVO = 4° to 0.0148 in GVO = 22°. Under each no-load condition, the amplitude first strengthens, then weakens, before peaking again thereafter, thus reaching the maximum at the downstream channel close to the runner inlet with the deepening of the channel, which is similar to the monitoring point RN4-2 in Figure 11. Nevertheless, the low-frequency components of the pressure pulsation become more complex and stronger with the increase of the GVO. This is due to the rotating stall frequency and its harmonic parts (i.e., $0.56f_n$~$0.7f_n$) at the runner inlet, to which is added the typical frequency $0.14f_n$ of the vortex rope, which is the dominant frequency at the runner outlet due to its proximity to the inlet of the conical section of the draft tube. Under the combined effects of high-frequency components caused by the RSI and low-frequency components induced by the rotating stall, the structural stress of the downstream location close to the runner inlet is significant and worth considering for reliability analysis.

4.3. Pressure Pulsation in the Spiral Casing and Draft Tube

The capture of pressure pulsation characteristics is helpful to master its propagation law. Hence, several monitoring points are arranged along the flow path of a spiral casing and draft tube in Figure 2a,c, respectively. Figures 12 and 13 show the frequency domain diagrams of pressure pulsation under no-load conditions with different GVO.

As observed, the amplitude of pressure pulsation in the spiral casing decreases gradually during upstream propagation when the GVO is small, and the low-frequency signal caused by the stall vortices attenuates significantly. In addition, the double blade-passing frequency $18f_n$ is still reflected, which indicates that the pressure pulsation caused by RSI is more powerful in the upstream propagation under low GVO conditions. When the GVO is larger, the dominant frequency and amplitude of pressure pulsation at each monitoring point are nearly equivalent (Figure 12c,d). Because of the complexity of the internal flow of the unit under no-load conditions with large GVO, the generated low-frequency signal gradually gains importance and surpasses the blade-passing frequency two times. The reason can be attributed to the amplitude of rotating stall frequency $0.7f_n$ being distinctly higher than RSI's $1818f_n$ along the upstream flow path of the spiral casing (SC6 to SC1), thereby lacking attenuation in the propagation process. Compared with the low GVO conditions, the stall vortices with $0.7f_n$ evolve continually in the runner channel and propagate upstream through the double-row cascade, exerting the effect in the spiral casing. Furthermore, this low-frequency signal is transmitted downstream into the draft tube.

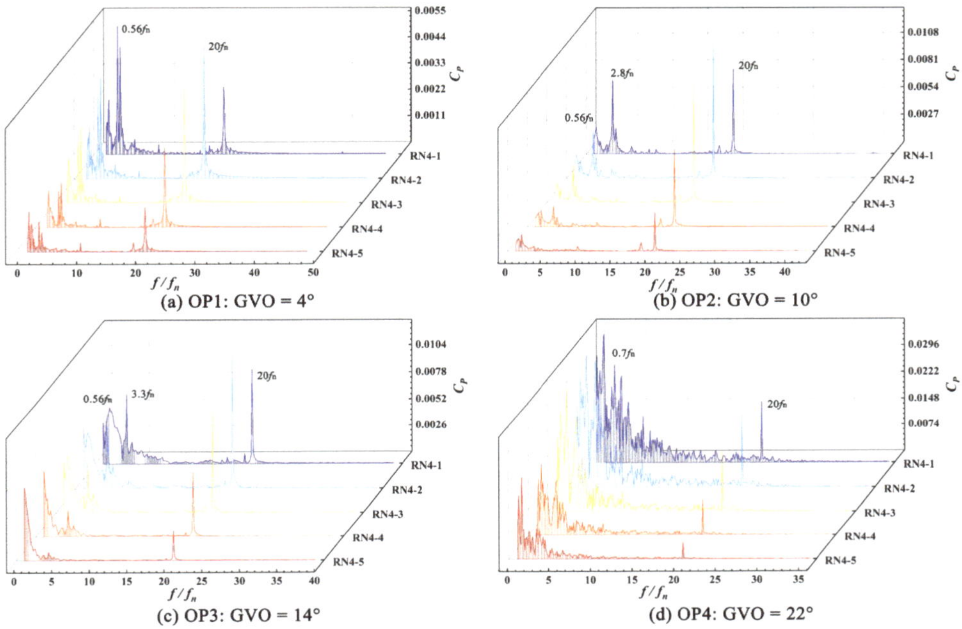

Figure 11. The frequency domain diagram of pressure pulsation in the runner channel (RN).

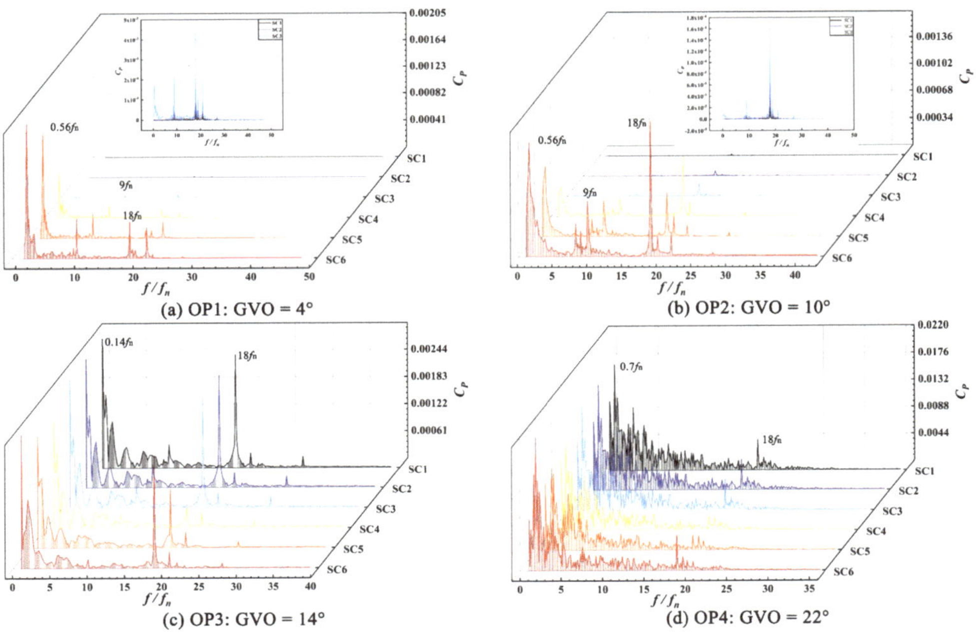

Figure 12. The frequency domain diagrams of pressure pulsation in the spiral casing (SC).

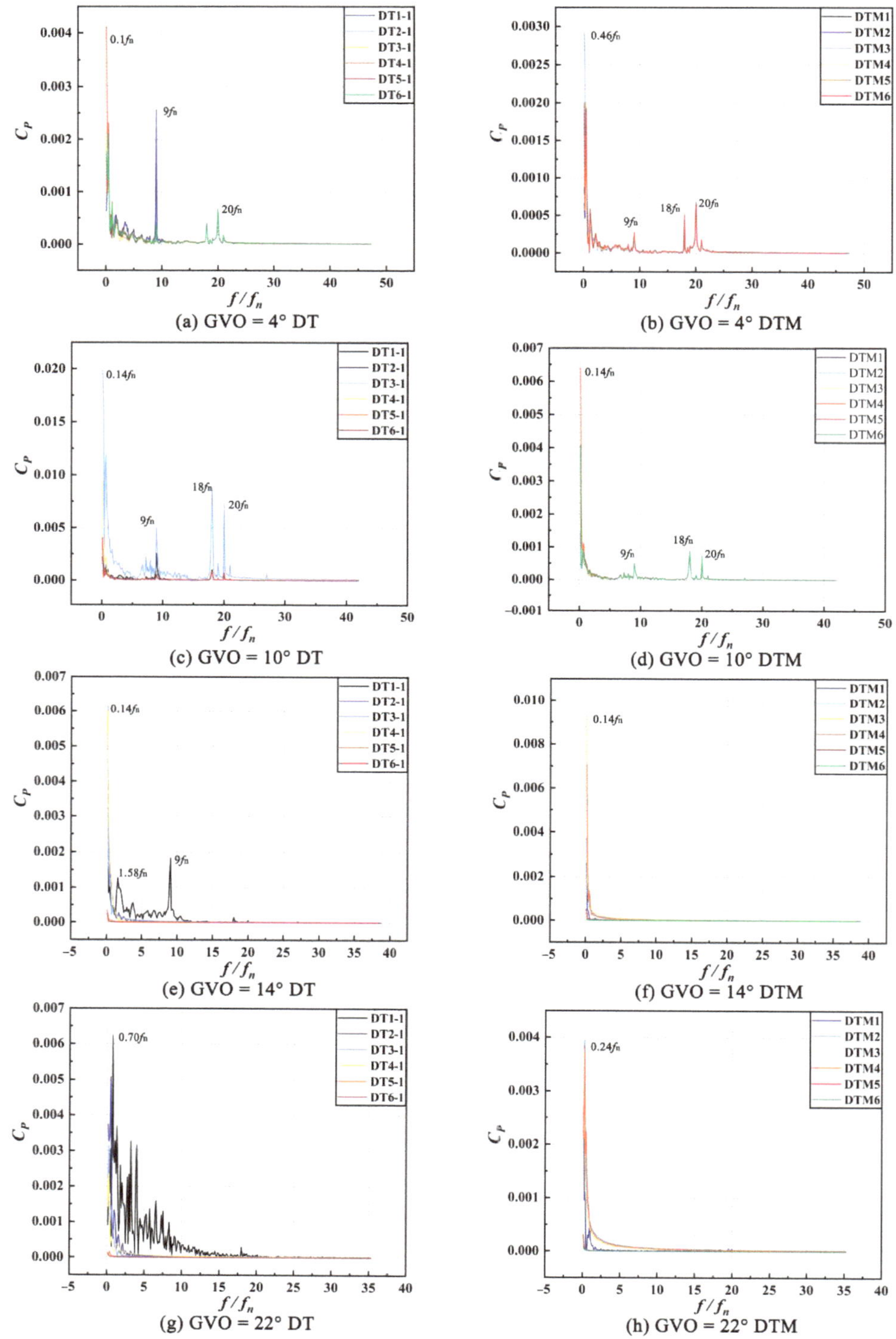

Figure 13. The frequency domain diagrams of pressure pulsation in the draft tube.

Figure 13 shows the changes in pressure pulsation signals in the frequency domain among the monitoring points in the near wall region (DT) and center of each section (DTM) of the draft tube. Under the no-load conditions, the flow in the draft tube is affected by the high-speed circulation at the outlet of the runner, resulting in the RSI frequencies (such as guide vane passing frequency $20f_n$ and blade-passing frequency $9f_n$ and its harmonic $18f_n$) shown in the near wall region (DT). In addition, the vortex rope forms easily from non-design conditions, and the pressure pulsation is dominated by low-frequency signals. As the vortex rope gradually develops downstream, due to the sudden change of movement direction after entering the elbow section, the flow state in this region is more complicated than that in the tapered pipe section. The amplitude of pressure pulsation reaches its maximum in the elbow section of the draft tube, corresponding to the monitoring points DT3-1 and DTM3 in Figure 13. However, the influence of the RSI on the pressure pulsation weakens gradually with the increase of GVO and then disappears in the draft tube. The low-frequency components represented by the rotating stall $0.7f_n$ are more obvious in the conical section, but they then develop and attenuate downstream. This in turn results in more complex low-frequency components, as shown in Figure 13g.

5. Analysis of Structure Stress

Through the analysis of internal flow characteristics and pressure pulsation, it can be seen that the flow state of the runner and guide vanes is complicated, and the pressure pulsation in the vaneless zone is more significant under the combined influence of RSI and the rotating stall. To gain a better understanding, the weak-coupling method [32] is applied for analyzing the structure stress of runner and guide vanes. The discretized form of the linear dynamic equilibrium equation can be manifested as follows:

$$[M]\{\ddot{x}\} + [C]\{\dot{x}\} + [K]\{x\} = \{F\} \tag{6}$$

where $[M]$, $[C]$, and $[K]$ denote the mass matrix, damping matrix, and stiffness matrix, respectively; $\{\ddot{x}\}$, $\{\dot{x}\}$, and $\{x\}$ represent acceleration, velocity, and displacement, respectively; and $\{F\}$ is the external load force acting on the nodes.

As shown in Figure 14b, the structural load appears on the surface of the runner blades and guide vanes, which is the imported pressure caused by water flow [33]. Moreover, several constraints are set on the structure domain. For instance, the hub flange of runner and the shaft head of the guide vane are given with fixed support, and the upper, middle, and lower shaft necks are restrained by cylindrical support. Considering the influence of gravity on the parts and the runner rotational motion, the gravitational acceleration and the rotation speed of the runner are erected as well. The tetrahedral mesh is used for dividing the structure domain (Figure 14c), and the grid numbers of the runner and guide vane are 3.6×10^6 and 1.4×10^6, respectively, which is verified by mesh independence.

Figure 15 presents the equivalent stress distribution of the runner structure. Under low GVO conditions, the uniform water ring with high rotational velocity is guided by the guide vanes, and the runner is subject to relatively weak hydraulic excitation due to the low discharge. Thus, the stress of the runner is regular in the circumferential direction and only occurs at the flange connected to the turbine shaft and the connections between the blades, as well as in the shroud and hub with higher values. The GVO increases the overall stress of the runner. Affected by the stronger unstable flow, the symmetry of the stress distribution of blades disappears gradually, leading to a more apparent stress difference on the blade surface. In addition, the main stress concentration points are still located in the "T" connection between the blade leading edge and the runner crown and the blade outlet edge and the lower ring. When the unit operates at the OP4 (GVO = 22°), the stress moves away along the blade's root, resulting in the local stress concentration of the weld between the blade and the shroud (Figure 15d), which becomes a potential factor for inducing the runner crack. This is in line with what is observed in the actual NO. 8 blade, as shown in Figure 16. The two cracks are located between the blade and lower ring, which are 30 mm and 15 mm long, respectively. In the long term, the reversible pump operation may increase the crack size, causing failure.

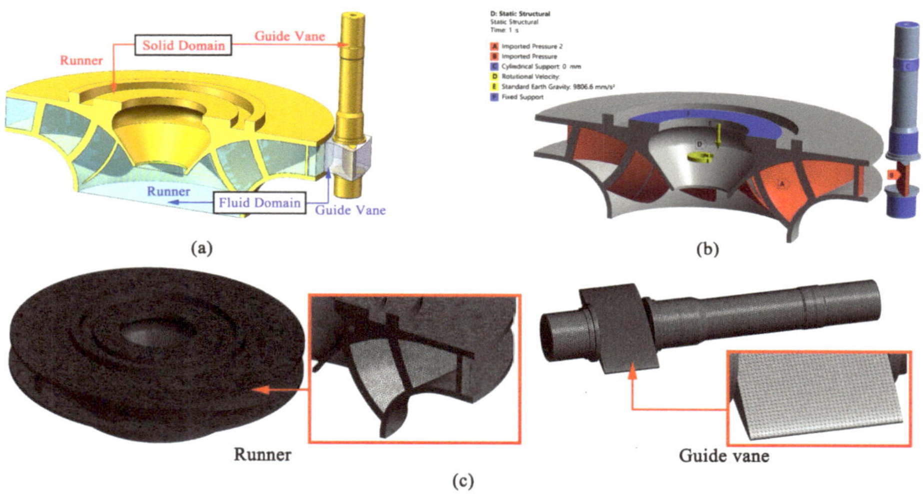

Figure 14. The illustration of fluid–structure interaction (FSI): (**a**) schematic diagram; (**b**) boundary conditions; (**c**) mesh of the runner and the guide vane.

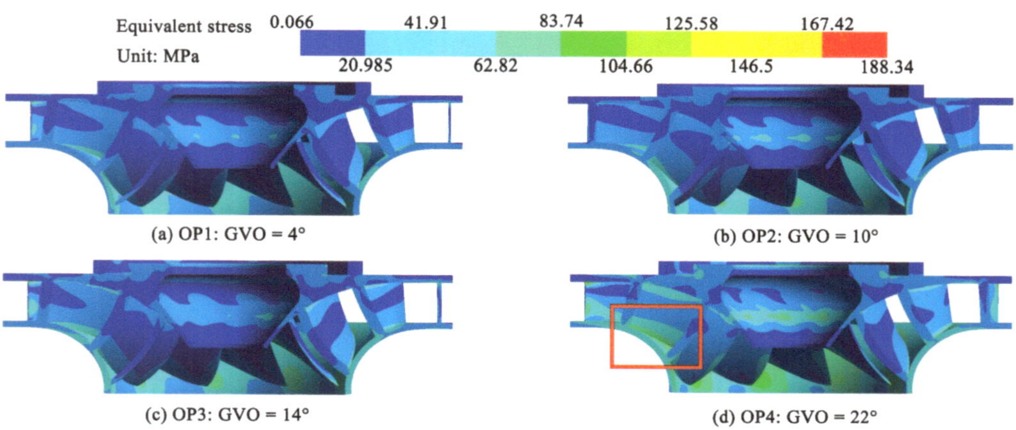

Figure 15. The equivalent stress distribution of the runner under different conditions. The stress concentration is marked with red rectangular boxes.

Figure 17 reveals that the maximum stress mainly aggregates at the joint of the guide vane profile and the middle or lower shaft neck under all selected no-load conditions. Likewise, the equivalent stress of the guide vane rises first, but then it decreases with the increase of the GVO. This reveals that the maximum value of stress increases from 45.6 MPa in GVO = 4° to 87.5 MPa in GVO = 10° and declines to 30.6 MPa in GVO = 22°. Meanwhile, the deformation of the guide vane starts at the trailing edge, and then it spreads to the head with GVO. After the analysis, it was found that when the GVO is small, the high-speed water ring in the vaneless zone leads to flow blockage, i.e., the water flow gathers at the tail of the guide vanes and the runner inlet (Figure 5a). The angle of discharge at the outlet of guide vanes also changes with the larger GVO, which also results in a more intense flow impact, causing the structure stress to rise as well (Figure 17b). However, when the GVO continues to increase, more discharge could flow into the runner channel after passing the guide vanes; thus, the runner blades are subject to more hydraulic pressure and stress, (Figure 15d) whereas the stress on the guide vanes is alleviated correspondingly (Figure 17d).

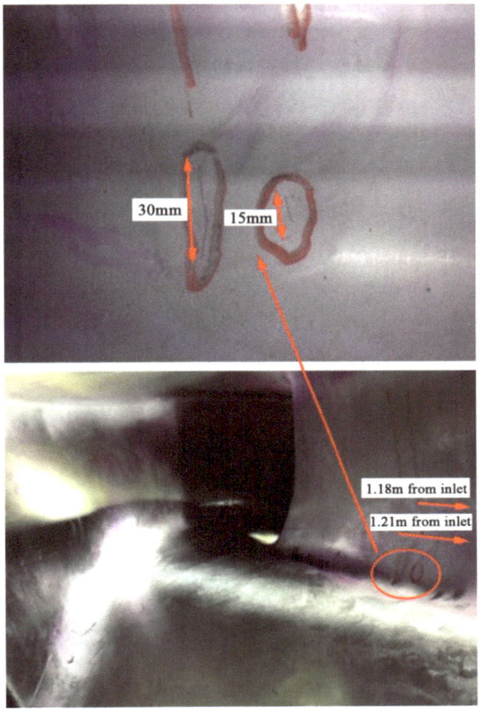

Figure 16. Blade cracks near the runner inlet.

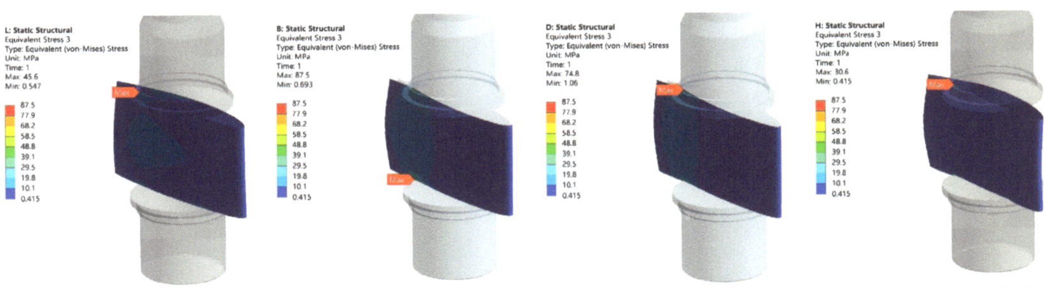

(a) OP1: GVO = 4° (b) OP2: GVO = 10° (c) OP3: GVO = 14° (d) OP4: GVO = 22°

Figure 17. The equivalent stress distribution of guide vanes under different conditions.

6. Conclusions

In this study, a prototype RPT was selected to investigate its internal flow characteristics. The propagation law of pressure pulsation under no-load conditions was simulated using the SST model, whereas the influence of hydraulic characteristics on the structural performance of the guide vane and the runner was analyzed based on the unidirectional FSI method.

1. Under the no-load conditions, the flow stability of the pump-turbine is closely related to the vortex evolution in the runner. By increasing the GVO, the channel vortices first gather at the suction side of the blades and then spread to the pressure side. Meanwhile, the reverse pressure turning points of the blade start to move downstream, and the circumferential symmetry of the high-speed water ring in the vaneless region is broken. As a result, the rotating stall with frequency $0.7f_n$ is more drastic than under low GVO conditions, which also threatens the structural safety of the runner.
2. During low GVO conditions, the pressure pulsation in the vaneless zone is dominated by the RSI, namely, the blade-passing frequency $9f_n$ and its harmonic $18f_n$. The propagation capacity of the RSI frequencies is more powerful than that caused by the rotating stall. With the increase of GVO, the circumferential propagation rate and appearance of low-frequency signals are represented by the rotating stall frequency $0.7f_n$ rise, which transmits upstream to the spiral casing and downstream to the elbow section of the draft tube. At the downstream channel, near the runner inlet, the amplitude of these low-frequency signals reaches its maximum. These phenomena align with the results of flow characteristics and have an effect on structural performance.
3. Under no-load conditions, the greater the GVO, the more uneven the circumferential distribution of the runner stress and the more significant the stress development of the blade root. That is, the stress concentrates at the weld between the blades and the lower ring, potentially forming and expanding blade cracks. This substantiates the cracks observed in the physical blade. Moreover, the structural stress of the guide vanes is alleviated with the increase of the GVO, whereas its deformation becomes enlarged under the combined effects of frequency and amplitude of the hydraulic excitation.
4. To improve the structural performance affected by hydraulic characteristics, the RPT should avoid operating under no-load conditions with large GVO as far as possible. In order to design or optimize the S-characteristics of the RPT, the guide vanes and runner could be considered as a whole to investigate their internal flow and structure characteristics.

Author Contributions: Conceptualization, Y.Z. (Yuquan Zhang) and E.F.-R.; methodology, C.Y. and Y.Z. (Yuquan Zhang); supervision, Y.Z. (Yuan Zheng) and Y.Z. (Yuquan Zhang); writing—original draft, S.R.; software, S.R. and B.L.; validation, S.R., Y.Z. (Yuan Zheng) and C.Y.; investigation, S.R., Y.Z. (Yuan Zheng), C.Y. and B.L.; writing—review and editing, E.F.-R. and Y.Z. (Yuquan Zhang). All authors have read and agreed to the published version of the manuscript.

Funding: This work was supported by the National Natural Science Foundation of China (52271275).

Data Availability Statement: Data are contained within the article.

Acknowledgments: Thanks to Zhang for his advice.

Conflicts of Interest: The authors declare no conflict of interest.

Nomenclature

P_r	Rated power of the unit, MW
n_r	Rated rotational speed of the runner, rpm
H_r	Rated head, m
D	Nominal diameter of the runner, m
f_n	Rotational frequency of the runner, Hz
Z_r	The blade number of runner
Z_g	The number of guide vanes
Z_s	The number of stay vanes
n_{11}	Unit speed
Q_{11}	Unit discharge
C_p	The pressure pulsation coefficient

Abbreviations

BEP	Best efficient point
CFD	Computational fluid dynamics
FSI	Fluid–structure interaction
GVO	Guide vanes opening
GGI	General grid interface
OP	Operation point
RSI	Rotor–stator interaction
RPT	Reversible pump-turbine
SST	Shear stress transport

References

1. Han, X.; Wang, X.; Zheng, H.; Xu, X.; Chen, G. Review of the development of pulsating heat pipe for heat dissipation. *Renew. Sustain. Energy Rev.* **2016**, *59*, 692–709. [CrossRef]
2. Der, O.; Alqahtani, A.A.; Marengo, M.; Bertola, V. Characterization of polypropylene pulsating heat stripes: Effects of orientation, heat transfer fluid, and loop geometry. *Appl. Therm. Eng.* **2021**, *184*, 116304. [CrossRef]
3. Rodríguez-Pérez, Á.M.; Pulido-Calvo, I.; Cáceres-Ramos, P. A computer program to support the selection of turbines to recover unused energy at hydraulic networks. *Water* **2021**, *13*, 467. [CrossRef]
4. Wang, R.; Yang, W.J.; Ji, L.T.; Li, X.D.; Zhang, S.S.; Jing, X.Y. Regulation intensity assessment of pumped storage units in daily scheduling for renewable energy consumption. *Sustain. Energy Technol. Assess.* **2023**, *56*, 103027. [CrossRef]
5. Zhao, X.R.; Xiao, Y.X.; Wang, Z.W.; Luo, H.Y.; Ahn, S.H.; Yao, Y.Y.; Fan, H.G. Numerical analysis of non-axisymmetric flow characteristic for a pump-turbine impeller at pump off-design condition. *Renew. Energy* **2018**, *115*, 1075–1085.
6. Zhang, Y.Q.; Sun, J.; Zheng, Y.; Ge, X.F.; Peng, X.Y.; Hong, Y.L.; Fernandez-Rodriguez, E. Unsteady characteristics of lubricating oil in thrust bearing tank under different rotational speeds in pumped storage power station. *Renew. Energy* **2022**, *201*, 904–915. [CrossRef]
7. Suh, J.W.; Yang, H.M.; Joo, W.G.; Park, J.; Choi, Y.S. Unstable S-shaped characteristics of a pump-turbine unit in a lab-scale model. *Renew. Energy* **2021**, *171*, 1395–1417. [CrossRef]
8. Casartelli, E.; Del Rio, A.; Mangani, L.; Schmid, A. Capturing the S-shape of pump-turbines by computational fluid dynamics simulations using an anisotropic turbulence model. *J. Fluids Eng.* **2021**, *144*, 021203. [CrossRef]
9. Lenarcic, M.; Bauer Ch Giese, M.; Jung, A. Prediction of S-shaped characteristics in reversible pump-turbines using different numerical approaches. *IOP Conf. Ser. Earth Environ. Sci.* **2016**, *49*, 042009. [CrossRef]
10. Zhao, Z.W.; Zhang, H.; Chen, D.Y.; Gao, X. No-load stability analysis of pump turbine at startup-grid integration process. *J. Fluids Eng.* **2019**, *141*, 081113. [CrossRef]
11. Yang, H.X.; He, Q.L.; Huang, X.X.; Yang, M.Q.; Bi, H.L.; Wang, Z.W. Experimental and Numerical Investigation of Rotor–Stator Interaction in a Large Prototype Pump–Turbine in Turbine Mode. *Energies* **2022**, *15*, 5523. [CrossRef]
12. Zhang, Y.L.; Wu, Y.L. A review of rotating stall in reversible pump turbine. *J. Mech. Eng. Sci.* **2017**, *231*, 1181–1204. [CrossRef]
13. Tanaka, H. Vibration behavior and dynamic stress of runners of very high head reversible pump-turbines. *Int. J. Fluid Mach. Syst.* **2011**, *4*, 289–306. [CrossRef]
14. Hu, J.H.; Yang, J.B.; He, X.H.; Zeng, W.; Zhao, Z.G.; Yang, J.D. Transition of amplitude-frequency characteristic in rotor-stator interaction of a pump-turbine with splitter blades. *Renew. Energy* **2023**, *205*, 663–677. [CrossRef]
15. Han, F.; Chen, X.H.; Yang, Y.; Wang, C. Numerical and experimental study on the effect of rotor-stator interaction strength within mixed-flow centrifugal pumps. *J. Mar. Sci. Eng.* **2022**, *10*, 1114. [CrossRef]
16. Deniz, S.; Von Burg, M.; Tiefenthaler, M. Investigation of the flow instabilities of a low specific speed pump turbine part 2: Flow control with fluid injection. *J. Fluids Eng.* **2022**, *144*, 071210. [CrossRef]
17. Chen, Z.M.; Jiang, Z.Y.; Chen, S.; Zhang, W.W.; Zhu, B.S. Experimental and numerical study on flow instability of pump-turbine under runaway conditions. *Renew. Energy* **2023**, *210*, 335–345. [CrossRef]
18. Fu, X.L.; Zuo, Z.G.; Chang, H.; Wang, H.J.; Wei, X.Z. Mechanism of low frequency high amplitude pressure fluctuation in a pump-turbine during the load rejection process. *J. Hydraul. Res.* **2021**, *59*, 280–297. [CrossRef]
19. Hu, D.L.; Cheng, Y.G.; Zhang, P.C.; Wang, X.; Ding, J.H.; Zhang, X.X. Distribution features of flow patterns and pressure pulsations of pump-turbine in five operating modes on the four-quadrant plane. *Front. Energy Res.* **2022**, *10*, 880293. [CrossRef]
20. Fu, X.L.; Li, D.Y.; Song, Y.C.; Wang, H.J.; Yang, J.L.; Wei, X.Z. Dynamic characteristics of a running away pump-turbine with large head variation: 1D+3D coupled simulation. *Eng. Appl. Comput. Fluid Mech.* **2023**, *17*, 15. [CrossRef]
21. Hu, J.H.; Yang, J.D.; He, X.H.; Yang, J.B. Improved S-shaped characteristic curve of pump-turbine based on numerical simulation. *J. Huazhong Univ. Sci. Technol. (Nat. Sci. Ed.)* **2020**, *48*, 109–114.
22. Li, D.Y.; Qin, Y.L.; Wang, J.P.; Zhu, Y.T.; Wang, H.J.; Wei, X.Z. Optimization of blade high-pressure edge to reduce pressure fluctuations in pump-turbine hump region. *Renew. Energy* **2022**, *181*, 24–38. [CrossRef]
23. Qin, Y.L.; Li, D.Y.; Wang, H.J.; Liu, Z.S.; Wei, X.Z.; Wang, X.H.; Yang, W.B. Comprehensive hydraulic performance improvement in a pump-turbine: An experimental investigation. *Energy* **2023**, *284*, 10. [CrossRef]

24. Rodríguez-Pérez, Á.M.; Rodríguez, C.A.; Márquez-Rodríguez, A.; Mancera, J.J.C. Feasibility analysis of installing a tidal turbine using fuzzy logic: Case study and design considerations. *Axioms* **2023**, *12*, 778. [CrossRef]
25. Zhang, Z.; Mom, X.; Liu, M.; Liu, M. Fuzzy comprehensive evaluation of site selection for offshore wind farms based on Delphi method. *Renew. Energy* **2016**, *86*, 1152–1159.
26. Mao, X.L.; Pavesi, G.; Chen, D.Y.; Xu, H.S.; Mao, G.J. Flow induced noise characterization of pump turbine in continuous and intermittent load rejection processes. *Renew. Energy* **2019**, *139*, 1029–1039. [CrossRef]
27. Zhang, C.Y.; Wang, W.J.; Pei, J.; Yuan, S.Q. Numerical simulation of unsteady flow characteristics in a centrifugal pump during runaway process caused by power failure. *J. Phys. Conf. Ser.* **2022**, *2217*, 01247. [CrossRef]
28. Pokharel, N.; Ghimire, A.; Thapa, B.; Singh Thapa, B.; Qian, Z.D.; Guo, Z.W. Numerical and experimental study of pump as turbine for sediment affected micro hydropower project in Nepal. *IOP Conf. Ser. Earth Environ. Sci.* **2021**, *774*, 012062. [CrossRef]
29. Xu, J.H.; Zhang, Y.Q.; Peng, B.; Zheng, Y.; Li, C.Y.; Zang, W.; Fernandez-Rodriguez, E. Study on the dynamics and wake characteristics of a floating tidal stream turbine with pitch motion under free surface. *Phys. Fluids* **2023**, *35*, 085101. [CrossRef]
30. Li, B.; Yang, W.J.; Li, J.; Jing, S.Y.; Chen, Y. Analysis of the spanwise vortex of open channel flows based on the Omega-Liutex vortex identification method. *J. Hydrodyn.* **2022**, *34*, 148–155. [CrossRef]
31. Trieu Tran, C.; Long, X.P.; Ji, B. Vortical structures in the cavitating flow in the Francis-99 draft tube cone under off-design conditions with the new omega vortex identification method. *J. Phys. Conf. Ser.* **2019**, *1296*, 012011. [CrossRef]
32. Li, X.Y.; Cao, J.W.; Zhuang, J.L.; Wu, T.M.; Zheng, H.Y.; Wang, Y.F.; Zheng, W.Q.; Lin, G.Q.; Wang, Z.W. Effect of operating head on dynamic behavior of a pump-turbine runner in turbine mode. *Energies* **2022**, *15*, 4004. [CrossRef]
33. Zhang, Y.Q.; Zang, W.; Zheng, J.H.; Cappietti, L.; Zhang, J.S.; Zheng, Y.; Fernandez-Rodriguez, E. The influence of waves propagating with the current on the wake of a tidal stream turbine. *Appl. Energy* **2021**, *290*, 116729. [CrossRef]

processes

MDPI

Article

Investigating the Impact of Operating Conditions on Relief Pressure Valve Flow through CFD and Statistical Analysis

Petrică Cană, Razvan George Ripeanu *, Iulian Pătîrnac, Alin Diniță and Maria Tănase *

Mechanical Engineering Department, Petroleum-Gas University of Ploiești, 100680 Ploiesti, Romania; cana_petrica@yahoo.co.uk (P.C.); pata_iulian@yahoo.com (I.P.); adinita@upg-ploiesti.ro (A.D.)
* Correspondence: rrapeanu@upg-ploiesti.ro (R.G.R.); maria.tanase@upg-ploiesti.ro (M.T.); Tel.: +40-721212250 (R.G.R.);+40-726555632 (M.T.)

Abstract: This paper presents a comprehensive computational fluid dynamics (CFD) analysis of air/water flow through a discharge valve, focusing on four different seat–valve distances and three adjustment nozzle positions. The study investigates the velocity distribution, pressure profiles, tangential stresses, and turbulent kinetic energy within the valve and analyzes its performance under various operating conditions. Notably, peak velocities of 3210 m/s were observed between the valve seat and valve, with significant variations for different nozzle positions. Extreme pressure values centered on the valve plate, reaching 4.3 MPa. Tangential stresses were highest on the chamfered plate surface and varied on the seat, turbulent kinetic energy (TKE) exhibited randomness. This study provides valuable information for enhancing the valve's efficiency in a wide range of industrial applications.

Keywords: pressure valve; numerical analysis; CFD; fluid flow; turbulent kinetic energy; shear stress

Citation: Cană, P.; Ripeanu, R.G.; Pătîrnac, I.; Diniță, A.; Tănase, M. Investigating the Impact of Operating Conditions on Relief Pressure Valve Flow through CFD and Statistical Analysis. *Processes* **2023**, *11*, 3396. https://doi.org/10.3390/pr11123396

Academic Editors: Krzysztof Rogowski and Ireneusz Zbicinski

Received: 31 October 2023
Revised: 4 December 2023
Accepted: 7 December 2023
Published: 9 December 2023

1. Introduction

Effective fluid flow control is a fundamental requirement in a wide array of industrial processes, where precision and efficiency are predominant. Within this context, discharge valves play a pivotal role in regulating fluid flow within pipelines and systems. A comprehensive understanding of airflow dynamics within these valves is essential for their optimal design and operation.

The stability of safety relief valves, essential for their reliable performance is influenced by various factors, such as the design of the valve, spring pre-deformation, the length of the inlet pipeline, valve opening length, the position of regulator rings and the volume of the pressure vessel [1].

Dasgupta and Karmakar [2] conducted research on the dynamics of a pilot-operated pressure relief valve using the Bondgraph simulation technique. They derived the governing equations of the system from their model and, in the process of numerically solving these equations, they considered the different pressure-flow characteristics of the valve ports. The simulation study identified specific critical design parameters that exerted a substantial impact on the transient response of the system. Similarly, Prescott and Ulanicki [3] created a straightforward highly precise model for analyzing the dynamic behavior of a pressure reducing valve. The primary contributions of their work included the formulation of multiple dynamic models, encompassing two phenomenological models, one behavioral model, and one linear model, all designed to effectively describe the functioning of pressure reducing valves.

Advancements in computational techniques have made it increasingly practical to analyze the intricate flow patterns within valves [4–6]. As a result, there has been a significant increase in research related to the functionality and design of valves [4–11].

The study [7] develops a numerical model using advanced computational fluid dynamics techniques to analyze a direct-operated safety relief valve. The model accurately simulates the entire process from valve opening to closure, providing insights into compressible fluid flow and the effects of design parameters.

A three-dimensional numerical analysis with CFD was performed in [1] to examine a safety relief valve's performance under various conditions, including different openings and inlet pressures. The study compared mass flow and discharge coefficients from simulations with ASME 2011a Section 1 standards and the model proved consistent for openings below 12 mm and closely matched the ASME standard. Additionally, CFD simulations revealed that the average normal disc force was approximately 19% lower than the theoretical ASME force, which could help prevent valve oversizing.

A simplified dynamic model (based on the equation of motion for a single degree of freedom system, with the assumption that damping effects are negligible due to their minimal impact on reclosing time) was developed in [12] to predict the dynamic behavior during both the reclosing and blowdown phases of a conventional pressure relief valve. The study introduces a novel approach by integrating static computational fluid dynamics (CFD) analysis results, calculating the lift force coefficient of the valve at various fixed lift positions using static CFD analysis and then incorporates these values into the simplified dynamic model as the inherent characteristics of the valve.

CFD simulation was also used by Erdődi and Hős [13] to explore the dynamic instability (flutter/chatter) of gas-service direct spring operated pressure relief valves, specifically focusing on the acoustic coupling between the valve body dynamics and the upstream piping. To enhance accuracy and enable analytical computations, the paper introduces two CFD-based methods. The first involves steady-state CFD runs to validate and improve the reduced-order models. The second employs dynamic CFD simulations with deforming mesh technology, providing a high-fidelity resolution of valve response and transient fluid–structure interactions. Comparing results from both approaches validates the use of reduced-order models for stability predictions.

The scientific work [14] focused on the design and modeling of a buffer relief valve using ANSYS 2021 R1 software, in order to achieve an optimized design for this valve. The findings revealed a high level of agreement between the results obtained from ANSYS simulations and the actual experimental results, affirming that the valve is suitable and safe for operation at a pressure rating of 0.14 MpPa.

Hos et al. (2014) [15] conducted a study that both summarized and expanded upon recent scientific investigations into the causes of instability in pressure relief valves. The primary objective was to obtain a deeper and more comprehensive understanding of the factors influencing the stability of pressure relief valves during their operation. The authors specifically focused on direct spring-loaded pressure relief valves used in gas applications, with a particular emphasis on how valve dynamics interact with acoustic pressure waves in the inlet pipe, considering their combined effect on valve stability.

The paper [16] offers a comprehensive analysis of experimental findings concerning the static and dynamic behavior of a hydraulic pressure relief valve, specifically focusing on the factors that influence valve instability. The study reveals that both signals exhibit similar frequency content, and the frequency of chatter remains relatively constant over various parameter ranges, such as fluid velocity and set pressure. The paper also delineates different motion patterns, including stable operation, free motion, impacting oscillations, and chaotic behavior, as they relate to the interplay between fluid velocity and set pressure.

A model that captures the dynamic response of a pressure relief valve in gas or vapor service is introduced by Darby [17], taking into account a range of influencing factors. The model allows the prediction of the valve disk's position over time, considering specific valve characteristics, operational conditions, and installation parameters. The model also incorporates the impact of these various parameters on the stability or instability of the valve disk's response. Part II of the investigation program [18] presents an experimental testing program designed to investigate the opening stability of direct-acting pressure relief valves used in gas/vapor service, testing three different valve sizes from three manufacturers at two set pressures to assess their opening characteristics, specifically the relationship between disk lift and time. The tests varied the length of inlet piping and considered the presence or absence of discharge piping to identify conditions leading to unstable valve operation. Part III [19] includes a comparative analysis between the model's predictions and the actual dynamic response data and also examines the details necessary for putting the model into practical use.

The article [10] aims to expand the understanding of self-excited oscillations in pressure relief valves through a combination of time history analysis, spectrogram post-processing, and experimental testing under various conditions. The study reveals the presence of low-frequency and high-frequency self-excited oscillations, identifying their potential causes, which include external vibrations and acoustic resonance in the vessel-pipe system, shedding light on the factors contributing to valve damage.

The scientific paper [20] gives important details regarding the utilization of Computational Fluid Dynamics (CFD) simulation techniques to analyze hydrodynamic phenomena within pipes, exploring various aspects related to the hydraulic efficiency of the pipe system.

On the other hand, external mechanical vibrations can disrupt the stability of valves when applied to their bodies. In such instances, vibrations are induced in the valve control element, such as the plug, ball, or spool [21]. Regarding this aspect, the work performed by Stosiak et al. [22] focuses on reducing the impact of external mechanical vibrations on hydraulic valves within various military hydraulic drive systems. The research explores the potential reduction of vibrations on the valve casing by incorporating flexibility into the valve installation on vibrating surfaces. This involves introducing materials with known stiffness and damping characteristics, such as steel spring packages or special cushions made of elastomer or oil-resistant rubber. Additionally, the study suggests that elastomer cushions within the valve casing, positioned between the casing and centering springs, can serve as a supplementary or alternative solution for dampening vibrations.

The primary focus of the present study centers on exploring the influence of four distinct seat–valve distances and three adjustment nozzle positions on flow characteristics within the valve. The novelty of this paper lies in its comprehensive investigation of the impact of seat–valve distances and adjustment nozzle positions on flow characteristics within the valve. The performed investigation explores parameters such as velocity distribution, pressure profiles, tangential stresses, and turbulent kinetic energy, providing a detailed understanding of how various operational conditions affect the valve's performance.The practical application of this research is significant, as it offers valuable insights into optimizing the design and operation of these valves, the obtained results can be used to make informed decisions regarding seat–valve distances and adjustment nozzle positions, ultimately improving the efficiency and reliability of valves in real-world applications.

2. Materials and Methods

2.1. Characteristics of the Relief Pressure Valve

Figure 1 shows the main components of the pressure valvestudied in this research (commonly used in the oil and gas industry).

Due to the fact that both the sealing surface of the valve and the surface of the seat are covered with hard materials, the valve consists of the valve body and plate, and the seat assembly consists of the seat body and adjusting nozzle. The assembly of these two components, characteristic of discharge valves, is shown in Figure 2.

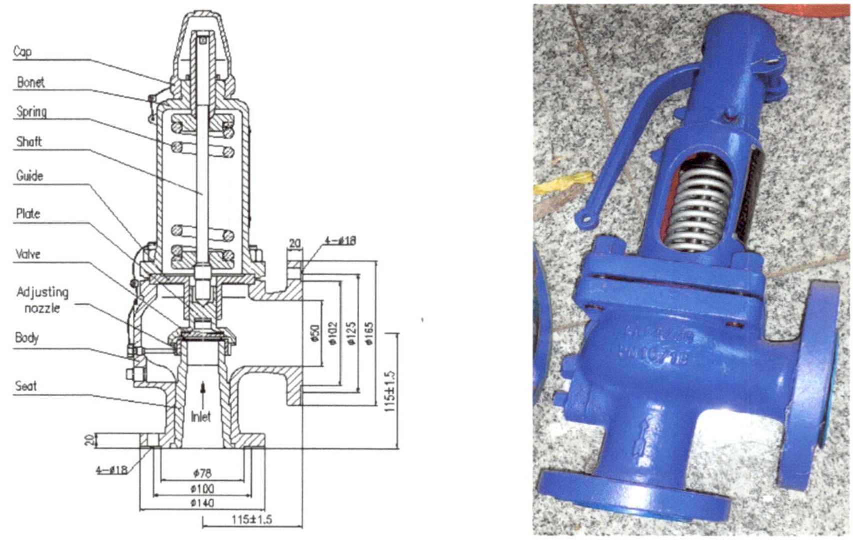

(**a**) (**b**)

Figure 1. Shape and dimensions of the analyzed pressure relief valve: (**a**)schematic representation; (**b**) real shape.

(**a**) (**b**)

Figure 2. Pressure valve subassemblies: (**a**)—the plate assembly; (**b**)—the seat assembly.

The geometric model of the fluid part was created in a three-dimensional design program, with small simplifications of its geometry being necessary, with the aim of eliminating possible errors within the CFD simulation. The geometric model of the fluid part is shown in Figure 3.

The technical characteristics of the valve considered in this analysis are presented in Table 1. The discharge pressure is the pressure corresponding to the maximum opening of the valve.

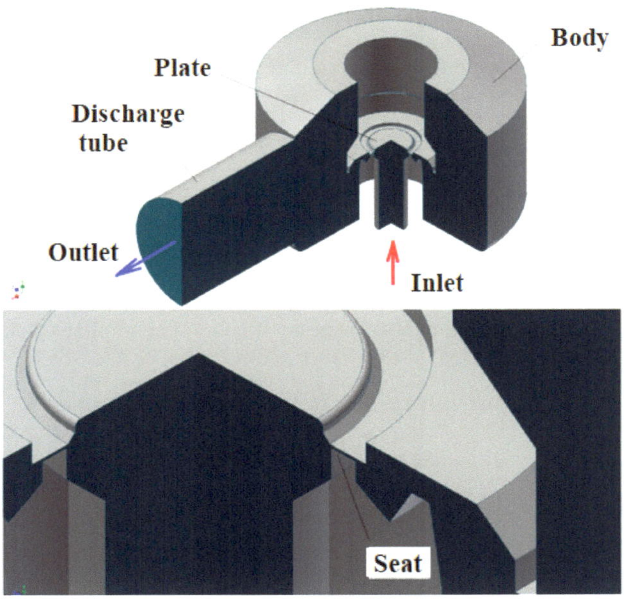

Figure 3. Fluid part of the3D model.

Table 1. Technical characteristics of the valve Dn25/50-Pn40.

Maximum distance seat–valve, [mm]	4
Adjustment pressure, [MPa]	4
Closing pressure, [MPa]	3.4
Discharge pressure, [MPa]	4.4
Pressure	PN40
Material	G20Mn5
Set pressure [MPa]	4
Spring range [MPa]	3.2–4
Maximum temperature [°C]	300
Rated coefficient of discharge	0.7
Applicable medium	Water, oil, gas etc.

2.2. Calculation Model

2.2.1. The Forces Acting on the Plate

Hydraulic force—is the main force that produces the displacement of the valve plate. This force can act both on the active face of the plate and on its dorsal surface, the analytical expression of this force is given by relation (1).

$$F_h = p \cdot A \tag{1}$$

where p—is the pressure acting on the plate; A—wetted surface area.

The weight force of the assembly of the movable part of the valve is given by the relationship:

$$F_g = mg \tag{2}$$

where m—is the total mass of the movable subassemblies in the valve composition; g—gravitational acceleration.

Spring force. During the valve opening process, the force generated by the spring increases due to its compression, given by Hooke's simple law:

$$F_{el} = -kx = -(F_0 + kx_e) = -k(x_0 + x_e) \quad (3)$$

where F_0 represents the compressive force of the spring when it is compressed by the amount x_0 and x_e—spring compression during valve operation.

Apart from these forces, the friction force, the viscous forces, as well as the damping force also act, but their effect can be neglected compared to the three forces described above [23]. In this case, the mass forces opposing the movement, given by the weight of the rod spring and its fixing elements, represent 9.49 N, so they were neglected. In addition, the rod is guided by a clearance adjustment, which is why the friction force was also neglected.

2.2.2. Establishing the Elastic Constant of the Spring

The determination of the elastic constant of the spring (k) from the analyzed valve composition was carried out both analytically and experimentally. The analytical determination of the elastic constant of the spring (stiffness coefficient) was carried out according to relation (4), used in the case of helical springs with a small pitch.

$$k = \frac{G \cdot d^4}{64 \cdot n \cdot R^3} \quad (4)$$

The terms presented in relation (4) are: G, the transverse modulus of elasticity of the spring material; d, diameter of spring wire; n, number of turns; R, mean arc radius. The geometric characteristics of the spring were determined by measurements and are presented in Table 2.

Table 2. Valve spring dimensions Dn25/50-Pn40.

Wire diameter, d [mm]	7.5
Mean radius, R [mm]	18
The number of turns, n	8.5
Pitch, p [mm]	12.5

The determination of the elastic constant of the spring was carried out on the Walter +BAI LF300 tensile testing machine, shown in Figure 4, and the characteristic diagram obtained following the compressive stress of the spring is shown in Figure 5.

The results obtained in the case of determining the elastic constant of the spring, both analytically and experimentally, are presented in Table 3. The value of the transverse modulus of elasticity was adopted as being: $G = 8.5 \cdot 10^4$ MPa.

Table 3. The values of the elastic constant of the spring, k.

Analytical determination	85.77 [N/mm]
Experimental determination	83.73 [N/mm]

It can be seen that the two results are very close, the error between these two determinations being 2.3%. To simplify the calculations, the value of the elastic constant of the spring of 84 N/mm will be adopted.

Figure 4. Determination of the elastic constant of the valve spring.

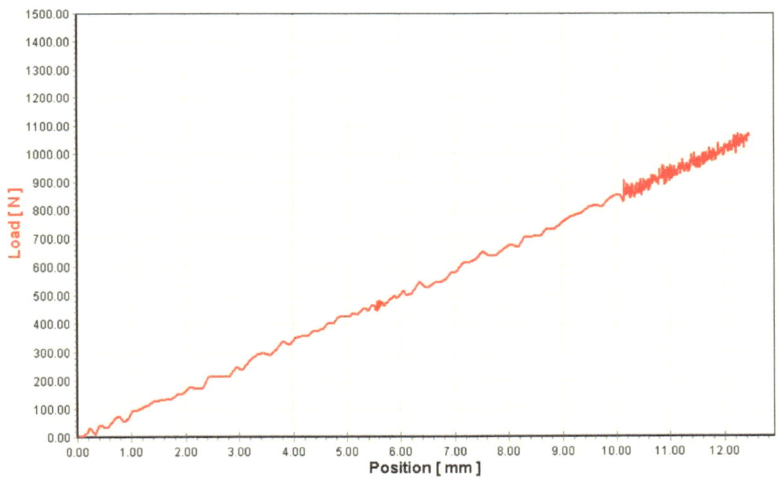

Figure 5. The elastic characteristic of the valve spring.

The CFD analysis of the relief valve aims to simulate the flow of air through the section between the seat and the valve, as well as the influence of its constructive elements on the flow. The simulation was carried out using Ansys-Fluent adopting a tetrahedral mesh obtaining approximately 47,394 nodes and 243,920 elements, as shown in Figure 6. The tetrahedral mesh was adopted after performing a sensitivity analysis making comparison between the results obtained with hexahedral mesh. It was observed that the results were very close, but the computational time was significantly reduced in case of tetrahedral mesh. The same mesh was used in similar works [1,4,23].

To study the behavior of the air flow through the discharge valve, two values of the distance (H) between the seat and the valve were adopted, adistance of 1mm and 4 mm, respectively (corresponding to the maximum opening of the valve), shown in Figure 7. The air pressure at the inlet in the valve is 4.1MPa for H = 1 mm and 4.4 MPa corresponding to the distance H = 4 mm.

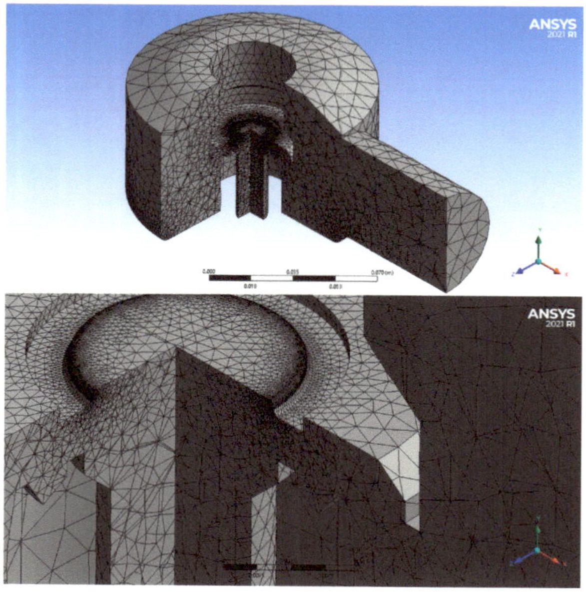

Figure 6. The mesh adopted in the CFD analysis of the valve.

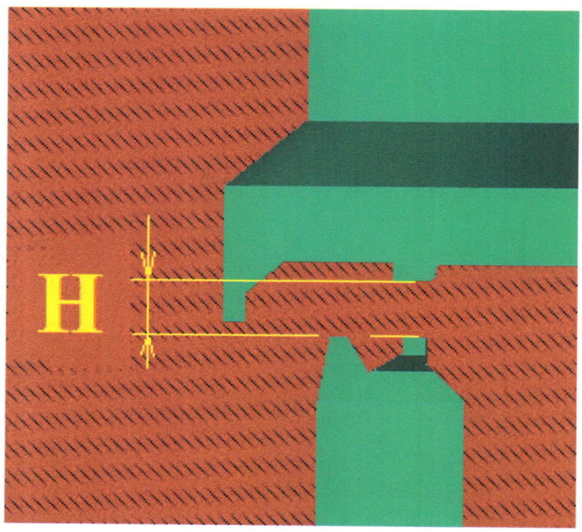

Figure 7. Distance (*H*) between seat and valve.

To analyze the influence of the adjusting nozzle on the flow through the discharge valve, three positions of it were analyzed: the first—the nozzle is at the same level as the valve seat; second—the nozzle is at the same level as the valve plate; the third—the nozzleis 1.5 mm below the level of the seat face. The latter position was adopted, taking into account certain aspects of the flow. These three situations are shown in Figure 8. To simplify the explanations, the first position of the nozzle was denoted by S0, the second by S1, and the third by S-1.

(**a**) (**b**) (**c**)

Figure 8. Position of the adjusting nozzle: position S0 (**a**), position S1 (**b**), position S-1 (**c**).

The flow model adopted in the simulation with ANYSYS Fluent, as well as the values of the adopted parameters are presented in Table 4. It was considered that the pressure at the valve inlet increases proportionally with the distance between the seat and the valve.

Table 4. The model adopted in the flow simulation.

The flow model	k-epsilon [1,8,23–26], Realizable, non-equilibrium wall function [23]
Environment: air/water	Air density: 1.204 kg/m^3 Water density: 998.2 kg/m^3 Air dynamic viscosity: 1.813·10^{-5} kg/(m·s) Water dynamic viscosity: 1.003·10^{-3} kg/(m·s)
Connection conditions	Inlet pressure: (4.1–4.4) MPa Outlet pressure: 0.1 MPa
Type of analysis	Steady, 300 iteration

Regarding the selection of turbulence models, the impact of different turbulence models on the accuracy of calculations is very important. The performed analysisinvolving the simulation of a pressure relief valve handling water and air as fluids, the choice of the k-epsilon turbulence model aligns well with common practices for such scenarios in CFD, being used in many scientific papers, such as [1,8,23–26]. Also, this model makes a balance between accuracy and computational efficiency, making it well-suited for engineering analyses [8].The choice to use non-equilibrium wall functions in the simulation, particularly in the context of analyzing air/water flow through a discharge valve, is typically motivated by the flow characteristics near the solid boundaries (such as the valve surface) and the computational efficiency considerations [23].

The concept of wall functions is commonly employed in simulations to model the near-wall behavior of the flow. Near the walls, the turbulence intensity is typically high, and resolving all the turbulent scales in this region would be computationally expensive. Wall functions allow the simulation to model the effects of turbulence close to the wall without explicitly resolving the near-wall region [27,28].The non-equilibrium wall function approach, in the context of the k-epsilon model, refers to a modification of the standard wall functions to account for non-equilibrium effects. In the standard equilibrium wall functions, it is assumed that the flow near the wall reaches a state of equilibrium, meaning that the production of turbulence (turbulent kinetic energy) is balanced by its dissipation. However, in some flows, especially those with strong adverse pressure gradients or complex geometries, the flow near the wall may deviate from equilibrium [28,29].The non-equilibrium wall function approach considers the departure from equilibrium in the near-wall region. It involves additional terms or corrections in the wall functions to account for the non-equilibrium effects on the turbulence quantities (k-epsilon) near the wall. These modifications aim to improve the accuracy of the simulation in capturing the turbulence behavior in challenging flow conditions [29–31].

The use of non-equilibrium wall functions in simulations, especially in the context of the k-epsilon turbulence model, can offer several advantages in capturing more realistic turbulent flows, particularly near walls: improved accuracy in complex flows; better representation of near-wall effects; applicability to challenging geometries; enhanced predictions of separated flows; reduced sensitivity to grid resolution; consistency with experimental data; enhanced physical realism.

The fluid flowing through the safety relief valve is assumed to be incompressible in the simulation, as in [7,23]. The statement from [32] emphasizes the practical considerations in fluid dynamics, acknowledging that, in many cases involving liquid service, incompressible flow assumptions are valid. It also acknowledges that for valves designed for gas service, there are situations where the effects of compressibility are minor and incompressible flow models may be applicable.

The convergence was monitored by examining the residual levels and it was decided to adopt 300 iterations, taking into account the computational efficiency and analyzing the residual plots shown in Figure 9, where it can be observed that theresidual showed a consistent trend of reaching a sufficiently low and steady state at 300 iterations.

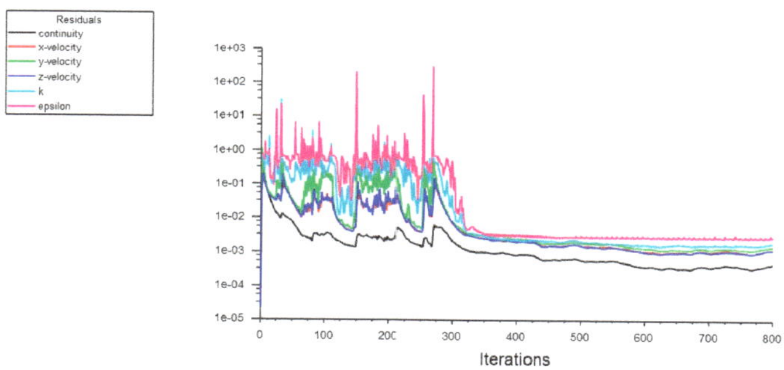

Figure 9. Residual plots.

3. Results and Discussion

3.1. CFD Analysis

The results obtained regarding the air/water speed for the three positions of the adjustment nozzle according to the distance between the seat and the valve, are presented in Table 5, and its variation graphs are described in Figures 10 and 11.

Table 5. Maximum air speed values, [m/s].

Position of Adjusting Nozzle	Seat-to-Valve Distance, *H* [mm]							
	H1		H2		H3		H4	
	Air	Water	Air	Water	Air	Water	Air	Water
S0	3080	107	3210	112	3060	108	3160	109
S1	2490	86.7	2940	104	3115	111	3020	105
S-1	2850	99.8	2900	104	3010	105	3010	106

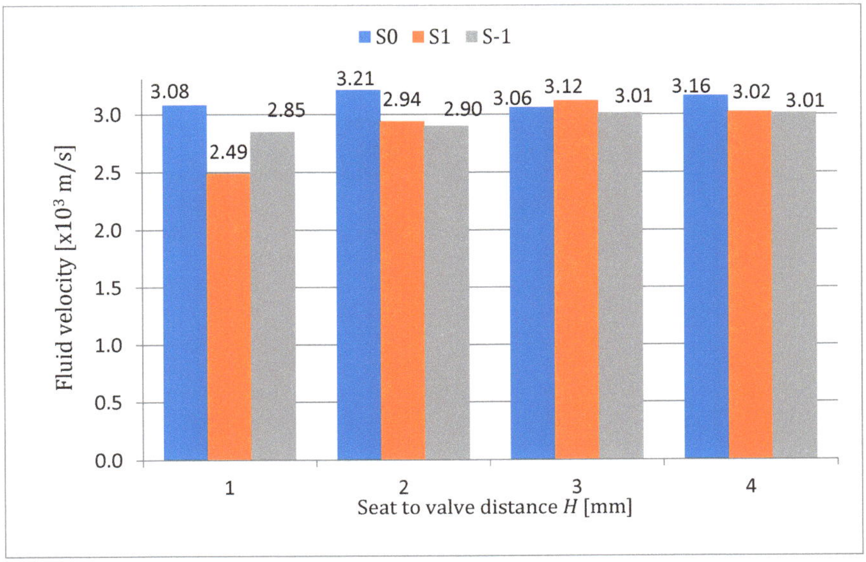

Figure 10. Air speed distribution.

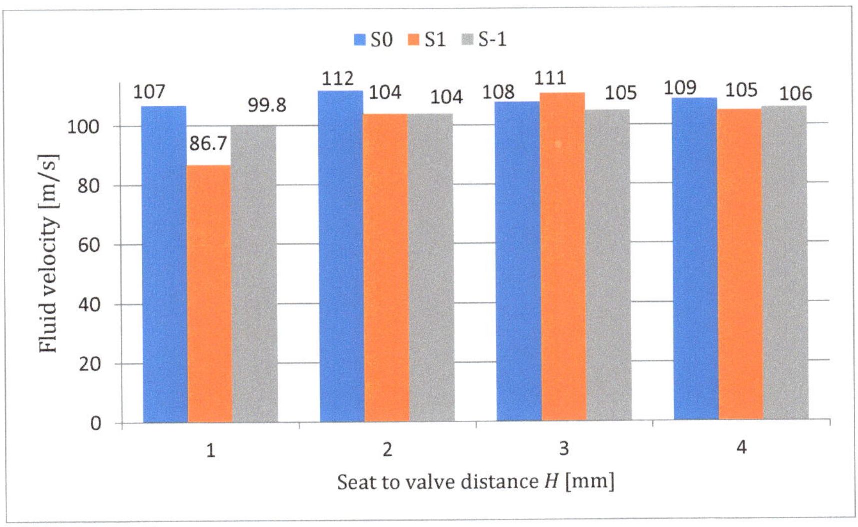

Figure 11. Water speed distribution.

The first significant observation is the impact of the nozzle position on maximum airspeed. Across all seat-to-valve distances (H), the S0 configuration consistently yielded higher airspeed values compared to S1 and S-1. This suggests that positioning the nozzle at the same level as the seat (S0) promotes higher airflow velocities.

Within each nozzle position category (S0, S1, and S-1), variations in seat-to-valve distance (H) also affect maximum airspeed values. For instance, in the S0 configuration, the highest airspeed of 3210 m/s was achieved at H2, while in the S1 configuration, the maximum airspeed of 2940 m/s occurred at H3. These findings imply that the choice of seat-to-valve distance is a critical factor in determining airflow velocity.

Notably, the S-1 configuration showed relatively consistent airspeed values across different seat-to-valve distances (H), with values ranging from 2850 m/s to 3010 m/s. This suggests that the S-1 configuration may be less sensitive to variations in seat-to-valve distance, providing stability in airflow performance.

In case of water environment, the results from Figure 11 show that the fluidvelocity is significantly influenced by the position of the adjusting nozzle. Notably, when the nozzle is in the S1 position, the fluid velocity is generally lower compared to the S0 and S-1 positions. For instance, the highest fluid velocity observed in the S0 position is 112.0 m/s, whereas the highest fluid velocity in the S1 position is 111.0 m/s. The lowest value of fluid velocity corresponds to the S1/H1 configuration. The seat-to-valve distance also plays a role in affecting the fluid velocity of water through the relief valve. In general, smaller seat-to-valve distances tend to result in lower water velocity's compared to larger distances. This trend indicates that increasing the distance between the seat and valve allows for increased water flow. Also, to obtain a higher distance between the seat and valve the overpressure was higher as specified in the standard for the safety valve analyzed.

Graphical representations of the results obtained from the CFD simulation regarding pressure, shear stresses and turbulent kinetic energy are depicted in Figures 12–17 for the plate and Figures 18–23 for the seat.

It is evident that pressure varies considerably across different configurations. In the S0 configuration, the pressure increases with seat-to-valve distance. In the S1 and S-1 configurations, pressure values generally increase until H3, after decrease. This suggests that nozzle position and seat-to-valve distance interact to influence pressure distribution within the valve.

Similar to pressure, shear stress exhibits variations across different combinations of seat-to-valve distances and nozzle positions. For all positions of adjusting nozzle, the maximum shear stress is recorded at H3 distance and is almost equal for all the analyzed configurations. For S1 and S-1 configurations, the minimum values of shear stresses correspond to minimum seat to valve distance (H1), while for S0 configuration it corresponds to the maximum seat to valve distance (H4). The variations in shear stress highlight the complex flow patterns and boundary interactions within the valve.

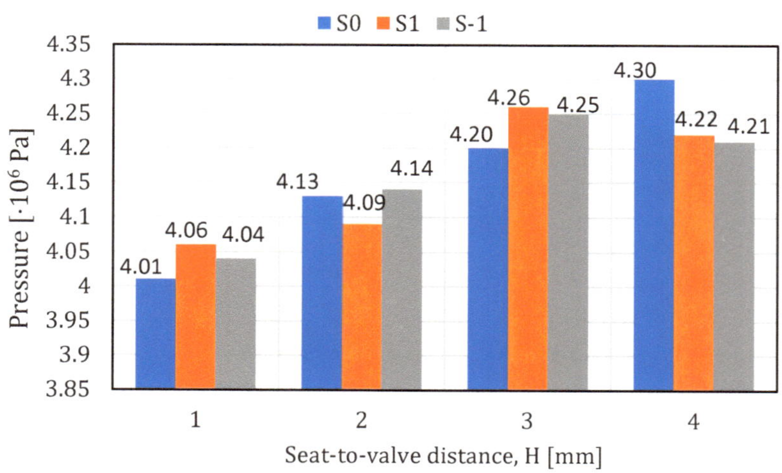

Figure 12. Air pressure variation in case of plate.

Turbulent kinetic energy varies significantly, reflecting the turbulent nature of the airflow. The highest value of turbulent kinetic energy is recorded at H2 distance for all the analyzed configurations. Some configurations, such as S-1-H2 and S-1-H4, exhibit notably higher turbulent kinetic energy compared to others. This suggests that specific nozzle-seat combinations may promote more turbulence within the valve.

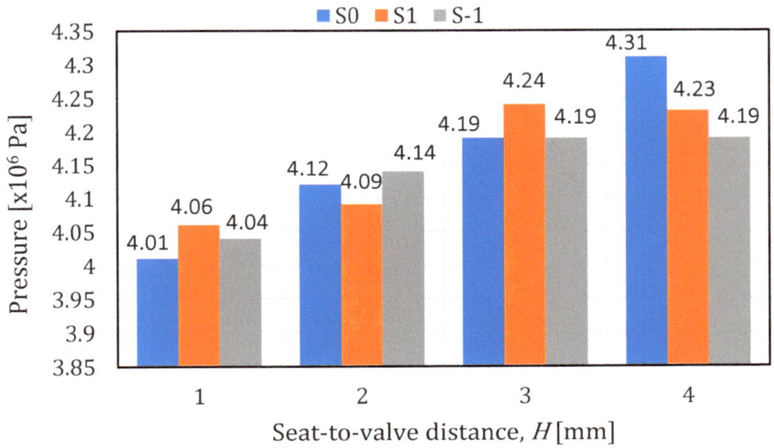

Figure 13. Water pressure variation in case of plate.

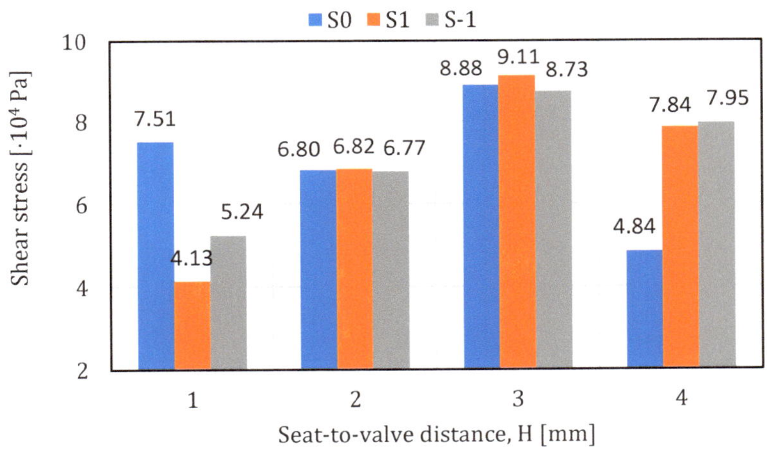

Figure 14. Shear stress variation in case of plate (air environment).

In the case of water environment, the shear stress is notably affected by nozzle position and seat-to-valve distance. Highest shear stress is observed in the S0 and S1 nozzle positions at 3 mm seat-to-valve distance. Smaller seat-to-valve distances generally result in higher shear stress values. It can be also observed that TKE varies significantly with nozzle position and seat-to-valve distance. Maximum TKE values are found in the S0 and S-1 positions at 2 mm seat-to-valve distance.

Pressure values vary considerably across different configurations. In general, the S0 and S1 configurations exhibit higher pressure values than the S-1 configuration. Additionally, within each configuration, pressure tends to decrease with an increase in seat-to-valve distance. These findings suggest that the choice of nozzle position and seat-to-valve distance significantly affects pressure distribution on the seat.

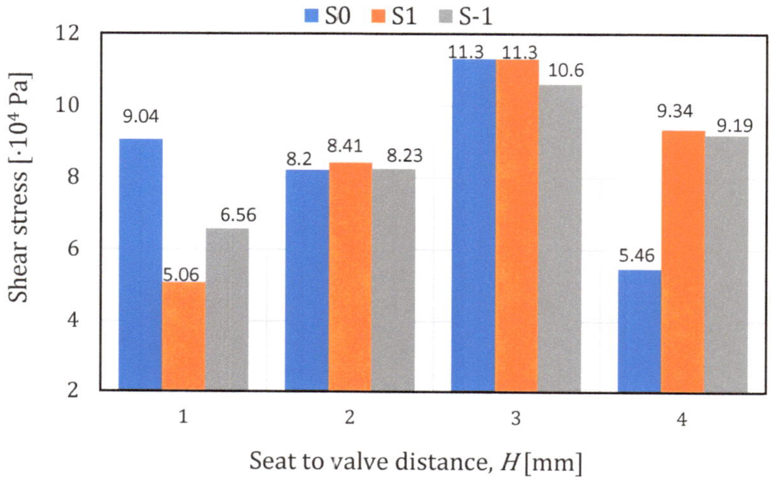

Figure 15. Shear stress variation in case of plate (water environment).

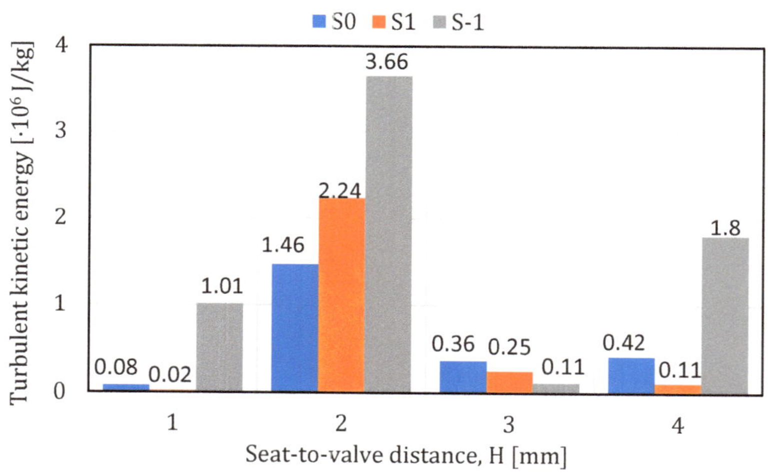

Figure 16. Turbulent energy variation in case of plate (air environment).

The shear stress values decrease with seat-to-valve distance, the highest values correspond to S1 configuration, except at H4 distance, where S0 presents the higher shear stress. These variations in shear stress highlight the complex fluid dynamics and boundary interactions occurring on the seat.

Turbulent kinetic energy varies significantly, reflecting the turbulence present in the airflow. In an air environment, generally, the turbulent kinetic energy decreases with seat-to-valve distance until H3. Some configurations, such as S0-H1 and S0-H2, show higher turbulent kinetic energy values compared to others. This suggests that specific nozzle-seat combinations can promote greater turbulence around the seat area.

For example, Figure 24a,b shows the current lines in the case of S0-H1 flow for the air and water environment, respectively. In Figures 25 and 26, the distributions of pressure, shear stress and turbulence kinetic energy for plate and seat, respectively, are presented.

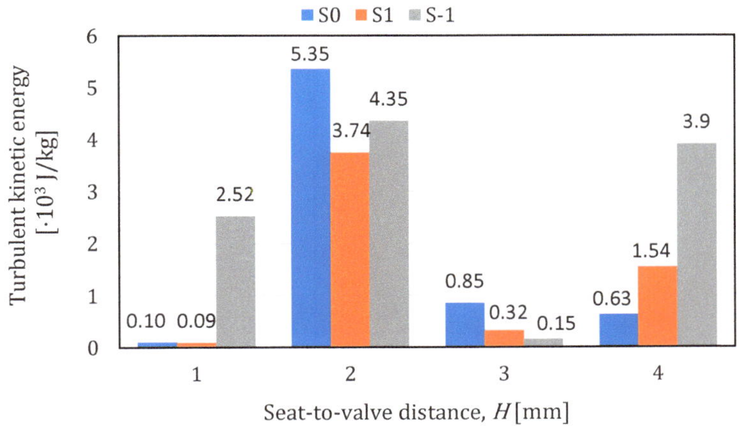

Figure 17. Turbulent energy variation in case of plate (water environment).

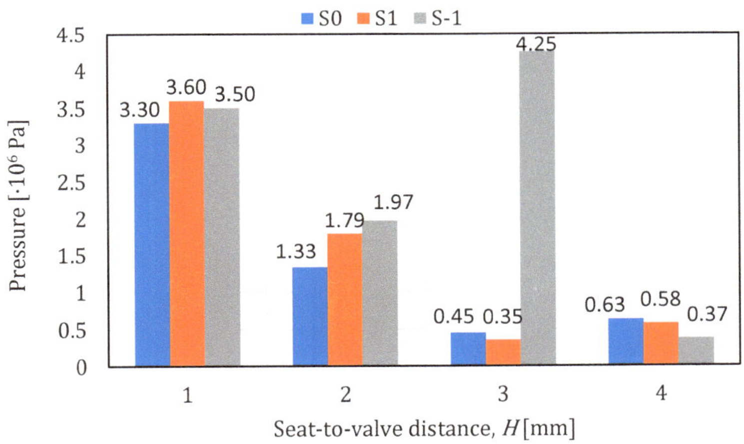

Figure 18. Air pressure variation in case of seat.

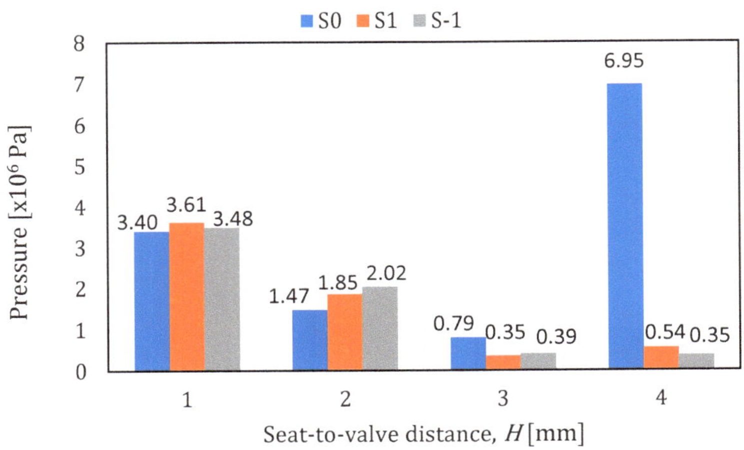

Figure 19. Water pressure variation in case of seat.

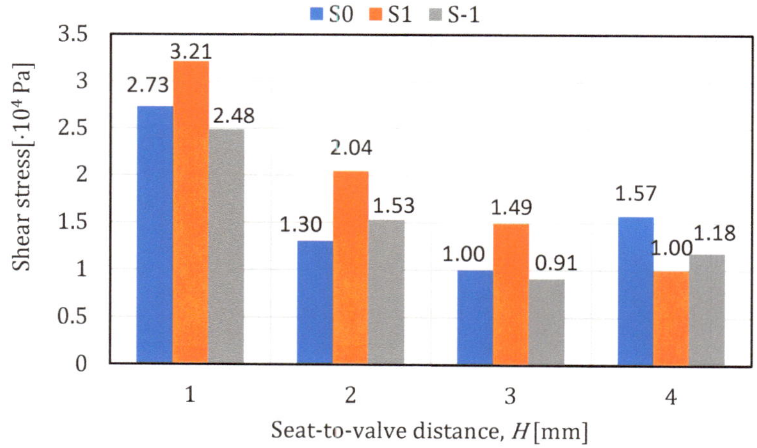

Figure 20. Shear stress variation in case of seat (air environment).

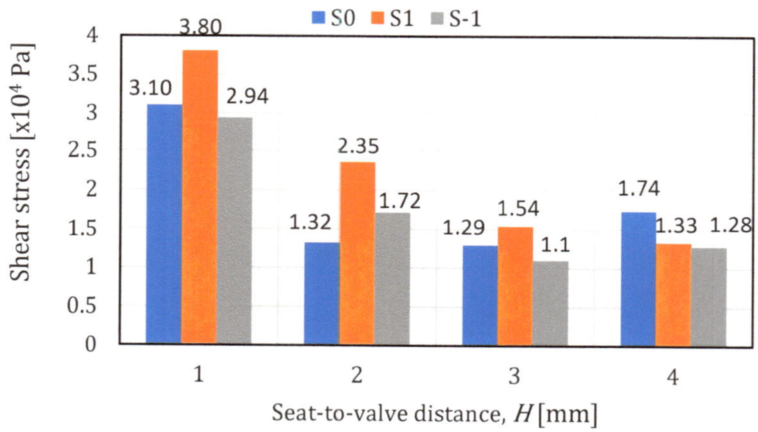

Figure 21. Shear stress variation in case of seat (water environment).

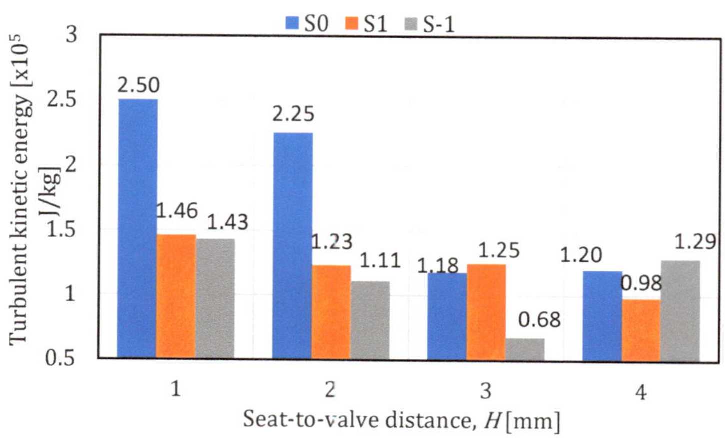

Figure 22. Turbulent energy variation in case of seat (air environment).

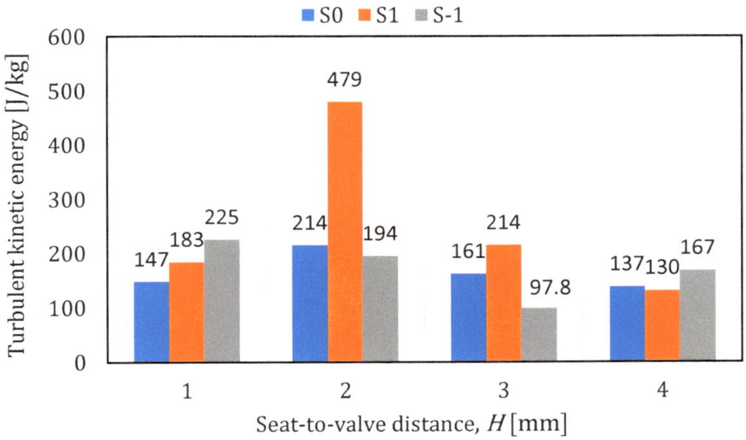

Figure 23. Turbulent energy variation in case of seat (water environment).

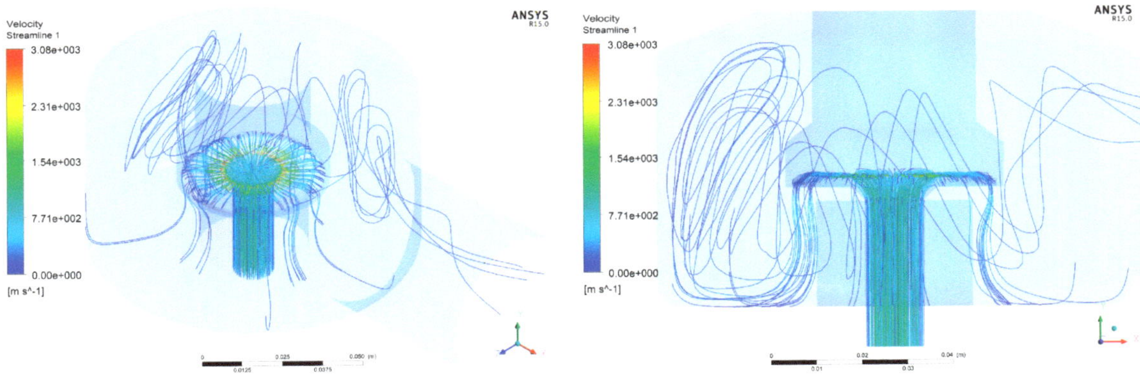

(a)

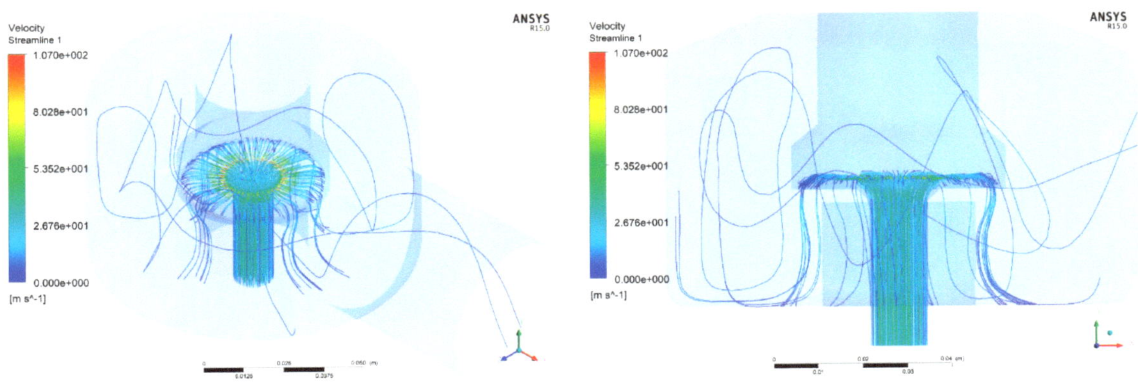

(b)

Figure 24. Airflow streamlines, case S0-H1: air (**a**); water (**b**).

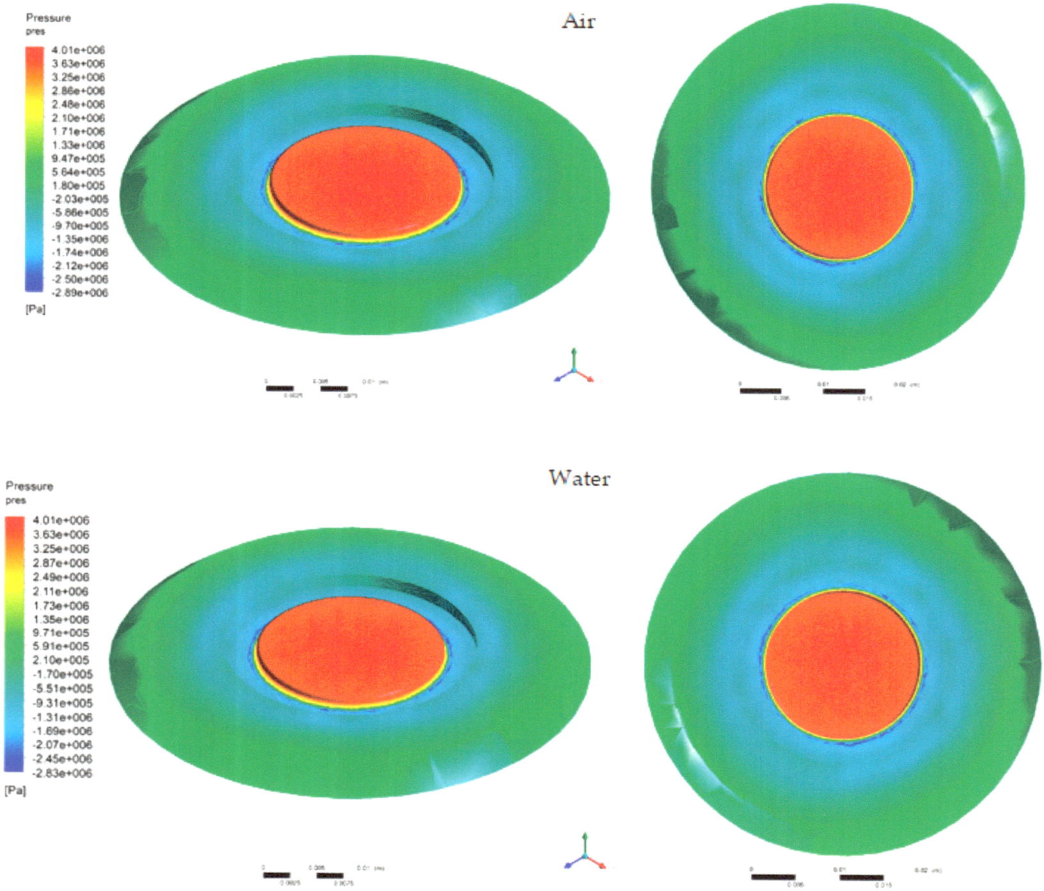

(a)

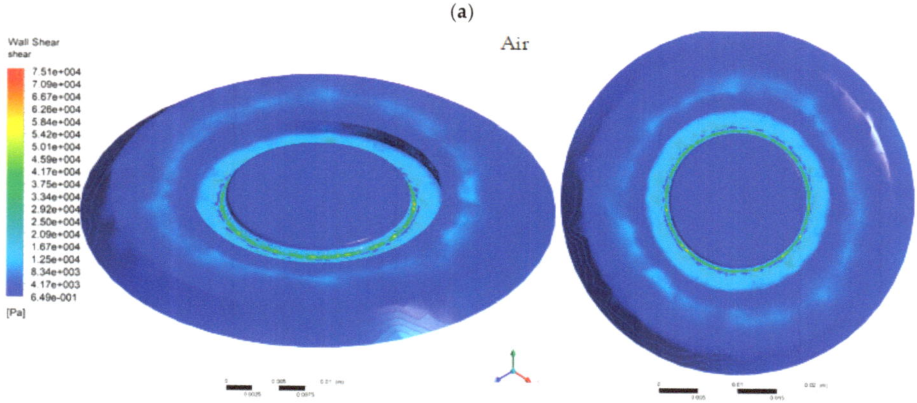

Figure 25. *Cont.*

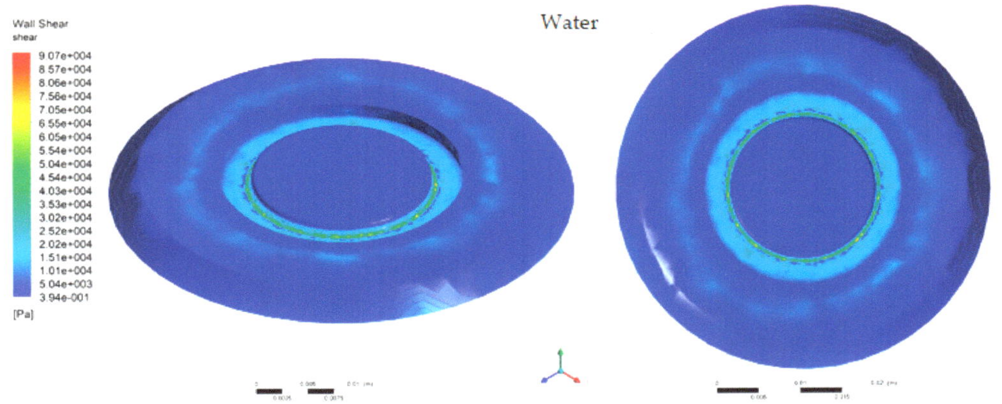

(b)

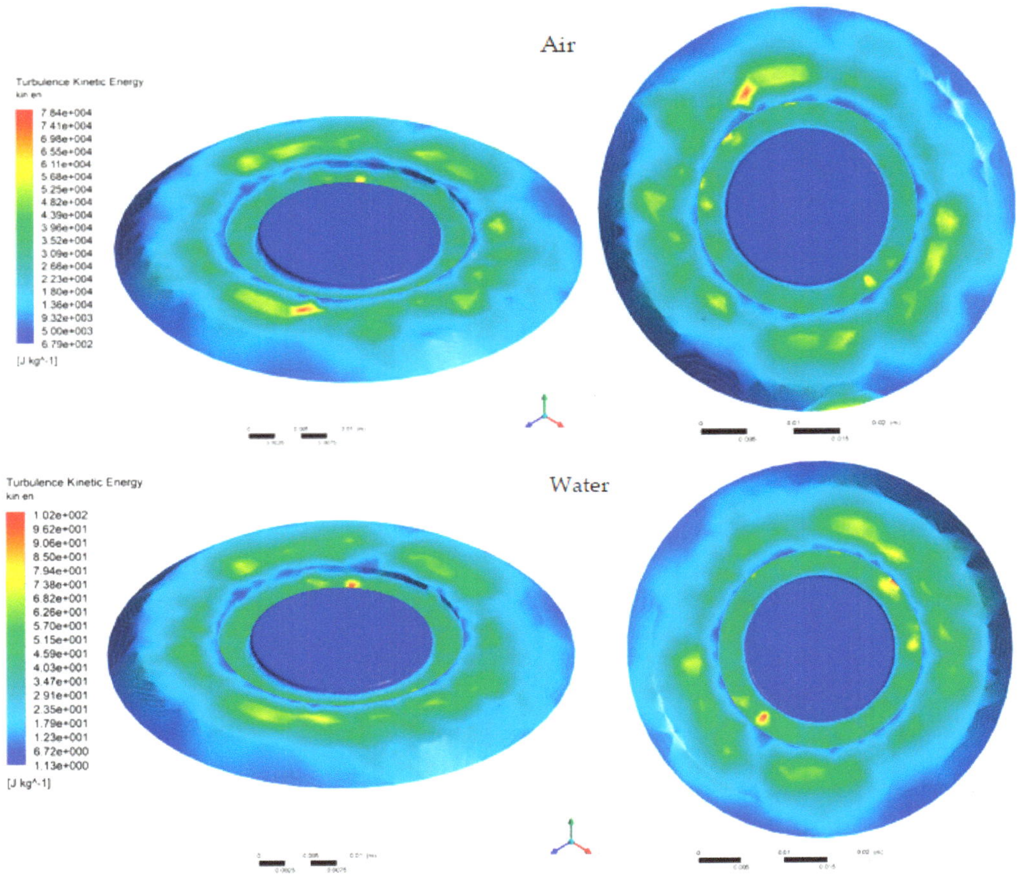

(c)

Figure 25. CFD results for plate, (case S0-H1): pressure (**a**), shear stress (**b**), turbulence kinetic energy (**c**).

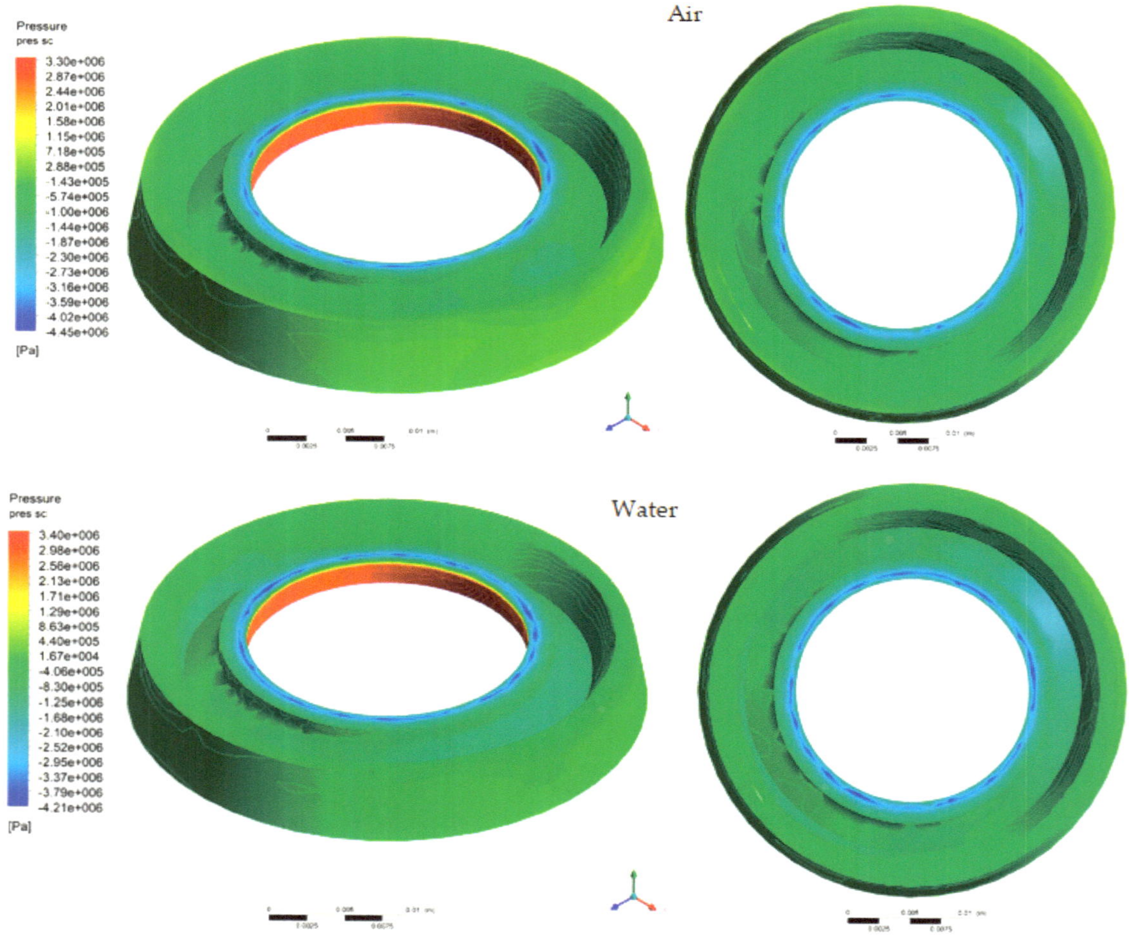

(a)

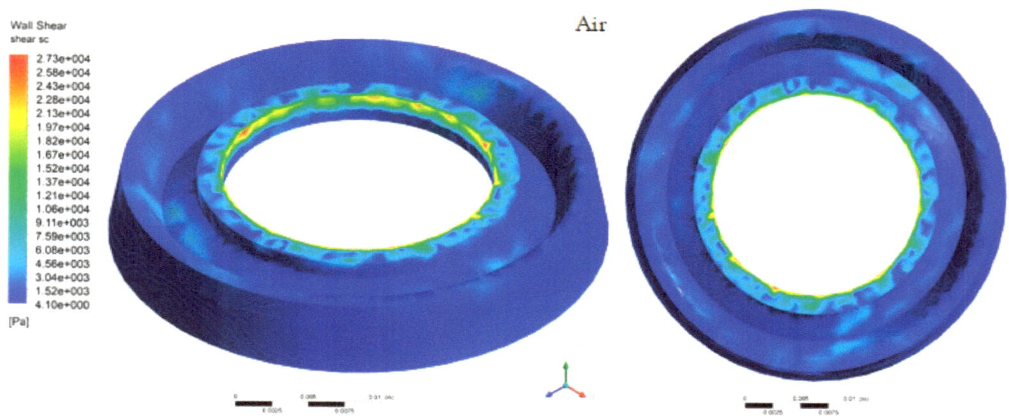

Figure 26. *Cont.*

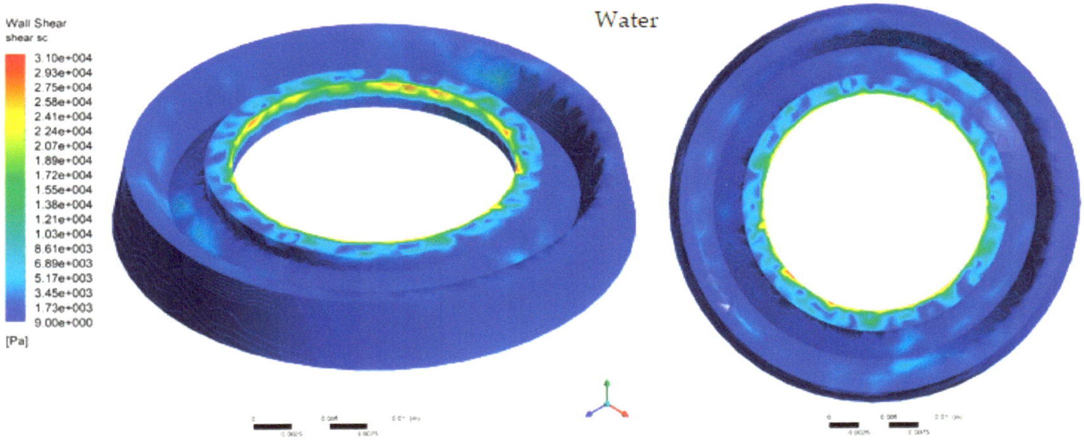

(**b**)

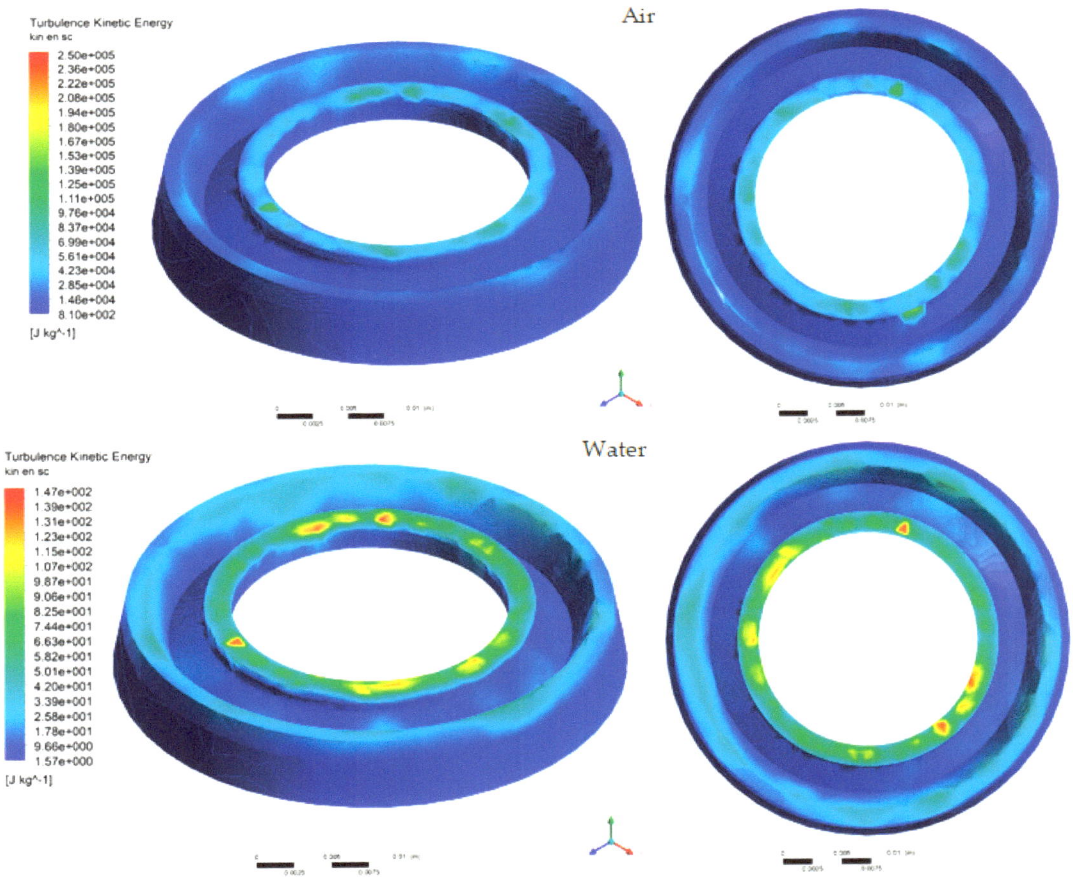

(**c**)

Figure 26. CFD results for seat, (case S0-H1): pressure (**a**), shear stress (**b**), turbulence kinetic energy (**c**).

3.2. Statistical Analysis

3.2.1. The Relative Significance of Input Parameters

The relative significance of input parameters (A—adjusting nozzle position, B—seat to valve distance), based on their impact, can be effectively illustrated using Pareto charts (Figures 27–29). The charts from Figure 27 indicate that, for fluid velocity, adjusting nozzle position emerges as the most critical factor in case of air, whereas in case of water, the seat-to-valve distance is the most significant factor. Regarding the shear stress and turbulent kinetic energy, for both fluids used in the analysis, the factor with the greatest significance is seat to valve distance, as indicated in Figures 28 and 29.

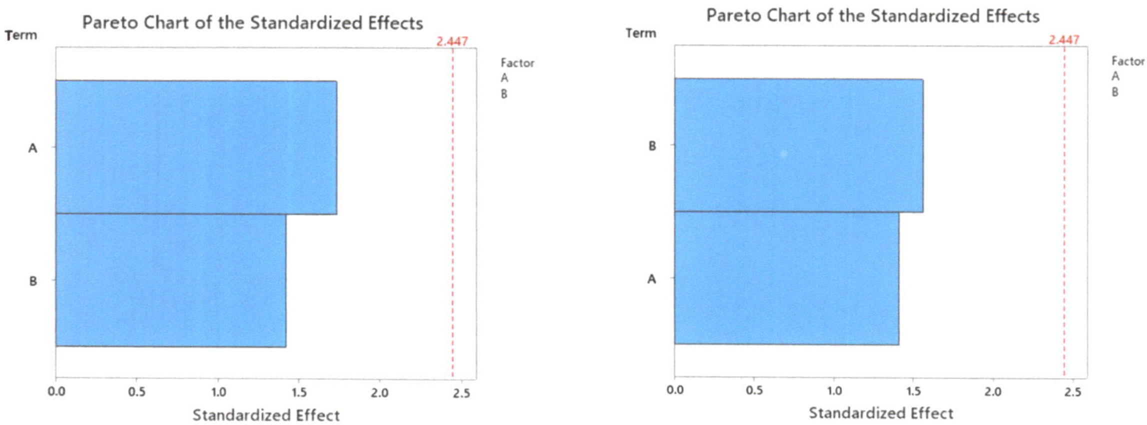

(a) (b)

Figure 27. Pareto Charts for fluid velocity: (**a**) air; (**b**) water.

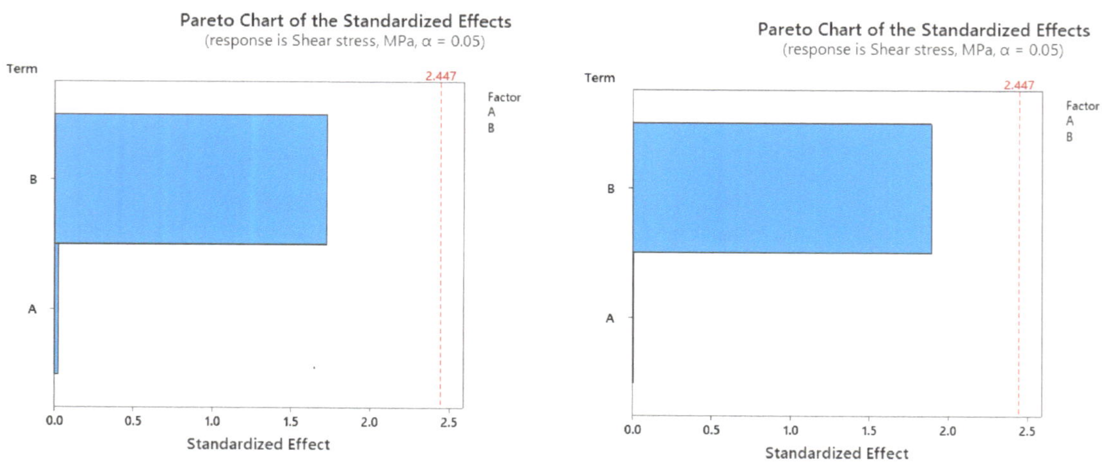

(a) (b)

Figure 28. Pareto Charts for shear stress: (**a**) air; (**b**) water.

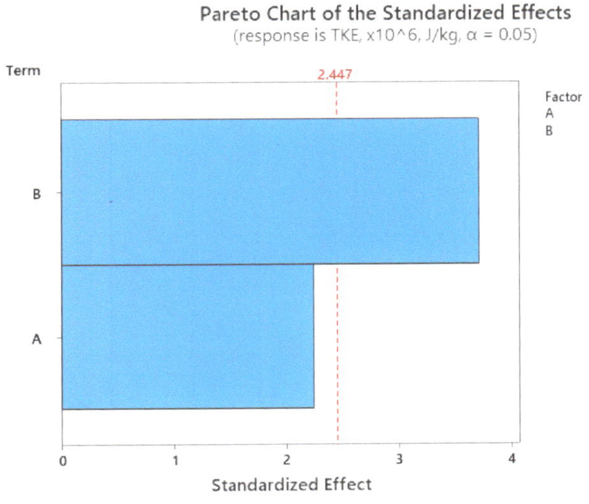

(a)

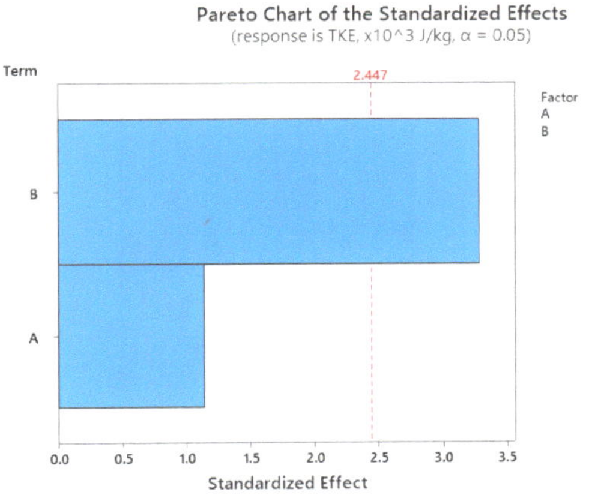

(b)

Figure 29. Pareto Charts for turbulent kinetic energy: (**a**) air; (**b**) water.

The main effect plot graphs, presented in Figures 30–32, reveal a relatively similar evolution of results for both air and water fluids, except for the fluidvelocity parameter when the maximum value corresponds to the S0-H4 configuration for air, while for water, it corresponds to the S0-H3 configuration. As in Figure 28, it can be also highlighted that, in the case of shear stress, adjusting the nozzle position has a very small influence, since the line is almost horizontal (see Figure 31), suggesting that the response mean is the same across all factor levels.

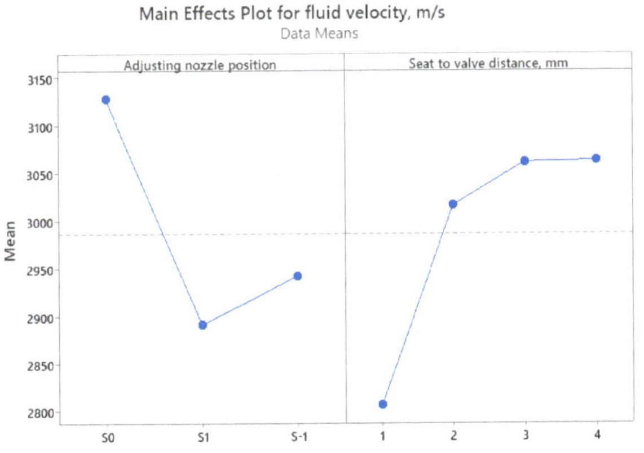

(a)

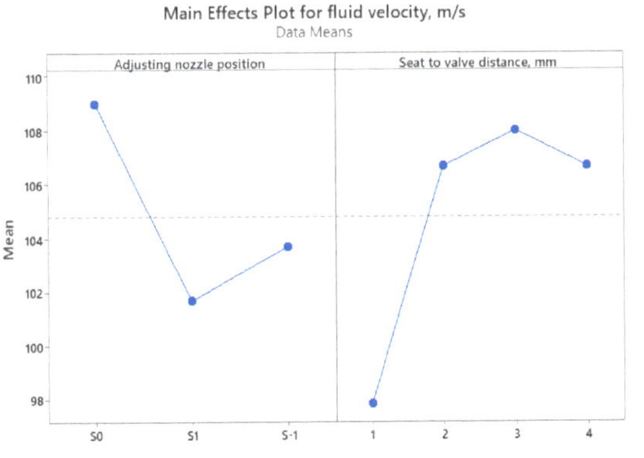

(b)

Figure 30. Main effects plot for fluid velocity: (**a**) air; (**b**) water.

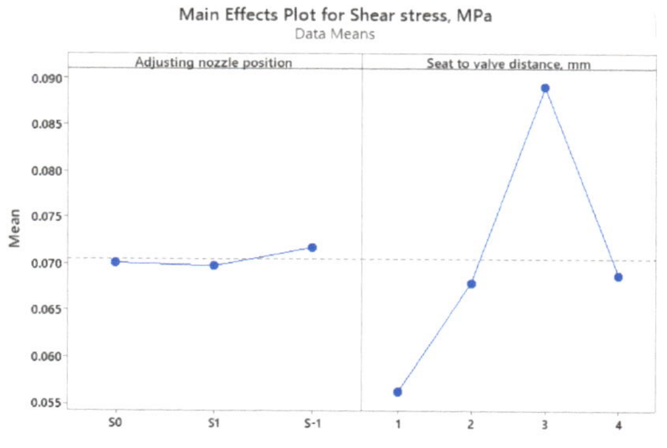

(a)

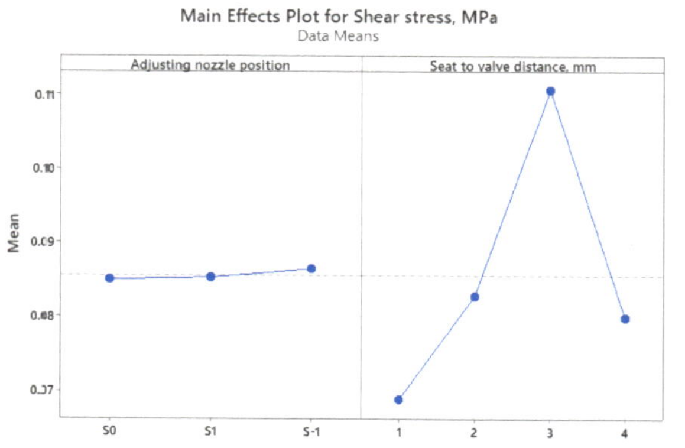

(b)

Figure 31. Main effects plot for shear stress: (**a**) air; (**b**) water.

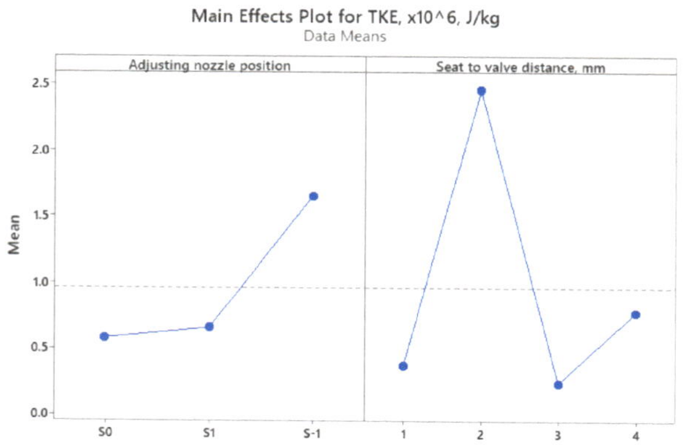

(a)

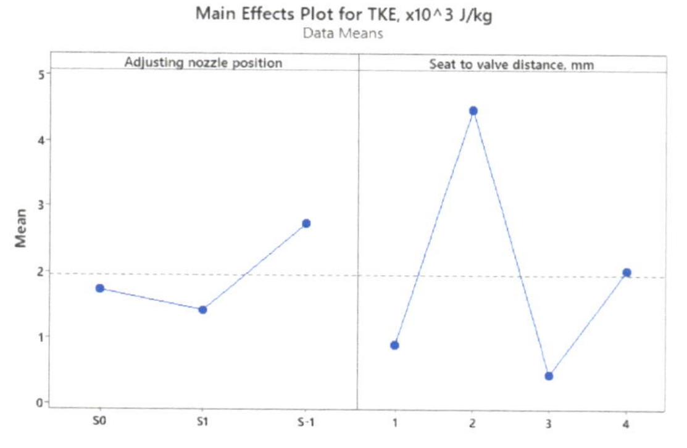

(b)

Figure 32. Main effects plot for turbulent kinetic energy: (**a**) air; (**b**) water.

3.2.2. Multi-Response Optimization Using Desirability Function

The optimization analysis was performed to establish the best combination of input parameters (adjusting nozzle position, seat to valve distance) leading to the simultaneously expected evolution of output parameters (fluid velocity, shear stress and turbulent kinetic energy). In the case of fluid velocity, the aim is to maximize it, and the shear stress expected solution is to minimize it, while for turbulent kinetic energy, two situations were investigated since TKE provides a measure of turbulence intensity. Higher TKE levels could be necessary to dissipate excess energy or prevent pressure buildup within the system. In certain applications, such as safety relief valves in high-pressure systems, larger TKE values might indicate that the valve is effectively handling the release of energy and maintaining system safety. In some cases, is preferred to minimize TKE to reduce turbulence within the valve or the downstream flow. Lower TKE values can lead to more stable and controlled flow conditions, which can be advantageous in situations where excessive turbulence could lead to issues like noise, vibration, or erosion of valve components. Smaller TKE values might be desirable when the primary objective is to achieve a more laminar and less turbulent flow.

The optimization was made utilizing the Desirability Function through Minitab 19 software. Desirability analysis assigns values on a scale from zero to one, where a score of one represents the utmost level of suitability.

For the case when the importance is the same for each response, the composite desirability D is calculated with the formula [33,34]:

$$D = (d_1 \cdot d_2 \cdot \ldots \cdot d_n)^{1/n} \quad (5)$$

where n is the number of responses, d_i represents the desirability for each individual response, calculated as [17]:

- when the goal is to maximize the response desirability:

$$\begin{aligned} d_i &= 0,\ if\ y_i < L_i \\ d_i &= \frac{(y_i - L_i)\cdot r_i}{(T_i - L_i)},\ if\ L_i \le y_i \le T_i \\ d_i &= 1,\ if\ y_i > T_i \end{aligned} \quad (6)$$

- when the goal is to minimize the response desirability:

$$\begin{aligned} d_i &= 0,\ \text{if}\ y_i > U_i \\ d_i &= \frac{(U_i - y_i)\cdot r_i}{(T_i - L_i)},\ \text{if}\ T_i \le y_i \le U_i \\ d_i &= 1,\ \text{if}\ y_i < T_i \end{aligned} \quad (7)$$

y_i, T_i, L_i, U_i represent the predicted value, target value, lowest and highest value, respectively, of the analyzed response.

Tables 6 and 7 present the rankings assigned to various configurations of input parameters for the air, respectively, and water environment. The configuration with the highest desirability score in the analysis (highlighted in Tables 6 and 7) is considered more suitable according to the criteria and objectives defined for the analysis.

The plots for optimization (Figures 33 and 34) visually depict the impact of individual factors (presented in columns) on the responses or composite desirability (displayed in rows). In these graphs, the red lines indicate the existing factor settings, and the red numbers at the top of each column represent the current levels of these factors. The horizontal blue lines, along with their associated numbers, show the responses linked to the current factor levels.

Table 6. Composite Desirability and Ranks in case of air environment.

Simulation No	Adjusting Nozzle Position	Seat to Valve Distance, mm	Greater Is Better		Smaller Is Better	
			Composite Desirability	Rank	Composite Desirability	Rank
1	S0	1	0.000	10	0.915	1
2	S0	2	0.825	2	0.832	5
3	S0	3	0.000	10	0.716	9
4	S0	4	0.681	7	0.905	2
5	S1	1	0.542	8	0.844	4
6	S1	2	0.810	3	0.790	8
7	S1	3	0.000	10	0.694	11
8	S1	4	0.682	6	0.865	3
9	S-1	1	0.772	4	0.828	7
10	S-1	2	0.839	1	0.704	10
11	S-1	3	0.425	9	0.594	12
12	S-1	4	0.769	5	0.831	6

The highlighted lines correspond to the optimal values (having Rank 1) of input parameters.

Table 7. Composite Desirability and Ranks in case of water environment.

Simulation No	Adjusting Nozzle Position	Seat to Valve Distance, mm	Greater Is Better		Smaller Is Better	
			Composite Desirability	Rank	Composite Desirability	Rank
1	S0	1	0.364	8	0.728	1
2	S0	2	0.720	1	0.462	7
3	S0	3	0.103	11	0.357	9
4	S0	4	0.550	5	0.703	2
5	S1	1	0.228	9	0.598	4
6	S1	2	0.620	3	0.443	8
7	S1	3	0.000	12	0.312	10
8	S1	4	0.455	6	0.639	3
9	S-1	1	0.436	7	0.577	5
10	S-1	2	0.701	2	0.190	12
11	S-1	3	0.158	10	0.245	11
12	S-1	4	0.582	4	0.572	6

The highlighted lines correspond to the optimal values (having Rank 1) of input parameters.

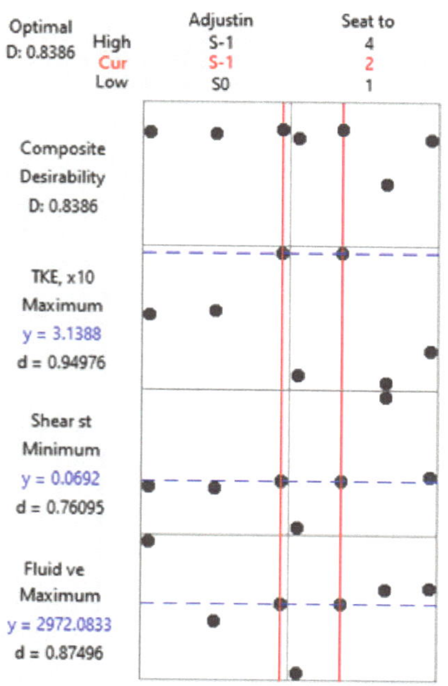

(a)

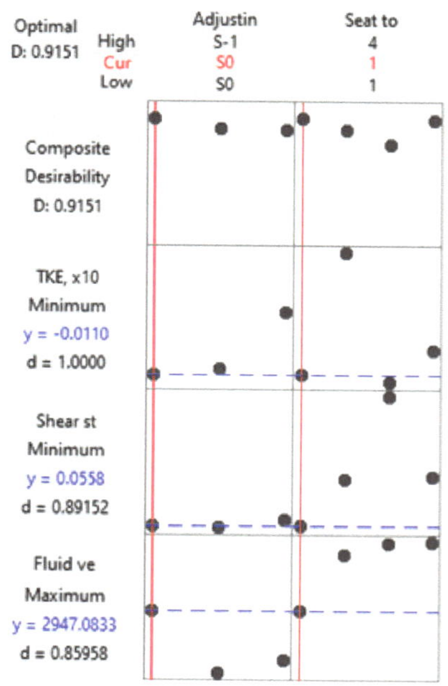

(b)

Figure 33. Multi-response optimization plots in case of air environment: (**a**) "greater is better" for TKE, (**b**) "smaller is better" for TKE.

The optimization process indicated the following combination of input parameters to achieve the expected outcomes:

- √ for the air environment: S-1/H2 for "greater is better" option of turbulent kinetic energy and S0/H1 for "smaller is better" option of turbulent kinetic energy;

√ for the water environment: S0/H2 for "greater is better" option of turbulent kinetic energy and S0/H1 for "smaller is better" option of turbulent kinetic energy.

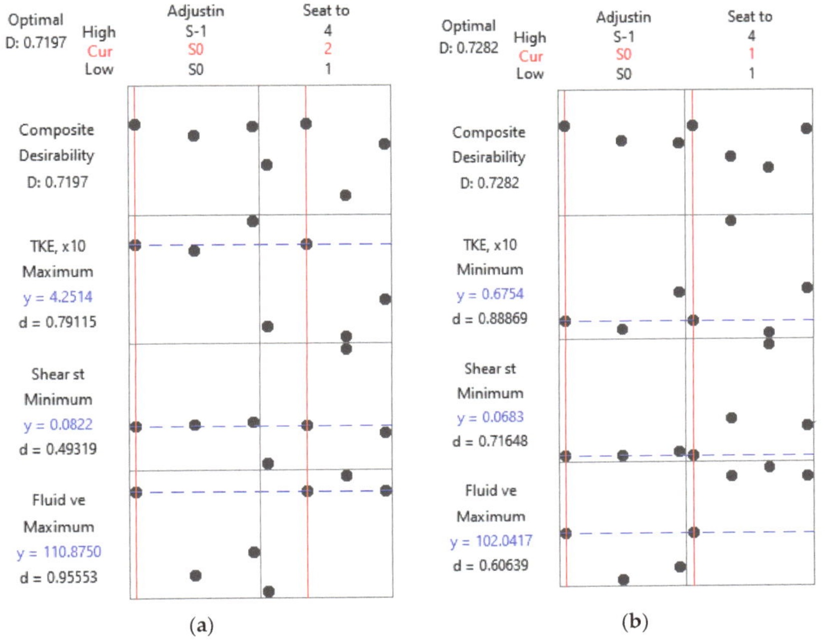

Figure 34. Multi-responseoptimizationplots in case of water environment: (**a**) "greater is better" for TKE, (**b**) "smaller is better" for TKE.

3.3. Synthesis of Results

√ In most of the analyzed cases, the maximum airspeed was obtained in the space between the seat and the valve, this having the value of 3210 m/s for the S0-H2 case. For cases where the adjusting nozzle is at or below the seat, the variation in air velocity is relatively small. The calculated velocity difference between the extreme values is 150m/s for position S0 and approximately 160m/s for S-1.

√ For nozzle position S1, the air velocity varies significantly compared to the other two, with the minimum value being 2490 m/s for H1. At high fluidvelocity, the vortices created inside the valve body are much more pronounced, but the streamlines respect its geometric configuration.

√ In all analyzed cases, the extreme pressure values were obtained on the circular surface in the center of the plate. The highest value of the pressure is 4.3 MPa (air) and 4.31 MPa (water), being evenly distributed on the plate surface. In the case of the seat, the pressure drops with valve lift for all three positions of the adjusting nozzle except S-1-H3 (in air environment) and S0-H4 (in water environment).

√ The maximum values of the tangential stresses on the plate appear in most cases on the surface of the 0.4mm chamfer, and in the case of the seat, they vary randomly on its surface. The maximum value of the tangential stresses on the plate is 0.0911 MPa (air) and 0.113 MPa (water), and in the case of the seat, the maximum values were obtained at H1, having a decreasing variation depending on the seat–valve distance.

√ In air environment, the turbulent kinetic energy has a random distribution over the surface of both the plate and the seat. In the case of the plate, the maximum value is 3660kJ/kg and was obtained in the S-1-H2 configuration. In the case of the seat, the turbulent kinetic energy has a relatively uniform variation for positions S1 and S-1, the highest value was obtained in the case of S0-H1, being 78.4kJ/kg.

- √ In most of the analyzed cases, the maximum water speed was obtained in the space between the seat and the valve, this having the value of 112 m/s for the S0-H2 case. For cases where the adjusting nozzle is at or below the seat, the variation in water velocity is relatively small. The calculated difference between the extreme values is 5 m/s for the nozzle position S0 and about 6 m/s for S-1.
- √ In the case of nozzle position S1, the watervelocity varies significantly compared to the other two, with the minimum value being 87 m/s for H1. At high fluid velocity, the turbulence created inside the valve body is much more pronounced, but the streamlines follow its configuration.
- √ In case of water, the turbulent kinetic energy has a random distribution over the surface of both the plate and the seat. In the case of the plate, the maximum value is 5350 J/kg and was obtained in the S0-H2 situation. In the case of the seat, the turbulent kinetic energy has a relatively uniform variation for positions S0 and S-1, the highest value was obtained in the case of S1-H2 and is 479 J/kg.

4. Conclusions

In this paper, the analysis of the air/water flow through the Dn25/50-Pn40 discharge valve was carried out, using CFD simulation and considering four values of the seat–valve distance and three positions of the adjustment nozzle. Standards and norms, used by the manufacturers and operators of safety valves, do not specify the position of adjusting nozzle. Also, there are no studies regarding the correct position of the adjusting nozzle. This work establishes the proper position of the adjusting nozzle considering the main safety valve function—to avoid overpressure in the working system as quick is possible—with a maximum fluid velocity, and with a minimum turbulent kinetic energy. The study, applied to a Dn25/50-Pn40 safety valve, shows the importance of safety valve construction, working fluid and pressure for correct positioning of the adjusting nozzle.

The study utilized Computational Fluid Dynamics (CFD) simulations to analyze the air/water flow through the Dn25/50-Pn40 discharge valve. This suggests a reliance on numerical methods for understanding the fluid dynamics involved.

The analysis took into account four values of the seat–valve distance and three positions of the adjustment nozzle. This indicates a comprehensive investigation, exploring the impact of different configurations on the valve's performance.

The focus on air/water flow implies a consideration of two-phase flow dynamics. Understanding how the valve behaves under different conditions with both air and water has implications for its applicability in diverse operational scenarios.

The variations in seat–valve distance and adjustment nozzle positions suggest a sensitivity analysis to evaluate how changes in these parameters affect the overall performance of the discharge valve. This information is crucial for optimizing the valve's functionality.

The use of a Pn40 pressure class indicates that the valve is designed to handle relatively high-pressure conditions. This choice of pressure class is likely relevant to the specific applications for which the valve is intended. Analyzing the pressure distribution from Figures 25a and 26a, some negative pressure contours can be observed, due the phenomenon of cavitation in the case of liquids, respectively, of compressibility; in the case of gases, phenomena do not interfere with the purpose of the article, namely to determine the optimal position of the adjusting nozzle depending on the value of the fluid velocity and turbulent kinetic energy.

The paper likely provides insights into the engineering aspects of discharge valves, offering valuable information for designers and engineers involved in fluid system design.

The performed statistical analysis showed that when dealing with air, the adjusting nozzle position stands out as the most critical factor regarding fluidvelocity, while for water, the seat–valve distance takes precedence. Moving on to the assessment of shear stress and turbulent kinetic energy, the analysis consistently highlights the seat–valve distance as the factor of greatest significance, for both air and water. These conclusions emphasize the essential role of specific variables in influencing fluid behavior and underline the importance of optimizing these factors to achieve desired outcomes in fluid flow and related parameters.

The optimization process has successfully identified the optimal combination of input parameters necessary to attain the desired outcomes in both air and water environments. For the air environment, the "greater is better" option for turbulent kinetic energy was achieved by setting S-1/H2, while the "smaller is better" option for turbulent kinetic energy was attained with the setting S0/H1. Similarly, in the water environment, the optimization process determined that the combination of S0/H2 results in "greater is better" turbulent kinetic energy, while S0/H1 is the ideal setting for "smaller is better" turbulent kinetic energy. These findings provide a clear and actionable roadmap for achieving the desired results in both environments, underscoring the importance of tailored parameter adjustments for specific conditions.

Therefore, this work provides valuable insights into the airflow dynamics within the examined valve, offering a comprehensive understanding of how different operational conditions influence its performance. The novelty of the performed research lies in the detailed examination of seat–valve distances and adjustment nozzle positions, revealing the complex interaction of variables within the valve, which has practical implications for valve optimization and design in various applications.

Author Contributions: Conceptualization, R.G.R. and A.D.; methodology, R.G.R., A.D. and P.C.; validation, R.G.R.; formal analysis, R.G.R.; investigation, R.G.R., A.D., P.C., I.P. and M.T.; resources, P.C.; writing—original draft preparation, I.P. and M.T.; writing—review and editing, M.T.; visualization R.G.R. and A.D.; supervision, R.G.R. All authors have read and agreed to the published version of the manuscript.

Funding: This research received no external funding.

Data Availability Statement: Data are contained within the article.

Conflicts of Interest: The authors declare no conflict of interest.

References

1. Scuro, N.L.; Angelo, E.; Angelo, G.; Andrade, D.A. A CFD Analysis of the Flow Dynamics of a Directly-Operated Safety Relief Valve. *Nucl. Eng. Des.* **2018**, *328*, 321–332. [CrossRef]
2. Dasgupta, K.; Karmakar, R. Dynamic Analysis of Pilot Operated Pressure Relief Valve. *Simul. Model. Pract. Theory* **2002**, *10*, 35–49. [CrossRef]
3. Prescott, S.L.; Ulanicki, B. Dynamic Modeling of Pressure Reducing Valves. *J. Hydraul. Eng.* **2003**, *129*, 804–812. [CrossRef]
4. Szpica, D.; Mieczkowski, G.; Borawski, A.; Leisis, V.; Diliunas, S.; Pilkaite, T. The Computational Fluid Dynamics (CFD) Analysis of the Pressure Sensor Used in Pulse-Operated Low-Pressure Gas-Phase Solenoid Valve Measurements. *Sensors* **2021**, *21*, 8287. [CrossRef]
5. Lisowski, E.; Filo, G.; Rajda, J. Analysis of Energy Loss on a Tunable Check Valve through the Numerical Simulation. *Energies* **2022**, *15*, 5740. [CrossRef]
6. Sun, Y.; Wu, J.; Xu, J.; Bai, X. Flow Characteristics Study of High-Parameter Multi-Stage Sleeve Control Valve. *Processes* **2022**, *10*, 1504. [CrossRef]
7. Song, X.; Cui, L.; Cao, M.; Cao, W.; Park, Y.; Dempster, W.M. A CFD Analysis of the Dynamics of a Direct-Operated Safety Relief Valve Mounted on a Pressure Vessel. *Energy Convers. Manag.* **2014**, *81*, 407–419. [CrossRef]
8. Alkhulaifi, K.; Alharbi, A.; Alardhi, M.; Alrajhi, J.; Almutairi, H.H. Comparative Analysis of the Performance Characteristics of Butterfly and Pinch Valves. *Processes* **2023**, *11*, 1897. [CrossRef]
9. Yang, L.; Wang, Z.; Dempster, W.; Yu, X.; Tu, S.-T. Experiments and Transient Simulation on Spring-Loaded Pressure Relief Valve under High Temperature and High Pressure Steam Conditions. *J. Loss Prev. Process Ind.* **2017**, *45*, 133–146. [CrossRef]
10. Makaryants, G.M. Fatigue Failure Mechanisms of a Pressure Relief Valve. *J. Loss Prev. Process Ind.* **2017**, *48*, 1–13. [CrossRef]
11. Yuan, G.; Wang, Y.; Fang, Y.; Ma, R.; Ning, K.; Tang, Y. High-Temperature and Pressure Downhole Safety Valve Performance Envelope Curve Study. *Processes* **2023**, *11*, 2525. [CrossRef]
12. Song, X.-G.; Park, Y.-C.; Park, J.-H. Blowdown Prediction of a Conventional Pressure Relief Valve with a Simplified Dynamic Model. *Math. Comput. Model.* **2013**, *57*, 279–288. [CrossRef]
13. Erdődi, I.; Hős, C. Prediction of Quarter-Wave Instability in Direct Spring Operated Pressure Relief Valves with Upstream Piping by Means of CFD and Reduced Order Modelling. *J. Fluids Struct.* **2017**, *73*, 37–52. [CrossRef]

14. Desai, S.; Desai, A.; Karande, V. Design and Weight Optimization of Buffer Relief Valve Using FEA with Experimental Validation. *Mater. Today Proc.* **2020**, *27*, 1466–1472. [CrossRef]
15. Hős, C.J.; Champneys, A.R.; Paul, K.; McNeely, M. Dynamic Behavior of Direct Spring Loaded Pressure Relief Valves in Gas Service: Model Development, Measurements and Instability Mechanisms. *J. Loss Prev. Process Ind.* **2014**, *31*, 70–81. [CrossRef]
16. Bazsó, C.; Hős, C.J. An Experimental Study on the Stability of a Direct Spring Loaded Poppet Relief Valve. *J. Fluids Struct.* **2013**, *42*, 456–465. [CrossRef]
17. Darby, R. The Dynamic Response of Pressure Relief Valves in Vapor or Gas Service, Part I: Mathematical Model. *J. Loss Prev. Process Ind.* **2013**, *26*, 1262–1268. [CrossRef]
18. Aldeeb, A.A.; Darby, R.; Arndt, S. The Dynamic Response of Pressure Relief Valves in Vapor or Gas Service. Part II: Experimental Investigation. *J. Loss Prev. Process Ind.* **2014**, *31*, 127–132. [CrossRef]
19. Darby, R.; Aldeeb, A.A. The Dynamic Response of Pressure Relief Valves in Vapor or Gas Service. Part III: Model Validation. *J. Loss Prev. Process Ind.* **2014**, *31*, 133–141. [CrossRef]
20. Karpenko, M.; Stosiak, M.; Šukevičius, Š.; Skačkauskas, P.; Urbanowicz, K.; Deptuła, A. Hydrodynamic Processes in Angular Fitting Connections of a Transport Machine's Hydraulic Drive. *Machines* **2023**, *11*, 355. [CrossRef]
21. Stosiak, M.; Karpenko, M.; Deptuła, A.; Urbanowicz, K.; Skačkauskas, P.; Deptuła, A.M.; Danilevičius, A.; Šukevičius, Š.; Łapka, M. Research of Vibration Effects on a Hydraulic Valve in the Pressure Pulsation Spectrum Analysis. *JMSE* **2023**, *11*, 301. [CrossRef]
22. Stosiak, M.; Karpenko, M.; Prentkovskis, O.; Deptuła, A.; Skačkauskas, P. Research of Vibrations Effect on Hydraulic Valves in Military Vehicles. *Def. Technol.* **2023**, in press. [CrossRef]
23. Budziszewski, A.; Thorén, L. CFD Simulation of a Safety Relief Valve for Improvement of a One-Dimensional Valve Model in RELAP5. Master's Thesis, Chalmers University of Technology, Gothenburg, Sweden, 2012.
24. Singh, R.; Ahmed, R.; Karami, H.; Nasser, M.; Hussein, I. CFD Analysis of Turbulent Flow of Power-Law Fluid in a Partially Blocked Eccentric Annulus. *Energies* **2021**, *14*, 731. [CrossRef]
25. Leutwyler, Z.; Dalton, C. A CFD Study of the Flow Field, Resultant Force, and Aerodynamic Torque on a Symmetric Disk Butterfly Valve in a Compressible Fluid. *J. Press. Vessel. Technol.* **2008**, *130*, 021302. [CrossRef]
26. Leutwyler, Z.; Dalton, C. A Computational Study of Torque and Forces Due to Compressible Flow on a Butterfly Valve Disk in Mid-Stroke Position. *J. Fluids Eng.* **2006**, *128*, 1074–1082. [CrossRef]
27. Nazif, H.R.; Basirat Tabrizi, H. Applying a Non-Equilibrium Wall Function in k–ε Turbulent Modelling of Hydrodynamic Circulating Flow. *Appl. Math. Model.* **2014**, *38*, 588–598. [CrossRef]
28. Parente, A.; Gorlé, C.; van Beeck, J.; Benocci, C. Improved k–ε Model and Wall Function Formulation for the RANS Simulation of ABL Flows. *J. Wind. Eng. Ind. Aerodyn.* **2011**, *99*, 267–278. [CrossRef]
29. Nazif, H.R.; Basirat Tabrizi, H.; Farhadpour, F.A. Comparative Analysis of the Boundary Transfer Method with Other Near-Wall Treatments Based on the—Turbulence Model. *Eur. J. Mech.—B/Fluids* **2014**, *44*, 22–31. [CrossRef]
30. Nazif, H.R.; Tabrizi, H.B. Comparison of Standard Turbulent Wall Function with a Non-Equilibrium Wall Model. *Inter. J. Fluid Mech. Res.* **2011**, *38*, 499–508. [CrossRef]
31. El Gharbi, N.; Absi, R. Effect of Different Near-Wall Treatments on Indoor Airflow Simulations. *JAFM* **2012**, *5*, 63–70. [CrossRef]
32. Francis, J.; Betts, P.L. Modelling Incompressible Flow in a Pressure Relief Valve. *Proc. Inst. Mech. Eng. Part E J. Process Mech. Eng.* **1997**, *211*, 83–93. [CrossRef]
33. D'Addona, D.M.; Raykar, S.J.; Singh, D.; Kramar, D. Multi Objective Optimization of Fused Deposition Modeling Process Parameters with Desirability Function. *Procedia CIRP* **2021**, *99*, 707–710. [CrossRef]
34. Zisopol, D.G.; Tănase, M.; Portoacă, A.I. Innovative Strategies for Technical-Economical Optimization of FDM Production. *Polymers* **2023**, *15*, 3787. [CrossRef] [PubMed]

processes

MDPI

Article

Influence of Blade Trailing-Edge Filing on the Transient Characteristics of the Centrifugal Pump during Startup

Hongchang Ding *, Fei Ge, Kai Wang and Fanyun Lin

College of Mechanical and Electronic Engineering, Shandong University of Science and Technology, Qingdao 266590, China

* Correspondence: dhchang@sdust.edu.cn; Tel.: +86-17852728669

Abstract: During the startup process of a centrifugal pump, the vibration and noise problems caused by unsteady flow are the focus of attention, and pressure pulsation is one of the main reasons for this problem. In the current research, a special impeller with blade pressure side trimming was proposed to reduce the strong pressure pulsation phenomenon during the startup process of centrifugal pumps. This article uses numerical simulation methods to simulate three typical blade trailing edges: original trailing edge (OTE), pressure side long linear (LLPS), and pressure side short linear (SLPS), and verifies them with experimental results. The results indicate that although the head of the centrifugal pump after filing has been reduced, its efficiency has been improved to a certain extent. Thirteen monitoring points were set up near the impeller outlet circumference and volute tongue to analyze the changes in pressure pulsation, verifying that blade trimming has a significant inhibitory effect on pressure pulsation during the startup of centrifugal pumps. The average maximum pressure pulsation amplitude of all monitoring points decreased by 32.23%, and the maximum pressure pulsation amplitude decreased by 56%. Blade trimming can affect the internal flow field distribution of centrifugal pumps. By analyzing the static pressure distribution, velocity streamline distribution, and vorticity distribution at the middle interface of three different impellers during startup, it was verified that there is a close relationship between pressure pulsation and unsteady flow structure during startup. The final conclusion is that blade trimming has a significant inhibitory effect on the pressure pulsation of the centrifugal pump during startup, and the impeller outlet vorticity is significantly reduced. The scheme proposed in this study has far-reaching prospects in the design of low noise centrifugal pumps.

Keywords: centrifugal pump; startup; blade trailing-edge filing; transient characteristics

Citation: Ding, H.; Ge, F.; Wang, K.; Lin, F. Influence of Blade Trailing-Edge Filing on the Transient Characteristics of the Centrifugal Pump during Startup. *Processes* **2023**, *11*, 2420. https://doi.org/10.3390/pr11082420

Academic Editors: Ambra Giovannelli and Jie Zhang

Received: 13 June 2023
Revised: 4 August 2023
Accepted: 8 August 2023
Published: 11 August 2023

1. Introduction

As a general fluid machinery, the centrifugal pump has been widely used in industrial and agricultural production [1]. Under normal working conditions, the centrifugal pump always maintains a stable state of operation. However, the pump will also encounter unstable conditions such as shutdown, variable speed adjustment, throttle adjustment, and speed fluctuation during operation, including at startup. In these cases, the flow characteristics of the centrifugal pump will change dramatically in a short time, resulting in strong vibration and noise, which will affect the safe operation of the pump and may cause damage to the pipeline and connecting equipment [2].

Considering the aforementioned circumstances, it becomes imperative to conduct an in-depth study on the transient characteristics of centrifugal pumps during startup to enhance their operational reliability. Tsukamoto et al. [3] first evaluated the transient characteristics of the pump by measuring the instantaneous speed, flow rate, and total pressure during the starting period. Duplaa et al. [4] established that low final flow rates usually enable the pump to reach low cavitating conditions at the end of startup, whereas increasing the final flow rate results in more developed cavitating conditions.

Chalghoum et al. [5] analyzed the dynamic characteristics of pumps using the characteristic analysis method, revealing that the pressure evolution during startup is influenced by the valve opening percentage and starting time. Li et al. [6] established the transient performance test system of a centrifugal pump. The startup process of a centrifugal pump was tested, and PIV technology was used to analyze the test. Wang et al. [7] studied the transient behavior of an ultra-low specific speed centrifugal pump when it was stopped and started in the process of rapid depression lowering at the inlet and found that the rapid reduction of pump inlet pressure would accelerate the formation of cavitation. Li et al. [8] used a close-loop numerical model to realize unsteady self-coupled computations during the starting period and found that the transient vortex evolution between blades in the mean reason leading to the transient head coefficient was lower than the steady state value. Meng [9] also introduced a coupling simulation method for a closed-loop pipe system including a pump.

Zhou et al. [10] analyzed the circulating pipeline system and found that, during the startup process, the radial force and pulsation amplitude of the double volute were lower than those of the single volute. Zou et al. [11] studied the valve closing startup process of a large double-suction centrifugal pump and put forward the transient formula of radial force evolution in the startup process for the first time. Li et al. [12] designed an inlet elbow structure with a visible window for axial-flow pumps and observed an instantaneous impact head phenomenon in the final stage of centrifugal pump startup, affecting the impeller's volume efficiency. Zhang et al. [13] analyzed that the difference between solid–liquid two-phase flow and pure water in the startup process is mainly reflected in the later stage. Zhang et al. [14] studied the startup of the pump by establishing a circulating pipeline system and found that the metastability flow calculation method was insufficient to predict the transient performance of the pump, especially in the early stage.

To reduce the vibration and noise generated during centrifugal pump startup, researchers have explored various blade structure design methods to minimize pressure pulsations. Dai [15] provided internal flow parameters for judging and predicting drag and noise reduction effects, and placed pits on the blade surface to reduce pressure pulsation and noise. Wang et al. [16] designed and optimized a bionic blade of equal thickness inspired by the C-shaped initial mid-arc of carp, and the unsteady interaction between blade wake and volute was significantly reduced, thus reducing noise. Kim et al. [17] found that under a small flow rate, the shape of the leading edge of the blade can be changed by increasing blade thickness to reduce eddy currents near the leading edge. Tao et al. [18] analyzed a semi-open centrifugal mud pump and obtained the blade trailing edge thickness with good wear characteristics and acceptable hydraulic loss. Based on the fluid-structure coupling method, Cui et al. [19] found that after cutting the trailing edge, the dominant frequency amplitude and low-frequency vibration displacement on the centrifugal pump impeller decreased. Qian et al. [20] utilized the NLPQL algorithm to optimize blade thickness, resulting in lower pressure pulsation intensity compared to the original impeller and similar performance to centrifugal pumps with shunt blades. Wu et al. [21] modified the pressure side profile of the blade to reduce the intensity of shedding vorticity at the trailing edge of the blade and the pressure pulsation in the volute. Furthermore, Wu et al. [22] redesigned the blade thickness of a five-blade centrifugal impeller based on force balance perpendicular to the streamline direction, effectively suppressing secondary flow development and improving energy efficiency.

Compared to the methods mentioned above, blade trailing edge filing is a straightforward and effective technique for mitigating operational vibration and noise in centrifugal pumps. Ryi et al. [23] found that a traditional sawtooth trailing edge has a better noise reduction effect on wind turbine rotors than an inclined sawtooth trailing edge. Gao et al. [24] studied the influence of five blade trailing edge shapes on the performance and pressure pulsation of low specific speed centrifugal pumps. The results showed that EPS (ellipse on pressure side) and EBS (ellipse on both sides) can both reduce the eddy current intensity at the trailing edge of the blade, significantly improve the efficiency of the pump, and reduce the pressure pulsation. Huang et al. [25] found that filing the trailing edge of the blade reduces the outlet angle of the blade, leading to an increase in head and a decrease in pressure pulsation.

In summary, blade trailing edge filing is a method suitable for practical engineering, but its impact on centrifugal pump startup is not clear, and research results are relatively limited. The article analyzes the internal flow field of a low specific speed centrifugal pump. To achieve this, A full channel model of three different trailing edge impellers of a centrifugal pump was established using 3D design software. Subsequently, both steady-state and unsteady simulations were conducted on the model by utilizing CFX 19.2 software. This paper analyzes the relationship between pressure distribution, velocity streamline distribution, vorticity distribution, and pressure pulsation phenomenon of centrifugal pumps with different blade trailing edges during startup, providing a reference for the vibration reduction of centrifugal pumps during startup. The main layout of this article is as follows: Section 2 introduces the pump model and grid division and describes the solver settings and experimental verification methods, Section 3 discusses the external and internal flow characteristics of the pump, and Section 4 presents the conclusions discussed above.

2. Numerical Setup

2.1. Computational Domain

The model pump designed in this paper is a low specific speed centrifugal pump with $n_s = 33$. The rated flow rate is 12.5 m^3h^{-1} and the head is 50 m. In order to ensure the flow stability and accuracy during the simulation and avoid the backflow phenomenon, a reasonable extension is carried out at the impeller inlet and volute outlet.

For the CFD simulation, the computational domain includes inlet, impeller, volute, and outlet. Figure 1 shows the 3D geometry model of each component, and the specific parameters of the centrifugal pump model are shown in Table 1.

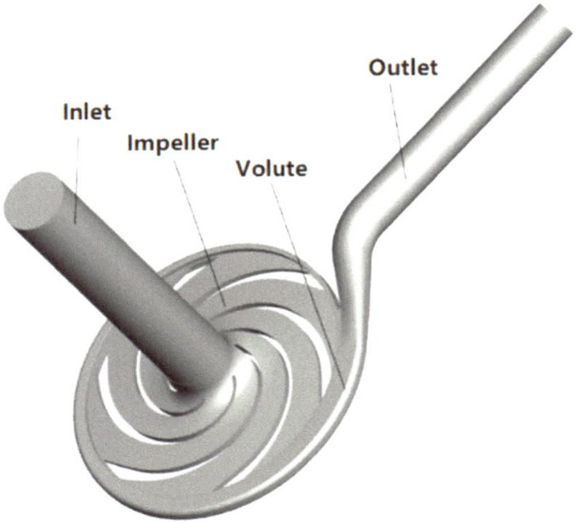

Figure 1. 3D model of centrifugal pump.

Due to the thickness of the pump blade, the fluid velocity and pressure in the wake area of the blade outlet are different from the mainstream, which will cause flow loss and reduce efficiency. However, changing the blade thickness and adjusting the blade outlet angle by trimming the blade trailing edge can reduce the shedding vortex, reduce the dynamic and static interference between the impeller and the tongue, reduce the wake loss at the blade outlet, and improve the non-uniformity of fluid flow direction and velocity distribution.

Table 1. Main structural parameters of the model pump.

Parameters	Value
Design flow rate/Q_d	12.5 m^3h^{-1}
Head/H_d	50 m
Rotating speed/n	2900 rpm
Impeller blade number/Z	6
Impeller inlet diameter/D_1	50 mm
Impeller outer diameter/D_2	198 mm
Blade outlet width/b_2	4 mm
Volute inlet diameter/D_3	205 mm
Volute outer diameter/D_4	32 mm
Volute inlet width/b_3	12 mm

By filing the trailing edge of the impeller blade, the problem that the performance parameters of the centrifugal pump in engineering do not meet the requirements of field use can be solved. Because the performance of the pump can be improved by changing less geometric parameters, the vibration of the pump system can be reduced and the cost of impeller modification can be reduced. Therefore, it is necessary to study the blade trailing edge filing.

Under the condition of keeping the main structural parameters in Table 1 unchanged, the pressure side of the centrifugal pump blade was filed, and the pressure side was linearly filed with two lengths, respectively. The trailing edge of the blade after filing is shown in Figure 2. Among them, the prototype centrifugal pump is recorded as OTE and the file pressure side thickness is all blade outlet circumference length (18 mm) and half blade outlet circumference length (9 mm) recorded, respectively, as LLPS and SLPS. The blade filing is retained at least 2 mm in the circumferential direction to ensure the impeller blade structural strength.

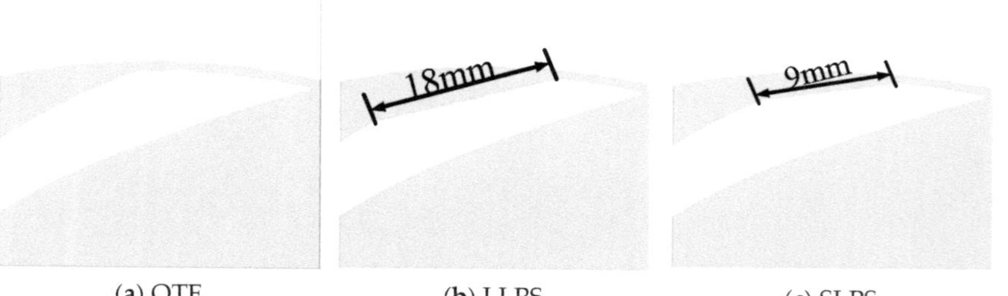

(**a**) OTE (**b**) LLPS (**c**) SLPS

Figure 2. Schematic diagram of blade outlet filing.

2.2. Meshes

The fluid domain of the centrifugal pump that needs to be calculated includes impeller, volute, inlet extension section, and outlet extension section. The calculation domain adopts a hexahedral grid; Figure 3 is the final mesh details used for the centrifugal pump. In order to improve the mesh quality and calculation accuracy of the computational domain, the mesh refinement is carried out in the regions with large curvature, such as blade, tongue, and impeller outlet, and the boundary layer is divided. The first layer has a thickness of 0.005 mm and a growth rate of 1.2. The mesh quality of all computational regions is above 0.5. ANSYS-ICEM-CFD 19.2 software was used to build the structured meshes associated with the block topology of Y-type and O-type, and appropriate mesh density was decided according to the gird independence test.

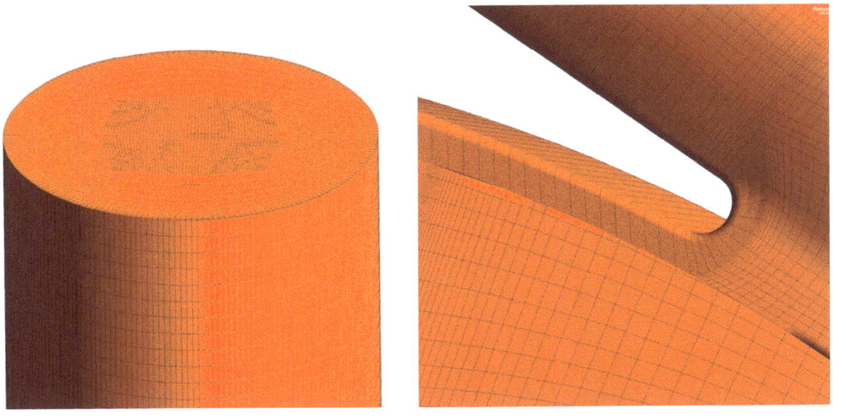

Figure 3. Grid diagram of centrifugal pump.

Aiming at the external characteristics of the pump, the interference of grid size is eliminated, and the grid independence verification results shown in Table 2 are obtained. Taking the head and efficiency of Test case 3 as the standard, when the number of grids is more than 3.6 million, increase the number of grids, and the change of head and efficiency of simulation is very small. The relative scatter is very small and the workload is greatly reduced, which further verifies the accuracy of the CFD calculation results. As long as it does not affect the predicted values of water head and mechanical power, the possible errors in calculating the flow field are acceptable [26]. Finally, it can be determined that the number of grids in impeller, volute, inlet section, and outlet section are 1,730,000, 990,000, 570,000, and 430,000, respectively, and the total number of grids is 3.6 million.

Table 2. Computational domain grid scheme.

Test Case	Impeller Mesh Elements	Volute Mesh Elements	Stator Mesh Elements	Outlet Mesh Elements	Total Mesh Elements	Relative Head	Relative Efficiency
1	973,794	638,384	312,900	301,800	2,216,918	1.00453844	1.00120579
2	1,448,054	824,976	478,500	362,160	2,995,770	1.00147528	1.00094366
3	1,731,326	999,268	572,100	434,592	3,603,964	1.00000000	1.00000000
4	1,933,406	1,477,016	672,900	521,516	4,389,610	0.99959429	0.99819130

2.3. Numerical Settings

ANSYS CFX 19.2 software was used to run the steady state calculation and transient calculation of the centrifugal pump startup process for this study, in which the SST k-ω turbulence model and the SIMPLEC algorithm was used to solve the discrete difference equations of the second-order upwind scheme. By controlling the change of rotational speed, n, to simulate the starting state of the centrifugal pump, centrifugal pump instantaneous speed, n, is defined as follows:

$$n = n_{\max} \cdot e^{-\frac{t}{t_0}} \quad (1)$$

where $n_{\max}$ is the rated speed of the centrifugal pump, t is the instantaneous time, and t_0 is rated acceleration time; in this paper, t_0 = 0.15 s.

The inlet boundary condition is set to Total Pressure (stable), which is set to 1 atm. The outlet boundary condition is set to Mass Flow Rate and the outlet flow is set to Q; Q can be obtained by the relationship between flow and speed through the similarity law. The instantaneous flow of the centrifugal pump, Q, is defined as follows:

$$Q = Q_0 \cdot \frac{n}{n_{max}} \quad (2)$$

where Q_0 is the rated flow rate of the centrifugal pump.

In the setting of transient numerical calculation, the turbulence intensity is set to be medium (Intensity = 5%), each surface of the impeller is set to a rotating wall, and the roughness of the wall is set to a smooth wall. The interface between the rotating area and the static area of the impeller is set to transient Rotor Stator, and the grid connection mode is set to GGI. The computational time interval was set to 0.0000556 s in the unsteady computations, so the impeller will rotate 1° in a time step; the convergence criterion was 10^{-4} in the simulation at all working conditions, and the number of internal cycles is set to 10.

When focusing on the pressure pulsation characteristics in some important regions of the centrifugal pump, some monitoring points are mounted on the mid-span plane of the centrifugal pump, as illustrated in Figure 4. Thirteen monitoring points (P0–P12) are evenly mounted on the diffuser outlet with an angle of 30°.

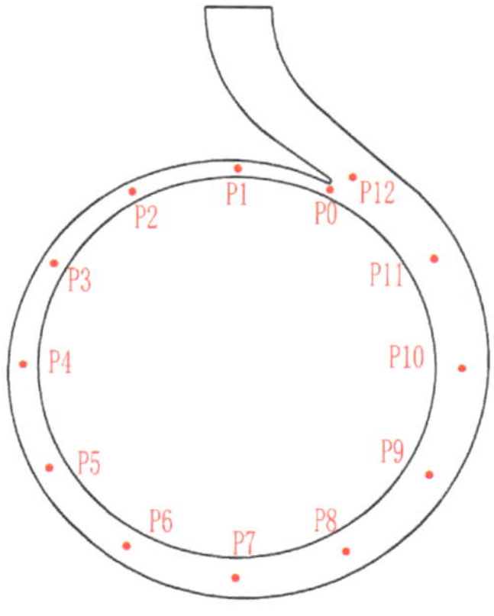

Figure 4. Locations of monitoring point.

2.4. Experimental Verification

As shown in Figure 5a, a closed experimental platform is established to verify the accuracy of the numerical simulation. The system includes valve, water tank, pressure sensor, data acquisition card, electromagnetic flowmeter, and centrifugal pump with the same model as the model pump. The measurement range of the electromagnetic flowmeter is 0.87 m^3h^{-1}~43.42 m^3h^{-1}, and its accuracy is less than 0.5% of the measured value. Two identical pressure sensors are installed on the inlet and outlet pipes of the pump, with a measurement range of -1~5 bar (accuracy $\pm$ 0.5%). In order to measure the power, the actual output power of the no-load test motor can be obtained by the loss analysis method, and the rotational speed of the centrifugal pump can be measured by a photoelectric tachometer. Figure 5b shows the layout of the experimental platform.

The methodology was as follows: Check the connection status of the centrifugal pump pipeline before the test and fill the pump with water. Run the centrifugal pump at low speed for 3 min with the outlet valve closed, and then open the outlet valve to run at the design flow for 20 min to check whether the platform is normal. If the system is normal, carry out 5 flow point tests from small flow to large flow. After the centrifugal pump runs for five minutes at each flow point, collect data every other minute for a total of 9 times, and then close the test bench.

(**a**) Enclosed test bench

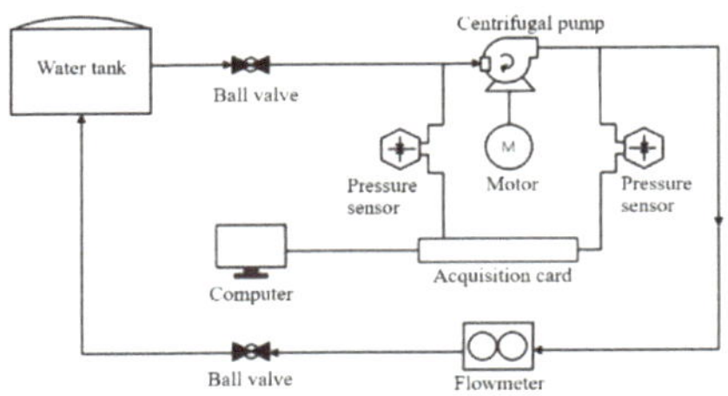

(**b**) Experimental platform schematic diagram

Figure 5. Experimental test platform.

The comprehensive uncertainty of the pump under five working conditions is calculated, and the comprehensive uncertainty is not more than ±1%. The measurement accuracy of each parameter meets the standard.

Figure 6 shows the comparison between the measured experimental performance curve and the numerical head of the OTE pump, which verifies the calculation accuracy of the numerical model adopted. For all relevant flow conditions, we note that the predicted head agrees well with the experimentally measured head curve. At nominal flow rate, the difference in head rise coefficient between the datasheet curve and numerical results is approximately 0.0374 and the average relative error of efficiency is 4.76%, both of which are less than 5%. Therefore, the calculation model is more accurate in predicting the external characteristics of the centrifugal pump, which verifies the validity of the numerical calculation method.

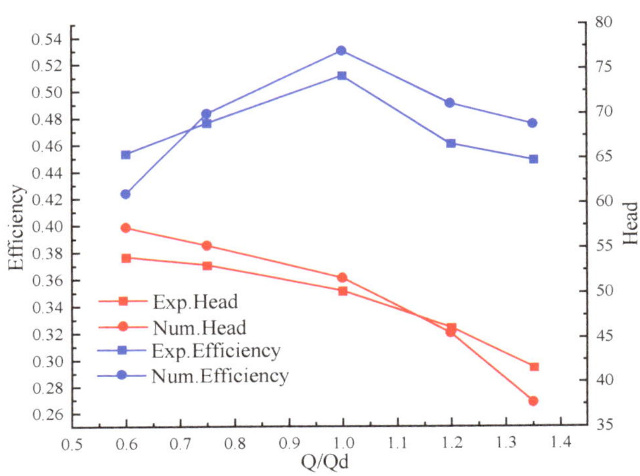

Figure 6. External characteristic curve of low specific speed centrifugal pump.

3. Results and Discussion

3.1. External Characteristic Analysis

The external characteristic curves of centrifugal pumps with different blade trailing edges during startup are shown in Figure 7.

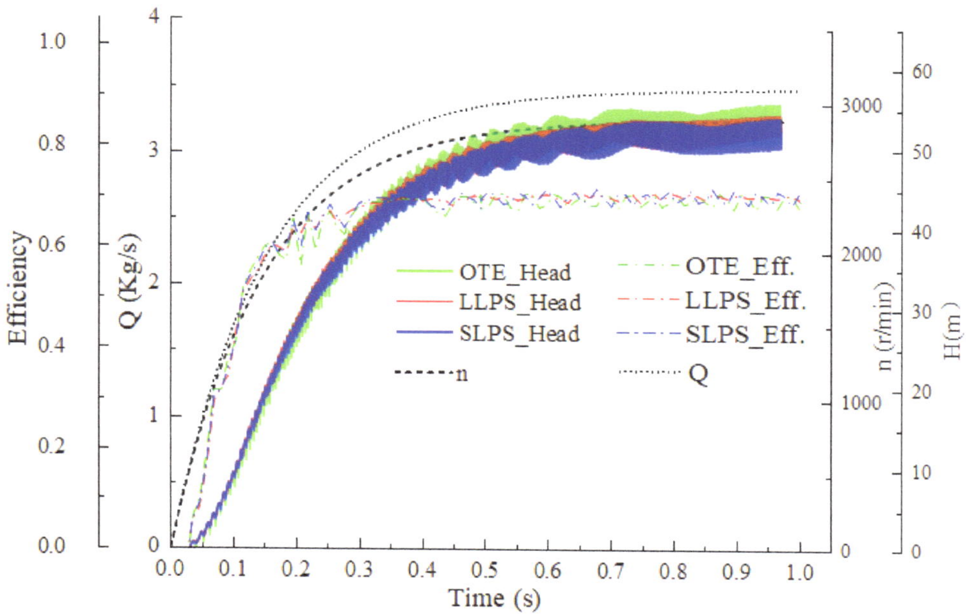

Figure 7. External characteristic curves of centrifugal pumps with different blade trailing edges during startup.

It can be seen from Figure 7 that the head and efficiency of the centrifugal pump show a fluctuating upward trend during the startup process, and the curves of the LLPS or SLPS are delayed, and the delay time is 0.0322 s. The reason may be that there is no effective flow field in the centrifugal pump at the initial startup stage, and there is no ability to pump the fluid, resulting in the head and efficiency being 0 before 0.0322 s. It can be seen from the efficiency–time curve that the curve becomes stable at 0.3 s. At 0.2~0.3 s, the efficiency fluctuates obviously and reaches a stable state quickly after the pulsation. It can be clearly seen that the average efficiency of OTE after stability is 69.05%, and the average efficiency of LLPS and SLPS is 69.68% and 69.63%, respectively, indicating that the cutting of the trailing edge of the pressure side blade can improve the efficiency of the centrifugal pump to a certain extent. This may be because the trailing edge filing prevents the separation of the boundary layer, thus reducing the friction and separation loss.

The centrifugal pump speed tends to be stable after 0.5 s, and the flow rate change is not obvious. Fast Fourier transform (FFT) was performed on the heads of three centrifugal pumps with different blade trailing edge filings after 0.5 s, and the frequency spectrum of the head in the starting process was obtained. Figure 8 shows that the change trend of the head of centrifugal pumps with different blade trailing edge filings is basically the same, but the head amplitude of the OTE pump is larger than that of the LLPS pump and SLPS pump. The head amplitudes of the OTE pump, LLPS pump, and SLPS pump are 1.58 m, 1.26 m, and 1.12 m, respectively. The head amplitude of the LLPS pump decreases by 29%, and the head amplitude of the SLPS pump decreases by 20%. Thus, the blade pressure side filing can effectively reduce the amplitude of the head, and the effect of the LLPS pump is better. It can be seen from the diagram that the main frequency of each type of centrifugal pump is 76.1411 Hz, which is 1.5 times the rotating frequency of the centrifugal pump.

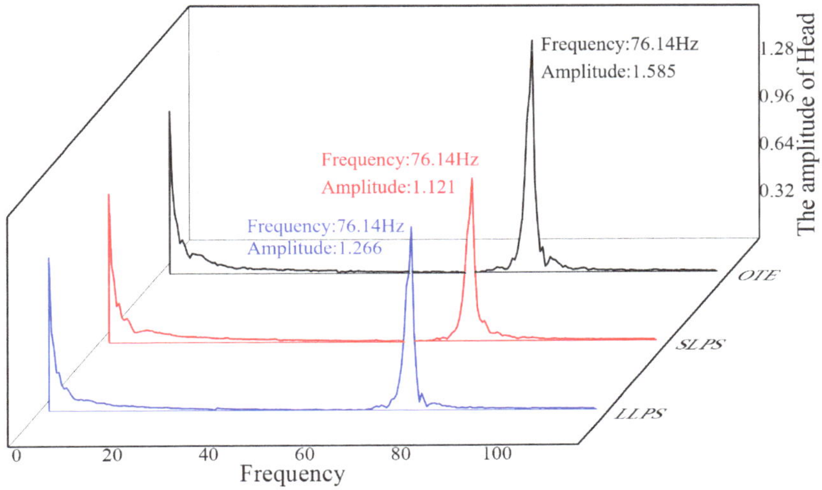

Figure 8. Head fluctuation frequency characteristics with different blade trailing edges.

3.2. Analysis of Pressure Fluctuation

The pressure fluctuation characteristic is an important indicator of the stability of the pump, while the representative fluctuation characteristic lies in the top point of the volute casing. Specifically, P0, P1, P4, P7, P10, and P12 are the monitoring points under study. In this paper, the amplitude of pressure fluctuation is expressed as the pressure coefficient, which is defined as:

$$C_P = \frac{\Delta p}{0.5\rho u_2^2} \tag{3}$$

where ρ is the density of the medium and u_2 is the circumferential velocity of impeller outlet.

Figures 9 and 10 show the time-domain diagrams of pressure fluctuation coefficients at P0, P1, P10, P4, P7, and P12 monitoring points, respectively. It can be seen that the change trend of centrifugal pumps with different blade trailing edge filings is basically the same at 0−1.0 s; pressure fluctuation shows a rising trend of continuous fluctuation, there are several throat areas in the fluctuation process, and the fluctuation range of pressure fluctuation is the smallest. The distance between adjacent pressure pulsation throats also increases with time. After the last throat is generated, the pressure pulsation tends to fluctuate steadily. There are oscillations at points P0, P1, and p12 because these monitoring points are located near the volute tongue. Due to the dynamic and static interference, the eddy current generated by the rotation of the impeller causes greater pressure pulsation.

It can be seen that the blade pressure fluctuation amplitude after filing has been significantly reduced and it is easier to reach a stable state. As shown in Figure 9d–f, the LLPS pump reaches the equilibrium state at 0.8526 s, the SLPS pump reaches the equilibrium state at 0.8974 s, and the OTE pump is close to the stable state at 0.9801 s. The pressure fluctuation amplitude of the OTE pump is significantly greater than that of the centrifugal pump after two kinds of file repair, and there is also the same trend in other monitoring points, indicating that the blade pressure side file repair has a significant inhibitory effect on the vibration of the centrifugal pump during the startup process.

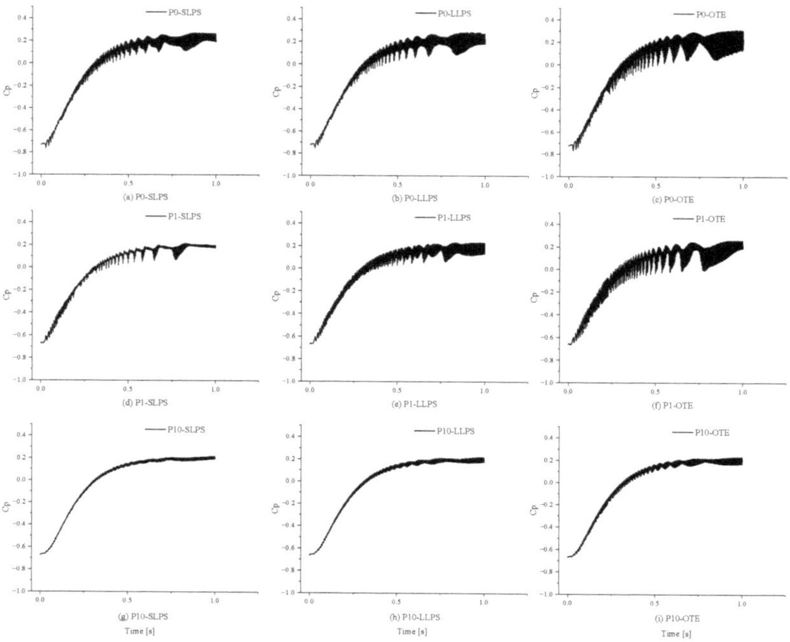

Figure 9. Time domain pressure pulsation at monitoring points in centrifugal pumps with different blade edges (P0, P1, P10).

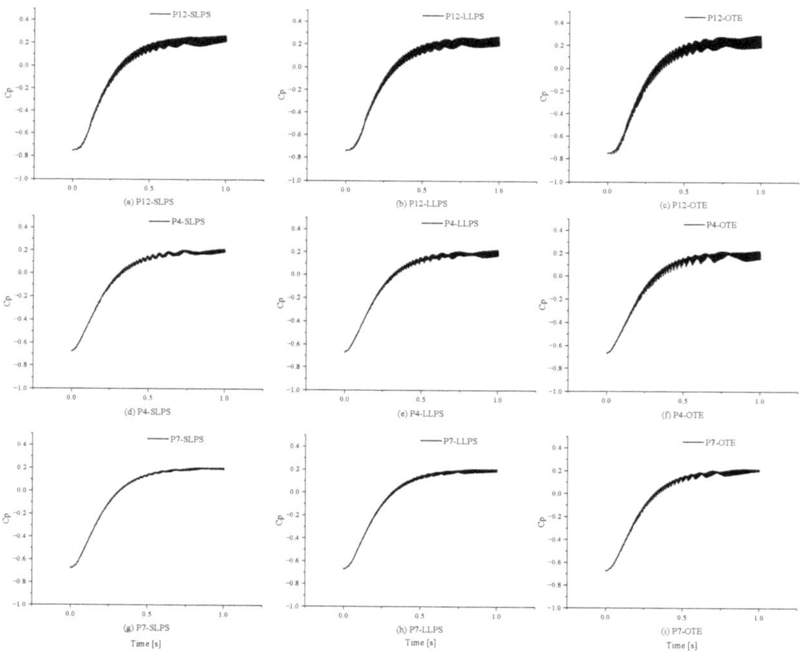

Figure 10. Time domain pressure pulsation at monitoring points in centrifugal pumps with different blade edges (P4, P7, P12).

In order to more intuitively display the effect of suppressing pressure pulsation after blade trailing edge filing, the time-domain and frequency-domain conversion of the pulsation signal is carried out by fast Fourier transform (FFT). Since the trends before 0.5 s are basically the same, this paper conducts fast Fourier transform for the time-domain diagrams of monitoring points P0, P1, P4, and P7 after 0.5 s. Figure 11 shows the frequency domain diagrams of pressure pulsation at the outlet of the impeller at each monitoring point. The vibration signals mainly include high frequency and a small amount of low frequency signals. The main frequency of each monitoring point is 3699.48 Hz, which is 76.5 times that of the centrifugal pump under stable working conditions (48.33 Hz). The pressure coefficient amplitude of the OTE pump is the largest at point P0, which is 0.053, while the amplitude decreases at P1, P4, and P7. This is because point P0 is located near the tongue, and the dynamic and static interference is serious. Moreover, the amplitude of the SLPS pump at each monitoring point is the smallest of the three pumps, and the minimum point is at P7, which is 0.0031. It can be seen that the pressure pulsation amplitude of the filing centrifugal pump is significantly reduced, and the inhibition degree of pressure pulsation by different filing methods at different monitoring points is also different. As shown in Figure 12, in order to better analyze the maximum amplitude of the main frequency, the maximum amplitudes of the centrifugal pumps with different filing methods at all monitoring points were compared and analyzed.

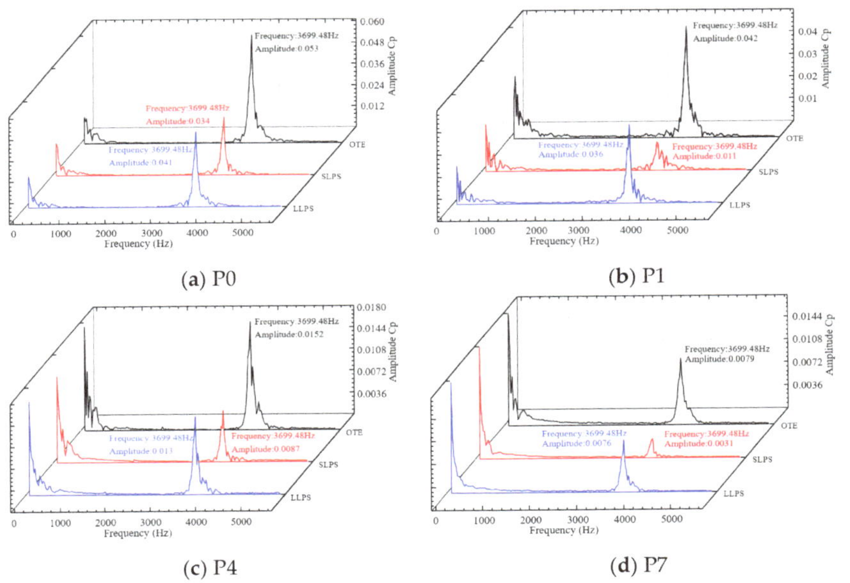

Figure 11. Flow channel outlet pressure pulsation frequency domain diagram.

As can be seen from Figure 12, the pressure pulsation amplitude of three different centrifugal pumps at all monitoring points shows the same change trend, and the maximum pressure pulsation amplitude is at P0 (volute tongue). The pressure pulsation amplitude of the monitoring points far from P0 showed a downward trend. At P6 and P7, which are farthest from P0, the amplitude of pressure pulsation reaches the minimum, which proves that the dynamic and static interference of blade and volute tongue is the main cause of high amplitude pressure pulsation, vibration, and noise. The centrifugal pump with two kinds of blades filed has the obvious ability to suppress pressure pulsation, and the LLPS centrifugal pump has the best inhibition effect. The maximum pressure pulsation amplitude of each monitoring point decreased by 32.23% on average, and the maximum pressure pulsation amplitude of the SLPS centrifugal pump decreased by 20.25% on average. The inhibition effect of centrifugal pumps is best when the water body of volute and impeller is narrow. Due to the strong interference between the liquid flowing out of the impeller and the static volute, which then causes the abnormal vibration of the centrifugal pump, the pressure pulsation and vibration energy can be effectively reduced by filing the centrifugal pump here. The maximum pressure pulsation amplitude of the LLPS pump at the P3 monitoring point is reduced by 56%.

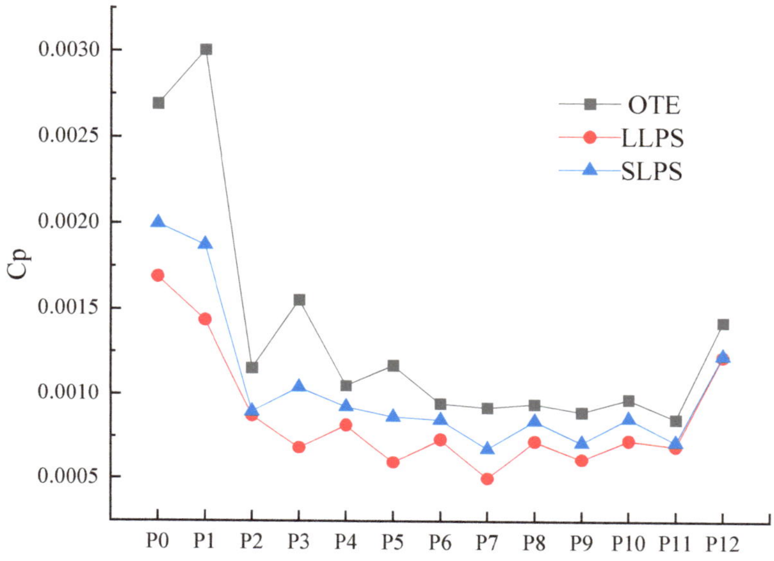

Figure 12. Distributions of pressure amplitude at volute channel.

3.3. Analysis of Internal Flow Field of Centrifugal Pumps with Different Filing Blades during Startup

Figure 13 is the static pressure distribution at the middle interface of the centrifugal pump during the startup period. At 0.02 s and 0.05 s, the pressure distribution inside the impeller is uneven, and the high-pressure area is mainly concentrated in the narrow area of the volute flow channel. In the initial stage of centrifugal pump startup, the flow development in the pump is insufficient, resulting in a large-scale uneven distribution of pressure. After 0.15 s, the low-pressure area appeared at the inlet of the impeller, the high-pressure annular distribution appeared around the outer circle of the volute, and the pressure gradient near the tongue decreased. This may be because the trailing edge of the blade after filing can improve the streamline of the fluid and reduce the generation of turbulence. Thus, the energy loss and pressure loss are reduced, and the pressure inside the volute is increased. It can be seen that there is an obvious small low-pressure area at the blade outlet of the OTE pump, which is prone to flow disorder. The low-pressure areas of the LLPS and SLPS pumps at the same position are significantly reduced, and the pressure distribution at the impeller outlet is more uniform. After 0.60 s, the flow tends to be stable. It can be seen that the low-pressure area at the outlet of the blade of the OTE pump persists, and the low-pressure zones of the LLPS and SLPS pumps have disappeared. When the centrifugal pump reaches the rated speed at 1.00 s, the static pressure of the volute of the LLPS and SLPS pumps is significantly greater than that of the OTE pump, indicating that the transport capacity of the centrifugal pump has been improved after blade filing.

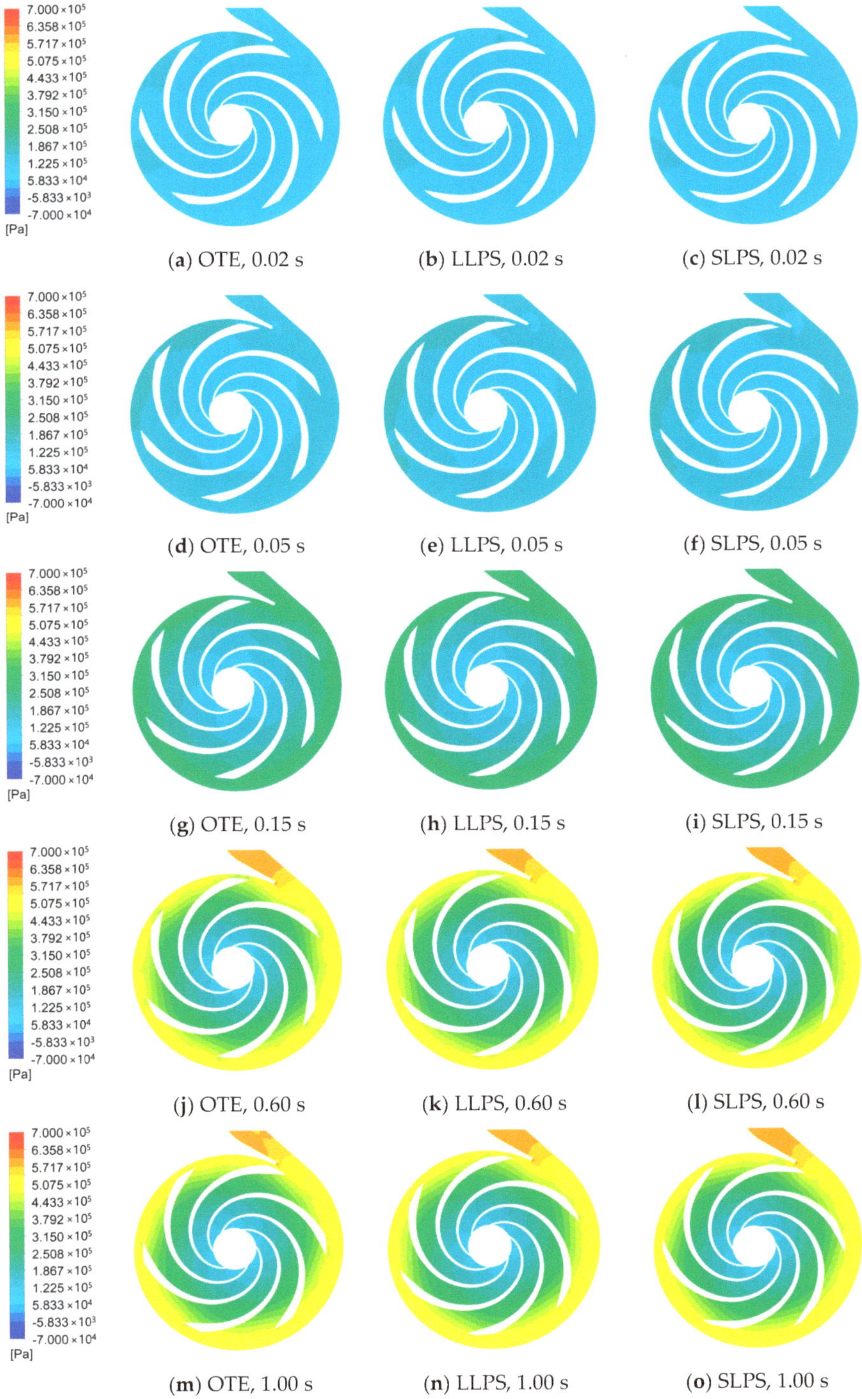

Figure 13. Static pressure distribution of middle section of centrifugal pump under startup condition.

Figure 14 shows the distribution diagram of velocity streamline at the central section of the centrifugal pump under starting condition. As can be seen from the figure, in the initial stage of starting the centrifugal pump, the internal flow of the centrifugal pump is relatively chaotic, and a large vortex area appears on the pressure side of the three flow channels of the impeller This vortex leads to flow blockage, which influences the performance of the centrifugal pump. This may also be the reason for the phenomenon that the head and efficiency are 0 before 0.0322 s. After 0.05 s, the vortex area in all channels decreased significantly, and some split into two vortices, and the core of the vortex moves to the impeller inlet. At 0.08 s, the vortex area in the impeller channel is further reduced, and the core of the vortex is mainly concentrated in the middle of the impeller channel. At 0.10 s, only the OTE pump still has a vortex area, and the streamlines of the LLPS pump and SLPS pump, whose blades have been filed, tend to be smooth, and there is no vortex. At 0.15 s, the rated acceleration time of the centrifugal pump is reached, the speed is from static to 63.2% of the rated speed, and the flow line in the impeller channel is basically stable.

(**a**) OTE, 0.02 s (**b**) LLPS, 0.02 s (**c**) SLPS, 0.02 s

(**d**) OTE, 0.05 s (**e**) LLPS, 0.05 s (**f**) SLPS, 0.05 s

(**g**) OTE, 0.08 s (**h**) LLPS, 0.08 s (**i**) SLPS, 0.08 s

Figure 14. *Cont.*

(j) OTE, 0.10 s (k) LLPS, 0.10 s (l) SLPS, 0.10 s

(m) OTE, 0.15 s (n) LLPS, 0.15 s (o) SLPS, 0.15 s

Figure 14. Velocity streamline distribution in the middle section of the centrifugal pump under starting state.

Based on the analysis of the distribution characteristics of the static pressure and velocity streamlines of the centrifugal pump at the startup period, this paper reveals the relationship between the pressure pulsation and the unsteady flow structure during the startup period from the law of vorticity, Ω, distribution, which is defined as:

$$\Omega = 2\omega_z = \left(\frac{\partial v_x}{\partial x} - \frac{\partial v_x}{\partial y}\right) \tag{4}$$

Figure 15 shows the vorticity distribution of the central section of the centrifugal pump under starting condition. As can be seen from the figure, at 0.02 s, only the high vorticity area appears near the volute tongue, which means that the flow in the pump is relatively slow at the initial stage of the centrifugal pump startup. The high vorticity area out of the tongue corresponds to the high-pressure gradient phenomenon in Figure 13. At 0.05 s, besides the tongue, there is also a large vorticity distribution area at the interface of the impeller volute, especially in the trailing edge of the blade pressure side. After 0.15 s, an obvious spatio-temporal evolution process of vortex structure can be seen: vortex is generated from the inlet of impeller and gradually spreads to the impeller suction side. With the impeller rotating periodically into the volute region, the trailing edge vortex structure of the suction side of the blade produces a shedding vortex. When the shedding vortex passes through the tongue, it is divided into two vortex structures, which develop towards the exit of the volute and the left side of the tongue. The vortex structure in the left region of the tongue is gradually reduced with the change of the volute channel. Figures 12 and 13 can verify this trend. In Figure 14, the maximum amplitude of the pressure fluctuation from P0 to P11 is also a gradually decreasing trend. In Figure 15, the left static pressure of the middle section is also significantly greater than that of the right. At 1.00 s, the internal flow of the centrifugal pump tends to be stable. It is obvious that the vorticity of the SLPS impeller is significantly lower than that of the other two schemes. This may be due to the fact that the short linear filing at the trailing edge of the blade reduces the turbulence loss at the impeller outlet, improves the flow of fluid at the trailing edge of the blade, slows down the vortex separation phenomenon, and then reduces the vorticity. However, after long linear filing, the flow rate further increases, which increases the strength of vortex formation and rotation. At the same time, the increase in filing degree leads to the decrease of blade interference and the increase in vorticity.

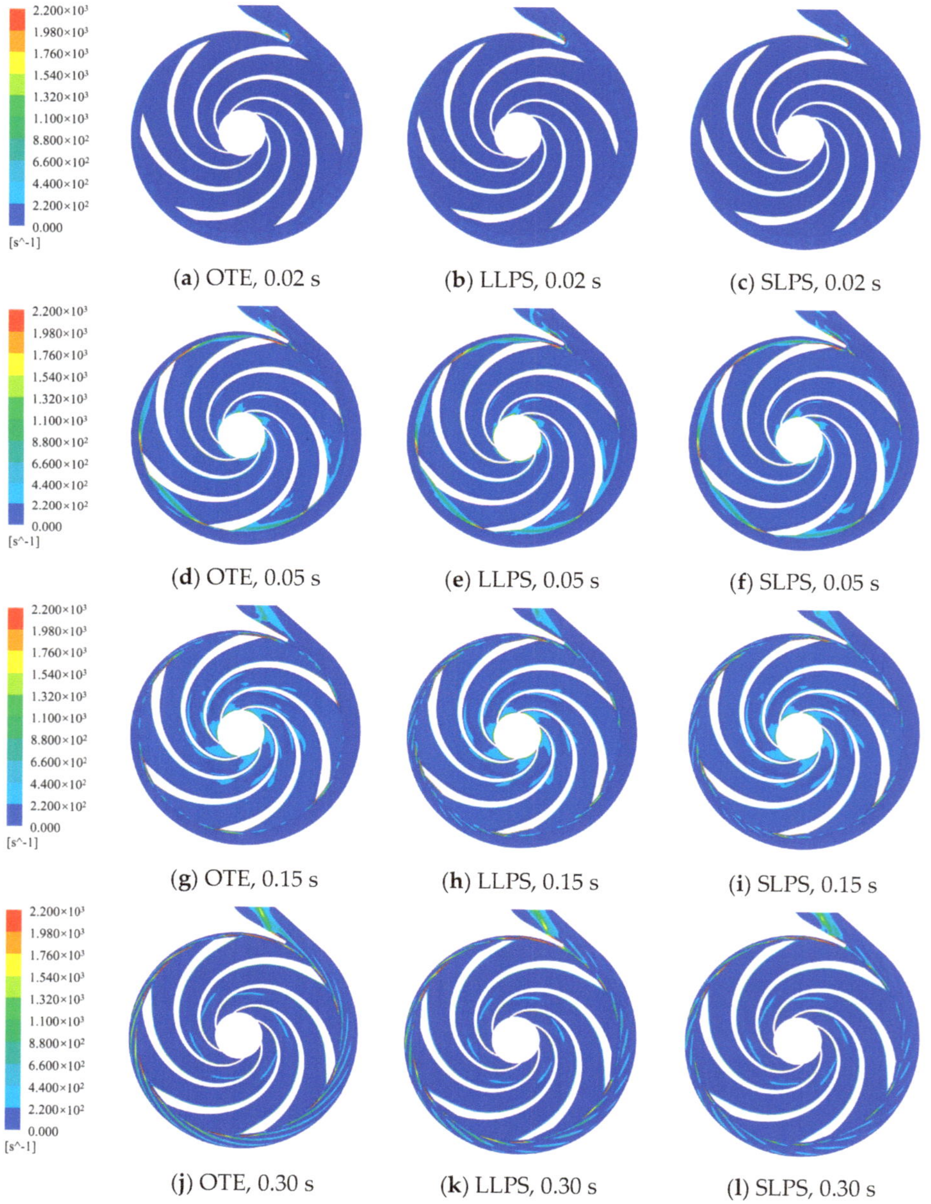

(**a**) OTE, 0.02 s (**b**) LLPS, 0.02 s (**c**) SLPS, 0.02 s

(**d**) OTE, 0.05 s (**e**) LLPS, 0.05 s (**f**) SLPS, 0.05 s

(**g**) OTE, 0.15 s (**h**) LLPS, 0.15 s (**i**) SLPS, 0.15 s

(**j**) OTE, 0.30 s (**k**) LLPS, 0.30 s (**l**) SLPS, 0.30 s

Figure 15. *Cont.*

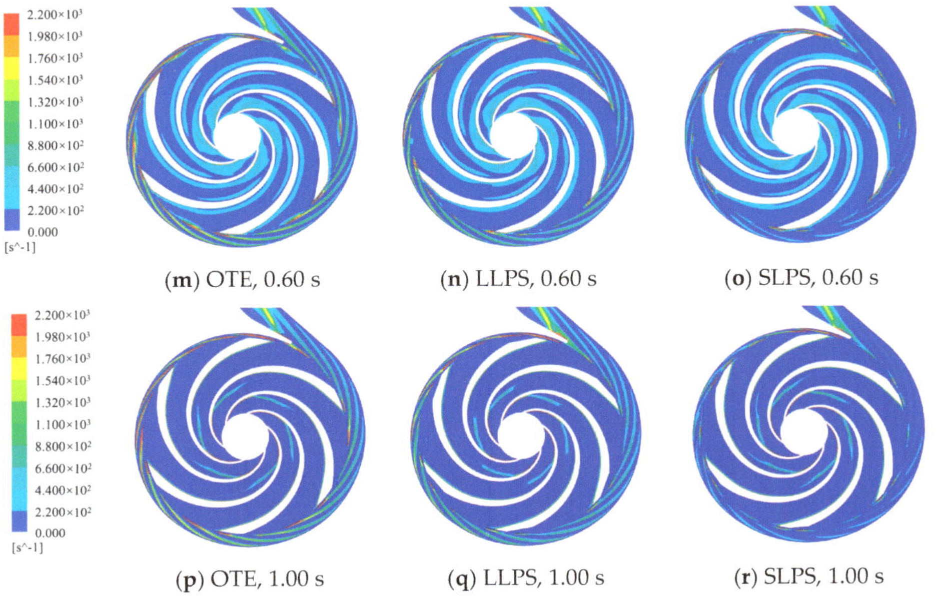

(**m**) OTE, 0.60 s (**n**) LLPS, 0.60 s (**o**) SLPS, 0.60 s

(**p**) OTE, 1.00 s (**q**) LLPS, 1.00 s (**r**) SLPS, 1.00 s

Figure 15. Vorticity distribution of the middle section of the centrifugal pump at the startup period.

4. Conclusions

In this paper, the startup process of the centrifugal pump tail edge filing scheme suitable for engineering practice was analyzed, the pressure fluctuation during the startup process was studied, and its internal flow mechanism was explored. The main findings are as follows:

1. During the startup period of the centrifugal pump, the external characteristic curves showed a fluctuating upward trend and have a lag phenomenon. The head of the centrifugal pump after filing has decreased, but the efficiency has improved. The blade pressure side filing can effectively reduce the amplitude of the head.
2. The blade pressure side file has a significant inhibitory effect on the pressure fluctuation during the startup process of the centrifugal pump, and the maximum pressure fluctuation amplitude at the monitoring point is reduced by 32.23% on average and the maximum pressure pulsation amplitude of the SLPS centrifugal pump decreased by 20.25% on average.
3. Through the analysis of static pressure distribution and velocity streamline, it was found that the small low-pressure area at the outlet of OTE pump blade is prone to flow disorder. The low-pressure area of the LLPS and SLPS pumps at the same location is significantly reduced, and the impeller outlet pressure distribution is more uniform. There is a large eddy current in the three flow channels in the impeller, which leads to flow blockage and makes the centrifugal pump lose pumping effect. Through vortex analysis, it was found that the vortex at the outlet of the centrifugal pump blade after trailing edge filing is significantly reduced, which effectively promotes the stability of the centrifugal pump and inhibits the pressure fluctuation.

This paper provides ideas and methods for reducing the vibration caused by pressure pulsation during the starting process of centrifugal pumps and provides theoretical support for application in practical engineering. However, this paper does not explore the energy loss and noise during the startup process, which can be further studied in the future.

Author Contributions: H.D., conceptualization, methodology, supervision; F.G., software, validation, writing—review and editing; K.W., investigation, formal analysis; F.L., data curation, software. All authors have read and agreed to the published version of the manuscript.

Funding: This research was funded by [Natural Science Foundation Project of Shandong province] grant number [ZR2019MEE068].

Data Availability Statement: No data were used for the research described in the article.

Conflicts of Interest: The authors declare that they have no known competing financial interests or personal relationships that could have appeared to influence the work reported in this paper.

References

1. Zhang, Y.L.; Zhu, Z.C.; Li, W.G. Experiments on transient performance of a low specific speed centrifugal pump with open impeller. *Proc. Inst. Mech. Eng. Part A J. Power Energy* **2016**, *230*, 648–659. [CrossRef]
2. Li, Z.F.; Wu, D.Z.; Wang, L.Q.; Dai, W.P.; Chen, F.Q. Experimental study on transient characteristics of centrifugal pump during start-up process. *J. Drain. Irrig. Mach. Eng.* **2010**, *28*, 389–393.
3. Tsukamoto, H.; Asakura, J.; Tominaga, N.; Nanba, H.; Tanaka, T. Dynamic response of a cavitating centrifugal pump to fluctuating rotational speed. In Proceedings of the International Conference on Fluid Machinery, Beijing, China, 9–12 September 1996; pp. 441–448.
4. Duplaa, S.; Coutier-Delgosha, O.; Dazin, A.; Bois, G. X-ray Measurements in a Cavitating Centrifugal Pump during Fast Start-Ups. *J. Fluids Eng.* **2013**, *135*, 41204. [CrossRef]
5. Chalghoum, I.; Elaoud, S.; Akrout, M.; Taieb, E.H. Transient behavior of a centrifugal pump during starting period. *Appl. Acoust.* **2016**, *109*, 82–89. [CrossRef]
6. Li, Z.F.; Wu, D.Z.; Dai, W.P.; Wang, L.Q. Experimental study on the transient characteristics and flows in a centrifugal pump during starting period. *J. Eng. Thermophys.* **2010**, *31*, 2019–2022.
7. Wang, Y.; Xie, L.; Chen, J.; Liu, H.-l.; Luo, K.; Zhang, Z.-l.; Cao, M.-h. Experimental Study on Transient Startup Characteristics of a Super Low Specific Speed Centrifugal Pump. *J. Chem. Eng. Jpn.* **2019**, *52*, 743–750. [CrossRef]
8. Li, Z.; Wu, P.; Wu, D.; Wang, L. Experimental and numerical study of transient flow in a centrifugal pump during startup. *J. Mech. Sci. Technol.* **2011**, *25*, 749–757. [CrossRef]
9. Meng, L.; Liu, M.; Zhou, L.; Wang, W.; Liao, C.; Zhao, L.; Li, T. Coupling simulation of the fast startup of a centrifugal pump with cavitation in a closed-loop pipeline system. *Eng. Comput.* **2018**, *35*, 2010–2024. [CrossRef]
10. Zhou, R.; Yang, J.; Liu, H.L.; Dong, L. Effect of Volute Geometry on Radial Force Characteristics of Centrifugal Pump during Startup. *J. Appl. Fluid Mech.* **2021**, *15*, 25–36. [CrossRef]
11. Zou, Z.C.; Wang, F.J.; Yao, Z.F.; Tao, R.; Xiao, R.F.; Li, H.C. Impeller radial force evolution in a large double-suction centrifugal pump during startup at the shut-off condition. *Nucl. Eng. Des.* **2016**, *310*, 410–417. [CrossRef]
12. Li, W.; Ji, L.L.; Shi, W.D.; Li, E.D.; Ma, L.L.; Yang, Z.Y. Particle image velocimetry experiment of the inlet flow field in a mixed-flow pump during the startup period. *Proc. Inst. Mech. Eng. Part A J. Power Energy* **2020**, *234*, 300–314. [CrossRef]
13. Zhang, Y.L.; Li, Y.; Zhu, Z.C.; Cui, B.L. Computational Analysis of Centrifugal Pump Delivering Solid-liquid Two-phase Flow during Startup Period. *Chin. J. Mech. Eng.* **2014**, *27*, 178–185. [CrossRef]
14. Zhang, Y.L.; Zhu, Z.C.; Dou, H.S.; Cui, B.L.; Li, Y.; Zhou, Z.Z. Numerical Investigation of Transient Flow in a Prototype Centrifugal Pump during Startup Period. *Int. J. Turbo Jet-Engines* **2017**, *34*, 167–176. [CrossRef]
15. Dai, C.; Guo, C.; Chen, Y.P.; Dong, L.; Liu, H.L. Analysis of the Influence of Different Bionic Structures on the Noise Reduction Performance of the Centrifugal Pump. *Sensors* **2021**, *21*, 15. [CrossRef]
16. Wang, J.; Liu, X.; Tian, C.; Xi, G. Aerodynamic performance improvement and noise control for the multi-blade centrifugal fan by using bio-inspired blades. *Energy* **2023**, *263*, 125829. [CrossRef]
17. Kim, S.-J.; Choi, Y.-S.; Cho, Y.; Choi, J.-W.; Kim, J.-H. Effect of Runner Blade Thickness on Flow Characteristics of a Francis Turbine Model at Low Flowrates. *J. Fluids Eng.* **2020**, *142*, 31104. [CrossRef]
18. Tao, Y.; Yuan, S.Q.; Liu, J.R.; Zhang, F.; Tao, J.P. Influence of Blade Thickness on Transient Flow Characteristics of Centrifugal Slurry Pump with Semi-open Impeller. *Chin. J. Mech. Eng.* **2016**, *29*, 1209–1217. [CrossRef]
19. Cui, B.L.; Li, W.Q.; Zhang, C.L. Effect of Blade Trailing Edge Cutting Angle on Unstable Flow and Vibration in a Centrifugal Pump. *J. Fluids Eng.-Trans. ASME* **2020**, *142*, 15. [CrossRef]
20. Qian, B.; Wu, P.; Huang, B.; Zhang, K.; Li, S.Y.; Wu, D.Z. Optimization of a Centrifugal Impeller on Blade Thickness Distribution to Reduce Hydro-Induced Vibration. *J. Fluids Eng.-Trans. ASME* **2020**, *142*, 12. [CrossRef]
21. Wu, C.S.; Zhang, W.Q.; Wu, P.; Yi, J.L.; Ye, H.J.; Huang, B.; Wu, D.Z. Effects of Blade Pressure Side Modification on Unsteady Pressure Pulsation and Flow Structures in a Centrifugal Pump. *J. Fluids Eng.-Trans. ASME* **2021**, *143*, 14. [CrossRef]
22. Chengshuo, W.; Kexin, P.; Changqin, L.; Peng, W.; Bin, H.; Dazhuan, W. Blade redesign based on secondary flow suppression to improve energy efficiency of a centrifugal pump. *Energy* **2022**, *246*, 123394. [CrossRef]
23. Ryi, J.; Choi, J.S. Noise reduction effect of airfoil and small-scale rotor using serration trailing edge in a wind tunnel test. *Sci. China-Technol. Sci.* **2017**, *60*, 325–332. [CrossRef]

24. Gao, B.; Zhang, N.; Li, Z.; Ni, D.; Yang, M.G. Influence of the Blade Trailing Edge Profile on the Performance and Unsteady Pressure Pulsations in a Low Specific Speed Centrifugal Pump. *J. Fluids Eng.-Trans. ASME* **2016**, *138*, 10. [CrossRef]
25. Huang, B.; Zeng, G.T.; Qian, B.; Wu, P.; Shi, P.L.; Qian, D.Q. Pressure Fluctuation Reduction of a Centrifugal Pump by Blade Trailing Edge Modification. *Processes* **2021**, *9*, 19. [CrossRef]
26. Fecarotta, O.; Messa, G.V.; Pugliese, F. Numerical assessment of the vulnerability to impact erosion of a pump as turbine in a water supply system. *J. Hydroinforma.* **2020**, *22*, 691–712. [CrossRef]

processes

MDPI

Article

Design of Electric Supercharger Compressor and Its Performance Optimization

Yiyang Wei [1], Binglin Li [1,*], Xiaomei Xu [1], Minxiang Wei [2] and Chengdong Wang [2]

1 College of Automobile and Transportation engineering, Nanjing Forestry University, Nanjing 210037, China; weiyiyang@njfu.edu.cn (Y.W.); xxm120480@126.com (X.X.)
2 College of Energy and Power Engineering, Nanjing University of Aeronautics and Astronautics, Nanjing 210016, China; weimx@nuaa.edu.cn (M.W.); wcd123@nuaa.edu.cn (C.W.)
* Correspondence: lblqx@njfu.edu.cn

Abstract: The performance of the centrifugal compressor, which is the main component of the electric supercharger, significantly impacts the engine's dynamics, economy, emissions, and responsiveness. The purpose of this paper is to enhance the aerodynamic performance of the centrifugal compressor of the electric supercharger for the two-stroke engine by optimizing the design of its impeller and diffuser parameters. The paper employs the numerical simulation method and applies the Spalart–Allmaras turbulence model to solve the RANS equations to analyze the impact of impeller-related parameters on the centrifugal compressor's performance. Subsequently, the paper optimizes the initial model parameters based on the simulation results and confirms its performance through an experiment. The findings indicate that enhancing the isentropic efficiency and pressure ratio of the compressor can be achieved by increasing the number of blades on the impeller, selecting an appropriate blade backward angle, and increasing the relative outlet width. After optimization, the compressor's efficiency can achieve 0.842, the pressure ratio can reach 1.49 with a working margin of 22%, and the efficiency is enhanced by 1.4%, while the pressure ratio is increased by 1.8% compared to the pre-optimization state. Moreover, the optimized model is experimentally validated to meet the design requirements.

Keywords: centrifugal compressor; electric supercharger; numerical simulation; optimized design; experimental verification

Citation: Wei, Y.; Li, B.; Xu, X.; Wei, M.; Wang, C. Design of Electric Supercharger Compressor and Its Performance Optimization. *Processes* **2023**, *11*, 2132. https://doi.org/10.3390/pr11072132

Academic Editor: Andrew Hoadley

Received: 25 June 2023
Revised: 15 July 2023
Accepted: 16 July 2023
Published: 17 July 2023

1. Introduction

With the continuous development of general aviation, small unmanned aerial vehicles (UAV) gradually emerging, two-stroke engines with low fuel consumption, low failure rate, low noise, and light mass are widely used [1]. Compared with the traditional turbocharger, the electric supercharger can improve the combustion quality of the engine and reduce the emission of smoke and fuel consumption rate, which has a certain effect on environmental protection. Moreover, it can improve the engine dynamics, especially in the low-speed and high-torque operating conditions [2–7].

Centrifugal compressors, as the main components of electric pressure intensifiers, are characterized by small size, simple structure, and high boost ratio, and their main structures are the diffuser, impeller, and worm housing [8]. The design of the diffuser and impeller has an essential impact on the performance of the pressurizer, so the current research on the pressurizer is mainly reflected in the analysis and optimization of its implementation [9–13]. The diffuser increases the pressure of the compressed medium and reduces the medium's velocity through the expansion of the flow channel. The bladeless diffuser is widely used in compressors because of its simple structure, good adaptability, and wide working range compared to the vane diffuser [14]. Seralathan and Roy analyzed the effect of the bladeless diffuser on the compressor performance by numerical simulation: when the diffuser flow angle is small, the flow path length is shorter, the impeller friction loss is reduced, and the compressor efficiency is increased [15]. Chen et al. used the method of appropriately increasing the bladeless area, which can significantly reduce the separation loss and make the flow velocity at the impeller outlet and diffuser inlet uniform, which improves the performance of the compressor [16]. Ma and Li reduced the solidity of the blade diffuser and slotted the hub side of the diffuser impeller, which can make the compressor more efficient [17]. Gunadal and Govardhan investigated the relationship between blade angle and compressor performance in a bladed diffuser, and, when the diffuser blade tilt was 6°, the stall margin and operating range were improved, and the compressor pressure ratio was also increased [18].

The impeller also has an important impact on the performance improvement of the compressor, so a lot of research work has been carried out by scholars in the design optimization of the compressor [19–22]. Li et al. studied a set of high-speed compressor impellers with different geometries and analyzed the secondary flow model, where the ratio of impeller inlet diameter to outlet diameter has an effect on the impeller flow uniformity, which, in turn, affects the compressor efficiency [23]. Yang et al. modified the hub alignment to investigate the effect of hub curvature on the performance of the compressor [24]. Hong et al., through numerical simulations of a centrifugal compressor for a hydrogen-powered fuel electric vehicle, found that the generation of leaf top leakage, hub separation, and horseshoe vortex structure determined the degree of the aerodynamic performance of the centrifugal compressor [25]. Jaatinen-Värri et al. studied that impellers with different top clearance and spreading width, and found that increasing the top clearance can significantly increase the reverse flow in the impeller, which will mix with the top clearance flow, leading to a reduction in impeller wake loss and an increase in compressor efficiency [26].

This article makes the following contributions: (1) Designing a centrifugal compressor for the electric turbocharger by exploring the internal flow field of the compressor through numerical simulations of the initial model; (2) Analyzing the impact of impeller blade number, relative outlet width, and impeller backbend angle on compressor performance and optimizing the initial model to determine whether optimized performance is improved over initial performance; (3) Experimentally verifying whether the optimized compressor performs according to the design requirements.

The main sections of this article are as follows. Section 2 introduces the theoretical foundation of numerical simulation and formulates the initial model of the compressor. Section 3 analyzes the effects of the impeller blade number, relative outlet width, and impeller backbend angle on the compressor's performance and flow field, evaluating the optimized compressor model. Section 4 outlines the experimental procedure, while Section 5 outlines the main conclusions of the research.

2. Centrifugal Compressor Modeling and Numerical Simulation

This chapter introduces the theoretical background of compressor numerical simulation and presents the procedures for designing a compressor model, grid division, discretization, and performance analysis. The factors affecting the compressor performance are investigated.

2.1. Numerical Simulation Theory

Numerical simulation methods [27] allow accurate and efficient analysis of the flow and heat transfer in the compressor model. In this paper, the numerical calculations are mainly performed using the Euranus solver in FineTurbo, and the Spalart-Allmaras equation turbulence model is selected to solve the RANS equations. The RANS equations are obtained by Re averaging the conventional mass conservation equation, momentum conservation equation, and energy conservation equation, where the mass conservation equation is expressed as:

$$\frac{\partial \rho}{\partial t} + \frac{\partial(\rho \overline{\mathbf{V}}_i)}{\partial x_i} = 0, \quad i = 1,2,3 \tag{1}$$

where ρ is the fluid density; t is the time; x_i is the spatial coordinate position; i is the three directions in space; and $\overline{\mathbf{V}}_i$ is the velocity in the direction.

The conservation of momentum equation is:

$$\frac{\partial \rho \overline{\mathbf{V}}_i}{\partial t} + \frac{\partial (\rho \overline{\mathbf{V}}_i \overline{\mathbf{V}}_j)}{\partial x_j} = -\frac{\partial \overline{p}}{\partial x_i} + \frac{\partial}{\partial x_j}(\rho v \frac{\partial \overline{\mathbf{V}}_i}{\partial x_j}) + \frac{\partial}{\partial x_j}(-\rho \overline{\mathbf{V}}_i{}' \overline{\mathbf{V}}_j{}'), i = 1,2,3, j = 1,2,3 \quad (2)$$

where $\overline{p}$ is the combined pressure force acting on the fluid microcluster; v is the kinematic viscosity coefficient of the fluid; m^2/s; $-\rho \overline{\mathbf{V}}_i{}' \overline{\mathbf{V}}_j{}'$ is the Reynolds stress term; x_i is the spatial coordinate position; j denotes the three directions of space; $\overline{\mathbf{V}}_i$, $\overline{\mathbf{V}}_j$ is the velocity of the fluid microcluster in the x_i, x_j direction, respectively; and $\overline{\mathbf{V}}'_i$, $\overline{\mathbf{V}}'_j$ is the velocity of the pulsating mass in the x_i, x_j direction, respectively.

The conservation of energy equation is:

$$\frac{\partial (\rho \overline{E})}{\partial t} + \frac{\partial (\rho \overline{E} \overline{\mathbf{V}}_i)}{\partial t} = \frac{\partial}{\partial x_i}(\lambda \frac{\partial \overline{T}}{\partial x_i}) + S_E, \quad i = 1,2,3 \quad (3)$$

where $\overline{E}$ is the total energy of the fluid microcluster; λ is the thermal conductivity; $\overline{T}$ is the temperature; and S_E is the energy source term. The finite difference method and finite volume method are used to discretize the control method.

The SA turbulence model contains turbulent viscosity, generation, and dissipation terms. The turbulent viscosity is:

$$v_t = \widetilde{v} f_{v1} \quad (4)$$

where $\widetilde{v}$ represents is the Spalart-Allmaras variable [28], with f_{v1} expressed as follows:

$$f_{v1} = \frac{\chi^3}{\chi^3 + c_{v1}} \quad (5)$$

where χ is the ratio between the turbulent fluctuation velocity $\widetilde{v}$ and the molecular viscosity v. The turbulent fluctuation velocity $\widetilde{v}$ is obtained from the following equation:

$$\frac{\partial \widetilde{v}}{\partial t} + \vec{\mathbf{V}} \cdot \Delta \widetilde{v} = \frac{1}{\sigma}\{\Delta \cdot [(v + (1 + c_{b2})\widetilde{v})\Delta \widetilde{v}] - c_{b2}\widetilde{v}\Delta \widetilde{v}\} + Q \quad (6)$$

where $\vec{\mathbf{V}}$ is the average velocity vector; Q is the source term, σ and c_{b2} is a constant. Q consists of a generating term and a dissipative term, and the expression is as follows:

$$Q = \widetilde{v}p(\widetilde{v}) - \widetilde{v}D(\widetilde{v}) \quad (7)$$

where

$$\widetilde{v}p(\widetilde{v}) = c_{b1}\widetilde{S}\widetilde{v} \quad (8)$$

$$\widetilde{v}D(\widetilde{v}) = c_{w1} f_w \left(\frac{\widetilde{v}}{d}\right) \quad (9)$$

$$\widetilde{S} = S f_{v3} + \frac{\widetilde{v}}{k^2 d^2} f_{v2} \quad (10)$$

$$f_{v2} = \frac{(1 + \chi f_{v1})(1 - f_{v2})}{\chi} \quad (11)$$

$$f_{v3} = \frac{(1 + \chi f_{v1})(1 - f_{v2})}{\chi} \quad (12)$$

where the expression of f_w, g, and r are as follows:

$$f_w = g\left(\frac{1 + c_{w3}^6}{g^6 + c_{w3}^6}\right)^{\frac{1}{6}} \tag{13}$$

$$g = r + c_{w2}(r^6 - r) \tag{14}$$

$$r = \frac{\tilde{v}}{\tilde{S}\kappa^2 d^2} \tag{15}$$

where the constant term is shown below:

$$c_{w1} = \frac{c_{b1}}{\kappa^2} + \frac{\sigma}{(1 + c_{b2})} \tag{16}$$

where $c_{w2} = 0.3$, $c_{w3} = 2$, $c_{v1} = 7.2$, $c_{v2} = 5$, $c_{b1} = 0.1357$, $c_{b2} = 0.619$, $\kappa = 0.41$, $\sigma = \frac{2}{3}$.

2.2. Compressor Modeling

As shown in Table 1, the design index of the centrifugal compressor is that the total pressure ratio of the compressor is more significant than 1.4, the isentropic efficiency is greater than 0.75, and the working margin is greater than 15% under the condition that the speed is 70,000 r/min and the air mass flow rate is 0.027 kg/s. The design needs to meet the above requirements.

Table 1. Impeller basic design parameters.

Parameters/Unit	Value
Total inlet temperature/kPa	101.325
Total inlet pressure/K	288.15
Design speed/r · min^{-1}	7×10^4
Design flow rate/kg · s^{-1}	0.027
Isentropic efficiency	0.75
Pressure ratio	1.4

This paper explores the design of the impeller and diffuser for centrifugal compressors. The impeller inlet design [29,30] considers various parameters, such as rim diameter and hub diameter. The impeller inlet rim diameter has an enormous impact on the performance of the compressor. At the same mass flow rate, a smaller rim diameter can reduce the hub diameter, which reduces the tip Mach number and plays a role in improving the downstream flow field of the impeller; if the hub diameter is too large, the local Mach number of the compressor will reach transonic velocity and generate surge phenomenon in the leading-edge region of the blade, which, in turn, will produce flow separation. Impeller outlet design [31,32] mainly considers impeller diameter, impeller outlet width, and outlet blade angle. Increasing the impeller diameter significantly enhances its work capacity and pressure ratio, but excessively enlarging it will increase the impeller's circumferential velocity, thereby affecting its safe operation. The impeller outlet width mainly affects the impeller import and export speed ratio, and, if the impeller outlet width is too small, it will lead to the top of the impeller and a significant increase in the impact of gap leakage. The impeller outlet blade angle significantly affects the impeller's aerodynamic performance. The backward curved impeller sufficiently reduces the impeller outlet flow rate, regardless of the subsequent stationary parts, which enhances its performance. Its conditions range widely compared to the forward curved and radial impellers, and it also weakens the strength of the secondary flow inside the impeller, which increases compressor efficiency. Hence, this paper adopts the backward curved impeller.

This paper presents a bladeless diffuser, with the main focus on the ratio of the diffuser diameter D_3 to the impeller diameter D_2, referred to as f_1. Studies indicate that a value of f_1 equal to 1 results in better diffuser performance; however, for this design, a value of 0.8 for f_1 is considered. Additionally, the ratio of the diffuser inlet width b_1 to the impeller outlet width f_2 is typically between 1 and 1.1. Thus, the value of f_2 is set to 1 in this design, resulting in b_1 being equal to the impeller outlet width for consistency.

According to the above requirements, the relevant geometric parameters of the compressor are designed, and the results are shown in Table 2, which are input into the compal to generate the meridional flow channel surface, as shown in Figure 1a. D_1 is the impeller inlet diameter, d is the impeller inlet hub diameter, D_2 is the impeller outlet diameter, b_1 is the impeller outlet width, and D_3 is the diffuser diameter. The designed 1D model is imported into Axcent to generate a 3D model, as shown in Figure 1b.

Table 2. Impeller geometry parameters.

Parameters/Unit	Value
Impeller inlet hub diameter d/mm	14
Impeller inlet diameter D_1/mm	19
Impeller outlet diameter D_2/mm	37
Leaf top clearance/mm	0.3
Blade back bending angle	45°
Number of blades	7 × 7
Width of diffuser inlet/mm	2.8173
Diffuser diameter D_3/mm	29

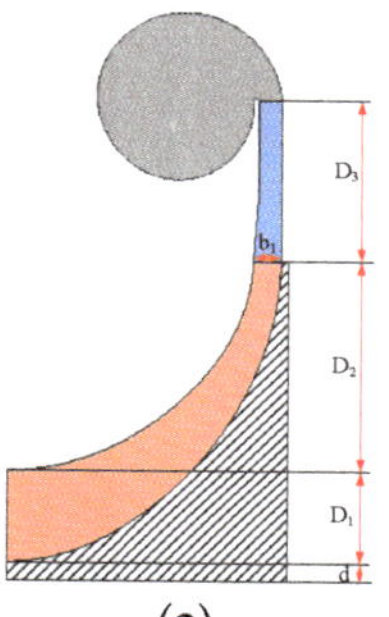

(a)

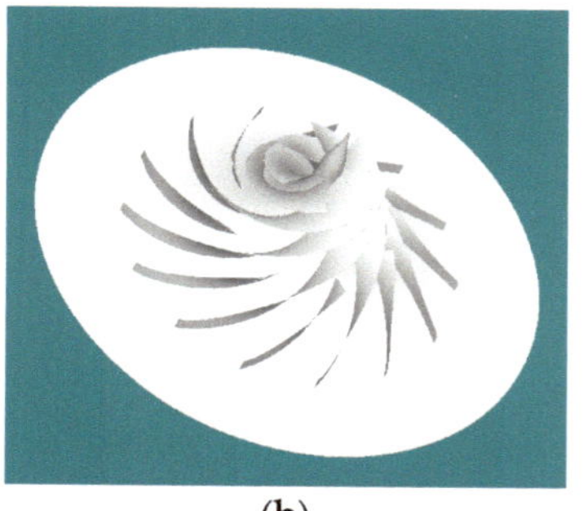
(b)

Figure 1. Compressor model. (**a**) The meridional flow channel surface of the compressor; (**b**) Compressor 3D model.

2.3. Simulation Analysis

Figure 2 displays the grid partitioning diagram for the compressor in the FINE/Turbo16.1 software. The boundary conditions are appropriately set, assuming the working fluid to be an ideal gas. The inlet total pressure is set at 101.3 KPa, and the inlet total temperature is set to 288.15 K. The turbulent viscosity is 0.00005 m^2/s. The wall condition is defined as adiabatic with no slip, and the compressor operates at a rotational speed of 7×10^4 r/min in a rotating coordinate system. The convergence criterion for residual values is set to 10^{-6}. Different turbulence models have distinct thresholds for the y+ value. For the S-A model, the y+ value must be less than 10 to directly solve the viscous sublayer. To ensure precise capturing of the fluid state near the boundary, the first-layer mesh near the wall has a size of 0.005 mm with a y+ value of 5. Three grid models of varying coarseness, composed of 2,121,904, 1,881,016, and 1,526,405 grid cells, respectively, are selected to verify grid independence. The following quality requirements must be met: possess no negative cells, exhibit minimum orthogonality greater than 20°, demonstrate maximum expansion ratio less than 5, and display a maximum aspect ratio less than 1000. Finally, the pressure ratio and efficiency errors for different grid models are maintained within 0.3%. Grid coarseness exhibits minimal influence on the calculation results, as indicated in Table 3.

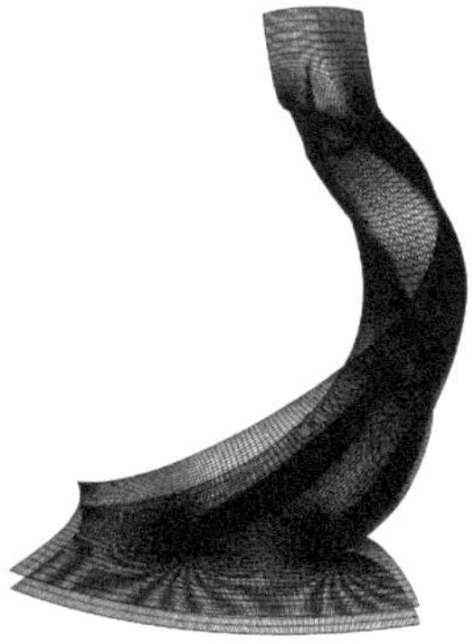

Figure 2. Centrifugal compressor single passage meshing.

Table 3. Grid independence verification.

Number of Grids	Mass Flow/(kg·s^{-1})	Pressure Ratio	Isentropic Efficiency
2,121,904	0.27	1.421	0.815
1,881,016	0.27	1.419	0.812
1,526,405	0.27	1.409	0.81

According to the above simulation conditions for the flow field simulation of the compressor model, its maximum flow rate is 0.0561, the wheezing flow rate is 0.0233, the adiabatic efficiency is 0.828, and the pressure ratio is 1.46, so the performance of the compressor meets the design requirements.

By analyzing the internal flow field of the centrifugal compressor, the deficiencies in the performance of the compressor can be accurately explored so that the design of the compressor can be optimized. Figure 3a displays the impeller meridional surface static pressure diagram, where the gas converts kinetic energy into pressure energy by overcoming the adverse pressure gradient in the diffuser flow. Figure 3b depicts the entropy increase of the meridional surface and highlights the tip clearance and bladeless diffuser as the primary regions of entropy gain. The tip clearance is mainly designed to prevent the collision between the compressor blade and the shroud. Due to the pressure difference between the blade suction and pressure surfaces, a portion of the fluid flows over the top of the blade, forming a leakage flow that leads to a loss of flow and an increase in the blade top entropy value. The bladeless diffuser has no blades, which eliminates impact loss. However, it incurs primary losses due to friction and separation loss. The separation occurs when the boundary layer lacks enough kinetic energy to convert pressure to overcome the adverse pressure gradient.

In Figure 4a, the entropy distribution of the compressor is presented in the 95% and 50% blade height directions. The comparison between Figures 4a and 3b shows that the larger entropy distribution is mainly observed on the suction surface of the main blade and splitter blade. This phenomenon is caused by the large hub turning angle where the airflow travels from the impeller inlet to the diffuser. During this process, the airflow needs to reduce its flow velocity and convert kinetic energy to pressure energy to overcome pressure gradients in the flow. However, a large pressure gradient results in a gradual decrease in flow velocity in both the main flow region and the boundary layer. The gas flow in the boundary layer, which has insufficient kinetic energy to overcome adverse pressure gradients, leads to stagnation, deceleration of gas, and the generation of large flow loss in the separation region. This results in reduced compressor efficiency. Figure 4b shows two separation regions on the suction surface of the blade with low gas flow velocities, leading to significant flow loss, lower impeller efficiency, and lower pressure expansion capacity.

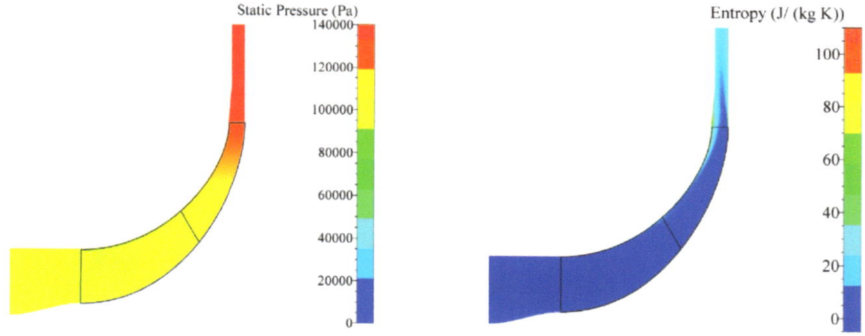

(a) (b)

Figure 3. The initial model of the compressor meridian surface runner surface parameters. (**a**) Meridian surface static pressure; (**b**) Meridian surface entropy distribution.

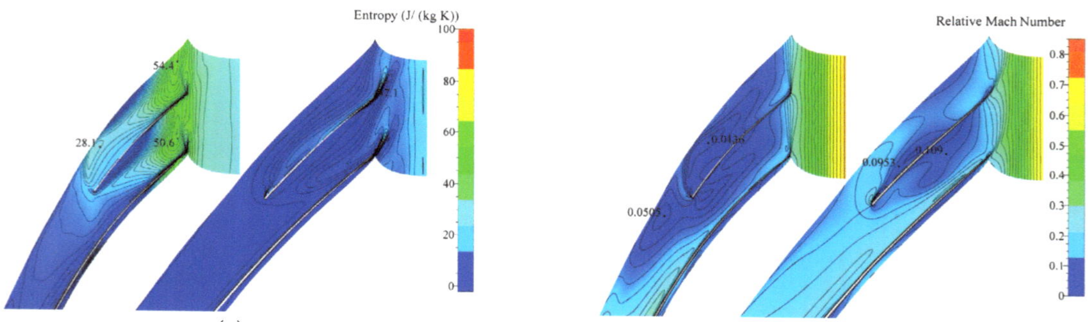

(a) (b)

Figure 4. Entropy distribution and relative Mach number distribution of the initial model of the compressor. (**a**) 95% and 50% blade height direction entropy distribution; (**b**) Relative Mach number distribution in 95% and 50% blade height directions.

3. Study of the Influence of Relevant Parameters on the Performance of Centrifugal Compressors

Several factors influence the performance of a centrifugal compressor, including the compressor's tip clearance, hub ratio, number of impeller blades, impeller inlet angle, circumferential position of the splitter blades, the impeller outlet width, and the backbend angle of the blades. This paper focuses on the main optimized parameters, namely, the number of impeller blades, the relative impeller outlet width, and the impeller backbend angle. By examining the effects of these parameters on the centrifugal compressor's isentropic efficiency, pressure ratio, and flow rate, the impeller is optimized by selecting the parameter values that enhance the compressor's performance. Ultimately, the optimized model is compared and analyzed with the preliminary model.

3.1. *Number of Impeller Blades*

The centrifugal compressor's impeller blades' number has a significant impact, albeit not in a clear linear relationship. Therefore, specific analysis is necessary. This study used 7 × 7 blades in the initial model, and the number of blades was increased and decreased in subsequent models, choosing two groups with 6 × 6 and 8 × 8 blades, respectively. Figure 5 illustrates the efficiency and pressure ratio as influenced by the impeller blades' number. From the figure, it is evident that the compressor with 8 × 8 blades performs better than the others in the compressor's working range, with an isentropic efficiency of up to 83% and a corresponding pressure ratio of 1.477.

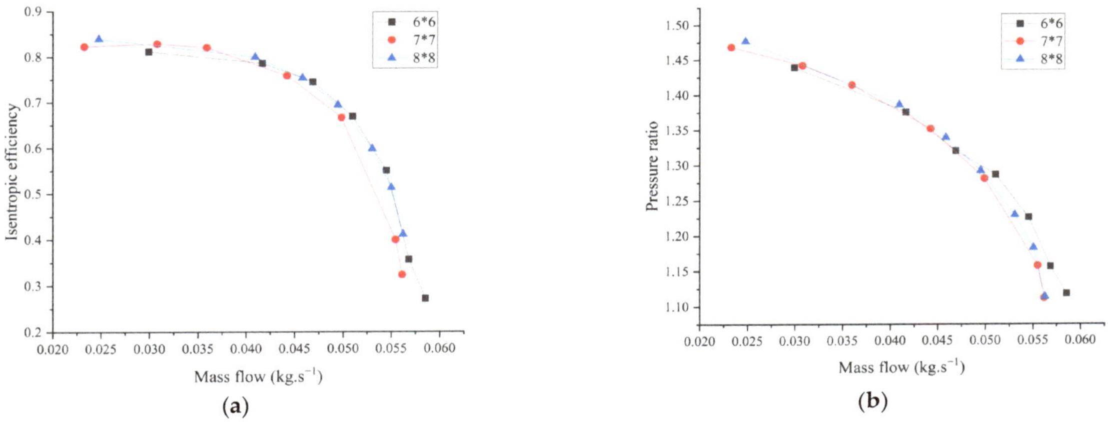

Figure 5. Isentropic efficiency and pressure variation curves for different number of blades. (**a**) Eefect on isentropic efficiency; (**b**) Effect on pressure ratio.

Figure 6 shows the distribution of entropy values at the impeller meridian surface for the 6 × 6 and 8 × 8 impeller blades when the compressor efficiency is highest. Comparing with Figure 3b, it can be seen that, when the number of blades is 8 × 8, the entropy increasement at the tip clearance of the impeller is the smallest, and the energy loss of the compressor compared with the other two cases is reduced, which makes the efficiency of the compressor improved.

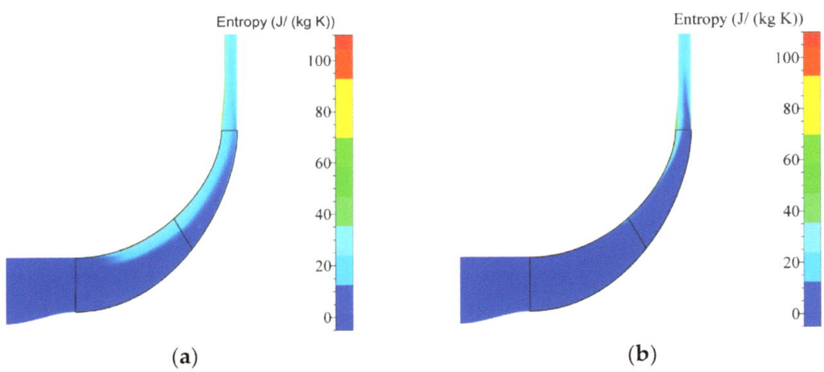

Figure 6. Meridional surface mass entropy distribution with different number of blades. (**a**) Meridional surface mass entropy (number of blades 6 × 6); (**b**) Meridional surface entropy distribution (number of blades 8 × 8).

Figure 7 illustrates the relative Mach number distribution at 95% and 50% of the blade height under different blade numbers. Specifically, the relative Mach number of the 8 × 8 blade configuration is higher than that of 6 × 6 and 7 × 7 configurations. By examining Figure 3b, it is clear that the impeller with the 6 × 6 blade configuration has a larger low-speed zone range than the impeller with the 7 × 7 configuration at the suction surface of the splitter blade at the 95% blade height direction. However, for the 8 × 8 blade configuration, the range of the low-speed zone is significantly reduced compared to the first two configurations, which indicates a decrease in separation loss. This decrease in separation loss is further supported by Figure 8, which demonstrates that as the number of blades increases, the separation loss decreases. Thus, the 8 × 8 blade configuration has a significantly reduced entropy increasement region compared to the other two impellers. Finally, Figure 5a shows an improvement in compressor efficiency for the 8 × 8 blade configuration.

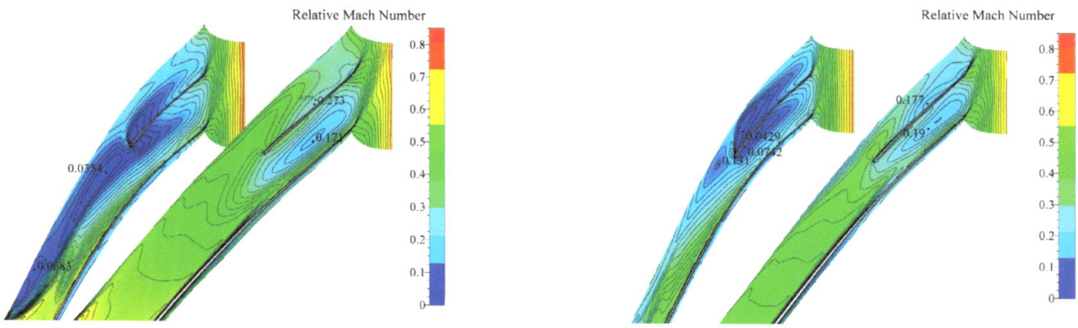

(**a**) (**b**)

Figure 7. Relative Mach number for different blade number at 95% and 50% blade height direction. (**a**) Relative Mach number distribution (number of blades 6 × 6); (**b**) Relative Mach number distribution (number of blades 8 × 8).

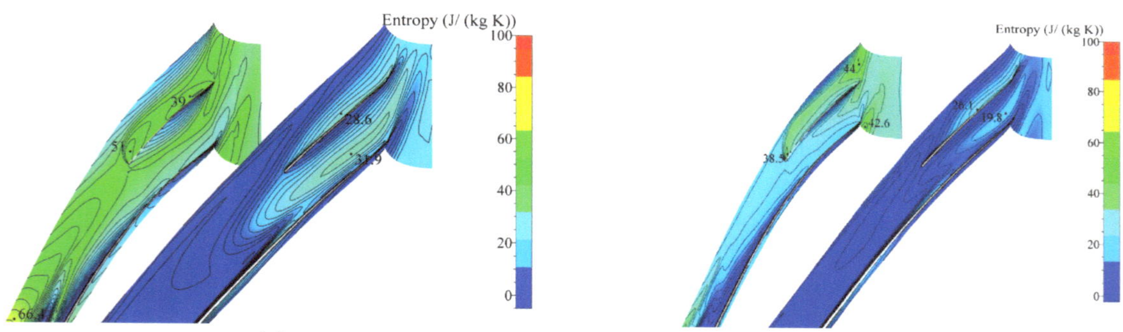

(**a**) (**b**)

Figure 8. Mass entropy distribution of different blade number at 95% and 50% blade height direction. (**a**) Entropy distribution (number of blades 6 × 6); (**b**) Entropy distribution (number of blades 8 × 8).

3.2. Impeller Outlet Relative Width

The impeller outlet relative width is defined as the ratio of impeller outlet width d to impeller outlet diameter D_2. In this study, we set the initial impeller outlet width d as 2.8173 mm and impeller outlet diameter D_2 as 19.627 mm, resulting in the initial impeller outlet relative width is 0.1435. By maintaining the impeller outlet radius, the impeller outlet width d was altered to 3.601 mm and 2.0313 mm, resulting in relative outlet widths of 0.1835 and 0.1035, respectively. The simulation results of the compressor performance are shown in Figure 9. An increase in the relative impeller outlet width corresponded to higher compressor efficiency and pressure ratio. The highest isentropic efficiency achieved was 0.81 when the relative outlet width was 0.1835. However, it should be noted that the isentropic efficiency is lower than that of the other two cases during low-flow conditions. On the other hand, the pressure ratio significantly improved when compared to the other two cases, as illustrated in Figure 9.

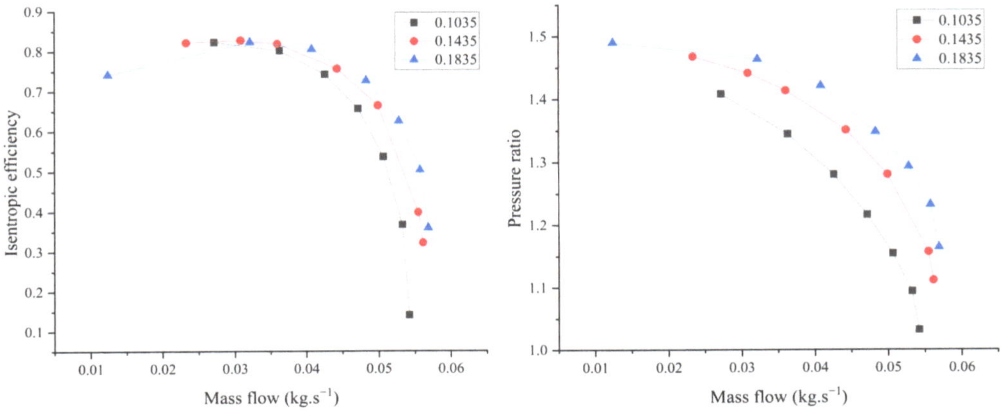

(a) (b)

Figure 9. Variation curves of isentropic efficiency and pressure ratio of impeller with different relative outlet width. (**a**) Effect on isentropic efficiency; (**b**) Effect on pressure ratio.

Impeller relative outlet width change will affect the impeller outlet relative velocity and impeller outlet airflow direction. As the relative outlet width increases, the impeller outlet relative velocity decreases, while the pressure expansion degree increases. Figure 10's static pressure diagram of the compressor meridian surface highlights this relationship. Compared with Figure 3a, it can be observed that an outlet width of 0.1835 results in the highest adverse pressure gradient on the impeller meridian surface. The high adverse pressure gradient reduces the relative flow velocity of the airflow in the diffuser, suppressing the radial flow of the vortex at the impeller's trailing edge. As a result, the pressure expansion capacity is increased. Therefore, the pressure ratio will be higher than in the other two cases.

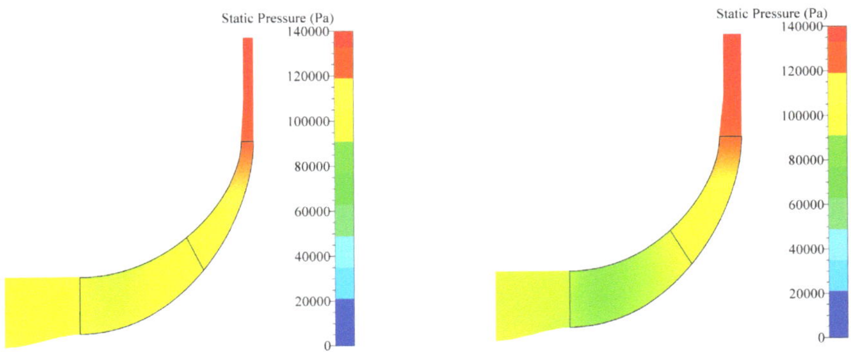

(a) (b)

Figure 10. Different relative outlet width impeller radial surface static pressure. (**a**) Meridional surface static pressure (relative exit width 0.1035); (**b**) Meridional surface static pressure (relative exit width 0.1835).

Figure 11 shows the relative Mach number distribution at the 95% and 50% blade heights of different impeller relative outlet widths. It can be observed that at both relative outlet widths, a lower relative Mach number region exists at the 95% blade height. The width of the impeller outlet affects the range of the low-speed zone, which is wider at the 0.1835 outlet width range than at the 0.1035 range with lower flow rate. At the 50%

blade height, a low-speed zone is also present on the suction surface of the splitter blade, although not noticeably. The impeller's outlet relative width affects the impeller's inlet and outlet relative velocity ratio, which increases as the outlet relative width increases. This results in premature blade suction surface separation and increases separation loss, ultimately reducing efficiency. However, an adverse pressure gradient can improve the pressure expansion capacity. Therefore, in order to maintain efficiency at the design value, the relative outlet width of 0.1835 is recommended.

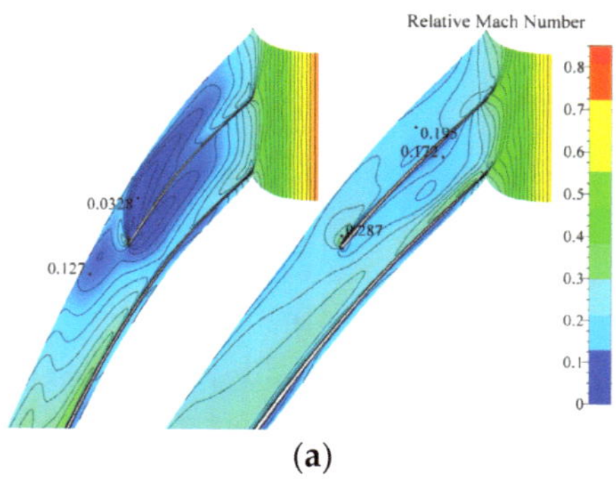

Figure 11. Relative Mach number at different relative exit widths of 95% and 50% blade height directions. (**a**) Relative Mach number distribution (relative exit width 0.1035); (**b**) Relative Mach number distribution (relative exit width 0.1835).

3.3. Blade Backbend Angle

The outlet backbend blade angle can be used to categorize impellers according to their orientation: forward impeller, backward impeller, or radial impeller. For this research, a backward impeller was used. Changing the blade backbend angle has an effect on the outlet airflow's deflection angle, which consequently affects the compressor's efficacy and pressure ratio. The initial blade backbend angle was set to 45°. The blade backbend angle was changed to 40° and 50°, and the impact on the compressor's pressure ratio and efficacy were investigated. The findings of this study are presented in Figure 12.

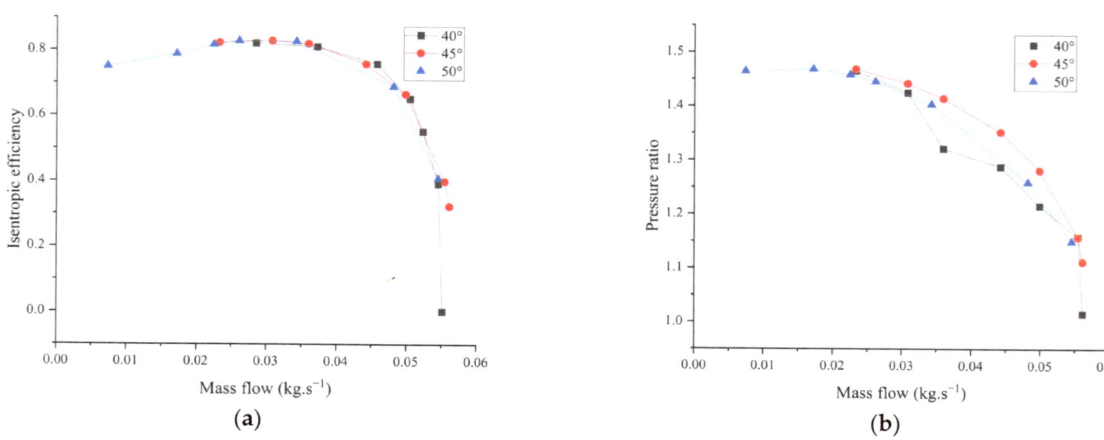

Figure 12. Isentropic efficiency and pressure ratio variation curves for different back-bending angles. (**a**) Effect on isentropic efficiency; (**b**) Effect on pressure ratio.

From Figure 12b, it is evident that the pressure ratio of the impeller initially rises, then falls, as the blade backbend angle increases. The compressor's pressure ratio is at its minimum when the blade backbend angle is 40°. An increase in the angle to 45° raises the pressure ratio, but as the impeller backbend angle advances, a decline in the pressure ratio is observed. Although the compressor efficiency remains relatively constant, the one with a blade backbend angle of 45° yields higher efficiency compared to the other two impellers.

Figure 13 shows the entropy map of the impeller meridian surface for different blade backbend angles. The entropy value is observed to be higher for a blade backbend angle of 40° when compared to that of 50°. As depicted in Figure 3b, a blade backbend angle of 45° results in the smallest meridian surface entropy value. Moreover, the regions with increased entropy values are primarily located at the tip clearance, blade trailing edge position, and diffuser internal.

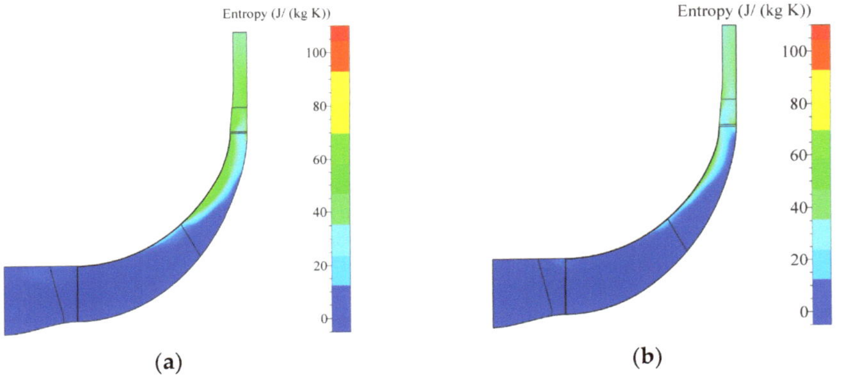

Figure 13. Impeller different back-bending angle meridian surface entropy distribution. (**a**) Meridional surface entropy distribution (impeller backbend angle 40°); (**b**) Meridional surface entropy distribution (impeller backbend angle 50°).

Figure 14 displays the entropy distribution at the 95% and 50% blade heights for different impeller backbend angles. A comparison with Figure 4a indicates that for three different backbend angles of the impeller, both pressure surface entropy values in the main and splitter blades increase more at a backbend angle of 40° compared with the other two impellers. The entropy increase range for the splitter blade suction surface is also wider, leading to more serious gas mixing and, consequently, lower compressor efficiency and pressure ratio. While the impeller with a blade backbend angle of 45° experiences a significant decrease in entropy value at 95% of the blade height and an improved gas mixing phenomenon, the impeller with a backbend angle of 50° experiences a significant increase in entropy value due to the gas changing from axial to radial. This change induces a vortex area close to the compressor shroud, in which the velocity is reduced, and the energy loss increases. As the blade backbend angle increases, the outlet pressure of the compressor gradually decreases, leading to more airflow separation in the flow channel, increased energy loss, and decreased compressor efficiency. Therefore, the impeller with a backbend angle of 45° was selected for this study, to optimize its efficiency.

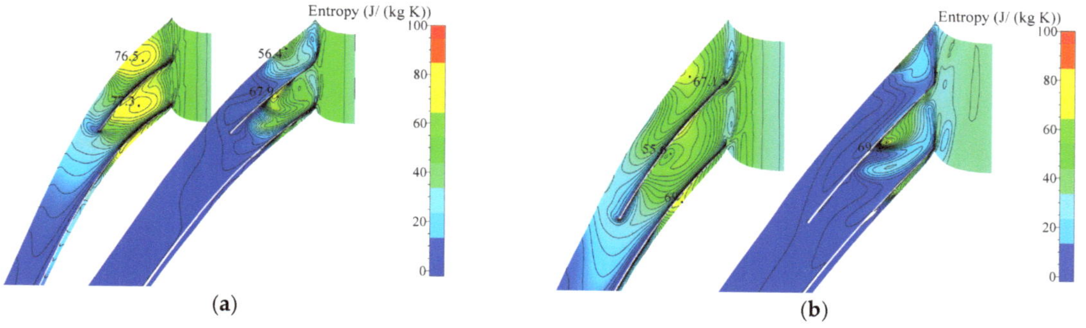

Figure 14. Distribution of 95% and 50% blade height entropy values for different impeller backbend angles. (**a**) Entropy distribution (impeller backbend angle 40°); (**b**) Entropy distribution (impeller backbend angle 50°).

3.4. Design Parameter Optimization

According to the aforementioned study, optimization of parameters from the initial model had a significant impact on the compressor performance. The number of impeller blades was increased from the original 7 × 7 to 8 × 8 while retaining the backbend angle of the impeller at 45°. The relative outlet width was also increased to 0.1835, which adjusts the impeller outlet width to 3.601. All other parameters remained constant.

The optimized model is meshed, and Figure 15 shows the performance curves of the compressor before and after optimization. The optimized efficiency and pressure ratio can reach 0.842 and 1.49, respectively, indicating a 1.4% and 1.8% improvement from their previous levels. Furthermore, the optimized isentropic efficiency and pressure ratio show higher readings at various flow rates compared to their previous values before optimization.

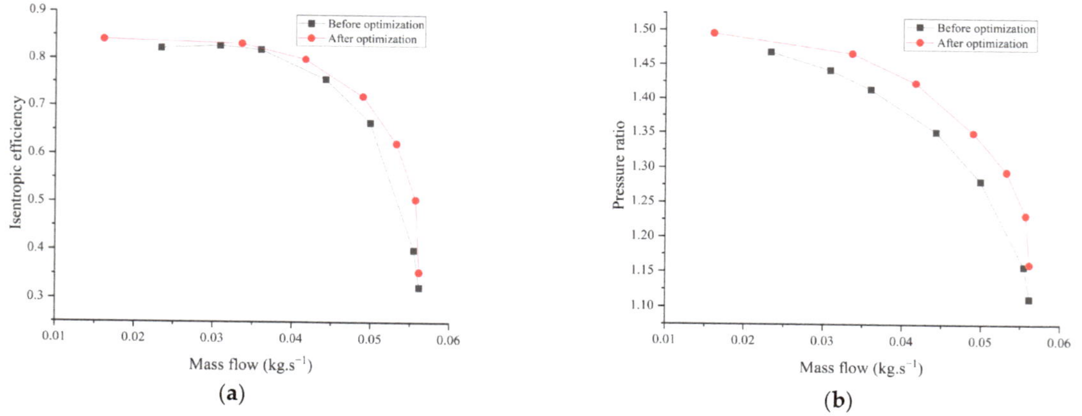

Figure 15. Isentropic efficiency and pressure ratio change curves before and after optimization. (**a**) Effect on isentropic efficiency;(**b**) Effect on pressure ratio.

Figure 16a presents the optimized impeller meridian static pressure and entropy distribution diagrams. A comparison with Figure 3 reveals that before optimization, the impeller meridian surface exhibits a small adverse pressure gradient. After optimization, the static pressure distribution is more even, characterized by a higher adverse pressure gradient, thereby lowering the relative flow velocity of airflow in the diffuser and improving the pressure expansion capacity. As Figure 16b depicts, the entropy increase on the optimized impeller meridian surface is reduced by a considerable amount, indicating the mitigation of gas confusion in the tip clearance and bladeless diffuser section, and the improved uniformity of airflow in the bladeless diffuser section. This improvement enhances the compressor performance.

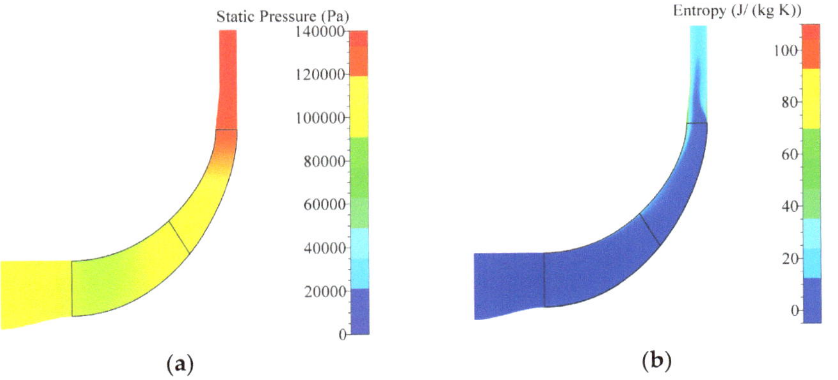

Figure 16. Optimized meridian flow path surface parameters for compressor. (**a**) Optimized meridian surface static pressure; (**b**) Meridian surface entropy distribution after optimization.

Figure 17a presents the relative Mach number distribution of the impeller at blade heights of 95% and 50%. The same low-speed region is shown in Figure 4b. However, compared to before optimization, the range of the low-speed region is significantly reduced, and the speed is also increased. The main area of the impeller entropy gain in Figure 17b overlaps with the low-speed region, where the presence of the low-speed region leads to the generation of separation loss. The entropy gain region is considerably smaller, and the entropy gain is significantly lower than before optimization. As a result of these changes, the efficiency and pressure ratio of the impeller are improved.

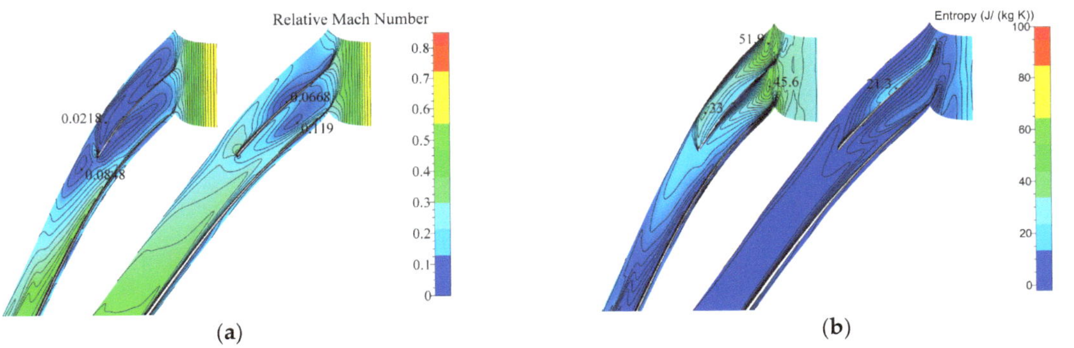

Figure 17. Relative Mach number and entropy distribution in f and 50% blade height directions after optimization. (**a**) Optimized relative Mach number distribution; (**b**) Optimized entropy distribution.

4. Experimentation

The experimental test bench consists of several key components: the INV9832-50 three-axis vibration acceleration sensor, an electric booster, an electronic control board, external water cooling, and tow atmospheric pressure sensors. The three-axis acceleration sensor is responsible for measuring speed variations in three dimensions, allowing for the monitoring of compressor vibrations during operation to ensure proper functionality. This sensor is capable of measuring within a range of 50 g, operates at temperatures between −50 °C and 120 °C, and has a maximum error of 0.05 g. The electronic control board functions primarily to manage compressor speed through a computer application program.

This capability enables measurement of the compressor's pressure ratio and efficiency at different speeds. To determine pressure changes, tow atmospheric pressure sensors are installed at both the inlet and outlet of the centrifugal compressor. This sensor records pressure fluctuations at different speeds, enabling the calculation of pressure ratio changes. The Honeywell HPB200F5DA atmospheric pressure sensor has a range of 500 to 1500 hPa, an operating temperature range of −40 °C to 110 °C, and an error range of ±0.4 hPa.

The physical model of the centrifugal compressor was developed based on the optimized impeller parameters, as shown in Figure 18a. The compressor's efficiency and pressure ratio experiments were carried out on the bench with an ambient temperature of 25.61 °C and pressure of 103.03 KPa. The flow rate, pressure ratio, and efficiency of the compressor were evaluated at different RPM levels from 40,000–70,000 RPM, in steps of 10,000 RPM due to an insignificant boost effect of the electric supercharger at 10,000–30,000 RPM, as illustrated in Figure 18b. Moreover, the overall vibration noise of the supercharger and the electric control board motor temperature was monitored. The experiment was conducted under the aforementioned experimental conditions, and the results are presented in Figure 19. The experimental and simulated compressor performances were compared, indicating a minor difference within the acceptable error range. The compressor met the design requirements. To establish the overall credibility of the experiment, Figure 20 depicts the engine model equipped with a propeller, matching it with an electric booster, primarily utilized on unmanned aerial vehicles.

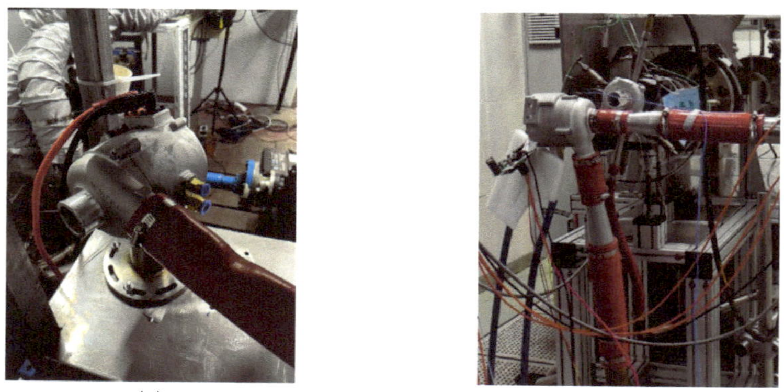

(**a**) (**b**)

Figure 18. Centrifugal compressor and experiment. (**a**) Centrifugal compressor; (**b**) Centrifugal compressor bench test.

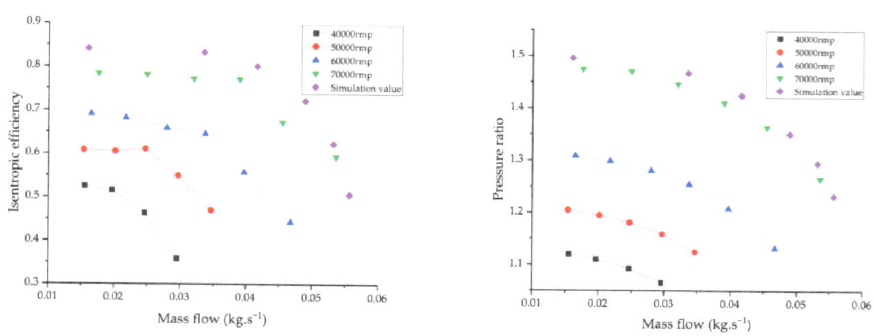

(**a**) (**b**)

Figure 19. Pressure ratio and actual efficiency. (**a**) Efficiency of centrifugal compressor; (**b**) Centrifugal compressor pressure ratio.

Figure 20. An engine equipped with an electric booster, used on UAV.

5. Conclusions

In this paper, the relevant operational conditions of the compressor were calculated according to design guidelines. The impeller and diffuser models were also designed, and the impact of these operational conditions on the compressor's performance were studied. The parameters were then optimized, and the conclusions obtained are as follows:

(1) The performance of a centrifugal compressor is markedly influenced by the number of impeller blades, the impeller outlet's relative width, and the impeller backbend angle. Increasing the number of impeller blades will decrease the low-speed zone range at the compressor exit, resulting in a decrease in entropy and an increase in compressor efficiency. An appropriate increase in the impeller's backbend angle enhances its efficiency and pressure ratio. However, a continuous rise of backbend angle will reduce the exit airflow deflection angle and intensify the airflow separation phenomenon in the flow channel. These factors result in an increase in inefficiencies. Increasing the relative width of the impeller exit decreases the relative speed of the impeller and increases pressure expansion, thereby amplifying the compressor's pressure ratio.

(2) By optimizing the above-mentioned parameters of the compressor, compared with the initial model, the performance of the centrifugal compressor can be effectively improved when the number of impeller blades is 8×8, the relative width of the impeller outlet is 0.01835, and the backbend angle of the impeller is 45°. The optimized compressor efficiency can reach 0.842, and the pressure ratio can reach 1.49 with a working margin of 22%. The optimized efficiency is 1.4%, and the pressure ratio is 1.8% higher than before.

(3) The experimental bench was utilized to evaluate the compressor's efficiency and pressure ratio. The compressor's electric supercharger was operated between 40,000 and 70,000 RPM to calculate both the pressure ratio and actual efficiency. The experimental results demonstrated, basically, consistency with the simulated values. Hence, validating the engineered model met the design requirements.

Author Contributions: Formal analysis, Y.W.; Funding acquisition, B.L.; Investigation, M.W.; Methodology, C.W.; Software, Y.W.; Writing—original draft, Y.W.; Writing—review & editing, X.X. and B.L. All authors have read and agreed to the published version of the manuscript.

Funding: This work was supported by the Industrial Proactive and Key Technology Program of Jiangsu Province (Grant number BE2022053-2).

Data Availability Statement: Not applicable.

Conflicts of Interest: The authors declare no conflict of interest.

References

1. Carlucci, A.P.; Ficarella, A.; Trullo, G. Performance Optimization of a Two-Stroke Supercharged Diesel Engine for Aircraft Propulsion. *Energy Convers. Manag.* **2016**, *122*, 279–289. [CrossRef]
2. Xu, B.; Li, D.; Ma, Z.; Zheng, M.; Li, Y. Thermodynamic Optimization of a High Temperature Proton Exchange Membrane Fuel Cell for Fuel Cell Vehicle Applications. *Mathematics* **2021**, *9*, 1792. [CrossRef]
3. Chen, Y.; Cheng, Z.; Qian, Y. Fuel Consumption Comparison between Hydraulic Mechanical Continuously Variable Transmission and Stepped Automatic Transmission Based on the Economic Control Strategy. *Machines* **2022**, *10*, 699. [CrossRef]
4. Lu, Q.; Zhou, W.; Zheng, Y. Regenerative Braking Control Strategy with Real-Time Wavelet Transform for Composite Energy Buses. *Machines* **2022**, *10*, 673. [CrossRef]
5. Alshammari, M.; Alshammari, F.; Pesyridis, A. Electric Boosting and Energy Recovery Systems for Engine Downsizing. *Energies* **2019**, *12*, 4636. [CrossRef]
6. Li, D.; Li, S.; Ma, Z.; Xu, B.; Lu, Z.; Li, Y.; Zheng, M. Ecological Performance Optimization of a High Temperature Proton Exchange Membrane Fuel Cell. *Mathematics* **2021**, *9*, 1332. [CrossRef]
7. Lee, W.; Schubert, E.; Li, Y.; Li, S.; Bobba, D.; Sarlioglu, B. Electrification of Turbocharger and Supercharger for Downsized Internal Combustion Engines and Hybrid Electric Vehicles-Benefits and Challenges. In Proceedings of the 2016 IEEE Transportation Electrification Conference and Expo (ITEC), Dearborn, MI, USA, 27–29 June 2016; pp. 1–6.
8. Li, Z.; Lu, X.; Han, G.; Zhu, J. Investigation on Flow Mechanism of an Advanced Transonic Centrifugal Compressor with Free-Form Impeller at Design and off-Design Speeds. *Eng. Appl. Comput. Fluid Mech.* **2022**, *16*, 1739–1760. [CrossRef]
9. Tang, Q.; Wu, H.; Li, J.; Lou, H.; Yang, C. Performance Optimization of Centrifugal Compressors Based on Throughflow Model. *Arab J. Sci. Eng.* **2022**, *47*, 16439–16450. [CrossRef]
10. Omidi, M.; Liu, S.-J.; Mohtaram, S.; Lu, H.-T.; Zhang, H.-C. Improving Centrifugal Compressor Performance by Optimizing the Design of Impellers Using Genetic Algorithm and Computational Fluid Dynamics Methods. *Sustainability* **2019**, *11*, 5409. [CrossRef]
11. Tang, X.; Gu, N.; Wang, W.; Wang, Z.; Peng, R. Aerodynamic Robustness Optimization and Design Exploration of Centrifugal Compressor Impeller under Uncertainties. *Int. J. Heat Mass Transf.* **2021**, *180*, 121799. [CrossRef]
12. Zhou, W.; Zheng, Y.; Pan, Z.; Lu, Q. Review on the Battery Model and SOC Estimation Method. *Processes* **2021**, *9*, 1685. [CrossRef]
13. Li, Z.; Lu, X.; Wu, Y.; Han, G. Quantitative Investigation of the Turbulence Model Effect on High-Pressure-Ratio Centrifugal Compressor Performance Prediction. *Int. Commun. Heat Mass Transf.* **2023**, *142*, 106644. [CrossRef]
14. Kong, D.; Song, S.J. Influence of Flow Coefficient on Unsteady Impeller Loading Induced by Impeller-Diffuser Interaction. *J. Mech. Sci. Technol.* **2022**, *36*, 2403–2413. [CrossRef]
15. Seralathan, S.; Hindustan Institute of Technology and Science; Hindustan University; Roy Chowdhury, D.G. Hindustan University Numerical Studies on the Effect of Diffuser Rotational Speeds on Low Pressure Ratio Centrifugal Compressor Performance. *JAFM* **2017**, *10*, 785–799. [CrossRef]
16. Chen, X.; Xu, W.; Yang, S.; Du, Y.; Dou, H.-S. Effects of Vaneless Region Length on the Performance of Centrifugal Compressor with Vaned Diffuser. *J. Braz. Soc. Mech. Sci. Eng.* **2021**, *43*, 227. [CrossRef]
17. Ma, C.; Li, W. The Influence of Low-Solidity and Slotting Combined Treatment Method of Diffuser Vanes on Performance of Centrifugal Compressor. *J. Braz. Soc. Mech. Sci. Eng.* **2019**, *41*, 68. [CrossRef]
18. Gunadal, S.M.; Govardhan, M. Improvement in Stable Operating Range of a Centrifugal Compressor with Leaned Diffuser Vanes. *J. Mech. Sci. Technol.* **2019**, *33*, 5261–5269. [CrossRef]
19. Li, C.; Ji, C.; Shi, M.; Sun, Q. Influence of Rotor–Stator Component Partition Structure on the Aerodynamic Performance of Centrifugal Compressors. *Eng. Appl. Comput. Fluid Mech.* **2019**, *13*, 1080–1094. [CrossRef]
20. Wang, Y.; Wang, K.; Lin, F.; Nie, C.; Engeda, A. Performance Analysis and Improvement of a High Flow Coefficient Centrifugal Compressor. *Sci. China Technol. Sci.* **2014**, *57*, 1647–1657. [CrossRef]
21. Zhang, Y.; Gao, F.; Zhao, F. Research on Path Planning and Tracking Control of Autonomous Vehicles Based on Improved RRT* and PSO-LQR. *Processes* **2023**, *11*, 1841. [CrossRef]
22. Moussavi Torshizi, S.A. Improving the Performance of Centrifugal Compressors by Flow Recirculation. *J. Appl. Fluid Mech.* **2019**, *12*, 1791–1799. [CrossRef]
23. Li, Z.; Lu, X.; Han, G.; Huang, E.; Yang, C.; Zhu, J. Numerical and Experimental Investigation of Flow Mechanism and Application of Tandem-Impeller for Centrifugal Compressor. *Aerosp. Sci. Technol.* **2020**, *100*, 105819. [CrossRef]
24. Yang, X.; Tang, M.; Li, S.; Li, W. Design of a High-Performance Centrifugal Compressor Impeller and an Investigation on the Linear Effects of the Hub. *AIP Adv.* **2022**, *12*, 025322. [CrossRef]
25. Hong, S.; Mugabi, J.; Jeong, J.-H. Numerical Study on Vortical Flow Structure and Performance Enhancement of Centrifugal Compressor Impeller. *Appl. Sci.* **2022**, *12*, 7755. [CrossRef]
26. Jaatinen-Värri, A.; Tiainen, J.; Turunen-Saaresti, T.; Grönman, A.; Ameli, A.; Engeda, A.; Backman, J. Centrifugal Compressor Tip Clearance and Impeller Flow. *J. Mech. Sci. Technol.* **2016**, *30*, 5029–5040. [CrossRef]
27. Tian, J.; Chen, Y.; Ma, Z. Numerical Simulation of Performance of an Air–Water Separator with Corrugated Plates for Marine Diesel Engines. *Processes* **2020**, *8*, 1617. [CrossRef]
28. Semlitsch, B.; Huscava, A. Shape Optimisation of Turbomachinery Components. In Proceedings of the 8th European Congress on Computational Methods in Applied Sciences and Engineering, Oslo, Norway, 5–9 June 2022.

29. Li, X.; Huang, N.; Chen, G.; Zhang, Y.; Zhao, Y.; Zhang, J.; Tong, D. Numerical Simulation on the Influence of Inlet Flow Characteristics on the Performance of a Centrifugal Compressor. *Energies* **2023**, *16*, 3869. [CrossRef]
30. Liu, Y.; Jiang, T.; Zheng, Y.; Tian, J.; Ma, Z. Multi-Scale Multi-Field Coupled Analysis of Power Battery Pack Based on Heat Pipe Cooling. *Processes* **2019**, *7*, 696. [CrossRef]
31. Tai, Y.; Li, P.; Zheng, Y.; Tian, J. Entropy Generation and Thermoelastic Damping in the In-Plane Vibration of Microring Resonators. *Entropy* **2019**, *21*, 631. [CrossRef]
32. Casey, M.; Robinson, C. Some Properties of the Exit Velocity Triangle of a Radial Compressor Impeller. *J. Turbomach.* **2022**, *145*, 051007. [CrossRef]

processes

MDPI

Article

Structure Size Optimization and Internal Flow Field Analysis of a New Jet Pump Based on the Taguchi Method and Numerical Simulation

Zhiliang Wang [1,2], Yu Lei [2,3], Zhenhua Wu [2,3], Jian Wu [2,3], Manlai Zhang [2,4] and Ruiquan Liao [1,2,*]

1 School of Petroleum Engineering, Yangtze University, Wuhan 430100, China
2 Lift Test Base of China National Petroleum Corporation, Hami 839009, China
3 Research and Development Center, Tuha Oilfield Company, China National Petroleum Corporation, Hami 839009, China
4 School of Mechanical Engineering, Yangtze University, Jingzhou 434023, China
* Correspondence: zhiliang-wang1996@foxmail.com

Abstract: Interlayer contradiction (high-pressure oil that prevents low-pressure oil from being extracted) has always been the main factor affecting the oil-recovery efficiency of the many oil-bearing series in shale oil wells in Eastern Shandong, China. If steps to deal with interlayer contradiction are not taken, Shengli Oilfield's oil-recovery efficiency will be significantly reduced after a certain period of exploitation. Furthermore, as the drilling depth increases, the formation-fluid supply capacity of Shengli Oilfield becomes worse and further increases the difficulty of oil recovery as well as production costs. In order to improve the oil-recovery efficiency of shale oil wells in Eastern Shandong and realize cost reductions and efficiency increases, we designed a new jet pump in this study. The pump can be used for oil recovery according to the principle of Venturi jet propulsion, as the required power fluid is not a high-pressure fluid injected from the ground, but rather high-pressure oil that is present in the formation. Through the analysis of the overall structure of the new jet pump, it was found that the pump could not only transform the existing interlayer contradiction (co-mining of high and low oil layers by utilizing interlayer contradiction), but also had the characteristics of a simple structure and low production costs. Since the structural dimensions of the jet pump and the physical characteristic parameters of the fluid have significant impacts on pump efficiency, we first analyzed the internal flow field of the jet pump by using numerical simulations and found that the throat–nozzle distance, area ratio, throat length–diameter ratio, diffuser angle, and flow ratio had the most significant impacts on pump efficiency. After obtaining the specific numerical range of the abovementioned structural parameters when the pump efficiency was as its maximum, an orthogonal array designed according to the Taguchi method was used to conduct experiments. According to a range analysis and an analysis of variance, at an unchanged flow ratio (0.3156), the new jet pump achieved the highest efficiency (31.26%) when the throat–nozzle distance was 2.62 mm, the throat length was 46 mm, the throat diameter was 6.8 mm, and the diffuser angle was 7.5°. In comparing its efficiency with that before optimization, we noticed that the efficiency was significantly improved by about 10%. These research results not only offer a new idea for the existing oil-recovery mode, but also introduce a new method for optimizing the structure of jet pumps.

Keywords: Taguchi method; numerical simulation; new jet pump; structural size optimization; comparison and analysis

Citation: Wang, Z.; Lei, Y.; Wu, Z.; Wu, J.; Zhang, M.; Liao, R. Structure Size Optimization and Internal Flow Field Analysis of a New Jet Pump Based on the Taguchi Method and Numerical Simulation. *Processes* **2023**, *11*, 341. https://doi.org/10.3390/pr11020341

Academic Editor: Alfredo Iranzo

Received: 18 December 2022
Revised: 10 January 2023
Accepted: 10 January 2023
Published: 20 January 2023

1. Introduction

With the continuous development of domestic drilling and production technology, China's onshore drilling depth has been advancing from 4500–6000 m (deep wells) to 6000–9000 m (ultra-deep wells) [1,2]. It has been proven that there are a large number of unconventional oil reservoirs in Eastern Shandong. These reservoirs require artificial lift technology [3,4] (mechanical devices sent down the shaft to replenish the energy of the fluid in the well by performing work on the fluid in the pipe) to aid in extracting oil from the wells because of the large buried depth and natural formation energy in the wells, which is insufficient to sustain their flow production [5]. We often use artificial lift systems with deep-well jet pumps [6,7] when extracting oil from unconventional reservoirs in the area. However, Shengli Oilfield has many oil-bearing series (multiple sections of oil layers with different levels of pressure in the wells), making interlayer contradiction more prominent. When extracting high-pressure oil and low-pressure oil in a well, high-pressure oil enters the low-pressure reservoir under the effect of its own pressure, preventing the recovery of low-pressure oil. In addition, after the extraction proceeds for some time, the oil pressure in the well significantly drops, which further impedes oil recovery and leads to a significant decline in the oil-lifting efficiency of the jet pump. Therefore, the following oil-recovery engineering problems in East China urgently require solutions: (1) the current oil-recovery efficiency of deep-well jet pumps needs to be improved, (2) the drilling and production costs need to be lowered, and (3) the amount of downhole safety accidents needs to be reduced.

To fully understand the basic theory [8,9], structural composition [10], and application effect of the jet pump [11], as well as to ensure safe on-site operations and to help decrease costs to increase efficiency, scholars at home and abroad have carried out a large amount of analysis and research on the jet pump. For example, Y.R. Reddy et al. [12] conducted a theoretical study on the efficiency of jet pumps. When the flow ratio was 1, they found that the maximum ideal pump effect of the jet pump was 50%, but the pump effect was only 39.4% after the structure of the jet pump was optimized using basic theory. When the appropriate material was chosen to reduce the friction resistance of the inner wall of the jet pump facing the fluid, the pump efficiency was able to reach more than 40%. K.E. Brown and H. Petrie jointly published *The Artificial Lift Method* [13], a book that systematically summarizes the principle, formula derivation, and application of the jet pump. However, their study was limited, as it assumed the two fluids in a jet pump have the same density. A. W. Grupping [14] elaborated on the working principle, the calculation formula, the curve of the working characteristic, the calculation program, and an example calculation of the oil-well jet pump, and they provided a set of theories and calculation methods combined with oil-production conditions. S.H. Winoto et al. [15] conducted a theoretical analysis and laboratory experiments on the efficiency of the jet pump and studied the impacts of the area ratio between the nozzle and throat pipe and the nozzle sections of different shapes on jet pump performance by using an experimental bench. Finally, they concluded that jet pumps with circular nozzles have the greatest pump efficiency. C.B. Wang [16] deduced the basic characteristic equation and efficiency of the jet pump using energy conservation and obtained an optimal parameter equation by choosing the multivariate function method to find the extreme value. After analyzing the friction loss coefficient and other parameters, he concluded that the friction loss coefficient directly affects the size of the optimal area ratio and indirectly affects the efficiency of the jet pump.

J. Thomson [11] was the first person to draw the structural diagram of the jet pump. Since then, more and more scholars have engaged in theoretical analysis, structural design, and research on the applications of jet pumps. O. B. Kwon [17] and other scholars used numerical simulations [18,19] and experiments to study the effects of the shape of the mixing chamber and different jet rates on the performance of jet pumps. After comparing simulation data with experimental data, it was found that the RNG k–ε model was more suitable to simulate the internal flow field of pumps than the standard k–ε model. If the flow rate is low, there will be some reverse flow in a pump. Y. Yamazaki [20] studied the influence of the shape of the nozzle (circular, notched, and flower-shaped) and the shape of the throat (traditional straight throat and diverging throat with a very gentle gradient) on the performance of jet pumps. The velocity distribution in the pipe and the local friction loss coefficient were experimentally analyzed, and the performance of the jet pumps was evaluated using these concrete values. During the experiment, the pump efficiency of all jet pumps could be improved when the diverging throat with a very gentle gradient was selected and only the nozzle shape was changed. K. Aldaş et al. [21] used CFD to compare the performance of jet pumps with different levels of surface roughness. They found that the SST model could best improve simulation accuracy: when the area ratio was 5.92 and the relative roughness was 0.05, the relative pump efficiency decreased by about 60%. K Xu et al. [22] used CFD to numerically simulate an annular jet pump; after establishing an accurate RBF neural network model and an NSGA-II algorithm, the multiobjective optimization of the annular jet pump was carried out. By comparing a preoptimized jet pump with an optimized jet pump, it was found that the head ratio of the former was increased by 30.46% and the efficiency was slightly improved. Regarding the application of jet pumps, S. Chen et al. [23] developed an integrated sand-removal system that used a jet pump to remove sand particles in wells. After optimizing the design of the jet pump and applying the system in the field, they found that the jet pump could effectively remove loose sand and the system could effectively avoid loss circulation. C. C. Gao [24] analyzed the application of jet pump technology for environmental protection and found that jet pump technology has played a very important role in sewage treatment, waste gas treatment, dust treatment, and other treatments. S. Sarshar [25] analyzed the application of jet pumps in oil and gas production and found that jet pump technology cannot only improve the production of oil and gas wells and increase low gas pressure, but is also a relatively economical and simple production method. B. Sun [26] studied the exploitation of coalbed methane with a jet pump by conducting an analysis of the basic theory of jet pumps and experiments on the solid–liquid two-phase flow. Thus, this study created new jet pump technology that can be used to extract coalbed methane and was applied to 116 gas wells with improved results.

To summarize, although jet pumps have been studied for many years [8,9,27], the basic performance equations of jet pumps are still not unified [11] and some basic performance equations are not universal (the equations are only applicable when the two fluid densities in a jet pump are the same) [11,16]. Although deep-well jet pumps [6,7,14] have been widely used in oil recovery engineering [7,28], the power fluid currently used in pumps is manually injected high-pressure fluid, which leads to high production costs. In addition, when there are multiple reservoirs with different levels of pressure in a well, interlayer contradiction (high-pressure oil that hinders low-pressure oil recovery [29,30]) further increases the difficulty and cost of oil recovery. In order to improve oil recovery efficiency, effectively exploit low energy and low production layers, and better realize cost reduction and efficiency increases, this paper designed a new jet pump after comprehensively considering the characteristics of strata, production cost, and oil-recovery efficiency in Eastern Shandong, and the basic characteristic equation of this new kind of jet pump was derived by using a momentum equation, an energy equation, and a continuity equation (when the two fluid densities in a jet pump are different, this equation is also applicable). Compared with traditional jet pumps [6,7,16,28], the new jet pump can effectively transform interlayer contradiction and further achieve lower costs and a higher efficiency. Since pump efficiency is subject to structural dimensions and other factors, we first used numerical simulations to understand the main structural parameters influencing pump efficiency and the intervals of the main structural dimensions at maximum pump efficiency. Then, we optimized the structural dimensions of the new jet pump using the Taguchi method. Based on range and signal-to-noise ratio analyses, the order of impact of the throat diameter, diffuser angle, throat–nozzle distance, and throat length on pump efficiency was obtained. The specific contribution (relative influence on pump efficiency) of each control factor was also derived using an ANOVA. Ultimately, the specific throat–nozzle distance, throat diameter, and throat length needed for maximum pump efficiency were calculated.

2. The Structure and Working Principle of the New Jet Pump

The conventional tube-string structure currently used for commingled production in an oil well containing multiple reservoir sections (e.g., high- and low-pressure reservoirs) is shown in Figure 1a. Although this method [8,9,31] can be used to extract high- and low-pressure oil, it presents a prominent contradiction between layers, causing the high-pressure oil to be pushed by its own pressure into the low-pressure reservoir, thus hindering the recovery of low-pressure oil. To reduce the difficulty of oil recovery and achieve lower costs and a higher efficiency, we propose an oil-recovery process with a new type of jet pump, whose overall tube-string structure is shown in Figure 1b. The new jet pump is installed between two packers such that the lower packer is located between the high-pressure reservoir and the low-pressure reservoir, whereas the upper packer is located above the new jet pump so that the low-pressure reservoir is separated from the other reservoirs.

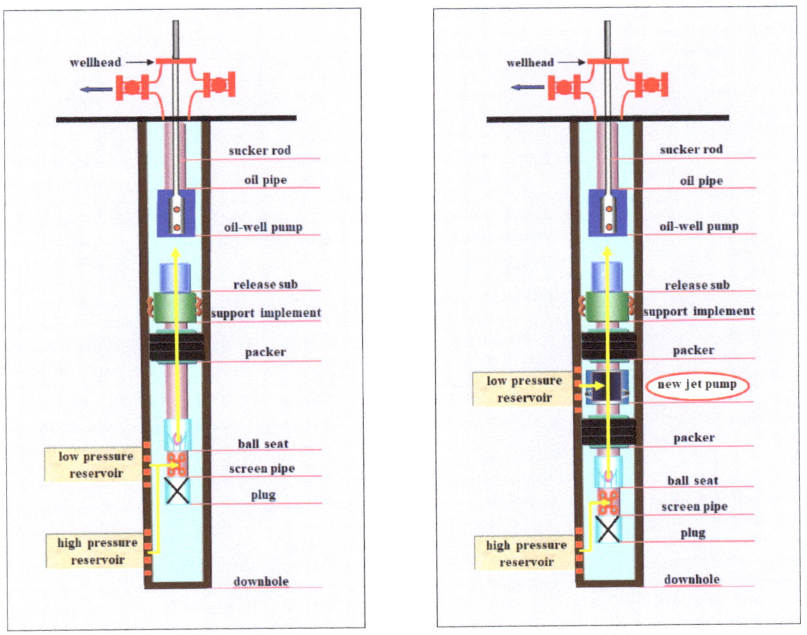

(a) (b)

Figure 1. Comparison between a conventional co-production string and a co-production string with a deep-well jet pump. (**a**) Schematic diagram of a conventional co-production string. (**b**) Schematic diagram of a co-production string with a new jet pump.

Figure 2 shows the overall structure of the new jet pump. The tool is assembled from 25 parts, with an overall simple structure [32]. Under its own pressure, when high-pressure oil enters the jet pump along the internal flow path of the tube string (the orange arrow in Figure 2 indicates the flow direction of the high-pressure oil in the jet pump), the high-pressure fluid generates an entrainment effect through the nozzle jet flow due to the small internal flow channel of the nozzle in the pump. This forms a negative-pressure zone near the nozzle outlet, which, combined with the pressure difference, causes low-pressure fluid (the green arrow in Figure 2 indicates the flow direction of the low-pressure fluid in the jet pump) to enter the tool through the small hole on the short joint of the shell and then pass through the check valve along the flow path to the negative-pressure zone. At this time, only a small amount of the two fluids is mixed. The full mixture is achieved in the throat. When the mixture (the red arrow in Figure 2 indicates its flow direction) enters the diffuser tube, part of its kinetic energy is converted by the expanding flow path (a gradually expanding path along the central flow channel of the diffuser tube) into pressure energy, thus effectively increasing the oil pressure in the low-pressure reservoir. When the mixture enters the oil-well pump along the connected oil-tube string, the plunger inside the oil-well pump lifts the mixture to the ground through an up-and-down reciprocating motion, achieving the simultaneous extraction of high- and low-pressure oil.

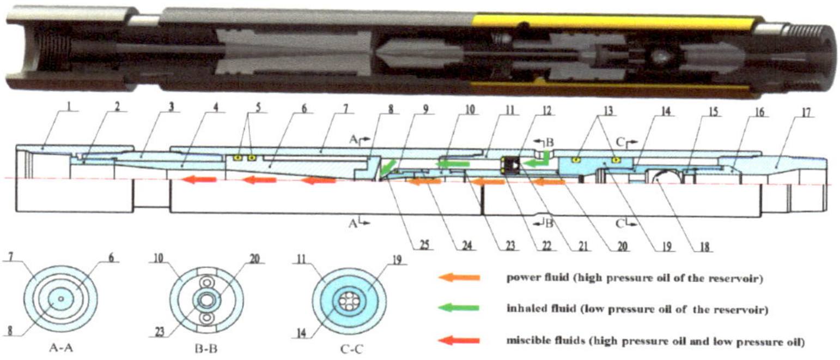

Figure 2. The structure and working principle of the new jet pump: 1. tubing clamp; 2. compression bolt; 3. tubing connector; 4. expanded pipe; 5. seal ring A; 6. diffuser tube; 7. pump housing; 8. throat; 9. seal ring B; 10. center connector; 11. joint of pump housing; 12. valve disk; 13. seal ring C; 14. check valve block; 15. check valve seat; 16. seat gland; 17. tail tube; 18. steel ball; 19. joint pipe; 20. sleeve; 21. check valve with spring; 22. straight pin; 23. joint of valve disk; 24. nozzle holder; 25. nozzle holder.

This is a new type of jet pump that was designed to address the shortcomings of the existing technology, and the strata series of Shengli Oilfield and the energy of high-pressure reservoirs was used to realize the simultaneous extraction of high- and low-pressure reservoirs based on the Venturi jet principle. Compared with traditional jet pumps [7,16], this jet pump not only solves the existing problems of difficult fluid extraction and low oil-recovery efficiency in low-pressure reservoirs, but also better transforms the existing contradiction between layers; additionally, it avoids the production cost of injecting high-pressure fluid into a well and, thus, allows for lower costs and a higher efficiency. Thanks to its simple structure, good practicality, and high reliability, it also reduces the risk of underground accidents.

3. Establishment of Calculation Method and Simulation Model

3.1. Basic Governing Equations of Fluids

When high-pressure oil and low-pressure oil in reservoirs enter the new jet pump, although there are two kinds of fluids in the pump, there is only a single liquid phase. As the high-pressure oil is sprayed through the nozzle in the jet pump, the fluid velocity significantly increases and the pressure is high. At this time, the flow field in the jet pump should be regarded as the incompressible turbulent flow field and the flow of oil in the pump should also follow the Navier–Stokes equation in fluid mechanics; thus, the following continuity equation can be established in the Euler coordinate system [33,34]:

$$\frac{\partial \rho}{\partial t} + \frac{\partial(\rho u_x)}{\partial x} + \frac{\partial(\rho u_y)}{\partial y} + \frac{\partial(\rho u_z)}{\partial z} = 0 \tag{1}$$

where ρ is the density of the fluid (kg/m^3); t is time (s); and u_x, u_y, and u_z are the velocity components of the fluid microelement along the X, Y, and Z axes, respectively (m/s).

The momentum equation [32]:

$$\begin{cases} \rho \frac{du}{dt} = -\frac{\partial P_0}{\partial x} + \frac{\partial \tau_{xx}}{\partial x} + \frac{\partial \tau_{yz}}{\partial y} + \frac{\partial \tau_{zx}}{\partial z} + \rho f_x \\ \rho \frac{dv}{dt} = -\frac{\partial P_0}{\partial y} + \frac{\partial \tau_{yy}}{\partial x} + \frac{\partial \tau_{zy}}{\partial y} + \frac{\partial \tau_{xy}}{\partial z} + \rho f_y \\ \rho \frac{dw}{dt} = -\frac{\partial P_0}{\partial z} + \frac{\partial \tau_{zz}}{\partial x} + \frac{\partial \tau_{yz}}{\partial y} + \frac{\partial \tau_{xz}}{\partial z} + \rho f_z \end{cases} \tag{2}$$

where P_0 is the static pressure (Pa); τ_{ij} is the stress tensor (i and j can be taken as x, y, and z) (Pa); and f_k is the volume force of gravity (k can be taken as x, y, and z) (N/m^3).

3.2. Choice of Turbulence Model

The oil flow in the new jet pump is turbulent flow. In order to select the most suitable turbulence model for the jet pump to improve the simulation accuracy of the flow field, the overall structure of the new jet pump and the applicable conditions of the turbulence model were carefully analyzed. Since the jet pump contains the nozzle jet flow and the boundary layer flow, the realizable k–ε turbulence model cannot only accurately predict the divergence ratio of flat and cylindrical jets, but it is also suitable for boundary layer flow and flow separation. Therefore, we ultimately selected the turbulence model as the realizable k–ε model in this paper (in case there is a need to learn about the process in order to directly verify the numerical model, [32] can be referred to for reference). The equation of the turbulence model is:

$$\frac{\partial}{\partial t}(\rho k) + \frac{\partial}{\partial x_i}(\rho k u_j) = \frac{\partial}{\partial x_i}\left[\left(\mu + \frac{\mu_t}{\sigma_k}\right)\frac{\partial k}{\partial x_j}\right] + G_k + G_b - \rho\varepsilon - Y_M + S_k \tag{3}$$

$$\frac{\partial}{\partial t}(\rho\varepsilon) + \frac{\partial}{\partial x_j}(\rho\varepsilon u_j) = \frac{\partial}{\partial x_j}\left[\left(\mu + \frac{\mu_t}{\sigma_k}\right)\frac{\partial \varepsilon}{\partial x_j}\right] + \rho C_\varepsilon S_\varepsilon - \rho C_2\frac{\varepsilon^2}{k + \sqrt{v\varepsilon}} + C_1\frac{\varepsilon}{k}C_3 G_b + S_\varepsilon \tag{4}$$

where k is the turbulence kinetic energy (m^2/s^2); u_j is the velocity of the fluid (m/s); ε is the turbulent dissipation rate (m^2/s^3); μ is the dynamic viscosity of turbulence (Pa·s); G_k is the turbulent kinetic energy caused by the velocity gradient of the laminar flow (m^2/s^2); G_b is the turbulent kinetic energy caused by buoyancy (m^2/s^2); Y_M is the term that contributes to turning the dissipation rate of the turbulent pulsating expansion into the global flow field in the compressible flow; C_1, C_2, and C_3 are three constants; σ_ε and σ_k are the Prandtl numbers; and S_k and S_ε are user-defined source items.

3.3. Calculation Formula of Pump Efficiency

The evaluation of jet pump performance is mainly reflected by pump efficiency. After reviewing the literature [31], we found that the key components of jet pumps are not standardized; if differently sized jet pump components are selected, their working characteristics will be different. In the analysis of their working characteristics, dimensionless parameters are selected to describe characteristics of a jet pump; therefore, the specific calculation formula of pump efficiency is:

$$\eta = q\frac{p}{1-p} \tag{5}$$

$$q = \frac{q_2}{q_1} \tag{6}$$

$$p = \frac{p_d - p_s}{p_n - p_s} \tag{7}$$

where η is the efficiency of the new jet pump; q is the flow ratio; p is the pressure ratio; q_1 is the volume flow rate of power fluid (m^3/s); q_2 is the volume flow rate of the intake fluid (m^3/s); p_d is the pressure at the outlet of the diffuser tube (Pa); p_s is the pressure at the inlet of the intake fluid (Pa); and p_n is the pressure at the inlet of the power fluid (Pa).

3.4. Basic Characteristic Equation of Jet Pump

The basic characteristic equation of a jet pump aims to determine the relationship between the pressure, flow rate, and geometric dimensions of the pump. This paper used a momentum equation, an energy equation (Bernoulli equation), and an equation of continuity to derive the basic characteristic equation of the new jet pump (the derivation process is tedious, and thus, it is omitted here, but it is described in detail in [32]). This equation provided an important basis for the theoretical research, structural design, and manufacturing of the current jet pump. Its specific form is:

$$N = \varphi_1^2 \left[\frac{2\varphi_2}{\mathrm{m}} + \left(2\varphi_2 - \frac{\mathrm{r}}{\varphi_4^2} \right) \frac{\rho q^2 \mathrm{r}}{\mathrm{m}^2} - \left(2 - \varphi_3^2 \right)(1 + \rho q) \frac{(1+q)}{\mathrm{m}^2} \right] \quad (8)$$

$$\rho = \frac{\rho_2}{\rho_1} \quad (9)$$

where N is the water head ratio, which is about equal to p; φ_1, φ_2, φ_3, and φ_4 are velocity coefficients [16] that are related to the surface roughness of the internal flow road in the jet pump (generally, the value of φ_1 is 0.975, the value of φ_2 is 0.975, the value of φ_3 is 0.925, and the value of φ_4 is 0.85) and whose values are determined empirically; m is the ratio between the area at the inlet of the throat and the area at the outlet of the nozzle in the new jet pump; r is the ratio between the area at the inlet of the throat and the area at the inlet of the throat minus the area at the outlet of the nozzle in the new jet pump; ρ is the ratio of the intake fluid density to power fluid density; ρ_2 is the density of the intake fluid; and ρ_1 is the density of the power fluid.

In order to verify the basic characteristic equation, the equation was compared with Zou's equation [35]. It can be seen in Figure 3 that the flow rate of our equation is smaller and that the fitted pump efficiency curves of the two equations almost coincide. As the flow ratio continues to increase, the deviation between the highest pump efficiency is always less than 5% and the overall trend of the two is almost the same, with certain deviations. Thus, the correctness of the basic characteristic equation in this paper was proven. The fundamental difference between the Zou equation and Equation (8) depends on the selection of the friction loss coefficient and different derivation formulas. If the flow ratio is small, the friction loss coefficient would have a slight influence on the results of the equation; thus, the calculated pump efficiency would be almost the same. Once the flow ratio increases, the pump efficiency from the equation with a large friction loss coefficient would be small, leading to a numerical difference.

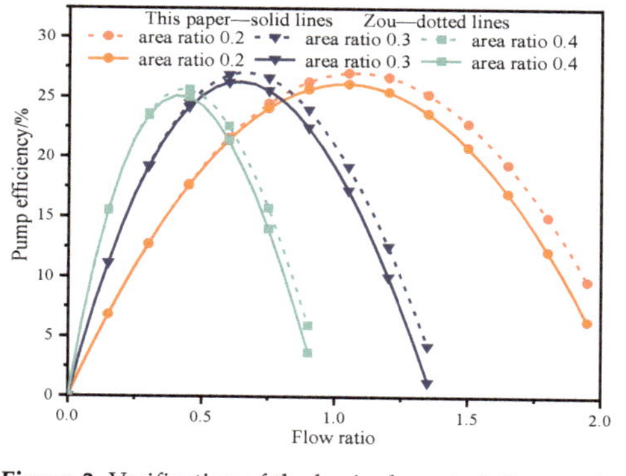

Figure 3. Verification of the basic characteristic equation.

After obtaining the basic characteristic equation of the new jet pump, we used MATLAB and the equation to draw the basic characteristic curve of the new jet pump. When the density of the power fluid and the intake fluid were the same (a density of 900 kg/m^3), the performance envelope curve of the new jet pump, as shown in Figure 4a, was obtained by changing the area ratio. It can be seen in the figure that in the case of a given area ratio, the pressure ratio gradually decreases with the increase in the flow ratio, whereas the pump efficiency first increases and then decreases, and there is a maximum efficiency value in each area ratio. With the increase in the area ratio, the maximum efficiency values of these different area ratios also first increase and then decrease. This conclusion is basically consistent with previous conclusions in the literature [35]. It can be seen that both the area ratio and the flow ratio have a significant influence on pump efficiency.

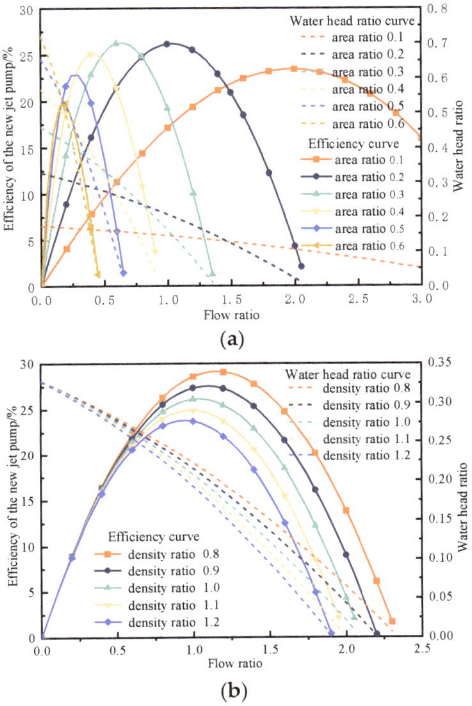

Figure 4. Performance envelope curve of the new jet pump. (**a**) Performance envelope curves of the new jet pump with different area ratios. (**b**) Performance envelope curves of the new jet pump at different density ratios.

The performance envelope curve of the new jet pump, as shown in Figure 4b, was obtained by changing the density ratio. It can be seen in the figure that in the case of a certain area ratio, the pressure ratio gradually decreases with the increase in the flow ratio, whereas the pump efficiency first increases and then decreases, and there is a maximum efficiency value at different density ratios. With the increase in the area ratio, the flow ratio corresponding to the highest efficiency value and the highest pump efficiency value of different density ratios gradually decreases, which indicates that the density ratio has a significant influence on the pump efficiency. This conclusion is also basically consistent with previous conclusions in the literature [16].

3.5. Establishment of a Simulation Model of the New Jet Pump

Due to the complex internal structure of the new jet pump, the three-dimensional modeling software SolidWorks and the numerical simulation software Fluent were used to establish the geometric model of the internal flow channel of the new jet pump. In order to improve the accuracy of the simulation results, the model was constructed according to a ratio of 1:1 (the whole process of modeling and simulation of the new jet pump is complicated, and thus, it is only briefly described here, but the whole process of the simulation was described in detail in [32]). The specific structure of the simulation model is shown in Figure 5; because subsequent meshing leads to poor mesh quality due to the complexity of the model, the simulation model was divided into two parts (simulation model a and simulation model b). It can be seen in the figure that a total of eight boundary conditions were set in the simulation model: three inlet boundaries (Inlet 1, Inlet 2, and Inlet 3), one outlet boundary (Outlet), two boundaries of the interface (Interface 1 and Interface 2; the interface was mainly used for processing the data transfer between interfacial surfaces in the multizone calculation model), and two wall boundaries (Wall 1 and Wall 2). During simulation, the two boundaries of the interface are in contact; Wall 1 refers to all the remaining surfaces that are not defined in simulation model b, and Wall 2 refers to all the remaining surfaces that are not defined in simulation model a.

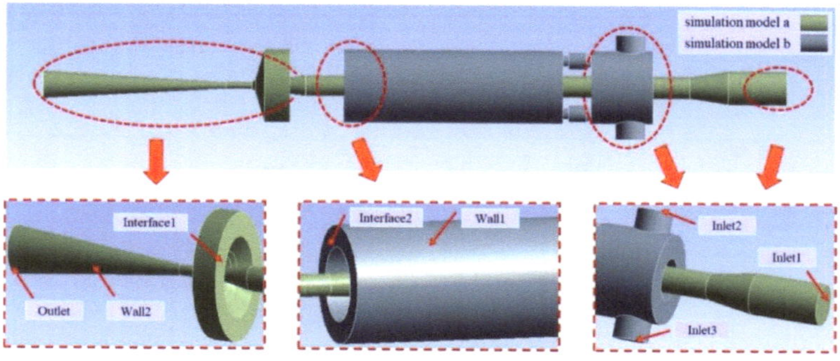

Figure 5. Simulation model of the new jet pump.

Based on actual conditions, our calculations showed that the inflow rate of power fluid was about 4.28 kg/s, the inflow rate of suction fluid was about 1.28 kg/s, and the outlet pressure of the jet pump was about 22.45 MPa. Therefore, we set Inlet 1 as the mass-flow inlet boundaries of the power liquid (4.28 kg/s), Inlet 2 and Inlet 3 as the mass-flow inlet boundary of the suction liquid (0.64 kg/s), and Outlet as the pressure outlet boundary of the jet pump (22.45 MPa). For the inlet boundary Inlet 1, we set the turbulence intensity as 5% and the hydraulic diameter as 5.22 mm. For the inlet boundaries Inlet 2 and Inlet 3, we set the turbulence intensity as 3% and the hydraulic diameter as 14 mm. For the outlet boundary Outlet, we set the turbulence intensity as 3% and the hydraulic diameter as 32 mm.

4. Meshing and Convergence Analysis of Simulation Model

When the simulation model of the new jet pump was meshed, in order to further improve the computational accuracy of the CFD (computational fluid dynamics) and effectively reduce the computational time of the numerical simulation, the influence of different mesh sizes on the simulation results was analyzed by changing the mesh size. (The simulation model was divided into mixed grids instead of single grids due to the complex simulation model. Therefore, the influence of the geometrical shape of single grids on the solution time and the accuracy of results are not discussed here. Because the two interfacial surfaces were circular, they were divided into hexahedral grids to process the data transfer between interfacial surfaces in the multizone calculation model.) It can be seen in Table 1 that a total of seven mesh partitioning methods were selected, and after comparing and analyzing the simulation results of these methods, it was found that the pressure value of the power fluid at Inlet 1 was almost the same when the grid sizes in the model were 1 mm and 2 mm, and the pressure values of these seven methods fluctuated between 36.89 MPa and 37.31 MPa. The simulation results not only verified the mesh independence of the model, but also showed that the convergence of the simulation results was good. In order to better improve the accuracy of the simulation results and minimize the computational time of the numerical simulation, the third mesh partition method was used to partition the simulation model.

Table 1. Grid independence verification of simulation model.

Meshing Method	Meshing Size (mm)	Number of Grid Nodes	Total Number of Units	Pressure of the Power Fluid at Inlet 1 (MPa)
#1	1	2,219,303	12,580,271	36.89
#2	1.5	670,295	3,693,996	37.01
#3	2	291,678	1,566,177	36.90
#4	2.5	154,344	807,660	37.08
#5	3	91,142	465,920	37.17
#6	3.5	58,893	294,429	37.26
#7	4	40,866	199,317	37.31

To verify the accuracy of the simulation results, the simulated pressure values of the power fluid inlet and two intake fluid inlets were extracted, and then the simulation results were compared with the theoretical calculation data one by one. It can be seen in Table 2 that the numerical difference between the simulated pressure value and the theoretical pressure value was only 2.88 MPa, and the relative error between them was only 8.4%. The simulated pressure value and the theoretical pressure value of the intake liquid inlets were also very close, and the average relative error between them was only 7.8%. The relative errors were all less than 10%, which not only indicated the correctness of the numerical simulation, but also further proved the accuracy of the simulation results.

Table 2. Verification of the accuracy of simulation results.

Locations	Theoretical Pressure Values (MPa)	Simulated Pressure Values (MPa)	Relative Errors (%)
Pressure of the power fluid at Inlet 1	34.02	36.90	8.4
Pressure of the intake fluid at Inlet 2	10	10.79	7.6
Pressure of the intake fluid at Inlet 3	10	10.84	8.0

5. Analysis of Influencing Factors of Pump Efficiency

The structure size and fluid physical characteristics have a significant influence on the efficiency of a jet pump. In order to better improve the efficiency of the new jet pump in the exploitation of high- and low-pressure reservoirs, based on the actual working conditions of the field, a numerical simulation analysis of the jet pump was carried out. The key structural parameters of the new jet pump are shown in Table 3. After the density of high-pressure oil was set to 900 kg/m^3 and that of low-pressure oil was set to 850 kg/m^3, the control variable method was used to analyze the influence of the throat–nozzle distance, the area ratio, and the length–diameter ratio of the throat on the efficiency of the jet pump and the flow field in the pump. Then, the optimal numerical intervals of the key structural parameters were obtained when the efficiency of the jet pump was at its maximum.

Table 3. Specific parameter values of key structures in the new jet pump.

Main Parameter Names	Units	Numerical Values
The number of the nozzle		1
Diameter of the nozzle outlet	mm	5.22
Diameter of the nozzle inlet	mm	15
Length of the throat at the nozzle outlet	mm	2.61
The convergence angle	°	14
Diameter of the throat	mm	6.65
Length of the throat	mm	40
Throat–nozzle distance	mm	10.44
Angle of the diffuser tube	°	8
Length of the diffuser tube	mm	167

5.1. Effect of the Throat–Nozzle Distance on Pump Efficiency

Low-pressure oil is carried by high-pressure oil between the nozzle outlet and the throat inlet; thus, the throat–nozzle distance has a significant influence on the suction range of high-pressure oil. After consulting the literature [8,21,36], it was found that the throat–nozzle distance is one of the main factors affecting pump efficiency. Therefore, the influence of different throat–nozzle distances on pump efficiency was studied by changing this distance. As shown in Table 4, when the throat–nozzle distances were 2.61 mm (0.5 times the diameter of the nozzle outlet) and 5.22 mm (the same as the diameter of the nozzle outlet), the pump efficiency was above 29%, and it can be seen in the figure that the range of the throat–nozzle distance that allowed for the maximum pump efficiency was between 2.61 mm and 5.22 mm. As the throat–nozzle distance continued to increase, the pump efficiency gradually and obviously decreased. When the throat–nozzle distance was 13.05 mm (2.5 times the diameter of the nozzle outlet), the pump efficiency was only about 20%, which further proved that the throat–nozzle distance was one of the main factors affecting the pump efficiency. Under the same working conditions, because the flow ratio did not change, the water head ratio curve showed the same trend as the pump efficiency curve (this conclusion can be obtained by analyzing Equation (5)).

Table 4. Effect of different throat–nozzle distances on pump efficiency.

Diameter of Nozzle Outlet (mm)	Distance from the Throat to Nozzle (mm)	Flow Ratio	Water Head Ratio	Pump Efficiency (%)
5.22	2.61	0.3156	0.4802	29.15
5.22	5.22	0.3156	0.4789	29.01
5.22	7.83	0.3156	0.4379	24.58
5.22	10.44	0.3156	0.4263	23.45
5.22	13.05	0.3156	0.3906	20.23
5.22	15.66	0.3156	0.3633	18.01

By extracting the velocity from the nozzle exit to the throat entrance of the new jet pump, as shown in Figure 6, we found that the maximum velocity at the nozzle outlet always fluctuated between 239 m/s and 247 m/s, the throat–nozzle distance gradually increased, and the size of the region with the highest fluid velocity (the red region) gradually decreased (this conclusion is basically consistent with the numerical simulation results in [35], which further proves the correctness of the numerical simulation in this paper). When the throat–nozzle distance was 3 times the diameter of the nozzle outlet, it could be clearly seen that only a small amount of high-speed fluid entered the throat. The reason for this result is that as the throat–nozzle distance gradually increased, the effective action distance for carrying the intake fluid became longer due to the power fluid and the velocity of the fluid was significantly reduced before it entered the throat.

5.2. Effect of Area Ratio on Pump Efficiency

The area ratio refers to the ratio of the inlet area of the throat to the outlet area of the nozzle. The area ratio should generally be greater than 1; if the area ratio is too small, it will cause serious wear to the inner wall of the jet pump, and if the area ratio is too large, there will be some liquid reflux. After consulting the literature [11,37], it was found that the area ratio is an important geometric parameter of jet pumps and that its value directly affects jet pump performance. Therefore, the influence of different area ratios on pump efficiency was studied by changing the diameter of the venture. Table 5 shows that when the diameter of the throat increased from 6.393 mm (area ratio of 1.5) to 7.382 mm (area ratio of 2), the efficiency of the jet pump gradually increased, but once the area ratio exceeded 2, the pump efficiency gradually decreased. It can be seen in the figure that the range of the diameter of the throat with maximum pump efficiency was between 6.39 mm and 8.25 mm. Under the same working conditions, the flow ratio did not change; thus, the water head ratio curve had the same trend as the pump efficiency curve.

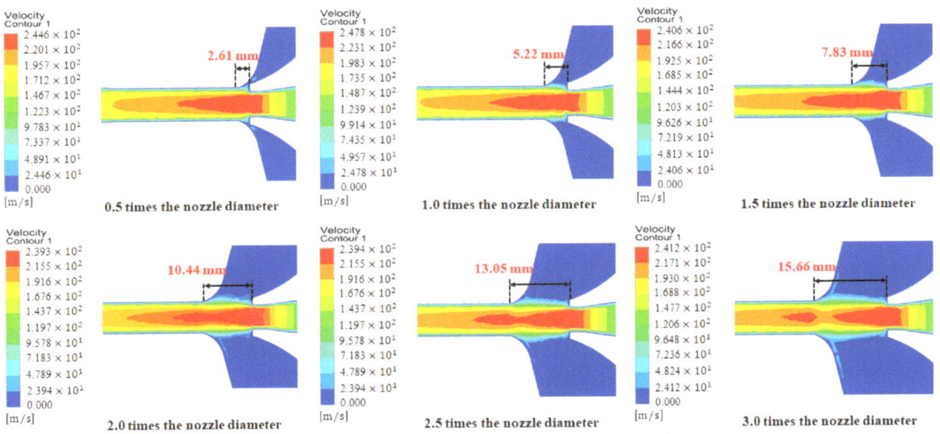

Figure 6. Velocity cloud diagrams from the nozzle outlet to the throat inlet with different throat–nozzle distances.

Table 5. Effect of different area ratios on pump efficiency.

Diameter of Nozzle Outlet (mm)	Diameter of Throat Inlet (mm)	Flow Ratio	Water Head Ratio	Pump Efficiency (%)
5.22	6.39	0.3156	0.4549	26.34
5.22	7.38	0.3156	0.4585	26.72
5.22	8.25	0.3156	0.4410	24.90
5.22	9.04	0.3156	0.404	21.39
5.22	9.77	0.3156	0.3599	17.74
5.22	10.44	0.3156	0.3215	14.95

By creating the velocity cloud diagrams of the throat outlet of the new jet pump with different throat diameters, as shown in Figure 7, it was found that when the throat diameter was 6.39 mm, the maximum velocity at the outlet was about 204 m/s and the velocity area was large, though there was a gradual increase in the diameter of the throat pipe. However, as the throat diameter gradually increased, the maximum velocity at the throat outlet gradually decreased, and as the area ratio gradually increased, the maximum velocity area gradually decreased (the conclusion of this numerical simulation was basically the same as that of the theoretical analysis in [11]). The reason for this result is that as the area ratio gradually increased, the sectional area at the throat outlet gradually increased and the contact area between the power fluid and the intake fluid in the throat also increased. These two parameters allow for a more adequate energy transfer in the throat tube, thus causing the maximum speed and the maximum speed area to gradually decrease with the increase in the area ratio.

5.3. *Effect of the Throat Length–Diameter Ratio on Pump Efficiency*

The throat is the main mixing area of the power and intake fluid, and the two carry out the main energy transfer in the throat; therefore, the throat length–diameter ratio has an important impact on the performance of a jet pump. After consulting the literature [11,38], it was found that the throat length–diameter ratio is one of the main factors affecting pump efficiency, and the throat length–diameter ratio of most jet pumps is greater than 4 and less than 10. Therefore, the influence of the throat length–diameter ratio on pump efficiency was studied by changing the length of the throat. As shown in Table 6, when the length of the throat was changed from 33.25 mm (where the throat length–diameter ratio was 5) to 46.55 mm (where the throat length–diameter ratio was 7), the pump efficiency gradually increased. However, once the throat length–diameter ratio exceeded 7, the pump efficiency gradually decreased. It can be seen in the figure that the range of the length of the throat at maximum pump efficiency was between 39.3 mm (where the throat length–diameter ratio was 6) and 53.2 mm (where the throat length–diameter ratio was 8).

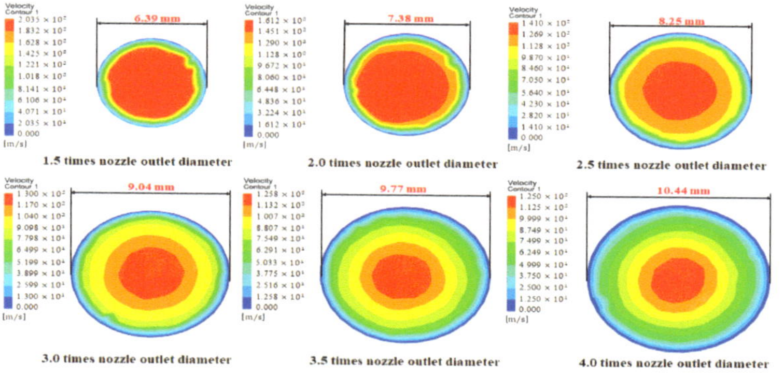

Figure 7. Velocity cloud diagrams of the throat outlet with different area ratios.

Table 6. Effect of different throat length–diameter ratios on pump efficiency.

Diameter of the Throat (mm)	Length of the Throat (mm)	Flow Ratio	Water Head Ratio	Pump Efficiency (%)
6.65	33.25	0.3156	0.4519	26.02
6.65	39.90	0.3156	0.4625	27.16
6.65	46.55	0.3156	0.4712	28.12
6.65	53.20	0.3156	0.4679	27.76
6.65	59.85	0.3156	0.4626	27.16
6.65	66.50	0.3156	0.4572	26.58

By extracting the velocity cloud diagrams of the throat outlet of the new jet pump with different throat length–diameter ratios, as shown in Figure 8, it was found that when the length of the throat was 33.25 mm, the maximum velocity at the outlet was about 134 m/s and the velocity area was small. However, as the length of the throat increased, the maximum velocity at the outlet of the throat gradually decreased, and when the length–diameter ratio gradually increased, the maximum velocity area gradually increased. The reason for this is that as the length–diameter ratio increased, the contact time between the power fluid and the intake fluid in the throat became longer, and thus, the power fluid and the intake fluid could allow for a more full energy transfer, causing the maximum velocity area to gradually increase with the increase in the length–diameter ratio.

5.4. Effect of Angle of the Diffuser Tube on Pump Efficiency

The function of the diffuser tube in a jet pump is to convert the kinetic energy of the mixed oil into pressure energy to lift the oil. After consulting the literature [11,39], it was found that the diffuser angle in the diffuser tube has a significant influence on the performance of jet pumps. Therefore, the influence of different diffuser angles on pump efficiency was studied by changing the diffuser angle. As shown in Table 7, when the diffuser angle was between 6° and 8°, the pump efficiency gradually increased with the increase in the diffuser angle. However, once the diffuser angle exceeded 8°, the pump efficiency gradually decreased with the increase in the diffuser angle. It can be seen in the figure that when the diffuser angle was 8°, the pump efficiency was at its maximum. The actual maximum pump efficiency was not reached at this point, but the range of the diffuser angle of the maximum pump efficiency was between 6° and 10°.

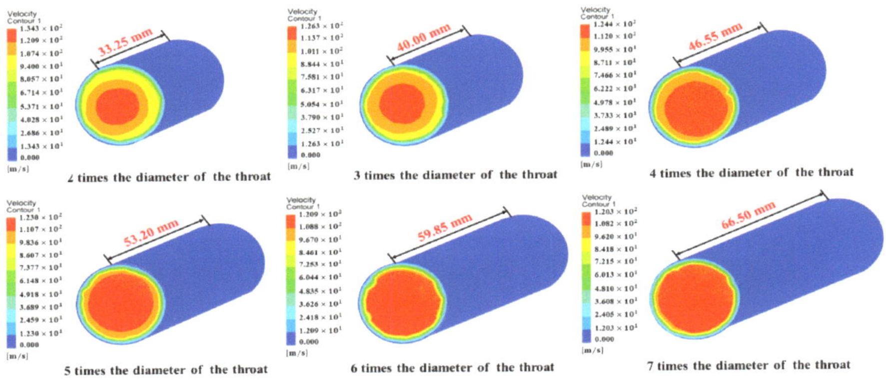

Figure 8. Velocity cloud diagrams of the throat outlet with different ratios of the length of the throat to its diameter.

Table 7. Effect of the different angles of the diffuser tube on pump efficiency.

Outlet Diameter of the Diffuser Tube (mm)	Angle of the Diffuser Tube (°)	Flow Ratio	Water Head Ratio	Pump Efficiency (%)
26.25	6	0.3156	0.4727	28.29
32.80	8	0.3156	0.4765	28.72
39.36	10	0.3156	0.4677	27.73
45.95	12	0.3156	0.4439	25.20
52.56	14	0.3156	0.4294	23.75
59.20	16	0.3156	0.4090	21.84

By extracting the velocity cloud diagrams from the diffuser tube of the new jet pump with different diffuser angles, as shown in Figure 9, it was found that the maximum velocity in the diffuser tube always fluctuated between 194 m/s and 196 m/s. When the diffuser angle was 6°, the miscible fluids could be fully developed for the entire section. The reason for this is that the fluid velocity was relatively fast, which increased the friction loss and caused the pump efficiency of the jet pump to decrease. As the diffuser angle increased, the velocity of miscible fluids decreased significantly less than the increase in the diffuser angle; thus, the velocity distribution surface in the diffuser tube did not fill the entire diffuser tube section. Thus, the performance of the new jet pump was significantly reduced (this conclusion is the same as that of the theoretical analysis in [35]).

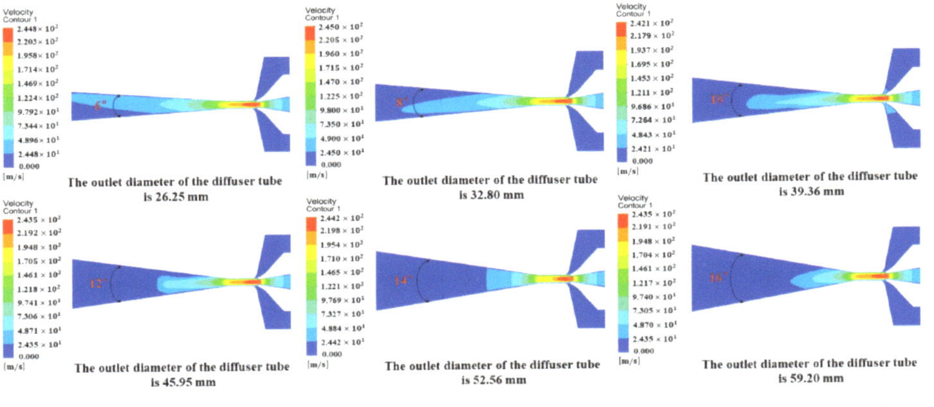

Figure 9. Velocity cloud diagrams of the diffuser tube with different diffuser angles.

5.5. Effect of the Flow Rate on the Pump Efficiency

Low-pressure oil is mixed with high-pressure oil in the throat to improve its speed and pressure; thus, the flow ratio is closely related to the mixing of the two fluids. After consulting the literature [11,40], it was found that the flow ratio is one of the main factors affecting pump efficiency. Therefore, the influence of the different flow rates on pump efficiency was studied by changing the volume flow of the power fluid. As shown in Table 8, when the power liquid flow was 0.45 (flow ratio of 0.2), the pressure ratio was at its maximum value, but the pump efficiency was less than 18%. As the flow ratio gradually increased, the pressure ratio began to decrease and the efficiency first increased and then decreased. It can be seen in the table that the range of the flow ratio of the maximum pump efficiency was 0.3 to 0.5.

Table 8. Effect of different flow rates on pump efficiency.

Flow of Intake Fluid (m^3/min)	Flow Ratio	Outlet Pressure of Diffuser Tube (MPa)	Water Head Ratio	Pump Efficiency (%)
0.09	0.2	22.45	0.4714	17.84
0.09	0.3	22.45	0.4449	24.04
0.09	0.4	22.45	0.3743	23.93
0.09	0.5	22.45	0.3224	23.79
0.09	0.6	22.45	0.2720	22.42
0.09	0.7	22.45	0.2210	19.86

By extracting the velocity cloud diagrams of the internal flow field in the jet pump with different flow ratios, as shown in Figure 10, it was found that when the flow ratio was 0.2, the maximum velocity of the power liquid at the nozzle outlet of the jet pump was 252 m/s, and as the flow ratio gradually increased, the maximum velocity at the nozzle outlet (the red area) gradually decreased (the conclusion of the numerical simulation is basically consistent with the theoretical analysis in [7]). The reason for this is that the volume flow of the intake fluid was a certain size; the volume flow of the power liquid gradually decreased with the increase in the flow ratio and the diameter of the nozzle outlet did not change; thus, the maximum velocity of the power fluid gradually decreased with the decrease in the volume flow of the power fluid.

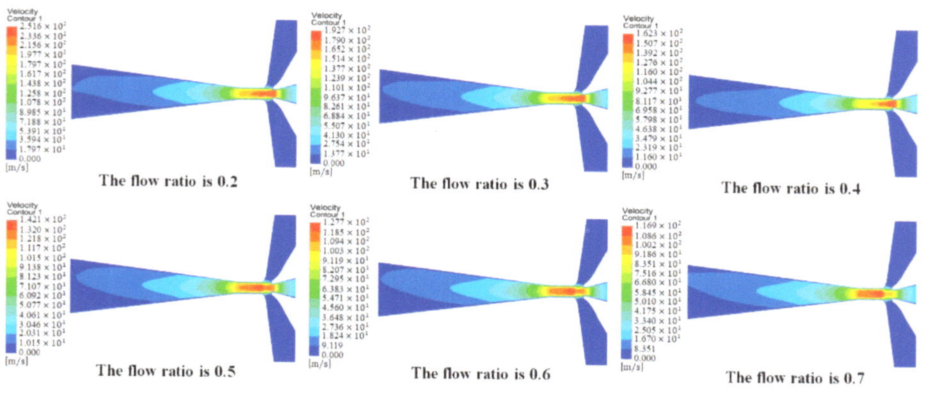

Figure 10. Velocity cloud diagrams of the flow field in the new jet pump with different flow ratios.

6. Structure Size Analysis of a New Jet Pump Based on the Taguchi Method

To achieve the lowest production costs and the shortest design and manufacturing time while maintaining customer satisfaction, Genichi Taguchi of Japan invented a robust design method that improved not only the quality of products, but also the research and development and design capabilities of domestic industries in Japan. This method is commonly called the Taguchi method [41]. The specific implementation process of Taguchi's method (where the contents of the red-dashed box are the main steps) is shown in Figure 11, where the first step is to determine the test index, which directly reflects the quality of the product. Then, the test parameters that have the greatest impact on the test index are selected. The test parameters comprise two parts: the determination of control factors (the range and level of all control factors need to be determined) and the determination of noise factors (the internal and external noise factors need to be determined). After the test parameters are determined, the number of control factors and the number of levels are used to arrange the test plan. At this time, an orthogonal table needs to be constructed (the construction of the orthogonal table requires the use of relevant knowledge, such as probability theory [41]). According to the determined orthogonal table, the test is carried out, and the results are entered into the table. Then, range and S/N ratio analyses are used to determine the relative importance of each test parameter, and the variance is used to judge the relative influence of each factor on the target value in order to determine the best parameter scheme.

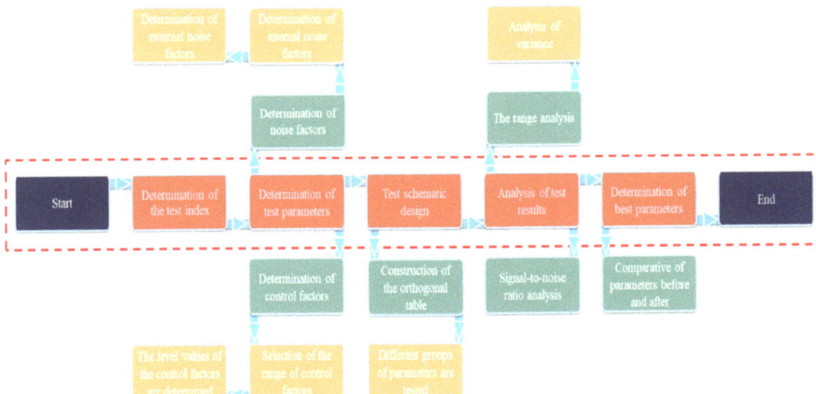

Figure 11. Implementation process of the Taguchi method.

6.1. Determination of the Test Index and Control Factors

The test index is the dependent variable in the test [42], and it mainly reflects the quality of the test results. The performance of a new jet pump is mainly reflected by the pump efficiency, and the value directly affects the production cycle and the current production cost. Here, the pump efficiency of the new jet pump was used as the test index.

Control factors are the independent variables in the test [42], and they must be considered when studying the test index. In general, a test index is mostly affected by multiple control factors; there are not only certain connections, but also significant differences between these control factors. By analyzing the abovementioned factors influencing pump efficiency, we found that the throat–nozzle distance, the diameter of the throat (area ratio), the length of the throat (throat length–diameter ratio), the diffuser angle, and the flow rate ratio have a significant influence on the pump efficiency because the power fluid flow and the intake fluid are not easy to control. Therefore, the throat–nozzle distance, the diameter of the throat, the length of the throat, and the diffuser angle were selected as control factors for Taguchi's method.

6.2. The Range and Level of the Control Factors

After the control factors were selected according to the optimal throat–nozzle distance interval, the diameter of the throat (area ratio) interval, the length of the throat (throat length–diameter ratio) interval, and the diffuser angle interval described above, the optimal range of the control factors from the Taguchi method was further determined. After the optimization of each control factor range was determined, it was necessary to design the number of levels of control factors and the specific value of each level. The optimal range and specific values of control factors are shown in Table 9; it can be seen in the table that five levels were selected for each control factor, and the five levels of each control factor were equally distributed within the set optimal range.

Table 9. The range and level design of control factors.

The Control Factors	The Range of Factors	Level 1	Level 2	Level 3	Level 4	Level 5
Distance from the throat to nozzle (mm)	2.62~5.22	2.62	3.27	3.92	4.57	5.22
Diameter of the throat (mm)	6.4~8.0	6.4	6.8	7.2	7.6	8.0
Length of the throat (mm)	44~52	44	46	48	50	52
The diffuser angle (°)	6~9	6	6.75	7.5	8.25	9

6.3. Determination of Noise Factors

The test process was carried out using simulation software, and thus, there were no external noise factors. Although factors such as environment temperature, human operation errors, production errors, and other factors can be excluded in the process of finite element simulation, if the same model is divided into different mesh sizes, the final test index will also have a certain error caused by mesh division that can be used as an internal noise factor. To prevent the internal noise factors from influencing the final analysis results and to improve the accuracy of the test results, two mesh sizes (1 mm and 2 mm) with better convergence and higher accuracy were selected as described above; then, the simulation model was divided into these two mesh sizes, and the numerical simulation was carried out.

6.4. Construction of Orthogonal Tables and Orthogonal Tests

An orthogonal table is an essential tool for the Taguchi method. Such a table usually has two characteristics. The first is that the number of factor levels occurs with the same frequency in each column, and the second is that for all test groups, the level in one column has the same frequency of occurrence as the level in the other column. When a table satisfies both of these characteristics, it can be called an orthogonal table [43]. An orthogonal table and the results of these tests are shown in Table 10. Because the number of control factors in this test was 4, the level number of each control factor was 5. Additionally, the internal noise factor caused by grid division was also considered; therefore, a total of 50 tests were needed.

Table 10. Orthogonal table and the results of these tests.

Number	Distance from the Throat to the Nozzle	Diameter of the Throat	Length of the Throat	The Diffuser Angle	Pump Efficiency	
	(mm)	(mm)	(mm)	(°)	η_1 (%)	η_2 (%)
1	2.62	6.4	44	6	27.144	27.142
2	2.62	6.8	46	6.75	31.007	31.010
3	2.62	7.2	48	7.5	30.532	30.514
4	2.62	7.6	50	8.25	28.368	28.348
5	2.62	8	52	9	25.454	25.452
6	3.27	6.4	46	4.5	30.406	30.389
7	3.27	6.8	48	8.25	29.351	29.327
8	3.27	7.2	50	9	28.939	29.060
9	3.27	7.6	52	6	28.798	28.803
10	3.27	8	44	6.75	24.540	24.524
11	3.92	6.4	48	9	27.086	27.189
12	3.92	6.8	50	6	30.641	30.646
13	3.92	7.2	52	6.75	30.426	30.426
14	3.92	7.6	44	7.5	27.752	27.755
15	3.92	8	46	8.25	25.668	25.668
16	4.57	6.4	50	6.75	27.857	27.937
17	4.57	6.8	52	7.5	29.610	29.556

Table 10. *Cont.*

Number	Distance from the Throat to the Nozzle	Diameter of the Throat	Length of the Throat	The Diffuser Angle	Pump Efficiency	
	(mm)	(mm)	(mm)	(°)	η_1 (%)	η_2 (%)
18	4.57	7.2	44	8.25	28.293	28.324
19	4.57	7.6	46	9	25.660	25.702
20	4.57	8	48	6	25.455	25.431
21	5.22	6.4	52	8.25	25.899	25.942
22	5.22	6.8	44	9	27.970	28.084
23	5.22	7.2	46	6	29.270	29.435
24	5.22	7.6	48	6.75	26.852	26.944
25	5.22	8	50	7.5	25.431	25.453

6.5. Signal-to-Noise Ratio Analysis of Test Results

In order to better analyze the test results, the data in Table 5 were imported into the Minitab statistical software [43]. Since the software includes the Taguchi test module, the operation was easy and the analysis results were reliable. Generally speaking, the larger the signal-to-noise ratio, the better the quality, and when the delta of a parameter was large, this parameter had a significant influence on pump efficiency. It can be seen in Table 11 that the influencing factors of pump efficiency, from strong to weak, were the throat diameter, the diffuser angle, the throat–nozzle distance, and the length of the throat (the results obtained with this signal-to-noise ratio analysis are basically consistent with those obtained by other theoretical methods in [35]). For pump efficiency, bigger is better, and it can be seen in Figure 12 that when the throat–nozzle distance was level 1, the diameter of the throat was level 2, the length of the throat was level 2, and the diffuser angle was level 3 (i.e., the throat–nozzle distance was 2.62 mm, the diameter of the throat was 6.8 mm, the length of the throat was 46 mm, and the diffuser angle was 7.5°), the pump efficiency was at its maximum value.

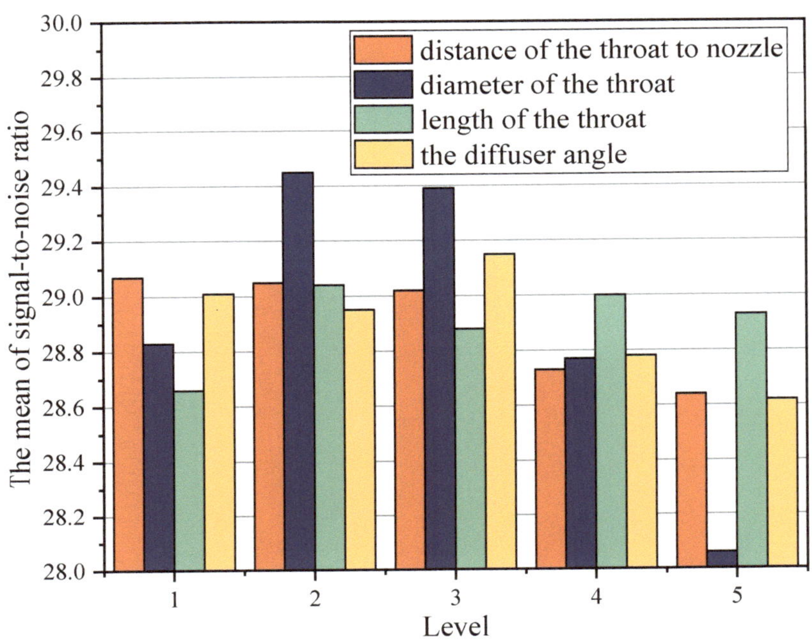

Figure 12. The main effect plot of the mean of the signal-to-noise ratio.

Table 11. S/N response table.

Level	Distance from the Throat to the Nozzle	Diameter of the Throat	Length of the Throat	The Diffuser Angle
	(mm)	(mm)	(mm)	(°)
1	29.07	28.83	28.66	29.01
2	29.05	29.45	29.04	28.95
3	29.02	29.39	28.88	29.15
4	28.73	28.77	29.00	28.78
5	28.64	28.06	28.93	28.62
Delta	0.43	1.39	0.38	0.53
Rank	3	1	4	2

6.6. Range Analysis of Test Results

After the Minitab statistical software was used to conduct the range analysis of the pump efficiency under each parameter combination in Table 10, a mean response table was obtained, as shown in Table 12. From this table, one can intuitively analyze the fluctuation between the pump efficiency and the level of each factor. By comparing the mean values and the rank of the mean values at different levels, it can be seen that the diameter of the throat had the greatest effect on pump efficiency, followed by the diffuser angle, the throat–nozzle distance, and the length of the throat. The optimal level of each factor was analyzed from the index effect, and it can be seen in Figure 13 that when the throat–nozzle distance of the new jet pump was level 1, the diameter of the throat was level 2, the length of the throat was level 2, and the diffuser angle was level 3, the pump efficiency reached its maximum value; this conclusion was consistent with the results of the signal-to-noise ratio analysis.

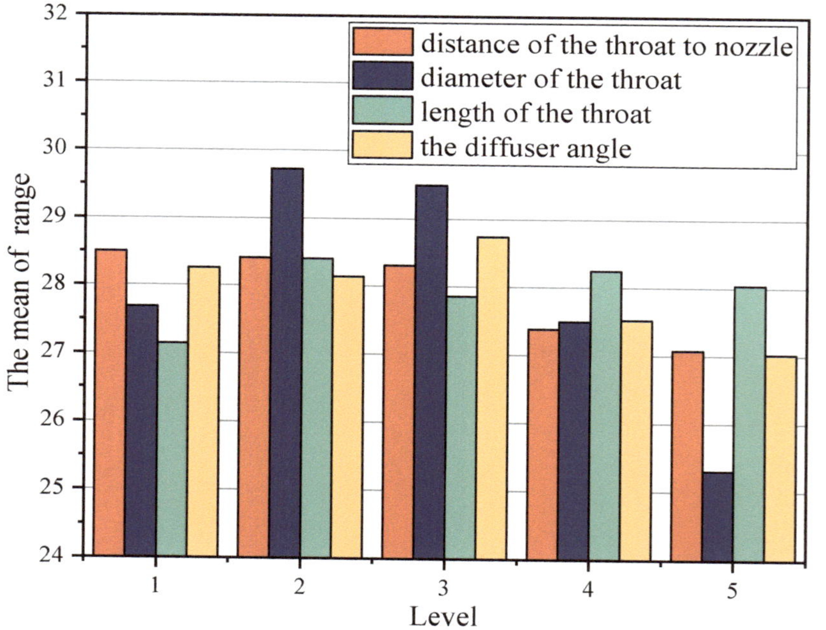

Figure 13. The main effect plot of the mean of the range.

Table 12. The mean response table.

Level	Distance from the Throat to the Nozzle	Diameter of the Throat	Length of the Throat	The Diffuser Angle
	(mm)	(mm)	(mm)	(°)
1	28.50	27.68	27.14	28.26
2	28.41	29.72	28.40	28.14
3	28.31	29.49	27.86	28.75
4	27.38	27.49	28.25	27.52
5	27.08	25.31	28.04	27.02
Delta	1.44	4.41	1.26	1.72
Rank	3	1	4	2

6.7. Variance Analysis of Test Results

In order to quantify the importance of the influence of the throat–nozzle distance, the throat diameter, the throat length, and the diffuser angle on pump efficiency, an analysis of variance was used to obtain the degree of freedom, the sum of squares, and the variance of each control factor. Then, the contribution degree of each control factor was calculated, and the relative influence degree of each control factor was determined. The results of the variance analysis of control factors are shown in Table 13; the degree of freedom in the table equals the total number of levels of each control factor minus 1. After determining the degree of freedom, the sum of squares, and other numerical values, we calculated the specific value of the variance and the F value (the value of F is the ratio of the variance of the control factor to the error variance). Then, each control factor's specific contribution to the pump efficiency of the new jet pump was obtained.

Table 13. Variance analysis of control factors.

Control Factors	Quadratic Sum	Degree of Freedom	Variance	Values of F	Contribution Degree
Distance from the throat to nozzle	0.766	4	0.192	1.721	9.12%
Diameter of the throat	6.378	4	1.595	14.293	75.93%
Length of the throat	0.446	4	0.112	1.004	5.31%
The diffuser angle	0.810	4	0.203	1.819	9.64%

It can be seen in Table 13 that the diameter of the throat contributed to a high degree, i.e., 75.93%, whereas the contribution of the other three was relatively small, and the degree of contribution of the throat–nozzle distance and the diffuser angle was similar. Comparing the two showed that the degree of influence of the diffuser angle on pump efficiency was slightly higher than that of the throat–nozzle distance. Among the four control factors, the contribution of the throat length was the lowest at 5.31%. Although the contributions of the throat–nozzle distance, the throat length, and the diffuser angle to the pump efficiency were not very high in this study, they are all important parameters that improve pump efficiency.

7. Comparative Analysis of the New Jet Pump before and after Optimization

The structure size, flow ratio, and pump efficiency of the new jet pump before and after optimization are compared in Table 14. It can be clearly seen that the throat–nozzle distance before optimization was much larger than that after optimization. The diameter and length of the throat after optimization were larger than those before optimization, but the diffuser angle after optimization was smaller than before optimization, and the difference between the two was only 0.5°. The new jet pump before optimization and the new jet pump after optimization were simulated under the condition of a certain flow ratio, and the pump efficiency of the new jet pump before and after optimization was calculated by using the pump efficiency calculation formula and the data derived from the numerical simulation. It was found that the pump efficiency before optimization was only 21.81%, the pump efficiency after optimization was 31.26% (at present, the highest efficiency of a deep-well jet pump is less than 36% in practical applications [43]), and the difference between the pump efficiencies of the two was about 10%.

Table 14. Comparative analysis of the structure size and pump efficiency of the new jet pump before and after optimization.

The New Jet Pump	Distance from the Throat to the Nozzle (mm)	Diameter of the Throat (mm)	Length of the Throat (mm)	The Diffuser Angle (°)	Flow Ratio	Pump Efficiency (%)
Before optimization	10.44	6.65	40	8.0	0.3156	21.81
After optimization	2.62	6.80	46	7.5	0.3156	31.26

By comparing the velocity cloud diagrams of the new jet pump before and after optimization in Figure 14, it can be clearly seen that the maximum velocities of the fluid in the new jet pump before and after optimization were almost the same. The effective action distance of the intake fluid and of the power fluid became longer, and it can be seen that only a small part of the high-speed fluid (the red area) in the new jet pump before optimization entered the throat pipe. Although the difference in the diffuser angles before and after optimization was small, the mixed fluid in the latter could be fully developed for the entire cross-section. A comprehensive analysis of the data in Table 14 and Figure 14 proves that the new jet pump after optimization had a significant improvement in performance compared with that before optimization, and it is further shown that using the Taguchi method allowed us to optimize the structure size and improve the pump efficiency of the jet pump.

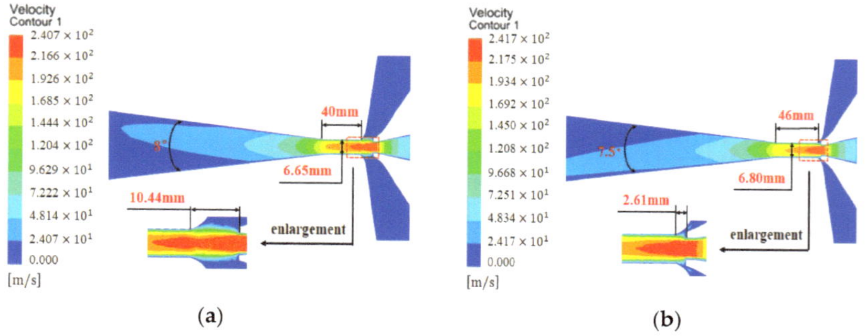

Figure 14. Velocity cloud diagram of the new jet pump before and after optimization. (**a**) Before optimization. (**b**) After optimization.

8. Conclusions

1. A new type of jet pump was designed to enhance the efficiency of oil recovery in wells in Eastern Shandong, China and to effectively exploit low-energy and low-yield zones. The jet pump simultaneously extracts oil from high- and low-pressure reservoirs by transforming the existing contradiction between layers in the well based on the Venturi jet principle. The method further lowers costs and enhances efficiency compared with traditional oil-recovery methods. Thanks to its simple structure, good practicality, and high reliability, the new jet pump also reduces the risk of underground safety accidents.
2. A new equation for the basic characteristics of the new jet pump was derived. A comparison of this new equation with the existing equations verified the correctness of this equation. According to the analysis of the envelope curve of the performance (drawn in the 2016 version of MATLAB using the derived equation of basic character-

istics) of the jet pump, the density ratio, flow rate ratio, and area ratio had a significant influence on pump efficiency; in addition, as the density ratio gradually increased, both the highest pumping efficiency among different density ratios and the flow ratio corresponding to the highest pumping efficiency gradually decreased.

3. According to an analysis of the internal flow field of the new jet pump, the increase in the throat–nozzle distance attenuated the flow core area because the effective action distance of the intake fluid being carried became longer due to the power fluid; thus, the velocity of the fluid was significantly reduced before it entered the throat. As the area ratio increased, both the maximum velocity at the throat outlet and the area of the maximum velocity gradually decreased (the reason for this is that as the area ratio gradually increased, the power fluid and the intake fluid carried out a more adequate energy transfer in the throat tube). An increase in the aspect ratio caused the maximum velocity area at the throat outlet to gradually increase (the reason for this is that the contact time between the power fluid and the intake fluid in the throat became longer, and therefore, the two could allow for a more full energy transfer). The increase in the diffuser angle resulted in the velocity distribution area in the diffuser tube not being able to cover the whole cross-section of the diffuser tube (the reason for this is that the velocity of miscible fluids decreased significantly less than the increase in the diffuser angle, which caused the velocity distribution surface in the diffuser tube to not fill the entire diffuser tube section).
4. To improve the pumping efficiency of the new jet pump in the extraction of high- and low-pressure reservoirs, its structural dimensions were optimized using the Taguchi method and numerical simulations (this is a new method for optimizing the structure of jet pumps). According to an analysis of the test results using signal-to-noise ratio and range analyses, the factors influencing pumping efficiency are, in descending order, the throat diameter, diffuser angle, throat–nozzle distance, and throat length. The contributions (relative influence on pumping efficiency) of these control factors were then obtained using an ANOVA. The throat diameter had the highest contribution (75.93%), followed by the diffuser angle (9.64%), throat–nozzle distance (9.12%), and throat length (5.31%).
5. The efficiency of the new jet pump was optimized with a combination of parameters including a level 1 throat–nozzle distance, a level 2 throat diameter, a level 2 throat length, and a level 3 diffuser angle (i.e., a throat–nozzle distance of 2.62 mm, a throat diameter of 6.8 mm, a throat length of 46 mm, and a diffuser angle of 7.5°). A comparison with the jet pump before optimization indicated that the pump efficiency was increased from 21.81% to 31.26%, further demonstrating a considerable improvement in the performance of the new jet pump.

Author Contributions: Methodology, Y.L.; software, J.W.; validation, Z.W. (Zhenhua Wu) and M.Z.; writing—original draft, Z.W. (Zhiliang Wang); funding acquisition, R.L. All authors have read and agreed to the published version of the manuscript.

Funding: This work was supported by the National Natural Science Foundation of China (No. 62173049).

Data Availability Statement: The datasets used and/or analyzed during the current study are available from the corresponding author upon reasonable request.

Conflicts of Interest: The authors declare no conflict of interest.

References

1. Wang, H.; Huang, H.; Bi, W.; Ji, G.; Zhou, B.; Zhuo, L. Deep and ultra-deep oil and gas well drilling technologies: Progress and prospect. *Nat. Gas Ind. B* **2022**, *9*, 141–157. [CrossRef]
2. Wu, X.; Wan, F.; Chen, Z.; Han, L.; Li, Z. Drilling and completion technologies for deep carbonate rocks in the Sichuan Basin: Practices and prospects. *Nat. Gas Ind. B* **2020**, *7*, 547–556. [CrossRef]
3. Bintarto, B.; Auliya, R.R.; Putra, R.A.M.; Pradipta, A.S.; Kurnia, R.A. Production Data Analysis and Sonolog for Determining Artificial Lift Design and Well Characteristic. *J. Pet. Geotherm. Technol.* **2020**, *1*, 28–35. [CrossRef]

4. Kolawole, O.; Gamadi, T.D.; Bullard, D. Artificial lift system applications in tight formations: The state of knowledge. *SPE Prod. Oper.* **2020**, *35*, 422–434. [CrossRef]
5. Clegg, J.; Bucaram, S.; Hein, N. Recommendations and Comparisons for Selecting Artificial-Lift Methods (includes associated papers 28645 and 29092). *J. Pet. Technol.* **1993**, *45*, 1128–1167. [CrossRef]
6. Holden, J. Fixing the Deep-Well Jet Pump. *Midwest Q.* **2009**, *50*, 357–359.
7. Zou, C.H.; Li, H.; Tang, P.; Xu, D.H. Effect of structural forms on the performance of a jet pump for a deep well jet pump. In Proceedings of the 17th International Conference on Computational Methods and Experimental Measurements, Opatija, Croatia, 5–7 May 2015.
8. Zhang, H.; Zou, D.; Yang, X.; Mou, J.; Zhou, Q.; Xu, M. Liquid–Gas Jet Pump: A Review. *Energies* **2022**, *15*, 6978. [CrossRef]
9. van der Lingen, T.W. A jet pump design theory. *J. Fluids Eng.* **1960**, *82*, 1128–1167. [CrossRef]
10. Mallela, R.; Chatterjee, D. Numerical investigation of the effect of geometry on the performance of a jet pump. *Proc. Inst. Mech. Eng. Part C J. Mech. Eng. Sci.* **2011**, *225*, 1614–1625. [CrossRef]
11. Lu, H.Y. *Theory and Application of Jet Technology*; Wuhan University Press: Wuhan, China, 2004.
12. Reddy, Y.R.; Kar, S. Theory and performance of water jet pump. *J. Hydraul. Div.* **1968**, *94*, 1261–1282. [CrossRef]
13. Cunningham, R.G. Gas compression with the liquid jet pump. *J. Fluids Eng.* **1974**, *96*, 203–215. [CrossRef]
14. Grupping, W.; Coppes, J.L.R.; Groot, J.G. Fundamentals of oil well jet pumping. *SPE Prod. Eng.* **1988**, *3*, 9–14. [CrossRef]
15. Winoto, S.H.; Li, H.; Shah, D.A. Efficiency of jet pumps. *J. Hydraul. Eng.* **2000**, *126*, 150–156. [CrossRef]
16. Wang, C.B. Research on the Jet Pumps Used in the Drain Sand of Petroleum Well. Ph.D. Thesis, Zhejiang University, Hangzhou, China, 2004. (In Chinese).
17. Kwon, O.B.; Kim, M.K.; Kwon, H.C.; Bae, D.S. Two-dimensional numerical simulations on the performance of an annular jet pump. *J. Vis.* **2002**, *5*, 21–28. [CrossRef]
18. Moghaddam, M.A.E.; Abandani, M.R.H.S.; Hosseinzadeh, K.; Shafii, M.B.; Ganji, G.G. Metal foam and fin implementation into a triple concentric tube heat exchanger over melting evolution. *Theor. Appl. Mech. Lett.* **2022**, *12*, 100332. [CrossRef]
19. Hosseinzadeh, K.; Moghaddam, M.A.E.; Asadi, A.; Mogharrebi, A.R.; Jafari, B.; Hasani, M.R.; Ganji, D.D. Effect of two different fins (longitudinal-tree like) and hybrid nano-particles (MoS2-TiO2) on solidification process in triplex latent heat thermal energy storage system. *Alex. Eng. J.* **2021**, *60*, 1967–1979. [CrossRef]
20. Yamazaki, Y.; Yamazaki, A.; Narabayashi, T.; Suzuki, J.; Shakouchi, T. Studies on mixing process and performance improvement of jet pumps (Effect of nozzle and throat shapes). *J. Fluid Sci. Technol.* **2007**, *2*, 238–247. [CrossRef]
21. Aldaş, K.; Yapıcı, R. Investigation of effects of scale and surface roughness on efficiency of water jet pumps using CFD. *Eng. Appl. Comput. Fluid Mech.* **2014**, *8*, 14–25. [CrossRef]
22. Xu, K.; Wang, G.; Zhang, L.; Wang, L.; Yun, F.; Sun, W.; Wang, X.; Chen, X. Multi-objective optimization of jet pump based on RBF neural network model. *J. Mar. Sci. Eng.* **2021**, *9*, 236. [CrossRef]
23. Chen, S.; Yang, D.; Zhang, Q.; Wang, J. An integrated sand cleanout system by employing jet pumps. *J. Can. Pet. Technol.* **2009**, *48*, 17–23. [CrossRef]
24. Gao, C.C.; Wang, X.H. Application of jet pump technology in environmental protection. *Environ. Prot. Sci.* **2009**, *35*, 11–14. (In Chinese)
25. Sarshar, S. The Recent Applications of Jet Pump Technology to Enhance Production from Tight Oil and Gas Fields. In Proceedings of the SPE Middle East Unconventional Gas Conference and Exhibition, Abu Dhabi, United Arab Emirates, 23–25 January 2012.
26. Sun, B. Study on Gas Production Technology of Coalbed Methane Jet Pump. Master's Thesis, Southwest Petroleum University, Chengdu, China, 2016. (In Chinese).
27. Chen, S.; Li, H.; Zhang, Q.; He, J.; Yang, D. Circulating usage of partial produced fluid as power fluid for jet pump in deep heavy-oil production. *SPE Prod. Oper.* **2007**, *22*, 50–58. [CrossRef]
28. Gazzar, M.E.; Meakhail, T.; Mikhail, S. Numerical study of flow inside an annular jet pump. *J. Thermophys. Heat Transf.* **2006**, *20*, 930–932. [CrossRef]
29. Kurkjian, A.L. Optimizing Jet-Pump Production in the Presence of Gas. *SPE Prod. Oper.* **2019**, *34*, 373–384. [CrossRef]
30. Samad, A.; Nizamuddin, M. Flow analyses inside jet pumps used for oil wells. *Int. J. Fluid Mach. Syst.* **2013**, *6*, 1–10. [CrossRef]
31. Sheha, A.A.A.; Nasr, M.; Hosien, M.A.; Wahba, E. Computational and experimental study on the water-jet pump performance. *J. Appl. Fluid Mech.* **2018**, *11*, 1013–1020. [CrossRef]
32. Wang, Z.L. Structure Design and Simulation Analysis of Jet Pump in High and Low Pressure Reservoir. Master's Thesis, Yangtze University, Jingzhou, China, 2022. (In Chinese).
33. Hosseinzadeh, K.; Moghaddam, M.A.E.; Asadi, A.; Mogharrebi, A.; Ganji, D. Effect of internal fins along with hybrid nano-particles on solid process in star shape triplex latent heat thermal energy storage system by numerical simulation. *Renew. Energy* **2020**, *154*, 497–507. [CrossRef]
34. Hosseinzadeh, K.; Mogharrebi, A.R.; Asadi, A.; Paikar, M.; Ganji, D. Effect of fin and hybrid nano-particles on solid process in hexagonal triplex latent heat thermal energy storage system. *J. Mol. Liq.* **2020**, *300*, 112347. [CrossRef]
35. Zou, M.M. Design and Analysis of Deep Well Jet Pump. Master's Thesis, Southwest Petroleum University, Chengdu, China, 2016. (In Chinese).
36. Lyu, Q.; Xiao, Z.; Zeng, Q.; Xiao, L.; Long, X. Implementation of design of experiment for structural optimization of annular jet pumps. *J. Mech. Sci. Technol.* **2016**, *30*, 585–592. [CrossRef]

37. Long, X.; Zhang, J.; Wang, Q.; Xiao, L.; Xu, M.; Lyu, Q.; Ji, B. Experimental investigation on the performance of jet pump cavitation reactor at different area ratios. *Exp. Therm. Fluid Sci.* **2016**, *78*, 309–321. [CrossRef]
38. Long, X.; Yan, H.; Zhang, S.; Yao, X. Numerical simulation for influence of throat length on annular jet pump performance. *J. Drain. Irrig. Mach. Eng.* **2010**, *28*, 198–201, 206. (In Chinese)
39. Liu, B.; Dong, S.; Tan, J.; Li, D.; Yang, J. Design and flow simulation of a micro steam jet pump. *Mod. Phys. Lett. B* **2022**, *36*, 2250018. [CrossRef]
40. Gan, J.; Zhang, K.; Wang, D. Research on Noise-Induced Characteristics of Unsteady Cavitation of a Jet Pump. *ACS Omega* **2022**, *7*, 12255–12267. [CrossRef] [PubMed]
41. Karna, S.K.; Sahai, R. An overview on Taguchi method. *Int. J. Eng. Math. Sci.* **2012**, *1*, 1–7.
42. Tsui, K.L. An overview of Taguchi method and newly developed statistical methods for robust design. *Iie Trans.* **1992**, *24*, 44–57. [CrossRef]
43. Mathews, P.G. *Design of Experiments with MINITAB*; ASQ Quality Press: Milwaukee, WI, USA, 2005.

processes MDPI

Article

Numerical Study of the Purge Flow's Effect on the Loss Mechanism of the Blocking and Shear Effects

Tao Bai [1,2,*], Qingzhen Yang [1] and Jian Liu [3]

1 School of Power and Energy, Northwestern Polytechnical University, Xi'an 710072, China
2 School of Aircraft Engineering, Xi'an Aeronautical Institute, Xi'an 710077, China
3 High Speed Aerodynamics Institute, China Aerodynamics Research and Development Center, Mianyang 621000, China
* Correspondence: 201407004@xaau.edu.cn

Abstract: The loss mechanism of shear and upstream blockage caused by the interaction of the purged flow and ingested gas needs to be systematically studied to optimize the flow near the rim. In order to study the causes and influence factors of blocking and shearing effects and quantify their losses reasonably, the three-dimensional unsteady numerical method validated by the experiment data was adopted to study the turbine with three kinds of sealing structures. The block and shear loss are quantified by integrating the dissipation coefficient in the volume where the specific aerodynamic loss occurs. The result indicated that there was a larger radial velocity and smaller tangential velocity of the purged flow relative to the main flow caused the blocking effect. Therefore, its loss is affected by the seal flow and seal structure. The shear effect is mainly affected by the tangential velocity gradient and the axial velocity gradient near the cavity exit. The contribution of the tangential velocity gradient to shear loss is increased with the enlarged sealing efficiency. Through research, it is clear that increasing the purge flow's tangential velocity is beneficial to reducing the shear loss and has a positive significance for weakening the blocking effect in the main flow channel. Furthermore, the influence of sealing structure on blocking and shear effect must be particularly considered since both are related to sealing efficiency.

Keywords: purge flow; seal structure; shear loss; blocking effect; sealing efficiency; tangential velocity; tangential velocity gradient

Citation: Bai, T.; Yang, Q.; Liu, J. Numerical Study of the Purge Flow's Effect on the Loss Mechanism of the Blocking and Shear Effects. *Processes* **2023**, *11*, 50. https://doi.org/10.3390/pr11010050

Academic Editor: Alfredo Iranzo

Received: 22 November 2022
Revised: 12 December 2022
Accepted: 13 December 2022
Published: 26 December 2022

1. Introduction

Introducing sealing air to the turbine cavity is an effective way to cool the high-temperature wall of the turbine cavity and prevent gas ingestion. However, the interaction between the purge flow and the mainstream leads to additional losses for the turbine's main annular, thus lowering the aerodynamic efficiency. Total-to-total efficiency would decrease by 0.6% when the sealing air rate *IR* is enlarged from −0.14% to 0.9% according to the research of Schuepbach et al. [1]. Reid et al. [2] pointed out that the turbine efficiency decreases by 0.56% with the purge flow rate of 1.0%.

Zlatinov et al. [3] quantified the loss of each part of the sealed air on the main flow through theoretical analysis and numerical calculation. It is pointed out that the loss caused by the sealed air can be attributed to three aspects: (1) the shear loss caused by the tangential velocity difference between the sealed air and the main flow; (2) the loss caused by the interaction between the sealing flow and the main secondary flow; (3) the loss induced by the blocking of the sealed flow to the main flow channel. They also pointed out that the blocking effect is independent of the swirl ratio and only affected by the sealing flow. The numerical studies by Jia Wei et al. [4] divide the losses associated with the purge flow into four types, emphasizing that they are not independent of each other.

Many studies have been carried out on the influence of purge flow on the secondary flow, and the significance of its influence has been confirmed [5–7]. The research in a

1.5-stage high-pressure turbine carried out by Schuepbach et al. [8] showed that the secondary flow in the hub would significantly develop in the radial direction with the purge flow. Ong et al. [9] pointed out that the negative angle of attack at the rotor inlet caused by the sealing air is the main reason for the enhancement and radial expansion of the rotor passage vortex. Regina et al. [10] focused on the influence mechanism of the sealing air on the rotor channel vortex. They believe the reason for changing the secondary flow intensity of the rotor is the increase of the flow coefficient at the hub rather than the increase of the negative angle of attack of the rotor. McLean et al. [11,12] found that the purge air not only significantly affects the secondary flow structure, but also causes apparent unsteady fluctuation of the main flow pressure and affects the development of the boundary layer at the hub. Schreiner and Mesny et al. [13,14] tracked the evolution mechanism of the interference of sealing air on the mainstream vortex structure through more advanced testing methods, making the research in this field more refined.

Compared to the influence on the interference of secondary flow, the research on shear and blocking loss has not been systematic. However, it also accounts for a large proportion of the interference loss. Reid et al. [2] found that severe mixing of the sealing air with the mainstream and its interference with the channel vortex are the leading causes of the losses. The losses caused by the two are equivalent. The experimental measurement also illustrated that the pre-swirl of the purge air would weaken the main flow aerodynamic loss caused by the increase of the sealing flow rate. Jenny et al. [15] obtained a similar conclusion. They also believed that the result of the reduction of turbine power capacity is that the turbine rotor works on the sealed air by accelerating the tangential direction and dissipating it through the hub vortex core region. Subsequently, Schrew et al. [16] focused on the mechanism of loss generation by analyzing the change law of sealing efficiency, inlet and outlet section Mach number, blade surface pressure coefficient, and hub pressure coefficient with sealing flow through the measured pressure on a two-stage low-pressure turbine. Yangfan [17–19] found that each part of the loss proposed by Zlatinov and Schrew et al. has prominent regional distribution characteristics. Based on this characteristic, the loss of each part is quantified. Similar to most previous studies, the conclusion is that shear loss is dominant in all kinds of losses. Therefore, the shear loss characteristics are further studied. It is found that the circumferential velocity gradient in the radial direction, the axial velocity gradient in the radial direction, and the tangential velocity gradient in the axial direction are the main factors causing viscous shear loss. They believe that the sealing flow and pre-swirl parameters at the inlet of the sealing flow significantly influence the shear loss. In contrast, other parameters, such as the sealing structure's angle and the sealing gap's position and width, can be ignored. Dario Barsi [20] quantified the loss caused by the interference of purged flow and main flow downstream of the vane and upstream of the rotor. The loss distributions indicated that the loss at the exit of the vane blade increased significantly; compared to the system without a cavity, the loss increased by 50%, below 50% of the blade span. Huang et al. [21] emphasized that the blocking caused by the sealed air reduces the loss of the upstream stator. Compared to the Chute sealing structure, the blocking of the radial sealing structure is more substantial. Hunter [22] concluded that the primary source of loss related to the sealing flow is the mixing loss caused by the difference in tangential momentum between the sealing air and the main flow. Zhang et al. [23] found that the Kelvin–Helmholtz instability near the turbine rim reduces the tangential velocity gradient of the main flow and the purge flow, thereby reducing the aerodynamic loss.

It is generally believed that the losses caused by the mixing shear due to the purge flow account for a relatively large proportion of all kinds of losses. In a typical front cavity structure, the loss associated with the purge air in the upstream stator region includes not only mixing shear but also blocking effect. Their influence mechanism has not been unified yet because of the complicated interference mechanism between ingested gas and purge air at the rim seal of the wheel space. In addition, the model assumptions are adopted in most quantitative studies on losses, which are not conducive to accurately assessing losses.

In this paper, the front cavity sealing structure was taken as the research object. The influence of sealing air on the upstream stator blocking and shear effect was analyzed in detail on account of the interaction between ingested gas and sealing air in the rim gap. In addition, the influence mechanism of sealing flow rate and sealing structure on blocking effect and shear effect was fully considered.

2. Numerical Method

2.1. Computational Models

A 1.5-stage Lisa turbine with a front cavity was adapted to investigate. The seal structures in this paper are axial, radial, and chute, with the angles of 30, 45, and 60. The turbine devices with five sealing structures were described as Axial, Radial, Chute_60, Chute_45, and Chute_30. The inclination angle is between the seal structure and the axial direction. The outlet of Chute_30 is closest to the axial direction; that is, the inclination is the largest. The model used in the paper is the same as that in the literature [24]. The Computational Model is shown in Figure 1. The r blade profile parameters are shown in Table 1. In addition, the turbine without cavity was also calculated as a comparative case, described as "No_purge" below.

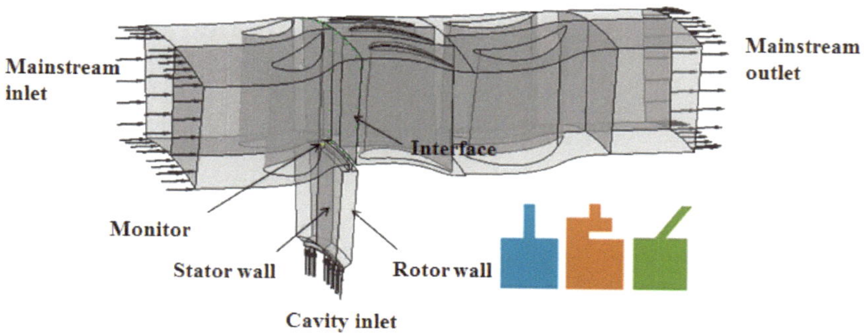

Figure 1. Computation model.

Table 1. Geometric and aerodynamic parameters of each blade row of a Lisa stage 1.5 turbine.

Parameter	S1	R1	S2
Blade numbers	36	54	36
Span/mm	70	70	70
Chord length/mm	80.88	59.72	85.50
Pitch/mm	63.7	42.5	63.7
Inlet angle/°	0.0	52.4	−35.4
Exit angle/°	72.0	−66.6	66.0
Exit Ma	0.54	0.50	0.48
Re/$\times 10^5$	7.1	3.8	5.1

2.2. Numerical Approach and Mesh

ANSYS CFX software(2020R2) was used to solve the Reynolds-averaged N–S equation, and the $k-\omega$ SST turbulence model was selected to solve the unsteady three-dimensional flow. To minimize the interpolation error, a matching grid interface should be defined. Thus, the mesh of the stator with the cavity was carried out in the ICEM software(2020R2), while the rotor and the second stator were undertaken using NUMECA AutoGrid5 software(8.2). The hexahedral structured grid is shown in Figure 2. The grids near the wall surface and the cavity gap were also densified. The y+ value of most areas near the wall surface was close to 1, and the maximum value was about 2. The total number of nodes was 8.0 million, and the detailed information referenced [24].

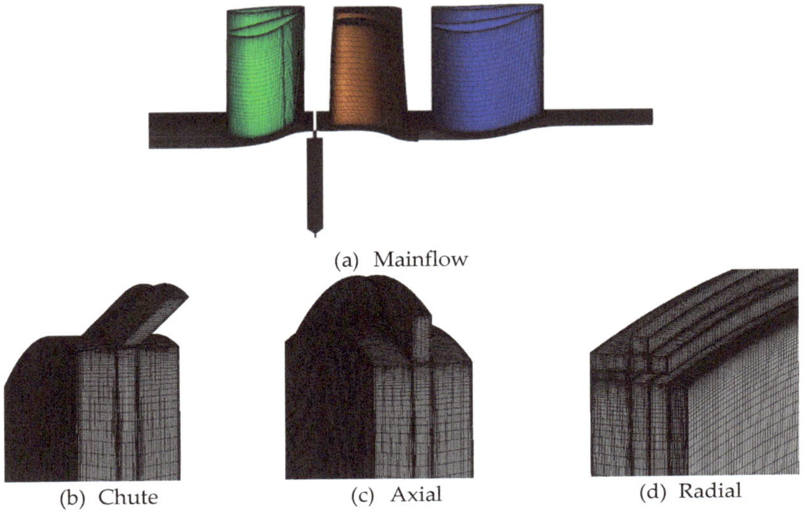

Figure 2. Computation mesh of a 1.5-stage turbine.

2.3. Boundary Conditions

The boundary condition is shown in Table 2. The working medium is a mixture of CO2 and ideal gas to distinguish the purge flow from the mainstream. The *IR* is defined as the ratio of sealing airflow to mainstream flow. The physical time step was 0.00978 us, which was obtained by dividing the blade into 42 steps when it passed through a rotor channel. After the convergence of unsteady calculation, which could be identified by the unsteady change of pressure in Figure 3, the data of 9 cycles was taken for the statistical average [24].

Table 2. Boundary condition [25].

Boundary Condition	Value
Total pressure in the mainstream inlet (Kpa)	140
Total temperature in mainstream inlet (K)	328
Total temperature in the cavity inlet(K)	308
Rim seal flow IR in cavity inlet	0%,0.5%,1.0%,1.5%
The concentration of CO_2 in mainstream inlet	0
The concentration of CO_2 in cavity inlet	1
The rotation speed of rotor wall(r/min)	2700

2.4. Validation of Numerical Method

The calculated total pressure coefficient and flow angle at the stator exit are compared with the experimental data to verify the numerical simulation method since the shear effect and blocking effect considered in this study occur in the stator domain. The total pressure coefficient in Figure 4 is defined as Equation (1), and the specific parameters in the equation are referenced [24]. The flow angle in Figure 5 is the angle between the flow and the axial direction.

$$c_{pt} = (p_t - p_{s,exit})/(p_{t,inlet} - p_{s,exit}) \tag{1}$$

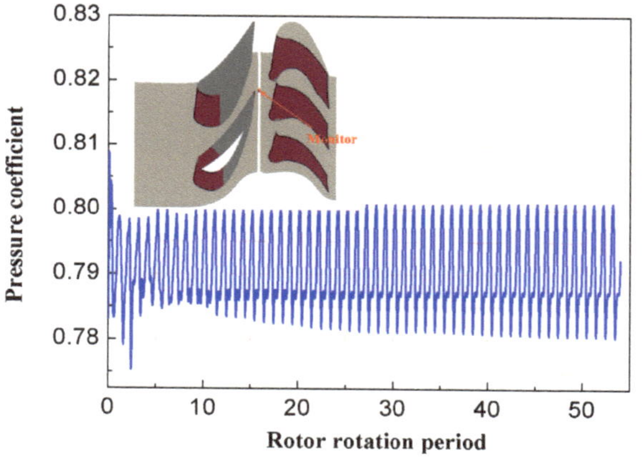

Figure 3. Pressure coefficient change at the monitoring point.

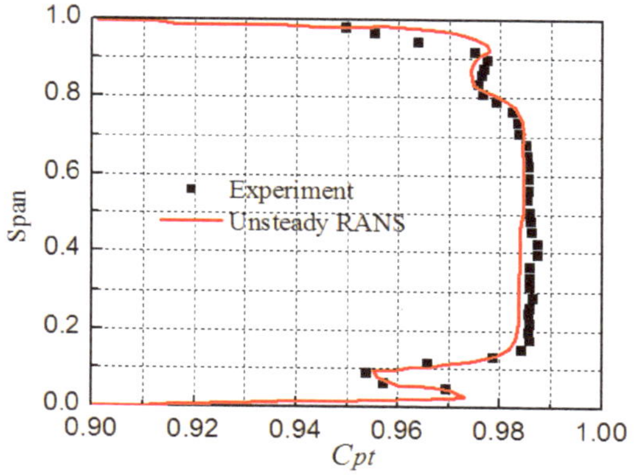

Figure 4. Total pressure coefficient at stator outlet.

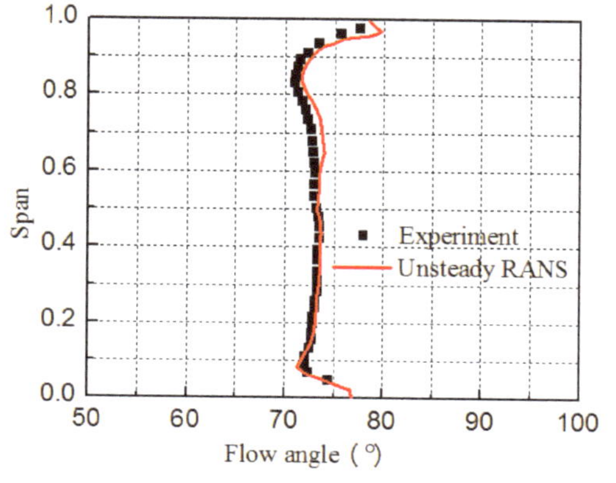

Figure 5. Flow angle at stator outlet.

It can be concluded that the position of the vortex at the end wall of the stator exit captured by the numerical method was consistent with the experimental results. The corresponding under-rotation and over-rotation position of airflow is also relatively consistent. The blocking and shear effects studied in this paper are related to the flow field near the stator hub. Figure 5 shows that the numerical results near the hub are in good agreement with the experimental results. Therefore, the selected numerical method is reasonable based on the research purpose of this paper.

3. Results and Discussion

3.1. Definition of Loss Parameters and Variation of Turbine Loss along Flow Direction

To evaluate the aerodynamic loss of turbine mainstream along the flow direction with or without the cavity, the entropy increase was introduced in this paper. The mainstream entropy increase of the turbine without a cavity is defined as Equation (2):

$$\Delta s_{No\_purge} = Cp\ln(\frac{T_s}{T_{t,in}}) - R_g\ln(\frac{P_s}{P_{t,in}}) \tag{2}$$

where $T_{t,in}$, $P_{t,in}$ are the total temperature and total pressure at the mainstream inlet, respectively, and T_s, P_s represent local static temperature and pressure, respectively.

Since there are two streams for the turbine with a cavity, the total pressure in the above formula is replaced by the work-averaged total pressure,$P_{t,in}^{wa}$ defined as Equation (3), which is based on the work-average of the purge flow and the mainstream, referred to in [26], while the total temperature is based on the mass average of the two streams.

$$P_{t,in}^{wa} = \left[\frac{\int T_t d\dot{m}}{\int (T_t/p_t^{\frac{\gamma-1}{\gamma}})d\dot{m}}\right]^{\frac{\gamma}{\gamma-1}} \tag{3}$$

The entropy increase of the turbine with a cavity obtained by using the work-averaged total pressure and the mass-averaged total temperature is defined as Equation (4):

$$\Delta s_{IR} = Cp\ln(\frac{T_s}{T_{t,in}{}^{ma}}) - R_g\ln(\frac{P_s}{P_{t,in}{}^{wa}}) \tag{4}$$

Therefore, the loss caused by the purge flow can be expressed as Equation (5):

$$\Delta s_{Purge=}\Delta s_{IR} - \Delta s_{No\_Purge} \tag{5}$$

The change of entropy increase along the flow direction can reflect the loss characteristics, thus helping to determine the key location in the streamwise direction. Figure 6a shows the distribution of entropy increase along the axial direction of the 1.5-stage turbine due to the interference between the purge air and the main flow for *IR* = *1.5%*. It can be seen that the purge flow increases the losses of the upstream stator, the downstream rotor, and the second-stage stator. The interference loss in the first-stage stator domain accounts for about 30% of the total loss of the 1.5-stage turbine.

The entropy increase along the streamwise direction for the first stator is shown in Figure 6b, and the cavity location is marked with a blue line (from 0.96 to 0.98). It can be seen that the loss slightly decreases from the stator inlet to the axial direction position of 0.90, but sharply decreases in the range of 0.9–0.93. Yangfan et al. [17] illustrated that the loss change was caused by the blocking of the sealed air on the upstream stator. It is called the blocking effect, and the corresponding loss reduction is called blocking loss in this paper. The loss significantly increases from the position of 0.93 to the interface of the rotor–stator, which is caused by the shear mixing of sealing air and main flow. It is called the shear effect, and the corresponding loss increase is called shear loss.

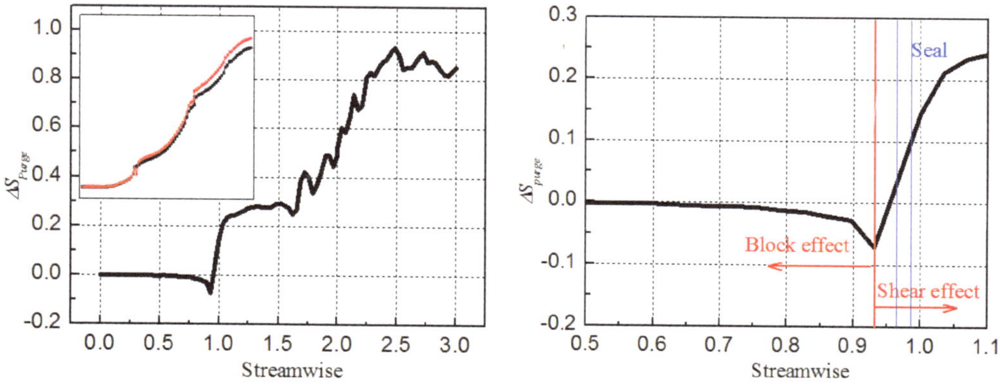

(**a**) 1.5 stage (**b**) stator

Figure 6. Distribution of entropy increase (Δs_{Purge}) along the flow direction (*IR* = *1.5%*).

Based on the above analysis, it can be determined that the blocking effect mainly occurs upstream of the cavity, i.e., before the flow direction position of 0.93. The shear mixing mainly occurs from 0.93 to the stator–rotor interface. Therefore, the loss distribution spatial region in the stator channel can be represented by Figure 7 schematically. The volume where blocking and shearing take place is defined as V_B, V_S. It should be noted that only the influence of sealing air on the main flow loss was considered in this paper, and variation loss caused by the interaction of main flow and secondary flow in the inner cavity was neglected.

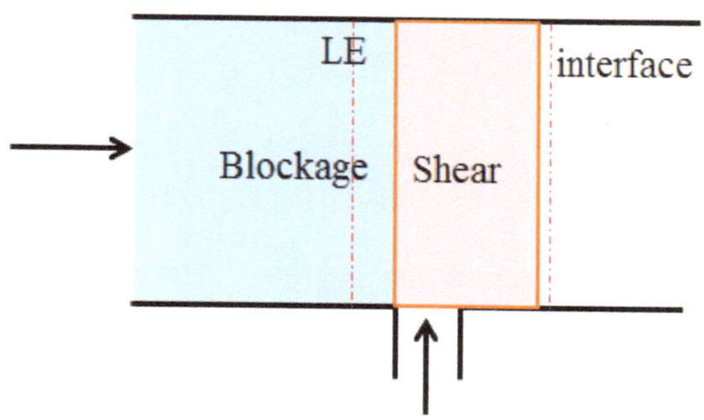

Figure 7. The loss area of turbine stator passage.

It is suitable to establish the regions where blocking and mixing occur using the entropy increase. However, there would be a relatively large error in quantifying the loss as the total inlet pressure and temperature in Equation (4) while adopting the model assumption when processing. Furthermore, the entropy increase is an integral quantity. Therefore, it seems inconvenient to quantitatively describe the local loss. This paper introduces the dissipation coefficient to characterize local loss and quantify the regional loss [3]. The dissipation coefficient is dimensionless by the stator inlet parameter, and its definition is Equation (6):

$$D = \frac{\begin{bmatrix} \left[\left(\frac{\partial V_\theta}{\partial r} + \frac{1}{r}\frac{\partial V_r}{\partial \theta}\right)^2 + \left(\frac{1}{r}\frac{\partial V_z}{\partial \theta} + \frac{\partial V_\theta}{\partial z}\right)^2 + \left(\frac{\partial V_r}{\partial z} + \frac{\partial V_z}{\partial r}\right)^2\right] + \\ 2\left[\left(\frac{\partial V_r}{\partial r}\right)^2 + \left(\frac{1}{r}\frac{\partial V_\theta}{\partial \theta} + \frac{\partial V_r}{r}\right)^2 + \left(\frac{\partial V_z}{\partial z}\right)^2\right] - \frac{2}{3}(\nabla \cdot V)^2 \end{bmatrix}}{\left(\rho V_{rel}^3 / 2C_{ax}\right)_{S_{in}}} \tag{6}$$

What needs to be further verified is that the dissipation coefficient and the entropic increase are consistent in describing the law of loss variation. Figure 8 shows the flow-based integral parameters of three axial sections (the streamwise position was 0.88, 0.93, and 0.98, respectively) with or without a cavity. The difference between the two represents the loss caused by sealing air at this axial position. It can be seen that the variation trend of the integral difference of dissipation coefficient along the flow direction is consistent with Figure 6b.

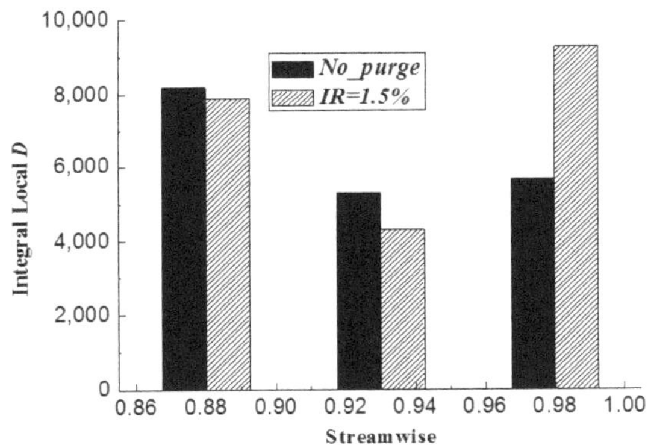

Figure 8. Dissipation coefficient integral.

3.2. Blocking Effect

Figure 9a shows the pressure coefficient distribution from the inlet to the outlet of the stator region. The pressure coefficient distribution on the blade surface is also shown in Figure 9b. The pressure coefficient is defined as Equation (7):

$$Cp = \frac{P}{P_1^*} \tag{7}$$

The blocking effect induced by the purge flow on the stator passage increases the pressure in the axial direction, especially upstream of the outlet of the cavity. The suction surface pressure coefficient increases remarkably after the maximum curvature of the suction surface of the blade, while the pressure surface side is almost unchanged. The change of the pressure gradient along the flow direction and tangential direction in the stator region will weaken the channel vortex and reshape the hub boundary layer. As a result, the position of the shedding vortex and the channel vortex lifted in the radial direction, and their size both reduced compared to the no purge flow situation, which is displayed in Figure 10. It should be responsible for the altered loss caused by the blocking effect. The cause and influence mechanism of the blocking effect this paper focuses on will be described in detail below.

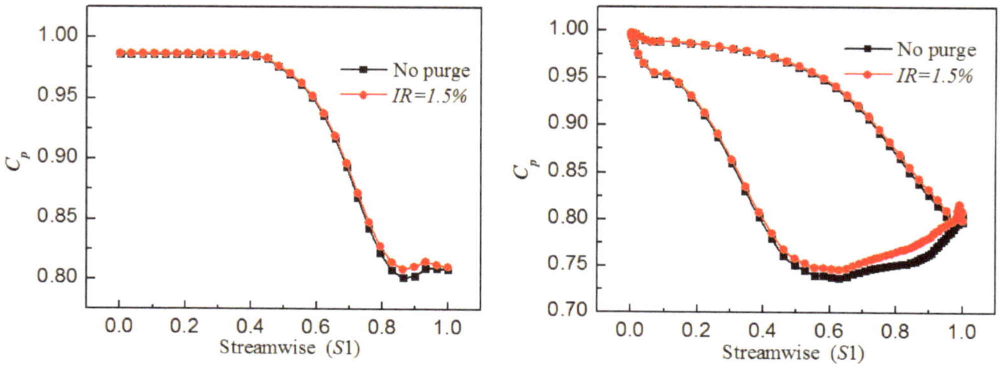

(**a**) From inlet to outlet (**b**) Blade surface

Figure 9. Time-averaged pressure coefficient distribution.

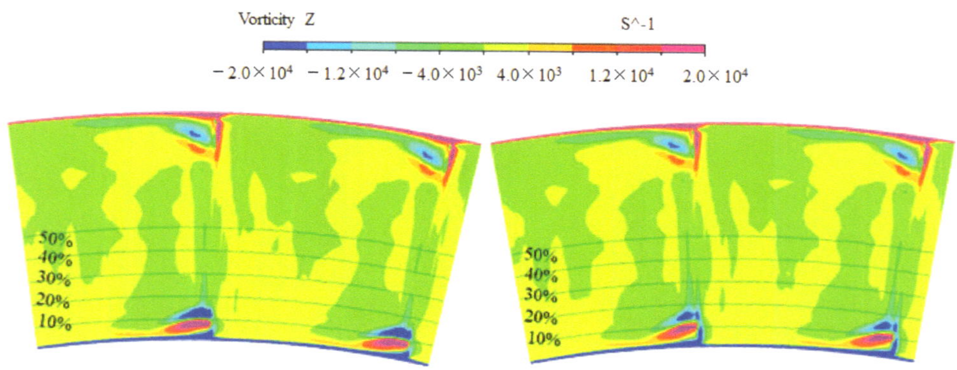

(**a**) No purge (**b**) IR = 1.5% (Axial)

Figure 10. Time-averaged vorticity Z contour (streamwise 0.92).

Blocking loss, expressed as ΔD_{Blok}, should be a negative value from the weakened vortex in Figure 10 with the existence of the purge flow. The volume integral of the dissipation coefficient in the blocking area of the turbine stator with or without cavity is carried out separately, and the difference between the volume integral of the dissipation coefficient is defined as the blocking loss, defined as Equation (8):

$$\Delta D_{Block} = \int_{V_{B,purge}} D - \int_{V_{B,No\_purge}} D \tag{8}$$

Blocking losses under different sealing flow rates and sealing structures are shown in Table 3. The blocking effect increases and blocking loss decreases significantly with the increase in sealing flow rate. The blocking effect of the chute seal structure is more evident than that of other seal structures. Yangfan et al. believe that the radial velocity of purge flow is the main reason for blocking, but it seems unreasonable to explain the effect of seal structure on the blocking effect.

Table 3. Blocking loss (dm^3).

ΔD_{Block}	Axial	Radial	Chute_60	Chute_45	Chute_30
0%	−0.003	—	—	—	—
0.5%	−0.009	—	—	—	—
1.0%	−0.018	−0.020	−0.021	−0.026	−0.028
1.5%	−0.026	—	—	—	—

To further analyze the key parameters causing the blocking effect, the sealing efficiency is listed in Table 4, and the tangential velocity coefficient ($\varphi_{V\theta}$) distribution in the cavity is displayed in Figure 11.

Table 4. Time-averaged sealing efficiency.

	Axial	Radial	Chute_60	Chute_45	Chute_30
0%	10.4%	—	—	—	—
0.5%	42.0%	—	—	—	—
1.0%	77.3%	82.3%	86.4%	94.1%	99.9%
1.5%	95.3%	—	—	—	—

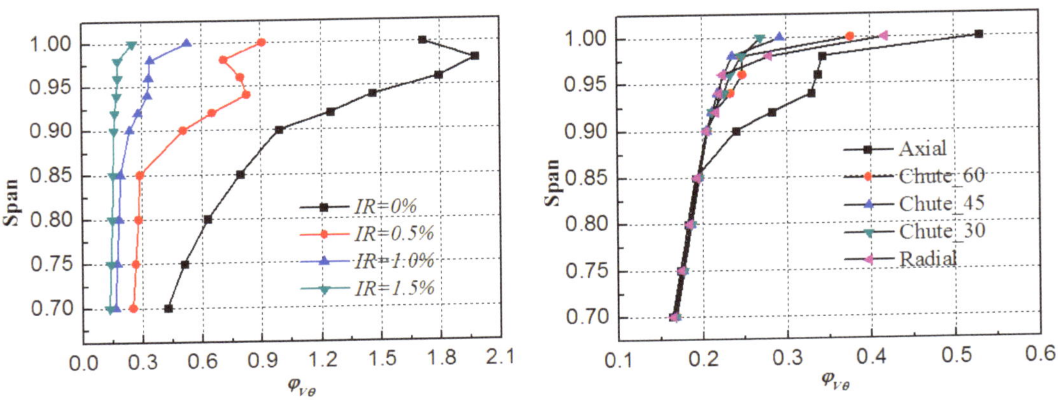

Figure 11. Time-averaged tangential velocity distribution.

The sealing efficiency is defined as Equation (9), where c_s is the local tracer gas (CO_2) concentration, c_a is the mainstream inlet tracer gas concentration, and c_0 is the tracer gas concentration of the purge flow in the cavity inlet.

$$\varepsilon_c = \frac{c_s - c_a}{c_0 - c_a} \tag{9}$$

The tangential velocity coefficient $\varphi_{V\theta}$ is the ratio of tangential velocity to rotational speed of the rotor hub.

The radial velocity at the outlet of the cavity increases with the enlarged seal flow rate for the axial seal structure according to the flow conservation, which forms a blocking effect on the upstream flow field. Therefore, the blocking effect becomes significant with the increase in flow rate. The radial velocity is indeed one of the factors causing the blocking effect from this point of view. It should be noted that the radial velocity for the

Chute structure should be reduced based on the above theory. Thus, the blocking effect weakens. However, the blocking effect becomes more remarkable with the inclination increase, illustrated by the lowered blocking loss in Table 3.

For exploring other factors affecting the blocking effect, the velocity triangle near the stator hub and downstream of the cavity is shown in Figure 12. Chilla et al. [27] pointed out that the tangential velocity difference between the main flow and the sealed flow at the cavity gap can reach 1.25 times the rim speed. That means the lower tangential velocity in the exit of the cavity promotes the decrease of the mainstream tangential velocity from C_{1u} to $C_{1u,j}$ in the velocity triangle. As a result, the effective flow area of the airflow decreases with the reduced turning angle of the airflow, thus helping to create a blocking effect on the stator passage. In other words, tangential velocity also affects the blocking effect. Next, the influencing factors of tangential velocity need to be found to reveal the blocking mechanism.

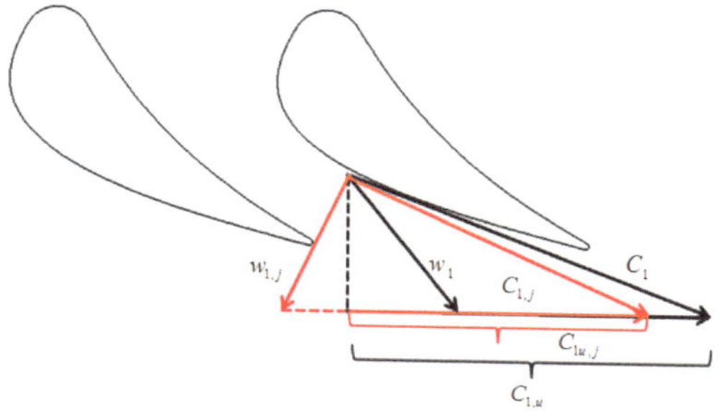

Figure 12. Velocity triangle distribution downstream of cavity exit.

As shown in Figure 11a, tangential velocity enlarged with the reduced sealing flow. Yang Fan et al. [18] believe that this is due to the increasing acceleration time of the rotating disc to the fluid in the cavity for the lower purge flow. What needs more attention is that the gas ingested from the mainstream will accelerate the tangential velocity of the sealed air, which can be verified by the variety in the tangential velocity of different seal structures. The tangential velocity should increase as the acceleration time increases for Chue_30 compared to Axial or Chute_60 at the same seal flow rate. However, the tangential velocity decreases from Figure 11b. Consequently, the main factor influencing the change of tangential velocity for the different seal strucuture is gas ingestion rather than disk acceleration time.

The blocking effect for chute seal structure is more vital than axial seal and becomes remarkable with the increased inclination. The lower tangential velocity caused by the weakened gas intrusion is responsible for the above situation. Compared to the axial sealing structure, the radial sealing structure has relatively higher sealing efficiency due to the damping effect of internal teeth on gas intrusion. Its blocking effect is between Axial and Chute_60. In addition, the gas ingested is also weakened with the increase of sealing flow, thus sealing efficiency increases and tangential velocity is reduced, which is another reason for the increasing blocking effect besides the higher radial velocity. In conclusion, blocking effect is negatively related to sealing efficiency.

The existing research and the velocity triangle in Figure 12 in this paper show that the blocking effect will not only affect the stator loss but also cause a negative angle of attack at the inlet of the downstream rotor, which leads to a significant secondary loss. Therefore, alleviating the upstream blocking effect has a positive effect in general.

3.3. Viscous Shear Effect

The time average dissipation coefficient and streamline diagram of the outlet section of the cavity for $IR = 1.5\%$ are displayed in Figure 13. A large shear region is formed near the hub due to the viscous shear of the main flow and the sealing flow. The dissipation coefficient is prominent in the shear region, and the local loss is enormous. According to the viscous shear theory, the velocity gradient is responsible for the viscous shear. To clarify the main velocity gradients affecting the viscous shear of the main flow and the sealed air, the velocity gradients of the above outlet sections are integrated based on the mass flow. The proportion of each velocity gradient is shown in Figure 14.The tangential velocity gradient along the radial direction $\partial V_\theta/\partial r$, the tangential velocity gradient along the axial direction $\partial V_\theta/\partial z$, and the axial velocity gradient along the radial direction $\partial V_z/\partial r$ account for 80% of the total. These three velocity gradients are the leading causes of viscous shear loss, which is consistent with the conclusion of Yang Fan et al. [18]

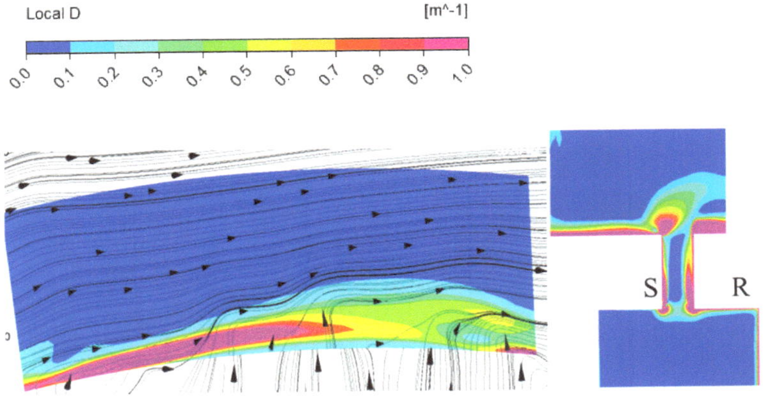

Figure 13. The dissipation coefficient distribution at the exit of disk cavity, *IR* = *1.5%*. (Left: axial direction, right: circumferential direction).

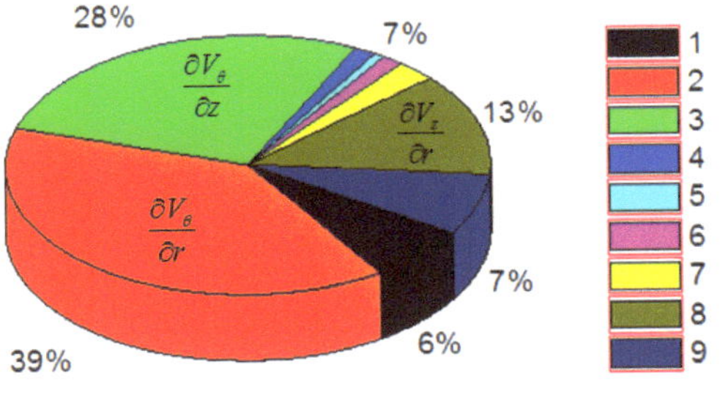

Figure 14. Proportion of velocity gradient based on mass flow integral, *IR* = *1.5%*.

It is still necessary to quantify the shear loss, similar to the blocking effect above. Yang Fan et al. [18] used the control volume to quantify the shear loss. This method needs to select the inlet section of the secondary flow and the main flow. Actually, the location of the inlet section will significantly impact the results because the complex interaction between purge flow and main flow makes the parameters in the gap more sensitive to the location, especially at the lower sealing flow rate. In addition, the model is mainly based on three main velocity gradients when quantifying the loss. In this paper, the difference between the volume integral of the dissipation coefficient of the turbine with or without a cavity in the mixing region (V_S), as shown in Figure 7, is the shear mixing loss, which is defined as Equation (10):

$$\Delta D_{\mathrm{Shear}} = \int_{V_{S,purge}} D - \int_{V_{S,No\_purge}} D \tag{10}$$

Table 5 shows the viscous shear loss under different sealing flows and sealing structures. The shear loss increases sharply with the enlarged sealing flow. The chute sealing structure is greater than that of the axial sealing structure, and the radial sealing structure is between the two for the constant sealing flow.

Table 5. Viscous shear loss (dm^3).

ΔD_{Shear}	Axial	Radial	Chute_60	Chute_45	Chute_30
0%	0.0467	—	—	—	—
0.5%	0.0534	—	—	—	—
1.0%	0.0670	0.0863	0.0734	0.0924	0.095
1.5%	0.0855	—	—	—	—

To further reveal the mechanism of viscous shear loss, the velocity gradient is carefully analyzed from the interaction of sealing air and the gas in the annular. The integral of each velocity gradient with different flow rates are shown in Table 6.

Table 6. Velocity gradient for variety *IR*.

Mass Average	0%	0.5%	1%	1.5%
$\frac{\partial V_\theta}{\partial r}$	121.0 (32.8%)	226.7 (36.2%)	319.7 (39.7%)	448.4 (39.1%)
$\frac{\partial V_\theta}{\partial z}$	23.14 (6.3%)	121.2 (19.4%)	204.2 (25.3%)	314.82 (27.4%)
$\frac{\partial V_z}{\partial r}$	117.0 (31.7%)	133.5 (21.3%)	119.7 (14.9%)	148.62 (13%)
Total proportion	70.8%	76.9%	79.9%	79.5%

The proportion of the main three gradients in the total gradient increases, from 70% to nearly 80%, with the increase of the *IR* from 0% to 1.5%. That is to say, it is much more accurate to use the model proposed by Wilcolx et al. [28] to evaluate the loss based on the control volume considering the main velocity difference for the high sealing flow.

The tangential velocity gradients $\partial V_\theta/\partial r$ and $\partial V_\theta/\partial z$ increased, and the proportion of the two enlarged from 39.1% to 66.5%, with the sealing flow increasing from 0% to 1.5%.To explain this phenomenon, the sealing efficiency contour and streamline are given in Figure 15. Based on the previous research work on gas ingestion, a rim seal vortex would form because of the interaction of the ingested gas from the rotating side of the cavity and the purge flow, and rotation direction was shown by a red arrow. The vortex area increases as the sealing flow rate decreases. The tangential acceleration effect on the purge flow becomes stronger with severe gas ingestion when the sealing rate is lower, such as *IR* = 0.5%. This is the reason for the decreased $\partial V_\theta/\partial r$ value as the sealing flow rate is reduced. Meanwhile, the rotating action of the rim seal vortex structure with high strength at low sealing flow rate speeds up the migration of the gas from the rotor wall side to the stationary side of the cavity, which will help to reduce the $\partial V_\theta/\partial z$.

In addition, the axial flow, which is opposite to the main flow, would form at the upper section of the gap as the rim seal vortex made the ingested gas flow from the rotor side to the stationary side in the upper rim clearance. It is the reason for the enlarged gradient $\partial V_z/\partial r$ with the lower sealing flow.

In conclusion, the interaction between gas and sealing flow near the outlet of the cavity is the main reason for the change in velocity gradient. The interference between gas and sealing air near the gap outlet diminishes with the increasing sealing efficiency, and the proportion of tangential velocity gradient gradually dominates. That is to say, the tangential velocity gradient plays a leading role in the viscous shear loss in the higher sealing efficiency situation.

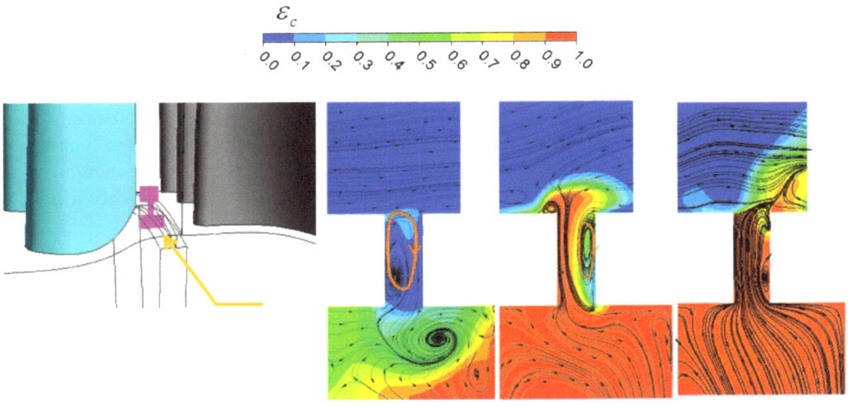

(**a**) 0.5% (**b**) 1.0% (**c**) 1.5%

Figure 15. The sealing efficiency and streamline (t = 0T).

The sum of the three gradients is almost the same as in Table 7, under the same sealing flow rate even with the different sealing structures. However, the chute sealing structure has a large tangential velocity gradient, especially for Chute_30; the tangential velocity gradient ratio reached 69.5%, while the axial and radial sealing structures were 65% and 59.9%, respectively. This is related to the higher sealing efficiency; of course, this weakens gas ingestion for the chute sealing structure at this flow rate. Combined with the sealing efficiency in Table 4, it can be concluded that the tangential velocity gradient is positively correlated with the sealing efficiency; therefore, the shear loss is related to sealing flow and sealing structure. According to the above research, it can be concluded that increasing the tangential velocity is of great significance for alleviating the upstream blocking effect and reducing the shear loss.

Table 7. Velocity gradient for a variety of sealing structures.

Mass Average	Axial	Radial	Chute_60	Chute_45	Chute_30
$\frac{\partial V_\theta}{\partial r}$	319.66 (39.7%)	265.95 (33.1%)	426.4 (46.3%)	440.9 (48.4%)	482.0 (51.7%)
$\frac{\partial V_\theta}{\partial z}$	204.214 (25.3%)	215.362 (26.8%)	183.2 (19.9%)	171.76 (18.9%)	165.4 (17.8%)
$\frac{\partial V_z}{\partial r}$	119.74 (14.9%)	154.05 (19.2%)	108.417 (11.8%)	98.65 (10.8%)	68.82 (7.4%)
Total proportion	79.9%	79.1%	78%	78.1%	76.9%

4. Conclusions

In this paper, the blocking and shear loss of different sealing structures are quantified based on the dissipation coefficient integral. The generation mechanism and influence factors of the loss are revealed from the gas ingestion and its interaction with sealing air. The main conclusions are as follows:

(1) The blocking effect of the purge flow on the upstream and the shear effect between the purge flow and the mainstream are responsible for the change of the stator loss. Both are sensitive to the sealing flow rate and sealing structure.

(2) The blocking effect is mainly affected by the radial and the tangential velocity at the exit of the cavity. The blocking effect is positively correlated with sealing efficiency. Thus, it is most apparent for the Chute_30 structure with a sealing efficiency of

99.9%. The tangential velocity, which is affected by gas ingestion, is the main factor in determining the blocking effect under the same sealing flow rate.

(3) The shear mixing loss is mainly affected by the tangential velocity gradient along the radial and the axial direction, and the axial velocity gradient along the radial. The rim shear vortex formed by the gas intrusion at the rim gap reduces the tangential velocity gradient and increases the axial velocity gradient for the gas transport from the rotor side to the stator side. With the increase in sealing efficiency, the proportion of tangential velocity gradient increases obviously, which accounts for 66.5% of the axial sealing structure with $IR = 1.5$, and 69.5% for the Chute_30 sealing structure with $IR = 1.0\%$. The tangential velocity gradient becomes the dominant factor of shear loss for the higher sealing efficiency situation. Thus, using the mixing shear model Sharon and Wilcox et al. proposed may be more accurate to evaluate the shear loss for the high sealing efficiency situation. In addition, the influence of the axial velocity gradient on the shear loss needs to be considered when the sealing efficiency is relatively low.

(4) Increasing the tangential velocity is an effective way to reduce the upstream blocking effect and shear loss. Although the blocking effect reduces the upstream stator loss, it is still significant to alleviate the blockage effect from the perspective of the turbine as a whole because the blockage will affect the upstream and downstream flow fields simultaneously.

Author Contributions: Conceptualization, Writing—Original Draft, Writing—Review and Editing, Data Curation, Methodology T.B.; Software, Supervision Q.Y.; Validation, Visualization, J.L. All authors have read and agreed to the published version of the manuscript.

Funding: This research received no external funding.

Data Availability Statement: Not applicable.

Conflicts of Interest: The authors declare no conflict of interest.

References

1. Schuepbach, P.; Abhari, R.S.; Rose, M.G.; Germain, T.; Raab, I.; Gier, J. Effects of suction and injection purge-flow on the secondary flow structures of a high-work turbine. *J. Turbomach* **2010**, *132*, 021021. [CrossRef]
2. Reid, K.; Denton, J.; Pullan, G.; Curtis, E.; Longley, J. The Effect of Stator-Rotor Hub Sealing Flow on the Mainstream Aerodynamics of a Turbine. In Proceedings of the ASME 51st Turbo Expo, Part A, Barcelona, Spain, 6–11 May 2006; Volume 6, pp. 789–798.
3. Zlatinov, M.B.; Tan, V.S.; Montgomery, M.; Islam, T.; Harris, M. Turbine Hub and Shroud Sealing Flow Loss Mechanisms. *J. Turbomach.* **2012**, *134*, 061027. [CrossRef]
4. Jia, W.; Liu, H. Numerical investigation of the interaction between upstream purge flow and mainstream in a highly-loaded turbine. In Proceedings of the ASME Turbo Expo 2014: Turbine Technical Conference and Exposition GT2014, Düsseldorf, Germany, 16–20 June 2014. ASME Paper GT2014-25501:1-15.
5. Gallier, K.D.; Lawless, P.B.; Fletter, S. Development of the unsteady flow on a turbine rotor platform downstream of a rim seal. In Proceedings of the ASME Turbo Expo 2004,Power for Land, Sea, and Air, Vienna, Austria, 14–17 June 2004; ASME Paper GT2004-53899. 2004.
6. Paniagua, G.; Denos, R.; Almeida, S. Effect of the Hub Endwall Cavity Flow on the Flow Field of a Transonic High-Pressure Turbine. *J. Turbomach.* **2004**, *126*, 578–586. [CrossRef]
7. Rehder, H.J.; Dannhauer, A. Experimental Investigation of Turbine Leakage Flows on the Three-Dimensional Flow Field and Endwall Heat Transfer. *J. Turbomach.* **2007**, *129*, 608–618. [CrossRef]
8. Schuepbach, P.; Abhari, R.S.; Rose, M.G.; Gier, J. Influence of Rim Seal Purge Flow on the Performance of an Endwall-Profiled Axial Turbine. *ASME J. Turbomach.* **2011**, *133*, 021011. [CrossRef]
9. Ong, J.H.P.; Miller, R.J.; Uchida, S. The Effect of Coolant Injection on the Endwall Flow of a High Pressure Turbine. *ASME J. Turbomach.* **2012**, *134*, 051003. [CrossRef]
10. Regina, K.; Kalfas, A.L.; Abhari, R.S. Experiment investigation of purge flow effects on a high pressure turbine stage. *J. Turbomach.* **2015**, *137*, 041006. [CrossRef]
11. McLean, C.; Camci, C.; Glezer, B. Mainstream Aerodynamic Effects Due to Wheelspace Coolant Injection in a High Pressure Turbine Stage: Part I—Aerodynamic Measurements in the Stationary Frame. *ASME J. Turbomach.* **2001**, *123*, 697–703. [CrossRef]
12. McLean, C.; Camci, C.; Glezer, B. Mainstream Aerodynamic Effects Due to Wheelspace Coolant Injection in a High Pressure Turbine Stage: Part II—Aerodynamic Measurements in the Rotational Frame. *ASME J. Turbomach.* **2001**, *123*, 687–696. [CrossRef]

13. Carvalho Figueiredo, A.J.; Schreiner, B.D.J.; Mesny, A.W.; Pountney, O.J.; Scobie, J.A.; Li, Y.S.; Cleaver, D.J.; Sangan, C.M. Volumetric Velocimetry Measurement of Purge-Mainstream Interaction in a one-Stage Turbine. *J. Turbomach.* **2021**, *143*, 041011. [CrossRef]
14. Mesny, A.W.; Glozier, M.A.; Pountney, O.J.; Scobie, J.A.; Li, Y.S.; Cleaver, D.J.; Sangan, C.M. Vortex Tracking of Purge-Mainstream Interactions in a Rotating Turbine Stage. *J. Turbomach.* **2022**, *144*, 041011. [CrossRef]
15. Jenny, P.; Abhari, R.S.; Rose, M.G. Unsteady rotor hub passage vortex behavior in the presence of purge flow in an axial low pressure turbine. *J. Turbomach.* **2013**, *135*, 051022. [CrossRef]
16. Schrewe, S.; Werschnik, H.; Schiffer, H.P. Experimental Analysis of the Interaction Between Rim Seal and Main Annulus Flow in a Low Pressure Two Stage Axial Turbine. *J. Turbomach.* **2013**, *135*, 051003. [CrossRef]
17. Yang, F.; Zhou, L.; Wang, Z. Investigation on Interaction Loss Mechanism Between Rim Seal Flow and Mainstream. *J. Propuls. Technol.* **2020**, *41*, 285–293.
18. Yang, F.; Zhou, L.; Wang, Z. Unsteady numerical investigation on viscous shear loss caused by rim seal purge flow. *J. Power Energy* **2019**, *233*. [CrossRef]
19. Yang, F.; Zhou, L.; Wang, Z. Unsteady numerical investigation on vortex interaction between rim seal flow and mainstream. *J. Propuls. Technol.* **2019**, *40*, 315–323.
20. Barisi, D.; Lengani, D.; Simoni, D.; Venturino, G.; Bertini, F.; Giovannini, M.; Rubechini, F. Analysis of the Loss Production Mechanism Due to Cavity-Main Flow Interaction in a Low-Pressure Turbine Stage. *J. Turbomach.* **2022**, *144*, 091004. [CrossRef]
21. Huang, J.W.; Fu, W.L.; Ma, G.J.; Wang, G.J.; Gao, J. Interaction between 1.5-stage turbine rim seal purge flow and mainstream and flow interface between rim seals affected by rim seal structure. *Acta Aeronaut. Astronaut. Sin.* **2021**, *42*, 124549.
22. Hunter, S.D.; Manwaring, S.R. Endwall Cavity Flow Effects on Gaspath Aerodynamics in an Axial Flow Turbine Part 1—Experimental and Numerical Investigation. In Proceedings of the ASME Turbo Expo 2000, Munich, Germany, 8–11 May 2000; ASME Paper 2000-GT-651. 2000.
23. Zhang, Z.; Zhang, Y.; Dong, X.; Qu, X.; Lu, X.; Zhang, Y. Flow mechanism between purge flow andmainstream in different turbine rim seal configurations. *Chin. J. Aeronaut.* **2020**, *33*, 2162–2175. [CrossRef]
24. Bai, T.; Yang, Q.; Liu, J.; Shi, Y.; Qiao, S. Numerical study of turbine rim seals performance with different sealing structures. *Métodos Numéricos Para Cálculo Diseño Ing. Rev. Int.* **2022**, *38*, 32. [CrossRef]
25. Behr, T. Control of Rotor Tip Leakage and Secondary Flow by Casing Air Injection in Unshrouded Axial Turbines. Ph.D. Thesis, Dresden University of Technology, Dresden, Germany, 2007.
26. Cumpsty, N.A.; Horlock, J.H. Averaging Nonuniform Flow for a Purpose. *ASME J. Turbomach.* **2006**, *128*, 120–129. [CrossRef]
27. Chilla, M.; Hodson, H.; Newman, D. Unsteady interaction between annulus and turbine rim seal flows. *J. Turbomach.* **2013**, *135*, 051024. [CrossRef]
28. Young, J.B.; Wilcock, R.C. Modeling the Air-Cooled Gas Turbine: Part 2-Coolant Flows and Losses. *ASME J. Turbomach.* **2002**, *124*, 214–221. [CrossRef]

processes

MDPI

Article

Gas–Liquid Interaction Characteristics in a Multiphase Pump under Different Working Conditions

Yuxuan Deng [1,*], Xiaodong Wang [2], Jing Xu [1], Yanna Li [1], Yanli Zhang [1] and Chunyan Kuang [1]

[1] BaiLie School of Petroleum Engineering, Lanzhou City University, Lanzhou 730071, China
[2] Pipechina West Pipeline Company Lanzhou Gas Transmission Branch Company, Lanzhou 730070, China
* Correspondence: dengyuxuan@lzcu.edu.cn; Tel.: +86-159-0815-5884

Abstract: In this study, we analyze gas–liquid interaction characteristics using a heterogeneous two-fluid model to investigate the influence of interphase force on multiphase pump performance. Two-phase transport platforms are used in oil and gas development to eliminate the need for separation equipment and reduce costs. Full-channel numerical calculations were conducted for an axial-flow multiphase pump based on different inlet gas void fractions (IGVFs) and flow rates. The results indicate that the interaction force of each phase is relatively large in the rotor–stator interference region, and the drag, lift, virtual mass, and turbulent dispersion forces increase with an increase in IGVF or when deviating from the design condition (Q = 50 m^3/h). The interphase forces (resistance, lift, virtual mass force, and turbulent dispersion) increase considerably in the impeller passage and minimally in the guide blade passage. Under the conditions of small and high flows, the force of each phase changes considerably in the impeller and diffuser passages, respectively. Furthermore, the turbulent kinetic energy in the flow passage corresponds to the change trend of the interphase force, indicating that the interphase force causes energy loss inside a multiphase pump. These results provide essential information for the optimization of the hydraulic design of multiphase pumps.

Keywords: multiphase pump; two-phase flow; inlet gas void fraction; interphase force; numerical calculation

Citation: Deng, Y.; Wang, X.; Xu, J.; Li, Y.; Zhang, Y.; Kuang, C. Gas–Liquid Interaction Characteristics in a Multiphase Pump under Different Working Conditions. *Processes* **2022**, *10*, 1977. https://doi.org/10.3390/pr10101977

Academic Editor: Alfredo Iranzo

Received: 15 July 2022
Accepted: 17 August 2022
Published: 1 October 2022

1. Introduction

Deep-sea oil and gas resources usually comprise a mixture of oil, gas, and water. Utilizing separation equipment is the traditional approach to separating oil and gas, which are then transported separately using a pump and compressor. As such, this process employs with complex piping systems and separation equipment, resulting in high investment costs [1,2]. In contrast, a two-phase transport platform eliminates part of the pipeline and separation equipment, saving 40% of infrastructure investment [3]. This approach has been widely used in oil and gas development [4,5]. As the core equipment of a two-phase transport platform, the axial-flow multiphase pump ensures smooth and efficient operation and plays a key role in the development and utilization of oil and gas resources.

Many scholars have conducted research on optimization design and internal flow field analysis to improve the performance of multiphase pumps. Cao et al. proposed a design method to determine blade placement angle based on a given velocity moment [6]. Their experimental results showed that an impeller designed using this method had a wide working range. Zhang et al. proposed a hydraulic design method for impeller blades of multiphase pumps under large design flows based on a meridional surface network [7]. Kim et al. conducted variable sensitivity analysis, whereby geometric parameters were selected as optimization variables, such as the inlet width, outlet width, hub ratio, and placement angle [8]. Moreover, the impeller and diffuser were simultaneously optimized by constructing an approximate function relation between the geometric parameters, pressure rise, and efficiency under pure water conditions. The results showed that the flow separation in the optimized pump was suppressed, and the pump efficiency increased by 3.5%. Liu et al. adopted a polynomial function to directly control the blade placement angle of a mixed-transport pump [9]. They used the blade placement angle as an optimization variable and conducted orthogonal optimization with five factors and four levels. The results showed that the pressure of the optimized pump increased by 6.69%.

Zhang et al. used numerical simulation results as samples to study the influence of inlet gas void fractions (IGVFs) on multiphase pump performance [10]. They used a backpropagation neural network to establish the response function relationship between optimization variables and optimization objectives. Finally, the optimization results of comprehensive performance under 0%, 10%, 30%, and 40% gas content conditions were obtained through multiobjective genetic algorithm optimization in variable space. Xu et al. conducted an experimental study on a spiral axial-flow gas–liquid multiphase pump [11]. They concluded that the pump could operate stably at 0–50% gas content. The head and efficiency gradually decreased with an increase in inlet gas content. Zhang et al. divided various flow forms into a bubble balloon and stratified flows according to the diameter and number of bubbles in the pump [12,13]. Saadawi et al. debugged a multiphase pump [14]. They found that increasing the speed extended the pump's operating range at 50% inlet gas. Alberto et al. observed, through visualization experiments, that increasing impeller speed resulted in more uniform gas–liquid two-phase mixing [15,16].

Some parameters cannot be obtained using the current test techniques. With the rapid development of technology in recent years, many scholars now use computational fluid dynamics software to conduct numerical research on multiphase pumps. Zhang et al. conducted a numerical analysis on an axial-flow multiphase pump under different inlet gas holdup conditions [17]. They found higher gas holdup in the area near the outlet hub of the impeller. The higher the gas content of the inlet, the higher the degree of its aggregation. Liu et al. carried out a numerical simulation on four groups of mixed over-flow media, obtaining various viscosity by mixing glycerol with different proportions of water [18]. Their analysis showed that the turbulent kinetic energy in the flow channel increased with an increase in viscosity. Zhang et al. conducted unsteady calculations of an axial-flow multiphase pump under different inlet gas holdup and flow conditions [19,20]. Their analysis showed that the main cause of pressure pulsation was rotor–stator interference and the maximum value appeared near the rotor–stator interference area. Zhang et al. also reached a similar conclusion [21]. Yu et al. preliminarily calculated and analyzed the magnitude and variation of gas–liquid interforce in a multiphase pump [22].

In summary, most of the studies on multiphase pumps have focused on performance optimization and internal flow field analysis, and few studies have been conducted on gas–liquid interaction characteristics. In particular, the mechanism of gas–liquid interaction is not well understood. In this study, we numerically calculated the internal flow field of the axial flow in a multiphase pump using a heterogeneous two-fluid model to investigate the influence of interphase force on the performance of the multiphase pump. The change rule of interphase force in a multiphase pump was studied under different IGVF (5%, 10%, 15%, 20%, and 25%) and flow rate (Q = 30, 40, 50, 60, and 70 m^3/h) conditions to illustrate their effects on interphase force.

2. Numerical Model

2.1. Computational Method

A heterogeneous two-fluid model was used in this study to calculate the two-phase flow in a multiphase pump. Reynolds time and homogeneous Navier–Stokes equations were used to solve the calculations. The governing equations for a steady incompressible fluid are as follows:

The continuity equation:

$$\nabla \cdot (\alpha_k \rho_k V_k) = 0 \tag{1}$$

The momentum equation:

$$\nabla\cdot(\alpha_k\rho_k V_k V_k) = -\alpha_k\nabla p + \nabla\cdot(\alpha_k\tau_{ij}) + M_k + f_k \quad (2)$$

where the subscript k represents any phase (l is liquid, and g is gas); α_k is the k-phase volume fraction, satisfying $\alpha_g + \alpha_l = 1$; ρ_k is the k-phase density; V_k is the k-phase absolute velocity; p is the pressure; τ is the viscous stress tensor; M_k is the interphase force on phase K; and f_k is the mass force.

ANSYS CFX 17.0 computational fluid dynamics commercial software was used to solve the three-dimensional steady flow inside an axial-flow pump. The discretization of governing equations was based on the finite volume method, which considers the control volume composed of the internal cell nodes. The shear stress transport (SST) k-ω model was selected as the turbulence model to more accurately predict flow separation under an adverse pressure gradient. Compared with the standard and renormalization group k-ε models, this model considers the viscosity of the inner wall of the model. Therefore, it provides better turbulent shear stress transmission, a more stable algorithm, and better flow simulation performance in a narrow space [23,24]. The governing equations are as follows:

$$\frac{\partial(\rho k)}{\partial t} + \frac{\partial(\rho k u_i)}{\partial x_j} = \frac{\partial}{\partial x_j}[(\mu + \frac{\mu_t}{\sigma_k})\frac{\partial k}{\partial x_j}] + P_k - \beta'\rho k\omega \quad (3)$$

$$\frac{\partial(\rho\omega)}{\partial t} + \frac{\partial(\rho\omega u_i)}{\partial x_j} = \frac{\partial}{\partial x_j}[(\mu + \frac{\mu_t}{\sigma_{\omega 3}})\frac{\partial \omega}{\partial x_j}] - \beta\rho\omega^2 + \alpha_3\frac{\omega}{k}P_k + 2(1-F_1)\rho\frac{1}{\sigma_{\omega 2}\omega}\frac{\partial k}{\partial x_j}\frac{\partial \omega}{\partial x_j} \quad (4)$$

2.2. Interphase Force Model

The main interphase forces involved in the calculation of flow fields for a multiphase pump are resistance, lift, virtual mass, and turbulent dispersion. Resistance is produced by the relative motion of a gas and liquid phase and is the main cause of friction loss. Lift causes the gas phase to move perpendicular to the mainstream, which interferes with the mainstream and increases resistance. Virtual mass force is produced by local acceleration, and its effect on bubbles is opposite to that of centrifugal force. We assumed that the mainstream is perpendicular to the gravity, causing the bubbles move toward the rim before gathering at the hub in the impeller. Turbulent dissipative force is generated by the momentum exchange caused by the turbulence pulsation between phases. Its magnitude depends on the volume fraction gradient of the two phases [25,26]. The interphase force is shown in Figure 1 [27]. The net force of the interphase forces is M_k. The expression of the force in each phase is expressed as:

$$F_{M,k} = F_{D,k} + F_{A,k} + F_{L,k} + F_{T,k} \quad (5)$$

where $F_{D,k}$, $F_{A,k}$, $F_{L,k}$, and $F_{T,k}$ are the resistance, virtual mass, lift, and turbulent dispersion forces, respectively. The resistance force can be expressed as:

$$F_{D,g} = F_{D,l} = \frac{3}{4}\alpha_g\rho_l\frac{C_D}{d}|U_g - U_l|(U_g - U_l) \quad (6)$$

where d is the bubble diameter; U_g and U_l are the absolute velocities in the gas and liquid phases, respectively; α_g is the volume fraction of the gas; ρ_l is the liquid density; and C_D is the resistance coefficient [28], which is expressed as:

$$C_D = \begin{cases} 24(1+0.15\text{Re}^{0.687})/\text{Re} & \text{Re} \le 1000 \\ 0.44 & \text{Re} > 1000 \end{cases} \quad (7)$$

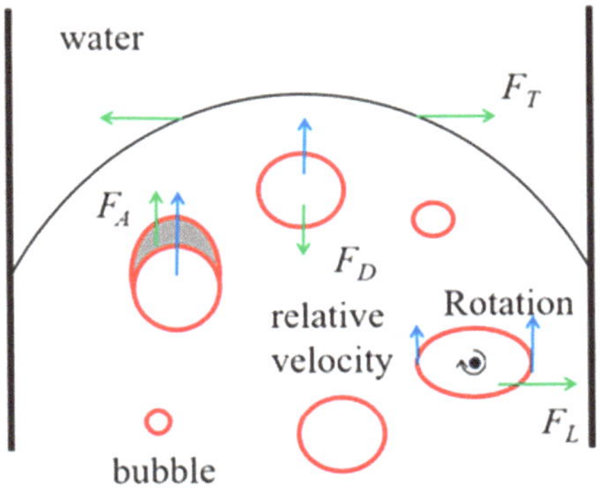

Figure 1. Interphase forces of gas–liquid flow in a multiphase pump [27].

The virtual mass force is expressed as:

$$F_{A,g} = F_{A,l} = \alpha_g C_A \rho_l \left(\frac{DU_g}{Dt} - \frac{DU_l}{Dt}\right) \tag{8}$$

where $C_A = 0.5$ is the virtual mass-force coefficient [26].

The lift force is expressed as:

$$F_{L,g} = F_{L,l} = C_L \alpha_g \rho_l (U_g - U_l) \times (\nabla \times U_l) \tag{9}$$

where $C_L = 0.5$ is the lift coefficient [27].

The turbulent dispersion force is expressed as:

$$F_{T,g} = F_{T,l} = -C_T \rho_l k \nabla \alpha_l \tag{10}$$

where $C_T = 0.1$ is the coefficient of turbulent dispersion force [20], and k is turbulent kinetic energy.

2.3. Computational Model

The axial-flow multiphase pump model includes an inlet pipe, outlet pipe, impeller, and diffuser, as shown in Figure 2. The main design parameters are as follows. The design head is H_d = 15 m, design flow is Q_d = 50 m^3/h, and rated speed is n = 2950 r/min. The structural parameters are listed in Table 1.

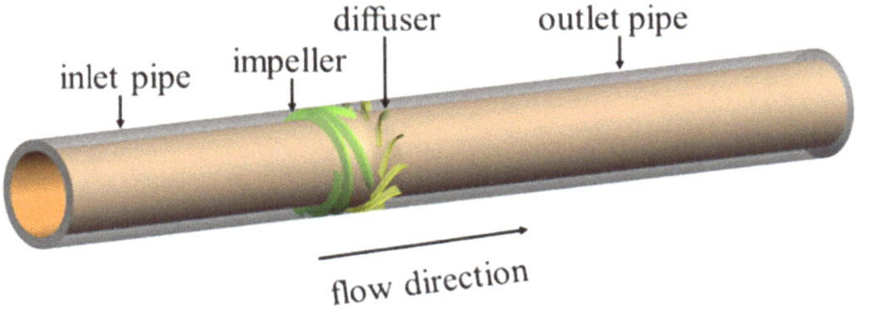

Figure 2. Computational model of an axial-flow multiphase pump.

Table 1. Structural parameters.

Parameter	Impeller	Diffuser
Blade number	4	11
Shroud radius (mm)	75	75
Inlet hub radius (mm)	58.96	67
Outlet hub radius (mm)	67	60
Axial length (mm)	55	65
Tip clearance (mm)	0.3	0

All flow components adopted a hexahedral structured grid, considering the full flow-channel structure and computer configuration (Intel Xeon W-2223 @ 3.6 ghz with 32 GB RAM). Integrated computer-aided engineering and manufacturing was adopted for the simple structure of the inlet and outlet pipes. The impeller and diffuser flow-channel structure was complex, so TurboGrid mesh generation software was used to divide the hexahedron structure grid. The target passage mesh size method was selected for global mesh partitioning. The wall area was encrypted with the proportional to mesh size method. The meshes of the impeller and diffuser are shown in Figure 3.

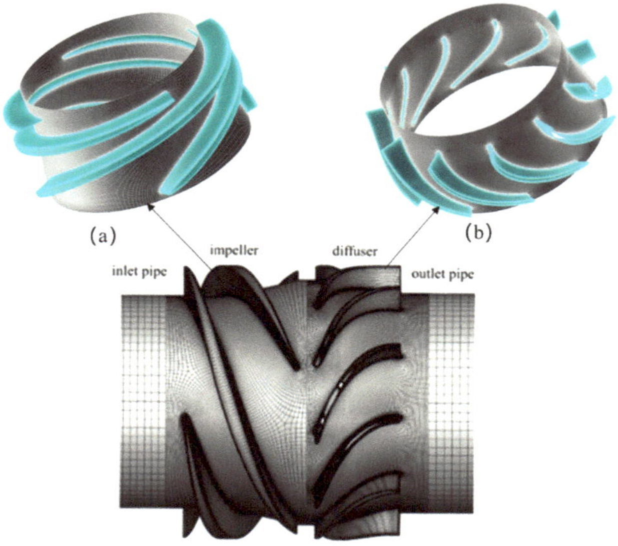

Figure 3. Computational meshes of the (**a**) impeller and (**b**) diffuser.

Grid independence verification analysis was conducted on the premise of ensuring grid quality. Five schemes with different grid numbers were verified and analyzed under a pure water condition (IGVF = 0%) with head and efficiency as the judgment criteria. Considering computing resources and accuracy, Scheme 3 had a grid number of 3,412,466 and was selected for numerical calculations. The results showed that the number of grids exceeded that of Scheme 3, and the head and efficiency changed gradually with the number of grids, as shown in Figure 4. The inlet section, impeller, diffuser, and outlet section had 143,658, 1,604,584, 1,478,546, and 185,678 grids, respectively.

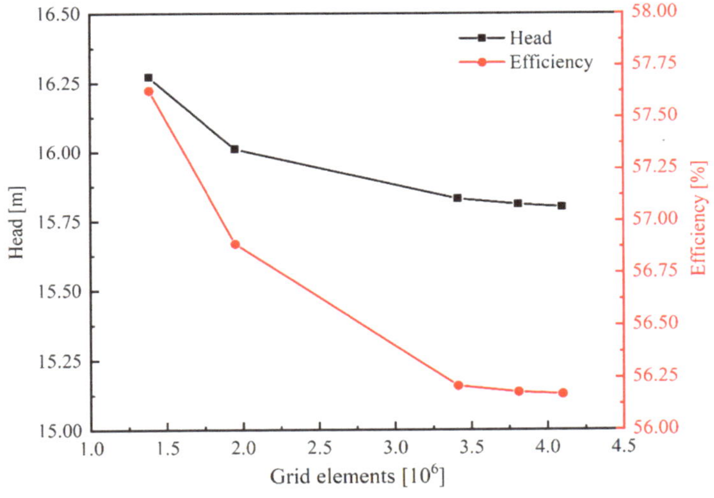

Figure 4. Verification of rid independence.

2.4. Boundary Conditions and Solution Settings

The high-resolution and second-order backward Euler models were selected because the SST k-ω turbulence model was used to solve the flow state inside the multiphase pump. An air–water two-phase mixed medium was selected as the flow medium. Water was the continuous phase, air was the discrete phase, and the bubble diameter was set as 0.1 mm [12]. The bubble diameter was assumed to be constant during the calculation. The volume fractions of air and water were determined according to working conditions. In all cases, they must add up to 1. The surface tension coefficient was set to 0.072 N/m, and interphase transfer was used as the particle model. The appropriate boundary conditions were set according to the operating conditions. The inlet was set to mass flow, and the outlet condition was the average static pressure outlet. The speed was set to 2950 r/min. The impeller was in the rotation domain, and the other regions were in the rest domain. The interfaces between the inlet pipe and impeller and between the impeller and diffuser adopted a frozen rotor interface. Other interfaces were set to general connection.

3. Numerical Results and Discussion

3.1. Analysis of External Characteristics

The influence of IGVF on the pressure rise performance of a multiphase pump was analyzed, as shown in Figure 5. The pressure rise was approximately linear with IGVF. When IGVF increased, the pressure rise decreased; that is, the pressurization capacity of the multiphase pump gradually decreased.

The gas-phase distributions in the flow passage of the impeller and diffuser are shown in Figure 6. The gas in the impeller runner gathered at the hub, mainly due to the density difference between gas and liquid. The gas was mainly concentrated in the hub area due to the high-speed rotation of the impeller. At the inlet of the diffuser, gas mainly collected at the hub. The gas gradually diffused toward the rim along the flow direction. The main reason for this is that the fluid maintained a velocity component in the circumferential direction due to inertia at the inlet. This left the fluid subject to centrifugal force, causing gas to accumulate at the hub. The inertial effect diminished as the fluid gradually moved toward the outlet. Subsequently, the fluid was less affected by centrifugal force, the degree of gas gathering at the hub gradually weakened, and the gas slowly diffused from the hub to the rim. In addition, the degree of gas accumulation in the impeller and diffuser passage gradually increased with an increase in IGVF. This is the main reason for the weakening of the supercharging capacity of the multiphase pump. The power capacity of the impeller and diffuser gradually decreased with the increasing degree of gas accumulation in the impeller and diffuser.

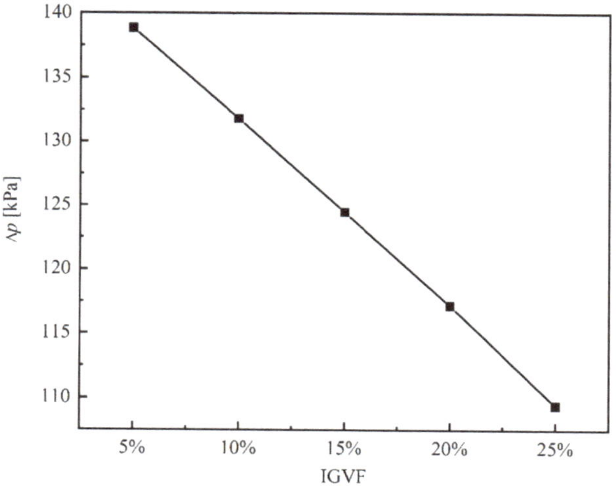

Figure 5. Influence of IGVF on pressure rise of a multiphase pump.

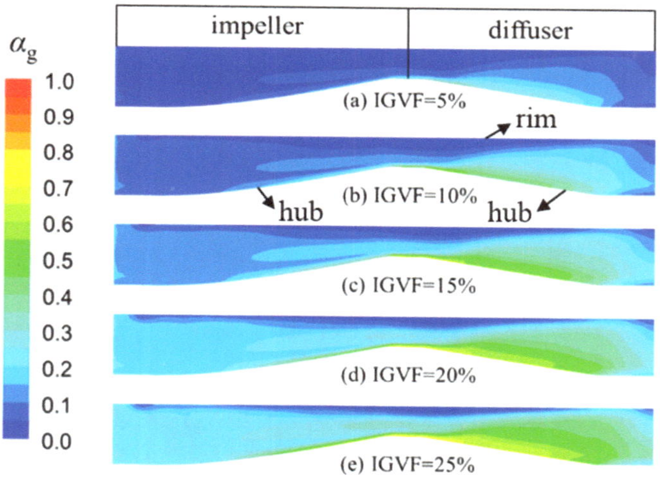

Figure 6. Distributions of gas void fractions in the impeller and diffuser under different IGVFs.

Next, the influence of flow rate on the pressure rise performance of the multiphase pump was analyzed. The pressure rise first increased and then decreased with an increase in flow rate, as shown in Figure 7. The reason for this is that the flow angle at the inlet of the diffuser was inconsistent with that of the blade, resulting in deviation from the design condition, as shown in Figure 8.

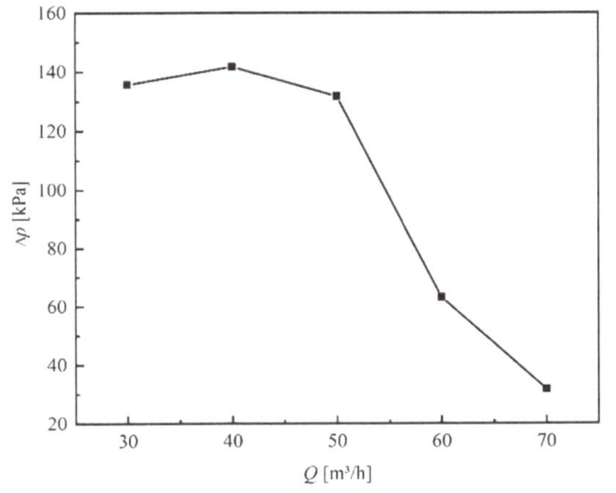

Figure 7. Influence of flow rate on pressure rise of a multiphase pump.

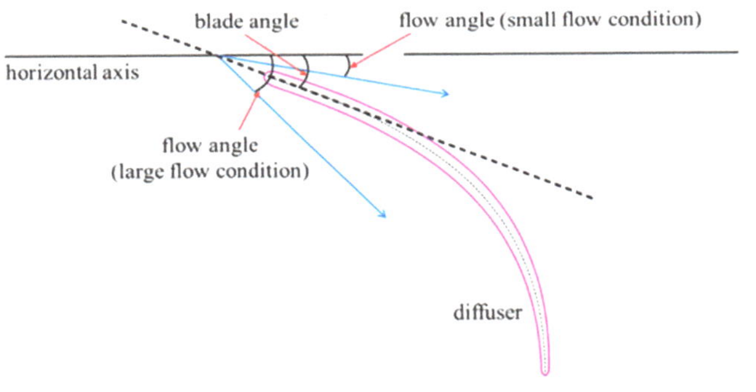

Figure 8. Diagram of flow angle.

The internal flow of the multiphase pump under different flow conditions was analyzed using a blade height of 0.5 as an example, as shown in Figure 9. The flow angle was smaller than the blade placement angle under the small low-flow condition. This tended to form shedding vortices on the blade suction surface, and the position of the shedding vortex was delayed as the flow increased. A shedding vortex easily formed on the blade pressure surface under a high flow condition because the streamline distribution was chaotic, as shown in Figure 9. Therefore, the flow angle at the diffuser inlet was inconsistent with the blade when deviating from the design condition. Subsequently, it was easy to form a shedding vortex and cause energy loss at the inlet of diffuser. Compared with low flow, the streamline distribution was more chaotic, and the energy loss was more serious under a high flow condition. Therefore, the pressurization capacity clearly decreased under the condition of high flow.

(a) Q=30 m³/h (b) Q=40 m³/h (c) Q=50 m³/h (d) Q=60 m³/h (e) Q=70 m³/h

Figure 9. Streamline distribution in the impeller and diffuser at span = 0.5 for different flow rates (Q = m^3/h).

3.2. Influence of IGVF on Interphase Force

The change rule of interphase force under different IGVF conditions (Q = 50 m^3/h) is shown in Figure 10. IGVF exerted a considerable influence on resistance and lift but had almost no influence on virtual mass and turbulent dispersion forces. The largest interphase force was drag force, followed by lift, virtual mass, and turbulent dispersion forces. Compared with other forces, the turbulent dispersion force was negligible. There were three peaks of interphase force along the flow direction, with a maximum value from the outlet of the impeller to the inlet of the diffuser; that is, the interaction force was the maximum in the area of static and static interference. The interphase force also reached its peak value in the impeller inlet and diffuser outlet areas. The main reason for this is that the interaction force was larger in the area of rotor–stator interference. In other regions, the IGVF was smaller, gas accumulation in the flow passage was not obvious, and the flow pattern was good. Therefore, the variation trend of interphase forces was also relatively orderly. With an increase in IGVF, the order of interphase forces was drag, lift, virtual mass, and turbulent dispersion. The variation trend of each force along the axial relative position remained basically unchanged, but the overall trend was upward. In particular, the range of change was large near the impeller inlet and diffuser outlet. The main reason for this is that the fluid in the impeller inlet entered the impeller runner from the inlet pipe. The high-speed rotation of the impeller led to fluid impingement on the leading edge of the blade, resulting in a chaotic flow field at the impeller inlet. A large amount of gas accumulated near the diffuser outlet to form an air sac, as shown in Figure 6. Resistance, virtual mass, and lift forces were positively correlated with gas holdup, according to the expression of the interphase force. Therefore, the interaction force was larger, and the change rule was more complex.

Contour plots of turbulent kinetic energy distribution in the impeller and diffuser passage under different IGVF conditions are shown in Figure 11. The turbulent kinetic energy in the guide-blade passage was significantly greater than that in the impeller passage, indicating that the flow pattern in the guide-blade passage was more chaotic. The turbulent kinetic energy in the impeller passage near the hub area increased with an increase in IGVF, whereas the change in turbulent kinetic energy in the diffuser passage was more obvious. This result is consistent with the streamline distribution in the flow channel shown in Figure 9. This indicates that the increase in IGVF led to an increase in turbulent kinetic energy and energy loss in the impeller and guide-blade runner. The interphase force also increased with an increase in IGVF according to the above analysis of interphase force. The increase in interphase force led to an increase in energy loss in the impeller and diffuser passage.

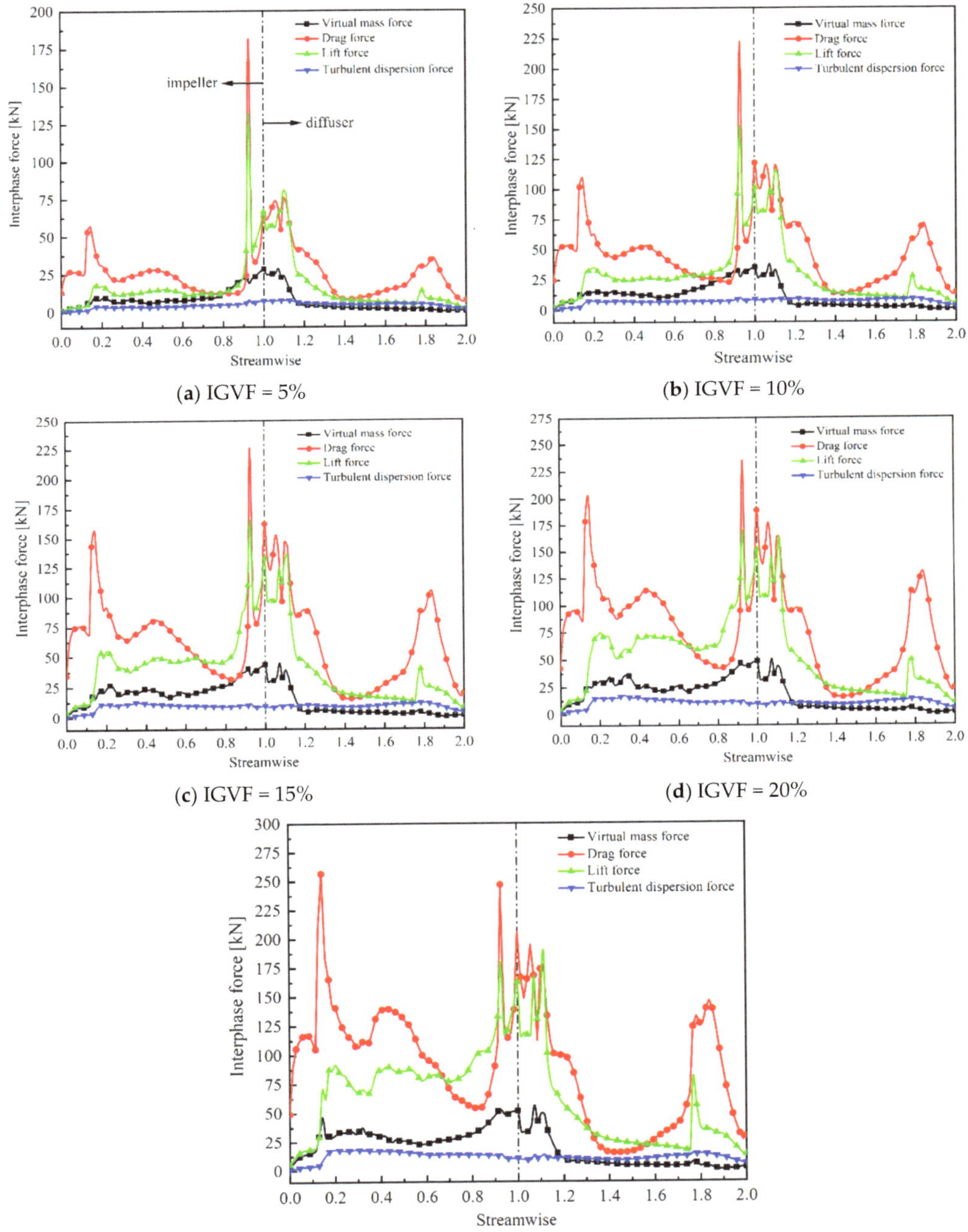

Figure 10. Analysis of interphase force under different IGVF conditions.

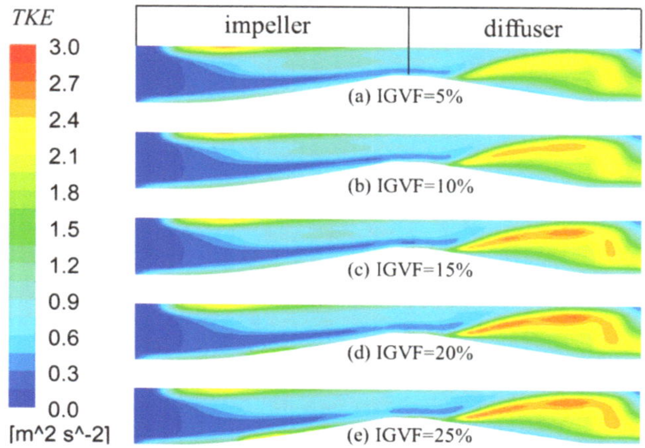

Figure 11. Distribution of turbulence kinetic energy in the impeller and diffuser under different IGVFs.

3.3. Influence of Flow on Interphase Force

The change rule of interphase force under different flow conditions with IGVF = 10% is shown in Figure 12. A gas holdup distribution of the impeller and diffuser passage under different flow conditions with IGVF = 10% is shown in Figure 13. Flow had a considerable influence on resistance and lift but almost no influence on virtual mass and turbulent dispersion forces, as shown in Figure 12. The order of the interphase forces was drag, lift, virtual mass, and turbulent dispersion. The resistance and lift reached the maximum near the impeller inlet when $Q = 30\ m^3/h$, followed by the region of rotor–stator interference. It was evident that resistance and lift were affected by the gas holdup in the flow passage. The gas content in the impeller passage was high when $Q = 30\ m^3/h$, resulting in significantly increased resistance and lift relative to other areas, as shown in Figure 13. The maximum resistance and lift forces were between the impeller outlet and guide-blade inlet when $Q \geq 40\ m^3/h$. This was affected by static and static interference, resulting in increased drag and lift. Additionally, the impeller inlet and diffuser outlet resistance and lift were relatively large. The interphase force was larger, mainly due to the collision between the high-speed rotation of the impeller and the fluid at the inlet of the impeller. The outlet position of the diffuser was affected by the inconsistency between the inlet flow angle of the diffuser and blade placement angle. This created a vortex at the outlet of the diffuser that caused the interaction force to be high. The gas content in the impeller and diffuser passage dropped significantly and tended to be evenly distributed when $Q \geq 40\ m^3/h$, with a minimal impact on the interphase force.

The designed flow rate in this study was $Q = 50\ m^3/h$, and IGVF was set to 10% when calculating each flow condition. The corresponding interphase force under the condition of IGVF = 10% and $Q = 50\ m^3/h$ was used as the benchmark to more clearly demonstrate the influence of flow rate on interphase force. The relative size of the interphase force under other working conditions was compared, namely for F/F (IGVF = 10%, $Q = 50\ m^3/h$). Here, the numerator F is the interphase force under different working conditions, and the denominator F (IGVF = 10%, $Q = 50\ m^3/h$) is IGVF = 10%, corresponding to an interphase force at $Q = 50\ m^3/h$. The relative forces of each phase under different flow conditions are shown in Figure 14. The virtual mass force was less affected by the flow rate in the impeller passage, as shown in Figure 14a. However, the virtual mass force was considerably affected by the flow in the diffuser passage, especially when it deviated from the design condition ($Q = 50\ m^3/h$). Additionally, it changed considerably under the condition of high flow rate ($Q = 70\ m^3/h$). The resistance in the impeller passage was considerably affected by the condition of low flow ($Q = 30\ m^3/h$), as shown in Figure 14b. The change was gradual under the other flow conditions, but the resistance clearly fluctuated in the diffuser passage when the flow exceeded that of the design condition. The variation law of lift was similar to that of resistance, as shown in Figure 14c. The variation of turbulent dispersion force was complex, as shown in Figure 14d. The turbulent dispersion force achieved the maximum value and gradually decreased along the flow direction at the middle and rear of the impeller and inlet of the diffuser under the condition of low flow rate (Q = 30 m^3/h). The turbulent dispersion force in the diffuser was considerably affected by the flow with a high flow rate.

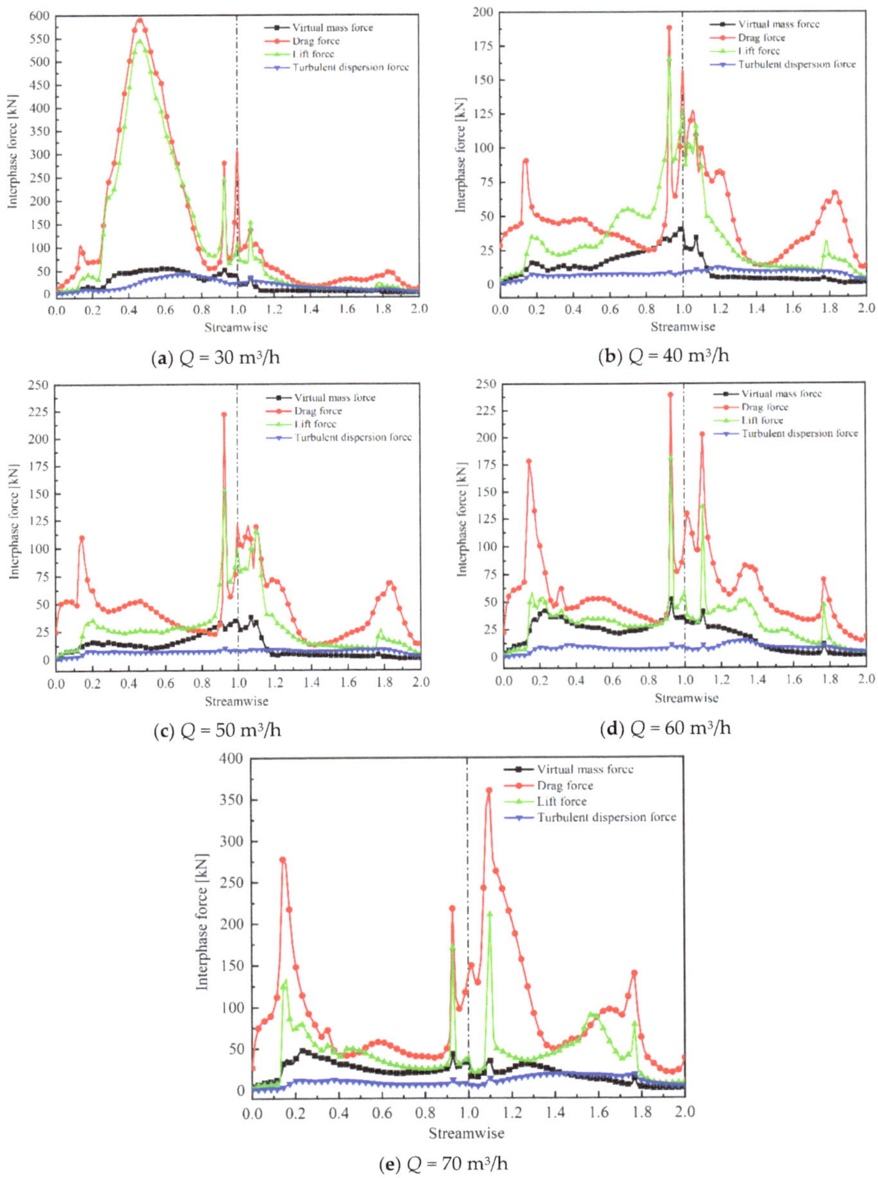

(**a**) Q = 30 m³/h (**b**) Q = 40 m³/h (**c**) Q = 50 m³/h (**d**) Q = 60 m³/h (**e**) Q = 70 m³/h

Figure 12. Analysis of interphase force under different flow rates (Q = m^3/h).

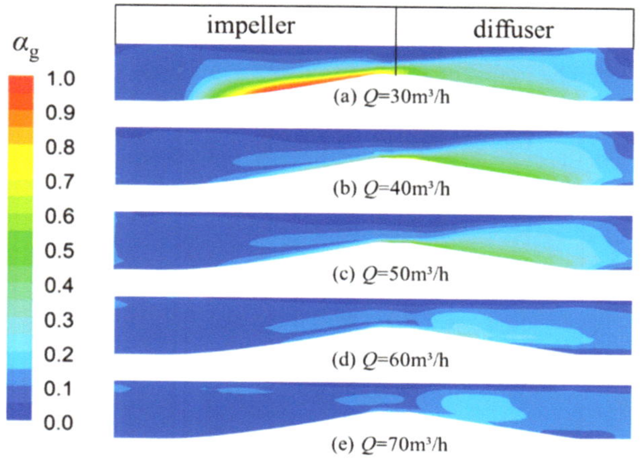

Figure 13. Distribution of gas void fraction in the impeller and diffuser under different flow rates (Q = m^3/h).

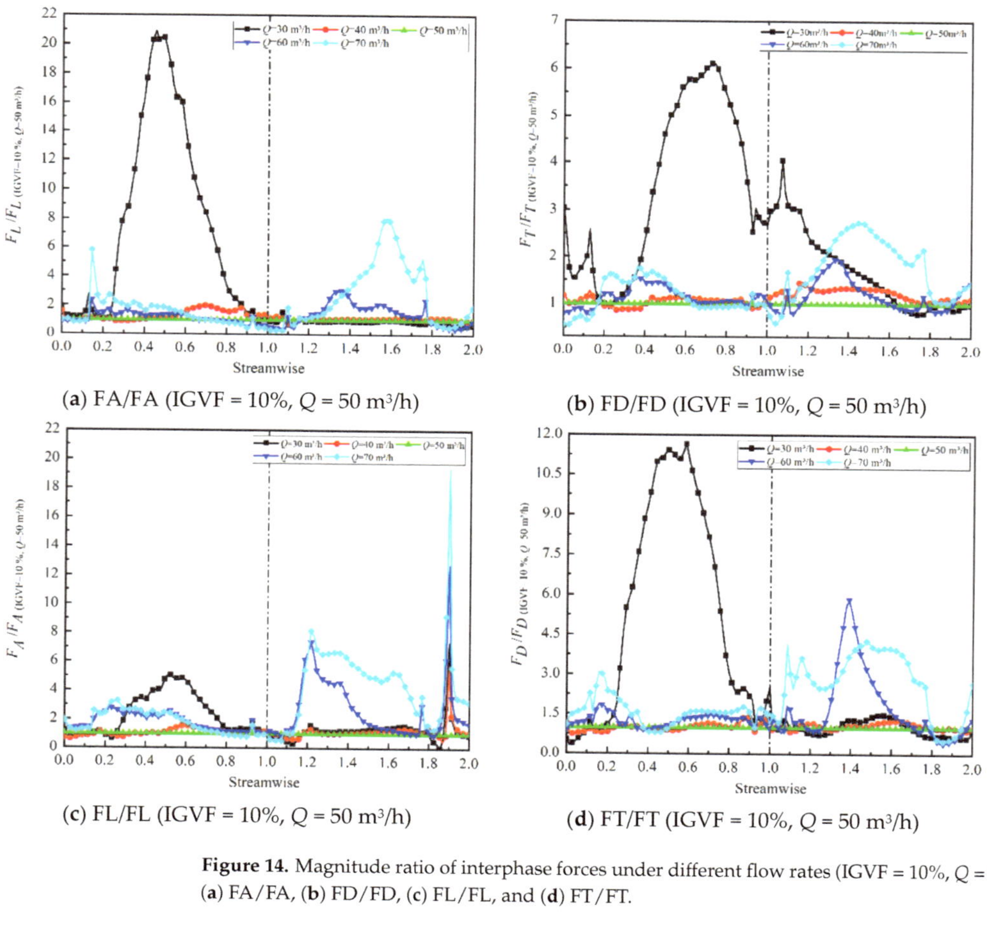

(**a**) FA/FA (IGVF = 10%, Q = 50 m³/h) (**b**) FD/FD (IGVF = 10%, Q = 50 m³/h)

(**c**) FL/FL (IGVF = 10%, Q = 50 m³/h) (**d**) FT/FT (IGVF = 10%, Q = 50 m³/h)

Figure 14. Magnitude ratio of interphase forces under different flow rates (IGVF = 10%, Q = 50 m^3/h): (**a**) FA/FA, (**b**) FD/FD, (**c**) FL/FL, and (**d**) FT/FT.

The distribution of turbulent kinetic energy in the impeller and diffuser passage under different flow conditions is shown in Figure 15. The turbulent kinetic energy generates a local high-intensity region at the hub in the middle of the impeller runner when $Q = 30\ m^3/h$, as shown in area A in the black dotted box in Figure 15. This region corresponds to the position where the resistance, lift, and turbulent dispersion forces in the middle rotation path increased, as shown in Figure 14. When $Q > 30\ m^3/h$, the turbulent kinetic energy in the impeller and diffuser passage decreased with an increase in flow rate, and the boundary was $Q = 50\ m^3/h$. The change in turbulent kinetic energy in the impeller passage was no longer obvious when $Q > 50\ m^3/h$, but the turbulent kinetic energy in the diffuser passage gradually increased. The turbulent kinetic energy occurred in the middle and rear segments of the guide blade when $Q = 70\ m^3/h$, as shown in B in the black dotted box in Figure 15. This corresponds to the position at which the virtual mass force increased, as shown in Figure 14. The region of turbulent kinetic energy enhancement corresponded to the position of interphase force increase. This demonstrates that the increase in interphase force led to an increase in energy loss in the impeller and diffuser passage.

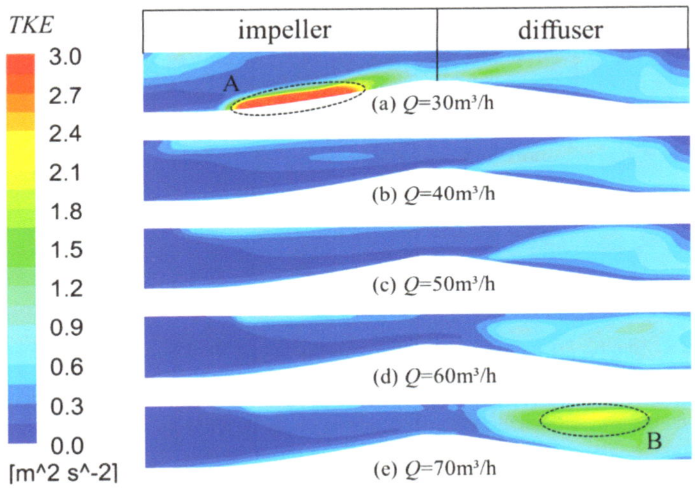

Figure 15. Distribution of turbulent kinetic energy in the impeller and diffuser under different flow conditions.

4. Conclusions

In this study, we analyzed gas–liquid interaction characteristics using a heterogeneous two-fluid model to investigate the influence of interphase force on multiphase pump performance. Full-channel numerical calculations were conducted, and the change rule of interphase force in an axial-flow multiphase pump was studied under different IGVF (5%, 10%, 15%, 20%, and 25%) and flow rate (Q = 30, 40, 50, 60, and 70 m^3/h) conditions to illustrate their effects on interphase force. The following conclusions can be drawn:

1. Under different IGVF conditions, the pressure rise decreased with an increase in IGVF. Under different Q conditions, the pressure rise first increased and then decreased with an increase in Q.
2. Under different IGVF conditions, the order of the interphase forces was drag, lift, virtual mass, and turbulent dispersion. The interaction force of each phase was relatively large in the region of static and static interference. The drag, lift, virtual mass, and turbulent dispersion forces all showed an increasing trend with an increase in IGVF. The interphase force increased considerably in the impeller passage but less in the guide-blade passage.

3. The largest interphase action order under different Q-flow conditions was drag, lift, virtual mass, and turbulent dispersion forces, which all tended to increase when deviating from the design condition. The force of each phase changed considerably in the impeller passage under the condition of low flow. Conversely, the force of each phase changed considerably in the diffuser passage under the condition of high flow rate.
4. The change trends of the interphase force and turbulent kinetic energy in the impeller and diffuser passage were basically constant with the changes in IGVF and flow conditions. This shows that the interphase force led to a change in turbulent kinetic energy, which affected the internal energy loss of the multiphase pump.

The results of this study provide important information for the optimization of the hydraulic design of multiphase pumps.

Author Contributions: Conceptualization, Y.D. and X.W.; methodology, Y.D.; writing—original draft preparation, J.X.; writing—review and editing, Y.L.; visualization, Y.Z.; supervision, Y.D.; software, C.K. All authors have read and agreed to the published version of the manuscript.

Funding: This research was funded by the Open Research Subject of the Key Laboratory of Fluid and Power Machinery, Ministry of Education (Grant No. LTDL2020-005), Doctoral Research Foundation of Lanzhou City University (Grant No. LZCU-BS2019-07).

Institutional Review Board Statement: Not applicable.

Informed Consent Statement: Not applicable.

Data Availability Statement: All relevant data are presented in the article.

Acknowledgments: The authors would like to thank all staff of the treatment plants for their contributions.

Conflicts of Interest: The authors declare no conflict of interest.

References

1. Zhang, J.; Cai, S.; Zhu, H.; Zhang, Y. Experimental investigation of the flow at the entrance of a rotodynamic multiphase pump by visualization. *J. Petrol. Sci. Eng.* **2015**, *126*, 254–261. [CrossRef]
2. Shi, G.; Liu, Z.; Liu, X.; Xiao, Y.; Tang, X. Phase distribution in the tip clearance of a multiphase pump at multiple operating points and its effect on the pressure fluctuation intensity. *Processes* **2021**, *9*, 556. [CrossRef]
3. Zhang, W.; Zhu, B.; Yu, Z. Numerical Analysis for the Influence of Flow Parameters on Phase Interaction in a Multiphase Rotodynamic Pump. *J. Eng. Thermophys.* **2020**, *8*, 1911–1916. (In Chinese)
4. Shi, G.; Liu, Z.; Xiao, Y.; Li, H.; Liu, X. Tip leakage vortex trajectory and dynamics in a multiphase pump at off-design condition. *Renew. Energy* **2020**, *150*, 703–711. [CrossRef]
5. Shi, G.; Yan, D.; Liu, X.; Xiao, Y.; Shu, Z. Effect of the gas volume fraction on the pressure load of the multiphase pump blade. *Processes.* **2021**, *9*, 650. [CrossRef]
6. Cao, S.; Peng, G.; Yu, Z. Hydrodynamic design of rotodynamic pump impeller for multiphase pumping by combined approach of inverse design and CFD analysis. *J. Fluids Eng.* **2005**, *127*, 330–338. [CrossRef]
7. Zhang, Y.; Zhang, J.; Zhu, H.; Cai, S. 3D blade hydraulic design method of the rotodynamic multiphase pump impeller and performance research. *Adv. Mech. Eng.* **2014**, *6*, 1–10. [CrossRef]
8. Kim, J.; Lee, H.; Kim, J.; Choi, Y.; Yoon, J.; Yoo, I.; Choi, W. Improvement of hydrodynamic performance of a multiphase pump using design of experiment techniques. *J. Fluids Eng.* **2015**, *137*, 081301. [CrossRef]
9. Liu, M.; Tan, L.; Cao, S. Design method of controllable blade angle and orthogonal optimization of pressure rise for a multiphase pump. *Energies* **2018**, *11*, 1048. [CrossRef]
10. Zhang, J.; Zhu, H.; Yang, C.; Li, Y.; Wei, H. Multi-objective shape optimization of helico-axial multiphase pump impeller based on NSGA-II and ANN. *Energy Convers. Manag.* **2011**, *52*, 538–546. [CrossRef]
11. Xu, Y.; Cao, S.; Sano, T.; Wakai, T.; Reclari, M. Experimental investigation on transient pressure characteristics in a helico-axial multiphase pump. *Energies* **2019**, *12*, 461. [CrossRef]
12. Zhang, J.; Cai, S.; Li, Y.; Zhu, H.; Zhang, Y. Visualization study of gas-liquid two-phase flow patterns inside a three-stage rotodynamic multiphase pump. *Exp. Therm. Fluid Sci.* **2016**, *70*, 125–138. [CrossRef]
13. Zhang, J.; Li, Y.; Cai, S.; Zhu, H.; Zhang, Y. Investigation of gas-liquid two-phase flow in a three-stage rotodynamic multiphase pump via numerical simulation and visualization experiment. *Adv. Mech. Eng.* **2016**, *8*, 1–13. [CrossRef]
14. Saadawi, H.N. Operating multiphase Helicoaxial pumps in series to develop a satellite oil field in a remote desert location. *SPE Proj. Facil. Constr.* **2008**, *3*, 1–6. [CrossRef]

15. Alberto, S.; Lars, E. Design of a multiphase pump test laboratory allowing to perform flow visualization and instability analysis. *Am. Soc. Mech. Eng.* **2015**, *23*, 857–878.
16. Alberto, S.; Lars, E. Flow visualization of unsteady and transient phenomena in a mixed-flow multiphase pump. In Proceedings of the ASME Turbo Expo: Turbomachinery Technical Conference & Exposition, Seoul, Korea, 13–17 June 2016; pp. 1–12.
17. Zhang, W.; Yu, Z.; Zahid, M.; Li, Y. Study of the gas distribution in a multiphase rotodynamic pump based on interphase force analysis. *Energies* **2018**, *11*, 1069. [CrossRef]
18. Liu, M.; Tan, L.; Cao, S. Influence of viscosity on energy performance and flow field of a multiphase pump. *Renew. Energy* **2020**, *162*, 1151–1160. [CrossRef]
19. Zhang, W.; Yu, Z.; Li, Y.; Yang, J.; Ye, Q. Numerical analysis of pressure fluctuation in a multiphase rotodynamic pump with air-water two-phase flow. *Oil Gas Sci. Technol.–Rev. d'IFP Energ. Nouv.* **2019**, *74*, 18. [CrossRef]
20. Zhang, W.W.; Zhu, B.S.; Yu, Z.Y. Numerical study on pressure fluctuation in a multiphase rotodynamic pump with different inlet gas void fractions. *IOP Conf. Ser. Earth Environ. Sci.* **2019**, *405*, 012029. [CrossRef]
21. Yu, Z.; Zhu, B.; Cao, S. Interphase force analysis for air-water bubbly flow in a multiphase rotodynamic pump. *Eng. Comput.* **2015**, *32*, 2166–2180. [CrossRef]
22. Li, X.; Li, Z.; Zhu, B.; Wang, W. Effect of tip clearance size on Tubular turbine leakage characteristics. *Processes* **2021**, *9*, 1481. [CrossRef]
23. Zhang, D.; Wu, S.; Shi, W.; Pan, D.; Yao, J.; Zhang, G. Application and experiment of different turbulence models for simulating tip leakage vortex in axial flow pump. *Trans. Chin. Soc. Agric. Eng.* **2013**, *29*, 46–53.
24. Xiao, W.; Tan, L. Review on research of rotodynamic multiphase pumps. *J. Hydroelectr. Eng.* **2020**, *39*, 108–120.
25. Yu, Z.; Zhu, B.; Cao, S.; Liu, Y. Effect of virtual mass force on the mixed transport process in a multiphase rotodynamic pump. *Adv. Mech. Eng.* **2014**, *6*, 958352. [CrossRef]
26. Wang, Q.; Yao, W. Computation and validation of the interphase force models for bubbly flow. *Int. J. Heat Mass Transf.* **2016**, *98*, 799–813. [CrossRef]
27. Pourtousi, M.; Sahu, J.N.; Ganesan, P. Effect of interfacial forces and turbulence models on predicting flow pattern inside the bubble column. *Chem. Eng. Process. Process Intensif.* **2014**, *75*, 38–47. [CrossRef]
28. Mohajerani, M.; Mehrvar, M.; Ein-Mozaffari, F. CFD analysis of two-phase turbulent flow in internal airlift reactors. *Can. J. Chem. Eng.* **2012**, *90*, 1612–1631. [CrossRef]

processes

Article

Wet Compression Study for an Aero-Thermodynamic Performance Analysis of a Centrifugal Compressor at Design and Off-Design Points

Hyun-Su Kang, Sung-Yeon Kim and Youn-Jea Kim *

School of Mechanical Engineering, Sungkyunkwan University, Suwon 16419, Korea; hskang1504@skku.edu (H.-S.K.); ksy203@skku.edu (S.-Y.K.)
* Correspondence: yjkim@skku.edu

Abstract: In this study, to analyze the effect of wet compression technology on the aero-thermodynamic performance of a centrifugal compressor, a numerical analysis study was conducted on the design point and off-design point. Wet compression technology sprays water droplets at the inlet of the compressor. During the compression process, water droplets evaporate, reducing the heat of compression and reducing the compression work, which improves the efficiency of the compressor. In wet compression technology, detailed research is needed for the application to compressors because the droplet behavior affects the internal flow. The main parameters for wet compression technology are the droplet size and injection rate selection, and the flow inside the compressor changed by the droplet behavior was analyzed. When the droplet size and injection rate were changed at the design point and the off-design point, it was confirmed that a small droplet size was effective in both areas, and it was confirmed that the performance improved as the flow rate increased. The internal flow changed greatly depending on the size of the droplet. As a result, the centrifugal compressor to which the wet compression technology was applied had a lower outlet temperature than dry compression at both the design point and the off-design point and had increases in the pressure ratio and efficiency. However, the surge margin decreased by about 2% in the surge region. The reason is that due to high-speed rotation, particles move in the outer diameter direction and are driven into a tip-leakage flow, and many stagnant flows occur without flowing into the main flow. Through the study results, it was possible to understand the effects of wet compression technology on the performance and efficiency increase of centrifugal compressors and the effects of particle behavior on the internal flow of the compressor at the off-design point.

Keywords: centrifugal compressor; wet compression; flow angle; two-phase flow; droplet evaporation

Citation: Kang, H.-S.; Kim, S.-Y.; Kim, Y.-J. Wet Compression Study for an Aero-Thermodynamic Performance Analysis of a Centrifugal Compressor at Design and Off-Design Points. *Processes* **2022**, *10*, 936. https://doi.org/10.3390/pr10050936

Academic Editor: Alfredo Iranzo

Received: 20 April 2022
Accepted: 3 May 2022
Published: 9 May 2022

1. Introduction

Compressors are used in a variety of industries. It is a very important component, especially in the energy field. Generally, compressors consume up to 40% of the energy consumed by the entire energy system or power plant. For this reason, lots of research has been done to improve compressor performance.

Mainly, research on improving the performance of the compressor has been conducted, with a lot of research on improving the pressure ratio or efficiency by redesigning the impeller or diffuser, which is a key element. However, most redesign methods have many engineering costs because aerodynamic performance and structural and vibration issues must be considered.

Wet compression technology is a way to compensate for these shortcomings. It does not require a separate component redesign and has proven to be effective just by injecting water droplets at the inlet. Especially in the power generation system, when the external temperature rises by 1 degree, the power of the power plant is reduced by up to 1%. The external environment cannot be controlled, but if water is injected to lower the compressor inlet temperature, the temperature of the compressor internal flow channel can be lowered, and a decrease in the temperature inside the flow channel has the effect of reducing compression work. Methods to lower the compressor inlet temperature include intercooling, evaporative cooling, inlet fog cooling, and wet compression. Among these methods, the most effective method is wet compression [1], where un-evaporated droplets flow into the compressor and evaporate and cool by the high-temperature, high-pressure environment of the internal flow path.

The wet compression technology was first proposed by Kleinschmidit [2] in 1940 and has been studied by many researchers. After that, in 1989, the wet compression technology was firstly installed in the gas turbine for the first time [3]. Jolly [4] experimentally applied wet compression technology to a gas turbine, resulting in a 9% power improvement. Bettocchi et al. [5] compared the performance curves of wet compression and conventional air compression and experimentally proved that the pressure ratio increased by 2% when the wet compression method was used. Neupert et al. [6] visualized the flow of droplets floating inside the compressor using a particle analyzer (Laser doppler anemometry, LDA, or Particle doppler anemometry, PDA). Recently, with the development of numerical analysis technology, it has become possible to study the interior of a compressor using wet compression technology in three dimensions. Furthermore, Sun et al. [7] advanced the numerical analysis of the physical interaction between droplets and blades.

Wet compression technology has been studied to have various advantages when applied to a gas turbine's axial compressor. However, research on applying wet compression technology to centrifugal compressors is relatively insufficient. In fact, there are more than 1000 cases of wet compression applied to a gas turbine [8], but it is still difficult to find cases applied to centrifugal compressors. Since a centrifugal compressor has different flow characteristics from axial compressors, additional studies are needed to apply wet compression. If wet compression technology is applied to the centrifugal compressor, wet compression technology can be effectively applied due to the advantage of the centrifugal compressor, which can be used in various operating conditions.

Since late 2010, several researchers have been conducting research to apply wet compression technology to a centrifugal compressor. Surendran et al. [9] applied a wet compression technique to a low-speed centrifugal compressor using numerical analysis and showed that the compression work decreased when the droplet inject amount was 3% or more. Sun et al. [10] numerically studied the effect of supercritical wet compression on the performance of centrifugal compressors used in compressed air energy storage systems.

Although the wet compression technique has been proven effective by various studies, it has problems, such as droplet collision and a reduction in the operating area [11]. In addition, the compression process changes due to droplets inside the compressor, and the flow angle changes. If the flow angle at the outlet of the impeller is largely changed, abnormal phenomena, such as a stall and surge, may occur along with the deterioration of the compressor performance, making it difficult to operate the compressor [12].

Wet compression technology has been proven to be effective by many studies, but in-depth research on droplet variables is required to apply wet compression due to complex compressor internal flow phenomena, such as droplet behavior. Therefore, in this study, a wet compression simulation was performed for not only the design point, but also the off-design point, and the internal thermodynamic performance analysis, as well as the internal flow angle and internal flow distribution, were considered. In addition, as a detailed goal, the variables in wet compression technology, diameter and injection rate, were calculated for the design point and surge area, and the effect of the variables on the flow and performance of the centrifugal compressor was studied.

2. Methodology

2.1. Governing Equations

2.1.1. Continuous Phase

In this paper, a numerical analysis was conducted to analyze the wet compression effect. For the numerical analysis, ANSYS CFX 2020R2, a commercial program, was used to solve the Reynolds-Averaged Navier–Stokes (RANS) equations. The RANS equation consists of a continuity equation, momentum equation, and an energy equation. The governing equations that describe the motion of continuous fluid are as follows:

Continuity equation:

$$\frac{\partial \rho}{\partial t} + \nabla \times (\rho \vec{v}) = S_M \quad (1)$$

where ρ is the fluid density, and $\vec{v}$ is the velocity field of the fluid. The right term (S_M) represents the source term of mass transfer from the dispersed phase to the continuous phase caused by the evaporation of droplets.

Momentum equation:

$$\frac{\partial(\rho \vec{v})}{\partial t} + [\nabla \times (\rho \vec{v}) \vec{v}] = -\nabla p + \nabla \times \vec{\tau} + S_F \quad (2)$$

where p is the pressure, and S_F represents the external body force. The external body force represents the momentum transfer between the dispersed phase and continuous phase. Furthermore, $\vec{\tau}$ means shear stress and is expressed as follows:

$$\vec{\tau} = \mu[(\nabla \vec{v} + \nabla \vec{v}^{\,t}) - \frac{2}{3}\nabla \times \vec{v} I] \quad (3)$$

Energy equation:

$$\frac{\partial(\rho e)}{\partial t} + \nabla \times (\rho \vec{v} e) = \nabla(\lambda \nabla t) + \rho\nabla \times \vec{v} + \nabla \times (\vec{v} \times \tau) + S_E \quad (4)$$

where λ is the thermal conductivity, and S_E is the energy source term, which represents the energy exchange between the dispersed phase and the continuous phase.

2.1.2. Dispersed Phase

The numerical analysis of wet compression is a two-phase flow, in which two different fluids flow, and is calculated using the Eulerian–Lagrangian method for calculating droplet information along with the RANS equations. Three equations are needed to exchange information on the motion and evaporation of droplets. The first is an equation to describe the motion of droplets, a heat transfer equation to describe the heat transfer effect of continuous and dispersed phases, and, finally, the mass transfer equation to calculate the interphase mass transfer.

- Particle transport equation

The particle transport equation is used to elucidate the motion of the droplet. The forces exerted on the droplet in the equation are forces by gravity, the centrifugal and Coriolis forces, the virtual mass force, and the pressure gradient force. However, in this study, only the forces by gravity and the centrifugal and Coriolis forces, which have a pronounced influence on the droplet movement, were used due to the high rotational speed of the fluid machine. The particle transport equation is written as follows:

$$m_d \frac{du_d}{dt} = F_D + F_R \quad (5)$$

where m_d is the droplet mass, and u_d is the droplet velocity. F_D and F_R are the forces by gravity and the centrifugal and Coriolis forces mentioned above and are expressed as follows:

$$F_D = \frac{1}{2} C_D \rho_{air} d^2 |u - u_d| (u - u_d) \tag{6}$$

$$F_R = m_d[-2\Omega \times u_d - \Omega \times (\Omega \times r)] \tag{7}$$

where d is the droplet diameter, u is the fluid velocity of continuous phase, Ω is the angular velocity, r is the position vector, and C_D is the drag coefficient, which is determined by the following empirical correlation [13]:

$$C_D = \frac{24\left(1 + 0.15 Re_{d_r}^{0.687}\right)}{Re_{d_r}},\ Re \leq 1000 \tag{8}$$

- Heat transport equation

The heat transfer between the continuous phase and dispersed phase is calculated from the following equation:

$$m_p C_d \frac{dT_d}{dt} = \pi d \lambda Nu (T - T_d) + h_l \frac{dm_p}{dt} \tag{9}$$

where C_d is the specific heat of water droplets, T_d is the temperature of droplet, and h_l is the latent heat of vaporization. Nu is the Nusselt number, which is the ratio of convective to conductive heat transfer. In this study, Nu was used through the Ranz–Marshall correlation, given by [14]:

$$Nu = 2 + 0.6 Re^{0.5} \left(\mu \frac{C_P}{\lambda}\right)^{\frac{1}{3}} \tag{10}$$

- Mass transport equation

The mass transport equation calculates the mass change that occurs in the process of phase change of droplets. In this study, two mass transfer correlations, according to the boiling point, were used. Mass transfer is decided by boiling point. The boiling point is determined through an Antoine equation given below [15]:

$$\log_{10} p_{sat} = A - \frac{B}{T + C - 273.15} \tag{11}$$

where p_{sat} is the saturation pressure, and A–C are the constant coefficients.

When the droplet temperature is above the boiling point, the mass transfer rate can be written as follows:

$$\frac{dm_d}{dt} = -\frac{\pi d \lambda Nu (T - T_p)}{h_{fg}} \tag{12}$$

Conversely, the mass transfer rate below the boiling point is as follows:

$$\frac{dm_d}{dt} = \pi d \rho_v D Sh \frac{M_v}{M} \log\left(\frac{1 - f_p}{1 - f}\right) \tag{13}$$

where ρ_v is the vapor density, and D is the diffusivity coefficient. Sh is the Sherwood number, and M and M_v are the air and water vapor molar mass, respectively. f and f_p are the mole fraction of the air–water mixture and the mole fraction of the water droplets, respectively.

2.2. Reference Model

The NASA CC3 model was used as an industrial air compressor to validate the effect of wet compression on the centrifugal compressor, and its performance and efficiency have been proven through various experiments [16,17]. The model impeller consists of 15 blades and 15 splitters, and the diffuser consists of 24 vanes. The design point of the analysis model is a 4.17 pressure ratio and 21,789 rpm, and the operating flow rate is 4.54 kg/s. Detailed dimensions are shown in Table 1 and Figure 1.

Table 1. Specification of the geometrical parameters.

Impeller	
Number of impeller blades	15
Number of splitter blades	15
Impeller LE radius at hub (r_{Hub})	41 mm
Impeller LE radius at shroud (r_{Shroud})	105 mm
Impeller TE radius (r_1)	215 mm
Diffuser	
Number of diffuser blades	24
Diffuser LE radius (r_2)	232 mm
Diffuser TE radius (r_3)	363 mm
Blade height (D_H)	17 mm

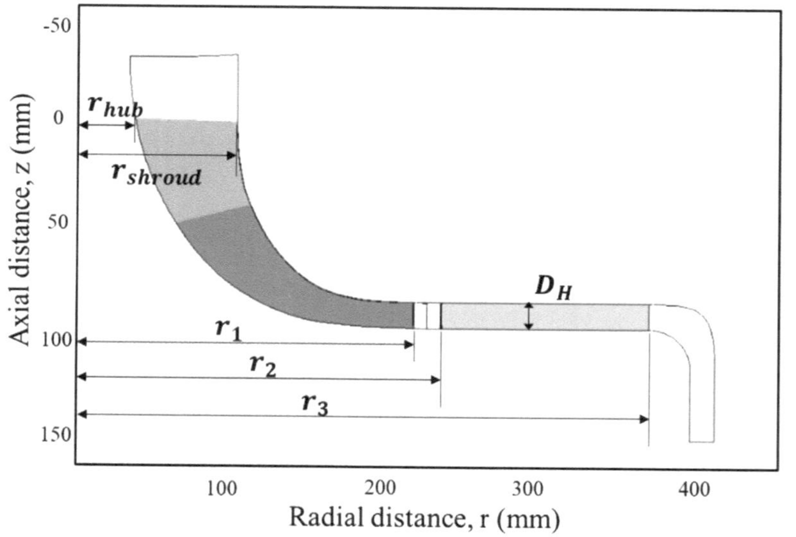

Figure 1. Specification of the geometrical dimensions.

2.3. Grid System

Securing a reliable grid system is one of the important processes in the CFD. In this study, structured grids for CFD were generated using ANSYS TurboGrid. Figure 2 shows the grid of the compressor, including the impeller and diffuser. For the blade flow path, an O-grid system was used for the surface near blades, and an H-grid system was used for other regions. In addition, a grid dependency test was conducted to minimize the influence of the grid (see Figure 3). The pressure ratio and isentropic efficiency were chosen as the evaluation index, and it was confirmed that the result value converged at about 1.2×10^6 elements. Therefore, a final 557,798 grids were used in the impeller region and 509,082 grids in the diffuser region, and a total of 1.2×10^6 grids were used.

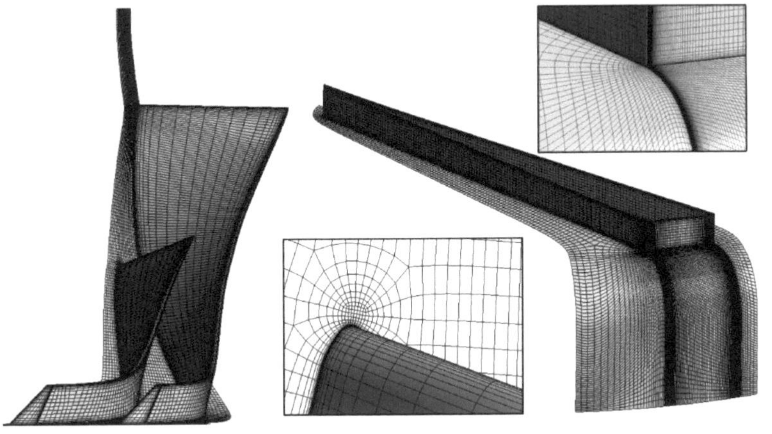

Figure 2. Grid systems of the numerical analysis.

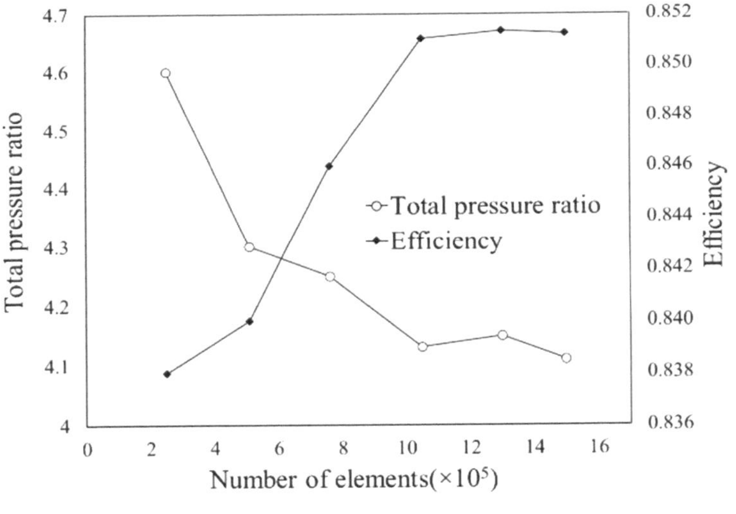

Figure 3. Performance results of the grid dependency test.

2.4. Boundary Conditions

In this study, a steady state analysis was performed to verify the performance curve. The MRF (moving reference frame) technique was used to analyze the rotation of the compressor. The SST (shear stress transport) model was used as the turbulence condition. Wet compression technology is a two-phase flow, in which droplets are injected with compressed air. For this reason, boundary conditions for fluids and droplets injection are required. The continuous phase fluid was calculated using the Eulerian method, and the injected droplets were calculated using the Lagrangian method. The calculation method through the Eulerian–Lagrangian method was calculated in a two-way coupled method, and the droplet information, such as temperature and trajectory, were calculated along with the flow. Detailed boundary conditions are described in Table 2.

Table 2. Boundary conditions of the centrifugal compressor.

Continuous Phase		
Rotating speed		21,789 [rpm]
Fluid		Air (ideal gas)
Inlet	Total pressure	101.325 [kPa]
	Temperature	288.15 [K]
Outlet	Average static pressure	410 [kPa]
Interface		Stage (mixing-plane)
Convergence criteria (RMS)		1.0×10^{-4}
Dispersed Phase		
Droplet		Water
Droplet temperature		288.15 [K]
Number of droplets		10,000
Injection type		Full cone

Wet compression technology exposes droplets to gas flow inside the compressor. Therefore, breakup occurs due to the difference in velocity between gas and liquid phases. In this study, the cascade atomization and breakup (CAB) model was used to consider the slip velocity. The CAB model was developed in the Taylor Analogy Breakup (TAB) model, in which the breakup of droplets is described as a system composed of a spring-mass system and damped resonance and is known to be suitable for liquid spray analysis. In addition, the Sommerfeld collision model [16], a probabilistic collision model, was used to consider the impact caused by collisions between droplets and blades.

3. Validation of Aero Performance and the Two-Phase Flow Model

3.1. Performance Curve Mapping

To validate the numerical analysis method, the CFD results were compared with existing published data [16–18]. For comparison, the pressure ratio and isentropic efficiency, which are the compressor performance index, were compared. The pressure ratio and isentropic efficiency are defined as follows:

$$\text{Pressure ratio(PR)} = \frac{P_{out}}{P_{inlet}} \tag{14}$$

$$\text{Isentropic efficiency(Eff)} = \frac{W_c}{W_{i,c}} \tag{15}$$

The comparisons of the CFD and experimental results are shown in Figure 4 and showed a good agreement.

3.2. Two-Phase Flow Model Validation

In this chapter, a two-phase flow model also needed to be verified. To validate the two-phase flow model, the NACA 0012 model was chosen. The experimental result [19] and CFD results were compared when NACA 0012 was rained on (refer to Figure 5). As a result of validation, it shows that an angle of attack (AOA) was performed on the airfoil from $-5°$ to $9°$, and all relative errors came within 7%. It was judged that the two-phase flow model used in this study is acceptable for this study.

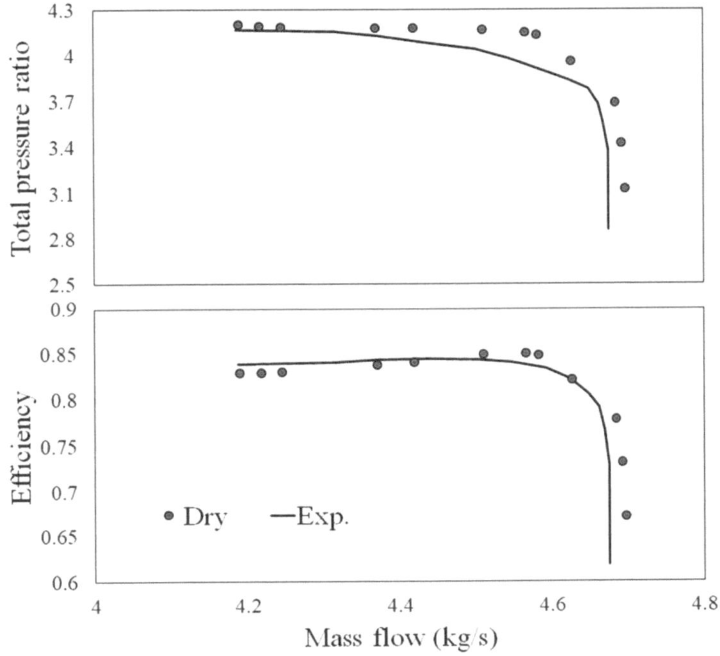

Figure 4. Comparison results of the total pressure ratio and efficiency between the CFD and experimental results.

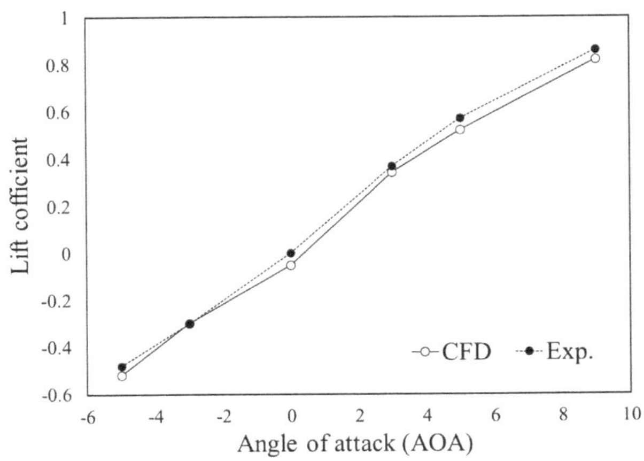

Figure 5. Comparison results of the lift coefficient curve for the NACA 0012 airfoil for rain conditions between the experiment [19] and CFD.

4. Results of Wet Compression at Design Point

4.1. Overall Performance Analysis

In this section, the influence of wet compression technology on aerodynamic performance at the design point was analyzed. To compare the effects of dry compression and wet compression, the wet compression technology injected droplets having a uniform size of 2 μm at 0.2% of the air flow rate. The actual droplet distribution consisted of various droplet sizes. The distribution of these droplets used a theoretical formula derived from an empirical formula. In general, the numerical analysis used the Rosin–Rammler (RR) distribution [20], which has few errors and is easy to apply. Accordingly, numerical analysis was performed by applying various droplet distributions along the RR distribution, not the uniform droplet size. The RR distribution equation is expressed as follows:

$$R = \exp\left[-\left(\frac{d}{d_e}\right)^{\gamma}\right] \tag{16}$$

where d_e is the RR mean diameter, and γ is the RR parameter. A general value of γ for the wet compression used is 1.5~3.0, but in this study, a value of 2 was used, referring to the literature on the existing wet compression research. Figure 6 compares the droplet trajectories when the uniform droplet size is 2 μm and when the average diameter is 2 μm according to the RR distribution. When looking at the droplet trajectory, the heterogeneous atomization condition had a droplet size of up to 3 μm, and thus, the trajectory was longer than that of the uniform droplet diameter. In terms of compression ratio, isentropic efficiency, and outlet temperature, there was no significant difference. Because the droplet size distribution in the two conditions differed, it was assumed that the differences did not have a substantial impact on performance, because the change in the droplet distribution was minor. Hence, to examine the compressor performance through the motion of a specific diameter, all analyzes that were conducted were performed using a homogeneous droplet diameter.

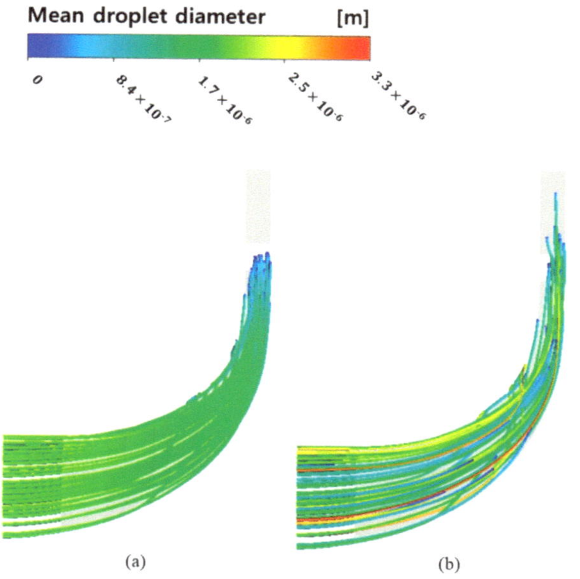

Figure 6. Droplet trajectories on the meridional plane under different injection conditions; (**a**) homogeneous injection and (**b**) heterogeneous injection.

The NASA CC3 model confirmed that when wet compression technology was applied to the design point, the pressure ratio and efficiency increased compared to the dry compression. At the design point, the pressure ratio improved by 0.12% compared to the existing dry compression. The mass flow rate of the air mixture increases because the temperature inside the compressor decreases by the latent heat of the droplet as the droplet is injected to increase the air density. The isentropic efficiency is an indicator of whether the temperature inside the compressor decreases. Therefore, the higher the efficiency, the clearer the heat absorption reaction induced by the latent heat of the droplet. The isentropic efficiency increased by 1.14% from air compression. The diffuser outlet temperature was 457.71 K in air compression, and it decreased by 3.6 K to 454.12 K when wet compression was applied.

Wet compression technology affects not only the thermodynamic aspects, but also the aerodynamic aspects, due to the droplets' motion. Figure 7 shows the velocity distribution of the impeller outlet, and Figure 8 shows the velocity angle distributions at same position as shown in Figure 7.

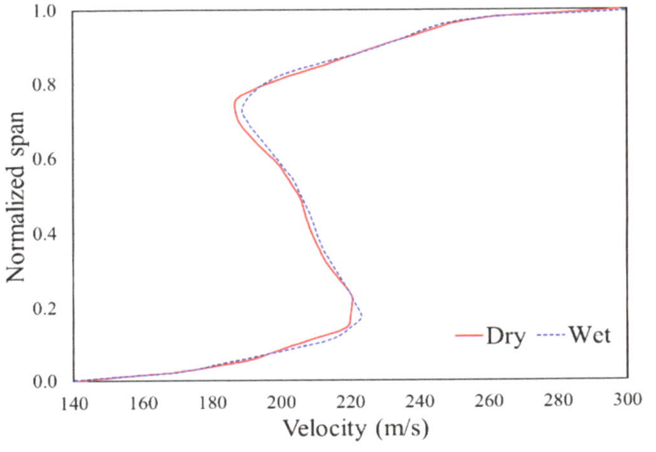

Figure 7. Impeller outlet velocity distribution of dry and wet compression.

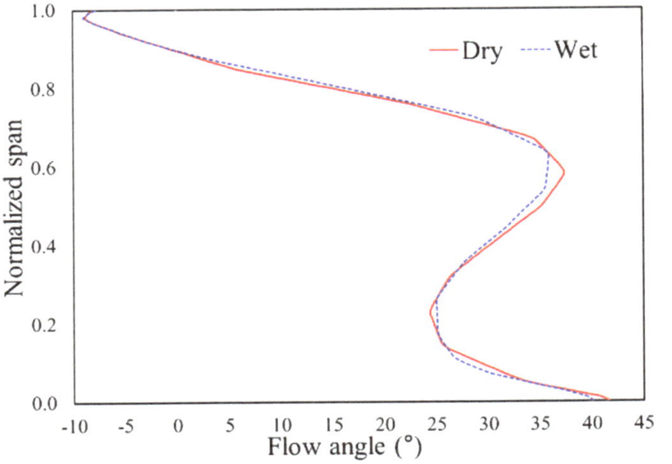

Figure 8. Impeller outlet flow angle distribution of dry and wet compression.

At the outlet of the impeller, the flow velocity differed by several m/s from about a range of 0.2~0.8, and overall, the speed was slower in the dry than the wet compression. Similarly, for the flow angle distribution, the flow angle difference occurred in the range of about 0.2~0.7. The presence or absence of droplets affects the flow speed and angle inside the impeller, and these changes affect not only the performance of the impeller, but also the diffuser, where kinetic energy is converted into pressure energy. From the next section, various flow rates and droplet diameters were analyzed for a detailed analysis of wet compression.

4.2. Effect of the Droplet Size on the Aerodynamic Performance of the Centrifugal Compressor at the Design Point

The droplet diameter is closely related to the evaporation rate and, thus, affects the performance. It is known that the appropriate droplet size in gas turbines is 5 μm to 10 μm. In addition, the smaller the droplet diameter, the less erosion is caused [21].

To scrutinize the effect of the droplet size on the centrifugal compressor performance, the flow rate, outlet temperature, pressure ratio, and isentropic efficiency were compared, varying the droplet diameter from 1~10 μm. The results were normalized for the convenience of comparison according to the droplet diameter and are shown in Figure 9. The isentropic efficiency had the highest values at the droplet size of 2 μm, and the values of other droplet diameters were normalized based on 2 μm. The outlet temperature was normalized with 6 μm and was calculated by the following equation:

$$\text{Normalized parameters} = \frac{\text{Value}}{\text{Value}_{\text{max}}} \tag{17}$$

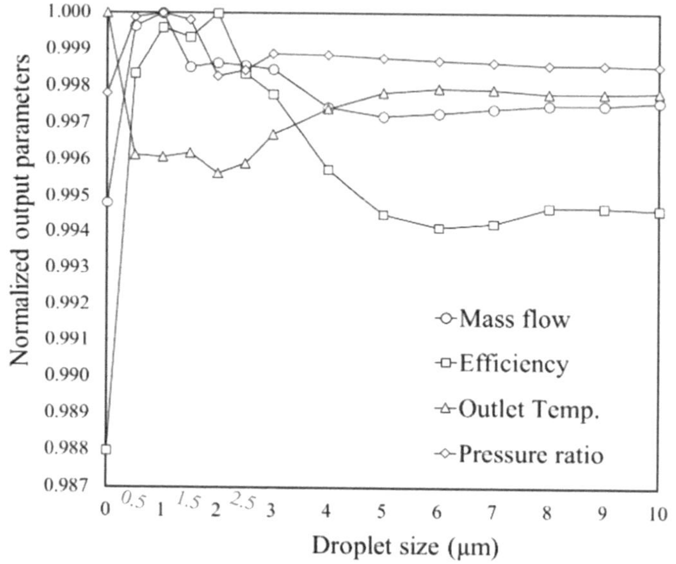

Figure 9. Effect of the different droplet sizes on the centrifugal compressor performance (injecting mass flow rate of 0.2%) at the design point.

To increase precision in the analysis results, the intermediate values of particle sizes having high efficiency intervals (1~3 μm) were additionally analyzed. When the droplet diameter was changed, the best aerodynamic performance was shown at 2 μm. On the contrary, the aerodynamic performance was lowest at 6 μm. The results of the droplet trajectory at 2 μm and 6 μm are compared in Figure 10.

In the case of 2 μm, all droplets were evaporated at the impeller outlet, and the particle diameter at the impeller outlet also decreased to 1 μm or less. On the other hand, when it was 6 μm, some droplets were delivered to the diffuser region without being evaporated inside the impeller. The droplet diameter at the impeller outlet showed a distribution of up to 5 μm or more. Additionally, if the diameter of the impeller is large, it is separated from the main flow upstream of the impeller, and it could be confirmed that a lot of it joins as a tip leakage flow.

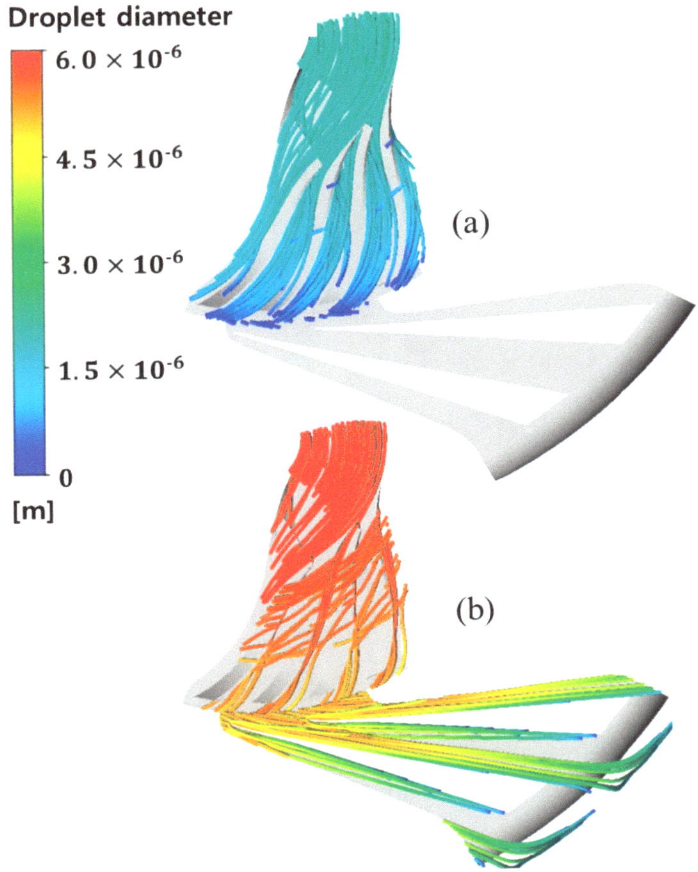

Figure 10. Comparison results of the droplet trajectories for the case of droplet diameter; (**a**) d_r = 2 μm and (**b**) d_r = 6 μm.

The droplets inside the centrifugal compressor are subjected to centrifugal force in the radial direction. The larger the droplet diameter, the more the centrifugal force is affected; hence, droplets flock toward the shroud. These results tend to be similar to previous studies on wet compression [22,23].

As the droplet diameter increases, the particle response time increases, which is greatly influenced by the centrifugal force. The particle response time refers to the time required for the particles to react to changes in fluid flow. Thus, droplets with a large diameter are not affected by the gas flow field; consequently, they deviate faster from the main flow. Figure 11 shows the mass fraction distribution of droplets with sizes of 2 μm and 6 μm in the flow direction. It was confirmed that a model with a droplet size of 6 μm was concentrated toward the gap by a large centrifugal force. Most of the droplets moved toward the tip clearance, resulting in more leakage flow, forming an unstable internal flow field. This was judged to be the reason that the performance was lower when the particle diameter was large. In wet compression technology, the larger the particle diameter, the particle response time increases to be longer, which leads to tip leakage. It was confirmed that the performance of the compressor was better when the particle diameter was small.

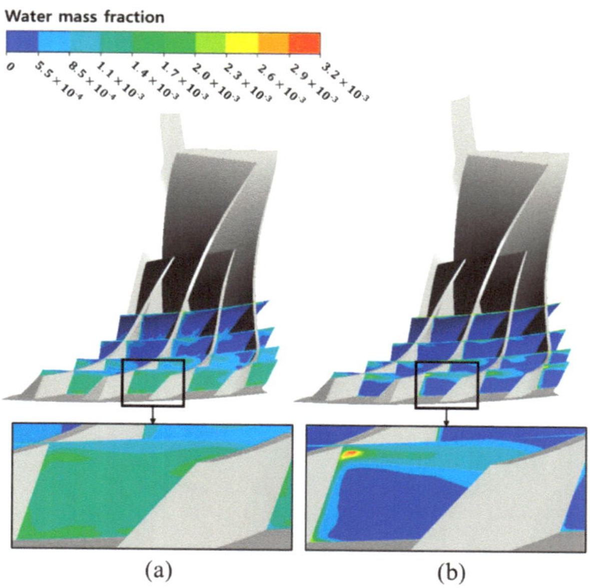

Figure 11. Comparison results of the mass fraction for the case of droplet diameter; (**a**) d_r = 2 μm and (**b**) d_r = 6 μm.

4.3. Effect of the Mass Flow of Water Injection on the Aerodynamic Performance of a Centrifugal Compressor at the Design Operating Condition

In this section, the calculation results, according to the injection rate, are discussed. The droplet diameter was fixed to 2 μm, having the highest efficiency at the design point, and the droplet flow rate was performed in the order of 0.2%, 0.4%, 0.6%, 0.8%, and 1.0% of the air flow rate. The increase in the amount of water injected indicates that the size of latent heat contained in water increases; hence, the cooling effect linearly improves performance. Figure 12 shows the results, and the internal temperature decreased as the flow rate increased. Figure 13 shows the temperature distribution by injection rate. The cooling effect clearly occurred in the diffuser region due to the evaporated latent heat. Accordingly, the outlet average temperature difference between 1% injection and 0.2% injection was 9.3 K. In addition, as the injection rate increases, the trajectory of the droplets inside the compressor increases. As the injection rate increases, more droplets enter the impeller region; thus, it is thought that the trajectory increases because relatively unevaporated droplets increase. Comparing the time and distance over which the actual droplet remains, the traveling time of the longest droplet was 0.004461 s when the flow rate was 1% and the traveling distance was 0.3883 m, and when it was 0.2%, it was 0.003275 s and 0.3746 m. From the design point, when water was injected at an injection rate of 0.2% of the air, most of the droplets evaporated at the end of the impeller region and in front of the diffuser region. As the injection rate increases, droplets exit to the diffuser region, but the traveling distance of the droplet is not different because most of the droplets with the longest traveling distance rotate in the centrifugal direction rather than the streamwise direction. The advantages of flow rate in wet compression are obvious, but if the amount exceeds a certain amount, a stall or flow instability might occur, so flow control is required, considering the operating schedule of the compressor [23].

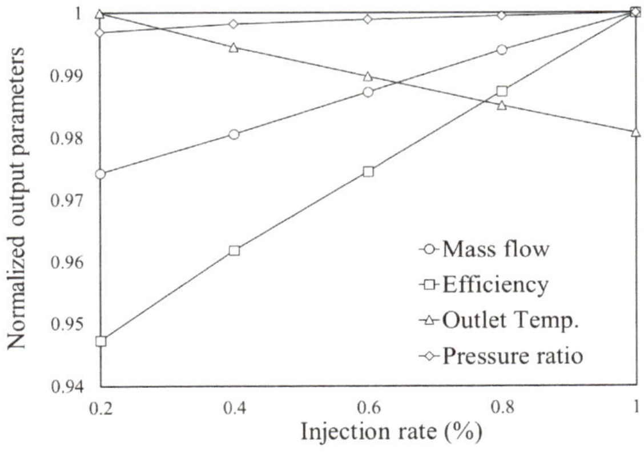

Figure 12. Effect of the different mass flow rates on the centrifugal compressor performance (droplet diameter, d_r = 2 µm) at the design point.

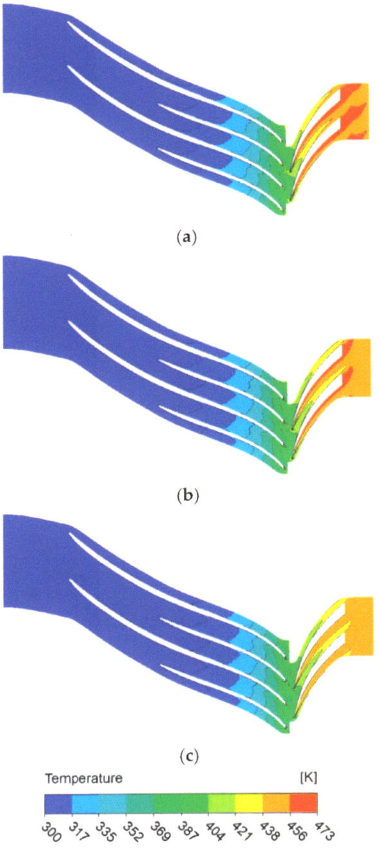

Figure 13. Comparison results of the temperature contour under different droplet mass flow rates at a span 50%. (**a**) 0.2% (0.00092 kg/s); (**b**) 0.6% (0.00276 kg/s); (**c**) 1% (0.0046 kg/s).

5. Results of Wet Compression at the Off-Design Region

5.1. Performance Analysis at the Off-Design Operating Condition

In this section, the effect of wet compression technology on a wide range of flow rates, from design flow rates to flow rates near surge, was analyzed. Figure 14 illustrates the dry and wet compression performance curves for the pressure ratio and efficiency. In the wet compression technology, based on the previous section, the size of the droplet was 2 μm, and the injection rate was 0.2% of the air flow rate. In wet compression, both the pressure ratio and the isentropic efficiency moved to the right, which means that more flow rates had passed compared to the same performance. When wet compression technology is applied, the surge condition moves toward a high flow rate. To confirm this, the stall margin was evaluated, which is expressed as follows:

$$SM = \left[1 - \frac{PR_{peak}}{PR_{stall}} \times \frac{m_{stall}}{m_{peak}}\right] \times 100\% \tag{18}$$

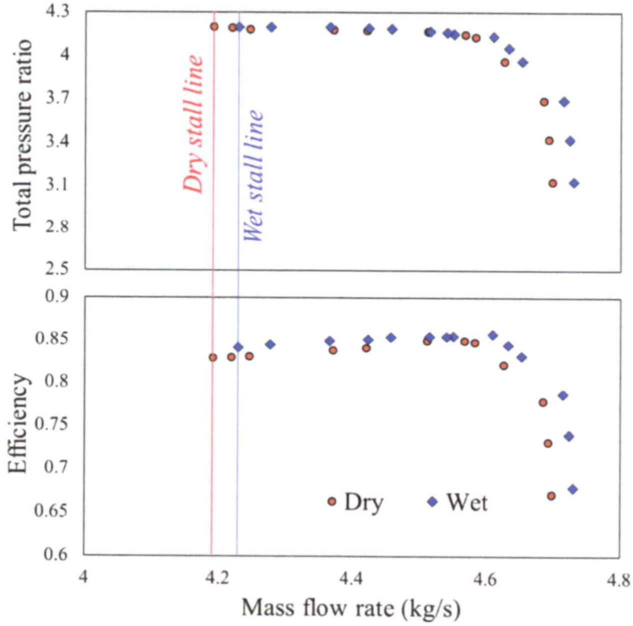

Figure 14. Comparison results of the performance curve between the dry and wet compression.

The SM value decreased by 2% in the wet compression compared to the dry compression. In general, the surge phenomenon occurs at a low flow rate point and occurs in the impeller and diffuser regions, causing an instability in the flow. Accordingly, the droplet motion changes the flow inside the impeller to reduce SM. Figure 15 shows the leakage flow occurring at the top of the tip clearance (red arrow) near the surge point. In the case of (a) dry compression, a few parts of the leakage flow generated from the top of the shroud went to the adjacent blade, but in the case of (b) wet compression, it can be noted that a large amount of the leakage flow generated from the top moved to the adjacent blade and formed a blockage (red circle). As a result, in wet compression, there is relatively more leakage flow that does not flow downstream and is stagnant.

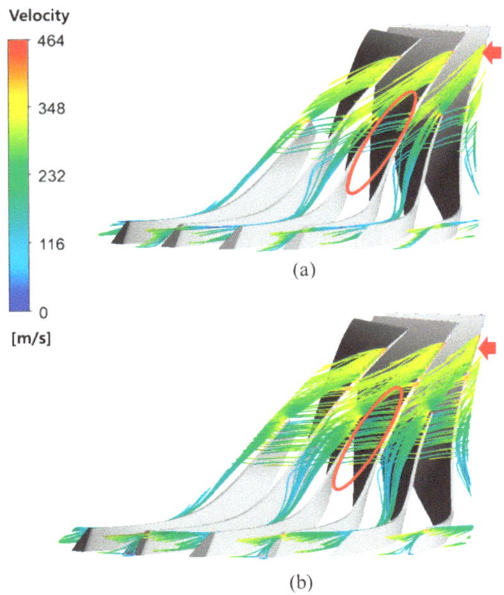

Figure 15. Comparison results of the leakage flow between the (**a**) dry and (**b**) wet compression.

Figures 16 and 17 show the eddy viscosity in the dry and wet compression at a span of 0.9 (normalized between the hub to shroud). Figure 16 shows that a relatively higher eddy viscosity occurred in (b) the wet compression compared to (a) the dry compression near the outlet of the impeller.

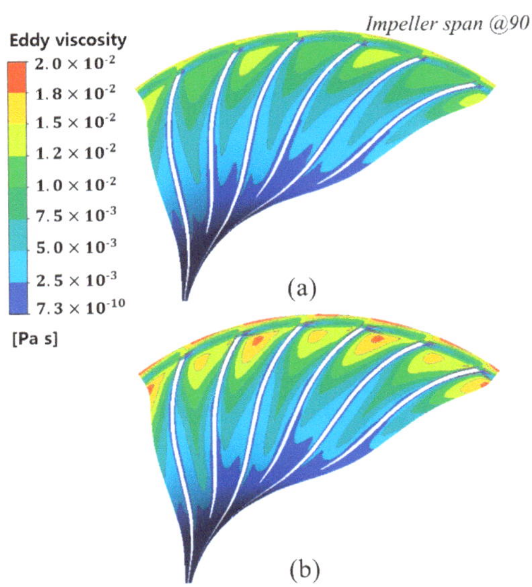

Figure 16. Eddy viscosity distribution between the (**a**) dry and (**b**) wet compression at the stall region.

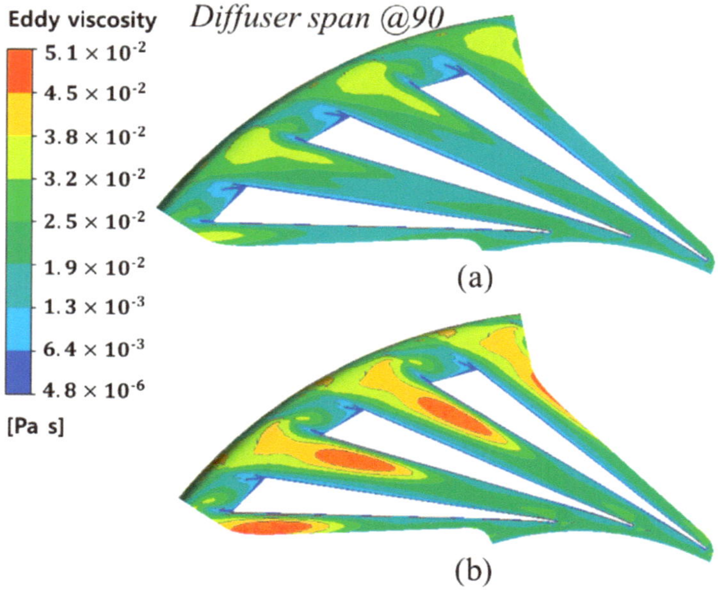

Figure 17. Diffuser eddy viscosity distribution between the (**a**) dry and (**b**) wet compression at the stall region.

Similarly, a higher eddy viscosity occurred near the exit of the diffuser in the wet compression, shown in Figure 17b. An unstable vortex field generated from the upstream was accumulated, and more unstable flow occurred in the wet compression. Blockages and strong vortices in the compressor internal flow path cause a numerical stall.

In the wet compression, the leakage flow developed from the upstream did not flow downstream as well as that of the dry compression, resulting in larger blockage near the span of 0.9 and a relatively larger eddy viscosity near the impeller and diffuser outlets. For this reason, the stall margin was lower in the wet compression than that in the dry compression.

5.2. Performance Analysis at the Off-Design Operating Condition

Figure 18 shows the performance according to the size of the droplet to analyze the performance according to the size of the droplet at the off-design point. As with the design point, the size of the droplet was changed from 1 to 10 μm to confirm the flow rate, outlet temperature, pressure ratio, and isentropic efficiency. Unlike the design point, it showed the best performance at 1 μm in the surge area. Results also showed that the performance decreases as the droplet diameter increases. When the droplet diameter grows, the number of droplets that have not evaporated inside the centrifugal compressor influences performance, just as it does with the design point. Figure 19 shows the average volume fraction in the direction of the streamwise, and the droplet diameter with 1 μm evaporated at the streamwise location of 0.6. On the other hand, when the droplet diameter was 10 μm, droplets existed at the end of the diffuser. As the droplet diameter increases, the point of evaporation is delayed. The latent heat size of the diffuser region cannot be used for cooling when the droplets leave the diffuser region in an evaporated state; hence, the smaller the droplets are injected, the higher the performance.

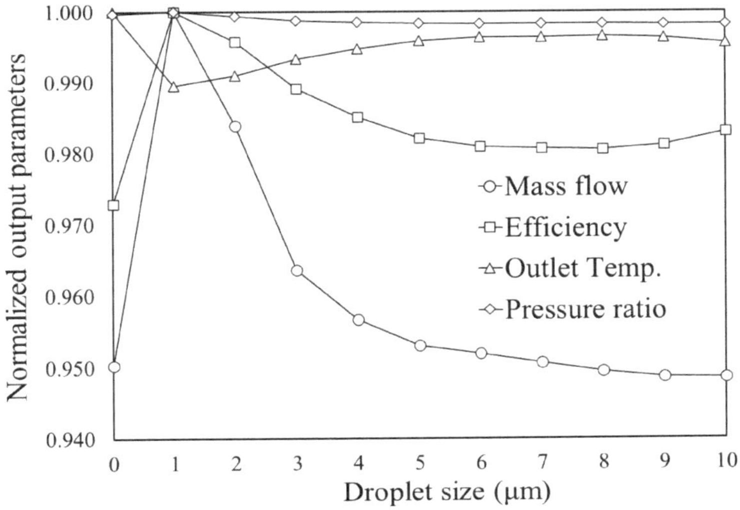

Figure 18. Effect of the different droplet sizes on the centrifugal compressor performance.

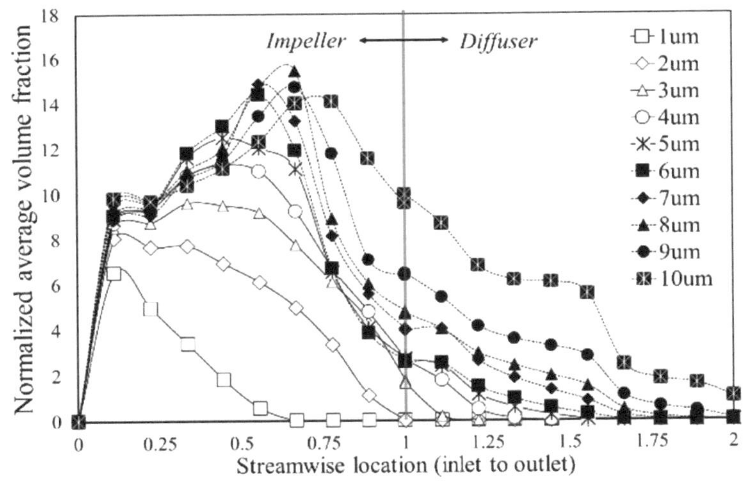

Figure 19. Averaged volume fraction along the streamline position with different droplet sizes.

5.3. Performance Analysis at the Off-Design Operating Condition

Figure 20 shows the performance change when the injection rate was changed by 0.2~1.0% in the surge area. As with the previous design point, the flow rate linearly increased. Figure 21 shows the average volume fraction for each streamwise location. As the flow rate increases, the mass fraction of water increases in the impeller region. However, as the injection rate increased, the evaporation position of all droplets did not change significantly. Increasing the flow rate increases the amount of latent heat to increase the cooling effect but does not change the evaporation position of the droplet. Consequently, the size of the droplet has the greatest influence on the evaporation of the droplet. Results also showed that the outlet temperature decreased to 458.239 K, 454.499 K, 454.865 K, 449.395 K, and 447.021 K, respectively, as the injection rate increased. Figure 22 shows the performance for the design point and the surge area to compare with the change in the injection rate. In both the surge area and the design point, the pressure ratio, outlet temperature, and isentropic efficiency showed similar trends, but the mass flow rate of air showed a steeper difference as the water injection rate increased. This is thought to have delivered more air compressed by the cooling effect in the surge region where the internal temperature is higher.

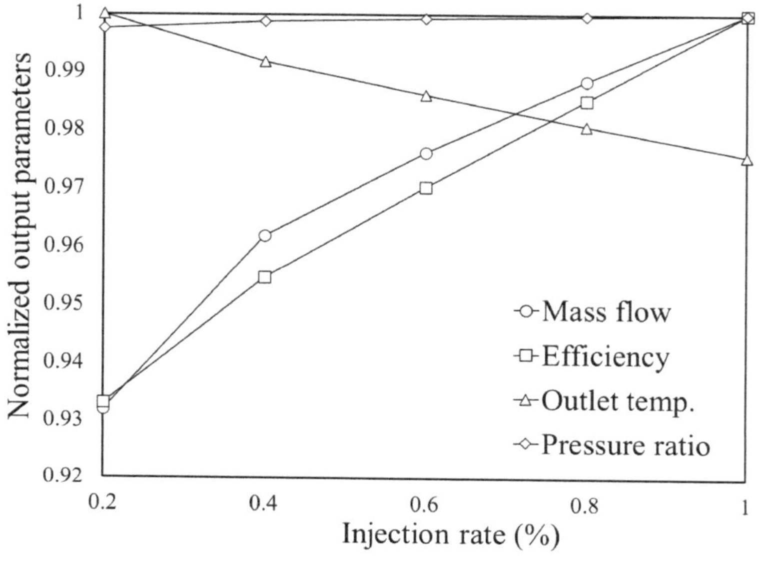

Figure 20. Effect of the different mass flow rates on the centrifugal compressor performance.

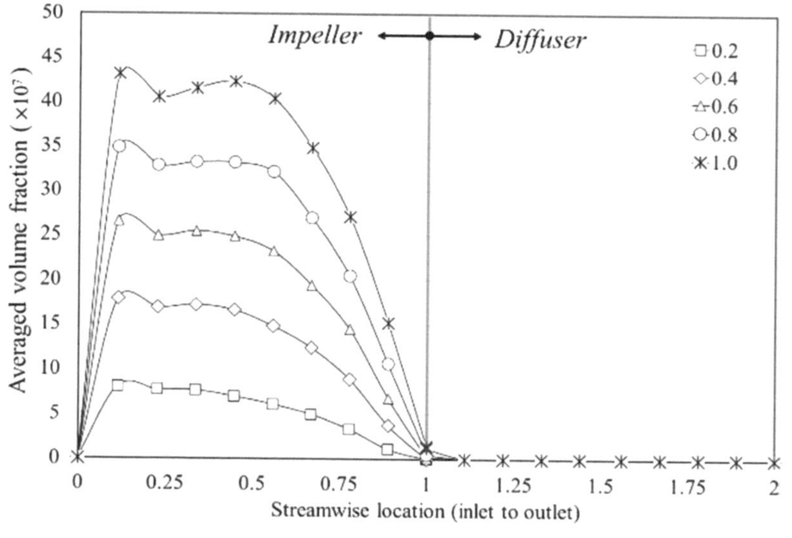

Figure 21. Averaged volume fraction along the streamline position with different water injection rates.

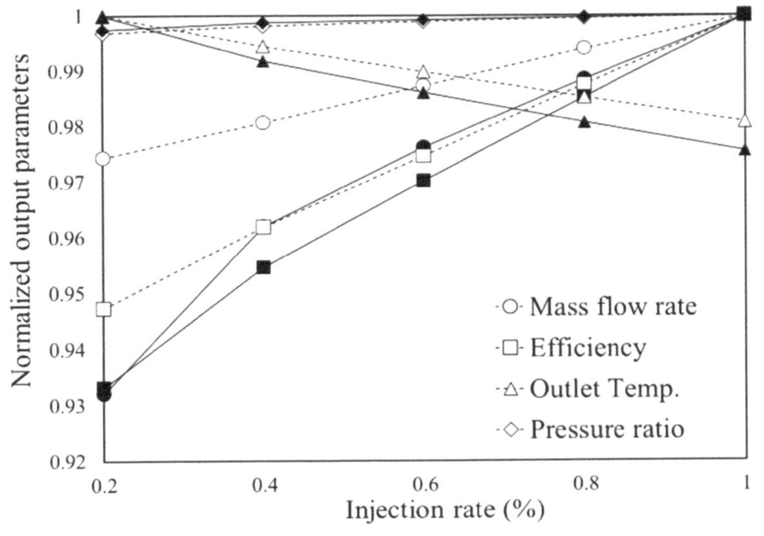

Figure 22. Effect of the different mass flow rate on the centrifugal compressor performance between the design point and surge region (Filled symbol: design point; open symbol: surge region).

6. Conclusions

In this study, the wet compression technology that improves compressor performance by injecting droplets to the inlet area was investigated. Through this, the effect on the aerodynamic performance of the centrifugal compressor was considered, and the results are summarized as follows:

(1) A droplet having a diameter 2 μm at the design point was injected into the compressor inlet area at 0.2% of the air flow rate (4.54 kg/s). At the design point, the pressure ratio increased by 0.12% and the isentropic efficiency by 1.14% compared to the air compression. The variable that has a dominant influence on the efficiency increases is the evaporation of the droplet. Droplet evaporation reduces the internal temperature of the compressor by the amount of latent heat and, consequently, increases the efficiency.

(2) At the design point, the performance was compared by changing the droplet diameter by 1 to 10 μm. The droplet diameter was best at 2 μm, and the performance deteriorated from the droplet diameter with a size of 6 μm or more. This is because the evaporation rate decreased as the surface area of the droplet was different while having the same mass flow rate. In addition, the droplet diameter with 6 μm quickly departs from the main flow due to its relatively large mass. As a result, it was established that most of the droplets moved toward the tip, generating more leakage flow and forming an unstable flow. The performance improves linearly as the water injection rate increases. Accordingly, the outlet average temperature difference between a 1% spray and 0.2% spray was 9.3 K.

(3) A numerical analysis was performed by applying wet compression technology to various operating points by changing the static pressure in the outlet area of the centrifugal compressor. Wet compression has been shown to affect not only the design point, but also the off-design point, including near the surge. In the surge region, wet compression generated more leakage flow according to droplet motion, and it reduced SM by 2% compared dry compression. The effective droplet size in the surge area was 1 μm. Like the design point, the small droplet size was effective, and as the droplet diameter increased, the unevaporated droplet increased. The water injection rate affected the performance in the surge region.

Author Contributions: Supervision, Y.-J.K.; writing original draft, H.-S.K. writing review and editing, H.-S.K. and S.-Y.K. All authors contributed to the manuscript. All authors have read and agreed to the published version of the manuscript.

Funding: This research was supported by the SungKyunKwan University and the BK21 FOUR (Graduate School Innovation) funded by the Ministry of Education (MOE, Korea) and National Research Foundation of Korea (NRF).

Conflicts of Interest: The authors declare no conflict of interest.

References

1. Sun, L.; Zheng, Q.; Li, Y.; Bhargava, R. Understanding effects of wet compression on separated flow behavior in an axial compressor stage using CFD analysis. *J. Turbomach.* **2011**, *133*, 031026. [CrossRef]
2. Kleinschmidt, R. Value of wet compression in gas-turbine cycles. *Mech. Eng.* **1947**, *69*, 115–116.
3. Nolan, J.; Twombly, V. Gas turbine performance improvement direct mixing evaporative cooling system American Atlas Cogeneration Facility Rifle, Colorado. In *Turbo Expo: Power for Land, Sea, and Air*; American Society of Mechanical Engineers: New York, NY, USA, 1990; Volume 79078, p. V004T11A021.
4. Jolly, S. *Wet Compression—A Powerful Means of Enhancing Combustion Turbine Capacity*; Power-Gen International: Orlando, FL, USA, 2002.
5. Bettocchi, R.; Morini, M.; Pinelli, M.; Spina, P.R.; Venturini, M.; Torsello, G. Setup of an experimental facility for the investigation of wet compression on a multistage compressor. *J. Eng. Gas Turbines Power* **2011**, *133*, 102001. [CrossRef]
6. Neupert, N.; Ober, B.; Joos, F. Experimental investigation on droplet behavior in a transonic compressor cascade. *J. Turbomach.* **2015**, *137*, 031009. [CrossRef]
7. Sun, L.; Zheng, Q.; Luo, M.; Li, Y.; Bhargava, R. On the behavior of water droplets when moving onto blade surface in a wet compression transonic compressor. *J. Eng. Gas Turbines Power* **2011**, *133*, 082001. [CrossRef]
8. Bhargava, R.; Meher-Homji, C. Parametric analysis of existing gas turbines with inlet evaporative and overspray fogging. In *Turbo Expo: Power for Land, Sea, and Air*; American Society of Mechanical Engineers: New York, NY, USA, 2002; Volume 36096, pp. 387–401.
9. Surendran, A.; Kim, H.D. Effects of wet compression on the flow behavior of a centrifugal compressor: A CFD analysis. In *Turbo Expo: Power for Land, Sea, and Air*; American Society of Mechanical Engineers: New York, NY, USA, 2014; Volume 45639, p. V02DT42A002.
10. Sun, J.; Hou, H.; Zuo, Z.; Tang, H.; Chen, H. Numerical study on wet compression in a supercritical air centrifugal compressor. *Proc. Inst. Mech. Eng. A J. Power Energy* **2020**, *234*, 384–397. [CrossRef]
11. Brun, K.; Kurz, R.; Simmons, H.R. Aerodynamic instability and life-limiting effects of inlet and interstage water injection into gas turbines. *J. Eng. Gas Turbines Power* **2004**, *128*, 617–625. [CrossRef]
12. Kang, J.S.; Cha, B.J.; Yang, S.S. Meanline Analysis of wet compression in a centrifugal compressor. *Aerosp. Eng. Technol.* **2005**, *4*, 7, 18–24.
13. Naumann, Z.; Schiller, L. A drag coefficient correlation. *Zeit. Ver. Deutsch. Ing.* **1935**, *77*, e323.
14. Ranz, W.E. Evaporation from drops, Parts I & II. *Chem. Eng. Prog.* **1952**, *48*, 141–146.
15. Reid, R.C.; Prausnitz, J.M.; Poling, B.E. *The Properties of Gases and Liquids*; McGraw-Hill Education: New York, NY, USA, 1987.
16. McKain, T.F.; Holbrook, G.J. *Coordinates for a High Performance 4: 1 Pressure Ratio Centrifugal Compressor*; NASA Technical Report No.19970024917; NASA: Washington, DC, USA, 1997.
17. Wernet, M.P.; Bright, M.M.; Skoch, G.J. An investigation of surge in a high-speed centrifugal compressor using digital PIV. *J. Turbomach.* **2001**, *123*, 418–428. [CrossRef]
18. Skoch, G.J.; Prahst, P.; Wernet, M.; Wood, J.; Strazisar, A. *Laser Anemometer Measurements of the Flow Field in a 4: 1 Pressure Ratio Centrifugal Impeller*; 97-GT-342; American Society of Mechanical Engineers: New York, NY, USA, 1997.
19. Douvi, E.C.; Margaris, D.P.; Lazaropoulos, S.D.; Svanas, S.G. Experimental and computational study of the effects of different liquid water content on the aerodynamic performance of a NACA 0012 airfoil at low Reynolds number. In Proceedings of the 5th International Conference on Experiments/Process/System Modeling/Simulation/Optimization, Athens, Greece, 3–6 July 2013; pp. 3–6.
20. Rosin, P. Laws governing the fineness of powdered coal. *J. Inst. Fuel* **1933**, *7*, 29–36.
21. Biglarian, M.; MomeniLarimi, M.; Ganji, B.; Ranjbar, A. Prediction of erosive wear locations in centrifugal compressor using CFD simulation and comparison with experimental model. *J. Braz. Soc. Mech. Sci. Eng.* **2019**, *41*, 106. [CrossRef]
22. Sun, J.; Zhou, X.; Liang, Q.; Zuo, Z.; Chen, H. The effect of wet compression on a centrifugal compressor for a compressed air energy storage system. *Energies* **2019**, *12*, 906. [CrossRef]
23. Yin, H.; Wu, H.; Li, Y.; Quan, J. Performance analysis of the water-injected centrifugal vapor compressor. *Energy* **2020**, *200*, 117538. [CrossRef]

processes

Article

Vortex Suppression and Flow Pattern Analysis of a Hydrofoil with Parallel Grooves

Yue Chen [1], Wei Zhang [1], Dehong Fang [1], Mingkang Sun [1,2], Jian Liu [1], Daoyuan Song [3,*] and Xiaoping Zhang [3,*]

1 State Key Laboratory of Hydroscience and Engineering, Department of Energy and Power Engineering, Tsinghua University, Beijing 100084, China; 13597567272@163.com (Y.C.); zhangw12@tsinghua.org.cn (W.Z.); ahdhfang@hotmail.com (D.F.); sunmingkang10121@163.com (M.S.); ljian8552@gmail.com (J.L.)
2 College of Mechanical Engineering and Automation, Fuzhou University, Fuzhou 350108, China
3 College of Electrical Engineering, Naval University of Engineering, Wuhan 430033, China
* Correspondence: hustsdy@126.com (D.S.); zxpbabymail01@163.com (X.Z.)

Abstract: As one of the typical negative factors affecting the vortex structure and flow characteristics of hydraulic machinery, the TLV has a non-negligible impact on the energy performance. In order to improve the utilization efficiency of hydraulic machinery in marine energy, the parallel grooves structure is proposed and applied to the NACA0009 hydrofoil. Subsequently, an exhaustive numerical analysis is carried out adopting the SST k-ω turbulence model, and the effects of the position and spacing on the suppression effect and flow characteristics are investigated. The presence of the parallel grooves leads to a decrease in the lift-to-drag ratio of the hydrofoil within 5%, but it can effectively suppress the development of TLV and reduce the area of TLV. The parallel grooves destroy the structure of PTLV and STLV, and the spacing and position have a greater influence on the flow characteristics of the hydrofoil. In order to take the TLV suppression effect and the energy performance of the hydrofoil into account, the L3T1 structure is recommended.

Keywords: marine energy; hydrofoil; tip clearance; parallel grooves; vortex suppression

Citation: Chen, Y.; Zhang, W.; Fang, D.; Sun, M.; Liu, J.; Song, D.; Zhang, X. Vortex Suppression and Flow Pattern Analysis of a Hydrofoil with Parallel Grooves. *Processes* **2021**, *9*, 816. https://doi.org/10.3390/pr9050816

Academic Editor: Alfredo Iranzo

Received: 28 March 2021
Accepted: 22 April 2021
Published: 8 May 2021

1. Introduction

Marine energy is an important source of energy for human development in the future, and the efficiency of utilization in marine energy affects the development of the energy industry, of which hydraulic machinery plays an important role in the exploitation and utilization of marine energy [1,2]. As a significant component of hydraulic machinery, the hydrofoil is not only one of the main topics in hydrodynamic engineering and research [3,4], but also gradually represents great impact in the development of marine energy. In the field of renewable energy, tidal energy relies on turbines to achieve the conversion from gravitational potential energy of the current to electrical energy, so the hydrofoil is widely adopted in the design of mechanical equipment such as turbines [5,6] and oil-gas multiphase pump [7]. Different structures and design schemes of hydrofoil will affect the hydrodynamic performance of hydraulic machinery due to the factors such as the interaction between ocean currents and marine equipment, which can produce uneven stress distribution [8,9], noise [10,11], and tip leakage vortex [12]. In order to improve the energy characteristics of hydraulic machinery and enhance the efficiency of marine energy utilization, it is necessary to explore how to eliminate the negative effects of hydraulic machinery.

For the purpose of investigating the impacts on energy characteristics caused by structure and operating conditions of hydrofoil, scholars have carried out studies around the effects brought by the shape of the hydrofoil [13], different attack angles and inlet flow velocity [14], and found that the tip clearance had an important impact on the energy characteristics of hydraulic machinery [15,16]. TLV is generated and developed in the blades of hydraulic machinery. In order to improve the energy performance of hydraulic machinery, many scholars have conducted in-depth research on the mechanism of the tip leakage vortex (TLV) generated from the tip clearance. It had been shown that the size of the tip clearance would affect the TLV generation [17], and a smaller tip clearance made less flow rate and generated a smaller TLV, resulting in less energy loss. In terms of numerical solutions, scholars have conducted calculation using RANS and LES methods [18], which meticulously demonstrated the flow structure from the gap between the tip of the NACA0009 hydrofoil and the wall, analyzed the moving trajectory of the TLV and the effect of gap widths on the generation of TLV. The unsteady RANS equation was also employed to study the TLV under unsteady incoming flow conditions [19], and it was found that the time-averaged entropy generation in the unsteady case was reduced by 18% compared with the steady incoming flow conditions, indicating that the flow stability also affected the energy characteristics of the hydrofoil. Wavelet analysis had also been applied to the analysis of TLV [20], and by analyzing the instabilities and the corresponding frequencies, it was possible to understand whether there existed irregularities in the flow field and explored the relevant factors affecting the flow pattern of hydrofoil. TLV leads to a reduction in efficiency [21], indicating that the tip clearance of hydrofoil and the TLV are issues worth focusing on.

Since TLV has a significant impact on the energy performance and efficiency of hydrofoil and hydraulic machinery, methods to suppress TLV need to be explored. In terms of experimental research, scholars have mainly adopted high-speed camera and instantaneous pressure measurement techniques to analyze the performance of hydraulic machinery [22], and the effect of TLV on the flow field had been systematically studied through the pressure pulsation. In recent years, scholars have found that a certain groove design for the hydrofoil can effectively improve its lift-to-drag ratio [23,24], and the intrinsic mechanism was that the groove changed the flow structure of the fluid and integrally improved the energy performance of the hydrofoil [25]. On this basis, special curves such as the bimodal Gaussian function curve were used to design and optimize the groove of the hydrofoil [26] in order to explore the solutions to improve the flow pattern and energy performance.

The Lagrangian method [27] can also be employed to evaluate the calculation method based on Helmholtz's concept of strong vortex dynamics. Similarly, Pereira et al. [28] established the pure Lagrangian model to simulate convection and proposed a method to simulate the diffusion of vorticity on the surface of research objects, in order to reveal the unsteady aerodynamic characteristics of a horizontal axis wind turbine. The establishment and verification of a time-accurate Lagrangian vortex wake model [29], and the Lagrangian numerical simulation method [30] can be employed to analyze the flow around a circular cylinder. Furthermore, a meshless three-dimensional flow simulation method [31] based on boundary integral and vortex method of Lagrangian differential equation was constructed. The propose of a new large eddy simulation method [32] can directly distinguish the sub grid motion of concentrated vortices.

In this paper, a kind of new parallel groove design scheme based on NACA0009 hydrofoil is proposed, numerical simulations are performed using the SST k-ω model, and the accuracy of the numerical method is validated. On this basis, the vortex structure and flow field of various groove configurations are compared, the mechanism of suppression on hydrofoil TLV by parallel groove is explored, and the analysis of energy characteristics and flow characteristics in hydrofoil is carried out to obtain the best optimized structure.

2. Computational Fluid Domain and Groove Design

2.1. Computational Fluid Domain

In this paper, the NACA0009 hydrofoil is selected for numerical simulation based on the previous experimental basis [33], and a three-dimensional model is established according to the experimental conditions, of which the model structure is shown in Figure 1. The basic parameters of the geometric model are as follows: the total length of the water tank l is 750 mm, the distance from the inlet to the leading edge of the hydrofoil is 200 mm, the distance from the outlet to the trailing edge of the hydrofoil is 450 mm, the width of the water tank w is 150 mm, the height of the water tank h is 50 mm, the chord length of the hydrofoil c is 100 mm, the span of the hydrofoil b is 146 mm, and the maximum thickness of the hydrofoil is 9.9 mm. The attack angle of the hydrofoil α is 10°, the incoming flow velocity W_∞ is 10 m/s, and the tip clearance δ is 0.4 mm.

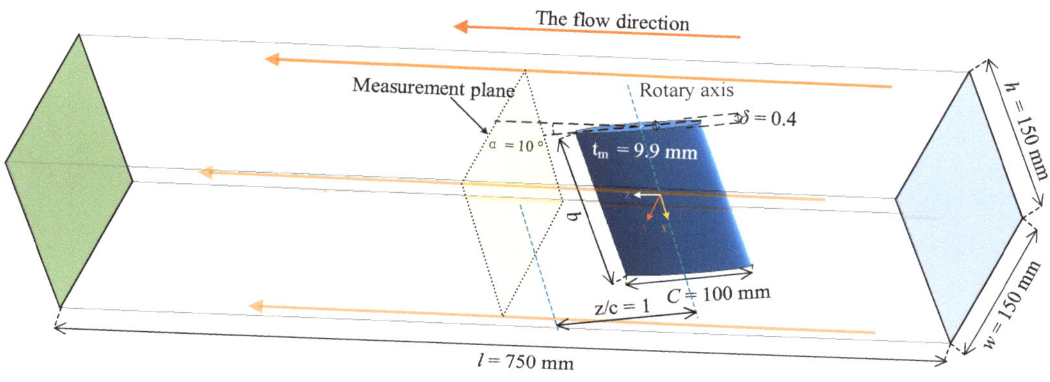

Figure 1. Computational fluid domain.

2.2. Design Scheme of Parallel Groove

Regarding the research on the optimal design of hydraulic machinery, previous authors adopted relevant methods such as multi-objective and multi-parameter design to optimize the flow pattern and enhance the energy characteristics [34]. As regard to the study of flow control in hydrofoil, Huang et al. [35] proposed a C-groove design, a structure that can reduce quadrupole noise as well as suppressing TLV. On this basis, the role of C-groove for hydrofoil energy performance enhancement was also investigated [36]. A new type of groove suspended at the tip of the hydrofoil was also proposed [37], which was composed of a transverse groove along the direction of the chord and multiple sets of oblique grooves, and the suppression effect on TLV of the grooves was validated by experiments and numerical simulations.

Due to the different locations of the primary tip leakage vortex (PTLV) and secondary tip leakage vortex (STLV) [38], a single groove can block only one kind of the TLV, while too many grooves can drastically affect the performance of the hydrofoil. Therefore, a parallel grooves structure is proposed in this paper, and the location of the groove is mainly concentrated at 5–30% of the chord length, the location where the tip leakage vortex is mainly generated. The shape of the single groove is a linear groove, the inclination angle of the groove is 45°, the depth of the groove δ_1 is half of the tip clearance, the vertical length of the groove d_1 is 3 mm, and the maximum length of the groove boundary d_2 is $\sqrt{2}d_1$, which is $1/23c$. The distance between the center of the first groove and the leading edge is x_1, and the distance between the center of the second groove and the leading edge is x_2.

In contrast, the structure of the parallel grooves proposed in this paper as shown in Figure 2 chose a double groove configuration, which can destruct the flow pattern of STLV while suppressing the development of PTLV, thus playing a significant role in suppressing the tip leakage vortex of the tip clearance. At the same time, the parallel grooves structure does not change the overall structure of the hydrofoil to a large extent, thus has relatively little negative effect on its hydrodynamic performance.

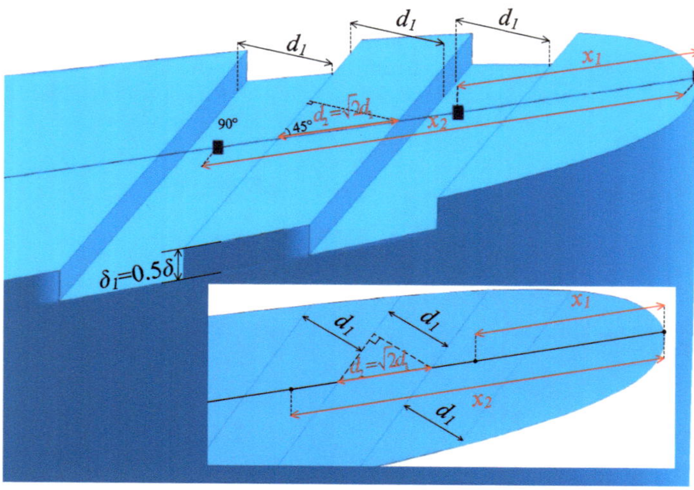

Figure 2. Structure parameter illustration of parallel groove.

In order to concentrate on the effect of parallel grooves on vortex suppression and flow characteristics, the qualitative and quantitative study is conducted on the location and spacing of parallel grooves. The distance between the center of the first groove and the leading edge x_1 is defined as the position, and three schemes are designed for the positions of the groove, which are located at 2/23, 3/23 and 4/23 of chord length, corresponding to L2, L3 and L4. Since the TLV is mainly located at the 0–30% chord length, to ensure the suppression effect of the grooves and also consider the interaction between the parallel grooves, the design of spacing is extremely important. The spacing of parallel grooves in this paper are mainly chosen as 1/23, 2/23, and 3/23 of chord length, corresponding to T1, T2, and T3. Overall, it is necessary to ensure that the parallel grooves have a suppressing effect on both PTLV and SPLV, and the vortex structures of the two grooves will not be sucked by each other to produce a larger vortex structure. The specific configurations are shown in Table 1.

Table 1. Design scheme of parallel grooves.

Location \ Spacing	1/23c	2/23c	3/23c
2/23c	L2T1	L2T2	L2T3
3/23c	L3T1	L3T2	L3T3
4/23c	L4T1	L4T2	L4T3

3. Mesh Generation and Numerical Method

3.1. Mesh Arrangement

The commercial software ANSYS ICEM 18.0 has been employed to generate structured meshes, and meanwhile, in order to ensure the calculation accuracy and reduce the calculation error, the mesh number irrelevance verification is carried out in this paper. Three sets of meshes are established, which are Mesh1, Mesh2 and Mesh3 with the mesh number 4.2×10^6, 5.2×10^6 and 6.2×10^6, respectively. The lift-to-drag force of the NACA0009 hydrofoil with different mesh number is shown in Table 2. The calculation results reveal that the value of lift-to-drag force remains stable when the mesh number reaches 5.2×10^6, which can be assumed that the mesh above 5.2×10^6 has reached the accuracy requirement. In order to capture the TLV in the tip clearance with as much detail as possible, Mesh3 has been chosen for subsequent calculations.

Table 2. Lift-to-drag under different mesh number.

Mesh Scheme	Mesh1	Mesh2	Mesh3
Mesh number	4.2×10^6	5.2×10^6	6.2×10^6
Lift	852.21	860.85	861.05
Drag	40.82	39.91	39.79
Lift-Drag	20.88	21.57	21.64

The details of node distribution of Mesh3 are represented in Figure 3. To improve the quality of the mesh, the surrounding grid of hydrofoil is treated with an O-block topology [39], and the mesh layer on the hydrofoil surface is 73, with y^+ values between 1–10 on the hydrofoil. Since the study in this paper focuses on the TLV and to capture the TLV structure as accurately as possible, the mesh of tip clearance is refined with 58 layers. The presence of the groove structure leads to the emergence of a nascent groove leakage vortex (GLV) inside the groove. In order to observe the development of the GLV, the interior mesh of the groove is refined with 20 layers. Furthermore, 138 layers are arranged along the chord length of the hydrofoil, and 78 layers are sped along the span of the hydrofoil.

3.2. Numerical Setting

The commercial software ANSYS CFX18.0 is adopted for numerical simulation, and the SST k-ω turbulence model is selected. In this case, Reynolds number is computed by Re = $\varrho cv/\mu$, where v is the inlet velocity, ϱ is the fluid density, μ is the viscosity, and c is the chord length of the hydrofoil, and the value of Re is 1×10^6. According to the experimental results, the boundary conditions of the numerical simulation are set as velocity inlet and static pressure outlet. In addition, a combination of high-resolution convective discretization and high-resolution turbulence numerical simulation is used in the calculation process. The convergence criterion is defined as the root mean square residual, which is less than 1×10^{-5}.

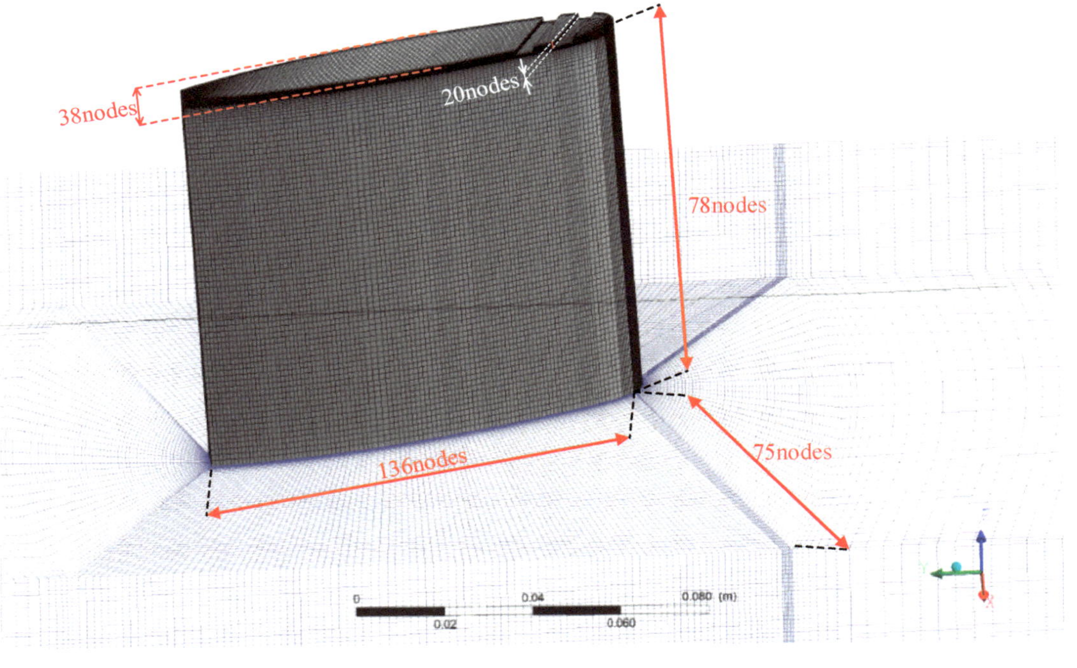

Figure 3. *Cont.*

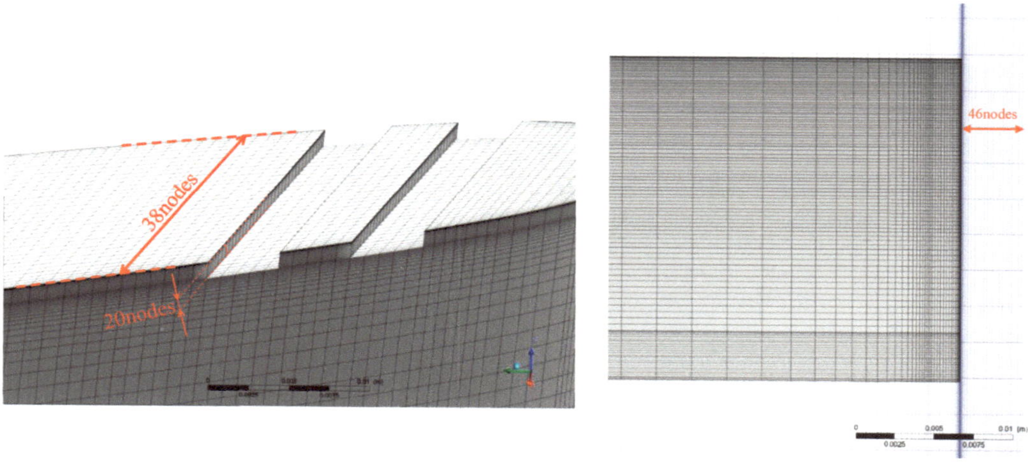

Figure 3. Mesh arrangement of hydrofoil.

4. Results and Discussion

4.1. Numerical Method Validation

In order to verify the accuracy of the numerical method, the calculated results are compared and analyzed with the experimental results in this paper, as displayed in Figures 4 and 5. The relevant parameters and working conditions for the comparison condition are as follows: 4 mm for tip clearance, 10 m/s for inlet velocity, and 10° for attack angle of the hydrofoil. The vortex defined by the Q criterion ($Q = 2 \times 10^7\ s^{-2}$) and the TLV trajectory portrayed in the experimental measurement can be seen in Figure 4. As the vortex in numerical simulation basically match with the experimentally obtained PTLV and SPLV trajectories, the numerical method is effective in predicting the location of the tip leakage vortex.

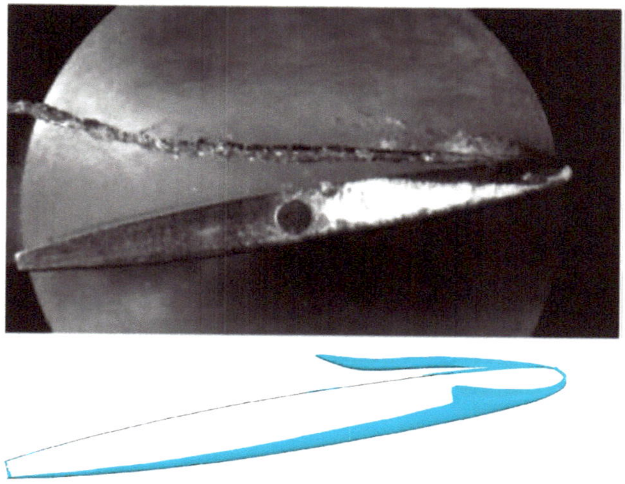

Figure 4. Comparison of vortex in experimental and numerical results.

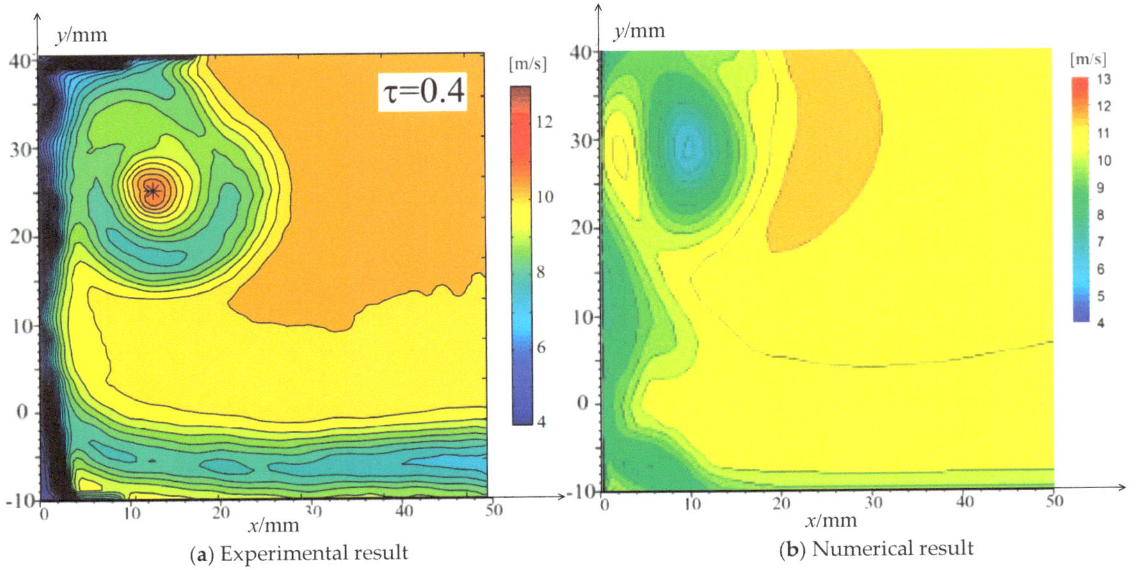

Figure 5. Comparison of velocity field in experimental and numerical results.

Meanwhile, in order to intuitively compare the error range in more detail, the velocity contour of the cross-section $z/c = 1$ are selected in this paper. As can be seen in Figure 5a,b, the numerical simulation is relatively successful in predicting the basic positions of the high-speed region and the low-velocity region. Although there is a certain error between the numerical simulation and the experimental results, which may be caused by the fact that the surface of the hydrofoil in the experiments is not absolutely smooth, but the error range is not large, therefore it is convinced that the accuracy of the SST k-ω turbulence model is verified.

4.2. Comparison of Groove Location and Spacing

In this paper, the effect of parallel grooves on the TLV is investigated from two aspects, the location and the spacing of the parallel grooves. Based on NACA0009 (M0), nine sets of different optimization schemes are designed, and Figure 6 represents the vortex structure for each set of optimization schemes under the definition of Q criterion ($Q = 2 \times 10^7\ s^{-2}$).

As revealed in the figure, there exists two main vortex structures generated by the hydrofoil in M0, the PTLV (primary tip leakage vortex) and the STLV (secondary tip leakage vortex). When there is a groove at the top of the tip clearance, the structure of PTLV is destructed and the vortex is cut off largely in the groove. Besides, the presence of the groove induces a GLV (groove leakage vortex), the GLV and the TLV are induced together to form BGLV (back groove leakage vortex). The interaction between GLV and PTLV also induces NPTLV (new primary tip leakage vortex). However, it is obvious from the figure that the GLV and BGLV are relatively small in diameter and length, while the area of NPTLV is reduced by half compared with PTLV due to the truncation effect of the grooves. Therefore, it can be concluded that the parallel groove structure has a good suppression effect on the tip leakage vortex.

As shown in configuration of L2T1, L2T2 and L2T3, the GLV and PTLV coil with each other and induce the generation of BGLV. When spacing between grooves increases, the diameter and length of BGLV gradually decreases and the vortex intensity becomes smaller. At the same time, as represented in cases L4T1, L4T2 and L4T3, the delayed grooves provide some space for the PTLV development, and entrainment with GLV to induce the generation of NPTLV. Furthermore, the BGLV can be generated only when the groove spacing is small. As revealed in configuration of L3T1, L3T2, and L3T3, the generation of PTLV is obviously suppressed, and its main vortex structure is produced within the co-development of GLV and STLV.

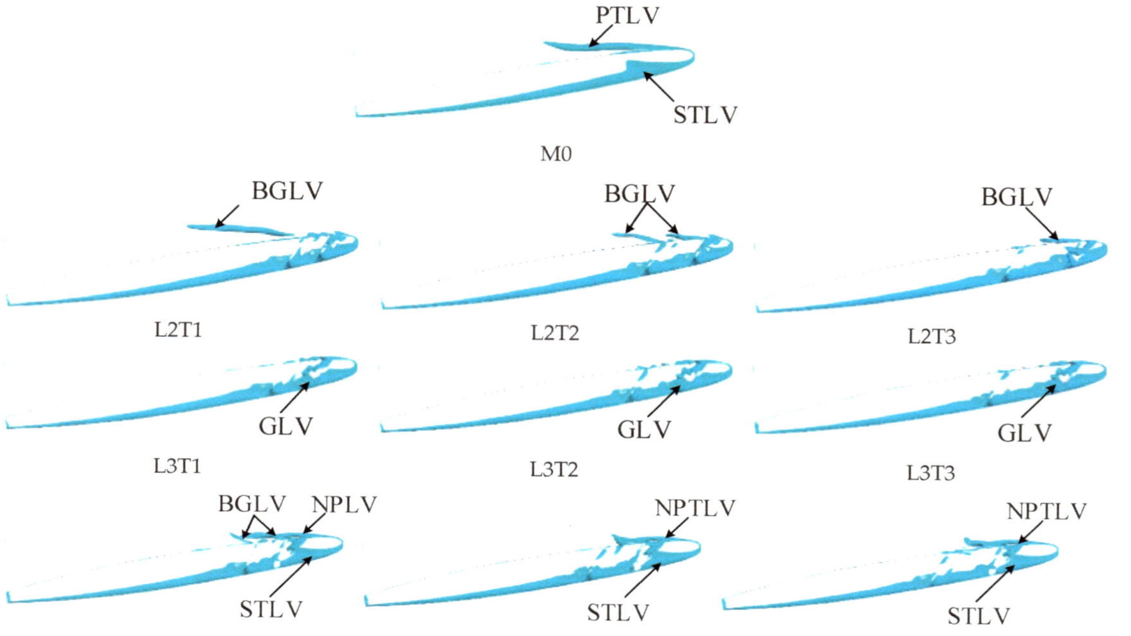

Figure 6. The vortex structure shown by Q criterion.

The vortex area and energy performance of different optimization schemes are listed in Tables 3 and 4. It can be indicated from Table 3 that the schemes with obvious suppression effect on TLV are L2T2, L2T3, L3T1, L3T2, and L3T3, among which the L3T1 scheme has the best suppression effect on TLV. It can be seen from Table 4 that although the configurations with the parallel grooves show a slightly decreasing trend in the lift-drag ratio, the reduction ranges are all within 5%. Moreover, the lift shows an increasing trend, and the lift-drag ratio decreases by only 2.35% in the L2T3, which has the best lift-drag ratio performance among various optimized schemes. Considering the effect of parallel grooves on the suppressing effect of the TLV and the energy performance of hydrofoil, the schemes with better performance are L2T2, L2T3, L3T1 and L3T3. This is because the parallel grooves structure can reduce the effect of TLV and improve the pressure gradient distribution in the pressure and suction sides of the hydrofoil, thus improving the flow pattern while causing less impact on the energy characteristics.

Table 3. Area of vortex for different schemes.

Scheme / Area (mm^2)	M0	L2T1	L2T2	L2T3	L3T1	L3T2	L3T3	L4T1	L4T2	L4T3
Area of vortex	2514	2566	2501	2484	2440	2480	2473	2550	2553	2547

Table 4. Energy performance for different schemes.

Item	M0	L2T1	L2T2	L2T3	L3T1	L3T2	L3T3	L4T1	L4T2	L4T3
Lift-drag	21.64	20.78	20.74	21.13	20.69	20.63	20.68	20.53	20.47	20.70
Change of Lift-drag (%)		−4.00	−4.15	−2.35	−4.38	−4.62	−4.42	−5.13	−5.04	−4.35
Drag	39.79	41.36	41.42	41.03	41.54	41.60	41.51	41.76	41.86	41.46
Lift	861.1	859.2	859.1	867.0	859.5	858.6	858.6	857.3	856.9	858.1
Change of Lift (%)		−0.22	−0.35	0.45	−0.53	−0.75	−0.87	−1.13	−1.29	−1.27

4.3. Analysis on Flow Characteristics and Vortex

Although four schemes represent better suppression effect on TLV, they also significantly affect the flow pattern around the hydrofoil. An undesirable structural design may induce unstable flow field, which reduces the energy performance of the hydrofoil. Therefore, it is necessary to investigate the flow characteristics of hydrofoils. As indicated in Figure 7, the three-dimensional streamline is released from the edge of the pressure surface of the hydrofoil. It can be seen from the figure that the starting position of PTLV is about 10–20%c in the original structure M0, while STLV is formed near the pressure side at 15%c, and the vortex intensity gradually decreases with the development of the TLV. Comparing to the configurations of L2T2, L2T3, L3T1 and L3T3, the starting position of the TLV tends to move forward significantly when the parallel grooves exist, the vortex intensity in the high-speed area decreases significantly, and the high-speed area becomes larger. For configurations of L2T3 and L3T3, when the spacing between the parallel grooves is certain, the tendency of the TLV position to move backward decreases when the groove positions move backward. The effect of increasing the area of the high-speed region is more obvious when the spacing increases. At the same time, the flow in the low-speed region is entangled with the PTLV in the high-speed region because of the stronger entrainment effect, which reduces the flow velocity.

Comparing the configurations of L2T2, L2T3 and L3T1, L3T3, increasing the spacing between the grooves affects the overall flow direction of the fluid at the tip clearance, with a relatively small effect on the entanglement strength of the PTLV, and the location where the fluid entangled into the PTLV is delayed. As for the M0 condition, the velocity of the fluid being entangled into the PTLV increases rapidly and then decreases gradually as the flow develops. The presence of the groove allows the fluid being entangled from the groove remains at low velocity.

In order to analyze the flow field inside the tip clearance and groove with more details, six cross-sections are selected at 10%c, 15%c, 20%c, 25%c, 30%c, and 40%c, respectively, and Figure 8 reveals the swirling strength of the cross sections. The swirling strength λ_{ci} is defined as $\lambda_{ci} = Im[eig(\nabla V)]$, where the imaginary part of eigenvalues of the velocity gradient tensor will be calculated. It can be clearly seen from the figure that the presence of parallel grooves enhances the developments of GLV compared with M0, but the swirling strength of the GLV is smaller, and the influence range of the vortex is concentrated inside the grooves. Moreover, the intensity of GLV is gradually weakened with the backward movement of the groove position when the groove spacing is certain comparing with L2T3 and L3T3.

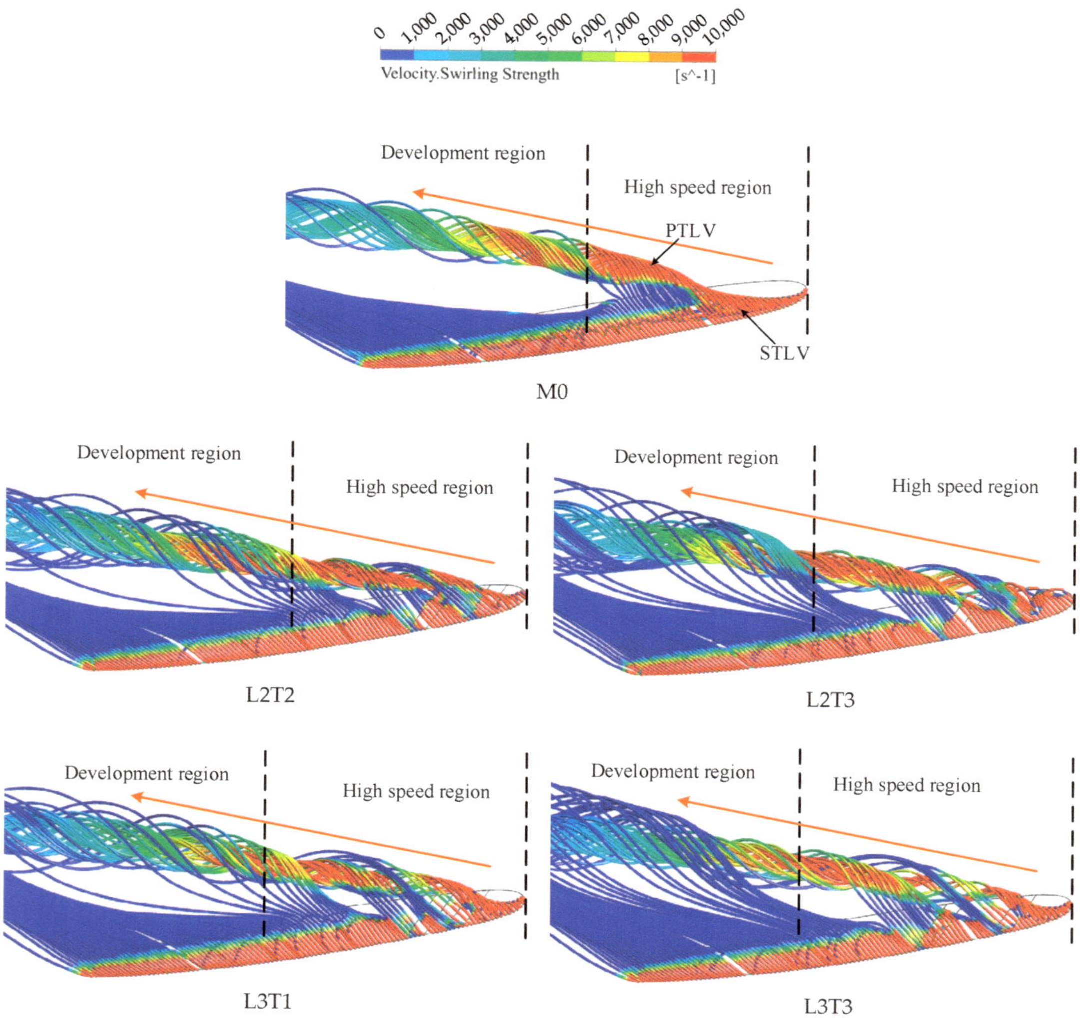

Figure 7. Three dimensional streamlines of different schemes.

Additionally, the vortex intensity of PTLV and STLV with parallel grooves is remarkably reduced both around the hydrofoil and tip clearance. The trajectory of the PTLV in M0 is marked by the orange dashed line in the figure, and it can be seen from L2T2, L2T3, L3T1, and L3T3 that the position of the vortex core in each cross-section is not significantly shifted, and the presence of the parallel grooves does not change the trajectory of the PTLV, but expands the vortex range. It can be assumed that the presence of parallel grooves makes the structures of PTLV and STLV diverge, thus suppressing the TLV.

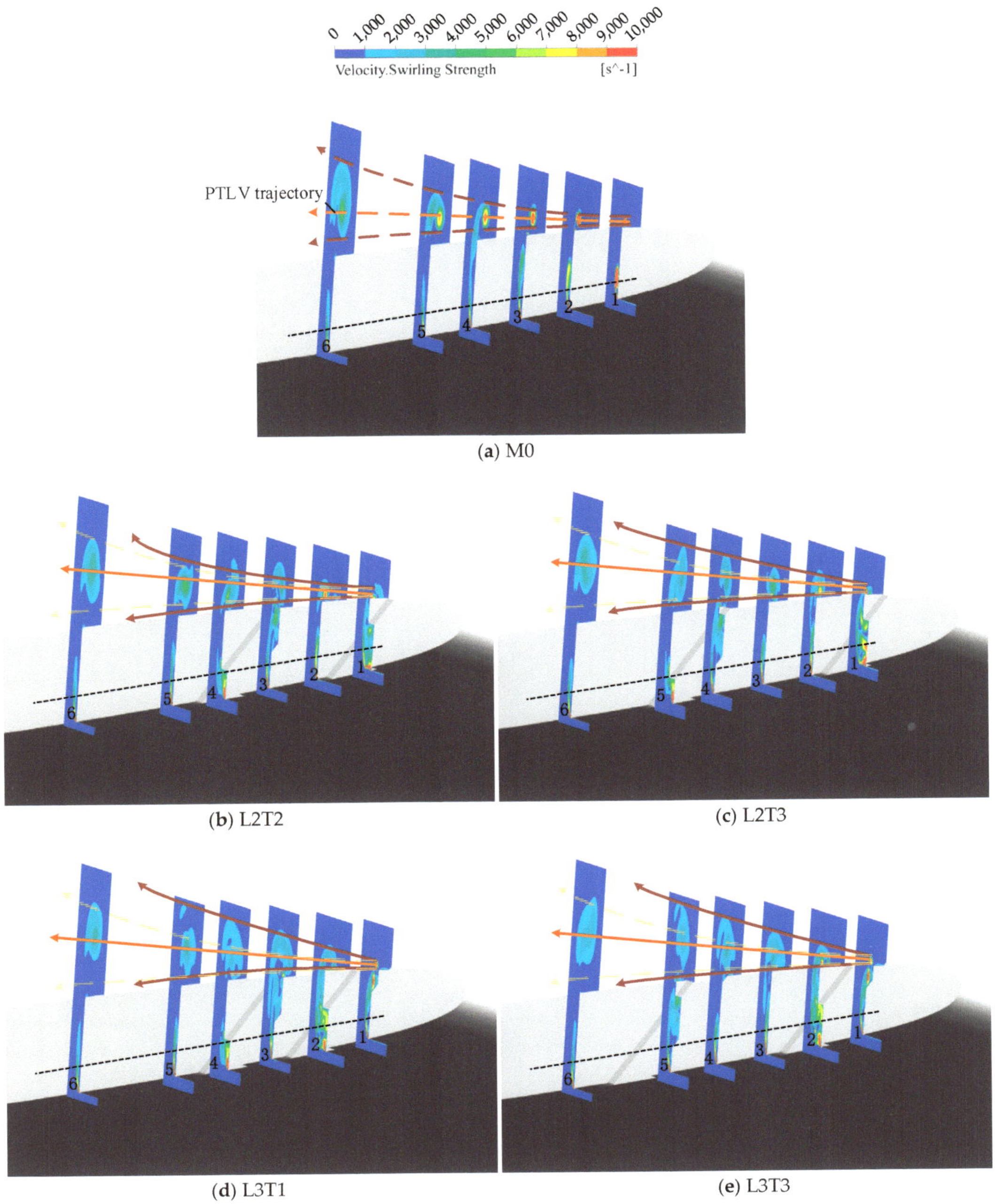

Figure 8. Distribution of velocity swirling strength on different sections.

Comparing M0 with the optimally designed configurations, the presence of the parallel grooves helps the development of the GLV, as shown in plane 4 of L2T2 and L2T3, and the swirling strength of the GLV is enhanced to some extent. With the backward movement of the groove, as shown in plane 5 of L2T3, the swirling strength of GLV is enhanced.

Comparing M0 with L3T1 and L3T3, the flow patterns shown in plane 1 are almost the same, indicating that the groove does not affect the fluid that is about to flow into the groove. Meanwhile, the presence of the groove makes the vortex intensity become lower, and the vortex intensity is significantly reduced after plane 3.

Figure 9 shows the swirling strength comparison of the vortex core in the six cross-sections within (b), (c), (d) and (e). It is obvious from the figure that the presence of the parallel grooves reduces the swirling strength of vortex core in each section compared with the M0. The scope of this effect is mainly focused on the front position of 30%c of the hydrofoil, among which the effect on Section 1 of L3T1 and L3T3 is the most prominent. The effect of the groove on the swirling strength is the smallest when the cross section is at 40%c position. Among the four configurations, the best comprehensive effect of vortex core swirling strength reduction is L3T1.

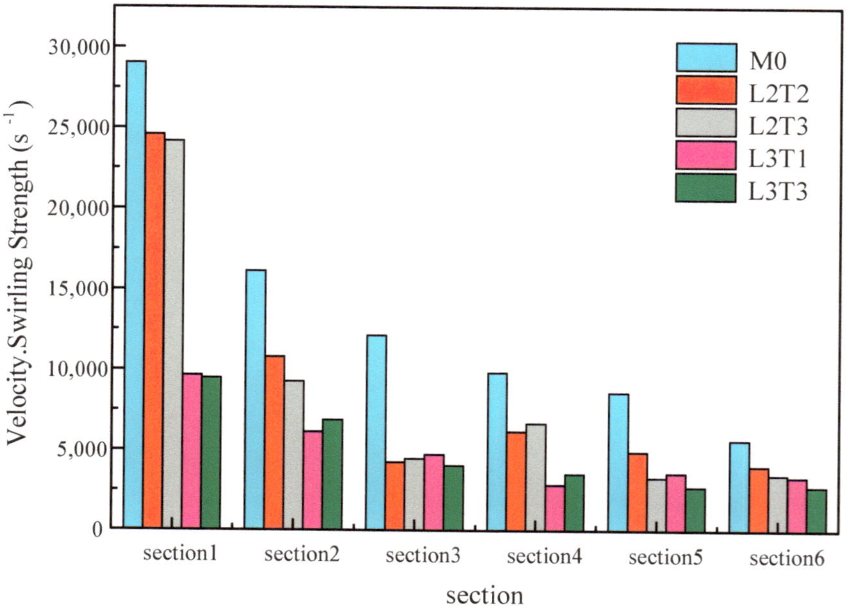

Figure 9. Swirling strength of vortex core comparison for different schemes.

In order to analyze the influence of parallel groove on the TLV in-depth, the central cross-section of the parallel groove is selected for further analysis. Figure 10 reveals the velocity distribution of the cross-section from (a) to (e), from which can be seen that an obvious high-speed zone at 0–30%c of the leading edge appears in M0, which is also the main distribution area of the TLV. The flow velocity of the TLV is high, and the velocity reaches peak value from the leading edge to the front part of the hydrofoil, and then gradually decreases but the existence of the parallel groove makes the area of high-velocity region reduced, the high-speed zone only exists at 10%c of the leading edge, and the flow velocity increases slowly after passing through the groove. A local low-velocity region appears inside the groove, and the flow velocity inside the second groove is lower compared with the first groove, therefore, the velocity of flowing through the grooves is significantly affected, which has a more prominent effect on the flow field around the hydrofoil.

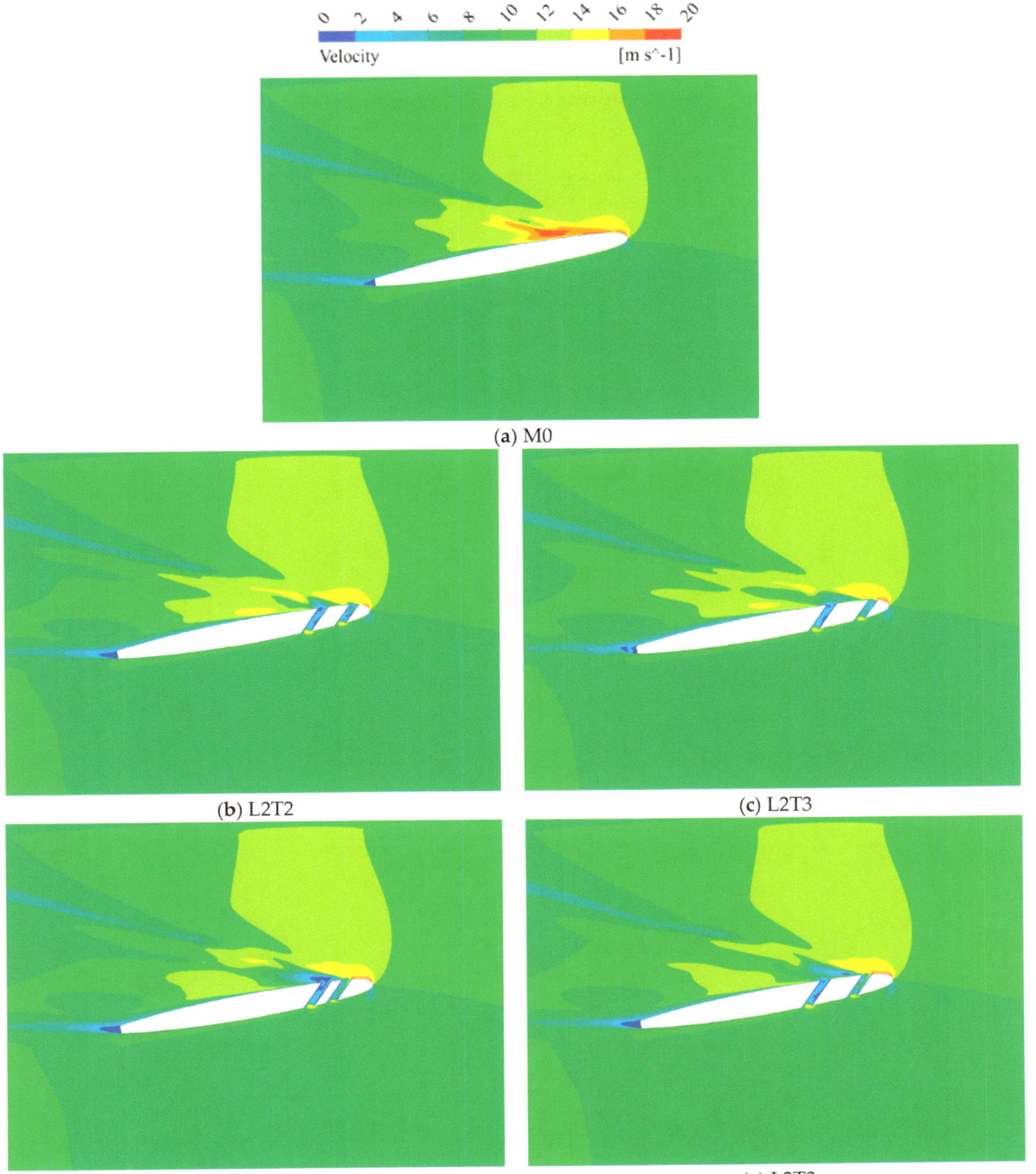

Figure 10. Velocity distribution for different schemes.

Comparing the L2T3 and L3T3, the area of the high-speed region shows a small increase to the end of the first groove when the groove moves backward. The influence of the low-speed region in the groove increases further, and a local low-speed region appears in the suction side. Compared with the L2T2 and L2T3, the grooves in the L3T1 and L3T3 are farther away from the leading edge of the hydrofoil, and the velocity in the grooves is smaller, so it is indicated that the local area with smaller velocity can be produced by appropriately moving the position of the first groove backward. The grooves change the velocity distribution around the hydrofoil, resulting in a change in the pressure difference between the suction side and pressure side, which affects the flow field around the hydrofoil.

Figure 11 represents the static pressure distribution in the center cross-section of the grooves, from which it can be seen that the parallel grooves have less influence on the pressure surface from (a) to (e). Only a small high-pressure area exists on the pressure side in the leading edge of the hydrofoil for M0. When the parallel grooves exist, a low-pressure area appears inside the groove, while a small high-pressure area appears below the grooves. As the spacing between grooves increases, the value in the low-pressure region inside the grooves increases slightly. The influence of the grooves on the suction side is greater, making the range and amplitude of the local low-pressure area at the leading edge decrease.

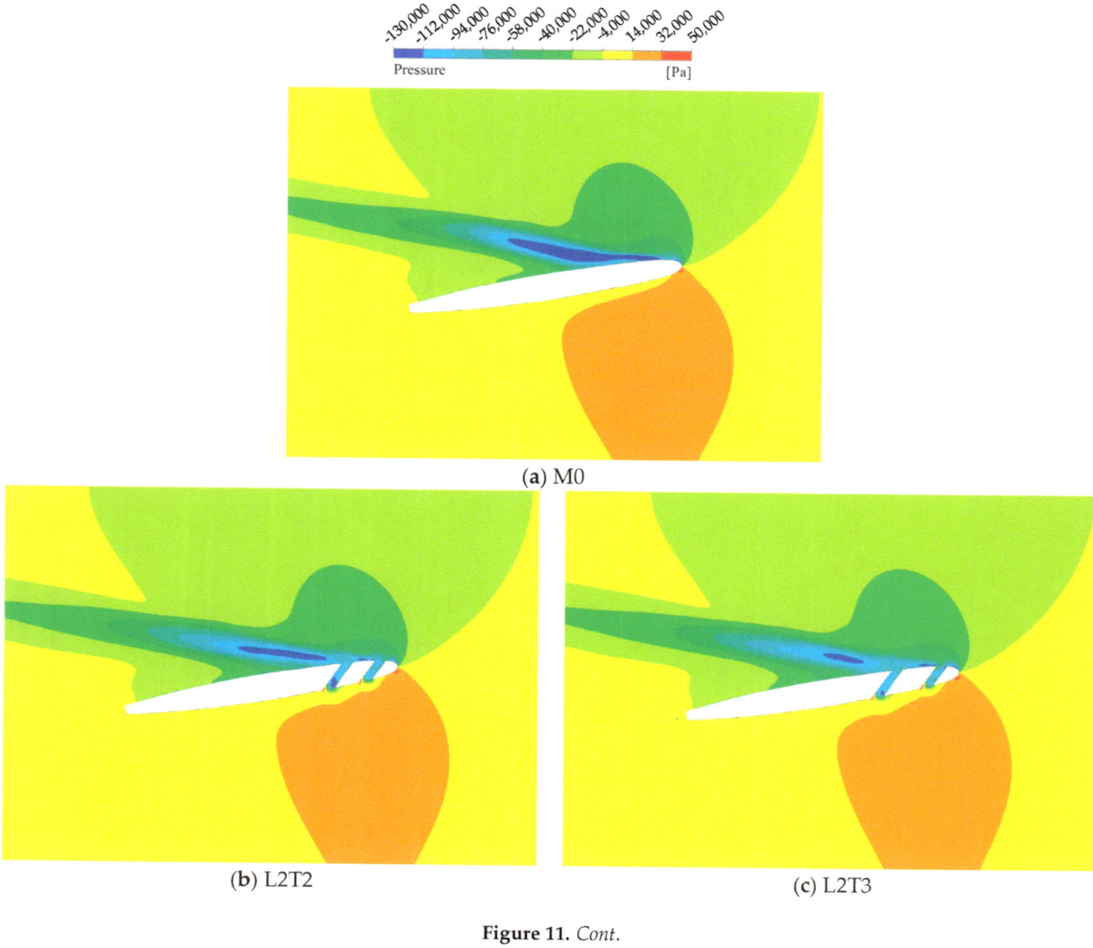

(**a**) M0

(**b**) L2T2

(**c**) L2T3

Figure 11. *Cont.*

(d) L3T1 (e) L3T3

Figure 11. Pressure distribution for different schemes.

The presence of the parallel grooves reduces the fluid velocity at the suction side, decreases the dynamic pressure in this region, and thus reduces the pressure difference between the pressure side and the suction side. The grooves decrease both the low-pressure region at the suction side and the high-pressure region at the pressure side, which reduces the static pressure difference between the two sides. Therefore, the existence of the parallel grooves reduces the total pressure difference to a certain extent, thus affecting the energy characteristics of the hydrofoil.

5. Expectation

In future research, the mechanism about the impact of the parallel grooves on the tip leaking vortex can be more systematically studied and analyzed, so as to provide a scientific basis for a more accurate understanding of the depth and position of the parallel groove. Secondly, a more detailed parametric study of the groove spacing can be carried out to explore the mechanism of groove–groove interaction in parallel grooves.

6. Conclusions

In response to the problem of performance degradation in hydraulic machinery due to the existence of tip clearance, a parallel grooves structure is proposed in this paper and applied to the NACA0009 hydrofoil. Subsequently, an exhaustive numerical analysis of parallel grooves is carried out using the SST k-ω turbulence model, and the impact of the position and spacing of the parallel grooves on the suppression effect and flow characteristics are investigated. The main conclusions are obtained as follows.

(1) Compared with the structure without parallel grooves, the presence of the grooves will lead to a decrease in the lift-to-drag ratio of the hydrofoil within 5%, but it can effectively suppress the development TLV and reduce the area of TLV. This is of great significance for improving the flow stability and studying the optimal design of hydraulic machinery.

(2) The location and spacing of the parallel grooves have a greater effect on the flow characteristics of the hydrofoil. With the backward movement of the groove location, the tendency of TLV to move backward slows down, and the vortex intensity of the GLV gradually decreases, which can produce a local area with smaller velocity. With the increase of the spacing, the effect of increasing the high-speed region is more obvious, and the flow in the low-speed region is mixed with the PTLV in the high-speed region.

(3) The parallel grooves destroy the structure of PTLV and STLV, and inhibits the development of TLV. In order to balance the TLV suppression effect and the energy performance of the hydrofoil, the L3T1 structure is recommended.
(4) The presence of parallel grooves leads to a local low-pressure area, which affects the pressure difference between the pressure side and the suction side, which in turn affects the work done by the hydrofoil and slightly reduces the lift-to-drag ratio.

Author Contributions: Conceptualization, Y.C. and W.Z.; methodology, D.S.; software, Y.C.; validation, D.F., M.S. and J.L.; formal analysis, Y.C., W.Z.; investigation, D.F.; resources, X.Z.; data curation, J.L.; writing—original draft preparation, Y.C., W.Z.; writing—review and editing, D.S., X.Z. All authors have read and agreed to the published version of the manuscript.

Funding: This work has been supported by the National Natural Science Foundation of China [52077218].

Conflicts of Interest: The authors declare no conflict of interest.

References

1. Zhang, W.; Xie, X.; Zhu, B.; Ma, Z. Analysis of phase interaction and gas holdup in a multistage multiphase rotodynamic pump based on a modified Euler two-fluid model. *Renew. Energy* **2021**, *164*, 1496–1507. [CrossRef]
2. Zhang, W.; Zhu, B.; Yu, Z. Characteristics of bubble motion and distribution in a multiphase rotodynamic pump. *J. Pet. Sci. Eng.* **2020**, *193*, 107435. [CrossRef]
3. Yan, H.; Su, X.; Zhang, H.; Huang, J.; Zhou, L.; Liu, Z.; Wang, Z. Design approach and hydrodynamic characteristics of a novel bionic airfoil. *Ocean Eng.* **2020**, *216*, 108076. [CrossRef]
4. Bai, X.; Cheng, H.; Ji, B.; Long, X. Spatial and spectral investigation of turbulent kinetic energy in cavitating flow generated by Clark-Y hydrofoil. *J. Hydrodyn.* **2020**, *32*, 175–178. [CrossRef]
5. Kinsey, T.; Dumas, G.; Lalande, G.; Ruel, J.; Mehut, A.; Viarouge, P.; Lemay, J.; Jean, Y. Prototype testing of a hydrokinetic turbine based on oscillating hydrofoils. *Renew. Energy* **2011**, *36*, 1710–1718. [CrossRef]
6. Kinsey, T.; Dumas, G. Computational fluid dynamics analysis of a hydrokinetic turbine based on oscillating hydrofoils. *J. Fluids Eng.* **2012**, *134*, 021104. [CrossRef]
7. Zhang, W.; Chen, Z.; Zhu, B.; Zhang, F. Pressure fluctuation and flow instability in S-shaped region of a reversible pump-turbine. *Renew. Energy* **2020**, *154*, 826–840. [CrossRef]
8. Zhang, P.; Duan, M.; Ma, J. A precise mathematical model for geometric modeling of wire rope strands structure. *Appl. Math. Model.* **2019**, *76*, 151–171. [CrossRef]
9. Zhang, P.; Ma, J.; Duan, M.; Yuan, Y.; Wang, J. A high-precision curvature constrained Bernoulli–Euler planar beam element for geometrically nonlinear analysis. *Appl. Math. Comput.* **2021**, *397*, 125986.
10. Dang, Z.; Mao, Z.; Tian, W. Reduction of hydrodynamic noise of 3d hydrofoil with spanwise microgrooved surfaces inspired by sharkskin. *J. Mar. Sci. Eng.* **2019**, *7*, 136. [CrossRef]
11. Zhang, L.; Duan, J.; Da, L.; Xu, G.; Sun, X. Vibroacoustic radiation and propagation properties of slender cylindrical shell in uniform shallow sea. *Ocean Eng.* **2020**, *195*, 106659. [CrossRef]
12. Bai, X.; Cheng, H.; Ji, B.; Long, X.; Qian, Z.; Peng, X. Comparative Study of different vortex identification methods in a tip-leakage cavitating flow. *Ocean Eng.* **2020**, *207*, 107373. [CrossRef]
13. Chen, F.; Zhang, L.; Huai, X.; Li, J.; Zhang, H.; Liu, Z. Comprehensive performance comparison of airfoil fin PCHEs with NACA 00XX series airfoil. *Nucl. Eng. Des.* **2017**, *315*, 42–50. [CrossRef]
14. Hoseinzadeh, S.; Bahrami, A.; Mirhosseini, S.M.; Sohani, A.; Heyns, S. A detailed experimental airfoil performance investigation using an equipped wind tunnel. *Flow Meas. Instrum.* **2020**, *72*, 101717. [CrossRef]
15. Miorini, R.L.; Wu, H.; Katz, J. The internal structure of the tip leakage vortex within the rotor of an axial waterjet pump. *J. Turbomach.* **2012**, *134*, 031018. [CrossRef]
16. Zhang, J.; Fan, H.; Zhang, W.; Xie, Z. Energy performance and flow characteristics of a multiphase pump with different tip clearance sizes. *Adv. Mech. Eng.* **2019**, *11*, 1687814018823356. [CrossRef]
17. Jung, Y.; Choi, M.; Park, J.Y.; Baek, J.H. Effects of recessed blade tips on the performance and flow field in a centrifugal compressor. *Proc. Inst Mech. Eng. Part A J. Power Energy* **2013**, *227*, 157–166. [CrossRef]
18. Decaix, J.; Balarac, G.; Dreyer, M.; Farhat, M.; Munch, C. RANS and LES computations of the tip-leakage vortex for different gap widths. *J. Turbul.* **2015**, *16*, 309–341. [CrossRef]
19. Zhou, K.; Zhou, C. Unsteady effects of vortex interaction on tip leakage vortex breakdown and its loss mechanism. *Aerosp. Sci. Technol.* **2018**, *82*, 363–371. [CrossRef]
20. Boudet, J.; Jacob M, C.; Caro, J.; Jondeau, E.; Li, B. Wavelet Analysis of a Blade Tip-Leakage Flow. *AIAA J.* **2018**, *56*, 3332–3336. [CrossRef]
21. Wang, L.; Lu, J.; Liao, W.; Zhao, Y.; Wang, W. Numerical simulation of the tip leakage vortex characteristics in a semi-open centrifugal pump. *Appl. Sci.* **2019**, *9*, 5244. [CrossRef]

22. Shen, X.; Zhang, D.; Xu, B.; Shi, W.; BPM, v.E. Experimental and numerical investigation on the effect of tip leakage vortex induced cavitating flow on pressure fluctuation in an axial flow pump. *Renew. Energy* **2021**, *163*, 1195–1209.
23. Seo, S.H.; Hong C, H. Performance improvement of airfoils for wind blade with the groove. *Int. J. Green Energy* **2016**, *13*, 34–39. [CrossRef]
24. Liu, Y.; Li, P.; He, W.; Jiang, K. Numerical study of the effect of surface grooves on the aerodynamic performance of a NACA 4415 airfoil for small wind turbines. *J. Wind Eng. Ind. Aerodyn.* **2020**, *206*, 104263. [CrossRef]
25. Yang, X.; Zhao, Q.; Liu, Z.; Feng, Z.; Simon, T. Controlling Leakage Flows Over a Rotor Blade Tip Using Air-Curtain Injection: Part I—Aero-Thermal Performance of Rotor Tips. In the proceedings of Turbo Expo: Power for Land, Sea, and Air. *Am. Soc. Mech. Eng.* **2019**, *58653*, V05BT18A002.
26. Jiang, S.; Chen, F.; Yu, J.; Chen, S.; Song, Y. Treatment and optimization of casing and blade tip for aerodynamic control of tip leakage flow in a turbine cascade. *Aerosp. Sci. Technol.* **2019**, *86*, 704–713. [CrossRef]
27. Sarpkaya, T. Computational methods with vortices—The 1988 Freeman scholar lecture. *J. Fluids Eng.* **1989**, *111*, 5–52. [CrossRef]
28. Pereira, L.A.A.; Hirata, M.H.; Silveira Neto, A. Vortex method with turbulence sub-grid scale modelling. *J. Braz. Soc. Mech. Sci. Eng.* **2003**, *25*, 140–146. [CrossRef]
29. Gupta, S. Development of a Time-Accurate Viscous Lagrangian Vortex Wake Model for Wind Turbine Applications. Ph.D. Thesis, University of Maryland, College Park, MD, USA, 2006.
30. Hirata, M.H.; Pereira, L.A.A.; Recicar, J.N.; Moura, W.H. High Reynolds number oscillations of a circular cylinder. *J. Braz. Soc. Mech. Sci. Eng.* **2008**, *30*, 304–312. [CrossRef]
31. Khatir, Z.; Lucey, A.D. A combined boundary integral and vortex method for the numerical study of three-dimensional fluid flow systems. *Int. J. Comput. Math.* **2012**, *89*, 1504–1524. [CrossRef]
32. Kornev, N.; Samarbakhsh, S. Large Eddy Simulation with direct resolution of subgrid motion using a grid free vortex particle method. *Int. J. Heat Fluid Flow* **2019**, *75*, 86–102. [CrossRef]
33. Dreyer, M.; Decaix, J.; Münch-Alligné, C.; Farhat, M. Mind the gap: A new insight into the tip leakage vortex using stereo-PIV. *Exp. Fluids* **2014**, *55*, 1–13. [CrossRef]
34. Fan, H.; Zhang, J.; Zhang, W.; Fharhat, M. Multiparameter and multiobjective optimization design based on orthogonal method for mixed flow fan. *Energies* **2020**, *13*, 2819. [CrossRef]
35. Huang, Z.; Huang, Z.; Fan, H. Influence of C groove on energy performance and noise source of a NACA0009 hydrofoil with tip clearance. *Renew. Energy* **2020**, *159*, 726–735. [CrossRef]
36. Hu, Z.; Huang, C.; Huang, Z.; Zhang, J. A Method of Bending Shrinkage Groove on Vortex Suppression and Energy Improvement for a Hydrofoil with Tip Gap. *Processes* **2020**, *8*, 1299. [CrossRef]
37. Cheng, H.; Long, X.; Ji, B.; Peng, X.; Farhat, M. Suppressing tip-leakage vortex cavitation by overhanging grooves. *Exp. Fluids* **2020**, *61*, 1–18. [CrossRef]
38. Legras, G.; Gourdain, N.; Trebinjac, I. Numerical analysis of the tip leakage flow field in a transonic axial compressor with circumferential casing treatment. *J. Therm. Sci.* **2010**, *19*, 198–205. [CrossRef]
39. Zhang, J.; Zhang, Q.; Wang, H.; Ye, P. Study on heat transfer performance of shaft seal in HTR-PM circulator. In Proceedings of the International Conference on Nuclear Engineering (ICONE), Tsukuba, Japan, 19–24 May 2019; p. 1776.

processes MDPI

Article

A Proper Shape of the Trailing Edge Modification to Solve a Housing Damage Problem in a Gas Turbine Power Plant

Thodsaphon Jansaengsuk [1], Mongkol Kaewbumrung [2], Wutthikrai Busayaporn [3] and Jatuporn Thongsri [1,*]

1 Computer Simulation in Engineering Research Group, King Mongkut's Institute of Technology Ladkrabang, College of Advanced Manufacturing Innovation, Bangkok 10520, Thailand; thodsaphon.jan@gmail.com
2 Department of Mechanical Engineering, Faculty of Engineering and Architecture, Rajamangala University of Technology Suvarnabhumi, Phranakhon Si Ayutthaya 13000, Thailand; mongkol.k@rmutsb.ac.th
3 Synchrotron Light Research Institute (Public Organization), Nakhon Ratchasima 30000, Thailand; wutthikrai@slri.or.th
* Correspondence: jatuporn.th@kmitl.ac.th

Abstract: To solve the housing damage problem of a fractured compressor blade (CB) caused by an impact on the inner casing of a gas turbine in the seventh stage (from 15 stages), modifications of the trailing edge (TE) of the CB have been proposed, namely 6.5 mm curved cutting and a combination of 4 mm straight cutting with 6.5 mm curved cutting. The simulation results of the modifications in both aerodynamics variables C_l and C_d and the pressure ratio, including structural dynamics such as a normalized power spectrum, frequency, total deformation, equivalent stress, and the safety factor, found that 6.5 mm curved cutting could deliver the aerodynamics and structural dynamics similar to the original CB. This result also overcomes the previous work that proposed 5.0 mm straight cutting. This work also indicates that the operation of a CB gives uneven pressure and temperature, which get higher in the TE area. The slightly modified CB can present the difference in the properties of both the aerodynamics and the structural dynamics. Therefore, any modifications of the TE should be investigated for both properties simultaneously. Finally, the results from this work can be very useful information for the modification of the CB in the housing damage problem of the other rotating types of machinery in a gas turbine power plant.

Keywords: aerodynamics; compressor blade; computational fluid dynamics; computer simulation; finite element analysis; gas turbine; housing damage; power plant; structural dynamics; trailing edge

Citation: Jansaengsuk, T.; Kaewbumrung, M.; Busayaporn, W.; Thongsri, J. A Proper Shape of the Trailing Edge Modification to Solve a Housing Damage Problem in a Gas Turbine Power Plant. *Processes* **2021**, *9*, 705. https://doi.org/10.3390/pr9040705

Academic Editor: Alfredo Iranzo

Received: 23 March 2021
Accepted: 12 April 2021
Published: 16 April 2021

1. Introduction

In 2019, Thailand's electricity production relied on natural gas turbine power plants for up to 64% of total production. Therefore, the country's energy can be pushed into crisis whenever any of these power plants have been paused for maintenance. Consequently, avoiding the problems that may obstruct production—and maintaining every power plant to be at the highest efficiency—plays an important role in the task of maintenance engineers. Recently, one type of damage has been reported by the maintenance engineers of a power plant concerning the compressor blade (CB). The damage on the CB then causes an impact on the inner casing. This damage is also known as "housing damage" and is shown in Figure 1. Investigation showed that this damage locally occurs at the trailing edge (TE). According to the previous report, the straight cutting of TE modification at 1.0, 5.0, and 10.0 mm was proposed [1]. The computational fluid dynamics (CFD) simulation shed light on the aerodynamics (AD), pressure ratio, and change of the lift and drag coefficients (C_l and C_d) from the modifications. The results led to the conclusion that the most suitable modification was the straight cutting at 5 mm at the specific stage which caused pressure ratio alterations less than 0.2%, and the C_l and C_d changed less than 3.7% and 8.7%, respectively. This modification also maintained the aerodynamics at the same level as the original CB. This method has been applied to gas turbine power plants for actual usage and showed steady efficiency of the power generator, as usual. However, this simple method of cutting can be improved to prolong the lifetime of the CB and also maintain the highest efficiency. In general, the original design of the CB without the modifications should be at the optimum conditions, as the production figured the structural dynamics (SD) and AD at each circumstance into the design. These parameters have been taken into account to obtain high durability and efficiency in the generator. Computer simulation has been used in an important role to study the properties of the SD, such as structural analysis, modal analysis, thermal analysis, fatigue analysis, and AD. In the details of the SD, the structural analysis can predict the profile of stress, strain, and other factors [2–7]. For modal analysis, it has been used to figure out the natural frequency, mode shape, and total deformation, among other properties [8,9]. Thermal analysis has been applied to investigate thermal stress, thermal strain, and heat transfer [7,10]. Finally, fatigue analysis has been employed to predict the damage, life, and safety factor of the CB [11,12]. For the AD, CFD can simulate the C_l, C_d, pressure, vector flow, and other variables in the compressor stage [1,13–17]. From previous computer simulations, it has been indicated that inappropriate modifications can lead to inefficiency of the generator and also shorten its lifetime. Furthermore, in [1–17], most researchers focused their study on such conditions (e.g., steady rotational speed, pressure, temperature, and flow rate). Hence, the AD study has been separated from the SD investigation, because simultaneous investigations of both the AD and the SD can be very complex and troublesome. However, various works mentioned that changes in the AD would certainly affect the SD [12,18]. In conclusion, the most suitable modification of the CB that can deliver full efficiency and sustain its level of operation requires investigations of both the AD and the SD. It is unlikely that the previous work proposed the appropriate simulation of the CB's modification, which applied these two calculations in their work.

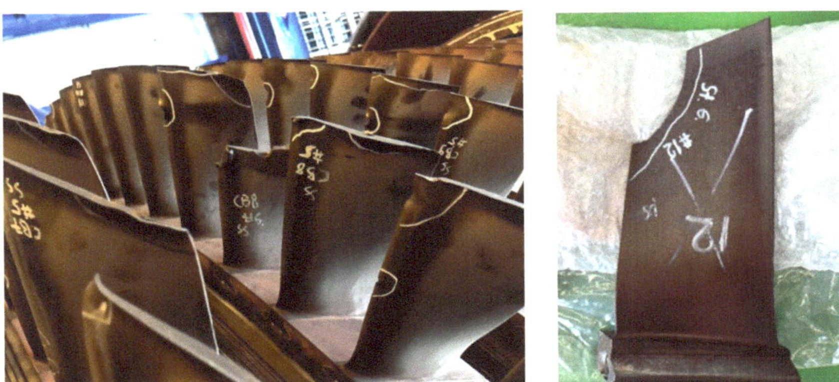

Figure 1. Housing damage on the compressor blades in a gas turbine power plant [1]. Copyright 2019 MDPI.

In this article, we propose new shapes for CB modification in stage seven, extending further than in [1]. After that, these CBs would be investigated for whether the AD were comparable to the original and the 5 mm straight cutting ones in the identical stage using ANSYS CFX R19.2 software in a transient state. The results of the AD will be consequently input into the investigation of the SD using ANSYS harmonic response analysis, structural analysis, thermal analysis, and fatigue analysis. The results of all simulations will be analyzed to figure out the most efficient modification. The feature of this work is the proposed modification, which will be investigated in terms of both the AD and SD together under the actual operating conditions of the gas turbine power plant.

2. Theoretical Background

When the gas flows into the compressor stage, the modifications of the CB will directly change the AD and its SD properties further from the factory's design. The employed software, theory, and calculation are mentioned below.

2.1. Aerodynamics

ANSYS CFX is CFD software developed via the finite volume method (FVM), based on unsteady Navier–Stokes equations. This package is widely used to solve the problems of rotating machinery in factories [1,17,19]. The AD properties of the fluid can be found by relying on the conservation of mass for Equation (1), momentum for Equation (2), and energy for Equation (3) [20] as follows:

$$\frac{\partial \rho}{\partial t} + \nabla \cdot \left(\rho \vec{U}\right) = 0 \tag{1}$$

$$\frac{\partial\left(\rho \vec{U}\right)}{\partial t} + \nabla \cdot \left(\rho \vec{U} \otimes \vec{U}\right) = -\nabla p + \nabla \cdot \tau + \vec{S}_M \tag{2}$$

where $\vec{U}$ is the mean velocity, ρ is the fluid density, p is the pressure, $\vec{S}_M$ is the source term of momentum, and τ is the stress tensor, all of which can be related to strain rate as follows:

$$\tau = \mu\left(\nabla \vec{U} + \left(\nabla \vec{U}\right)^T - \frac{2}{3}\delta \nabla \cdot \vec{U}\right)$$

and

$$\frac{\partial(\rho h_{tot})}{\partial t} - \frac{\partial p}{\partial t} + \nabla \cdot \left(\rho \vec{U} h_{tot}\right) = \nabla \cdot (\lambda \nabla T) + \nabla \cdot \left(\vec{U} \cdot \tau\right) + \vec{U} \cdot \vec{S}_M + \vec{S}_E \tag{3}$$

where $\vec{S}_E$ is the source term of energy, λ is the thermal conductivity of the fluid, and h_{tot} is the total enthalpy, which can be related to static enthalpy $h(T,p)$ as follows:

$$h_{tot} = h(T,p) + \frac{1}{2}\vec{U}^2$$

For the turbulence equation, the shear stress transport (SST) k-ω turbulence model was chosen, as in [1,20–23]. The SST k-ω equation can be written as

$$\frac{\partial}{\partial x_j}(\rho U_j k) = \frac{\partial}{\partial x_j}\left(\left(\mu + \frac{\mu_t}{\sigma k_3}\right)\frac{\partial k}{\partial x_j}\right) + P_k - 0.09 p k \omega \tag{4}$$

$$\frac{\partial}{\partial x_j}(\rho U_j \omega) = \frac{\partial}{\partial x_j}\left(\left(\mu + \frac{\mu_t}{\sigma \omega_3}\right)\frac{\partial \omega}{\partial x_j}\right) + 1.71(1 - F_1)\frac{\rho}{\omega}\frac{\partial k}{\partial x_j}\frac{\partial \omega}{\partial x_j} + \alpha_3 \frac{\omega}{k} P_k - \beta_3 \rho \omega^2 \tag{5}$$

where μ is the molecular dynamics viscosity, P_k is the shear production of turbulence, μ_t is the eddy viscosity, ω is the specific dissipation rate, k is the turbulent kinetic energy, F_1 is the blending function, and α, β, σ are the specific coefficients for the SST k-ω turbulence model.

Further details and definitions of the parameters can be found out in [1,20–23]. The ANSYS CFX software can calculate the AD, C_l, C_d, and pressure, via Equations (1)–(5). After that, the results will be used as the input for the SD for calculation later.

2.2. Structural Dynamics

Using the results for the AD from Section 2.1, which relate to the structure of the CB, this work examines the effect on the structure by using harmonic response analysis, structural analysis, thermal analysis, and fatigue analysis based on the finite element method.

In harmonic response analysis, the AD naturally causes external frequencies which act on the CB, which can be calculated by Equation (6) [24,25] as follows:

$$\left(-\omega_f^2[M] + j\omega_f[C] + [K]\right)\{u\} = \{F\} \tag{6}$$

where ω_f is the angular frequency, $[M]$ is the structural mass matrix, $[C]$ is the structural damping matrix, $[K]$ is the structural stiffness matrix, $\{u\}$ is the nodal displacement vector, and $\{F\}$ is the load vector.

In structural analysis and fatigue analysis, the AD also causes pressure on the CB in the transient state. The structural properties, such as the total deformation, equivalent stress, and safety factor, can be calculated by Equation (7) [26]:

$$[M]\{\ddot{u}\} + [C]\{\dot{u}\} + [K]\{u\} = \{F\} \tag{7}$$

where $\{\ddot{u}\}$ is the nodal acceleration vector and $\{\dot{u}\}$ is the nodal velocity vector.

In thermal analysis, the AD also causes the temperature changes in the CB, which can be expressed as follows [27]:

$$[K(T)]\{T\} = \{Q(T)\} \tag{8}$$

where T is the temperature and $[K(T)]$ and $\{Q(T)\}$ are the thermal conduction matrix and conduction vector, respectively, which are dependent on the temperature.

3. Methodology

3.1. CB Modification

The schematic diagram of all 15 compressor stages is shown in Figure 2. This figure is not in real proportions. In the simulation, the actual CB in stage seven from the gas turbine power plant was 3D scanned (FARO Edge Arm) and modified to be a solid model as in Figure 3 (left). This solid model was then modified to be as it is in Figure 3 (right) with (a) no modification, (b) 5.00 mm straight cutting, (c) 6.50 mm curve cutting, and (d) 6.50 mm curve cutting with 4.00 mm straight cutting models. These solid models were used to investigate the AD and SD properties with the software mentioned above in Sections 2.1 and 2.2 to find out the most suitable modification in the next section.

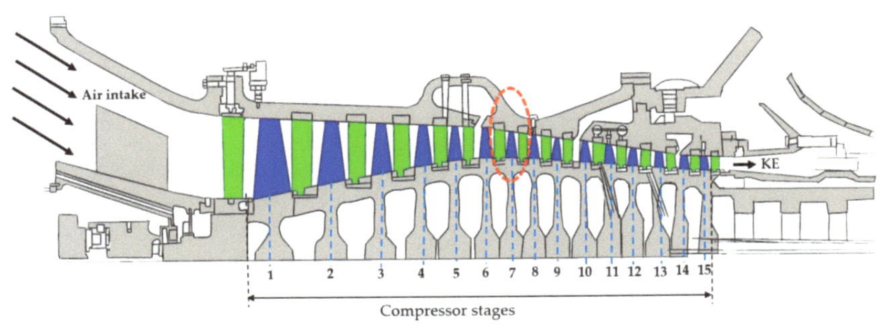

Figure 2. Schematic diagram of all 15 CBs in the compressor stages of the gas turbine power plant.

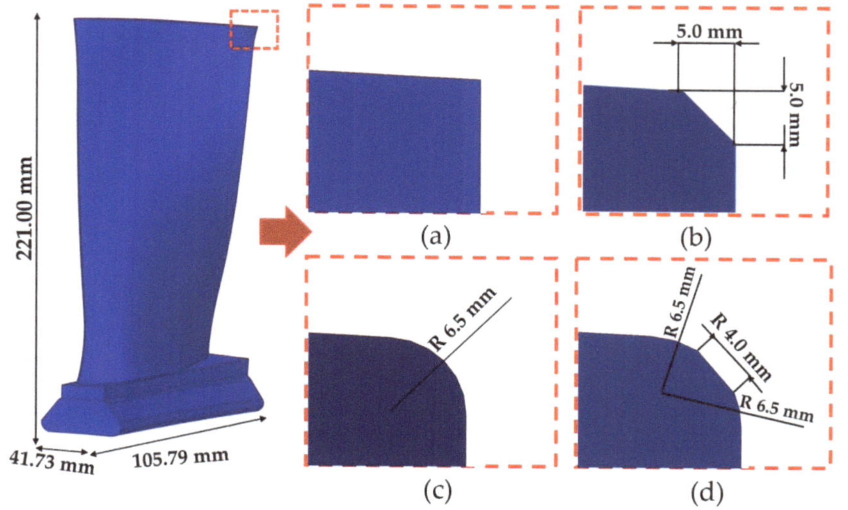

Figure 3. Solid model of the compressor blade in stage 7 from the gas turbine power plant, showing (**a**) the original design, (**b**) 5.0 mm straight cutting, (**c**) 6.5 mm curve cutting, and (**d**) 4.0 mm straight cutting with 6.5 mm curve cutting.

3.2. Models

From the models in Figure 3, the mesh model of the CB in the seventh stage was built, being composed of S6, R7, and S7. According to the complex shape of the CB, the 3D mesh model was set to be tetrahedral mesh in both the fluid's and the solid's domains. At the boundary of the two domains, an inflation was identified to be 20 layers with a y^+ of 1; the first layer thickness started at 1.41×10^{-3} mm and had a growth rate of 1.10. The inflation created wedge and prism mesh types. After that, the mesh model of the fluid domain was examined via mesh independent analysis to achieve the suitable mesh model. This model was finally adopted to build a structural mesh model later. In the independent analysis process, the maximum size of the mesh in the fluid domain was adapted to be 28–35 mm at a growth rate of 1.15. This setting caused the total elements of the mesh in both domains to be about 7.05–9.95 million elements, with the number of nodes at 2.93–3.20 million. The C_l and C_d were set to be the indicators to determine a suitable mesh model. The numbers of the elements and nodes of each model are illustrated in Table 1. In Figure 4, the mesh model of (a) the fluid's domain, (b) the interface's domain, and (c) the solid's domain of model A are shown. This model was processed with mesh independent analysis; hence, it already was suitable for further simulations. Please note that the mesh around the TE was defined with a higher number of elements than other areas because this TE was the area where the AD property would be investigated. For other models from B–D, they are similar to the model A except only on the numbers of elements and nodes. More details of the mesh independent analysis can be found in Supplementary Materials (Mesh_independent_analysis.pdf). The results of the C_l and C_d at each model exhibited differences at the third decimal places, which will be explained in Section 4.1. With the mesh independent analysis process, these indistinguishable C_l and C_d values expressed a confidential level of the mesh used in this simulation.

Table 1. Numbers of elements and nodes of each model after the mesh independent analysis (MIA) process.

Model	Fluid Domain		Solid Domain		Total	
	Element	Node	Element	Node	Element	Node
A	7,223,803	2,355,794	568,389	856,710	7,792,192	3,212,504
B	8,484,301	2,919,733	598,095	901,796	9,082,396	3,821,529
C	9,068,565	3,122,478	546,224	823,878	9,614,789	3,946,356
D	7,059,840	2,315,809	555,843	839,031	7,615,683	3,154,840

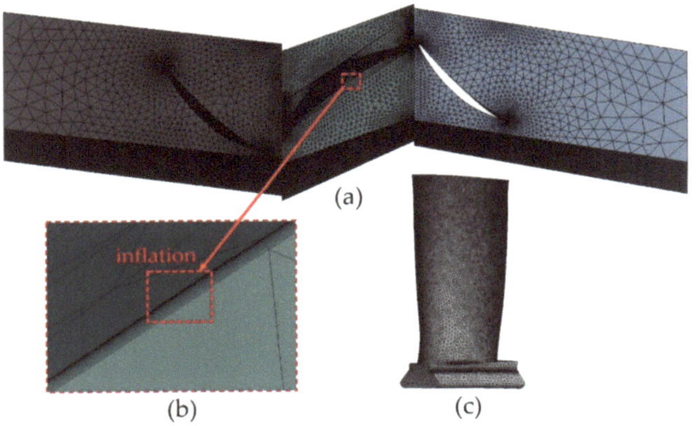

Figure 4. Mesh model of the (**a**) fluid's domain, (**b**) interface's domain, and (**c**) solid's domain of model A.

3.3. Boundary Conditions and Software Settings

The parameters to determine the boundary conditions were taken from the actual power plant and are shown in Figure 5. The gray blade is a stator, while the blue blade is a rotor. The yellow area indicates the investigating plane, which will explain the average pressure, the average temperature, and the pressure ratio for validation of the simulation in Section 4.1. The mass flow inlet was set to be 610 kg/s, and the temperature was set to be 174.15 °C. The pressure outlet was 553 kPa, while the rotation was set to 3000 rpm. Other parameters for the ANSYS CFX R19.2 software and static structural calculation are listed in Tables 2 and 3, respectively, being provided by a vendor. In the fatigue analysis, the CB was assigned to be the 422 stainless steel operating at 315 °C with the S-N curve reported in [28]. First, the calculation was in a steady state. Then, the obtained results were transferred to the transient calculation as initialized values to calculate the AD property. Finally, the AD results were applied as loads to calculate the SD property. The computational resource for simulation was 16 cores of an Intel Xeon 2.20 GHz processor with 64 GB of RAM, which required 24 h for CFD in a steady state, 120 h for CFD in a transient state, and 1.5 h for structural calculation per case. In total, it required 145.5 h per case.

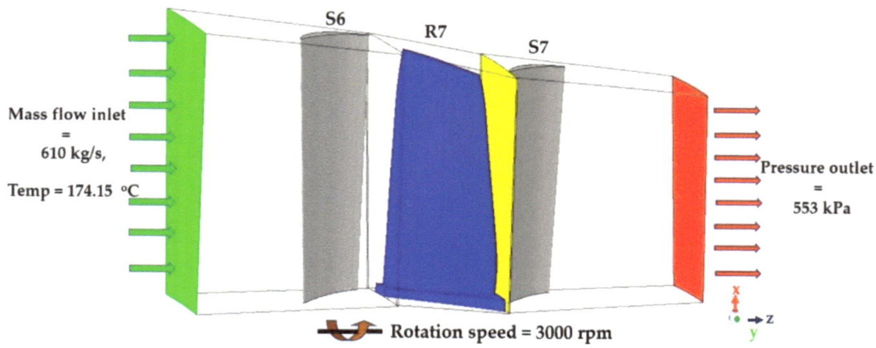

Figure 5. Boundary conditions. The positions of each blade and the observation's plan in the calculation show the (gray) stator, (blue) rotor, and (yellow) investigating plane.

Table 2. Settings and parameters for CFX calculation.

Numerical Parameters	Setting
Solvers	Pressure-based
Spatial Discretization	High-resolution scheme for the advection term High-resolution scheme for turbulence quantities
Convergence Control	Maximum Iteration 1000
Convergence Criteria	1.0×10^{-4}
Time Scale Control	Auto Timescale
Length Scale Option	Conservative
Time Scale Factor	Auto Timescale
Fluid	Air Ideal Gas
Heat Transfer	Total Energy
Turbulence	SST k-ω
Wall Function	High-speed (compressible) wall heat transfer model
Transient Blade Row Model	Profile Transformation
Frame Change or Mixing Model	Transient Rotor Stator
Pitch Change	Automatic

Table 3. Material's properties at 315 °C for the setting of the static structural calculation.

Property	Value
Density	7805 kg/m^3
Poisson's Ratio	0.195
Shear Modulus	79,300 MPa
Young's Modulus	1.896 × 105 MPa
Bulk Modulus	1.037 × 105 MPa
Tensile Yield Strength	1000 MPa
Tensile Ultimate Strength	1580 MPa
Thermal Conductivity	
100 °C	23.9 W/m·°C
350 °C	26.0 W/m·°C
Coefficient of Thermal Expansion	
100 °C	1.120 × 105 °C^{-1}
350 °C	1.147 × 105 °C^{-1}
S-N Curve	422 Stainless Steel [28]

4. Results and Discussion

4.1. Aerodynamics

To confirm the results of the simulation, the average pressure and the average temperature calculated at the investigating plane mentioned in Section 3.3 were compared to the actual pressure and temperature from the vendor's measurements. The results are shown in Table 4.

Table 4. Comparison of calculated and actual pressure and temperature at the investigating plane.

Parameter	Measurement	Simulation	Error
Pressure	570,000 ± 15% Pa	619,909 Pa	8.76%
Temperature	197.0 ± 10% °C	199.3 °C	1.17%

From Table 4, the results from the simulation were consistent with the actual ones, in which the errors of the pressure and temperature were 8.76% and 1.17%, respectively. These errors are still in the confidential level of statistics. Note that the errors may have occurred because only 1.5 stages of the CB were taken into the simulation without the influence of the previous stage. In contrast, the actual measurement was probed, which took the influences of stages 1–6 into account. If the simulation included the influence of the previous stage, the errors would drastically decrease. However, by focusing on the difference in the modification of the CB, the simulation with only 1.5 stages would be sufficient. The consistency of simulation and the actual measurement confirmed the confidential level of the results.

To compare the pressures occurring in models A–D, Figure 6 has been plotted to present the pressure ratio on the investigating plane, normalized to 1 atm. The results reveal that the areas around the root and the TE of models C and D could express the pressure ratio at a similar level to model A. Contrary to this, model B had the higher pressure ratio in a larger area than the other models. Figure 6 also depicts that the modifications of the CB were strongly influential on the pressure around the root area, as reported in [1,15].

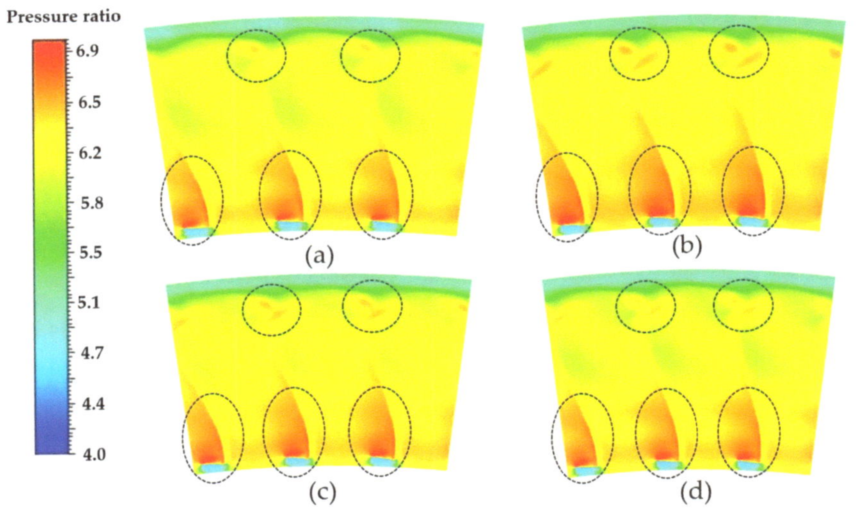

Figure 6. Pressure ratio on the investigating areas, normalized to 1 atm, for each model (**a**–**d**).

For further analysis, the results in Figure 6 were normalized with the pressure ratio of model A at the same positions and then reported in Figure 7 for blade spans of (a) 95.5%, (b) 95.8%, (c) 96.0%, and (d) 96.5%. The plots in Figure 7 have the x-axis representing the positions on the blades, while the y-axis represents the normalized pressure ratio (NPR). For the higher blade span, the position was from −0.03 m to −0.015 m, which means this position was situated close to the TE cutting. The results are also consistent with Figure 6, as the closer the positions to the TE cutting, the higher the level of the NPR found, especially in the models B–D. Models C and D obtained comparable NPRs to model A. In contrast, model B provided the most different NPR to model A, as expected. It is noted that when the positions were closer to the TE cutting, the pressure ratio exhibited larger changes. We also found that the larger the cutting of modification, the higher the difference of the NPS, but the results did not report this here.

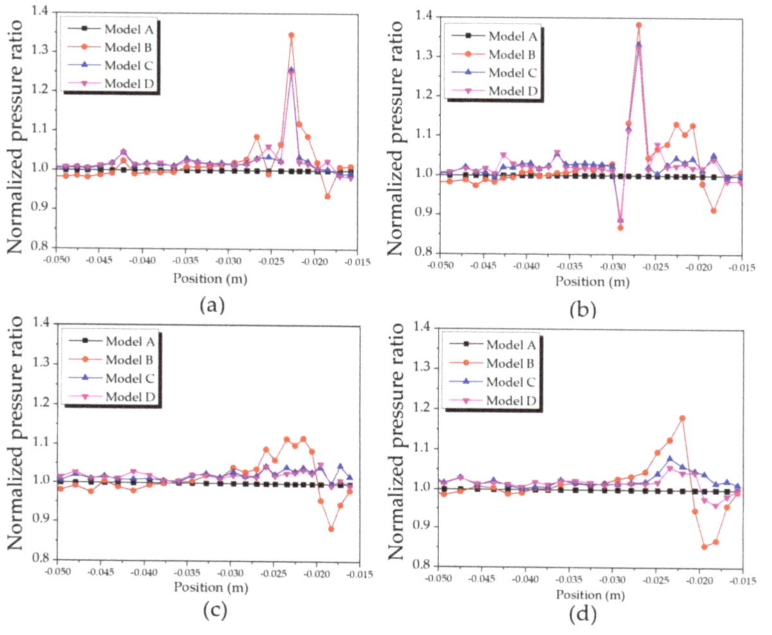

Figure 7. Plots of the normalized pressure ratio compared to blade spans of (**a**) 95.5%, (**b**) 95.8%, (**c**) 96.0%, and (**d**) 96.5%.

The AD results were used to calculate the C_l and C_d, shown in Table 5. The numbers in parentheses represent the percentage changes compared with the results of model A. The C_l and C_d could be altered up to 3.67% and 2.18%, respectively. It might be concluded that the modifications proposed in this work had an insignificant effect on the changes of the C_l and C_d. Model C led to C_l and C_d values more similar to those in model A than the other models, corresponding to the pressure ratio and the NPR reported in Figures 6 and 7.

Table 5. Result of the C_l and C_d for each model and the necessary parameters for calculation.

Model or Parameter	F_z (N)	F_y (N)	Area (m^2)	C_l	C_d
A	762.945	712.282	0.0415014	0.245	0.229
B	740.563	702.746	0.0414735	0.234 (−3.67%)	0.227 (−0.87%)
C	744.547	701.989	0.0414946	0.240 (−2.04%)	0.226 (−1.31%)
D	766.729	717.477	0.0414903	0.250 (+2.04%)	0.234 (+2.18%)

To determine the effect of the AD on the structure, Figure 8 shows the distributions of the (a) pressure load and (b) temperature load for model A, calculated from the AD and the boundary conditions mentioned in Section 3.3. In the actual operation of the CB, the pressure and the temperature increased in every higher stage. However, the results show that the pressure and the temperature of the CB in each position were not identical and became higher at the TE. In Figure 8a, the average pressure on the blade was about 687,574 Pa, while the pressure around the TE could be up to 1 MPa. Similarly, Figure 8b exhibits that the average temperature on the CB was approximately 234 °C, but the TE area presented a temperature much higher at 280 °C. Models B–D could provide a similar trend to the results. The pressure and temperature loads from the AD are important data and should be of concern.

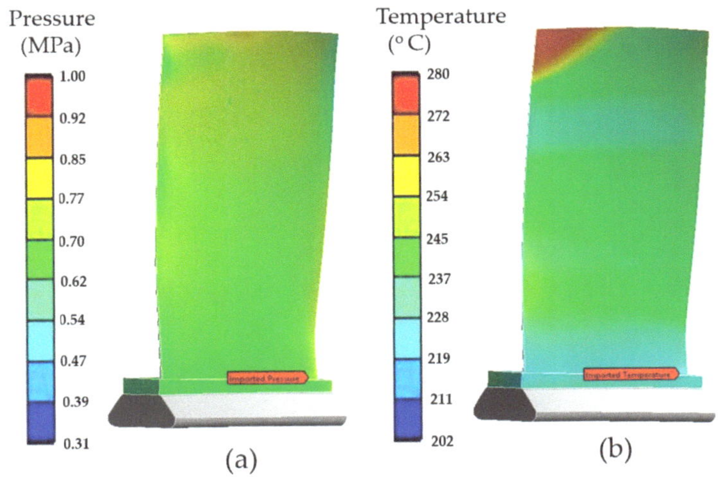

Figure 8. Distribution of the (**a**) pressure load and (**b**) temperature load of model A, calculated from the AD and the boundary conditions.

4.2. Structural Dynamics

To investigate the structural properties using harmonic response analysis, the pressure loads of models A–D in Figure 8a were averaged and then plotted in terms of the time domain, as shown in Figure 9a, from 0.00 to 0.02 s. Please note that the times later than 0.02 also exhibited similar results. After that, the results in Figure 9a would be calculated using a fast Fourier transform (FFT) and then normalized with the maximum pressure to obtain the normalized power spectrum (NPS) as in Figure 9b. This NPS could be the indicator to precisely investigate the frequency, as mentioned in [29]. From the results in Figure 9a, it was confirmed that the shapes of the modifications affected the average pressure. In Figure 9b, the pressure caused by the AD could be distinguished in terms of frequencies as 50, 100, 150, 200, 250, 300, and 350 Hz, corresponding to the value of the NPS. There was no frequency higher than 350 Hz that would react on the CB. These frequencies were caused by the rotational movement of the gas turbine at 3000 rpm, which was dominated by a 50 Hz fundamental frequency and composed of other frequencies as the harmonic frequencies. The CB's modification rarely affected the frequency. To analyze the total deformation, the 50 Hz frequency was taken into account using harmonic response analysis. The results can be seen in Figure 10. It was demonstrated that the maximum of the total deformation could be up to 1.12 mm. Moreover, most of the positions in every model showed an insignificant amount of deformation, which means the modifications had no influence on the harmonic frequency. For the higher frequencies, the effect of deformation also decreased.

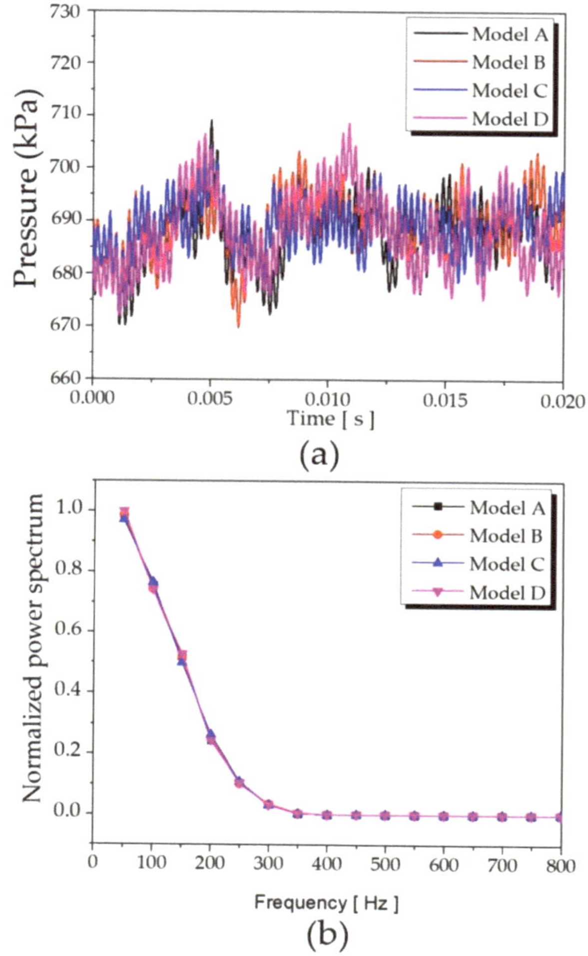

Figure 9. Plots of the (**a**) pressure load in the time domain and (**b**) NPS after the FFT and normalized by the maximum pressure.

To investigate the effect of the AD on the structure, Figure 11 shows the (a) total deformation (TD) and (b) equivalent von Mises stress (ES) from the pressure load and temperature of the AD, calculated using thermal analysis. In Figure 11a, the TD occurred at a higher rate near the upper part of the CB, while the ES occurred near the root of the CB. These results are also consistent with [30], which reported only the results in the structural simulation without loads from the AD. Furthermore, our work took the actual industrial parameters into account and provided more details than [30]. In addition, the simulation showed that with the loads that were applied, the TD and ES would be increased compared with the cases without the loads. The ES with loads appeared to be higher in the root area than the CB without loads, as expected. Our suggestion also mentions the computer simulation, as it should apply both the AD and the SD into the simulation for precise prediction, as in this work. Figure 12 reveals the TD in various models, considering only the area around the TE. We found that the TD had a similar pattern for all models. The modification did not affect the TD of the CB. From Figures 11 and 12, the maximum thermal expansion was discovered to be 3.70 mm at the leading edge, compared with the expansion at the TE of 2.62 mm. From Figure 2, the clearance between the inner casing and the TE was 5.20 mm, while the clearance between the inner casing and the leading edge was 6.80 mm. These expansions caused gaps between these parts, reduced to be 3.10 mm and 2.58 mm at the leading edge and the TE, respectively. If other factors had been included in the calculation, such as the vibration and expansion caused by the harmonic frequencies mentioned in Figure 10, the clearance, especially at the TE, would not be sufficient and lead to housing damage. Therefore, the larger gaps from the modification could dispute this problem at the leading edge. Figure 13 displays the ES in various models. It can be found that in every model, the ES had insignificant differences; hence, the modification of the TE had no effect on the ES at the edge area, only playing an important role at the root. As seen in Figure 2, the dimension of each CB was different. Hence, the length of the TE's cutting may be varied. From the results of the calculations, it can be confirmed that curved cutting was the proper choice for CB modification.

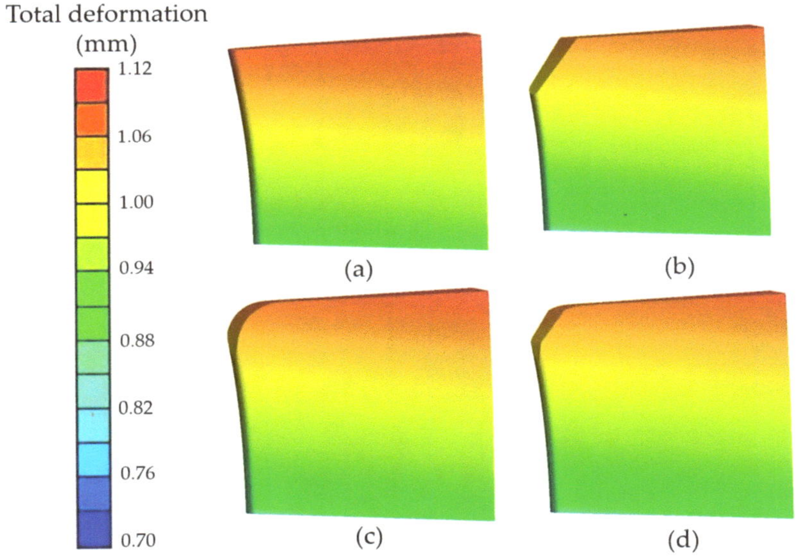

Figure 10. Total deformation on the trailing edge of (**a**) model A, (**b**) model B, (**c**) model C, and (**d**) model D at a 50 Hz frequency using harmonic response analysis calculation.

Figure 14 demonstrates the safety factors of the CB, calculated by fatigue analysis, for models A–D. The picture to the left (right) shows the plot of the safety factor for the unmodified CB (various modified CBs). As in the CB operation, a high fatigue cycle occurred. However, this work found that the modifications had no influence on the safety factor. On the other hand, every model showed the lowest safety factor around the root area. The reason for this is that the root area was the area where the CB had been plugged into the socket of the generator (see Figures 1 and 2). Therefore, the combinations of stress from rotation, compression from the CB attached to other components, thermal expansion, and temperature loads reacted to this area. These could cause the highest chance of damage happening. The safety factor around the root thus exhibited the low values of 0–1 at the surface. For further analysis into the axis, the safety factor drastically increased up to 15, which indicated that the deeper area had no chance of damage. This could be consistent with the actual results in Figure 1 (right picture) and the simulation results in Figure 16 in [12]. In more detail, the right section of Figure 1 clearly demonstrated the traces, scratches, and erosions around the root area. All of these agreements confirmed the consistency of the calculated safety factor and indicated that the modifications could be applied to the CB and had no effect on its operational lifetime.

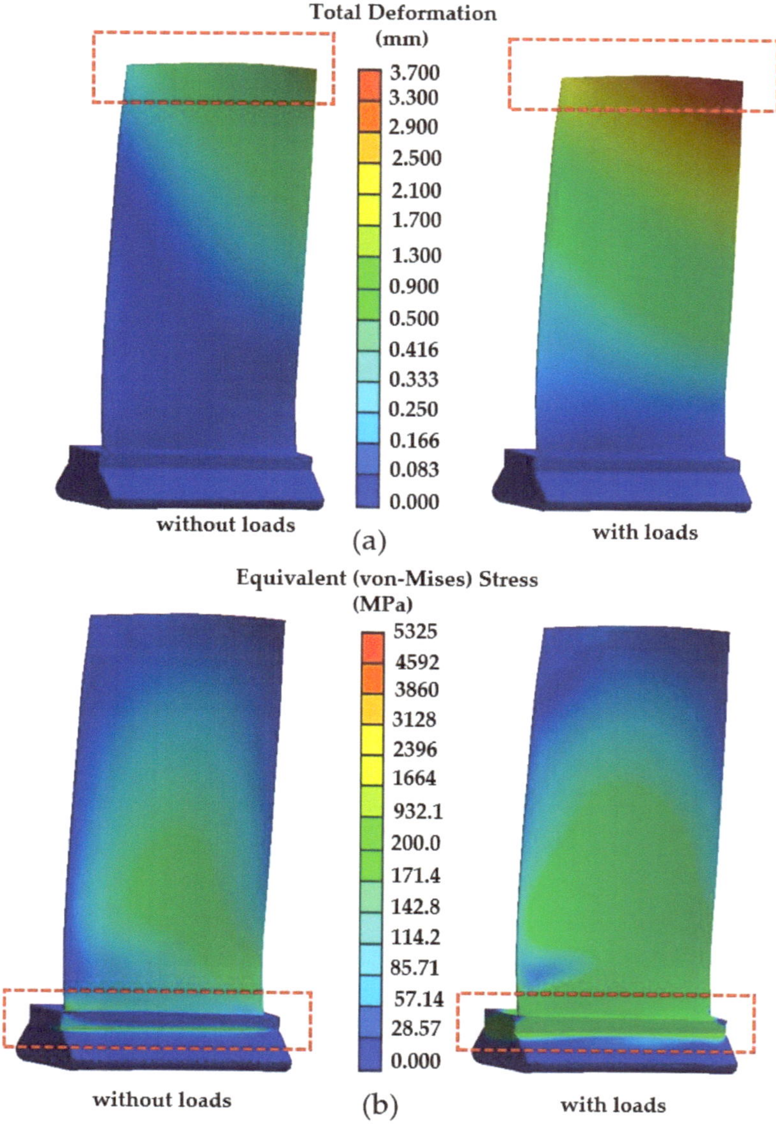

Figure 11. (**a**) Total deformation (TD) and (**b**) equivalent (von Mises) stress with and without pressure and temperature loads.

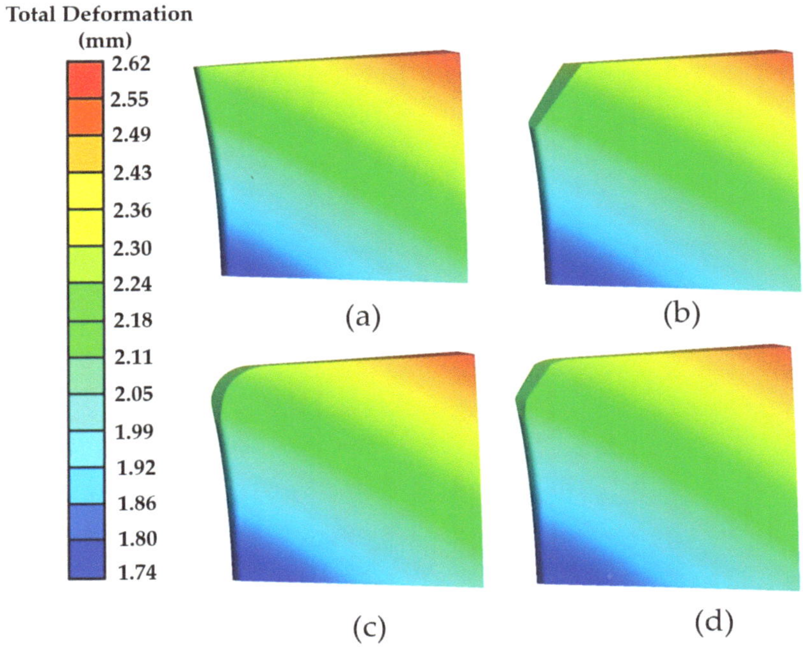

Figure 12. Total deformation on the trailing edge of (**a**) model A, (**b**) model B, (**c**) model C, and (**d**) model D, considering only the trailing edge area.

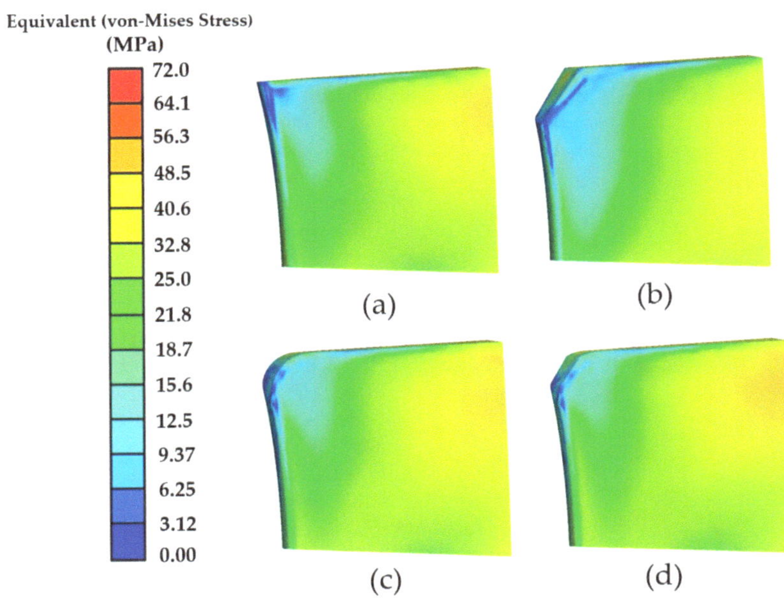

Figure 13. Equivalent von Mises stress in (**a**) model A, (**b**) model B, (**c**) model C, and (**d**) model D.

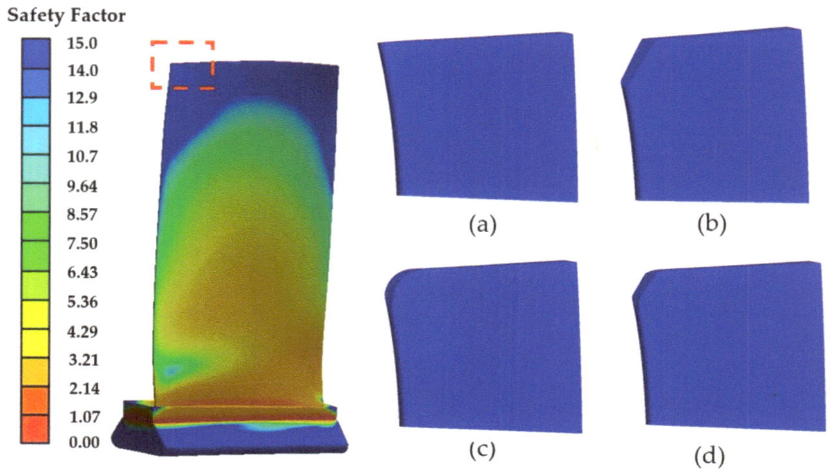

Figure 14. Safety factor on the trailing edge of (**a**) model A, (**b**) model B, (**c**) model C, and (**d**) model D.

From all the calculations of the AD and SD, the results led to the conclusion that models B, C, and D could be the solution to the housing damage. Model C obtained similar results to model A, considering the AD parameters such as the C_l, C_d, and pressure ratio. Moreover, considering the SD properties, such as the TD, ES, NPS, and safety factor, all models exhibited similar results. On the other hand, model B provided more unpleasant results in the AD than models C and D. In conclusion, model C was shown to be the most suitable model, considering the factor of uncomplicated modification and maintenance. Recently, this result has been applied for real use in a gas turbine power plant.

5. Conclusions

To solve the housing damage problem that happened in stage seven of the compressor stage in the gas turbine, this research proposed decent modifications of a CB composed of 6.5 mm curve cutting and the combination of 4 mm straight cutting with 6.5 mm curve cutting. These two models were simulated for their AD and structural properties. Using ANSYS CFX R19.2 with the SST k-ω turbulence model in the transient state, the AD results showed that the 6.5 mm curve cutting CB delivered results for the pressure ratio, C_l, C_d, and normalized power spectrum similar to the original CB. This modification also presented better results than the previous 5 mm straight cutting model and the combination of the 4 mm straight cutting with 6.5 mm curve cutting model. The consistency of all pressure and temperature values from the simulations and the measurements confirmed the confidence of the work. After that, the results of the AD were taken into account for the structural calculations, such as the structural analysis, harmonic response analysis, thermal analysis, and fatigue analysis using ANSYS structural analysis. The simulations revealed the increase of the pressure and temperature from the boundary conditions, especially at the upper edge of the CB. Therefore, in any modifications to the design and development of rotating machinery, the trailing edge area and upper edge should be investigated for both their AD and SD properties simultaneously. Then, the pressure from the aerodynamics calculations were used with a FFT and harmonic response analysis. The calculations found that the modifications had an insignificant influence on the harmonic frequency, total deformation, and equivalent stress. Then, the results from the AD were transferred to the thermal analysis and fatigue analysis. The results showed that the modification did not affect the SD properties, such as the total deformation, equivalent stress, and safety factor. A proper model with 6.5 mm curve cutting was proposed to be the most suitable model to be applied to actual modifications in the factory. This model has been proven to be the solution for the housing damage problem.

Supplementary Materials: The following are available online at https://www.mdpi.com/article/10.3390/pr9040705/s1. The Mesh_Independent_Analysis.pdf used to describe the mesh independent analysis is included within the Supplementary Materials.

Author Contributions: Conceptualization, T.J. and J.T.; methodology, T.J. and J.T.; software, T.J.; validation, T.J. and J.T.; formal analysis, J.T.; investigation, J.T.; writing—original draft preparation, T.J.; writing—review and editing, J.T., M.K. and W.B.; project administration, J.T.; funding acquisition, J.T. All authors have read and agreed to the published version of the manuscript.

Funding: The funding was sponsored by the Electricity Generating Authority of Thailand (EGAT) under a grant number 62-F205000-11-IO.SS03F3008494.

Institutional Review Board Statement: Not applicable.

Informed Consent Statement: Not applicable.

Data Availability Statement: The additional figures used to support the findings of this research are included within the supplementary materials.

Acknowledgments: This research was supported by the College of Advanced Manufacturing Innovation, King Mongkut's Institute of Technology Ladkrabang, and Electricity Generating Authority of Thailand (EGAT).

Conflicts of Interest: The authors declare that there are no conflicts of interest for this article.

Abbreviations

i, j	1, 2, and 3 correspond to x, y, and z, respectively
AD	aerodynamics
ω_f	angular frequency (rad/s)
F_1	blending function
μ_t	eddy viscosity (m^2/s)
ρ	density (kg/m^3)
$\{F\}$	load vector (N)
$\vec{U}$	mean velocity (m/s)
μ	molecular dynamics viscosity (Pa s)
$\{\ddot{u}\}$	nodal displacement vector (m)
$\{u\}$	nodal displacement vector (m)
$\{\dot{u}\}$	nodal velocity vector (m/s)
p	pressure (Pa)
P_k	shear production of turbulence
$\vec{S}_E$	source term of energy (N/m^2 s)
$\vec{S}_M$	source term of momentum (N/m^3)
α, β, σ	specific coefficient for SST k-ω
ω	specific dissipation rate (1/s)
τ	stress tensor (Pa)
$[C]$	structural damping matrix (N s/m)
SD	structural dynamics
$[M]$	structural mass matrix (kg)
$[K]$	structural stiffness matrix (N/m)
T	temperature (°C)
$[K(T)]$	thermal conduction matrix (W/°C)
λ	thermal conductivity (W/m K)
$\{Q(T)\}$	thermal conduction vector (W)
h_{tot}	total enthalpy (J)
k	turbulent kinetic energy (J/kg)

References

1. Kaewbumrung, M.; Tangsopa, W.; Thongsri, J. Investigation of the Trailing Edge Modification Effect on Compressor Blade Aerodynamics Using SST k-ω Turbulence Model. *Aerospace* **2019**, *6*, 48. [CrossRef]
2. Amoo, L.M. On the design and structural analysis of jet engine fan blade structures. *Prog. Aerosp. Sci.* **2013**, *60*, 1–11. [CrossRef]

3. Heidari, M.; Amini, K. Structural modification of a steam turbine blade. *IOP Conf. Ser. Mater. Sci. Eng.* **2017**, *203*, 012007. [CrossRef]
4. Pugachuk, A.S.; Muller, D.V.; Fominyh, N.K. Development of impellers based on prototypes in microturbine plants. *AIP Conf. Proc.* **2019**, *2141*, 030031.
5. Zhu, M. Design and analysis of steam turbine blades. *J. Phys. Conf. Ser.* **2019**, *1300*, 012056. [CrossRef]
6. NithinKumarK, C.; Tandon, T.; Silori, P.; Shaikh, A. Structural Design and Analysis of Gas Turbine Blade using CAE tools. *Int. J. Eng. Res. Technol.* **2014**, *3*, 469–474.
7. Ujade, G.; Bhambere, M. Review of Structural and Thermal Analysis of Gas Turbine Blade. *Int. J. Mech. Eng. Robot. Res.* **2014**, *3*, 347.
8. Singh, H.P.; Rawat, A.; Manral, A.R.; Kumar, P. Computational analysis of a gas turbine blade with different materials. *Mater. Today Proc.* **2020**. [CrossRef]
9. Boudounit, H.; Tarfaoui, M.; Saifaoui, D. Modal analysis for optimal design of offshore wind turbine blades. *Mater. Today Proc.* **2020**, *30*, 998–1004. [CrossRef]
10. Boyaraju, G.; Rajasekhar, S.; Sridhar, A.; Rao, J. Thermal analysis of a gas turbine rotor blade. *Int. J. Sci. Eng. Adv. Technol.* **2015**, *3*, 1181–1187.
11. Ngoret, J.K.; Kommula, V.P. Comprehending Occurrence of Premature Failure in Compressor Turbine (CT) Blades for Short-Haul Aircraft Fleet. *Int. J. Eng. Res. Afr.* **2019**, *42*, 10–23. [CrossRef]
12. Kou, H.J.; Lin, J.S.; Fu, X. Dynamic and fatigue compressor blade characteristics during fluid structure interaction: Part I-Blade modelling and vibration analysis. *Eng. Fail. Anal.* **2017**, *76*, 80–98. [CrossRef]
13. Zhu, F.-w.; Ding, L.; Huang, B.; Bao, M.; Liu, J.-T. Blade design and optimization of a horizontal axis tidal turbine. *Ocean Eng.* **2020**, *195*, 106652. [CrossRef]
14. Cui, B.; Zhang, C.; Zhang, Y.; Zhu, Z. Influence of Cutting Angle of Blade Trailing Edge on Unsteady Flow in a Centrifugal Pump Under Off-Design Conditions. *Appl. Sci.* **2020**, *10*, 580. [CrossRef]
15. Lu, J.L.; Chu, W.; Zhang, H. Influence of blade tip cutting on axial compressor aerodynamic performance. *Proc. Inst. Mech. Eng. Part G J. Aerosp. Eng.* **2009**, *223*, 19–29. [CrossRef]
16. Möller, D.; Jüngst, M.; Holzinger, F.; Brandstetter, C.; Schiffer, H.-P.; Leichtfuß, S. In Numerical Investigation of Tip Clearance Flow Induced Flutter in an Axial Research Compressor. In *ASME Turbo Expo 2016: Turbomachinery Technical Conference and Exposition*; American Society of Mechanical Engineers Digital Collection: New York, NY, USA, 2016.
17. Halder, P.; Kumar, P.M. Coupled CAD-CFD automated optimization for leading and trailing edge of an axial impulse turbine blade. *Ocean Eng.* **2020**, *213*, 107769. [CrossRef]
18. ANSYS, Inc. Lecture 01: Overview of FSI in Workbench. In *ANSYS Fluent Fluid-Structure Interaction (FSI) with ANSYS Mechanical*; ANSYS Inc.: Canonsburg, PA, USA, 2016.
19. ANSYS, Inc. ANSYS 17.1. In *ANSYS CFX Introduction*; ANSYS Europe Ltd.: Canonsburg, PA, USA, 2017.
20. ANSYS, Inc. *CFX-Solver Theory Guide R19.2*; ANSYS Inc.: Canonsburg, PA, USA, 2019.
21. Menter, F.R. Two-equation eddy-viscosity turbulence models for engineering applications. *AIAA J.* **1994**, *32*, 1598–1605.22. [CrossRef]
22. Puangburee, L.; Busayaporn, W.; Kaewbumrung, M.; Thongsri, J. Evaluation and Improvement of Ventilation System Inside Low-Cost Automation Line to Reduce Particle Contamination. *ECTI Trans. Electr. Eng. Electron. Commun.* **2020**, *18*, 35–44. [CrossRef]
23. Santati, S.; Thongsri, J.; Sarntima, P. Modified small-volume jet nebulizer based on CFD simulation and its clinical outcomes in small asthmatic children. *J. Healthc. Eng.* **2019**, *2019*, 2524583. [CrossRef]
24. Tangsopa, W.; Thongsri, J. Development of an industrial ultrasonic cleaning tank based on harmonic response analysis. *Ultrasonics* **2019**, *91*, 68–76. [CrossRef] [PubMed]
25. Tangsopa, W.; Thongsri, J. A Novel Ultrasonic Cleaning Tank Developed by Harmonic Response Analysis and Computational Fluid Dynamics. *Metals* **2020**, *10*, 335. [CrossRef]
26. ANSYS, Inc. ANSYS 2020 R1. In *ANSYS Mechanical APDL Theory Reference*; ANSYS Europe Ltd.: Canonsburg, PA, USA, 2020.
27. ANSYS, Inc. Module 07: Modal, Thermal, and Multistep Analysis. In *Introduction to ANSYS Mechanical*; ANSYS Inc.: Canonsburg, PA, USA, 2016.
28. Briggs, J.Z.; Parker, T.D. The super 12% Cr Steels. In *Source Book on Materials for Evaluated-Temperature Applications*; Elihu, F.B., Ed.; American Society for Metals: Geauga County, OH, USA, 1979; p. 121.
29. Tangsopa, W.; Thongsri, J. A Dual Frequency Ultrasonic Cleaning Tank Developed by Transient Dynamic Analysis. *Appl. Sci.* **2021**, *11*, 699. [CrossRef]
30. Kumar, R.R.; Pandey, K.M. Static Structural and Modal Analysis of Gas Turbine Blade. *IOP Conf. Ser. Mater. Sci. Eng.* **2017**, *225*, 012102. [CrossRef]

processes

Article

Large Eddy Simulation of Film Cooling Involving Compound Angle Holes: Comparative Study of LES and RANS

Seung Il Baek and Joon Ahn *

School of Mechanical Engineering, Kookmin University, 77 Jeongneung-ro, Seongbuk-gu, Seoul 02707, Korea; greenjet50@kookmin.ac.kr
* Correspondence: jahn@kookmin.ac.kr

Abstract: A large eddy simulation (LES) was performed for film cooling in the gas turbine blade involving spanwise injection angles (orientation angles). For a streamwise coolant injection angle (inclination angle) of 35°, the effects of the orientation angle were compared considering a simple angle of 0° and 30°. Two ratios of the coolant to main flow mass flux (blowing ratio) of 0.5 and 1.0 were considered and the experimental conditions of Jung and Lee (2000) were adopted for the geometry and flow conditions. Moreover, a Reynolds averaged Navier–Stokes simulation (RANS) was performed to understand the characteristics of the turbulence models compared to those in the LES and experiments. In the RANS, three turbulence models were compared, namely, the realizable k-ε, k-ω shear stress transport, and Reynolds stress models. The temperature field and flow fields predicted through the RANS were similar to those obtained through the experiment and LES. Nevertheless, at a simple angle, the point at which the counter-rotating vortex pair (CRVP) collided on the wall and rose was different from that in the experiment and LES. Under the compound angle, the point at which the CRVP changed to a single vortex was different from that in the LES. The adiabatic film cooling effectiveness could not be accurately determined through the RANS but was well reflected by the LES, even under the compound angle. The reattachment of the injectant at a blowing ratio of 1.0 was better predicted by the RANS at the compound angle than at the simple angle. The temperature fluctuation was predicted to decrease slightly when the injectant was supplied at a compound angle.

Keywords: film cooling; large eddy simulation (LES); reynolds averaged navier–stokes simulation (RANS); adiabatic film cooling effectiveness; compound angle

Citation: Baek, S.I.; Ahn, J. Large Eddy Simulation of Film Cooling Involving Compound Angle Holes: Comparative Study of LES and RANS. *Processes* **2021**, *9*, 198. https://doi.org/10.3390/pr9020198

Received: 28 December 2020
Accepted: 19 January 2021
Published: 21 January 2021

1. Introduction

Gas turbines, which are composed of compressors, combustors, and turbines, are mainly used as aircraft engines and prime movers for natural gas power generation. Ensuring cycle efficiency is critical in both these applications. Although additional devices such as reheaters or regenerators can be installed to increase cycle efficiency, this technique cannot be adopted in the case of aircraft engines due to the limitations of the weight and frontal area. Therefore, in general, to increase cycle efficiency, the turbine inlet temperature is raised above the allowable temperature limit of the materials by cooling the blade through the air from the compressor [1].

The cooling air extracted from the compressor is supplied to the internal cooling passage of the turbine blade to cool the blade. After cooling the inside of the blade, the air, injected through the blade surface, flows on the surface, thereby creating an insulating film to protect the blade from the hot combustion gases. To generate a uniform insulation film, it is advantageous to inject the coolant through a slot. However, to ensure the mechanical strength, cylindrical holes are usually arrayed in the spanwise direction, as shown in Figure 1a [2].

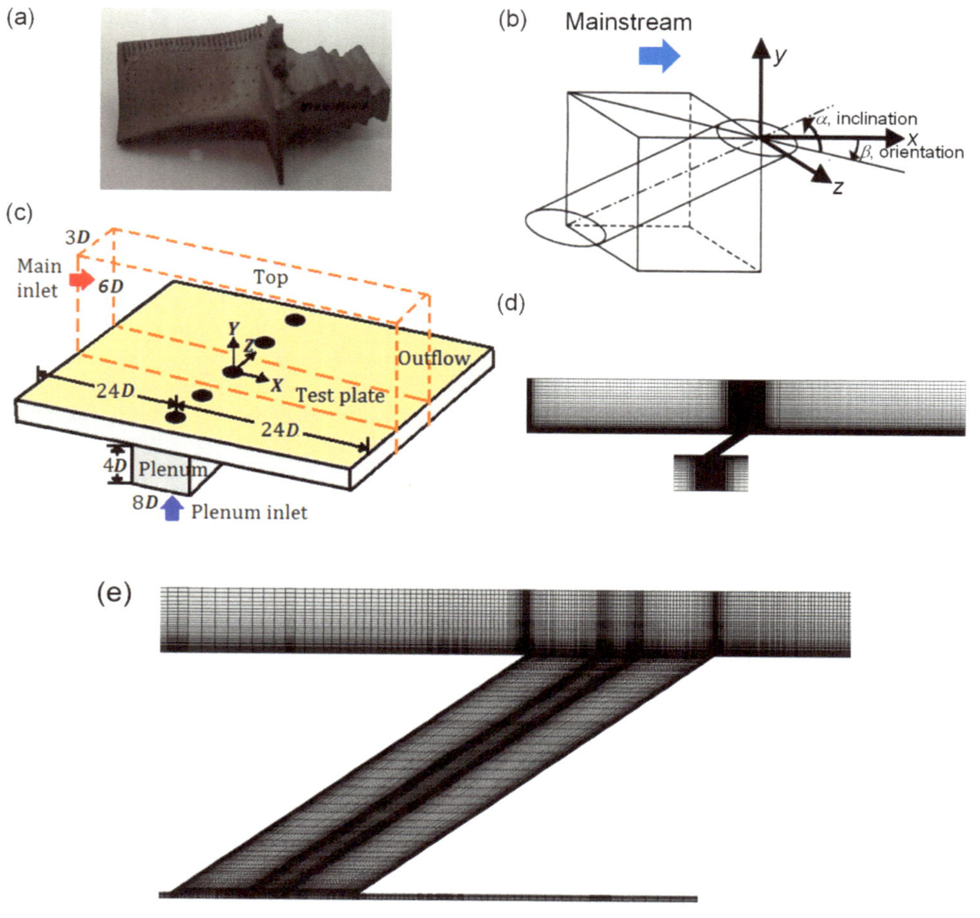

Figure 1. Film cooling geometry and computational domain: (**a**) Gas turbine blade with film cooling holes; (**b**) injection geometry of film cooling; (**c**) computational domain; (**d**) grid system; and (**e**) close-up of the mesh near the hole.

Over 50 years, numerous experiments and numerical studies have been conducted by many researchers to understand the physics of film cooling. Goldstein, Eckert, and Ramsey obtained the film cooling effectiveness on a flat plate with discrete cylindrical holes, experimentally [3]. They studied the effects of changing Reynolds number for the main flow and the inclination angle of the cooling holes on film cooling effectiveness and concluded that coolant spreading was almost the same regardless of the Reynolds number. They stated that increasing the coolant flow rate made an increase in film cooling effectiveness at low blowing ratios. Pedersen et al. investigated the effects of changing the density ratio of the coolant on film cooling effectiveness experimentally [4]. They also used a flat plate with cylindrical holes and stated that an increasing density ratio resulted in a higher film cooling effectiveness except for the narrow field near the cooling hole due to the coolant jet lifts off and reattachment to the test wall further downstream. Sinha et al. displayed the centerline and spanwise-averaged effectiveness for the various blowing ratios and density ratios experimentally [5]. They showed that increasing the coolant mass flow rate decreased the film cooling effectiveness by the coolant jet lift off. They stated that if the coolant jet to the main flow momentum flux ratio was between 0.3 and 0.7, the coolant jet detachment would occur in the narrow field near the hole and the jets would reattach downstream. If the momentum ratio was more than 0.7, a complete detachment of the coolant jets would be generated. They found that the spanwise-averaged film cooling effectiveness was strongly dependent on the spanwise spreading of the coolant jet.

Owing to structural reasons, the coolant cannot be injected parallel to the surface and is instead injected at an angle to the main flow, with inclination angle α (see Figure 1b) being approximately 30°. In this case, the film cooling jet lifts from the blade surface and interacts with the main flow to generate a counter-rotating vortex pair (CRVP). This flow structure renders the cooling effect uneven and deteriorates the insulation effect owing to the enhanced mixing with the main flow [6,7].

In general, film cooling jets are injected in the streamwise direction to avoid disturbing the main flow however in certain cases, the jets may be injected at a compound angle, with orientation angle β (see Figure 1b) to the main flow. Usually, a compound angle is adopted at the leading edge of the blade to ensure a certain inclination on the curved surface. Moreover, a compound angle is often adopted at the mid-chord to improve the uniformity of the film cooling.

When implementing a compound angle, the film cooling performance changes as the CRVP changes to a single vortex [8,9]. Lee et al. [10] measured the change in the film cooling jet flow for an α value of 35°, with β ranging from 15° to 90°. The CRVP of the film cooling jet exhibited strong asymmetry at an orientation angle of 15° and changed to a single vortex at an orientation angle of 30°. Jung and Lee [11] measured the change in the film cooling effectiveness with the orientation angle. When the compound angle was adopted, film cooling effectiveness increased from 20% to 80% depending on the orientation angle and blowing ratio [11].

Furthermore, flow field information is required to clarify the mechanism of film cooling performance enhancement owing to the compound angle injection or to determine the optimum injection geometry. Lee et al. [10] evaluated the film cooling flow involving a compound angle by using a five-hole Pitot tube however, the measurement did not clarify the turbulent quantities or velocity near the wall. Jung and Lee [11] measured the temperature field, which is a scalar quantity, instead of the velocity vectors. However, the relevant information near the wall was not clarified and the temperature field represents only indirect information regarding the flow field. Abhari et al. [12,13] used the particle image velocimetry (PIV) technique to minimize the disturbance of the flow field pertaining to the sensor and measure the flow field of the film cooling involving a compound angle nevertheless, the velocity near the wall could not be clarified.

The computational fluid dynamics (CFD) technique represents a promising tool to obtain three-dimensional flow field information, including the velocity near the wall. Among the relevant techniques, the most economical approach is the Reynolds averaged Navier–Stokes simulation (RANS), which can accurately reflect the basic flow structures however, the turbulence model cannot effectively predict the mixing of the main flow and film cooling jet, resulting in an overprediction of film cooling effectiveness [14]. The k-ε, k-ω, and Reynolds stress models (RSM) are frequently used in the RANS analysis of film cooling problems. Harrison and Bogard [15] applied the aforementioned three turbulence models and compared the prediction results with the experimentally obtained results. It was noted that the realizable k-ε can best predict the effectiveness of centerline film cooling.

In recent years, film cooling problems were attempted to be solved using the hybrid RANS/large eddy simulation (LES) [16] or the LES [17], which exhibits a higher accuracy but considerably larger computational cost compared to that of the RANS. In the simulation of the film cooling through the hybrid LES or LES, the film cooling effectiveness was better predicted than that in the RANS [16,17]. Moreover, using the instantaneous flow field obtained by the LES, the flow structures that were not visible in the RANS could be identified [18–20]. Nevertheless, most of the LES studies on film cooling were based on the use of a simple angle (β = 0°).

The LES results for film cooling involving a compound angle primarily pertained to the case with the leading edge application [21]. Since the leading edge has different film cooling characteristics than those of the mid-chord due to the influence of the pressure gradient and curvature [22,23], it is difficult to directly compare film cooling for simple and compound angle cases by using the data at the leading edge. To clarify the effect of the compound angle injection, the LES for compound angled film cooling applied to a flat plate must be realized.

The LES for a flat plate subjected to a compound angled film cooling was reported [24]. The researchers simulated a two-row film cooling with opposite orientation angles [25]. Nevertheless, the film cooling characteristics with the compound angle could not be identified because the film cooling jets exhibited a unique flow structure due to the strong interaction between the upstream and downstream jets. To determine these characteristics, Li et al. [26] performed an LES for single row compound angle film cooling and clarified the effect of the supply geometry and hole length to diameter ratio (L/D). McClintic et al. [27] studied the effects of internal cross-flow on film cooling effectiveness for a row of cylindrical holes with compound angles of 45° experimentally. They reported that a higher effectiveness was obtained for the cross-flow directed counter to the spanwise direction of the hole than for the cross-flow directed to the spanwise direction of the hole. Stratton et al. [28] investigated the effects of the internal cross-flow on the film cooling performance for a cylindrical hole with a compound angle of 45° by using RANS models and LES. They found that unsteady vortical structures were formed inside the hole and RANS results did not show a good match to an experimental data while LES predicted the film cooling effectiveness with reasonable accuracy.

Overall, although several LES results have been published for the film cooling involving a compound angle on the flat plate geometry, the discussion remains insufficient. The main point of this paper is to show LES results that have remained undiscovered in the past. The flow, thermal fields, and spatial distribution of the film cooling effectiveness obtained by LES and RANS were compared with experimental results. The predictive performance of the three turbulence models for RANS compared for compound angled film cooling, and information regarding the flow and temperature fields near the wall, which could not be determined experimentally, was discussed based on the LES data. In addition, by analyzing the instantaneous flow field obtained from the LES, the flow characteristics of film cooling with a compound angle that could not be determined through the time-averaged analyses, were clarified.

2. Numerical Methods and Validation

2.1. Computational Domain and Boundary Conditions

The computational domain and boundary conditions were set to be comparable with those in the experiments of Jung and Lee [11]. The boundary conditions in the computational domain are specified in Table 1. The domain involved the main flow, test plate, film cooling hole, and plenum, are indicated by the orange box in Figure 1c. The inclination angle of the film cooling jet was 35° and the L/D ratio was 4. The film cooling holes were arranged in a row in the spanwise direction, and the pitch was three times the diameter. The width of the computational domain was set as $3D$, according to the pitch and periodic conditions [29,30].

Table 1. Boundary conditions of the computational fluid dynamics (CFD) domain.

Surface	Boundary Condition
Main inlet	Velocity inlet (u = constant)
Plenum inlet	Velocity inlet (u = constant)
Top	Symmetry ($\frac{\partial u}{\partial y} = 0$, $\frac{\partial w}{\partial y} = 0$, $v = 0$)
Test plate	Adiabatic wall (u = v = w = 0)
Outflow	Pressure outlet
Main sides	Periodic (u (x, y, z, t) = u (x, y, z + P, t), ΔP = 0)
Sides of plenum	Wall (u = v = w = 0)
Tube wall	Wall (u = v = w = 0)

The main flow velocity was 10 m/s and the diameter of the film cooling hole was 1.5 cm. In the experiment of Jung and Lee [11], the flow at the exit of the film cooling hole was a turbulent boundary layer flow and the boundary layer thickness was almost 24 mm (1.6 D). In the CFD simulation, the velocity profile was set to be the same with the experiment. The length of the inlet section was determined as 24 D to ensure that the boundary layer, as in the experiment, was generated at the exit of the hole. To obtain a result not influenced by the boundary conditions up to $x/D = 15$, which corresponded to the film cooling effectiveness distribution in the work of Jung and Lee [11], the size of the computational domain downstream of the hole was set as 24 D and the pressure outlet conditions were imposed. In LES calculation, the vortex method was adopted at the main inlet and a fluctuation was added on a mean velocity profile by a perturbation of the vorticity field. The vortex points were convected randomly and the vortex method was based on the Lagrangian form of the evolution equation of the vorticity and the Biot–Savart law [31]. On the other hand, in RANS simulation, the turbulent intensity and turbulent length scale were set as 0.2% and 0.1 × height of the inlet, respectively. The turbulent intensity at the main inlet was set as the same in the experiment operating condition.

Simulations were performed considering two blowing ratios of 0.5 and 1.0. In the work of Jung and Lee [11], the film cooling jet penetrated from a high blowing ratio of 1.0 to $y/D = 2.0$ thus, in this work, the height of the computational domain in the wall-normal direction was determined to be 6 D. Symmetric conditions were imposed on the top surface of the computational domain [29,30]. As shown in Figure 1c, the flow separation occurring at the hole entrance and the interaction of the main flow and injectant at the hole exit were reflected by including the film cooling hole and plenum in the computational domain. The velocity boundary condition at the plenum entrance was imposed to satisfy the blowing ratio requirements.

Jung and Lee [11] compared four orientation angles of 0°, 30°, 60°, and 90°. The experimental apparatus involved five film cooling holes in a row and side walls were present in the test duct. Therefore, in the case of large orientation angles (i.e., 60° and 90°), complete periodicity did not appear in the spanwise direction. Correspondingly, in this study, the simulation cases were considered as those pertaining to 0° (simple angle) and 30°, in which the periodic distribution in the spanwise direction was well reflected in the experimental data.

In terms of the temperature boundary conditions, as in the experiment, the temperature of the main flow and film cooling jet was set as 293 K and 313 K, respectively. A no-slip condition was imposed as the velocity boundary condition on all the walls. Moreover, adiabatic conditions were considered, as in the experiment, to ensure that the adiabatic film cooling effectiveness could be obtained from the surface temperature of the test plate.

The grid system was composed of 2.56 million hexahedrons, based on the results of a sensitivity test. In the streamwise direction, the mesh spacing values range from 6 near the cooling hole to 35 near the end of the domain in x+ units. In the spanwise direction, the mesh spacing values are about 20 throughout the domain in z+ units. As shown in Figure 1d, a non-uniform grid, in which the grid was concentrated near the wall, was used, and 25 grids were stacked within $y^+ = 30$ on the wall and the value of y^+ of the first cell above the wall was set as about 1. Figure 1e illustrates the close-up of the mesh near the injection region of the hole.

2.2. Governing Equations and Turbulence Models

The CFD calculations were performed in ANSYS Fluent v. 19, and the meshes were generated in Pointwise v. 18, as shown in Figure 1d [29,30]. Fluent was based on cell-centered finite volume method and the whole computational domain was discretized into a finite number of control volumes. Pointwise was a commercial mesh generation software and structured, unstructured, and hybrid meshes were generated with precise control over points placement. The governing equations included the incompressible Navier–Stokes equation and energy equation. The numerical method used for the cases are a simple scheme. The convection term is solved via second order upwind while the diffusion term is solved by central differencing. In the LES, the governing equation was transformed to a grid filtered equation, as follows [31].

Conservation of Mass (Continuity equation):

$$\frac{\partial \rho}{\partial t} + \frac{\partial}{\partial x_i}(\rho \overline{u}_i) = 0. \tag{1}$$

Conservation of Momentum:

$$\frac{\partial(\rho \overline{u}_i)}{\partial t} + \frac{\partial}{\partial x_j}(\rho \overline{u}_i \overline{u}_j) = \frac{\partial}{\partial x_j}\left[\mu\left(\frac{\partial \overline{u}_i}{\partial x_j} + \frac{\partial \overline{u}_j}{\partial x_i}\right) - \frac{2}{3}\mu \frac{\partial \overline{u}_l}{\partial x_l}\delta_{ij}\right] - \frac{d\overline{p}}{dx_i} + \frac{\partial \tau_{ij}}{\partial x_j}. \tag{2}$$

Conservation of Energy:

$$\rho C_p \frac{\partial \overline{T}}{\partial t} + \rho C_p \frac{\partial}{\partial x_j}(\overline{u}_j \overline{T}) = \frac{\partial}{\partial x_j}\left(\kappa \frac{\partial \overline{T}}{\partial x_j}\right) + \frac{\partial q_j}{\partial x_j}. \tag{3}$$

In the equations, the overbar represents the grid filtered values. In Equation (2), τ_{ij}, the subgrid-scale stress was modeled using the Boussinesq hypothesis:

$$\tau_{ij} - \frac{1}{3}\tau_{kk}\delta_{ij} = -\mu_t\left(\frac{\partial u_i}{\partial x_j} + \frac{\partial u_j}{\partial x_i}\right) \tag{4}$$

where μ_t is the subgrid-scale viscosity. The Smagorinsky–Lilly model was adopted based on a validation test [29,30]:

Subgrid-scale viscosity in the Smagorinsky-Lilly model:

$$\mu_t = \rho {L_s}^2 \sqrt{\frac{1}{2}\left(\frac{\partial \overline{u}_i}{\partial x_j} + \frac{\partial \overline{u}_j}{\partial x_i}\right)\left(\frac{\partial \overline{u}_i}{\partial x_j} + \frac{\partial \overline{u}_j}{\partial x_i}\right)}. \tag{5}$$

In 1877, Boussinesq stated that the momentum transfer generated by turbulent eddies could be modeled with the eddy viscosity. In Equation (3), q_j refers to the subgrid-scale heat flux and can be expressed as follows by introducing the subgrid-scale diffusivity α_t.

Subgrid-scale heat flux:

$$q_j = -\rho C_p \alpha_t \frac{\partial \overline{T}}{\partial x_j} \tag{6}$$

μ_t was considered as α_t, assuming that the subgrid-scale turbulent Prandtl number is unity [31].

In the LES, the time step was fixed as 6.25×10^{-6} s, and the number of steps was 400 considering that the mainstream flowed as much as the hole diameter in this time interval. The film cooling flow is inherently unsteady and the current study performed the statistically steady state analysis by post-processing. A total of 20,000 time steps were required from the initial condition to reach steady state and 30,000 integrated time steps were considered to obtain the average flow field and statistics. The complete process required approximately one month of computation time on a 20-core-cluster computer. The RANS analyzed the time-averaged continuity, Navier–Stokes, and energy equations, as shown in Equations (7)–(9), respectively.

Conservation of Mass (Continuity equation):

$$\frac{\partial}{\partial x_j}(\rho U_j) = 0. \tag{7}$$

Conservation of Momentum:

$$\rho U_j \frac{\partial U_i}{\partial x_j} = -\frac{dP}{dx_i} + \rho g \delta_{i3} + \frac{\partial}{\partial x_j}\left(\mu \frac{\partial U_i}{\partial x_j} - \rho \overline{u_i u_j}\right) \tag{8}$$

Conservation of Energy:

$$\rho c_p U_j \frac{\partial \overline{T}}{\partial x_j} = -\frac{\partial}{\partial x_j}\left(-\kappa \frac{\partial \overline{T}}{\partial x_j} + \rho c_p \overline{u_j T'}\right). \tag{9}$$

Six turbulent stress components, $-\rho\overline{uu}$, $-\rho\overline{uv}$, $-\rho\overline{uw}$, $-\rho\overline{vv}$, $-\rho\overline{vw}$, and $-\rho\overline{ww}$ from Equation (8) and three turbulent heat flux components $\overline{T'u}$, $\overline{T'v}$, and $\overline{T'w}$ from Equation (9) were required to be modeled. The RSM, realizable k-ε model, and k-ω shear stress transport (SST) model were compared. Detailed equations for each of these models can be found elsewhere, such as the work of Tannehill et al. [32]. The RANS computations converged to the 10^{-6} level for all the equations in approximately 10 h on a 20-core-cluster computer.

2.3. Code Validation

Harrison and Bogard [15] performed a RANS by using three models and as shown in Figure 2a, the centerline film cooling effectiveness prediction of the realizable k-ε and RSM models was more accurate than that of the k-ω. Considering the experimental conditions adopted by Sinha et al. [5], the LES was performed on four subgrid-scale models, as shown in Figure 2b and the findings were compared with two sets of LES data in the open literature [17,33]. The LES results for the effectiveness corresponded to a higher correlation with the experimental data than the RANS data, as shown in Figure 2a.

Among the subgrid-scale models, the predictions of the wall adapting local eddy-viscosity model, Smagorinsky–Lilly model, and kinetic energy transport model (excluding those of the wall modeled LES) were equivalently correlated with the experimental values. WMLES (Wall-Modeled Large Eddy Simulation) and WALE (Wall-Adapting Local Eddy-Viscosity) models did not show a good match with the experimental data in the near-hole region and in the far field region, respectively. Considering these findings, the Smagorinsky–Lilly model was adopted in the present study.

Subsequently, a grid sensitivity test was conducted through the LES under the geometry and flow conditions considered by Jung and Lee [11] to compare the data in this study (see Figure 2c). The comparison for the centerline effectiveness at a blowing ratio of 0.5 indicated that similar results were obtained after 2 million grids. The results obtained using 2.56 million and 3.4 million grid points were completely collapsed and the series of simulations in the present study was thus performed using 2.56 million cells. A series of RANS for film cooling was performed using the three turbulence models in the grid system that passed the grid sensitivity test. As shown in Figure 2d, the RANS data overpredicted the overall film cooling effectiveness compared to those of the LES. For the downstream region after $x/D = 7$, similar characteristics to those reported by Harrison and Bogard [15] were obtained, as shown in Figure 2a. In other words, the film cooling effectiveness predictions of the realizable k-ε and RSM were equivalent and satisfactory, whereas the k-ω SST overpredicts the effectiveness. For the upstream region with x/D less than 5, the predictions of the RSM and k-ω SST were similar, whereas the realizable k-ε considerably overpredicted film cooling effectiveness.

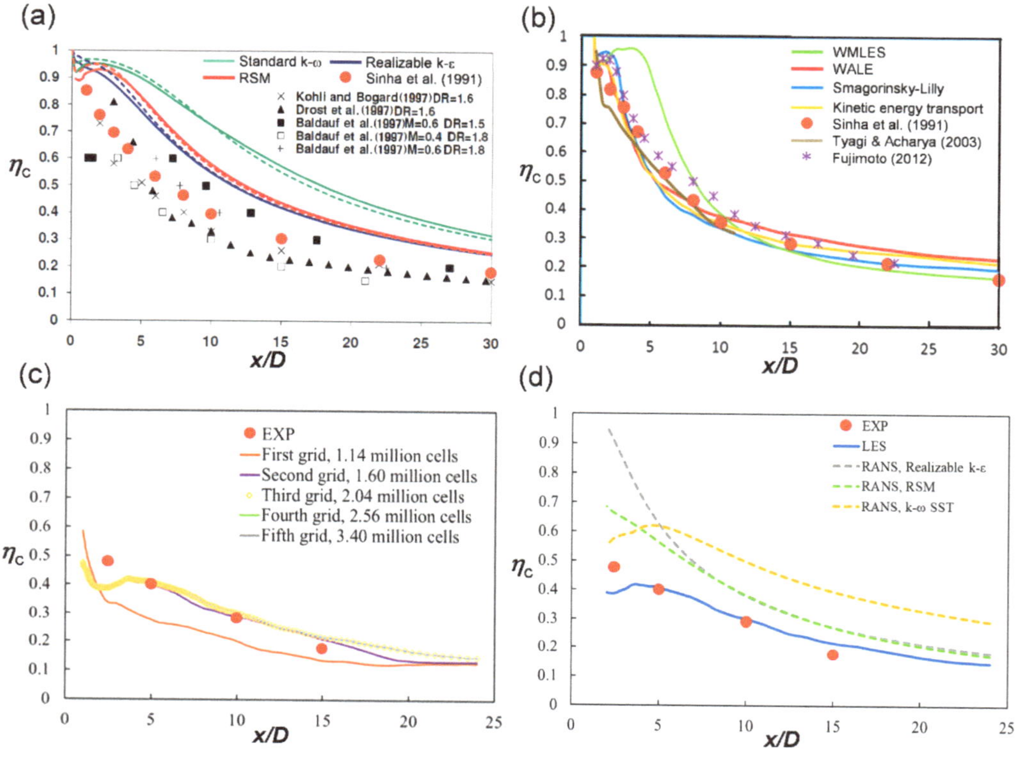

Figure 2. Code validation based on the centerline adiabatic film cooling effectiveness: (**a**) Turbulence model validation by Harrison and Bogard [15]; (**b**) subgrid scale model validation for the large eddy simulation (LES); (**c**) grid sensitivity test; and (**d**) code validation test compared to the data by Jung and Lee [11].

3. Results and Discussion

3.1. Time-Averaged Flow and Thermal Fields

The time-averaged flow and thermal fields obtained by the LES and RANS in a streamwise normal plane were compared with the experimental data of Burd et al. [34] and Jung and Lee [11]. As shown in Figure 2, the film cooling effectiveness differed considerably with the turbulence model however, no significant difference was observed between the time-averaged flow and thermal fields. Therefore, the results of the RANS with the realizable k-ε model were compared with those of the experiments and LES as a representative case. When the orientation angle was 0°, a kidney vortex occurred symmetrically in the film cooling jet and both the LES and RANS predict this phenomenon well, as shown in Figure 3a.

The flow near the wall, which affected the heat transfer characteristics, indicated that the flow of the CRVP created a parallel flow moving along the wall. The two parallel flows collided at the symmetric plane $z/D = 0$, and rose above the wall. In the experiment, the upward motion started to occur at approximately $z/D = 0.5$, and the LES predicted this phenomenon well. In the RANS, the v component was observed from approximately $z/D = 0.2$, closer to the plane of symmetry than in the experimental data. In terms of the streamwise velocity distributions, as shown in Figure 3b, both the LES and RANS exhibited distributions that were similar to the experimental data. Considering the central part of the film cooling jet ($z/D = 0$, $y/D = 0.4$) with a dimensionless velocity of 0.7 or less, indicated in blue, it can be noted that the area for the LES data is similar to that of the experiment however, the corresponding area for the RANS data is slightly larger, likely because the RANS predicts that the film cooling jet blocks the main flow more strongly than it actually does. According to the high-speed region marked in red, the LES prediction of the characteristics of the high velocity in the streamwise direction from the kidney vortex position is more similar to the experiment values than those obtained using the RANS.

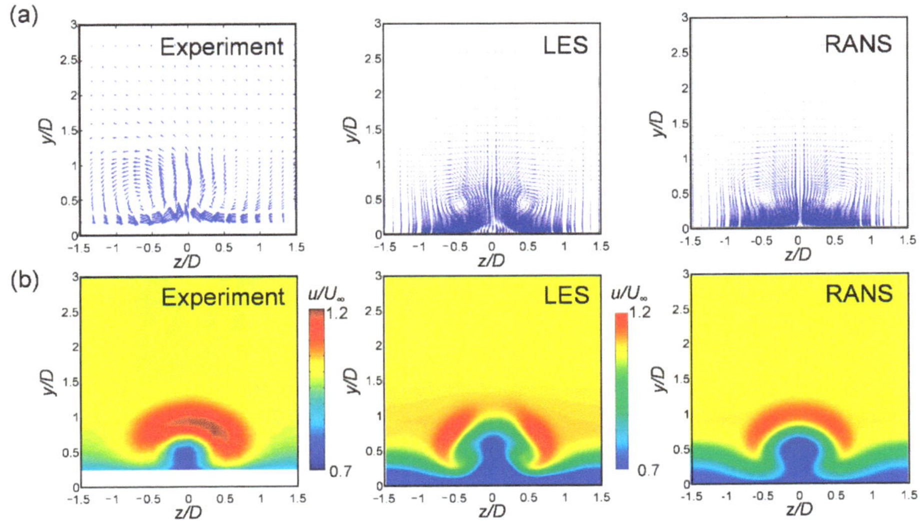

Figure 3. Time-averaged flow field at $x/D = 2.5$ for the streamwise film cooling when $M = 1.0$, compared with data from Burd et al. [35]: (**a**) Cross-stream velocity vectors and (**b**) streamwise velocity contour.

The CFD data provided the velocity information below $y/D = 0.2$, which could not be measured experimentally. Both the LES and RANS results indicated that the largest and smallest boundary layer thickness occurred at $z/D = 0$, the plane of symmetry, and at approximately $z/D = 0.5$, respectively. Therefore, the maximum heat transfer coefficient was expected to occur at approximately $z/D = 0.5$. The maximum film cooling effectiveness was expected to occur on the plane of symmetry without a strong jet lift-off and this aspect could be confirmed considering the data presented in the next section. Figure 4 shows the structure of the vortex generated in the injectant in a streamwise normal plane in comparison with the experimental data. Since the orientation angle considered in the results shown in Figure 3 was 0°, a left and symmetrical vortex pair appeared. In contrast, when the compound angle was adopted, an asymmetric or single vortex flow occurred. The results obtained using a five-hole Pitot tube [10] (Figure 4a) or PIV [12,13] (Figure 4b) indicated that when the orientation angle was 15°, the counter-rotating vortex remained weak. For the orientation angle of 30° or 45°, the vortex became a single vortex. The particle image velocimetry (PIV) technique minimized the disturbance of the flow field pertaining to the sensor, however, the velocity near the wall could not be clarified because of the measurement method and its limitations.

For the simple angle at a blowing ratio of 0.5 (Figure 4c), the CRVP was observed in both LES and RANS data, and the lift-off of the jet was predicted to a similar degree. Compared to the LES data, the upward flow predicted using the RANS was closer to the symmetry plane, the central temperature was higher, and the hot area marked in red was narrower. When the injection ratio reached 1.0 (Figure 4d), the prediction of the lift-off of the jet of the LES was larger than that of the RANS. In addition, the LES results clearly indicate that the high-temperature region becomes mushroom-shaped owing to the influence of the CRVP.

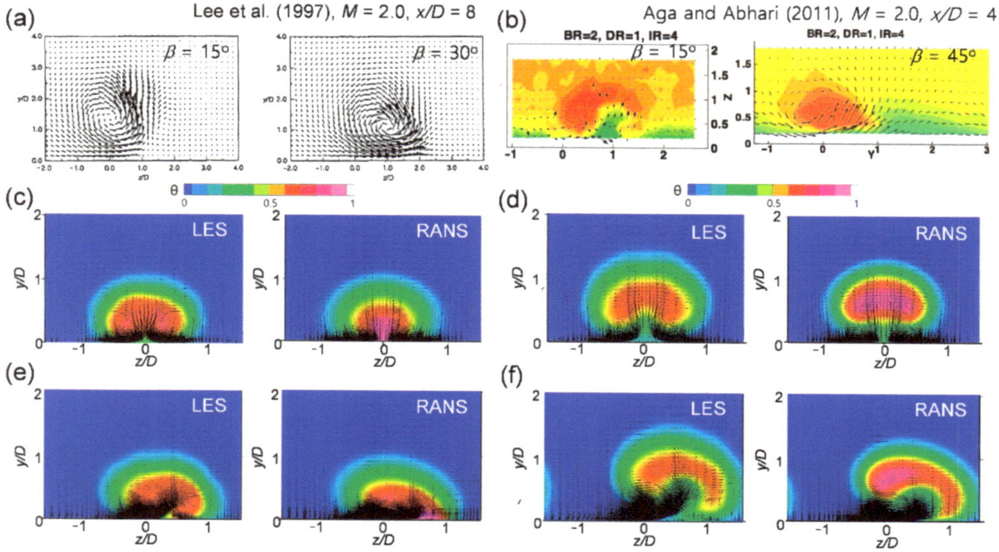

Figure 4. Time-averaged cross-stream velocity vectors with temperature contours at $x/D = 2.5$: (**a**) Cross-stream velocity vectors reported by Lee et al. [10]; (**b**) cross-stream velocity vectors with streamwise velocity contours, reported by Aga and Abhari [13]; (**c**) $M = 0.5$, $\beta = 0°$; (**d**) $M = 1.0$, $\beta = 0°$; (**e**) $M = 0.5$, $\beta = 30°$; and (**f**) $M = 1.0$, $\beta = 30°$.

When the compound angle was adopted, the flow patterns predicted through the LES and RANS at the blowing ratio of 0.5 (Figure 4e) were different. In the LES result, the counter-rotating vortex remained weak however, in the RANS, the vortex changed to a single vortex. The point at which the upward flow started was predicted to be near $z/D = 0$ and $z/D = 0.4$ by the LES and RANS, respectively. Lee et al. [10] reported that the CRVP changed to a single vortex at the orientation angle of 30° or more for a single jet. However, for the considered array jet, the transition to a single vortex was delayed at a low blowing ratio.

At a blowing ratio of 1.0 with the compound angle (Figure 4f), the CRVP transformed to a single vortex. The main flow was entrained along the upward flow of the vortex and the hot region was crescent-like. This aspect was common between the LES and RANS however, the LES predicted a higher lift-off than that pertaining to the RANS and the starting point of the upward flow was predicted to be closer to the plane of symmetry.

Figure 5 shows a comparison of the boundary layer temperature obtained through the CFD techniques at a blowing ratio of 0.5 with the experimental data presented by Jung and Lee [11] at three streamwise positions. At $x/D = 2.5$ with a simple angle (Figure 5a), the area of the high-temperature region (displayed in red) predicted by the LES was more similar to the experiment value. The RANS-predicted high-temperature region with a dimensionless temperature θ of 0.5 or higher was smaller than that in the experiment. Even at the downstream position, $x/D = 5.0$ (Figure 5b), the RANS underpredicted the intensity of mixing with the main flow compared to that obtained in the LES.

In the experimental data, the lift-off was unclear at the upstream locations (Figure 5a,b) because no data near the wall could be obtained through the cold wire measurement. At $x/D = 10$, as shown in Figure 5c, the center of the injectant appeared near $y/D = 0.5$ and lift-off could be observed in this part. The LES predicted that the lift-off occurred more intensely than that in the experiment, whereas the RANS did not predict any lift-off.

When the compound angle was adopted, the thermal field became asymmetric. This change in the temperature distribution was well predicted by both the LES and RANS (Figure 5d,f). However, in the case of the compound angle, the mixing characteristics could not be determined simply by considering the area of the hot part, as in the case of the simple angle. Nevertheless, the characteristics could be clarified considering the spacing of the isotherms. In particular, the isotherm spacing in the RANS was wider than that in the experiment.

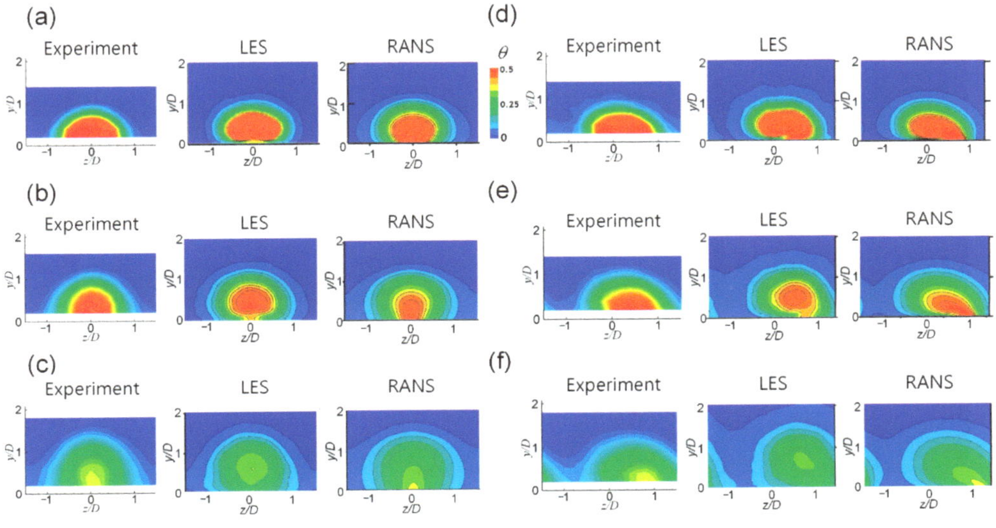

Figure 5. Boundary layer temperature distributions in the streamwise normal planes at $M = 0.5$: (**a**) $x/D = 2.5$, $\beta = 0°$; (**b**) $x/D = 5.0$, $\beta = 0°$; (**c**) $x/D = 10.0$, $\beta = 0°$; (**d**) $x/D = 2.5$, $\beta = 30°$; (**e**) $x/D = 5.0$, $\beta = 30°$; and (**f**) $x/D = 10.0$, $\beta = 30°$.

At a blowing ratio of 0.5, the LES predicted a lift-off and the RANS did not. At $x/D = 5.0$ (Figure 5b), the point at which the highest temperature occurred was measured to be $y/D = 0.3$, and the LES predicted the point as $y/D = 0.5$. The RANS predicted that the highest temperature would occur near the wall even in the downstream location, $x/D = 10$.

When the blowing ratio reached 1.0, lift-off of the injectant occurred in the case of both the simple and compound angle injection (Figure 6). In Figure 6a, which shows the temperature distribution at $x/D = 2.5$ for the simple angle, the area marked in red, pertaining to the CRVP, appeared to have an inverted heart shape. As shown in Figure 3, the LES predicted the location of the upward flow near the wall, closer to that in the experiment, and the shape of the high-temperature region obtained using the LES was similar to that in the experiment as well.

As the film cooling jet flowed downstream, it mixed with the main flow, thereby reducing the dimensionless temperature and changing the shape of the temperature distribution. In the case of a simple angle injection, the isotherm shape became similar to a concentric circle as the jet flowed downstream and both the LES and RANS predicted this phenomenon (Figure 6b,c). At $x/D = 10$ (Figure 6c), the maximum dimensionless temperature was approximately 0.4, observed at a height of approximately $y/D = 1$, and both the LES and RANS predicted this phenomenon to a similar extent.

In the case of the compound angle, at a blowing ratio of 1.0, the CRVP changed to a single vortex, and the high-temperature region exhibited a crescent shape at $x/D = 2.5$ (Figure 6d). In the case of the compound angle injection, the LES predicted the lift-off of the injectant to be 10% higher than that in the experimental data, whereas the RANS prediction was similar to the experiment value. The temperature distributions at the downstream locations for the RANS data were more similar to the experiment than those in the LES (Figure 6e,f).

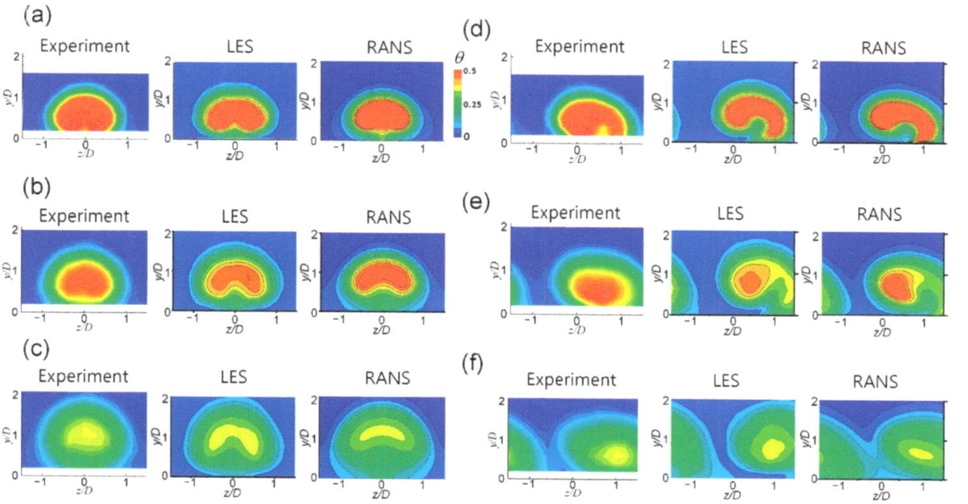

Figure 6. Boundary layer temperature distributions in the streamwise normal planes at $M = 1.0$: (**a**) $x/D = 2.5$, $\beta = 0°$; (**b**) $x/D = 5.0$, $\beta = 0°$; (**c**) $x/D = 10.0$, $\beta = 0°$; (**d**) $x/D = 2.5$, $\beta = 30°$; (**e**) $x/D = 5.0$, $\beta = 30°$; and (**f**) $x/D = 10.0$, $\beta = 30°$.

3.2. Adiabatic Film Cooling Effectiveness

For the local distributions of the adiabatic cooling effectiveness, the LES data and RANS results obtained using three turbulence models were compared with the experimental data presented by Jung and Lee [11]. Figure 7 shows comparisons of the distributions at a blowing ratio of 0.5. In the case of both the simple and compound angles, the LES-obtained distribution was closer to that obtained experimentally than that of the RANS. In the upstream region within $x/D = 2$, the realizable k-ε and RSM overpredicted the film cooling performance and thus a region with high effectiveness (red contour) appeared, which did not appear in the experimental data. The k-ω SST did not severely overpredict the film cooling effectiveness in the upstream region however, the diffusion in the downstream, compared with the other models, was predicted to be smaller than the experiment value.

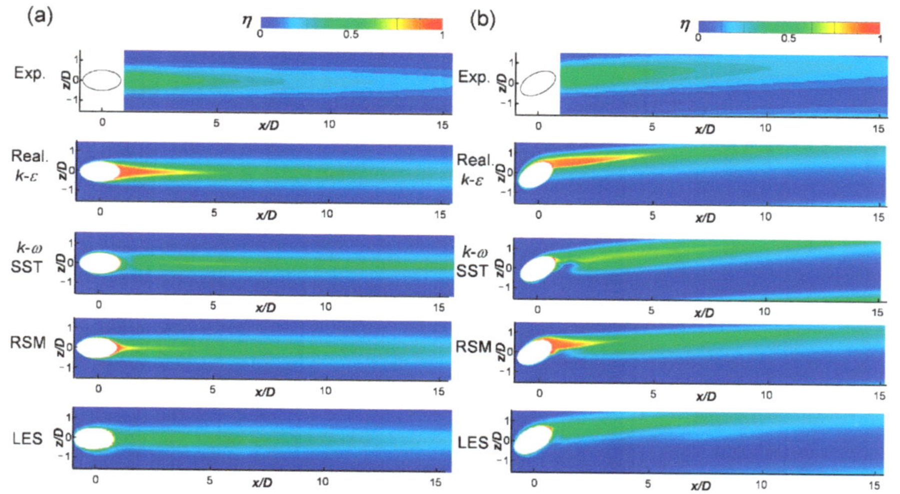

Figure 7. Local adiabatic film cooling effectiveness distributions at $M = 0.5$: (**a**) $\beta = 0°$ and (**b**) $\beta = 30°$.

In the simple angle case (Figure 7a), the CFD techniques predicted that the film cooling lasted longer than that in the experiment in the downstream region after $x/D = 10$. Moreover, the CFD results indicated that the spread of the film cooling effect between the holes was delayed downstream compared to that in the experiment. This phenomenon likely occurred because of the influence of the smaller prediction of the CFD for the mixing of the main flow and injectant and the inability to impose artificial perfect insulation similar to that in the CFD in the experiments. By observing the film cooling effectiveness around the hole, which could not be obtained experimentally, the effectiveness around the rim could be attributed to the presence of the horseshoe vortex clearly highlighted in the LES.

In the case of the compound angle (Figure 7b), the region with a high film cooling effectiveness spread widely and alleviated the degradation in the film cooling performance during the downstream flow. These characteristics were better predicted by the LES than by the RANS. The k-ω SST and RSM predicted that a region with low film cooling effectiveness would occur as the main flow was introduced to the surface with the downward flow of the main vortex generated in the film cooling jet in the downstream near the hole. This phenomenon was only slightly reflected in the LES and experiment. Khojasteh et al. [35] reported that the reattachment of the cooling jet occurred at approximately $x/D = 1.5$ in the simple angle case. Although this phenomenon was not clearly visible at $M = 0.5$, it could be noted in the experimental results for $M = 1.0$, as shown in Figure 8.

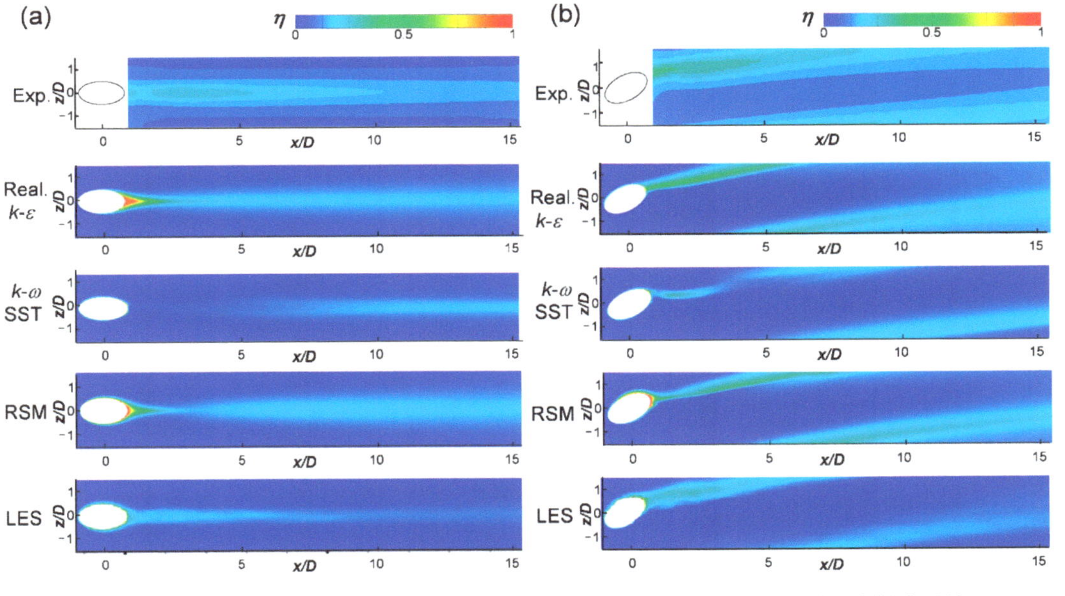

Figure 8. Local adiabatic film cooling effectiveness distributions at $M = 1.0$: (**a**) $\beta = 0°$ and (**b**) $\beta = 30°$.

Reattachment occurred in both the simple and compound angle cases and was more evident in the simple angle case. In the case of the simple angle, the area with high film cooling effectiveness due to the reattachment of the injectant could be observed only in the LES among the CFD results (Figure 8a). The RANS results indicated that the film cooling effectiveness increased after $x/D = 5$ however, this phenomenon was likely the effect of diffusion by mixing with the main flow rather than the reattachment. As shown in Figure 6, after $x/D = 2.5$, the height of the center of the injectant did not decrease even when the injectant moved downstream. In the case of the composite angle, both the LES and RANS predicted the occurrence of reattachment. Among the turbulence models, the realizable k-ε model obtained the distribution most similar to that in the experiment. At the blowing ratio of 1.0, the CFD techniques predicted film cooling effectiveness to be lower than that in the experiment and consequently, the distribution shown in Figure 8 involves more blue regions. Overall, the difference in the experimental data and simulation results was greater for the compound angle than for the simple angle.

To examine the film cooling performance between the holes and film cooling characteristics in the spanwise direction in the composite angle case, Figure 9 shows the spanwise variation of the film cooling effectiveness at $x/D = 2.5$. The LES predicted the spanwise diffusion of the film cooling effectiveness better than the RANS. When the blowing ratio increased (Figure 9c,d), the CFD techniques could not predict the spanwise spread accurately. Among the turbulence models, the k-ω SST and RSM obtained the best predictions for the film cooling performance at the blowing ratio of 0.5 and 1.0, respectively. In the case of the simple angle, the realizable k-ε and RSM could not predict the lift-off, and therefore, the centerline film cooling effectiveness was severely overestimated immediately downstream of the hole. The k-ω SST predicted the lift-off well at $M = 0.5$, albeit incorrectly at $M = 1.0$, corresponding to the largest difference from the experimental values. All the turbulence models predicted the lift-off more effectively in the case of the composite angle injection than the simple angle injection cases.

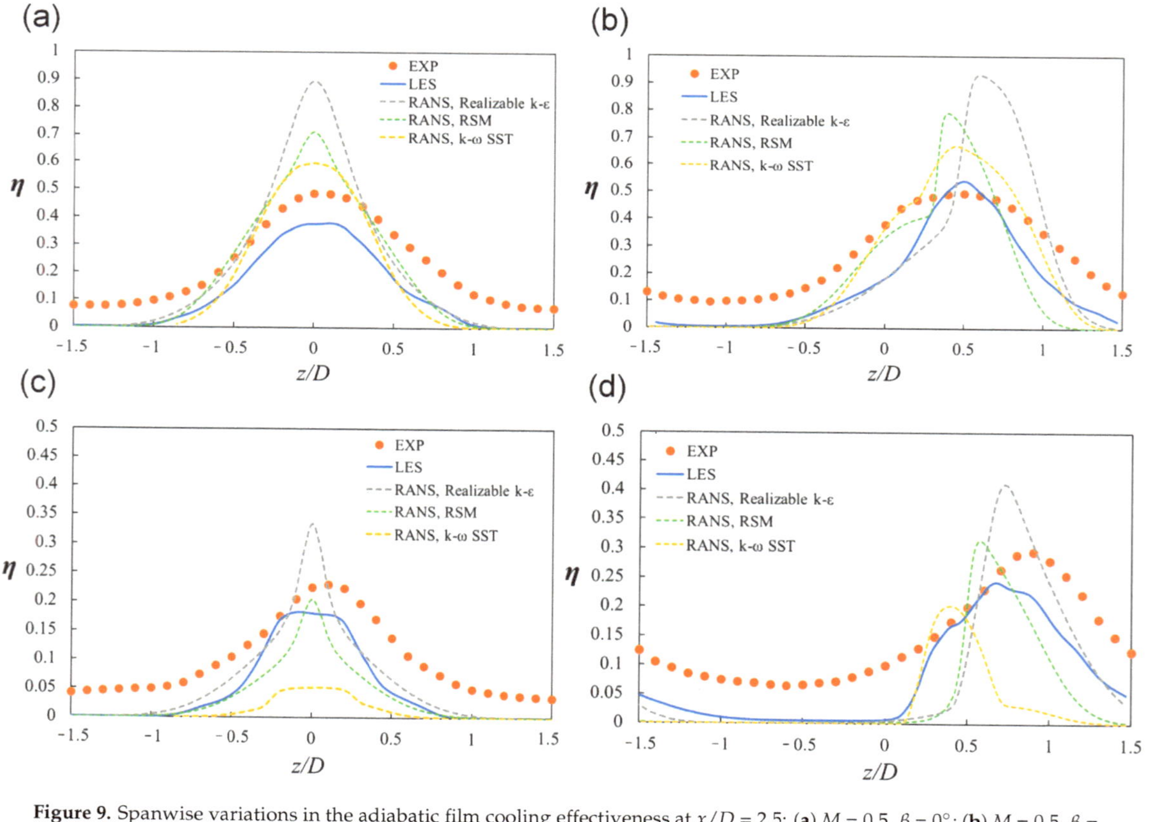

Figure 9. Spanwise variations in the adiabatic film cooling effectiveness at $x/D = 2.5$: (**a**) $M = 0.5$, $\beta = 0°$; (**b**) $M = 0.5$, $\beta = 30°$; (**c**) $M = 1.0$, $\beta = 0°$; and (**d**) $M = 1.0$, $\beta = 30°$.

3.3. Turbulence Statistics and Instantaneous Flow Fields

Figure 10 shows the turbulence intensity for each component obtained by the LES, in comparison with the experiment of Burd et al. [34]. As the RANS severely underpredicted the turbulence intensity, the maximum values of all the components for the three models did not exceed 10% and thus did not appear within the contour range in Figure 10. In the simple injection angle ($\beta = 0°$) case, at $M = 1.0$, the LES-obtained distribution was similar to that in the experiment under the same geometry and blowing ratio. Compared with that in Figure 3, the turbulence intensity of all the components was high in the part in which upward flow occurred in the CRVP.

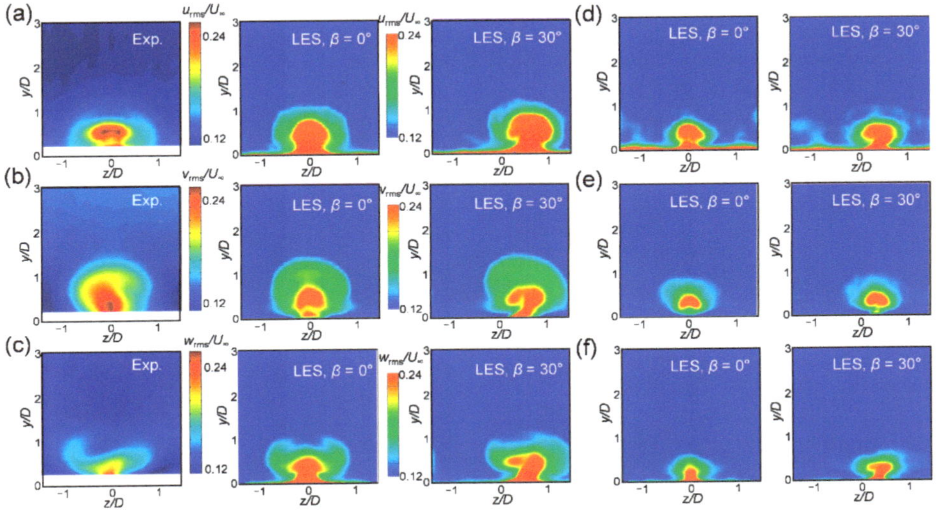

Figure 10. Turbulence intensities at $x/D = 2.5$ compared to those in the experiment of Burd et al. [32]: (**a**) u_{rms}, $M = 1.0$; (**b**) v_{rms}, $M = 1.0$; (**c**) w_{rms}, $M = 1.0$; (**d**) u_{rms}, $M = 0.5$; (**e**) v_{rms}, $M = 0.5$; and (**f**) w_{rms}, $M = 0.5$.

The three components exhibit a difference in the turbulence intensity below $y/D = 0.2$, which could not be measured experimentally. The u_{rms} and w_{rms} components not blocked by the wall appeared high along the wall, whereas the v_{rms} component converged to zero at the wall. In the case of the compound angle ($\beta = 30°$), the overall trend was similar to that of the simple angle. However, the CRVP changed to a single vortex and consequently, the contours became asymmetric, and the distribution moved in the +z direction.

At the blowing ratio of 0.5, the overall distributions appeared to be smaller than those at $M = 1.0$ (Figure 10d,f). Under the orientation angle of 30°, the movement in the z direction was smaller than that at the blowing ratio of 1.0. The u_{rms} component was larger than that when the blowing ratio was 1.0 near the wall.

Figure 11 shows the temperature fluctuations. The RANS predicted extremely small temperature fluctuations to be fitted within the contour range, and thus, only the LES results are shown in the figure. θ_{rms} was large in the region surrounding the area with large velocity fluctuation, as shown in Figure 10. In the same range of the contours, when $\beta = 0°$, the value was slightly higher than that when $\beta = 30°$. This finding supports the result of the high film cooling effectiveness under a compound angle as the injectant flowed downstream.

Figure 12 shows the isosurface of the second invariant of the velocity gradient tensor, which is known to reflect the vortex [36]. A horseshoe vortex, Kelvin–Helmholtz vortex, hanging vortex, rear vortex, and hairpin vortex could be observed in all the cases. The CRVP could be observed in the simple angle case, and at the blowing ratio of 1.0 (Figure 12c), the Kelvin-Helmholtz vortex appeared to be stronger with narrow intervals than that at the blowing ratio of 0.5 (Figure 12a).

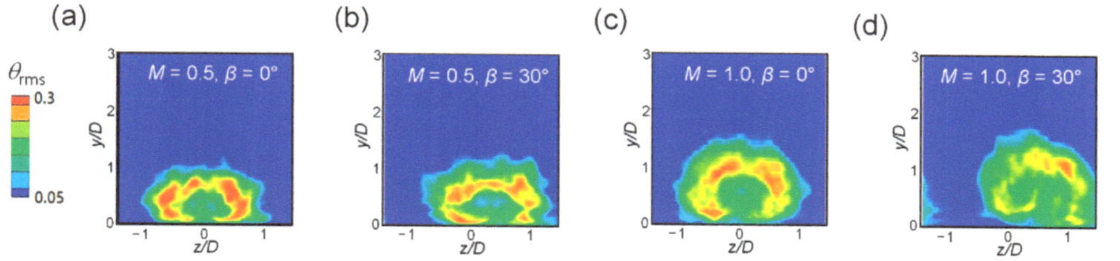

Figure 11. Temperature fluctuations obtained from the LES at x/D = 2.5: (**a**) M = 0.5, β = 0°; (**b**) M = 0.5, β = 30°; (**c**) M = 1.0, β = 0°; and (**d**) M = 1.0, β = 30°.

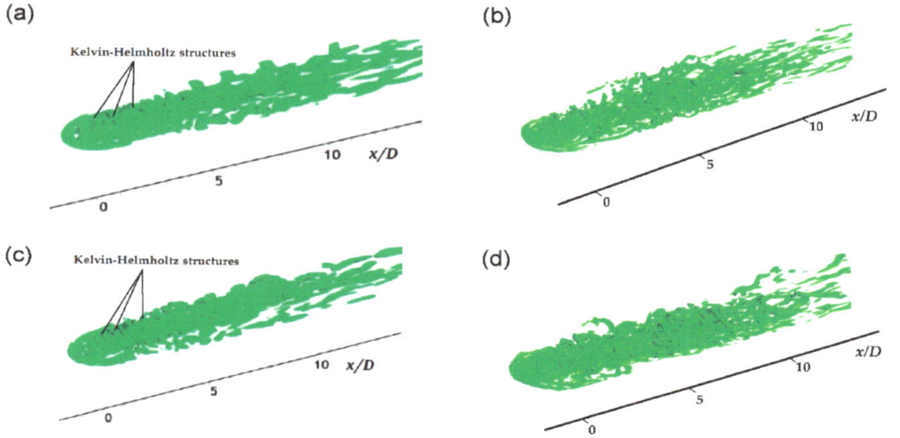

Figure 12. Isosurface of the second invariant of the velocity gradient tensor (Q): (**a**) M = 0.5, β = 0°; (**b**) M = 0.5, β = 30°; (**c**) M = 1.0, β = 0°; and (**d**) M = 1.0, β = 30°.

In the case of the compound angle (Figure 12b,d), the horseshoe vortex became asymmetric and the intensity of the Kelvin–Helmholtz vortex weakened. A single vortex was observed, with more complex vortices generated on the windward side. The vortical structures between the blowing ratios of 0.5 and 1.0 did not exhibit a distinct difference, and at the blowing ratio of 1.0, the injectant moved further in the spanwise direction.

Figure 13 shows the instantaneous temperature field at the wall obtained through the LES. t^* is the dimensionless time defined by the main flow velocity and diameter of the hole, such that t^* = 1 is the time that the main flow covers a distance equivalent to the hole diameter.

As the adiabatic condition was imposed on the wall, the instantaneous temperature field exhibited the instantaneous adiabatic film cooling effectiveness. When the blowing ratio was 0.5, the injectant flowed along the wall and traces of the injectant, indicated in light green on the contour, extended from the hole to the downstream region (Figure 13a,b). When the blowing ratio reached 1.0, lift-off and reattachment occurred intermittently. As shown in Figure 13c, when t^* = 2, traces of the injectant appeared from the hole however, when t^* = 4, the traces were eliminated owing to the influence of the near wall vortices, such as the rear vortex shown in Figure 12.

In the case of the simple angle (Figure 13a,c), the film cooling appeared around the rim of the hole due to the influence of the horseshoe vortex and hanging vortex. Under a compound angle, the film cooling effectiveness was reduced around the leeward side of the rim (Figure 13b) or eliminated (Figure 13d). This phenomenon was attributed to the main flow being entrained in the shear layer sweeps while the CRVP changed to a single vortex. In the case of the simple angle injection at the blowing ratio of 1.0, most of the insulating film was eliminated after x/D = 10 (Figure 13c) however, in the case of the compound injection angle, the injectant flowed along the wall and maintained the insulating film (Figure 13d).

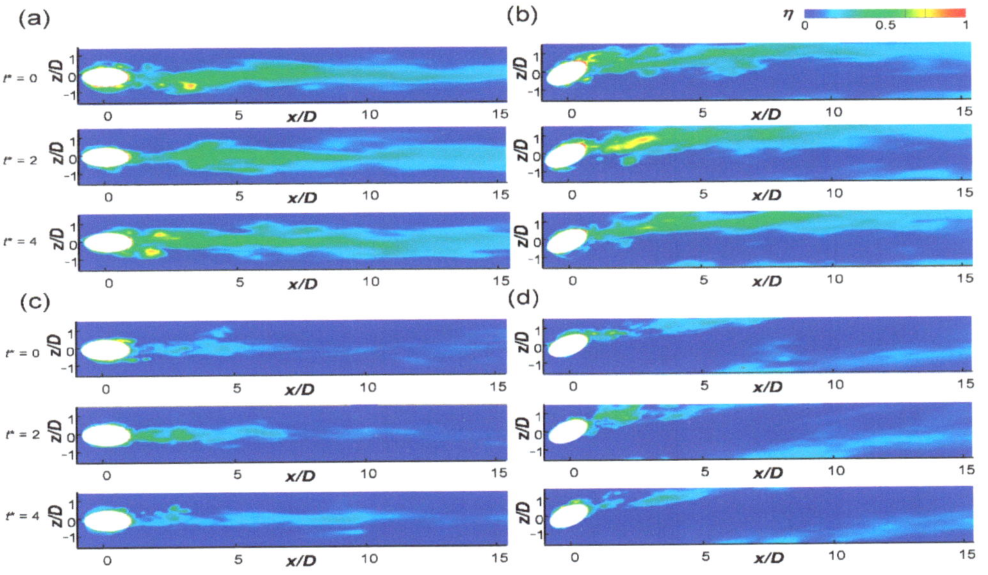

Figure 13. Instantaneous adiabatic film cooling effectiveness distributions: (**a**) M = 0.5, β = 0°; (**b**) M = 0.5, β = 30°; (**c**) M = 1.0, β = 0°; and (**d**) M = 1.0, β = 30°.

4. Conclusions

LES was performed for film cooling with simple and complex angles. Moreover RANS, with three turbulence models, namely, the realizable k-ε, k-ω SST, and RSM, was performed and the spatial distributions of the flow and temperature fields were compared with the experimental data presented at a high resolution. Certain information that was not well predicted with the RANS was clarified and the information that could not be measured experimentally was observed in the LES results. The following conclusions were derived:

1. In the time-averaged flow field, the RANS data exhibited a difference from the experiment and LES in terms of the rising point of the CRVP as the vortices collided with each other on the wall. When the injection ratio was 0.5 and the orientation angle is 30°, the LES predicted that the counter-rotating vortex remained weak and the RANS predicted that the vortex completely changed to a single vortex;
2. The RANS did not accurately predict the lift-off of the injectant or mixing with the main flow, and thus, it could not accurately predict the film cooling performance. The corresponding predictions obtained using the LES were better. The reattachment of the injectant at the blowing ratio of 1.0 was better predicted by the RANS in the compound angle case than that in the case of the simple angle;
3. The turbulence intensity was large in the region in which the upward flow of the vortex in the injectant was generated and the temperature fluctuation was large at the boundary of the turbulent intensity peak. The temperature fluctuation slightly decreased when the injectant was supplied at a compound angle;
4. In the compound angle case, the insulation film was eliminated near the leeward rim of the film cooling hole due to the influence of the single vortex however, at the

injection ratio of 1.0, the injectant flowed along the wall more smoothly than that in the simple injection angle, thereby enhancing the downstream film cooling performance.

Author Contributions: S.I.B. carried out the simulations, analyzed the CFD data, and helped to write the paper. J.A. supervised the research, analyzed the CFD data, and helped to write the paper. All authors have read and agreed to the published version of the manuscript.

Funding: This research was funded by the UAV High Efficiency Turbine Research Center program of Defense. The APC was funded by the UAV High Efficiency Turbine Research Center program of Defense.

Data Availability Statement: Data is contained within this article.

Acknowledgments: This work was supported by the UAV High Efficiency Turbine Research Center program of Defense.

Conflicts of Interest: The authors declare that there are no conflict of interest.

Nomenclature

C_s	Smagorinsky constant
C_p	Specific heat [J/(kgK)]
D	diameter of a single hole [mm]
d	wall distance [mm]
L	delivery tube length [mm]
L_s	mixing length of subgrid scales = $\min(\kappa d,\ C_s\Delta)$
M	blowing ratio = $(\rho_C U_C)/(\rho_G U_G)$
P	pitch of the holes [mm]
q_j	heat flux [W/mm^2]
T	emperature [K]
t	time [s]
t^*	non-dimensional time = $U_\infty\, t/D$
U	time-averaged flow velocity [m/s]
U_∞	freestream velocity [m/s]
u	streamwise velocity [m/s]
v	wall-normal velocity [m/s]
w	panwise velocity [m/s]
x	streamwise coordinate
y	wall-normal coordinate
z	spanwise coordinate
Greek symbols	
α	thermal diffusivity [m^2/s]
η	adiabatic film cooling effectiveness = $\frac{(T_G - T_{aw})}{T_G - T_C}$
η_C	centerline film cooling effectiveness
κ	thermal conductivity [W/(mK)]
ρ	density [kg/m^3]
τ_{ij}	subgrid-scale turbulent stress = $\rho\overline{u_i u_j} - \rho\overline{u}_i\overline{u}_j$
μ_t	subgrid-scale turbulent viscosity [kg/(m·s)]
Δ	local grid scale
θ	dimensionless temperature = $\frac{(T_G - T)}{T_G - T_C}$
Subscripts	
aw	adiabatic wall
c	centerline
C	coolant
G	mainstream gas
rms	root mean square value

References

1. Lakshminarayana, B. *Fluid Dynamics and Heat Transfer of Turbomachinery*; Wiley: Hoboken, NJ, USA, 1996; pp. 315–322.
2. Goldstein, R.J. Film Cooling. In *Advances in Heat Transfer*; Elsevier: Amsterdam, The Netherlands, 1971; Volume 7, pp. 321–379.
3. Goldstein, R.J.; Eckert, E.R.G.; Ramsey, J.W. Film Cooling with Injection through Holes: Adiabatic Wall Temperatures Downstream of a Circular Hole. *J. Eng. Power* **1968**, *90*, 384–393. [CrossRef]
4. Pedersen, D.R.; Eckert, E.R.G.; Goldstein, R.J. Film Cooling with Large Density Differences between the Mainstream and the Secondary Fluid Measured by the Heat-Mass Transfer Analogy. *J. Heat Transf.* **1977**, *99*, 620–627. [CrossRef]
5. Sinha, A.K.; Bogard, D.G.; Crawford, M.E. Film-Cooling Effectiveness Downstream of a Single Row of Holes with Variable Density Ratio. *J. Turbomach.* **1991**, *113*, 442–449. [CrossRef]
6. Lee, S.W.; Lee, J.S.; Ro, S.T. Experimental Study on the Flow Characteristics of Streamwise Inclined Jets in Crossflow on Flat Plate. *J. Turbomach.* **1994**, *116*, 97–105. [CrossRef]
7. Coletti, F.; Benson, M.; Ling, J.; Elkins, C.J.; Eaton, J. Turbulent transport in an inclined jet in crossflow. *Int. J. Heat Fluid Flow* **2013**, *43*, 149–160. [CrossRef]
8. Schmidt, D.L.; Sen, B.; Bogard, D.G. Film Cooling with Compound Angle Holes: Adiabatic Effectiveness. *J. Turbomach.* **1996**, *118*, 807–813. [CrossRef]
9. Ligrani, P.M.; Lee, J.S. Film Cooling from a Single Row of Compound Angle Holes at High Blowing Ratios. *Int. J. Rotating Mach.* **1996**, *2*, 259–267. [CrossRef]
10. Lee, S.W.; Kim, Y.B.; Lee, J.S. Flow Characteristics and Aerodynamic Losses of Film-Cooling Jets with Compound Angle Orientations. *J. Turbomach.* **1997**, *119*, 310–319. [CrossRef]
11. Jung, I.S.; Lee, J.S. Effects of Orientation Angles on Film Cooling over a Flat Plate: Boundary Layer Temperature Distributions and Adiabatic Film Cooling Effectiveness. *J. Turbomach.* **1999**, *122*, 153–160. [CrossRef]
12. Aga, V.; Rose, M.; Abhari, R.S. Experimental Flow Structure Investigation of Compound Angled Film Cooling. *J. Turbomach.* **2008**, *130*, 031005. [CrossRef]
13. Aga, V.; Abhari, R.S. Influence of Flow Structure on Compound Angled Film Cooling Effectiveness and Heat Transfer. *J. Turbomach.* **2011**, *133*, 031029. [CrossRef]
14. McGovern, K.T.; Leylek, J.H. A Detailed Analysis of Film Cooling Physics: Part II—Compound-Angle Injection with Cylindrical Holes. *J. Turbomach.* **2000**, *122*, 113–121. [CrossRef]
15. Harrison, K.L.; Bogard, D.G. Comparison of RANS Turbulence Models for Prediction of Film Cooling Performance. In Proceedings of the ASME Turbo Expo 2008, Berlin, Germany, 9–13 June 2008. GT2008-51423.
16. Yu, F.; Yavuzkurt, S. Near-Field Simulations of Film Cooling with a Modified DES Model. *Inventions* **2020**, *5*, 13. [CrossRef]
17. Tyagi, M.; Acharya, S. Large Eddy Simulation of Film Cooling Flow from an Inclined Cylindrical Jet. *J. Turbomach.* **2003**, *125*, 734–742. [CrossRef]
18. Ziefle, J.; Kleiser, L. Numerical Investigation of a Film-Cooling Flow Structure: Effect of Crossflow Turbulence. *J. Turbomach.* **2013**, *135*, 041001. [CrossRef]
19. Sakai, E.; Takahashi, T.; Watanabe, H. Large-eddy simulation of an inclined round jet issuing into a crossflow. *Int. J. Heat Mass Transf.* **2014**, *69*, 300–311. [CrossRef]
20. Dai, C.; Jia, L.; Zhang, J.; Shu, Z.; Mi, J. On the flow structure of an inclined jet in cross flow at low velocity ratios. *Int. J. Heat Fluid Flow* **2016**, *29*, 1–17.
21. Rozati, A.; Tafti, D. Large-eddy simulations of leading edge film cooling: Analysis of flow structures, effectiveness, and heat transfer coefficient. *Int. J. Heat Fluid Flow* **2008**, *29*, 1–17. [CrossRef]
22. Ahn, J.; Schobeiri, M.; Han, J.-C.; Moon, H.-K. Effect of rotation on leading edge region film cooling of a gas turbine blade with three rows of film cooling holes. *Int. J. Heat Mass Transf.* **2007**, *50*, 15–25. [CrossRef]
23. Lakehal, D.; Theodoridis, G.; Rodi, W. Three-dimensional flow and heat transfer calculations of film cooling at the leading edge of a symmetrical turbine blade model. *Int. J. Heat Fluid Flow* **2001**, *22*, 113–122. [CrossRef]
24. Graf, L.; Kleiser, L. Large-Eddy Simulation of double-row compound-angle film cooling: Setup and validation. *Comput. Fluids* **2011**, *43*, 58–67. [CrossRef]
25. Ahn, J.; Jung, I.S.; Lee, J.S. Film cooling from two rows of holes with opposite orientation angles: Injectant behavior and adiabatic film cooling effectiveness. *Int. J. Heat Fluid Flow* **2003**, *24*, 91–99. [CrossRef]
26. Li, W.; Li, X.; Ren, J.; Jiang, H. Large eddy simulation of compound angle hole film cooling with hole length-to-diameter ratio and internal crossflow orientation effects. *Int. J. Therm. Sci.* **2017**, *121*, 410–423. [CrossRef]
27. McClintic, J.; Klavetter, S.; Winka, J.; Anderson, J.; Bogard, D.; Dees, J.; Laskowski, G.; Briggs, R. The effect of internal cross-flow on the adiabatic effectiveness of compound angle film cooling holes. *J. Turbomach.* **2015**, *137*, 071006. [CrossRef]
28. Stratton, Z.; Shih, T.; Laskowski, G.; Barr, B.; Briggs, R. Effects of Crossflow in an Internal-Cooling Channel on Film Cooling of a Flat Plate through Compound-Angle Holes. In Proceedings of the ASME Turbo Expo Proceeding, Montreal, QC, Canada, 15–19 June 2015.
29. Baek, S.I.; Yavuzkurt, S. Effects of Flow Oscillations in the Mainstream on Film Cooling. *Inventions* **2018**, *3*, 73. [CrossRef]
30. Baek, S.I.; Ahn, J. Large Eddy Simulation of Film Cooling with Triple Holes: Injectant Behavior and Adiabatic Film-Cooling Effectiveness. *Processes* **2020**, *8*, 1443. [CrossRef]

31. ANSYS Fluent Theory Guide Version 19. Available online: https://www.ansys.com/products/fluids/ansys-fluent (accessed on 1 December 2020).
32. Tannehill, J.; Anderson, D.; Pletcher, R. *Computational Fluid Mechanics and Heat Transfer*, 2nd ed.; Taylor & Francis: New York, NY, USA, 1997.
33. Fujimoto, S. Large Eddy Simulation of Film Cooling Flows Using Octree Hexahedral Meshes. In Proceedings of the ASME Turbo Expo 2012, Copenhagen, Denmark, 11–15 June 2012. GT2012-70090.
34. Burd, S.W.; Kaszeta, R.W.; Simon, T.W. Measurements in Film Cooling Flows: Hole L/D and Turbulence Intensity Effects. *J. Turbomach.* **1998**, *120*, 791–798. [CrossRef]
35. Khojasteh, A.R.; Wang, S.F.; Peng, D.; Yavuzkurt, S.; Liu, Y. Structure analysis of adiabatic film cooling effectiveness in the near field of a single inclined jet: Measurement using fast-response pressure-sensitive paint. *Int. J. Heat Mass Transf.* **2017**, *110*, 629–642. [CrossRef]
36. Kolář, V. Vortex identification: New requirements and limitations. *Int. J. Heat Fluid Flow* **2007**, *28*, 638–652. [CrossRef]

processes

MDPI

Article

Numerical Investigation on the Transient Flow of a Boiler Circulating Pump Based on the Shear Stress Transport Turbulence Model

Fei Zhao [1,2], Fanyu Kong [1,*], Xiaohui Duan [3], Huiyuan Wu [2] and Jun Wang [2]

1 National Research Center of Pumps, Jiangsu University, Zhenjiang 212013, China; zhaof@wxit.edu.cn
2 Wuxi Institute of Technology, School of Mechanical Technology, Wuxi 214121, China; wuhy@wxit.edu.cn (H.W.); wangjun@wxit.edu.cn (J.W.)
3 NR Electric Power Electronics Co., Ltd., Changzhou 213025, China; duanxh_br@nrec.com
* Correspondence: kongfy2918@sohu.com

Received: 23 June 2020; Accepted: 11 August 2020; Published: 13 October 2020

Abstract: Based on the shear stress transport (SST) turbulence model, the influence of different outlet pipe angles on the head and efficiency of a boiler circulating pump was analyzed. When the outlet pipe angle changed from 115° to 130°, the head and efficiency of the pump reduced significantly. The boiler circulating pump with 115° outlet pipe angle was selected as the further research object, and the reliability of the numerical simulation was verified by experiments. The transient flow of the prototype pump under the design flow rate condition ($1.0Q_d$) and four other flow rate conditions ($0.6Q_d$, $0.8Q_d$, $1.2Q_d$, and $1.4Q_d$) was studied. The results show that, under the conditions of design flow and large flow rate ($1.0Q_d$, $1.2Q_d$, and $1.4Q_d$), the centrality and regularity of radial force distribution are obviously better than those of small flow rate ($0.6Q_d$, $0.8Q_d$). The leakage of the rear seal ring is less than that of the front seal ring under five flow rate conditions. As the flow rate increases, the leakage of front and rear seal rings decreases, and the leakage ratio of front and rear seal rings increases. The energy loss of the rear cover plate is greater than the energy loss of the front cover plate under five flow rate conditions. With the increase in flow rate, the total loss energy of the prototype pump decreases first and then increases, and the energy loss of the disc becomes larger and larger.

Keywords: SST; boiler circulating pump; outlet pipe angle; radial force; leakage of seal ring; energy loss of disc

1. Introduction

At present, electric energy mainly comes from thermal power, nuclear power, hydropower, wind power, and solar power, among which thermal power still occupies the leading position [1]. The boiler circulating pump is an important medium of circulating equipment in supercritical thermal power stations [2–4], which is used for circulating high-temperature and high-pressure water. Its stable operation directly affects the safety of the generating unit. For a long time, the boiler circulating pumps used in China mainly depended on imports, and the operation and maintenance costs were high. It is of great significance to carry out research on boiler circulating pumps to improve the level of localization.

Scholars around the world carried out research on boiler circulating pumps, mainly focusing on structural optimization, the hydraulic model, and flow characteristics. Baumgarten et al. [5] re-designed the impeller and diffuser in a boiler circulating pump, analyzed the internal flow of the diffuser and casing, and presented the structure of a hemispherical pump casing and acute angle outlet pipe, which could significantly reduce energy loss and improve the efficiency of the pump. Sha et al. [6] explored the function of a balance hole in the impeller of a high-temperature and high-pressure centrifugal pump, and they found that the axial force decreased by 15% after increasing the balance hole, which effectively improved the performance of the pump. Zuo et al. [7] optimized the casing of a boiler circulating pump based on computational fluid dynamics (CFD) and finite element method (FEM) analyses. Different combinations of casing materials, casing wall thickness, and casing sizes were numerically simulated to evaluate its safety and economic efficiency. Ni et al. [8] studied pressure fluctuation in the spherical shell of a nuclear reactor coolant pump based on large eddy simulation (LES), and pointed out that unsteady vortex flow is one of the main causes of pressure fluctuation in the pump. Su et al. [9] took a nuclear reactor coolant pump as the research object and investigated the change rule of pressure fluctuation during the starting process of the pump; they pointed out that the pressure fluctuation intensity of the guide vane passage was stronger than that of the impeller passage, while the pressure fluctuation intensity of the impeller and the pressure chamber was weaker.

Numerical simulations are widely used in the flow analysis of pumps [10–13]. The analysis of radial force distribution, seal ring leakage, and disc loss is helpful to improve the performance of pumps and optimize their hydraulic design.

Zhao et al. [14] studied the internal radial force of a deep-sea lifting pump; they pointed out that the dynamic and static interferences of the impeller and diffuser are important reasons for the unbalanced radial force in the pump, while the radial force and pressure in the pump have periodic fluctuation characteristics. Qian et al. [15] investigated the influence of export gland on the radial force in a multistage pump, and the results showed that the amplitude and main frequency of radial force decreased significantly after adding the export gland. Cheng et al. [16] analyzed the influence of the static and rotor clearance ratio on the hydraulic performance of a nuclear main pump. The results showed that the change in clearance ratio has great influence on the hydraulic loss of the guide vane and the water chamber. Gao and Yang [17–19] performed numerical simulations to study the effect of clearance on the flow field of a centrifugal pump. The results showed that, with the increase in clearance, the amplitude of pressure fluctuation and the radial force increased, whereas the head and efficiency decreased. Liu et al. [20–22] verified and modified the existing disc loss calculation method based on numerical calculations, and they finally obtained a prediction model with high accuracy. Dong et al. [23,24] combined experimental and numerical calculations to verify their disc loss prediction model, and they put forward the shortcomings and basis of the prediction model.

Most of the above studies were based on common centrifugal pumps, while there were few investigations on the radial force of a rotor, the leakage of a seal ring, and the disc loss of a boiler circulating pump.

In this paper, with the help of Pro/Engineer 5.0® software, the three-dimensional (3D) model of a boiler circulating pump was built. Based on the shear stress transport (SST) turbulence model, the influence of the boiler circulating pump with different outlet pipe angles on the head and efficiency of the boiler circulating pump was compared with the help of ANSYS 14.5® software. According to the analysis of different outlet pipe angles, a boiler circulating pump prototype with an outlet pipe angle of 115° was manufactured, performance experiments were carried out, and the experimental results of the boiler circulating pump prototype were compared with the results of a numerical simulation. The radial force of the rotor, the leakage of the seal ring, and the disc loss characteristics of the pump under different flow rate conditions were studied further. The conclusions of this paper can provide a reference for the design and optimization of boiler circulating pumps.

2. Materials and Methods

2.1. Pump Structure and Design Schemes of Pump Casing

The design parameters of boiler circulating pump are shown in Table 1. The transport medium of boiler circulating pump is high temperature and high-pressure water. The hemispherical structure shell can effectively balance the problem of asymmetric circumferential deformation, so that the pump casing has a strong compressive capacity. The structure of the boiler circulating pump is shown in Figure 1.

Table 1. Pump design parameters.

Parameters	Unit	Parameter Value
Design flow rate (Q_d)	m^3/h	966
Hydraulic head (H)	m	133
Design rotation speed (n)	r/min	2950
Specific speed (n_s)	–	142
Net Positive Suction Head Required ($NPSHr$)	m	17
Design pressure (P)	MPa	37
Design temperature (T)	°C	360

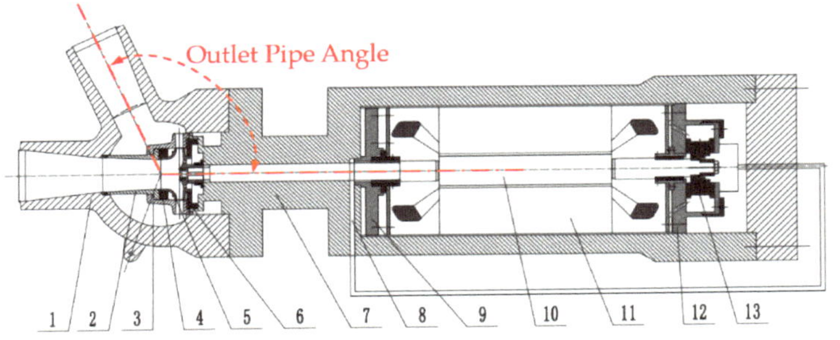

Figure 1. Structure of boiler circulating pump. 1 Pump casing; 2 connecting body; 3 guide vane; 4 front seal ring; 5 impeller; 6 rear seal ring; 7 motor housing; 8 cooling circuit; 9 front bearing seat; 10 rotor; 11 stator; 12 rear bearing seat; 13 thrust disk.

Different outlet pipe angles on the pump casing (as shown in Figure 1) will affect the utilization of the space around the boiler water circulating pump and the performance of the pump. Baumgarten S. et al. [5] of KSB group pointed out that the outlet pipe angle of the boiler circulating pump should be obtuse, but did not specifically analyze the impact of the outlet pipe angle on the performance of the pump. In order to analyze the influence of outlet pipe angle on the performance of boiler circulating pump, four different forms of pump casing have been designed by Pro/Engineer 5.0® software (Parametric Technology Corporation, Needham, MA, USA), as shown in Figure 2; the corresponding outlet pipe angles are 90°, 100°, 115°and 130°, respectively.

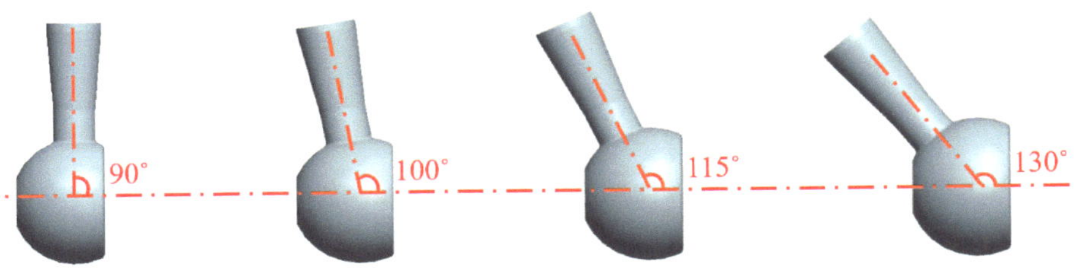

Figure 2. Pump Casings with different outlet pipe angles.

2.2. Mesh Generation and Independence Verification

ICEM 14.5® (ANSYS, Inc., Canonsburg, PA, USA) software was used to divide the hexahedral grid in the convection field. The grid at the blades were densified, and the quality of each water grid was controlled to be greater than 0.35. Figure 3 shows the flow field grid of the boiler circulating pump.

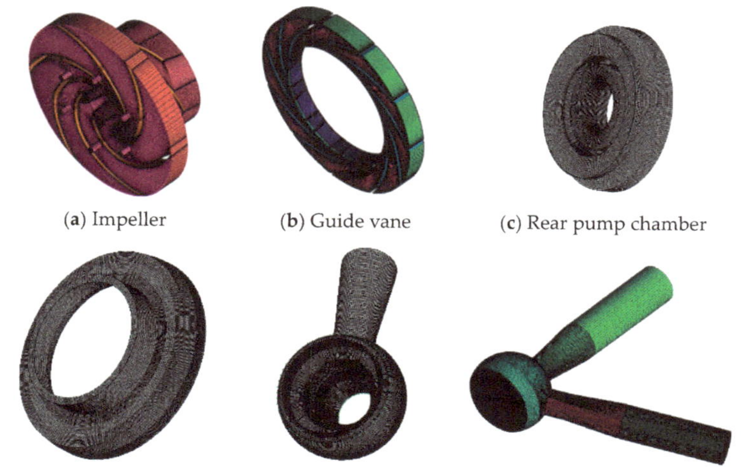

(a) Impeller (b) Guide vane (c) Rear pump chamber

(d) Front pump chamber (e) Volute (f) Grid assembly drawing

Figure 3. Flow field grid of boiler circulating pump.

Taking head and efficiency as the evaluation indexes, grid independence analysis was carried out, and the results are shown in Table 2. As can be seen from Table 2, FEM solution developed is convergent and the change of the solution is small when grid number is greater than 2.9×10^6. Considering the limitation of computing time and computer hardware, the 2.9×10^6 grid is selected for numerical simulation.

Table 2. Grid independence analysis.

Grid Number	Head/m	Efficiency/%
2.0×10^6	137.9	80.0
2.4×10^6	139.5	80.9
2.9×10^6	141.2	81.4
3.2×10^6	141.1	81.4
4.0×10^6	141.3	81.4

2.3. Turbulence Model and Simulation Setting

The mass conservation and momentum conservation are the basic laws that must be satisfied by the flow system, and the two equations are shown as follows.

$$\frac{\partial}{\partial x_i}(\rho u_i) = 0 \tag{1}$$

$$\frac{\partial}{\partial x_j}(\rho u_i u_j) = -\frac{\partial p}{\partial x_i} + \frac{\partial}{\partial x_j}[\mu(\frac{\partial u_i}{\partial x_j} + \frac{\partial u_j}{\partial x_i})] + \frac{\partial}{\partial x_j}(-\rho\overline{u_i' u_j'}) + F_i \tag{2}$$

where ρ is the density of liquid, μ is the dynamic viscosity, u means the flow velocity, and F_i stands for source item [25].

The shear stress transport (SST) turbulence model is used in numerical simulation of the boiler circulating pump. The SST turbulence model is a mixture of turbulence model $k-\varepsilon$ and turbulence model $k-\omega$, where k means the turbulence kinetic energy, ε means the dissipation of turbulence kinetic energy, ω means the turbulence frequency. The SST turbulence model not only has the accuracy of the model $k-\varepsilon$ in calculating the free flow in the boundary layer edges and free-shear layers, but also has the reliability of the model $k-\omega$ in calculating the flow in the near wall layers [26,27]. The control equations of SST turbulence model are as follows [28,29].

$$\frac{\partial}{\partial t}(\rho k)+\frac{\partial}{\partial x_i}(\rho u_i k)=\widetilde{P}_k-\beta^*\rho k\omega+\frac{\partial}{\partial x_i}[(\mu+\sigma_k\mu_t)\frac{\partial k}{\partial x_i}] \tag{3}$$

$$\frac{\partial}{\partial t}(\rho\omega)+\frac{\partial}{\partial x_i}(\rho u_i\omega)=\alpha\frac{1}{v_t}\widetilde{P}_k-\beta\rho\omega^2+\frac{\partial}{\partial x_i}[(\mu+\sigma_\omega\mu_t)\frac{\partial\omega}{\partial x_i}]+2(1-F_1)\rho\sigma_{\omega 2}\frac{1}{\omega}\frac{\partial k}{\partial x_i}\frac{\partial\omega}{\partial x_i} \tag{4}$$

$$v_t=\frac{a_1 k}{\max(a_1\omega,SF_2)} \tag{5}$$

$$S=\sqrt{2S_{ij}S_{ij}} \tag{6}$$

$$P_k=\mu_t\frac{\partial u_i}{\partial x_j}(\frac{\partial u_i}{\partial x_j}+\frac{\partial u_j}{\partial x_i})\rightarrow\widetilde{P}_k=min(P_k,10\cdot\beta^*\rho k\omega) \tag{7}$$

$$F_1=\tan\mathrm{h}\{\{min[max(\frac{\sqrt{k}}{\beta^*\omega y},\frac{500v}{y^2\omega}),\frac{4\rho\sigma_{\omega 2}k}{CD_{k\omega}y^2}]\}^4\} \tag{8}$$

$$F_2=\tan\mathrm{h}\{[max(2\frac{\sqrt{k}}{\beta^*\omega y},\frac{500v}{y^2\omega})]^2\} \tag{9}$$

$$CD_{k\omega}=max(2\rho\sigma_{\omega 2}\frac{1}{\omega}\frac{\partial k}{\partial x_i}\frac{\partial\omega}{\partial x_i},10^{-10}) \tag{10}$$

$$\alpha=\alpha_1F_1+\alpha_2(1-F_1) \tag{11}$$

where, ρ means the density of liquid, k means the turbulence kinetic energy, u_i means the flow velocity, ω means the turbulence frequency, y means the distance to the nearest wall, S means the invariant measure of the strain rate. F_1 and F_2 are the blending functions. The constants in the above equations are: $\beta^*=0.09,\alpha_1=5/9,\beta_1=0.075,\sigma_{k1}=0.85,\sigma_{\omega 1}=0.5,\alpha_2=0.44,\ \beta_2=0.0828,\ \sigma_{k2}=1.0,\sigma_{\omega 2}=0.856$.

The constant calculation used the inlet mass flow and outlet opening setting. Figure 4 shows the boundary conditions of the boiler circulating pump in the numerical simulation.

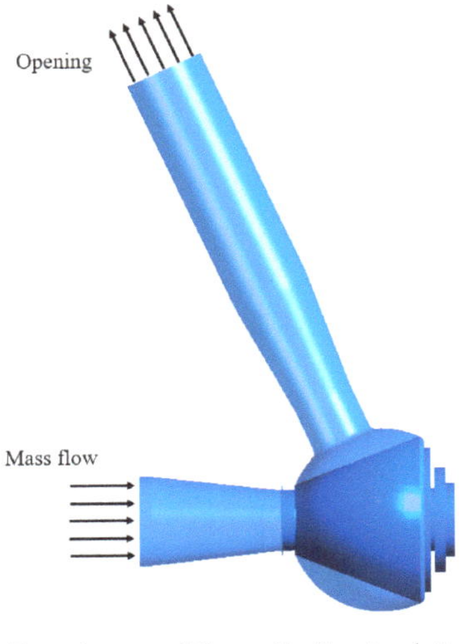

Figure 4. Boundary conditions of boiler circulating pump.

The outer wall of the impeller was set as the rotating wall, and the rotating speed was 2950 r/min. The solid wall adopted the non-slip boundary condition. According to the actual situation, the wall roughness was set to 25 μm, and the near wall area was treated by the standard wall function. The interface connection between the dynamic and static areas was set to the frozen rotor mode, and the calculation convergence residual was set to 10^{-5}.

The setting of time step in the unsteady calculation was based on the Courant number criterion [30]:

$$C_0 = \frac{|\overline{v}|\Delta t}{l} \leq 50 \tag{12}$$

where, $\overline{v}$ stands for estimate average speed (m/s), l stands for the minimum size of grid (m). Set the time step to 1.69492×10^{-4} s, and rotate 120 steps for one cycle.

2.4. Test Bench for Boiler Circulating Pump

The external characteristic experiment was carried out at the boiler circulating pump test bench in Wanhua Electrical Machinery Tech. Development Co., Ltd. Schematic diagram of the boiler circulating pump test bench structure and the scene of pump external characteristic experiment are shown in Figure 5a,b, respectively.

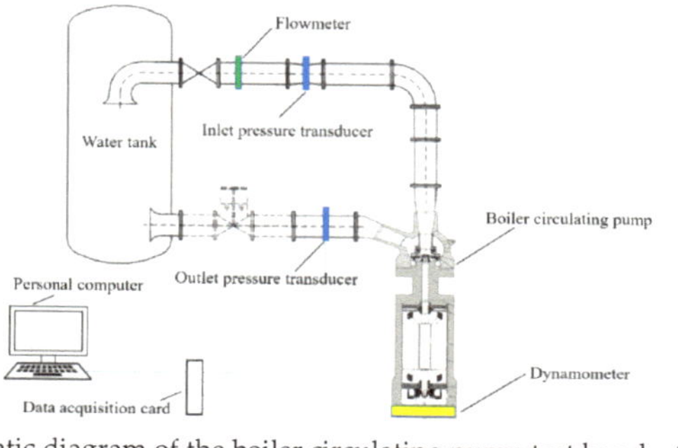

(**a**) Schematic diagram of the boiler circulating pump test bench structure.

(**b**) The scene of pump external characteristic experiment.

Figure 5. Test bench of boiler circulating pump.

3. Results and Discussion

3.1. Influence of Outlet Pipe Angle on Pump Head and Efficiency

The external characteristic curve of numerical simulation for different outlet pipe angles (90°, 100°, 115°, 130°) under different flow rates ($0.6Q_d$, $0.8Q_d$, $1.0Q_d$, $1.2Q_d$, and $1.4Q_d$) is shown in Figure 6.

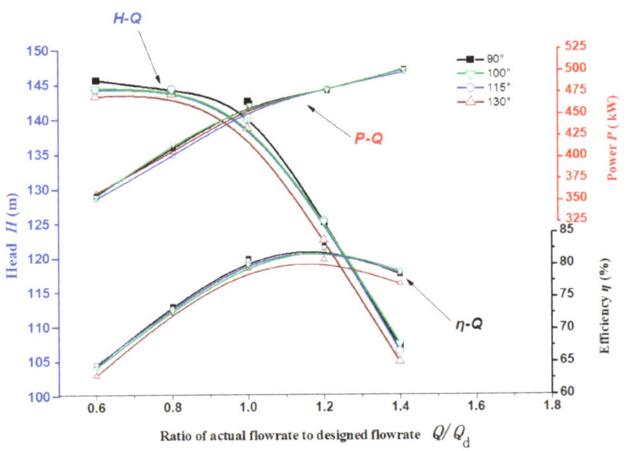

Figure 6. External performance curve of different outlet pipe angles. H-Q: Head (m)-Flowrate curve (m^3/h); P-Q: Power (kW)-Flowrate (m^3/h); η-Q: Efficiency (%)-Flowrate (m^3/h).

It can be seen from Figure 6 that the pump head, efficiency and shaft power curves corresponding to the outlet pipe angles of 90°, 100° and 115° are basically the same. When the outlet pipe angle is 130°, compared with the other three outlet pipe angles, the maximum head drop is about 2.8 m, and the maximum efficiency drop is about 2.1%. It can be seen from the above that in a certain range of outlet pipe angle (90° to 115°), the change of head and efficiency is very small. When the outlet pipe angle is 130°, the pump head and efficiency reduce significantly.

3.2. Experimental Verification

According to the analysis in the previous section, a boiler circulating pump prototype with an outlet pipe angle of 115° was manufactured; the experimental results of the boiler circulating pump prototype were compared with the results of numerical simulation. The uncertainty estimation of the experiment was conducted according to the ISO 9906:2012(E) international standard [31].

The experimental overall uncertainty e is given by:

$$e = \sqrt{e_R^2 + e_S^2} \tag{13}$$

where e_R is random uncertainty, and e_S stands for system uncertainty.

The random uncertainty e_R is given by:

$$e_R = \frac{100ts}{\overline{x}\sqrt{n}}\% \tag{14}$$

when the reading number n is eight, according to reference [31], the value of t is 2.32. The arithmetic mean $\overline{x}$ of a set of repeated observations x_i (i = 1, 2, ..., n) is calculated by the following formula.

$$\overline{x} = \frac{1}{n}\sum x_i \tag{15}$$

where n stands for the number of readings. The observation x_i could be the pump head H, flowrate Q or pump efficiency η.

The standard deviation s of the observations is given by:

$$s = \sqrt{\frac{1}{n-1}\sum (x_i - \overline{x})^2} \tag{16}$$

The system uncertainty e_S mainly depends on the uncertainty of the measuring instrument. The flowrate was measured by electromagnetic flowmeter with the uncertainty of ±0.25%. The inlet pressure and outlet pressure were measured by pressure transducers with the uncertainty of ±0.3%. The input electric power is measured by the dynamometer with the uncertainty of ±0.3%. Based on above formulas, the uncertainties of flowrate, pump head, and pump efficiency are calculated, as shown in Table 3.

Table 3. Experiment overall uncertainty.

Performance Characteristics	Flowrate	Pump Head	Pump Efficiency
Uncertainty (%)	±0.29	±0.34	±1.43

The head and efficiency at five flow rate conditions ($0.6Q_d$, $0.8Q_d$, $1.0Q_d$, $1.2Q_d$, and $1.4Q_d$) were calculated and experimented, respectively, and the corresponding external characteristic curve was drawn. The comparison between the experimental results and the computational fluid dynamics calculation results is shown in Figure 7.

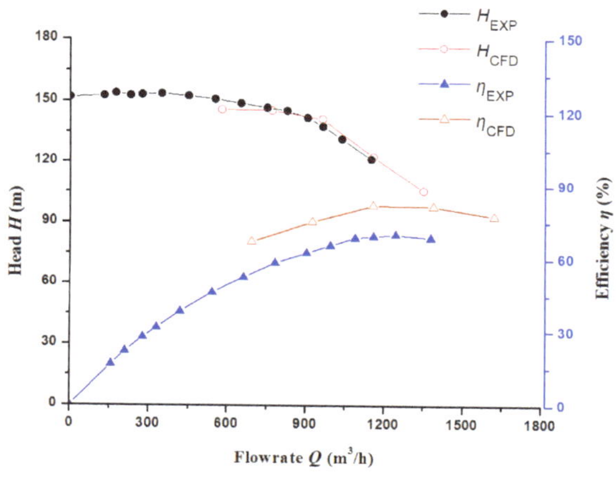

Figure 7. Comparison of experimental and simulated external performance curves.

As can be seen from Figure 7, the numerical calculation head is basically the same as the experimental head, while the numerical calculation efficiency is higher than the experimental efficiency. The loss of efficiency between the simulated and the experimental is shown in the Table 4.

Table 4. The loss of efficiency between the simulation and the experiment.

Flowrate (m^3/h)	Simulated Efficiency (%)	Experimental Efficiency (%)	Loss of Efficiency (%)
579.1	67.1	54.3	12.8
772.1	75.2	63.3	11.9
965.1	81.9	69.4	12.5
1149.7	81.5	68.7	12.8

The main reason for the efficiency loss between the simulation and the experiment is that the boiler circulating pump is an integral structure pump, and the shaft power cannot be measured directly. During the experiment, the input electric power is measured, and the loss of the motor converting electric energy to mechanical energy has not been deducted, so the experimental efficiency is low. Therefore, the numerical results are reasonable and reliable.

3.3. Analysis of Rotor Radial Force

In the vane pumps, because of the interference between the moving and stationary blades, the flow field inside the impeller is asymmetric, which causes the rotor to be affected by the transient radial force. The radial force of the rotor will affect the stability of the pump operation. The analysis of the radial force distribution of the boiler circulating pump at different flow rate conditions is helpful to improve and optimize the design of the pump.

Through the analysis of the transient calculation results, the radial force vector distribution of the rotor at different flow rate conditions is obtained, as shown in Figure 8. It can be seen from Figure 8 that the distribution regularity and track centricity of radial force vector are poor under the flow rate conditions of $0.6Q_d$ and $0.8Q_d$, and the regularity is stronger and the track centricity is better under flow rate conditions of $1.0Q_d$, $1.2Q_d$ and $1.4Q_d$. Under the conditions of small flow rate, the regularity and centricity are improved with the increase in flow rate, and under the conditions of design and large flow rate, the regularity is enhanced. With the increase in flow rate, the distribution range of the radial force vector decreases gradually.

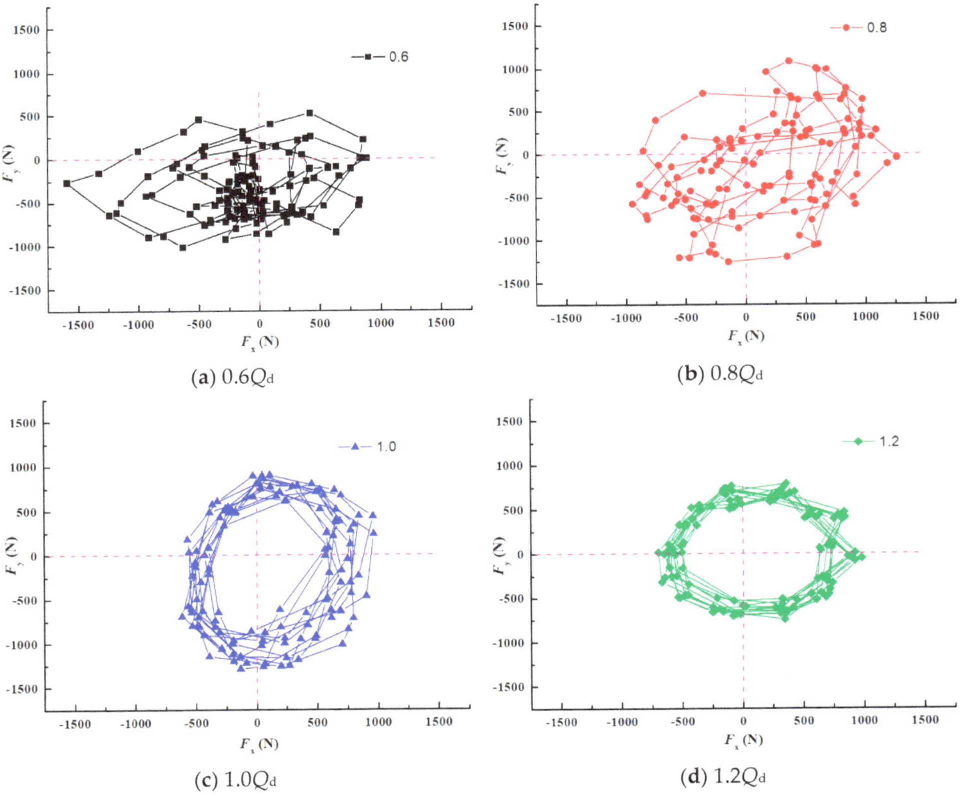

Figure 8. *Cont.*

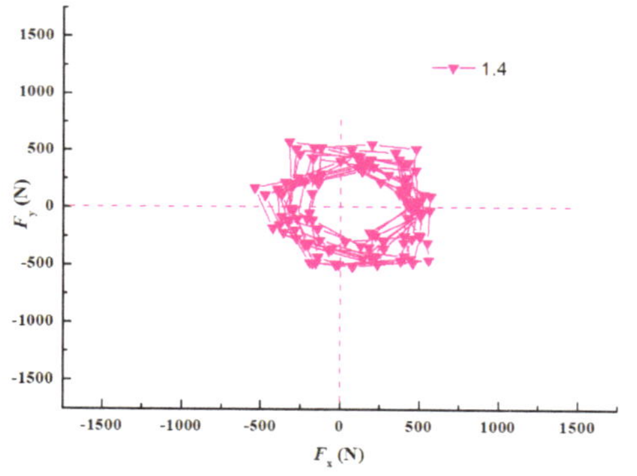

(e) $1.4Q_d$

Figure 8. Rotor radial force vector distribution at different flow rate conditions.

The guide vane has a certain interference effect on the flow field in the fluid domain of the impeller. Its direct effect is that the asymmetric interference makes the flow field in the impeller circumferentially asymmetric, and the asymmetric flow field will cause the unbalanced force on the impeller in the fluid domain. Figure 9 shows the circumferential pressure distribution at the impeller outlet under different flow rate conditions, showing certain periodicity under five flow rate conditions, with 11 fluctuations, consistent with the number of guide vanes.

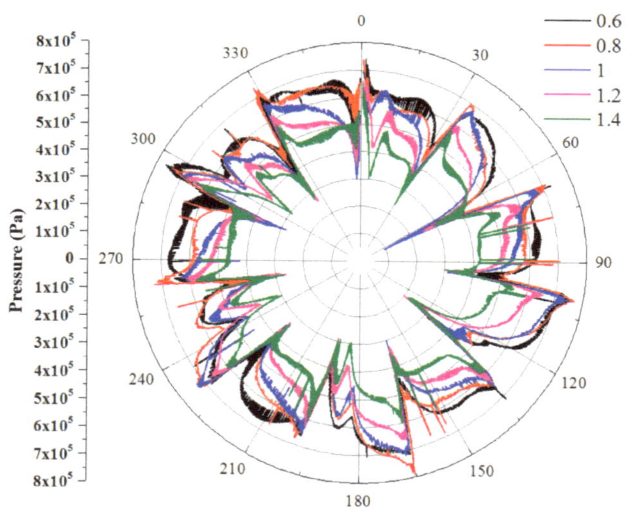

Figure 9. Distribution of pressure at the impeller outlet under different flow rate conditions.

It can be seen from Figure 9 that although the pressure distribution at the outlet of the impeller is periodic, the circumferential direction is not completely symmetrical, so it interferes with the internal flow field of the impeller, resulting in the hydraulic imbalance of the internal flow field of the impeller and the radial force described in Figure 8.

In Figure 10, areas A and B are circumferentially symmetrical. Under five flow rate conditions, the pressure distribution in the circumferentially symmetrical area and inside the flow passage of the impeller appears to be asymmetrical, resulting in the circumferential asymmetry of the pressure acting on the wall surface of the flow passage of the impeller. According to Figures 9 and 10, with the increase in flow rate, the pressure at the outlet of impeller decreases, and the pressure difference between guide vane and impeller decreases, which reduces the radial force of impeller caused by asymmetry to a certain extent. The pressure distribution in impeller and guide vane reveal that pressure fluctuation of the flow field is more periodic with the increase in flow rate, which to some extent controls the asymmetry of the flow field in the impeller.

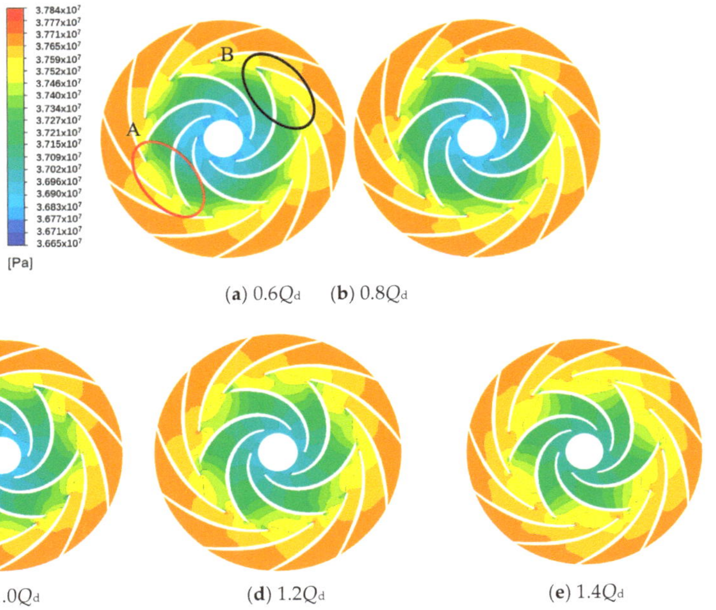

Figure 10. Pressure Distribution in Impeller and Guide Vane.

3.4. Leakage Analysis of Front and Rear Seal Ring

The dynamic and static cooperation between impeller and pump casing forms the clearances at the front and the rear ring, which prevent the medium in the high-pressure area from leaking to the low-pressure area under the joint action of the dynamic and static rotors. The clearances at the seal rings are very small, at the same time, there is relative rotation between the rotor and the stator, which makes the flow resistance of the structure larger and can effectively control the leakage. Under different flow rate conditions, the pressure distribution of the flow field is different, which makes the leakage of the front and rear seal ring change. The mathematical expressions of leakages are shown as follows [32].

$$Q_d = Q + q \tag{17}$$

$$q = q_f + q_r \tag{18}$$

$$\Delta m = \Delta m_f + \Delta m_r \tag{19}$$

$$\Delta m_f = \rho \cdot q_f \tag{20}$$

$$\Delta m_r = \rho \cdot q_r \tag{21}$$

where, Q is the actual flow rate of pump, q is the leaked flow rate of pump, Δm is the total leakage, Δm_f is the leakage of front seal ring, Δm_r is the leakage of rear seal ring, ρ is the density of liquid, q_f is the leaked flow rate of front seal ring, q_r is the leaked flow rate of rear seal ring.

Figure 11 describes the statistical chart of the leakage of seal rings and the pressure difference of the pump chamber, wherein the pressure difference of the pump chamber is the pressure difference between the inlet of the pump chamber and the inlet of the pump impeller. According to the overall distribution of the statistical chart, with the increase in the flow rate, the pressure difference at the inlet and outlet of the pump chamber, the leakage of the seal rings and the total leakage as a percentage of total flow all decrease. The variation trend of the pressure difference between the front and rear pump chambers is the same. The pressure difference at the inlet of the pump chamber decreases with the increase in the flow rate, which is in line with the characteristics of the pump. The absolute value of front and rear seal ring leakage changes little under five flow rate conditions, so the total leakage as a percentage of total flow decreases with the increase in flow rate.

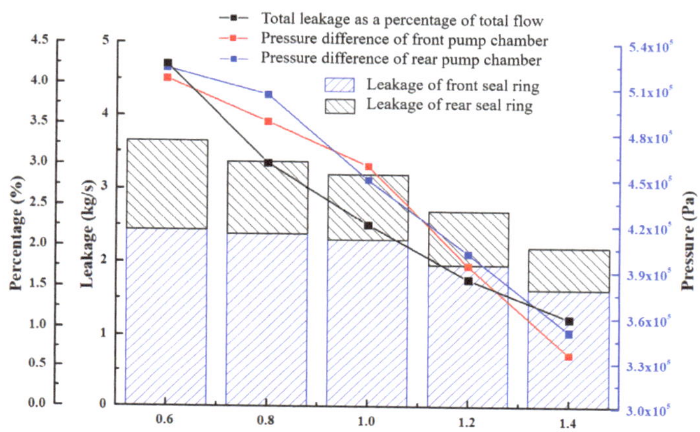

Figure 11. Leakage and pump chamber pressure difference statistics.

The results show that the pressure differences between the front and rear pump chambers at 0.6 Q_d and 1.4Q_d are 180,620.1 Pa and 172,596.3 Pa, respectively. Under five flow rate conditions (0.6Q_d, 0.8Q_d, 1.0Q_d, 1.2Q_d, and 1.4Q_d), the leakage values of the front seal ring are 2.43 kg/s, 2.38 kg/s, 2.29 kg/s, 1.95 kg/s and 1.63 kg/s, and the leakage values of the rear seal ring are 1.22 kg/s, 0.98 kg/s, 0.89 kg/s, 0.74 kg/s and 0.56 kg/s. The leakage ratio values of the front seal ring and the rear seal ring are 1.95, 2.48, 2.56, 2.64 and 2.88. With the increase in the flow rate, the leakage of the front seal ring and rear seal ring decreases. The main reason is that the pressure at the inlet of the pump chamber decreases with the increase in the flow rate. The absolute value of the leakage of the front seal ring is higher than that of the rear seal ring, mainly because the structure of the rear pump chamber is more complex than that of the front pump chamber, resulting in greater flow resistance. Therefore, the leakage of the rear pump chamber is less than that of the front pump chamber, and the leakage ratio of the front and rear seal ring increases with the increase in the flow rate.

3.5. Disc Loss Analysis

Disc loss is an important part of pump energy loss. Analyzing the characteristics of disc loss at different flow rate conditions can effectively guide the design and optimization of the pump. The mathematical expressions of disc losses are shown as follows [32].

The impeller energy consumption P_{total} is:

$$P_{total} = P_{uesful} + P_{loss} = N \cdot \omega \tag{22}$$

The useful energy P_{uesful} is:

$$P_{uesful} = \rho g Q_d H \tag{23}$$

The total energy loss P_{loss} is:

$$P_{\text{loss}} = P_{\text{total}} - P_{\text{uesful}} = M \cdot \omega - \rho g Q_{\text{d}} H \tag{24}$$

The energy loss of disc P_{dl} is:

$$P_{\text{dl}} = P_{\text{fdl}} + P_{\text{rdl}} = M_{\text{f}} \cdot \omega + M_{\text{r}} \cdot \omega \tag{25}$$

Percentage of front cover plate loss to total energy loss η_{fl} is:

$$\eta_{\text{fl}} = P_{\text{fdl}} / P_{\text{loss}} = M_{\text{f}} \cdot \omega / (M \cdot \omega - \rho g Q_{\text{d}} H) \tag{26}$$

Percentage of rear cover plate loss to total energy loss η_{rl} is:

$$\eta_{\text{rl}} = P_{\text{rdl}} / P_{\text{loss}} = M_{\text{r}} \cdot \omega / (M \cdot \omega - \rho g Q_{\text{d}} H) \tag{27}$$

Percentage of energy loss of disc to total energy loss η_{dl} is:

$$\eta_{\text{dl}} = P_{\text{dl}} / P_{\text{total}} = (M_{\text{f}} \cdot \omega + M_{\text{r}} \cdot \omega) / (M \cdot \omega) \tag{28}$$

Equation (28) can be simplified as:

$$\eta_{\text{dl}} = (M_{\text{f}} + M_{\text{r}}) / M \tag{29}$$

where, M is the torque of impeller, ω is the angular velocity of rotation, M_{f} is the torque of front cover plate, M_{r} is the torque of rear cover plate, Q_{d} is the design flow rate of pump, H is the hydraulic head of pump.

The numerical results show that the total energy loss values under five flow rate conditions are 52,754.1 W, 40,539.4 W, 27,163.1 W, 28,609.6 W and 42,177.7 W, respectively. The total energy loss decreases first and then increases with the increase in flow rate, and reaches the minimum at $1.0Q_{\text{d}}$. Figure 12 shows the statistical chart of disc loss. The disc loss energy increases with the increase in flow rate, which is in line with the actual design and operation conditions. As the total loss energy decreases first and then increases, the disc loss increases with the flow, so the disc loss percentage increases first and then decrease. The energy loss of the rear cover plate is higher than that of the front cover plate under five flow rate conditions. The main reason is that the length of the boundary line of the rear cover plate is longer than that of the front cover plate, and the volume of the rear pump chamber is larger than that of the front pump chamber, so the energy loss of the front cover plate is smaller than that of the rear cover plate.

The change of flow rate has a limited impact on the absolute value of energy loss of the front cover plate, and a greater impact on the rear cover plate. With the increase in flow rate, the absolute value of energy loss of rear cover plate increases. The structure of rear pump chamber is more complex, and the flow change has a great impact on the flow in the pump chamber. Thus, the energy loss of the rear cover plate changes greatly.

Figure 13 shows the shear stress distribution of the front and rear cover plates. Under different flow rate conditions, the distribution trend of shear stress on the wall of front and rear cover plates is the same, and the difference at different conditions is small. The shear stress increases at the abrupt change of the clearance and the wall. Therefore, the abrupt change of the cover plate wall should be avoided in the design process. At the same time, the reasonable control of the clearance size will also reduce the shear strain rate, thereby reducing the disc energy loss.

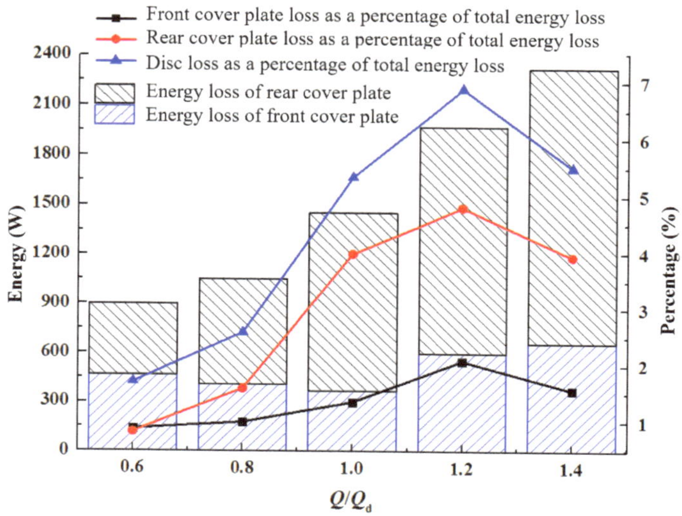

Figure 12. Statistical chart of disc loss.

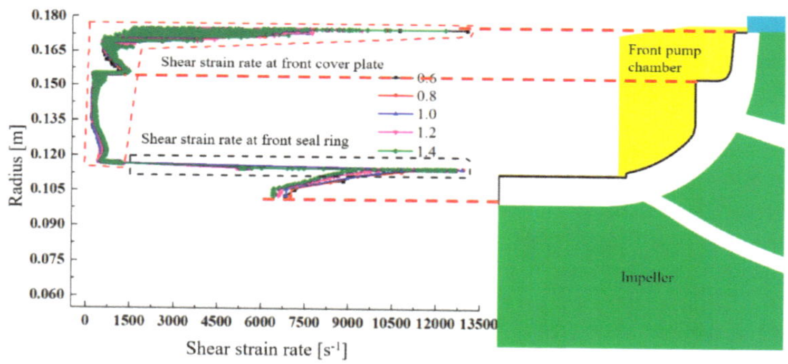

(**a**) Shear strain rate of front cover plate

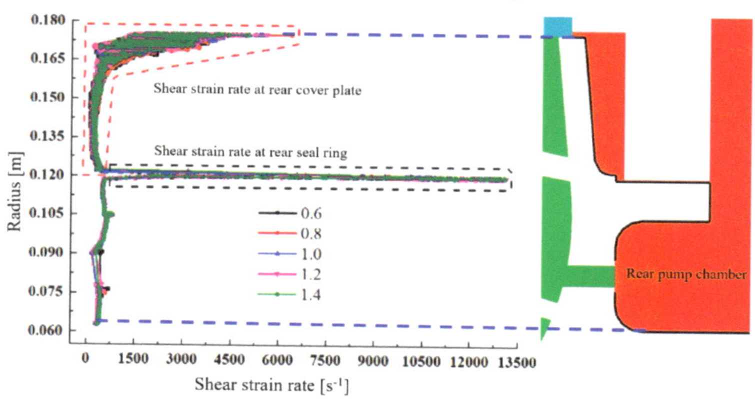

(**b**) Shear strain rate of rear cover plate

Figure 13. Shear strain rate distribution of cover plate.

4. Conclusions and Future Work

ANSYS-CFX is used to analyze the influence of different outlet pipe angles on the head and efficiency of the boiler circulating pump. The transient flow of the prototype pump with 115° outlet pipe angle under different flow rate conditions are analyzed. The conclusions are as follows.

(1) In a certain range of outlet pipe angle (90° to 115°), the change of head and efficiency is very small. When the outlet pipe angle is 130°, compared with the other three outlet pipe angles, the maximum head drop is about 2.8 m, and the maximum efficiency drop is about 2.1%.
(2) The centricity and regularity of rotor radial force distribution are poor under the small flow rate conditions ($0.6Q_d$, $0.8Q_d$). With the increase in flow rate, the centrality and regularity are obviously improved, and the distribution range of radial force vector gradually reduces. The reason is that with the increase in flow rate, the symmetry of flow field increases, the static pressure of the flow field decreases and the interference intensity decreases.
(3) Under different flow rate conditions ($0.6Q_d$, $0.8Q_d$, $1.0Q_d$, $1.2Q_d$, and $1.4Q_d$), the leakage ratio values of the front seal ring and rear seal ring are 1.95, 2.48, 2.56, 2.64 and 2.88, respectively. The leakage ratio of the front seal ring and the rear seal ring increases with the increase in the flow rate. With the increase in the flow rate, the leakage of the front and the rear seal ring decreases, which is mainly due to the decrease in the pressure at the inlet of the pump chamber. The absolute value of the leakage of the front seal ring is higher than that of the rear seal ring, mainly because the structure of the rear pump chamber is more complex than that of the front pump chamber, resulting in greater flow resistance.
(4) The total energy loss values of the boiler circulating pump under five flow rate conditions are 52,754.1 W, 40,539.4 W, 27,163.1 W, 28,609.6 W, and 42,177.7 W, respectively. With the increase in flow rate, the total loss energy of the pump first decreases and then increases, with the minimum loss energy at the design flow rate condition.
(5) Under all flow rate conditions, the energy loss of the front cover plate is less than that of the rear cover plate, which is mainly caused by the larger rotating surface and volume of the rear pump chamber than those of the front pump chamber, the viscous friction loss and the medium rotation in the pump chamber need more energy.
(6) Through the analysis of the energy loss and the wall shear strain rate of the disk, it is found that the shear strain rate increases at the clearance and the wall mutation. Therefore, in the design process of a boiler circulating pump, the sudden change of the cover plate wall should be avoided. At the same time, the reasonable control of the clearance size will also reduce the shear strain rate, thus reducing the disc energy loss.

This paper analyses the characteristics of the radial force of the rotor, the leakage of the seal ring and the disc loss in a boiler circulating pump, which can provide reference for the design and optimization of this kind of pump. Based on the research content of this paper, in the near future, the next research goal is to build mathematical model by applying artificial neural network models, use the traditional gradient-based optimization algorithm to improve the rotor force and reduce the disk loss.

Author Contributions: F.Z. and F.K. designed the experiments. F.Z. and X.D. made the numerical simulation and performed the experiment. F.Z., H.W. and J.W. analyzed the data. F.Z. wrote the paper. All authors have read and agreed to the published version of the manuscript.

Funding: This research was supported by: the Science and Engineering Research Project of Wuxi Institute of Technology, grant number ZK202001; the Industry-University-Research Cooperation Project of Jiangsu Province, grant number BY2019043; the Scientific Research Project of Wuxi Institute of Technology, grant number BT2018-01; the Excellent scientific and technological innovation team of Jiangsu Higher Education Institutions (Intelligent Manufacturing Equipment Design and Engineering Application); Qing Lan Project; the Scientific Research project of professor and doctor in Wuxi Institute of Technology, grant number BT2018-03; The Natural Science Foundation of the Jiangsu Higher Education Institutions of China, grant number 19KJD470006; and the Priority Academic Program Development (PAPD) of Jiangsu Higher Education Institutions.

Conflicts of Interest: The authors declare no conflict of interest.

References

1. Yang, Q.P.; Lin, W.J.; Wang, Y.M.; He, Y. Industry Development and Frontier Technology Roadmap of Thermal Power Generation. *Proc. Chin. Soc. Electr. Eng.* **2017**, *37*, 3787–3794.
2. Yao, J.C. Optimization and Application of Domestic BCP of 1050 MW Ultra-supercritical Efficient Once Reheat Unit. *Biol. Technol.* **2017**, *48*, 62–66.
3. Zhang, H.; Liu, X.; Kong, X.; Lee, K.Y. Stacked Auto-Encoder Modeling of an Ultra-Supercritical Boiler-Turbine System. *Energies* **2019**, *12*, 4035. [CrossRef]
4. Wang, C.; Feng, Q.; Lv, Q.; Zhao, L.; Du, Y.; Wang, P.; Zhang, J.; Che, D. Numerical Investigation on Co-firing Characteristics of Semi-Coke and Lean Coal in a 600 MW Supercritical Wall-Fired Boiler. *Appl. Sci.* **2019**, *9*, 889. [CrossRef]
5. Baumgarten, S.; Fritz, J.; Knierim, C.; Müller, T. Numerical simulation of the flow in a boiler circulating pump. *Proc. Inst. Mech. Eng. Part A* **2001**, *215*, 793–800. [CrossRef]
6. Sha, Y.J. Three-Dimensional Turbulence Numerical Simulation and Flow Characteristics of High Temperature and High Pressure Centrifugal Pump. Ph.D. Thesis, Tsinghua University, Beijing, China, 2011.
7. Zuo, Z.; Liu, S.; Fan, Y.; Wu, Y. Optimization of a Centrifugal Boiler Circulating Pump's Casing Based on CFD and FEM Analyses. *Adv. Mech. Eng.* **2014**, *2014*, 1–10. [CrossRef]
8. Ni, D.; Yang, M.G.; Gao, B.; Li, Z.; Li, Y.T.; Lu, S.; Jiangsu, H.P.M. The internal correlations between unsteady flow and pressure pulsations in a nuclear reactor coolant pump. *J. Eng. Thermophys.* **2017**, *38*, 1676–1682.
9. Su, S.Z.; Wang, P.F.; Xu, Z.B. Study on Pressure Fluctuation and Radial Force during Startup of Reactor Coolant Pump. *Nucl. Power Eng.* **2017**, *38*, 110–114.
10. Ohashi, H. Case Study of Pump Failure Due to Rotor-Stator Interaction. *Int. J. Rotat. Mach.* **1994**, *1*, 53–60. [CrossRef]
11. Wang, L.; Liu, H.; Wang, K.; Zhou, L.; Jiang, X.; Li, Y. Numerical Simulation of the Sound Field of a Five-Stage Centrifugal Pump with Different Turbulence Models. *Water* **2019**, *11*, 1777. [CrossRef]
12. Zhao, F.; Kong, F.; Zhou, Y.; Xia, B.; Bai, Y. Optimization Design of the Impeller Based on Orthogonal Test in an Ultra-Low Specific Speed Magnetic Drive Pump. *Energies* **2019**, *12*, 4767. [CrossRef]
13. Lin, P.; Li, Y.; Xu, W.; Chen, H.; Zhu, Z. Numerical Study on the Influence of Inlet Guide Vanes on the Internal Flow Characteristics of Centrifugal Pump. *Processes* **2020**, *8*, 122. [CrossRef]
14. Zhao, H.; Liu, S.J.; Hu, X.Z. Study on unsteady fluid radial force in deep-sea lifting pump. *J. Cent. South Univ. (Sci. Technol.)* **2019**, *50*, 829–836.
15. Qian, C.; Yang, C.X. Influence of High Pressure side outflow condition on radial force in multistage pump. *J. Xi'an Jiaotong Univ.* **2019**, *53*, 106–113.
16. Cheng, X.R.; Wei, Y.Q.; Wu, C.; Liu, H. The effect of static and rotor clearance ratio on hydraulic performance of nuclear main pump. *J. Drain. Irrig. Mach. Eng.* **2019**, *5*, 1–8.
17. Gao, B.; Wang, Z.; Yang, L.; Du, W.Q.; Wu, C.B. Analysis and test of performance and hydraulic excitation characteristics of centrifugal pump with different seal ring clearances. *Trans. Chin. Soc. Agric. Eng.* **2016**, *32*, 79–85.
18. Gao, B.; Wang, Z.; Yang, L.; Du, W.; Li, C. Effect of wear-ring clearance on performance and flow characteristics of centrifugal pump. *J. Drain. Irrig. Mach. Eng.* **2017**, *35*, 13–17.
19. Yang, C.X.; Qiang, P.; An, S.; Xu, N.; Liu, J. Effects of wear-ring clearance on performance of high-speed centrifugal pump. *J. Drain. Irrig. Mach. Eng.* **2017**, *35*, 18–24.
20. Liu, H.L.; Tan, M.G.; Yuan, S.Q. Calculation of disk friction loss of centrifugal pumps. *Trans. Chin. Soc. Agric. Eng.* **2006**, *22*, 107–109.
21. He, X.J.; Lao, X.S. Accuracy evaluation of several calculation formulas of disc friction loss power of low specific speed centrifugal pump. *Water Pump Technol.* **2010**, *10*, 16–20.
22. Yang, C.X.; Qian, C. Calculation of disc loss in centrifugal pump with low specific speed. *J. Lanzhou Univ. Technol.* **2012**, *38*, 56–60.
23. Dong, W.; Chu, W.L. Analysis of Flow Characteristics and Disc Friction Loss in Balance Cavity of Centrifugal Pump Impeller. *Trans. Chin. Soc. Agric. Mach.* **2016**, *47*, 29–35.
24. Feng, J.J.; Luo, X.Q.; Wu, G.K.; Zhu, G.J. Influence of clearance flow on efficiency prediction of Francis turbines. *Trans. Chin. Soc. Agric. Eng.* **2015**, *31*, 53–58.

25. Wang, F.J. *Principle and Application of CFD Software for Computational Fluid Dynamics Analysis*; Tsinghua University Press: Beijing, China, 2004; pp. 110–130.
26. Li, Y.B.; He, H.; Fan, Z.J.; Li, J.Z. Effect of blade tip clearance on cavitating flow in mixed-flow pump. *J. Drain. Irrig. Mach. Eng.* **2020**, *38*, 224–229.
27. Menter, F.R. Zonal two equation k-w turbulence models for aerodynamic flows. *AIAA Pap.* **1993**, *7*, 2906.
28. Menter, F.R. Ten years of experience with the SST turbulence model. *Turbul. Heat Mass Transf.* **2003**, *4*, 625–632.
29. Menter, F.R. Review of the shear stress transport turbulence model experience from an industrial perspective. *Int. J. Comput. Fluid* **2009**, *23*, 305–316. [CrossRef]
30. *ANSYS CFX-Solver Theory Guide*; ANSYS, Inc.: Canonsburg, WA, USA, 2011.
31. Technical Committee ISO/TC. *ISO 9906 Rotodynamic Pumps—Hydraulic Performance Acceptance Tests—Grades 1, 2 and 3*; International Standardization Organization: Geneva, Switzerland, 2012.
32. Guan, X.F. *Modern Pumps Theory and Design*; China Aerospace Press: Beijing, China, 2011; pp. 26–51.

processes

MDPI

Article

Investigation on Vortex Characteristics of a Multi-Blade Centrifugal Fan near Volute Outlet Region

Zhehong Li [1,2], Xinxue Ye [3] and Yikun Wei [2,*]

1 Faculty of Aeronautical Engineering, Taizhou University, Taizhou 318000, China; zhehongli@tzc.edu.cn
2 National-Provincial Joint Engineering Laboratory for Fluid Transmission System Technology, Zhejiang Sci-Tech University, Hangzhou 310018, China
3 Zhejiang Yilida Ventilator Co., Ltd., Taizhou 318056, China; yexinxue@yilida.com
* Correspondence: yikunwei@zstu.edu.cn

Received: 29 July 2020; Accepted: 27 September 2020; Published: 2 October 2020

Abstract: The origins and effects of the complex vortex structure near the volute outlet of a multi-blade centrifugal fan are investigated in this paper. Due to the wide blade and short blade channel, the airflow maintains a large radial velocity during the blade channel. This continuous radial partial velocity causes vortices to be generated at the region of volute outlet. Then, the secondary flow close to the impeller generate from the center to the sides in volute. It is obtained that the streamlines are divided into two parts (backflow and outflow) at volute outlet. Although the vortices near volute outlet region are complex, the main features of flow behavior caused by the vortex are understandable.

Keywords: centrifugal multi-blade fan; vortex; backflow; volute outlet

1. Introduction

The multi-blade centrifugal fans are widely used in heating, ventilation, air conditioning, and other fields. Kinds of centrifugal fan with small size coefficient and low noise mainly possess a great deal of outstanding advantages [1,2]. It is of great significance to improve the performance of centrifugal fan. Ballesteros et al. [3] investigated the unsteady flow conditions of the flow field of the multi-blade fan inlet flow, the intra-blade flow, and the outlet flow channel, and summarized their respective characteristics under different working conditions, respectively. Velarde et al. [4] pointed out the strong non-uniformity of the flow state in the blade channel of the multi-blade fan. The blade channel has obvious flow separation points and separation areas, the strong pressure pulsation at the volute, and the pressure distribution in the volute are asymmetric.

Adachi et al. [5–9] studied the influences of blades on the flow field and the performance of multi-blade centrifugal fans. Wang [10] studied the influence of blade trimming on the performance of multi-blade centrifugal fan. They argued that the blade trimming with proper trimming parameter enables the multi-blade fan to get an increasing fan performance. Kind et al. [11] proposed a method to predict the aerodynamic performance and flow state of the forward multi-blade centrifugal fan. This method decomposes the whole fan into three parts, namely the inlet part, the blade part, and the volute part. The main target should be placed on the blade and the gap. However, the vortices at the outlet extension of the multi-blade centrifugal fan also possess a significant influence on the aerodynamic performance of fan, so the optimization of the outlet extension should also be paid attention to. Velarde et al. [12] measured the velocity fluctuation at the outlet of a forward multi-blade fan by using the hot wire method and found that the outlet of the fan presented a strong asymmetry, especially at the volute tongue. Wen et al. [13] proposed two volute design methods to overcome the space limitation of the volute. The effects of turbulent flow structure on the performance of a centrifugal fan were studied numerically and experimentally [14,15]. The numerical studies [16–20] demonstrated that numerical simulations play an increasingly important role in fan performance design. Wang et al. [21] proposed that the representation technique is capable of effectively getting the nonuniform circumferential flow conditions at both the outlet and inlet of fan impeller.

The above researches show that the region between the impeller and volute plays an important role in the fan performance. Complex vortices are also mainly caused by the Coriolis force and unsteady separated flow. Interestingly, several complex vortex structures are easily formed and captured near the junction of the volute outlet and the outlet channel extension section, which blocks the pneumatic conveying of centrifugal fan. This phenomenon greatly suppresses the aerodynamic performance of multi-blade fan. There are great differences between the actual flow field features and the theoretical derivation, especially the influence of the vortex structure at the outlet of the volute on the performance of the fan. The volute plays a key role in effective work of the centrifugal fan and directly affects the performance of fan. Complex vortex structure near volute outlet is caused by Coriolis force and the unsteady flow in narrow blade channel. The aim of this paper is to provide some fundamental understanding of the complex vortex structure on the volute outlet by numerical simulations and to provide some physical insight into the control of unsteady flow field, rather than the performance of fan and results for specific applications.

2. Experimental Test of Aerodynamic Performance for Multi-Blade Fan

2.1. Key Parameters of Multi-Blade Fan Model

The several important parameters of multi-blade fan are shown in Figure 1 and Table 1. Figure 1 shows the physical picture of the centrifugal fan, which includes a volute and an impeller. As illustrated in Table 1, it is seen that the number of the impeller blade is 40, the impeller width (b) is 200 mm, the impeller diameter (D_2) is 150 mm, the angle of blade inlet (β_1) is 90°, the outlet angle of blade (β_2) is 27°, respectively.

(a)

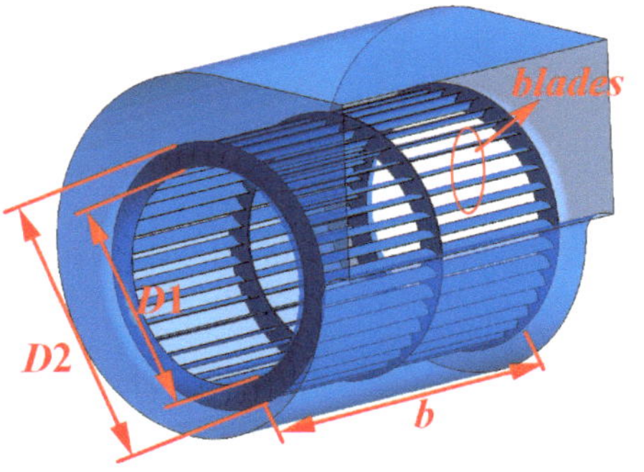

(b)

Figure 1. Three-dimensional model of a multi-blade fan: (**a**) the multi-blade fan manufacture; (**b**) schematic diagram of the multi-blade fan.

Table 1. Design dimensions for the multi-blade fan.

Parameter	Dimension
Inlet diameter of impeller (D_1)	132 mm
Outlet diameter of impeller (D_2)	150 mm
Width of impeller (b)	200 mm
Inlet angle of blade (β_1)	90°
Outlet angle of blade (β_2)	27°
Blade number	40

2.2. Laboratory Testing

The aerodynamics performance of the multi-blade fan was tested in the Zhejiang Yilida aerodynamics laboratory. The aerodynamics performance of a smulti-blade fan is usually obtained by total pressure P_t (Pa)-the flow rate of volume Q (m^3/h). The total pressure is the sum of dynamic pressure and static pressure. The total pressure of multi-blade fan is the difference between total pressure at the inlet and outlet of fan, its expression is followed as:

$$P_t = P_{st} + P_d \tag{1}$$

$$\eta = \frac{P_t \times Q}{1000N} \tag{2}$$

where P_{st} denotes the static pressure, P_d denotes the dynamic pressure, η is the efficiency of total pressure and N is the power. The aerodynamic performance test consists of negative pressure chamber, the positive pressure chamber and gas tunnel. Figure 2 illustrates the experimental system, and Figure 3 describes the photograph of the negative pressure chamber in the laboratory schematic. Each air chamber has four pressure taps which are located on the two side walls, the top surface and the ground in the chamber. The results of performance test include the total pressure P_t, the total pressure efficiency η. In this paper, the rated flow of multi-blade fan is 504 m^3/h (Q = 504 m^3/h), the flow coefficient is φ = 0.79, and n = 1000 r/min.

$$\psi = \frac{P_t}{\frac{1}{2}\rho u_2^2} \tag{3}$$

$$\phi = \frac{Q}{\frac{\pi}{4}D_2^2 u_2} \tag{4}$$

where Ψ is the total pressure coefficient and φ is the flow coefficient, respectively.

As shown in Figure 2, the flow rate was measured by the difference static pressure between the pressure taps before and after the nozzle the nozzle. All the pressure were measured by FCO332 (pressure transmitter, East Sussex, UK). By measuring the power of multi-blade fan, the total pressure efficiency of the multi-blade fan was obtained.

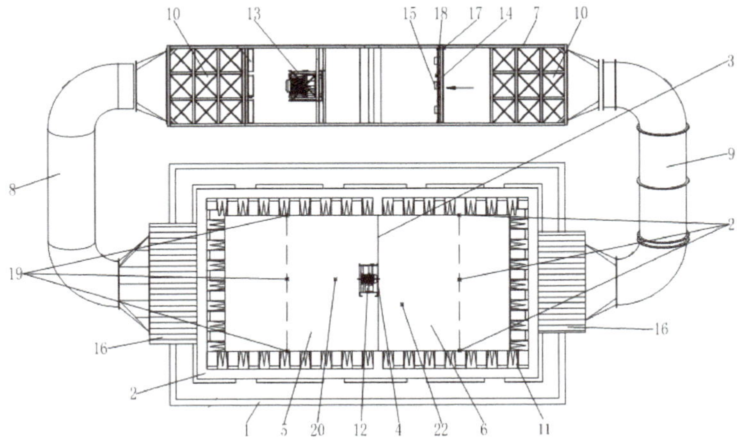

Figure 2. Sketch of the test installation. 1: Testing room, 2: Soundproof room, 3: Reflective wall, 4: Fan installation location, 5: Negative pressure chamber, 6: Positive pressure chamber, 7: Flow return room, 8: Inlet duct, 9: Outlet duct, 10: Silencer, 11: Muffler spike, 12: Testing fan, 13 Auxiliary fan, 14 Flow adjustment device, 15: Multiple nozzles, 16: Chip muffler, 17: Pressure taps in front of nozzle, 18: Pressure taps position behind the nozzle, 19: Pressure taps, 20: Noise taps, 21: Pressure taps, 22: Noise taps.

Figure 3. Negative pressure chamber.

3. Governing Equations and Numerical Method

3.1. Dynamic Model of Incompressible Fluid

The computational fluid dynamics methods are based on the basic governing equations of fluid mechanics: continuous equation, momentum equation, and energy equation. The turbulence models of incompressible Reynolds-averaged Navier-Stokes (RANS) are solved by the finite volume method. The mass and kinetic equations for the internal flow in the multi-blade fan are followed as [22,23]:

$$\frac{\partial \rho}{\partial t} + \frac{\partial}{\partial x_i}(\rho u_i) = 0 \tag{5}$$

$$\frac{\partial \rho u_i}{\partial t} + \frac{\partial}{\partial x_j}\left(\rho u_i u_j\right) = -\frac{\partial P^*}{\partial x_i} + \frac{\partial}{\partial x_j}\left[\mu_e\left(\frac{\partial u_i}{\partial x_j} + \frac{\partial u_j}{\partial x_i}\right)\right] + S_i \tag{6}$$

where ρ denotes the density of fluid, x_i and x_j are the components x, y, and z in the coordinate system of Cartesian in corresponding directions, P^* is the translation pressure, S is the sum of the volume force source terms, and μ is the dynamic viscosity coefficient. The formula of turbulence model Re-normalization group (RNG) k-ε is given as [22]

$$\frac{\partial(\rho k)}{\partial t} + \frac{\partial}{\partial x_i}(\rho u_i k) = \frac{\partial}{\partial x_i}\left[\left(\mu + \frac{\mu_t}{\sigma_k}\right)\frac{\partial k}{\partial x_i}\right] + P_k - \rho\varepsilon + P_{kb} \tag{7}$$

$$\frac{\partial(\rho \varepsilon)}{\partial t} + \frac{\partial}{\partial x_i}(\rho u_i \varepsilon) = \frac{\partial}{\partial x_i}\left[\left(\mu + \frac{\mu_t}{\sigma_\varepsilon}\right)\frac{\partial \varepsilon}{\partial x_i}\right] + \frac{\varepsilon}{k}(C_{\varepsilon 1}P_k - C_{\varepsilon 2}\rho\varepsilon + C_{\varepsilon 1}P_{\varepsilon b}) \tag{8}$$

where $C_{\varepsilon 1} = 1.44$, $C_{\varepsilon 2} = 1.92$, $\sigma_k = 1.0$, $\sigma_\varepsilon = 1.3$, $P_{\varepsilon k}$, $P_{\varepsilon b}$ represent the energy terms generated by the buoyancy, and P_k is used to generate the turbulent kinetic energy caused by viscous forces. The pressure boundary condition is used for the inlet of multi-blade fan and the mass flow boundary condition is implemented for the outlet of multi-blade fan. The upwind discretization is implemented in convection terms and the schemes of central difference are implemented in terms of diffusion [23].

3.2. Grid Independence Verification

The grid independence verification is relatively essential to obtain the high accuracy of numerical simulation and determining the amount of the calculation grid of the whole calculation area. Figure 4 shows the computational domain. Figure 4a illustrates two inlet domains, one volute domain, one outlet domain, and a rotating impeller domain. Figure 4b illustrates coordinate and origin of the impeller rotation. The size of every domain in Figure 4a should be presented here. The inlet boundary condition is the static pressure boundary condition and the outlet is the flow rate boundary condition. Different flow coefficients are obtained by varying the flow rate of outlet. A pair of dynamic and static interfaces are used for data transfer between the rotating part (impeller) and non-rotating part (volute). The Multiple Reference Frame (MRF) is used to set the rotating part (impeller). The rotating speed and torque of impeller are implemented to obtain the simulation power of the multi-blade fan.

To verify the grid independence of numerical simulations, four models with different grid resolutions, including 4.98×10^6, 7.40×10^6, 1.05×10^7, and 1.43×10^7, are implemented at flow rate condition (Q = 321 m^3/h). Figure 5 describes the relation between the total pressure and grid numbers. It is clearly seen that the total pressure of numerical simulation approaches the experimental results with the increment of the grids number. When the grid number reaches 1.05×10^7, the relative difference of total pressure between the numerical simulation results and the experimental results is only 1%, which proves that the grid resolution is fine enough for numerical results. Finally, to consider the accuracy and cost, a grid division method (topology method and grid node number setting) with a total grid of 1.05×10^7 (the computational time is about 11.5 h with 20 processes) is implemented instead of the 1.43×10^7 one (the computational time is about 24 h with 20 processes) in all numerical simulations.

Figure 6 shows the topological block structure and mesh region of the fan. The mesh density is increased towards the walls. As the inside flow of impeller is the very complicated, the mesh of impeller blade is refined in Figure 6a. The grid of volute region is refined in Figure 6b.

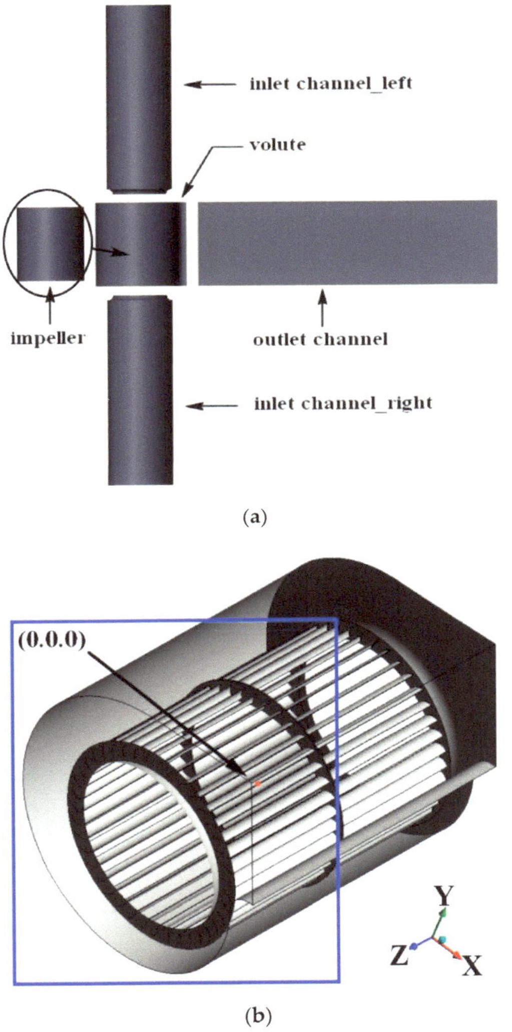

Figure 4. Computational domain and coordinate: (**a**) computational domain of fan; (**b**) coordinate and origin.

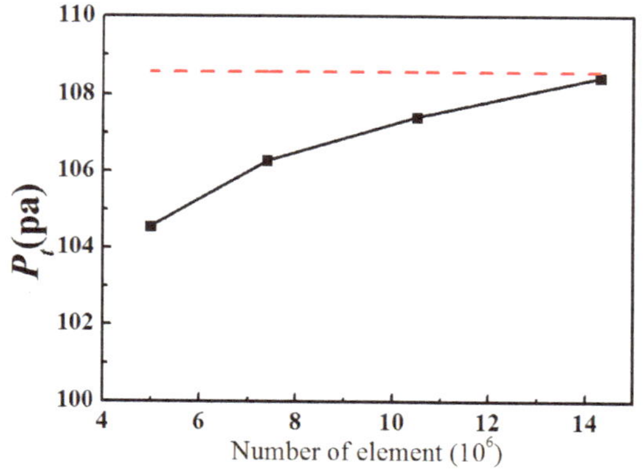

Figure 5. Total pressure with grid number.

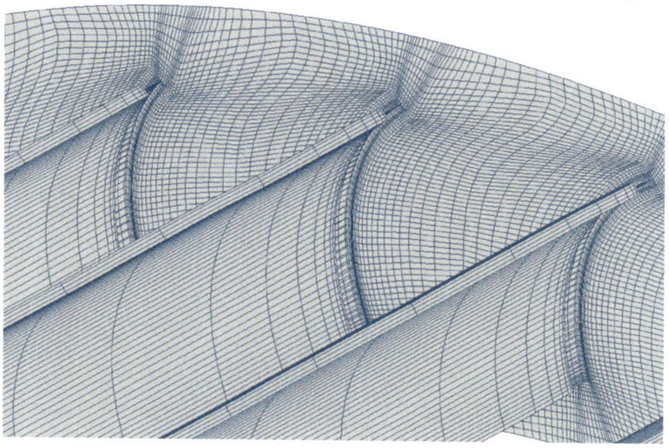

(**a**)

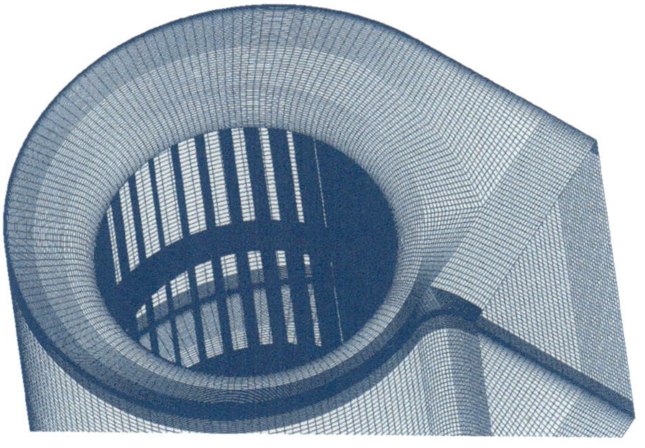

(**b**)

Figure 6. The grid of fan: (**a**) the grid of impeller; (**b**) the grid of volute.

4. Numerical Results and Some Discussions

In the following section, the complex internal flow fields of multi-blade fan will be presented and discussed. The vortex structure between outlet channel and volute will be studied. The origins and impacts of the complex flow structure on the volute outlet are further investigated.

4.1. Verification of Numerical Results

In order to prove the accuracy of numerical simulations, the reference model is used to study the total pressure and the efficiency of total pressure at various rate of flow. Figure 7 describes the comparison of numerical and experimental results about the total pressure and the efficiency of total pressure. The red solid line denotes the experimental results and the black line represents the numerical results, respectively. As described in Figure 7, it can be observed that the pressure coefficient gradually decreases with increasing flow coefficient, the pressure coefficients of numerical simulations are excellent consistent with that of experiment, and the total pressure efficiency of numerical simulations well agree with the experimental results in a range of medium and large flow rates. The numerical efficiency does not match the experimental results well at the low mass flow conditions. This mainly due to the error of motor efficiency at low power condition. As illustrated in Figure 7, one clearly observes that the total pressures of numerical simulations are in good agreement with the experimental results, which demonstrates that the method of numerical simulation is reliable in this paper [24,25].

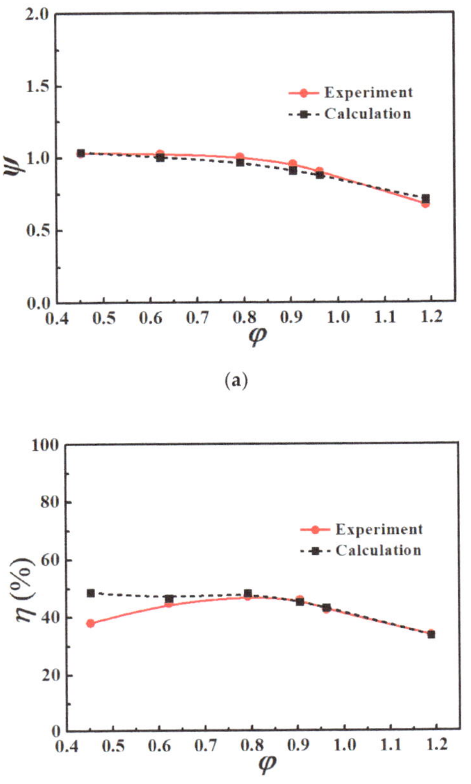

(a)

(b)

Figure 7. Comparison of experimental and numerical data under different flow rates: (**a**) total pressure; (**b**) efficiency.

The origins and impacts of the complex flow structure on the volute outlet are investigated. In order to demonstrate the detail of vortex near volute outlet, the streamline diagram with different flow coefficient will be presented and discussed in the following subsection.

4.2. Effect of Volute Outlet Vortex on Flow

Figure 8 shows the three-dimensional streamline distribution in the volute and the outlet of the volute. As illustrated in Figure 8a, it is obviously observed that the streamline is equally divided into two parts (backflow and outflow), where $\varphi = 0.435$ is a small flow condition, $\varphi = 0.609$ is a high efficiency condition, and $\varphi = 1.175$ is a high-flow condition, respectively. As shown in Figure 8a, it is clearly obtained that the flow field near the volute side wall of the multi-blade fan near the volute exit is relatively chaotic, and an obvious vortex can be observed at small flow conditions. A large area at the exit of the volute has a backflow phenomenon. From Figure 8a, it can be observed that most of the airflow from the impeller exit does not leave the fan by the volute exit but flows back to the volute through the volute. At this time, the airflow exits the volute and leaves the fan mainly through a limited passage near the upper wall of the volute. As shown in Figure 8b, one can see that a vortex appears in the junction between the volute exit and the fan outlet extension at high-efficiency conditions. The vortex goes down to the volute and up to the volute. At the half-height position of the exit, it is close to the volute side in the axial direction. Plotted in Figure 8c, at high flow conditions, the vortex area at the exit of the volute is slightly smaller than that at higher efficiency conditions. With increasing flow rate of fan, the volute exit flow rate is bound to increase, and the increasingly volute exit air velocity compresses the volute exit vortex area. As shown in Figure 8b, one volute exit vortex can be observed at the left and right side walls of the volute, and the two vortex structures are similar analysis. At the flow point of high efficiency working condition of fan, a large area of vortex structure appears in the exit of the volute. The analysis and study of this vortex structure are very helpful to understand the flow optimization of the multi-blade fan.

In order to investigate the flow characteristics of the vortex at the outlet of the volute, the velocity distribution of the impeller outlet is studied. Figure 9 illustrates the Line B-Outlet of the impeller. Plotted in Figure 9, the 'B-Outlet-Z25' means "the line near blade trailing at the cross-section $Z = 25$ mm" (where $Z = 0$ mm is located at the impeller center). The distribution position of 0° to 360° is illustrated in Figure 9.

From Figures 10–17, all the simulation results are obtained at the high efficiency condition and the flow coefficient (φ) is 0.609. Figure 10 illustrates the absolute velocity distribution of the line 'B-Outlet-Z'. As illustrated in Figure 10, one can see that the fluctuation of velocity due to multiple blades is very obvious at the area of 90° to 360°. As described in Figure 10d, one can observe that the absolute velocity at Z50 and Z75 is smaller than that of Z25, while the velocity at Z50 increases from 60° to 90°. It is further demonstrated that the airflow near the center of the impeller is almost not affected by the volute outlet vortex, but the flow from the Z50 to the side of the volute is greatly affected by the volute outlet vortex. The flow is blocked in the area from Z50 to the side of the volute due to the big volute outlet vortex. It is further obtained that the absolute velocity distribution with the increase of line B-Outlet-Z, which reveals that the flow obstruction appears in the region from Z25 to the side of the impeller. With the increase of Z, one can see that the flow obstruction is increasingly formed in a range of Z25 to Z75, which indicates that the performance of multi-blade fan is increasingly affected.

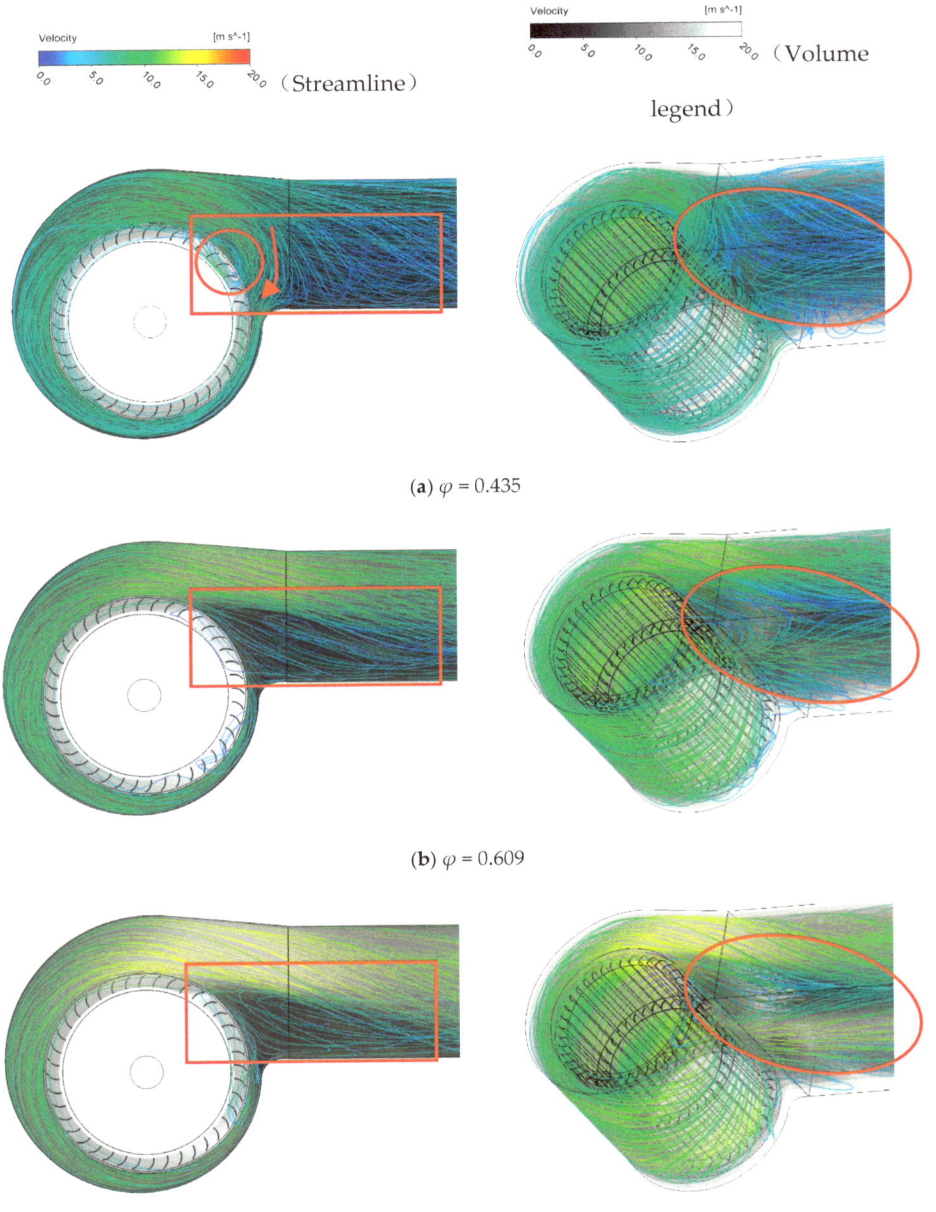

(**a**) $\varphi = 0.435$

(**b**) $\varphi = 0.609$

(**c**) $\varphi = 1.175$

Figure 8. Streamline diagram with different flow coefficient: (**a**) $\varphi = 0.435$; (**b**) $\varphi = 0.609$; (**c**) $\varphi = 1.175$.

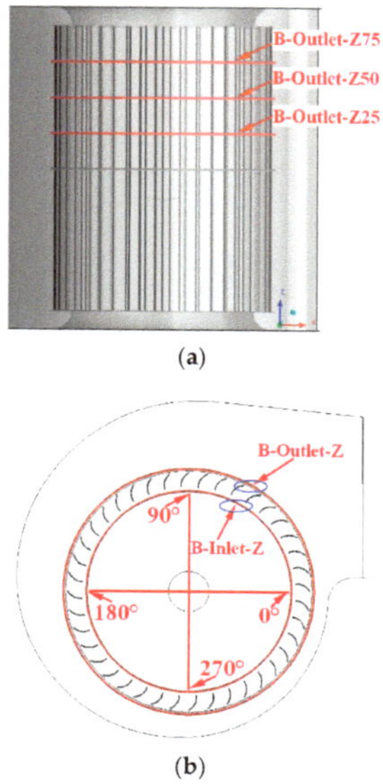

Figure 9. Line B-Outlet in impeller: (**a**) top view; (**b**) side view.

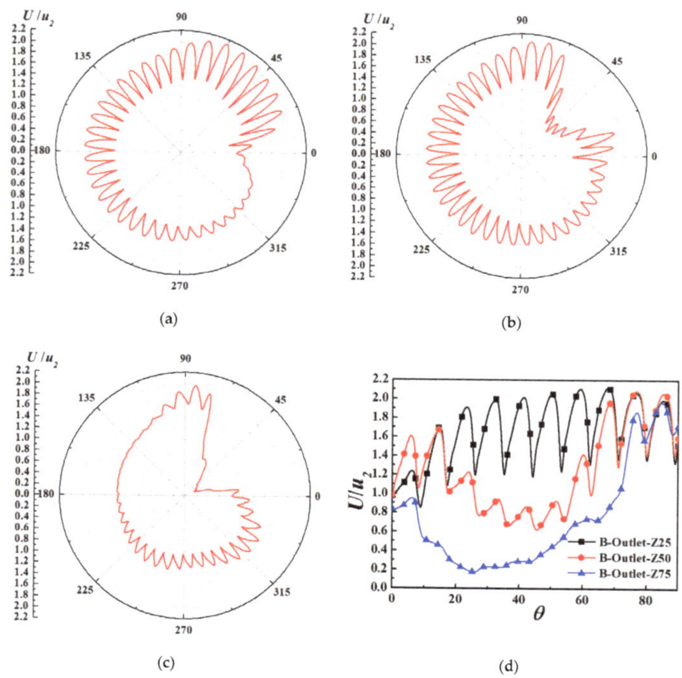

Figure 10. Absolute velocity distribution on-line B-Outlet-Z: (**a**) B-Outlet-Z25; (**b**) B-Outlet-Z50; (**c**) B-Outlet-Z75; (**d**) B-Outlet-(0°–90°).

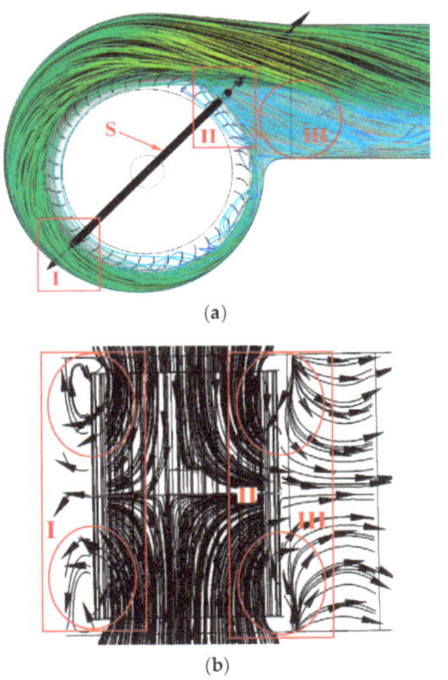

Figure 11. Streamline of Section S in multi-blade fan: (**a**) section S in fan; (**b**) streamline on section S.

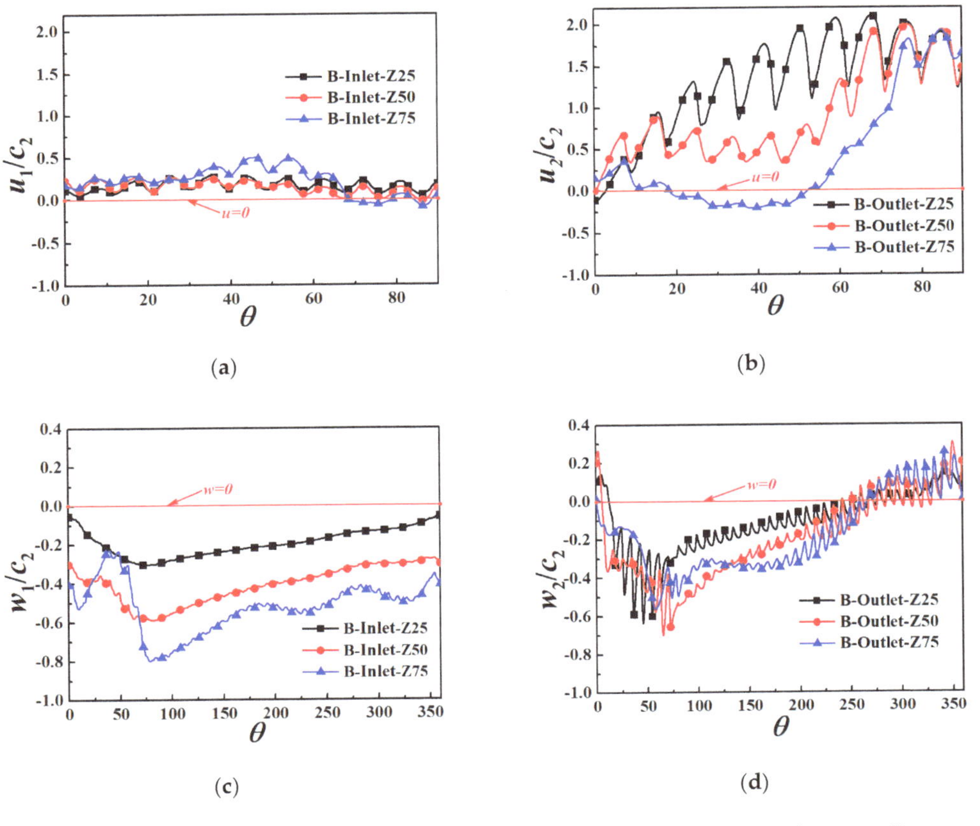

Figure 12. Velocity distributions on-line B-Inlet and B-Outlet: (**a**) Velocity u_1 distribution on-line B-Inlet-Z (0°–90°); (**b**) Velocity u_2 distribution on-line B-Outlet-Z (0°–90°); (**c**) Velocity w_1 distribution on-line B-Inlet-Z; (**d**) Velocity w_2 distribution on-line B-Outlet-Z.

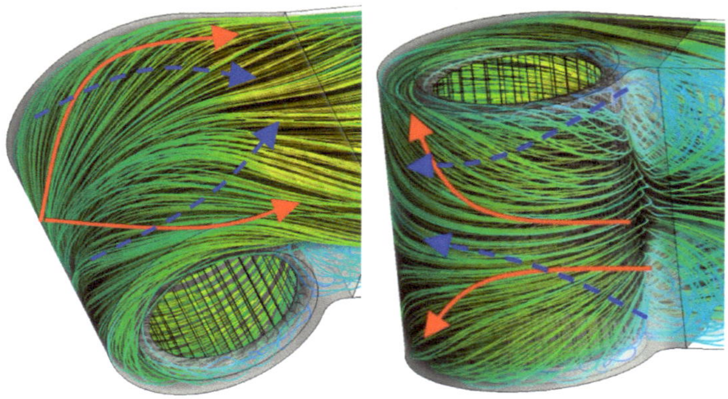

Figure 13. Streamline in volute.

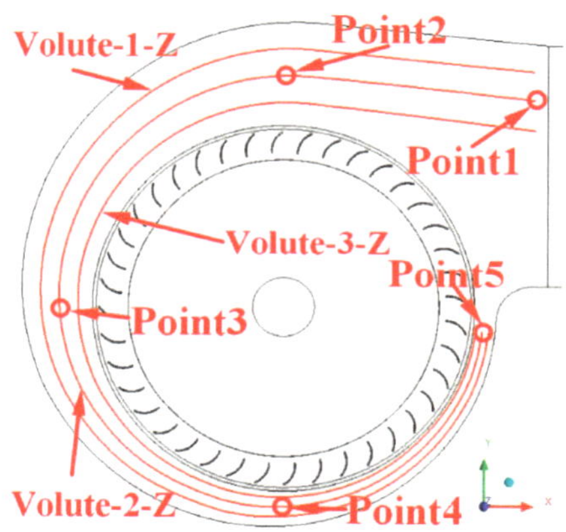

(a)

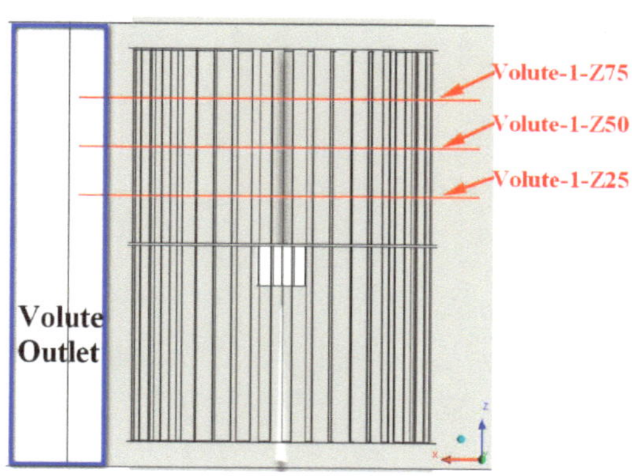

(b)

Figure 14. Line volute-Z in volute: (**a**) side view; (**b**) top view.

Interestingly, the flow turns 90 degrees from the volute inlet to the impeller outlet in multi-blade fan. The blade parameters of the multi-blade fan have little change along the axial direction, which shows that a large axial velocity before and after the airflow enters the blade flow channel (the velocity direction is from the volute inlet to the impeller center). Figure 11 shows the streamline of section S in multi-blade fan. As shown in Figure 11, some obvious secondary flows (region I) and back-flows (region II) obviously occur in the volute inlet. The restriction of the volute walls prevent back-flow in of the volute. However, the air flow with a large axial velocity in the blade passage due to the pressure gradient, secondary flows are formed. Moreover, the back-flow is caused by the high static pressure in region II, which directly cause the appearance of the vortex between the impeller outlet and volute outlet. A stepped structure is generated and captured at the blade passage outlet due to the regional back flows. Then the volute outlet vortex in region III is generated due to the flow stepped structure and the mechanism is similar to the stepped vortex. However, the volute outlet vortex is more slender and inclined than the stepped vortex.

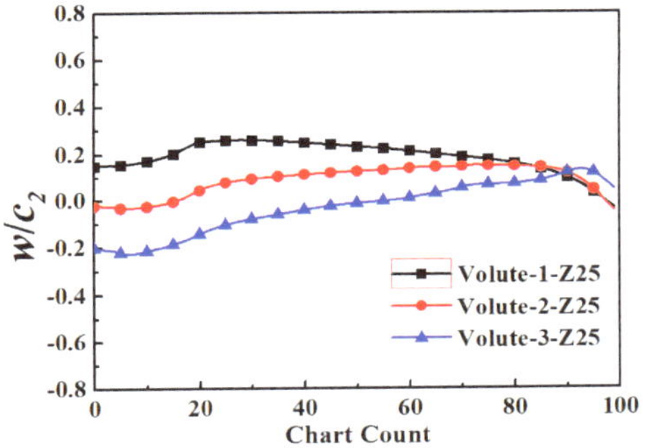

(a)

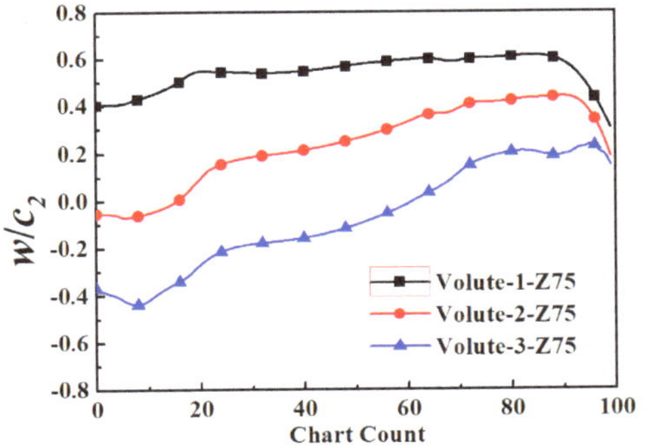

(b)

Figure 15. Velocity w distribution on-line Volute-Z: (**a**) won-line Volute Z25; (**b**) won-line Volute Z75.

Figure 12 displays the distribution of u and w of at blade passage inlet and outlet, where minus w means that the velocity direction is from the volute inlet to the impeller center and c_2 means the linear velocity of impeller outlet, respectively. It can be seen that back-flow is at the outlet of blade passage from 20° to 50° region (B-Outlet-Z75). As illustrated in Figure 12c,d, the axial velocity of the airflow does not change much along the blade passage. The main reason for this phenomenon is that the blade passage of the forward multi-blade fan is very short, the airflow passes through the blade passage rapidly. It is further found that the large axial velocity appears when the air flow goes by the blade flow channel. Interestingly, the airflow is far away from the lobes near the inlet side of the volute, and a significant backflow and step-like barrier structure appear in this area, which leads to the appearance of the vortex at the volute outlet. The vortex at the volute outlet causes the airflow of the impeller close to the volute to return to the blade channel. The secondary flow close to the impeller flows from the central to the sides of the volute.

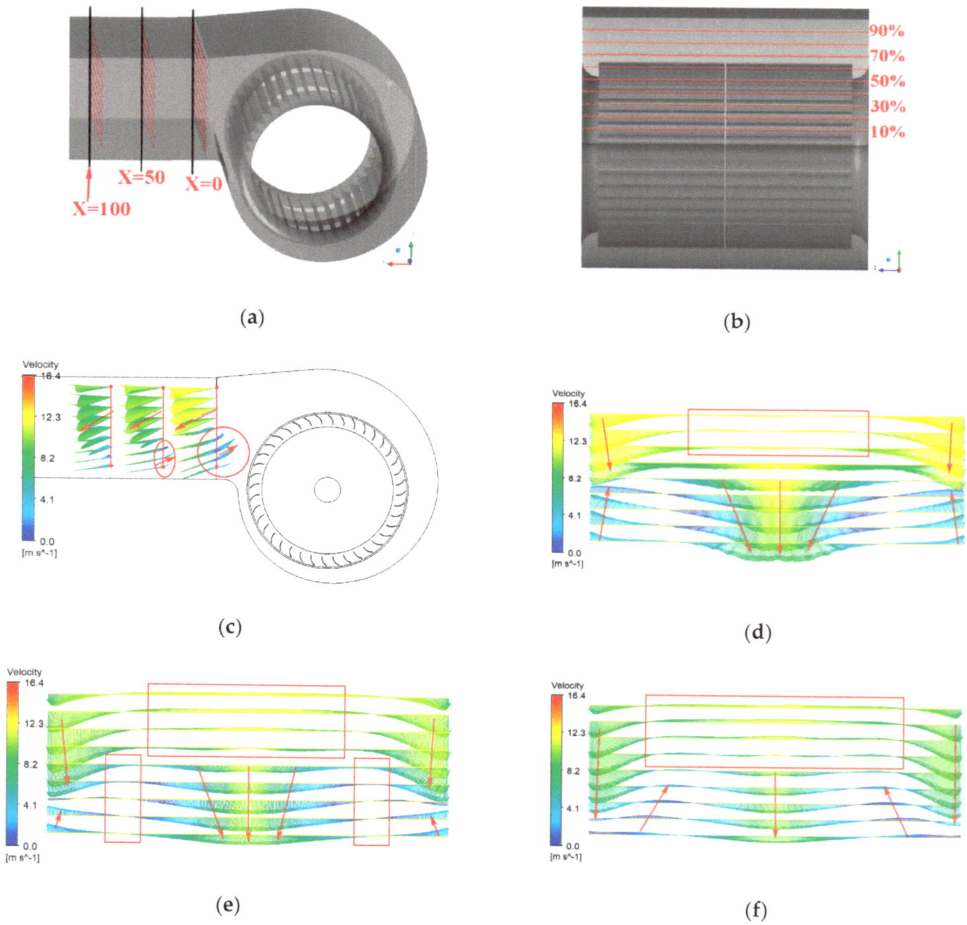

Figure 16. Velocity vector on-line X: (**a**) Line X sketch; (**b**) different height with X; (**c**) Velocity vector near volute outlet; (**d**) Velocity vector on-line X = 0; (**e**) Velocity vector on-line X = 50; (**f**) Velocity vector on-line X = 100.

4.3. Flow in the Volute

The flow in the volute is obstructed by the volute outlet vortex. Figure 13 illustrates the staggered flow in the volute. As illustrated in Figure 13, it is obviously observed that the airflow near the upper wall of the volute (solid line) flows from the center to both sides of the volute, while the flow near the impeller (dashed line) shows the opposite trend. This phenomenon is due to the large axial velocity of the flow in blade passage and the low velocity in volute outlet vortex.

As shown in Figure 14a, the flow in the volute is divided into four parts by three curves. The curves are located in the volute between the volute and the impeller. The axial position of the curves are shown in Figure 14b.

Figure 15 describes the distribution of w along the line in volute. Chart Count means that the line segment is divided into 100 equal parts, where 0 is near the exit of the volute and 100 is near the tongue. The velocity w at the Z25 section is smaller than that at the Z75, and there is no great difference in the velocity direction. This shows that the vortex near volute outlet region keeps the axial velocity direction of the flow near the volute tongue consistent.

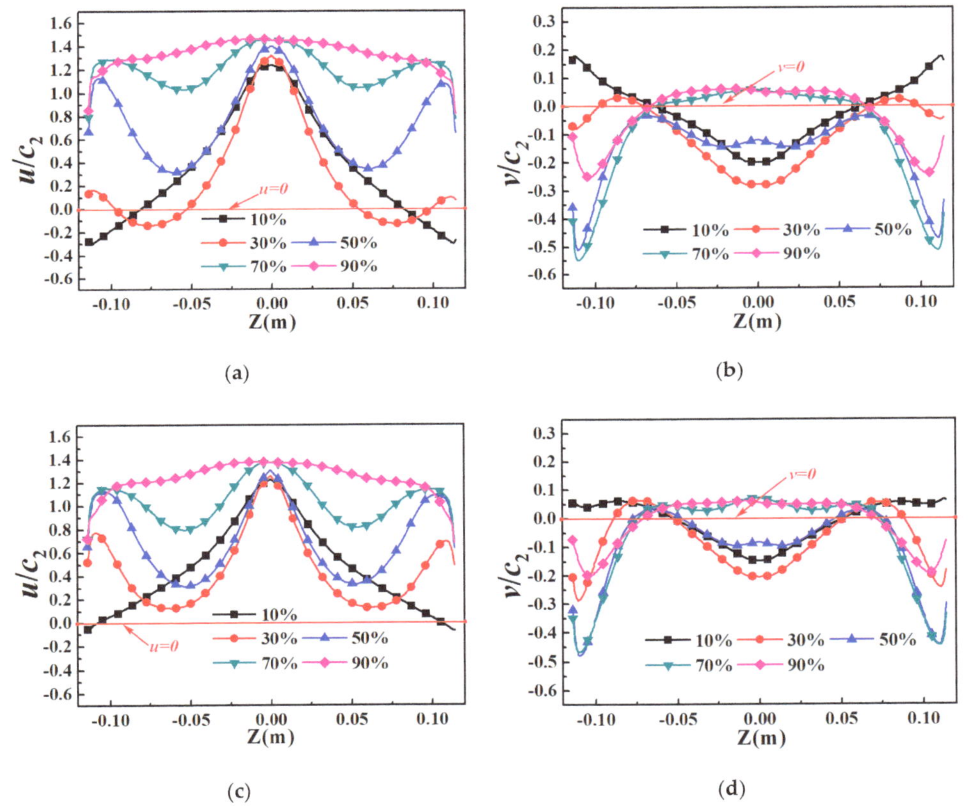

Figure 17. Velocity distribution on-line: (**a**) Velocity u distribution on-line (X = 50); (**b**) Velocity v distribution on-line (X = 50); (**c**) Velocity u distribution on-line (X = 100); (**d**) Velocity v distribution on-line (X = 100).

4.4. Outlet Extension Flow Structure

Figure 16 shows the velocity vector from volute outlet to the outlet extension of fan. There are three sections in the outlet extension where X = 100 (X100) means the section is 100 mm away from the volute outlet. As described in Figure 16c, one can see that a number of backflows occur from X = 0 to X = 50 at 10% to 40% height, and the flow from top to bottom can be seen at X = 0 to X100. It is also found that the flow with velocity component in the *Y*-axis direction (v) is blocked by the volute outlet vortex in Figure 16c. However, it is observed that not all positions have a large v-component velocity at the X = 0 in Figure 16d. It is obviously observed that the air flow converges towards the center of the passage below 50% height in Figure 16e. This mainly indicates that the effect of axial velocity has been continuous from impeller to the volute, which causes the shear friction. It is further obtained that the backflow decreases gradually with increasing x near the volute outlet.

Figure 17 displays the velocity distribution on the line of X = 50 and X = 100. As illustrated in Figure 17, it is obviously observed that the back-flow below occurs at X = 0 50% of the volute outlet height, however, the back-flow disappears at the X = 100. It can be observed that the velocity distribution at 90% height is relatively uniform, but from 70% height all the way to the bottom of the outlet extension section, the air flow is uneven. The velocity at both ends of the line are well consistent with the higher speed zone in the middle of the line segment, which indicates that the volute outlet vortexes are not close to the wall of the extended section. The flow is not the farther away from the volute outlet the better, but the entire area where the vortex outlet vortex exists, the air flow is very chaotic due to the existence of the vortex outlet vortex.

5. Conclusions

In this paper, the complex vortex structure at the volute outlet of a multi-blade fan is investigated by numerical simulation. It is clearly observed that the vortex structure between outlet channel and volute. The effect of channel on the internal flow and performance of multi-blade fan cannot be ignored. The main conclusions are as follows:

First of all, it can be clearly obtained that the streamlines are divided into two parts (backflow and outflow) at the volute outlet. The air flow has a large axial velocity before and after entering the blade flow channel. The airflow is far away from the lobes near the inlet side of volute, and an obvious backflow and step-like barrier structure mainly occur in this area, which leads to the appearance of the vortex at the outlet of the volute.

In addition, the flow vortex at the outlet of the volute causes the airflow of the impeller close to the volute to return to the blade channel. The secondary flow close to the impeller flows from the center to the sides of the volute.

What's more, it can be observed that the air flow converges towards the center of the passage below 50% height. This mainly indicates that the effect of axial velocity is continuous from impeller to the volute and causes the shear friction. It is further obtained that the backflow gradually decreases with increasing x near the volute outlet.

Finally, the origins and effects of the complex flow structure on the volute outlet are mainly investigated in this work. To benefit more from the vortex structure on the volute outlet, the volute and tongue can be further studied to achieve better flow quality and higher aerodynamic performance of multi-blade centrifugal fan.

Author Contributions: The following statements could be used Y.W. and Z.L. conceived and designed the experiments; X.Y. performed the experiments; Y.W. and Z.L. analyzed the data; Z.L. wrote the paper. All authors have read and agreed to the published version of the manuscript.

Funding: This works were mainly supported by the National Natural Science Foundation of China (11872337, 51906223) and Fundamental Research Funds of Zhejiang Sci-Tech University (2019Y004), the Key Research and Development Program of Zhejiang Province (2020C04011) and Public Projects of Zhejiang Province (LGG20E060001), respectively.

Conflicts of Interest: The authors declare no conflict of interest

References

1. Kind, R.J.; Tobin, M.G. Flow in a Centrifugal Fan of the Squirrel-Cage Type. *J. Turbomach.* **1990**, *112*, 84–90. [CrossRef]
2. Ohta, Y.; Outa, E.; Tajima, K. Evaluation and Prediction of Blade-Passing Frequency Noise Generated by a Centrifugal Blower. *J. Fluids Eng.* **1996**, *118*, 597–605. [CrossRef]
3. Ballesteros, T.R.; Francisco, G.C.; Sandra, V.S. Numerical Model for the Unsteady Flow Features of a Squirrel Cage Fan. In Proceedings of the ASME 2009 Fluids Engineering Division Summer Meeting, Vail, CO, USA, 2–6 August 2009.

4. Velarde, S.S.; Ballesteros, T.R.; González, P.J. Numerical simulation of the unsteady flow patterns in a smallmulti-blade fan. *Am. Soc. Mech. Eng.* **2006**, *47519*, 297–304.
5. Lin, S.C.; Huang, C.L. An integrated experimental and numerical study of forward-curved centrifugal fan. *Exp. Therm. Fluid Sci.* **2002**, *26*, 421–434. [CrossRef]
6. Wang, K.; Ju, Y.P.; Zhang, C.H. Design of multi-blade centrifugal fan based on grouping model and bionic volute tongue. *J. Eng. Thermophys.* **2017**, *38*, 1671–1675.
7. Adachi, T.; Sugita, N.; Yamada, Y. Study on the performance of a sirocco fan (optimum design of blade shape). *Int. J. Rotating Mach.* **2001**, *7*, 405–414. [CrossRef]
8. Adachi, T.; Sugita, N.; Yamada, Y. Study on the performance of a sirocco fan (flow around the runner blade). *Int. J. Rotating Mach.* **2004**, *10*, 415–424. [CrossRef]
9. Darvish, M.; Frank, S.; Paschereit, C.O. Numerical and experimental study on the tonal noise generation of a radial fan. *J. Turbomach.* **2015**, *137*, 101005. [CrossRef]
10. Wei, Y.; Ying, C.; Xu, J.; Cao, W.; Wang, Z. Effects of Single-arc Blade Profile Length on the Performance of a Forward multi-blade Fan. *Processes* **2019**, *7*, 629. [CrossRef]
11. Wang, K.; Ju, Y.; Zhang, C. Experimental and numerical investigations on effect of blade trimming on aerodynamic performance of squirrel cage fan. *Int. J. Mech. Sci.* **2020**, *177*, 105579. [CrossRef]
12. Kind, R.J. Prediction of flow behavior and performance of multi-blade centrifugal fans operating at medium and high flow rates. *J. Fluids Eng.* **1997**, *119*, 639–646. [CrossRef]
13. Velarde, S.S.; Ballesteros, T.R.; Santolaria, M.C. Unsteady flow pattern characteristics downstream of a forward-curved blades centrifugal fan. *J. Fluids Eng.* **2001**, *123*, 265–270. [CrossRef]
14. Wen, X.; Mao, Y.; Yang, X. Design method for the volute profile of a squirrel cage fan with space limitation. *J. Turbomach.* **2016**, *138*, 081001. [CrossRef]
15. Lun, Y.X.; Lin, L.M.; He, H.J.; Zhu, Z.C.; Wei, Y.K. Effects of Vortex Structure on Performance Characteristics of a multi-blade Fan with Inclined tongue. *Proc. Inst. Mech. Eng. Part A J. Power Energy* **2019**, *233*, 1007–1021. [CrossRef]
16. Yang, H.; Yu, P.Q.; Xu, J.; Ying, C.L.; Wen, B.C.; Wei, Y.K. Experimental investigations on the performance and noise characteristics of a forward-curved fan with the stepped tongue. *Meas. Control* **2019**, *52*, 1480–1488. [CrossRef]
17. Heo, M.W.; Kim, J.H.; Seo, T.W. Aerodynamic and aeroacoustic optimization for design of a forwardcurved blades centrifugal fan. *Proc. Inst. Mech. Eng. Part A J. Power Energy* **2016**, *230*, 154–174. [CrossRef]
18. Kim, J.S.; Jeong, U.C.; Kim, D.W. Optimization of sirocco fan blade to reduce noise of air purifier using a metamodel and evolutionary algorithm. *Appl. Acoust.* **2015**, *89*, 254–266.
19. Huang, C.H.; Hung, M.H. An optimal design algorithm for centrifugal fans. Theoretical and experimental studies. *J. Mech. Sci. Technol.* **2013**, *27*, 761–773. [CrossRef]
20. Khalkhali, A.; Farajpoor, M.; Safikhani, H. Modeling and multi-objective optimization of forward-curved blade centrifugal fans using CFD and neural networks. *Trans. Can. Soc. Mech. Eng.* **2011**, *35*, 63–79. [CrossRef]
21. Wang, K.; Ju, Y.; Zhang, C. Numerical investigation on flow mechanisms of squirrel cage fan. *Proc. Inst. Mech. Eng. Part A J. Power Energy* **2018**. [CrossRef]
22. Meneveau, C.; Lund, T.S.; Cabot, W.H. A Lagrangian Dynamic Subgrid Scale Model of Turbulence. *J. Fluid Mech.* **1996**, *319*, 353–385. [CrossRef]
23. Piomelli, U. High Reynolds Number Calculations Using The Dynamic Subgrid Scale Stress Model. *Phys. Fluids* **1993**, *5*, 1484–1490. [CrossRef]
24. Wang, C.; Hu, B.; Zhu, Y.; Wang, X.; Luo, C.; Cheng, L. Numerical study on the gas-water two-phase flow in the self-priming process of self-priming centrifugal pump. *Processes* **2019**, *7*, 330. [CrossRef]
25. Wang, Z.D.; Wei, Y.K.; Qian, Y.H. A bounce back-immersed boundary-lattice Boltzmann model for curved boundary. *Appl. Math. Model.* **2020**, *81*, 428–440. [CrossRef]

processes

MDPI

Article

Research on Flow Characteristics of Straight Line Conjugate Internal Meshing Gear Pump

Hongqiang Chai [1,2,*], Guolai Yang [1,2,*], Guoguo Wu [1,3,*], Guixiang Bai [1] and Wenqi Li [4]

1 College of Energy and Power Engineering, Lanzhou University of Technology, Lanzhou 730050, China; bai04guixiang@126.com
2 State Key Laboratory of Fluid Power and Mechatronic Systems, Zhejiang University, Hangzhou 310027, China
3 School of Intelligent Manufacturing Engineering, Chongqing University of Arts and Sciences, Chongqing 402160, China
4 Lanzhou Ls Group Co., Ltd., Lanzhou Ls Energy Equipment Engineering Research Institute Co., Ltd. Gansu Intelligent Control & Key Laboratory of Metal Plasticity Forming Equipment, Lanzhou 730314, China; li04wenqi@126.com
* Correspondence: chaihq35@163.com (H.C.); yanggl@lut.cn (G.Y.); wuguoguo@cqwu.edu.cn (G.W.)

Received: 9 January 2020; Accepted: 21 February 2020; Published: 27 February 2020

Abstract: The improvement of the overall performance of hydraulic pumps is the basis of intelligent hydraulics. Taking the straight line conjugate internal meshing gear pump as the research object, the theoretical flow rate and the geometric flow pulsation rate equations are established in this study through the volume change method. The change laws of the gear pair's geometric parameters on the theoretical flow rate and the geometric flow pulsation rate are studied. The simulation model of the internal flow channel is established, and the influence factors and the influence degree of the flow pulsation and average flow rate are analyzed. The high-pressure positive displacement pump test system is also designed and built. The performance evaluations are conducted, and the experimental results are analyzed. The results show that the periodic change of the meshing point position is the root cause of the geometric flow pulsation. The theoretical flow rate and the geometric flow pulsation rate are 103.71 L/min and 1.76%, respectively. To increase the theoretical flow rate whilst decreasing the geometric flow pulsation rate, the tip circle radius of the external gear should be increased as much as possible within the allowable range of the design calculation. Amongst the three influencing factors that produce flow pulsation, the oil compressibility has no effect on the flow pulsation. The uneven internal leakage is the main factor, and the geometric flow pulsation only accounts for a small proportion. The internal leakage reduces the simulated flow rate by 3.59 L/min. The difference between the experimental and simulated flow rates is less than 2%. Within the allowable speed range, the rotation speed of the external gear should be increased as much as possible to increase the average flow rate and the volumetric efficiency.

Keywords: straight line conjugate internal meshing gear pump; average flow rate; flow pulsation; gear pair's geometric parameters; simulation analysis; performance evaluations

---

1. Introduction

With the continuous development of hydraulic technology, digital and intelligent hydraulics have become hotspots in the industry. The advancement of the body part and the improvement of the hydraulic pump's overall performance (e.g., high pressure, low noise, low pulsation, and large and variable displacements) are also an objective of intelligent hydraulics [1–5]. The average flow rate, flow pulsation and internal flow field distribution are collectively referred to as the flow characteristics of gear pumps, which are the key factors affecting the large displacement, low pulsation, and low noise. These flow characteristics should be accurately determined for the research and development of high-performance gear pumps.

The straight line conjugate internal meshing gear pump is a gear pump with a novel design and has an advanced international level. The related patent appeared in 1970 [6]. The core components include an external gear with a straight tooth profile. Its internal ring gear has a high-order circular tooth profile and is conjugate to the straight tooth profile of the external gear. Given unfavourable factors such as the technical protection of foreign companies, open and complete standardised design systems are scarce. Numerous domestic companies and scholars have researched and optimized the design of straight line conjugate internal meshing gear pumps, mainly focusing on derivation of various equations and parameter optimization. Few studies on the performance of the entire pump have been published.

The tooth profile of the gear pair determines the key performance of gear pumps [7]. Ample research has been performed on the derivation of the tooth profile curve and meshing line equations and the optimization of the tooth profile's basic parameters using relevant constraints. Some scholars used the tooth profile normal method to derive the conjugate tooth profile curve and meshing line equations [8–12]. Xu et al. utilized the complex vector method to achieve similar objectives, whereas Wei et al. and Yang et al. employed the general tooth profile normal inversion and meshing angle function methods, respectively [13–15]. Some scholars established the mathematical models for multi-objective optimization and different interferences. By setting the continuous transmission conditions, meshing angles, absence of interference of tooth profile overlap, tooth tip thickness, and tip circle limit radius as the constraint functions, the researchers have investigated the influence of the gear pair's main parameters on the objective function, and optimized the relevant parameters [16–19]. However, the research on multi-objective optimization only focuses on a few sets of data, and does not analyze the optimal value of the objective function in a given interval.

Tooth surface contact transmission is an important source of mechanical vibration noise, which affects tooth surface strength and life [20–22]. Therefore, the analysis of tooth surface meshing and tooth surface contact strength are reported in numerous studies in the literature. Some scholars have established the sliding coefficient equation based on the polar coordinate method, developed the meshing efficiency equation using the tangential polar coordinate method, and used the finite element method to simulate the contact strength of tooth surface and pump's body structure, respectively. The influence of the relevant parameters on the objective function has been analyzed [23–26]. Some scholars established the coincidence degree equations on the basis of the basic law of tooth profile meshing and discussed the corresponding influence of relevant parameters [27,28].

The noise induced by the oil trap phenomenon and the pressure shock noise caused by the flow pulsation are the main noise sources of gear pumps [29–32]. Therefore, research on the mechanism of gear pump vibration and noise reduction is not only a key topic in the field, but also the interest of internationally renowned manufacturers that heavily invest in research. Wang et al. established the instantaneous oil trap volume equation based on the changes in the oil trap volume, and analyzed the influence of relevant parameters [33]. Some scholars established the instantaneous flow rate equations using the gear meshing principle, and derived the approximate equations of flow pulsation rate using the theoretical flow rate approximation equations [34–36]. However, the instantaneous flow rate equation is a general equation and does not change based on the distance from the node to the meshing point and the rotation angle of the external gear. In addition, error exists between the approximate and the exact equations of theoretical flow rate and flow pulsation rate, resulting in an inaccurate reflection of the true value. With respect to the magnitude of the error, no simulation analysis and experimental verification have been performed according to the parameters of the actual model. Similarly, whether the gear pair's geometric parameters change the theoretical flow rate and geometric flow pulsation rate has not been investigated. The development and design of the products should be equipped with the corresponding process equipment for computer-aided design [37]. With regard to the research on the machining method of gear pairs, Zhang et al. proposed a forming method for large-modulus internal gears. The external gear meshing with the internal gear was matched according to the conjugate principle [38].

None of the above-mentioned studies have performed the numerical simulation of the internal flow channel simulation model to obtain the flow characteristics of the internal flow field, nor analyzed the influence factors and influence degree of the outlet flow pulsation and average flow rate. Moreover, the performance evaluations of straight line conjugate internal meshing gear pumps are not executed using a high-pressure positive displacement pump test system, and the difference between the experimental and simulation results is not investigated.

The flow characteristics of a straight line conjugate internal meshing gear pump should be analyzed to provide reliable data to the designer for an accurate calculation of theoretical flow rate and geometric flow pulsation rate, and provide theoretical foundation to increase the geometric displacement, decrease the flow pulsation to reduce the pressure shock noise and improve the overall performance of gear pumps comprehensively. Based on these, this paper is conducted from three aspects. In terms of theoretical analysis, a general mathematical model of the theoretical instantaneous flow rate is established using the volume change method. By using the number-shape combination method, the theoretical instantaneous flow rate equations based on changes in the distance from the node to the meshing point and the rotation angle of the external gear, the theoretical flow rate equation, and the geometric flow pulsation rate equation are analyzed and determined. Taking the straight line conjugate internal meshing gear pump as the research object, the specific values of the theoretical flow rate and the geometric flow pulsation rate are calculated and analyzed. The design variables affecting theoretical flow rate and geometric flow pulsation rate and the corresponding value ranges are determined, and the change laws of the gear pair's geometric parameters on theoretical flow rate and geometric flow pulsation rate are studied. In terms of numerical simulations, the simulation model of the internal flow channel is developed. The numerical simulation results are analyzed and compared, and the influence factors and the influence degree of the flow pulsation and the average flow rate at the outlet of the gear pump are investigated. In terms of experimental researches, a high-pressure positive displacement pump test system is also designed and built. The performance evaluations for the test pump are implemented, and the experimental results are analyzed and compared with the simulation results.

2. Modeling of the Theoretical Flow Rate and Geometric Flow Pulsation Rate of a Straight Line Conjugate Internal Meshing Gear Pump

2.1. Tooth Profile Curve of the External Gear

The tooth profile curve of the external gear is a left-right symmetric straight line, which is established at the center O_1 of the external gear as the coordinate origin, as shown in Figure 1. The equation of l can be obtained from the geometric relationship of Figure 1.

$$l = r_1 \sin\left(\frac{\theta}{2} + \beta\right) \tag{1}$$

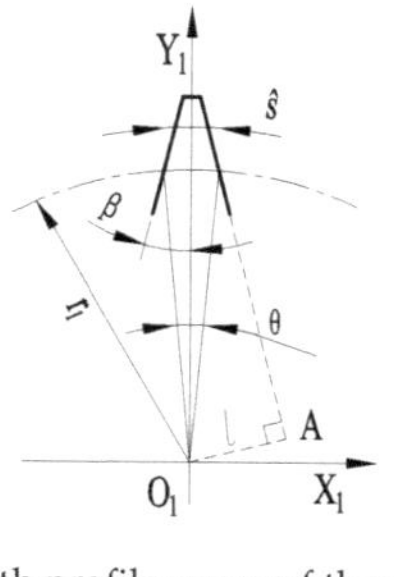

Figure 1. Tooth profile curve of the external gear.

2.2. General Mathematical Model Analysis of Theoretical Instantaneous Flow Rate

The theoretical instantaneous flow rate is established using the volume change method. The working dynamic diagram of straight line conjugate internal meshing gear pump is shown in Figure 2, where k is the meshing point. During the meshing process of the external gear and internal gear ring, the tooth profiles m, n, g, k of the external gear and the tooth profiles k, g', n', m' of the internal gear ring enclose an oil discharge chamber. When the external gear rotates, the full tooth profiles mn and $m'n'$ compress the oil in the oil discharge chamber to make the volume of oil discharge chamber smaller, while the other part of meshing tooth profiles gk and $g'k$ expand the volume of oil discharge chamber during continuous rotation. However, the rest tooth profiles that are completely surrounded by the high-pressure oil do not participate in the work and have no effect on the oil discharge.

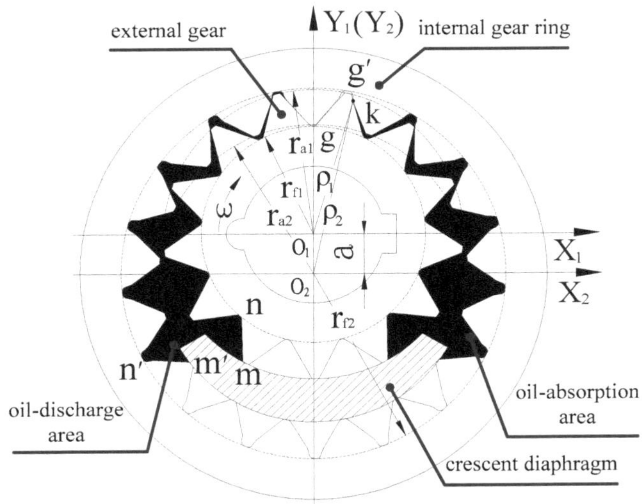

Figure 2. Working dynamic diagram of the straight line conjugate internal meshing gear pump.

Firstly, the sweep area method is used to calculate the volume change of oil discharge chamber. According to the above description, and given that the transmission ratio of the internal meshing gear pair is a fixed value, after the rotation angles of the external gear and the internal gear ring are $\Delta\varphi_1$ and $\Delta\varphi_2$, respectively, it can be obtained from Figure 2 that the full tooth profiles mn and $m'n'$ reduce the volume of oil discharge chamber.

$$\Delta V_1 = \frac{B}{2}\left[\left(r_{a1}^2 - r_{f1}^2\right)\Delta\varphi_1 + \left(r_{f2}^2 - r_{a2}^2\right)\Delta\varphi_2\right] \quad (2)$$

Meanwhile, due to the rotation of the meshing tooth profiles gk and $g'k$, it can be obtained from Figure 2 that the meshing tooth profiles gk and $g'k$ expand the volume of oil discharge chamber.

$$\Delta V_2 = \frac{B}{2}\left[\left(\rho_1^2 - r_{f1}^2\right)\Delta\varphi_1 + \left(r_{f2}^2 - \rho_2^2\right)\Delta\varphi_2\right] \quad (3)$$

According to the meshing principle of tooth profile, the position of the meshing point k constantly changes during working process. In other words, the values of ρ_1 and ρ_2 are not fixed, and they periodically change with the meshing point position. As a result, ΔV_2 also changes periodically, which is the root cause of the geometric flow pulsation. Therefore, in theory, the pulsation of the output instantaneous flow of gear pump is unavoidable.

The volume change of the oil discharge cavity can be obtained by subtracting Equations (2) and (3), which is expressed by Equation (4):

$$\Delta V = \frac{B}{2}\left[\left(r_{a1}^2 - \rho_1^2\right)\Delta\varphi_1 + \left(\rho_2^2 - r_{a2}^2\right)\Delta\varphi_2\right] \tag{4}$$

Equation (5) can be obtained from the basic law of the tooth profile meshing:

$$i_{12} = \frac{\omega_1}{\omega_2} = \frac{\Delta\varphi_1}{\Delta\varphi_2} = \frac{Z_2}{Z_1} = \frac{r_2}{r_1} \tag{5}$$

Equation (5) is substituted into the volume change Equation (4), and the theoretical instantaneous flow rate equation of straight line conjugate internal meshing gear pump can be obtained by derivation and simplification:"

$$q_{sh} = \frac{B\omega_1}{2}\left[r_{a1}^2 - \rho_1^2 + \left(\rho_2^2 - r_{a2}^2\right)\frac{Z_1}{Z_2}\right] \tag{6}$$

From the above analysis, it can be obtained that gear pumps with different tooth profiles exhibit different flow pulsation characteristics due to the different change laws of meshing point position. Further, the change laws can be determined based on the distance from the node to the meshing point and the rotation angle of the external gear.

2.3. Theoretical Instantaneous Flow Rate Equation Based on the Distance Change from the Node to the Meshing Point

According to the meshing principle of the gear pair, the schematic diagram of meshing point's position change is established in the internal meshing gear pair, as shown in Figure 3. By expressing ρ_1 and ρ_2 from the geometric relationship in Figure 3, the theoretical instantaneous flow rate equation based on the distance change from the node to the meshing point can be obtained:

$$q_{sh} = \frac{B\omega_1}{2}\left[r_{a1}^2 - i_{12}^{-1}r_{a2}^2 + (i_{12} - 1)r_1^2 - \left(1 - i_{12}^{-1}\right)f^2\right] \tag{7}$$

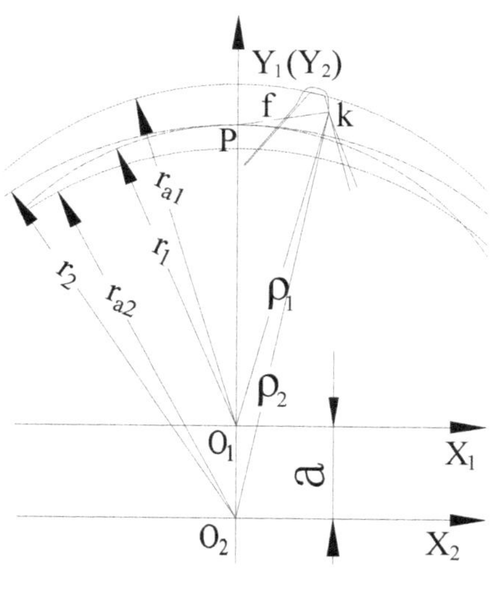

Figure 3. Meshing point's position change.

From the continuous contact transmission's conditions of the tooth profile, it can be known that transmission ratio $i_{12} > 1$ in the internal meshing gear pair, that is, the value of expression $1 - i_{12}^{-1}$ is a positive number less than 1. Therefore, when f is 0, that is, when the node coincides with the meshing point, the instantaneous flow rate reaches the maximum; when f is maximum, it is at the minimum. At this time, the meshing point may be located at the position where the gear pair is about to enter or exit the meshing. The specific location is discussed in detail in the following section.

2.4. Theoretical Instantaneous Flow Rate Equation Based on the Rotation Angle Change of the External Gear

The schematic diagram of the rotation angle change of the external gear is established according to the definitions of tooth profile parameters, as shown in Figure 4. According to the basic theorem of the tooth profile meshing (i.e., the common normal of two conjugate tooth profiles at the contact point passes through the instantaneous velocity center), in other words, the line between the meshing point and the node is perpendicular to the external gear's profile curve. Based on this and according to the geometric relationship in Figure 4, the equation sets about ρ_1 and ρ_2 can be obtained:

$$\left.\begin{aligned} \rho_1^2 &= l^2 + r_1^2\cos^2(\beta-\varphi) \\ \rho_2^2 &= [a\sin(\beta-\varphi)+l]^2 + r_2^2\cos^2(\beta-\varphi) \end{aligned}\right\} \tag{8}$$

By substituting the Equation (8) into Equation (6), the theoretical instantaneous flow rate equation based on the rotation angle change of the external gear can be obtained:

$$\begin{aligned} q_{sh} = {} & \tfrac{B\omega_1}{2}\Big[\big(r_1^2 + i_{12}^{-1}a^2 - i_{12}r_1^2\big)\sin^2(\beta-\varphi) + 2i_{12}^{-1}al\sin(\beta-\varphi) + r_{a1}^2 + i_{12}r_1^2 + i_{12}^{-1}l^2 - \ldots \\ & i_{12}^{-1}r_{a2}^2 - r_1^2 - l^2\Big] \end{aligned} \tag{9}$$

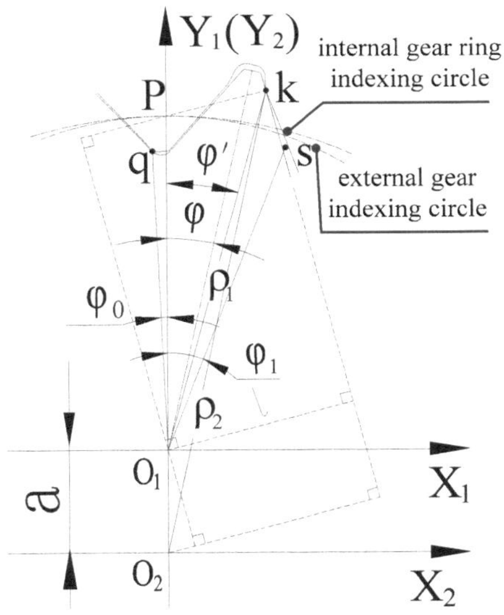

Figure 4. Rotation angle change of the external gear.

2.5. Theoretical Flow Rate Equation of the Straight Line Conjugate Internal Meshing Gear Pump

2.5.1. Theoretical Coincidence Degree Equation and Theoretical Oil Trap Time Equation Based on the Rotation Angle Change of the External Gear

From the meshing process of gear pair [39], it can be known that the starting point when the gear pair enters the meshing is the intersection of the tip circle of internal gear ring and the meshing line, corresponding to the initial meshing angle is φ_0. When the meshing reaches the intersection of the tip circle of the external gear and the meshing line, the gear pair are about to exit the meshing, which corresponds to the ending meshing angle is φ_1. Based on this and combined with Equation (8), the Equation (10) can be obtained:

$$\left.\begin{aligned} r_{a1}^2 &= l^2 + r_1^2\cos^2(\beta - \varphi_1) \\ r_{a2}^2 &= [a\sin(\beta - \varphi_0) + l]^2 + r_2^2\cos^2(\beta - \varphi_0) \end{aligned}\right\} \tag{10}$$

Equation (11) about the meshing angles φ_0 and φ_1 can be obtained from Equation (10):

$$\left.\begin{aligned} \varphi_0 &= \beta - \arcsin\frac{al + \sqrt{a^2l^2 - (i_{12}^2 r_1^2 - a^2)(r_{a2}^2 - l^2 - i_{12}^2 r_1^2)}}{i_{12}^2 r_1^2 - a^2} \\ \varphi_1 &= \beta - \arccos\frac{\sqrt{r_{a1}^2 - l^2}}{r_1} \end{aligned}\right\} \tag{11}$$

According to the definition of the theoretical coincidence degree in the meshing process of gear pair, the theoretical coincidence degree Equation (12) can be obtained:

$$\varepsilon_\alpha = \frac{\arcsin\frac{al + \sqrt{a^2l^2 - (i_{12}^2 r_1^2 - a^2)(r_{a2}^2 - l^2 - i_{12}^2 r_1^2)}}{i_{12}^2 r_1^2 - a^2} - \arccos\frac{\sqrt{r_{a1}^2 - l^2}}{r_1}}{2\pi/z_1} \tag{12}$$

Furthermore, the theoretical oil trap time Equation (13) can be obtained:

$$T_\alpha = \frac{\arcsin\frac{al + \sqrt{a^2l^2 - (i_{12}^2 r_1^2 - a^2)(r_{a2}^2 - l^2 - i_{12}^2 r_1^2)}}{i_{12}^2 r_1^2 - a^2} - \arccos\frac{\sqrt{r_{a1}^2 - l^2}}{r_1}}{\pi n_1/30} \tag{13}$$

2.5.2. Derivation of the Theoretical Flow Rate Equation

It can be known from the Figure 4 that each pair of gear teeth in the gear pair enters the meshing from the starting point q and exits the meshing from the ending point S. That is, in the rotation process of the external gear from the initial to the end meshing angles, one oil discharge cycle is formed. Assuming that the oil discharge volume is $\Delta\Sigma$ during this oil discharge cycle, and according to the definition of instantaneous flow rate, the mathematical model of the oil discharge volume through a single tooth can be obtained:

$$\Delta\Sigma = \int q_{sh}dt \tag{14}$$

By substituting the instantaneous flow rate Equation (9) and integrating Equation (14), the equation of the oil discharge volume by each pair of gear teeth in one oil discharge cycle can be obtained:

$$\begin{aligned} \Delta\Sigma = {} & \tfrac{B}{2}\Big\{\big(r_1^2 + i_{12}^{-1}a^2 - i_{12}r_1^2\big)\big[\tfrac{1}{2}(\varphi_1 - \varphi_0) - \tfrac{1}{4}\sin 2(\varphi_1 - \beta) + \tfrac{1}{4}\sin 2(\varphi_0 - \beta)\big] + 2i_{12}^{-1}al[\cos(\varphi_1 - \beta) - \cos(\varphi_0 - \beta)] + \ldots \\ & \big(r_{a1}^2 + i_{12}r_1^2 + i_{12}^{-1}l^2 - i_{12}^{-1}r_{a2}^2 - r_1^2 - l^2\big)(\varphi_1 - \varphi_0)\Big\} \end{aligned} \tag{15}$$

According to the definition of gear pump's displacement (i.e., the displacement is the oil discharge volume when the gear pairs rotate for one cycle), it is assumed that there are a total of teeth participating in meshing drainage within one cycle, thus the theoretical displacement equation can be obtained:

$$\Sigma = \Delta\Sigma z_1 \tag{16}$$

The equation of the theoretical flow rate of straight line conjugate internal meshing gear pump is obtained by combining the initial and ending meshing angles from Equation (11):

$$\begin{aligned} q_t = & \frac{z_1 n_1 B}{8}\Bigg\{\left(r_1^2+i_{12}^{-1}a^2-i_{12}r_1^2\right)\Bigg[2\left(\arcsin\frac{al+\sqrt{a^2l^2-(i_{12}^2r_1^2-a^2)(r_{a2}^2-l^2-i_{12}^2r_1^2)}}{i_{12}^2r_1^2-a^2}-\arccos\frac{\sqrt{r_{a1}^2-l^2}}{r_1}\right)+\sin 2\left(\arccos\frac{\sqrt{r_{a1}^2-l^2}}{r_1}\right)-\ldots \\ & 2\left(\frac{al+\sqrt{a^2l^2-(i_{12}^2r_1^2-a^2)(r_{a2}^2-l^2-i_{12}^2r_1^2)}}{i_{12}^2r_1^2-a^2}\right)\cos\left(\arcsin\frac{al+\sqrt{a^2l^2-(i_{12}^2r_1^2-a^2)(r_{a2}^2-l^2-i_{12}^2r_1^2)}}{i_{12}^2r_1^2-a^2}\right)\Bigg]+\ldots \\ & 8i_{12}^{-1}al\left[\frac{\sqrt{r_{a1}^2-l^2}}{r_1}-\cos\left(\arcsin\frac{al+\sqrt{a^2l^2-(i_{12}^2r_1^2-a^2)(r_{a2}^2-l^2-i_{12}^2r_1^2)}}{i_{12}^2r_1^2-a^2}\right)\right]+\ldots \\ & 4\left(r_{a1}^2+i_{12}r_1^2+i_{12}^{-1}l^2-i_{12}^{-1}r_{a2}^2-r_1^2-l^2\right)\left(\arcsin\frac{al+\sqrt{a^2l^2-(i_{12}^2r_1^2-a^2)(r_{a2}^2-l^2-i_{12}^2r_1^2)}}{i_{12}^2r_1^2-a^2}-\arccos\frac{\sqrt{r_{a1}^2-l^2}}{r_1}\right)\Bigg\} \end{aligned} \quad (17)$$

2.6. Geometric Flow Pulsation Rate Equation of the Straight Line Conjugate Internal Meshing Gear Pump

2.6.1. Maximum and Minimum Analysis of the Theoretical Instantaneous Flow Rate

The equation about φ' can be obtained from the geometric relation in Figure 4:

$$\varphi' = \arcsin\frac{l}{\rho_1} - (\beta - \varphi) \quad (18)$$

From the aforementioned analysis, we can know that when the meshing point coincides with the node, that is, when φ' is equal to 0, the theoretical instantaneous flow rate reaches the maximum, thus the equation of the maximum instantaneous flow rate can be obtained:

$$q_{sh\max} = \frac{B\omega_1}{2}\left[r_{a1}^2 - r_1^2 + \left(i_{12}^2 r_1^2 - r_{a2}^2\right)i_{12}^{-1}\right] \quad (19)$$

To obtain the minimum of the theoretical instantaneous flow rate, Equation (9) is derived with respect to the rotation angle φ, and make the derivative function be equal to 0 to solve the equation sets about φ:

$$\left.\begin{aligned} \varphi &= \beta + \left(\tfrac{\pi}{2} + k\pi\right) \\ \varphi &= \beta - \arcsin\frac{i_{12}^{-1}al}{i_{12}r_1^2 - r_1^2 - i_{12}^{-1}a^2} + k\pi \end{aligned}\right\} \quad (20)$$

Remove the inappropriate solution and determine the rotation angle of the external gear when the instantaneous flow rate reaches the limit:

$$\varphi_{\lim} = \beta - \arcsin\frac{i_{12}^{-1}al}{i_{12}r_1^2 - r_1^2 - i_{12}^{-1}a^2} \quad (21)$$

By comparison, it is found that Equation (21) is completely consistent with Equation (11) after deformation. Therefore, when the rotation angle of the external gear is the same as the ending meshing angle, the theoretical instantaneous flow rate is at the minimum, further, the minimum instantaneous flow rate equation can be obtained:

$$\begin{aligned} q_{sh\min} = & \frac{B\omega_1}{2}\Big[\left(r_1^2 + i_{12}^{-1}a^2 - i_{12}r_1^2\right)\sin^2(\beta - \varphi_1) + 2i_{12}^{-1}al\sin(\beta - \varphi_1) + r_{a1}^2 + i_{12}r_1^2 + i_{12}^{-1}l^2 - \ldots \\ & i_{12}^{-1}r_{a2}^2 - r_1^2 - l^2\Big] \end{aligned} \quad (22)$$

2.6.2. Determination of the Geometric Flow Pulsation Rate Equation

According to the definition of the geometric flow pulsation rate (i.e., it is the ratio of the difference between the maximum and minimum instantaneous flow rates to the theoretical flow), which is expressed by an equation:

$$\delta_q = \frac{q_{sh\max} - q_{sh\min}}{q_t} \quad (23)$$

The equations of the maximum and minimum instantaneous flow rates, and the theoretical flow rate are substituted into Equation (23), respectively. Thereby, the equation of the geometric flow pulsation rate of straight line conjugate internal meshing gear pump can be obtained:

$$\begin{aligned}
\delta_q = \tfrac{8\pi}{z_1}\Bigg[&\left(1-i_{12}^{-1}\right)l^2-2\left(r_1^2+i_{12}^{-1}a^2-i_{12}r_1^2\right)\frac{\sqrt{r_{a1}^2-l^2}}{r_1}\sin\left(\arccos\frac{\sqrt{r_{a1}^2-l^2}}{r_1}\right)-2i_{12}^{-1}al\sin\left(\arccos\frac{\sqrt{r_{a1}^2-l^2}}{r_1}\right)\Bigg]/\ldots\\
\Bigg\{&\left(r_1^2+i_{12}^{-1}a^2-i_{12}r_1^2\right)\Bigg[2\left(\arcsin\frac{al+\sqrt{a^2l^2-\left(i_{12}^2r_1^2-a^2\right)\left(r_{a2}^2-l^2-i_{12}^2r_1^2\right)}}{i_{12}^2r_1^2-a^2}-\arccos\frac{\sqrt{r_{a1}^2-l^2}}{r_1}\right)+\sin 2\left(\arccos\frac{\sqrt{r_{a1}^2-l^2}}{r_1}\right)-\ldots\\
&2\left(\frac{al+\sqrt{a^2l^2-\left(i_{12}^2r_1^2-a^2\right)\left(r_{a2}^2-l^2-i_{12}^2r_1^2\right)}}{i_{12}^2r_1^2-a^2}\right)\cos\left(\arcsin\frac{al+\sqrt{a^2l^2-\left(i_{12}^2r_1^2-a^2\right)\left(r_{a2}^2-l^2-i_{12}^2r_1^2\right)}}{i_{12}^2r_1^2-a^2}\right)\Bigg]+\ldots\\
&8i_{12}^{-1}al\left[\frac{\sqrt{r_{a1}^2-l^2}}{r_1}-\cos\left(\arcsin\frac{al+\sqrt{a^2l^2-\left(i_{12}^2r_1^2-a^2\right)\left(r_{a2}^2-l^2-i_{12}^2r_1^2\right)}}{i_{12}^2r_1^2-a^2}\right)\right]+\ldots\\
&4\left(r_{a1}^2+i_{12}r_1^2+i_{12}^{-1}l^2-i_{12}^{-1}r_{a2}^2-r_1^2-l^2\right)\left(\arcsin\frac{al+\sqrt{a^2l^2-\left(i_{12}^2r_1^2-a^2\right)\left(r_{a2}^2-l^2-i_{12}^2r_1^2\right)}}{i_{12}^2r_1^2-a^2}-\arccos\frac{\sqrt{r_{a1}^2-l^2}}{r_1}\right)\Bigg\}
\end{aligned} \quad (24)$$

3. Simulation Analysis of the Flow Characteristics of a Straight Line Conjugate Internal Meshing Gear Pump

Taking the straight line conjugate internal meshing gear pump as the research object, the specific values of the theoretical flow rate and the geometric flow pulsation rate are calculated and analysed according to gear pair's geometric parameters, and the change laws of gear pair's geometric parameters on the theoretical flow rate and geometric flow pulsation rate are studied. Then, in the case of considering the clearance leakage of the axial and radial, the dynamic grid technology is used to develop the simulation model of the internal flow channel, the distribution of internal flow field is analyzed, and the influence factors and the influence degree of the flow pulsation and the average flow rate at the outlet of the gear pump are investigated.

3.1. Analysis of the Theoretical Flow Rate Curves of a Straight Line Conjugate Internal Meshing Gear Pump

Given the gear pair's basic parameters of the straight line conjugate internal meshing gear pump, the gear pair's geometric parameters can be calculated. The calculation results are shown in Table 1.

Table 1. Geometric parameters of the gear pair.

Name	External Gear	Internal Gear Ring
Number of teeth	13	17
Profile half angle of the external gear (°)	28	
Reference circle diameter (mm)	65	85
Tip circle diameter (mm)	74.5	75.5
Root circle diameter (mm)	54.5	95.5
Center distance (mm)	10	
Actual coincidence degree	1.0288	

The gear pair's geometric parameters in Table 1 are substituted into the theoretical instantaneous flow rate Equation (9) and the initial and ending meshing angles equation sets (11). When the rotation speed is 2000 r/min, the theoretical flow rate curves corresponding to the theoretical coincidence degree can be obtained by programming [40] based on the M language and running the program, as shown in Figure 5.

It can be known from Figure 5 that the theoretical instantaneous flow rate curve changes continuously and periodically, and the curve intersects and distributes in a zigzag shape. The change period is the difference between the ending and initial meshing angles (i.e., the cycle is 0.5419 rad). The instantaneous flow rates corresponding to the ending meshing angle is much smaller than that corresponding to the initial meshing angle. The theoretical average flow rate is distributed horizontally in a straight line. According to Equation (12), the theoretical coincidence degree of the gear pump is 1.1212. However, the actual coincidence degree is 1.0288. Therefore, when the gear pair exits the meshing, the meshing point is not at the intersection of the tip circle of the external gear and the meshing line, but in advance. Based on this, the theoretical flow rate curves corresponding to actual coincidence degree can be obtained, as shown in Figure 6.

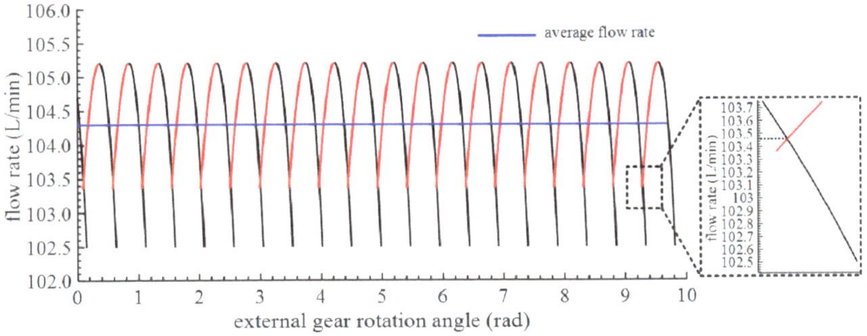

Figure 5. Theoretical flow rate curves corresponding to theoretical coincidence degree.

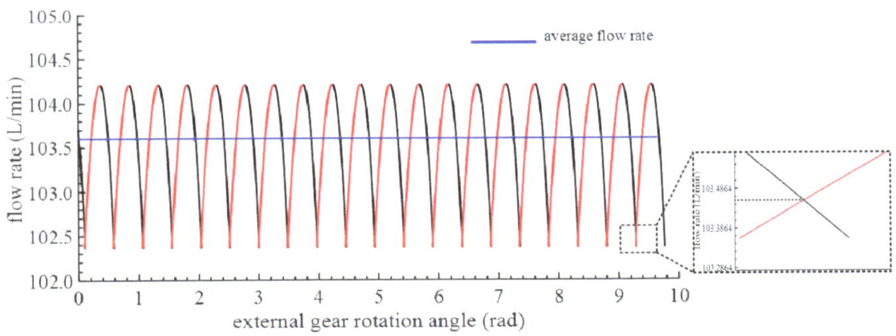

Figure 6. Theoretical flow rate curves corresponding to the actual coincidence degree.

Comparing Figure 5 with Figure 6, it can be known that the theoretical instantaneous flow rate curves correspond to the actual and theoretical coincidence degrees, respectively, are basically the same. The only difference is that the actual ending meshing angle is less than the theoretical, and the corresponding instantaneous flow rates are larger. The instantaneous flow rates corresponding to the initial and ending meshing angles are basically the same. According to Figure 6, the theoretical flow rate and the geometric flow pulsation rate are 103.61 L/min and 1.78%, respectively, whereas the theoretical flow rate and the geometric flow pulsation rate calculated by Equations (17) and (24) are 103.71 L/min and 1.76%, respectively. The difference rates are only 0.1% and 1%, respectively. Such a small error can be ignored. Since the oil compressibility and the internal leakage of the gear pump are not taken into account, the curve of the theoretical instantaneous flow rate is smooth and the geometric flow pulsation rate is low.

3.2. Study on Change Laws of Gear Pair'S Geometric Parameters to the Theoretical Flow Rate and Geometric Flow Pulsation Rate

3.2.1. Design Variables and the Corresponding Value Range

According to the theoretical flow rate Equation (17) and the geometric flow pulsation rate Equation (24), the gear pair's geometric parameters affecting the theoretical flow rate and the geometric flow pulsation rate are gear width, number of the external gear, reference circle radius of the external gear, center distance, transmission ratio, profile half angle of the external gear, center angle corresponding to tooth thickness of the external gear reference circle, tip circle radius of the external gear, and tip circle radius of internal gear ring, respectively. Given that the gear number combinations in gear pairs with the small gear difference (usually three or four) are several common groups, thus, the gear number should not be used as a design variable. When the reference circle radius of the external gear is determined, the center angle and the center distance are also determined. The effect of the gear width on the flow characteristics is obvious. Therefore, the identified design variables are as follows: reference circle radius of the external gear, tip circle radius of the external gear, tip circle radius of internal gear ring, and the profile half angle of the external gear.

Before simulation calculation, the value ranges of the design variables are determined according to constraint conditions. The corresponding constraint conditions include the limitation of the tooth thickness in the gear tip circle, the continuous transmission condition of gear pair, the limitation of meshing limit point, the limitation of tooth profile overlap interference, etc. The specific results are as follows: the reference circle radius of the external gear is 32.5–38 mm, the tip circle radius of the external gear is 33.25–37.25 mm, the tip circle radius of the internal gear ring is 33.75–40.75 mm, and the profile half angle of the external gear is 28–38°.

3.2.2. Influence of the Geometric Parameters on the Theoretical Flow Rate

The theoretical flow rate is programmed according to Equation (17) in MATLAB, and the change laws of the theoretical flow rate with the corresponding geometric parameters are obtained by running the program. The calculated results are shown in Figure 7.

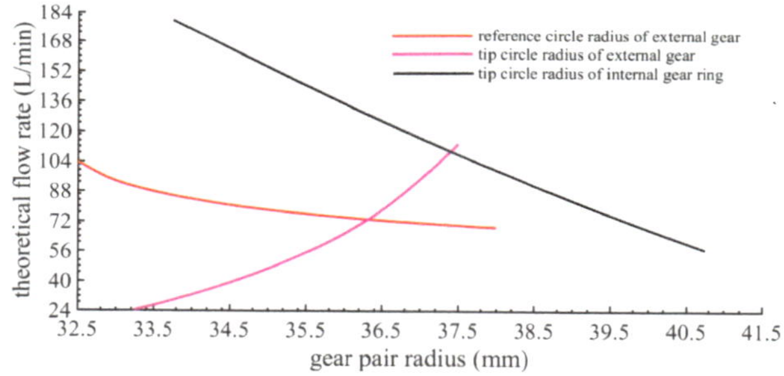

(a) With gear pair radius.

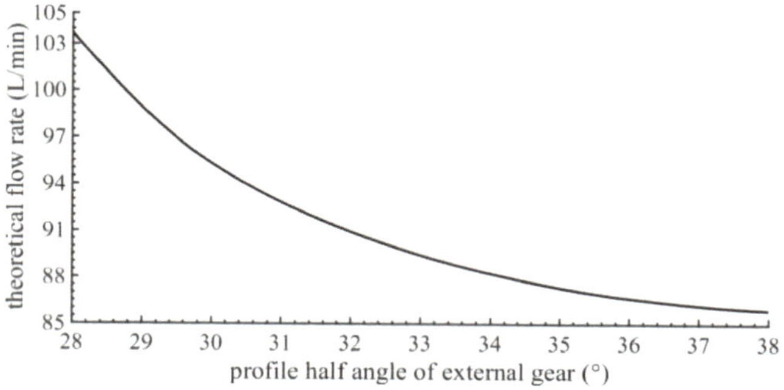

(b) With profile half angle of the external gear.

Figure 7. Change laws of the theoretical flow rate.

It can be known from Figure 7 that the theoretical flow rate curves linearly decrease with the increase of the reference circle radius of the external gear, the tip circle radius of internal gear ring and the profile half angle of the external gear, and linearly increases with the increase of the tip circle radius of the external gear. When the tip circle radius of internal gear ring increases 7 mm, the theoretical flow rate decreases by 122 L/min, and the change of the curve is relatively steep. When the reference circle radius of the external gear increases 5.5 mm, the theoretical flow rate decreases by 35 L/min, and the change of the curve is gentle. When the profile half angle of the external gear increases 10°, the theoretical flow rate decreases by 18 L/min, and the change of the curve is very gentle. When the tip circle radius of the external gear increases 4 mm, the theoretical flow rate increases by 79 L/min, and the change of the curve is steep. Therefore, the design variable that exerts the greatest influence on the theoretical flow rate is the tip circle radius of the external gear, and the tip circle radius of internal gear ring is the second.

3.2.3. Influence of the Geometric Parameters on the Flow Pulsation Rate

The geometric flow pulsation rate is programmed according to Equation (24) in MATLAB, and the change laws of the geometric flow pulsation rate with the corresponding geometric parameters are obtained by running the program. The calculated results are shown in Figure 8.

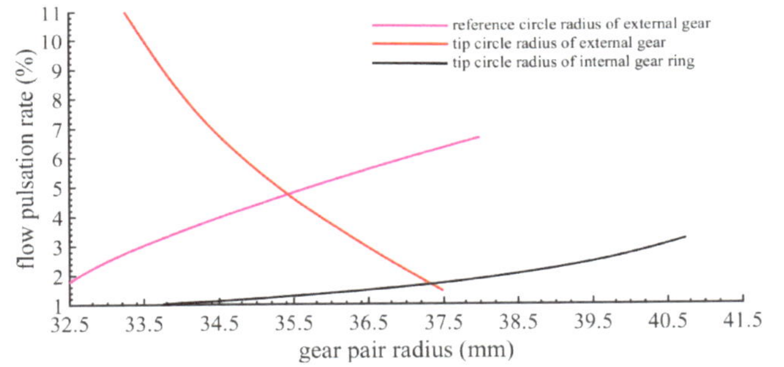

(**a**) With gear pair radius.

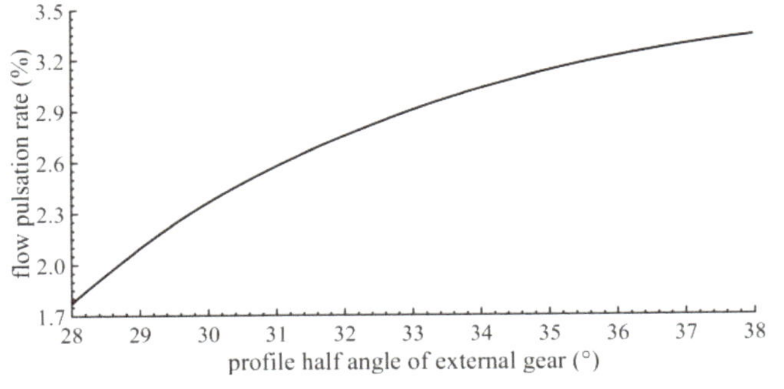

(**b**) With profile half angle of the external gear.

Figure 8. Change laws of the geometric flow pulsation rate.

It can be known from Figure 8 that the geometric flow pulsation rate increases linearly with the increase of the reference circle radius of the external gear, the tip circle radius of internal gear ring and the profile half angle of the external gear, and decreases linearly with the increase of the tip circle radius of the external gear. When the tip circle radius of internal gear ring increases 7 mm, the geometric flow pulsation rate increases by 2.2%, and the change of the curve is very gentle. When the reference circle radius of the external gear increases 5.5 mm, the geometric flow pulsation rate increases by 4.9%, and the change of the curve is gentle. When the profile half angle of the external gear increases 10°, the geometric flow pulsation rate increases by 1.5%, and the change of the curve is very gentle. When the tip circle radius of the external gear increases 4 mm, the geometric flow pulsation rate decreases by 9.2%, and the change of the curve is steep. Therefore, the design variable that exerts the greatest influence on the geometric flow pulsation rate is the tip circle radius of the external gear.

After analysis and comparisons, it can be concluded that the design variable that exerts the greatest influence on the theoretical flow rate and the geometric flow pulsation rate is the tip circle radius of the external gear. As this radius increases, the theoretical flow rate sharply increases, and the geometric flow pulsation rate sharply decreases. The tip circle radius of internal gear ring only influences the theoretical flow rate, which constantly decreases with the increase in this radius. The remaining variables exert no significant effect on the theoretical flow rate and the geometric flow pulsation rate.

Therefore, to increase the theoretical flow rate whilst decreasing the geometric flow pulsation rate, the tip circle radius of the external gear should be increased as much as possible within the allowable range of the design calculation.

3.3. Establishment of the Simulation Model of Internal Flow Channel

According to the model of the straight line conjugate internal meshing gear pump, the corresponding main characteristic parameters on the sample [41] are listed, as shown in Table 2.

Table 2. Main characteristic parameters.

Displacement (ml/r)	Mximum Speed (r/min)	Outlet Pressure (bar)		Mineral Oil (mm^2/s)
		rating	peak	viscosity range
50.3	2600	125	160	10–100

The schematic diagram of main components in the straight line conjugate internal meshing gear pumpis establish, as shown in Figure 9.

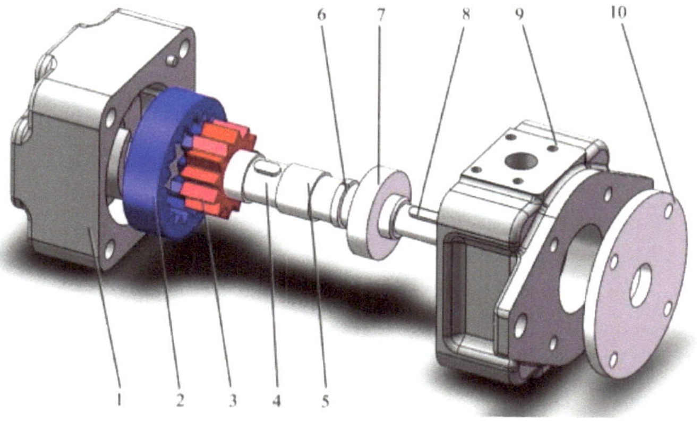

Figure 9. Three-dimensional structure diagram of straight line conjugate internal meshing gear pump: 1–pump body; 2–internal gear ring; 3–external gear; 4–pump shaft; 5–sliding bearing; 6–bearing retaining ring; 7–rolling bearing; 8–ordinary flat key; 9–pump cover; 10–bearing cover.

The sealed cavity in the straight line conjugate internal meshing gear pump is composed of the pump body end face, the internal wall surface, the pump cover end face, the crescent diaphragm side, the external gear bearing surface, and a pair of meshing gear pairs. The internal meshing gear pair is the moving part of gear pumps. The oil transportation function is realized through the meshing transmission of the gear pair.

The internal flow channel model is created and is divided into five parts: oil absorption area, oil discharge area, rotor area, oil distribution area on the absorption side, and oil distribution area on the discharge side, as shown in Figure 10.

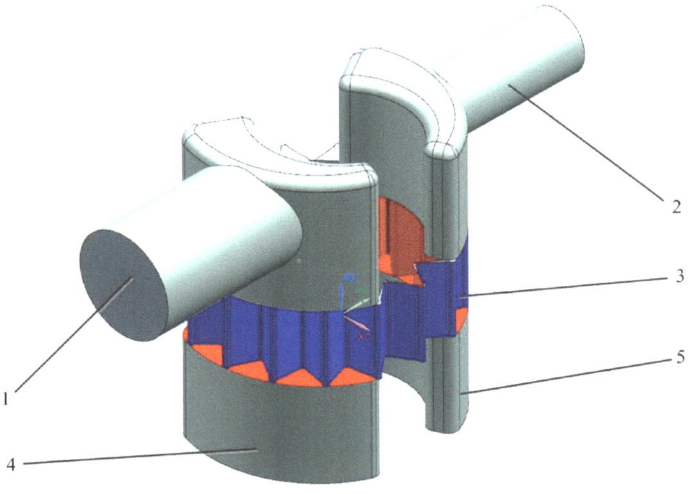

Figure 10. Flow channel model of the straight line conjugate internal meshing gear pump: 1–oil absorption area; 2–oil discharge area; 3–rotor area; 4–oil distribution area on the absorption side; 5–oil distribution area on the discharge side.

In the working process of the gear pump, the friction pairs' surfaces are separated by a certain thickness of oil film to prevent the friction pairs from directly contacting. There are four pairs of friction pairs in the gear pump, respectively, the gear end face and the side plate, the tooth tip, and the shell inner wall surface, between the meshing tooth surfaces, and the bearing and the hub. According to the friction pair design theory of the hydraulic pump and motor [42], the oil film thickness is set as follows: the clearances between the tip circle of the external gear and the crescent diaphragm, and between the tip circle of internal gear ring and the crescent diaphragm are all 20 μm; the gear side clearance is 4 μm; the axial clearance, the radial clearances between the internal ring gear and the shell inner wall surface, and between the bearing and the hub are all 30 μm. The specific oil film thickness is set in the grid model, which is not mentioned here.

The internal flow channel grid model of the gear pump is established [43], that is, the static region uses the geometry conformal adaptive binary-tree algorithm to generate Cartesian grids, which can well adapt to curves and surfaces. The built-in grid generation template is used to generate structured dynamic grid in the dynamic region. The boundaries of different fluid regions (e.g., dynamic regions and static regions) are set as a pair of interfaces by using the mismatched grid interface technology. In this way, each fluid variable passes through the moving surface automatically and implicitly. The mesh quality is aimed at non-broken surface, and the mesh in the rotor area is refined. The grid model is shown in Figure 11.

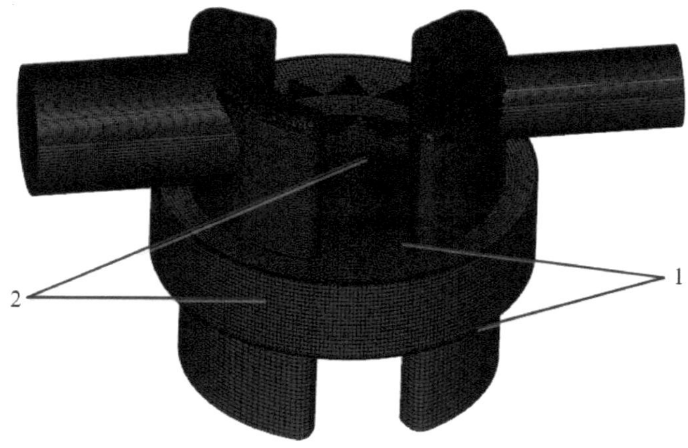

Figure 11. Flow channel grid model of the straight line conjugate internal meshing gear pump: 1–axial clearance; 2–radial clearance.

According to the structural conditions of the straight line conjugate internal meshing gear pump, the crescent internal gear pump model is selected. The rotation speed of the external gear is set to 2000 r/min and the remaining parameters are defined. The turbulence model is selected according to flow conditions. According to the main physical and chemical properties of the oil, the equilibrium dissolved gas model is selected in the cavitation model and the remaining parameters in the model are set.

For definition of boundary condition, since the maximum output flow rate is to be obtained, the inlet and outlet conditions are set to the pressure inlet and pressure outlet, respectively. The pressure values are all defined as 101,325 Pa. Given that the heaters and the heat exchangers are installed in the hydraulic system during normal operation, the temperature range of hydraulic oil is generally controlled at 20–50 °C. Due to short working time during the following experimental research, the temperature difference caused by the system heating can be ignored. In addition, the following experiment was performed under no-load conditions, so the system pressure has no effect on the physical and chemical properties of the oil. Therefore, for 46# mineral oil used in the simulation model, its medium properties are unchanged in the process of calculation, and the specific physical and chemical properties are shown in Table 3.

Table 3. 46# Mineral oil medium properties.

Viscosity Grade	Density	Dynamic Viscosity	Air Solubility	Bulk Elastic Modulus
	(kg/m^3)	(Pa/s)		(GPa)
46	843.7	0.039	1×10^{-5}	1.6

After the pre-processing is defined, the grid is first verified for irrelevance, and verification results are shown in Table 4.

Table 4. Comparison of simulation results under different grid numbers.

Grid Number	Node Number	Average Flow of Outlet (L/min)
454,430	792,420	97.552
567,169	1,045,980	99.975
735,362	1,312,560	100.067

It can be known from Table 4, when grid number is 454,000 and 567,000, the difference of the outlet average flow rate is 2.423 L/min, and difference rate is 2.5%. When grid number is 567,000 and 735,000, respectively, the difference of average flow rate is 0.092 L/min, and the difference rate is less than 0.1%. This deviation can be ignored. To speed up calculation speed, the grid model with grid number of 567,169 is selected for the numerical simulation.

3.4. Analysis of the Numerical Simulation Results

According to the relevant settings of aforementioned results, the simulation calculation of the flow channel model is performed. After the numerical simulation converged, we obtain the calculation results in the eighth cycle, such as the pressure cloud diagram on the symmetrical cross-section and the rotor section, the oil distribution area when absolute pressure is lower than 3000 Pa, the velocity vector diagram, the instantaneous and average flow rates curves, as shown below.

It can be known from Figure 12 that the oil films between the friction pairs completely separate the oil absorption cavity from the oil discharge cavity. The pressure distribution between the high and low pressure areas is uniformly transitioned. Thus, the applicability of the simulation model and the correctness of the simulation method are verified. In addition, when the rotation speed of the external gear is 2000 r/min, the corresponding minimum oil absorption pressure is 2259 Pa. This value is much less than atmospheric pressure, which indicates that the oil absorption process is very sufficient at this rotation speed. In addition, the corresponding maximum oil discharge pressure is 205,918 Pa, and this value is greater than the set pressure at the boundary, which will cause a certain degree of internal leakage.

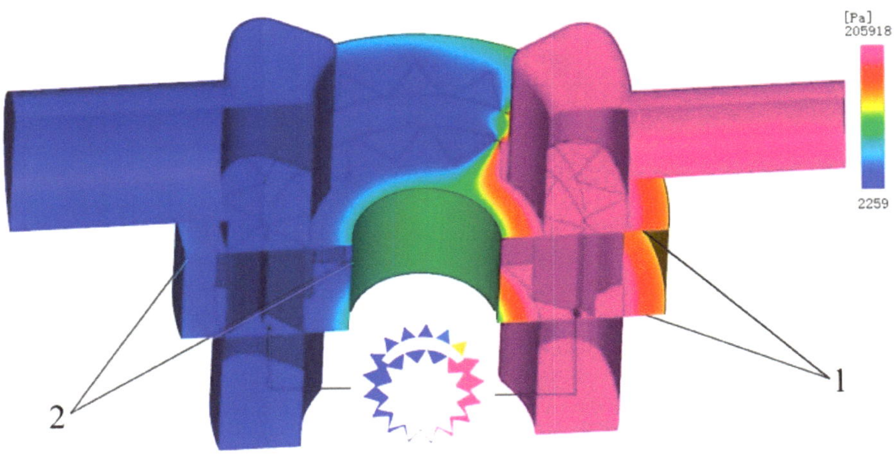

Figure 12. Pressure cloud diagram of the internal flow channel: 1–axial clearance; 2–radial clearance.

It can be known from Figure 13, when the absolute pressure is lower than 3000 Pa, the oil is all distributed in the oil absorption area, and concentrates in small area. This area is located after the front pair of gears withdraws the meshing, and far away from the oil absorption port. The oil is not sufficiently filled, so the established oil pressure is low in this area. In addition, due to the faster rotation speed of the external gear, there are also very small areas in the gap between the tooth tip circle and the crescent diaphragm.

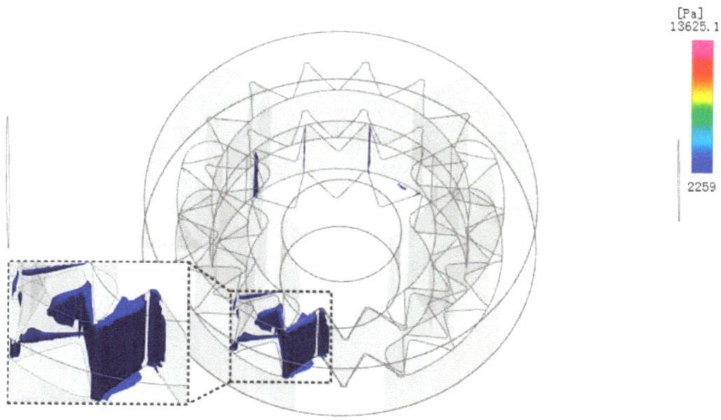

Figure 13. Oil distribution area when the absolute pressure is lower than 3000 Pa.

However, under the condition of oil working temperature is set at 30 °C and a certain amount of air is dissolved, the air separation pressure of the oil is about 1500 Pa. The saturated vapor pressure at this temperature is about 500 Pa [44]. Thus, the lowest pressure in the whole flow channel is higher than the air separation pressure of the oil. In addition, given that the highest oil pressure in the oil discharge area is only two atmospheric pressures. Therefore, the gas dissolved in the oil will not separate out, and the effective bulk modulus of elasticity will not change. Further, the oil compressibility has no effect on the flow pulsation and the average flow rate at outlet of the gear pump.

It can be known from Figure 14, under the meshing transmission of gear pair, the oil of the oil absorption cavity is continuously transported to the oil discharge cavity through interdental cavity. The oil velocity in the oil absorption area is relatively uniform, while the motion of fluid particles in the rotor area is complex and irregular. The oil velocity in the gap between the external gear, the internal gear ring and the crescent diaphragm, and between the gear surfaces is about 11 m/s, while the oil velocity in the rest is obviously lower. However, the high-speed oil has a great impact on the parts' surfaces, so it is necessary to implement the surfaces material modification of the gear surface and the surface of the crescent diaphragm. The oil distribution area is used to buffer the oil velocity in the oil absorption and oil discharge areas. The oil close to the oil absorption and oil discharge cavities has the gradually decreasing velocity, and the rest oil is in a static state. The oil in the oil discharge area first moves vertically upwards and then flows out horizontally along the outlet. Due to the large upward speed (approximately 7 m/s), it is difficult for the oil to make a smooth transition to the horizontal direction. Most oil passes along the upper wall of the outlet channel. Hence, increasing the wall thickness of the upper wall surface is necessary.

It can be known from Figure 15 that the instantaneous flow rate changes continuously and periodically, and there are 13 fluctuating pulsations. This is determined by the number of teeth on the driving gear (13). Secondly, unlike the theoretical instantaneous flow rate curve, the simulation instantaneous flow curve is not a smooth transition. The reason is that there is a certain gap between the surfaces of all friction pairs in the gear pump, and the gear side clearance changes constantly with the rotation of the gear pair. Therefore, the internal leakage from the clearances is inevitable. Furthermore, the non-uniformity of the internal leakage causes the gear pump to output a pulsating flow rate, and the pulsating flow rate changes periodically with the movement of the driving gear. Finally, according to the aforementioned analysis, under the working conditions applied in this study, the oil compressibility will not affect the flow pulsation at the outlet of the gear pump. In summary, the combined effect of the geometric flow pulsation and the uneven internal leakage leads to the instantaneous flow rate curve as shown in Figure 15.

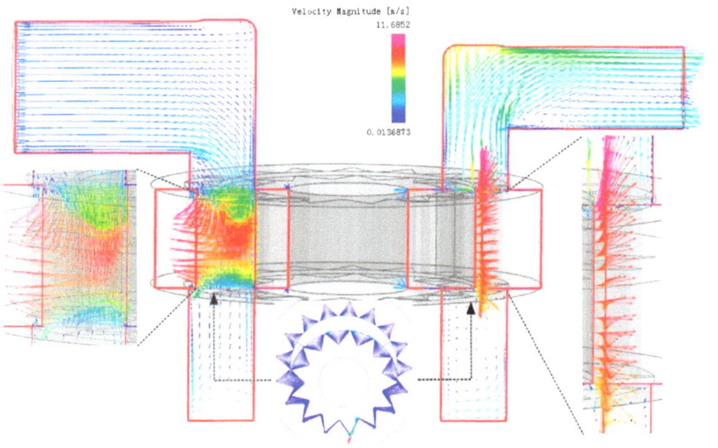

Figure 14. Velocity vector diagram of the cross-section.

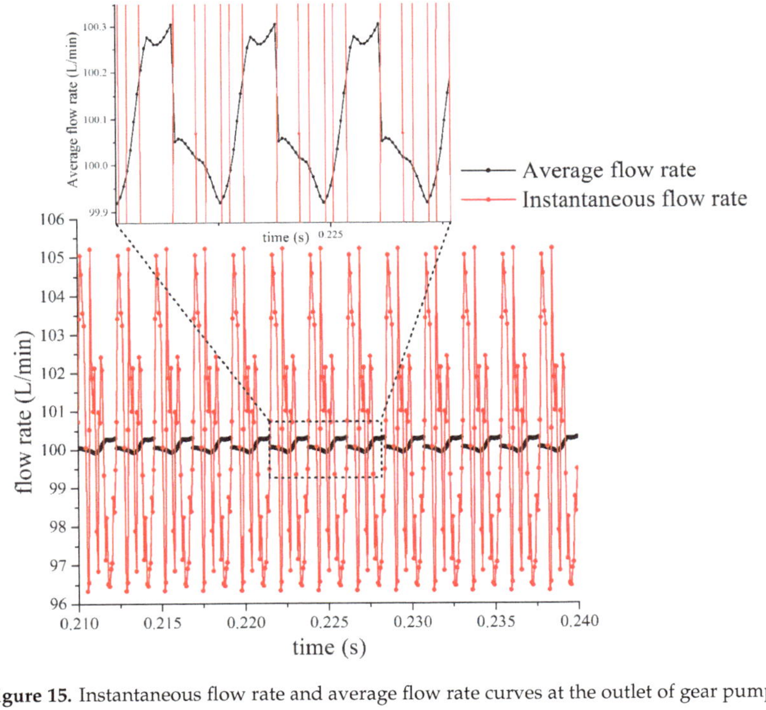

Figure 15. Instantaneous flow rate and average flow rate curves at the outlet of gear pump.

In addition, unlike the theoretical average flow rate curve, due to the pulsation of the instantaneous flow rate, the simulation average flow rate curve is also non-linear and changes periodically. However, the change of the average flow rate curve is very small (less than 0.4 L/min), which can be ignored compared with the instantaneous flow rate pulsation amplitude.

To further analyze the proportion of the flow pulsation caused by the geometric flow pulsation and the uneven internal leakage, the maximum, minimum, and average flow rates in Figure 15 are obtained, and the flow pulsation rate and the volume efficiency are calculated. In this way, the flow characteristic data of the straight line conjugate internal meshing gear pump under no-load conditions are obtained, as shown in Table 5.

Table 5. Flow characteristic data under no-load conditions.

Maximum Flow Rate	Minimum Flow Rate	Average Flow Rate	Pulsation Rate	Volumetric Efficiency
(L/min)	(L/min)	(L/min)	(%)	(%)
105.21	96.34	100.12	8.86	99.52

It can be known from Table 5 that the total flow pulsation rate under no-load conditions is 8.86%. From the aforementioned results, the geometric flow pulsation rate is 1.76%, accounting for 19.9% of the total flow pulsation rate. In this way, the flow pulsation rate caused by the uneven internal leakage is 7.1%, accounting for 80.1%. Therefore, the uneven internal leakage is the main factor affecting the flow pulsation at the outlet of the gear pump.

In addition, the average flow rate calculated by the formula is 103.71 L/min. However, the average flow rate calculated by simulation is 100.12 L/min. The difference between the two is 3.59 L/min, and difference rate is 3.46%. The reason is that the internal leakage of oil is considered in the simulation calculation, and the leakage leads to the reduction of the simulation average flow rate.

4. Experimental Research on Flow Characteristics of a Straight Line Conjugate Internal Meshing Gear Pump

To further study the flow characteristics of the straight line conjugate internal meshing gear pump, a high-pressure (32 MPa) positive displacement test system is designed according to the test outline. According to the system schematic diagram, the required components are accurately selected and the physical objects are reasonably arranged. The operation of entire test bench is controlled by a computer, and all data is processed by professional software.

4.1. Overall Design of Test System

To realize the digital and automation requirements of data collection, the overall design scheme of the test system is arranged as shown in Figure 16.

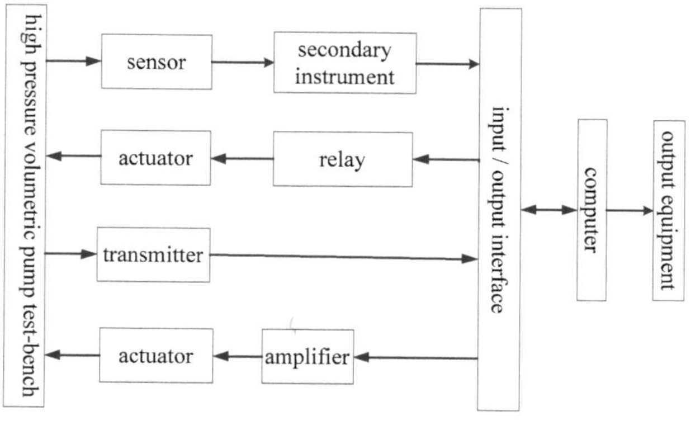

Figure 16. Overall design principle of test system.

It can be known from Figure 16 that the operation process of the test system is as follows: after the system is started, the signal is sent by the computer, and the test pump starts to run under the drive of variable frequency speed regulation three-phase asynchronous motor. The analog signal from the test pump is transmitted to the general input interface through each sensor and secondary instrument, and directly or after conversion by the A/D converter is input to the computer, which is processed by the computer and stored in the database. The computer outputs the digital signal according to the program instructions, and then converts it to the analog signal after D/A converter. The analog signal is converted into a current signal by a proportional amplifier, which drives the relevant electronic control components of the test bench to operate. After the test is completed, the computer outputs the test results through the output device.

The hydraulic system design principle of the high-pressure positive displacement pump test bench in the test system is shown in Figure 17.

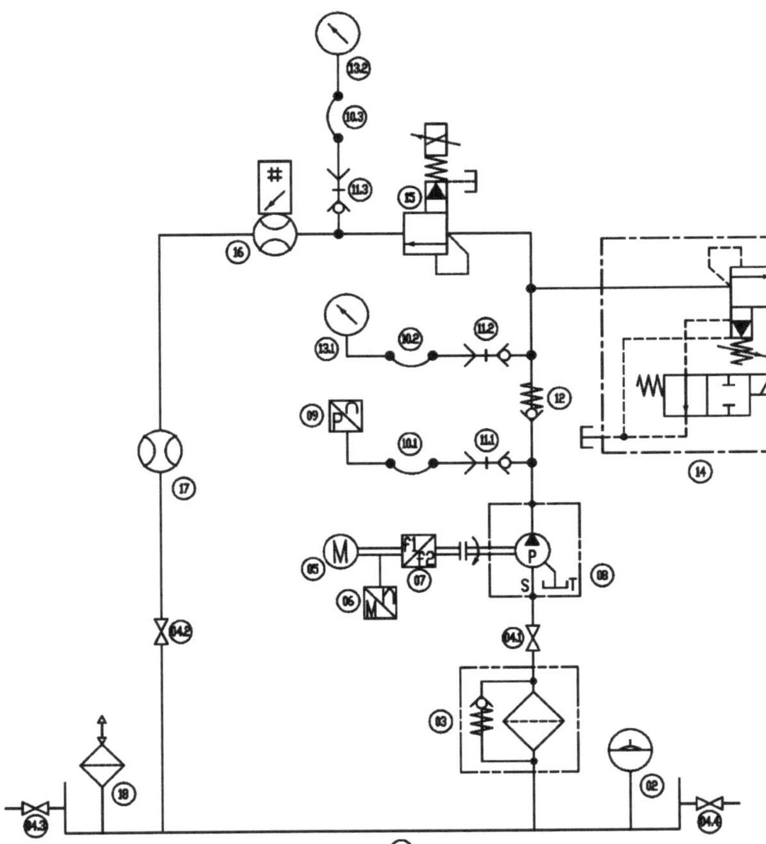

Figure 17. Design schematic diagram of hydraulic system: 01–tank; 02–level gauge; 03–suction filter; 04–shut-off valve; 05–variable frequency motor; 06–torque speed sensor; 07–frequency converter; 08–test pump; 09–pressure sensor; 10–pressure measuring hose; 11–pressure measuring connector; 12–check valve; 13–pressure gauge; 14–electromagnetic spill valve; 15–proportional relief valve; 16–digital flow meter; 17–flow meter; 18–air filter.

It can be known from Figure 17 that the hydraulic system is composed of power components, control components and hydraulic accessories, in which the power components are mainly composed of variable frequency motor (YVF2-55-2B35), test pump (QX41-050R), frequency converter (PR-F740-55K-CH), torque speed sensor (JC2C), torque meter (JW-3), and other components. The control components are mainly composed of check valve (TVP25-10), electromagnetic spill valve (DBW20B-1-50/31.5W220-50NZ5L), proportional relief valve (EBG-06-H), digital flowmeter

(FBLZJ-40-165J0), pressure sensor (FB3351GP0S22M3), and pressure gauge (YN-100-IV). The hydraulic accessories are mainly composed of level gauge (YWZ-350T), suction filter (TF-400 × 10L-Y), and shut-off valve (KH-G1/2″). The electromagnetic spill valve is used as a safety valve, and the system is in the unloading state before the electromagnet is energized. The proportional relief valve is installed in series on the main oil circuit, and the pressure of the test system is adjusted linearly according to the input current. The pressure sensor monitors total pressure at the outlet of the test pump in real-time. The two pressure gauges are arranged one behind the other of the proportional relief valve to accurately obtain the pressure loss of the system. Two digital flowmeters are installed in series in the main oil circuit, and the function of latter flowmeter is to check the measurement results of the previous flowmeter. The overall layout of the test system is shown in Figure 18.

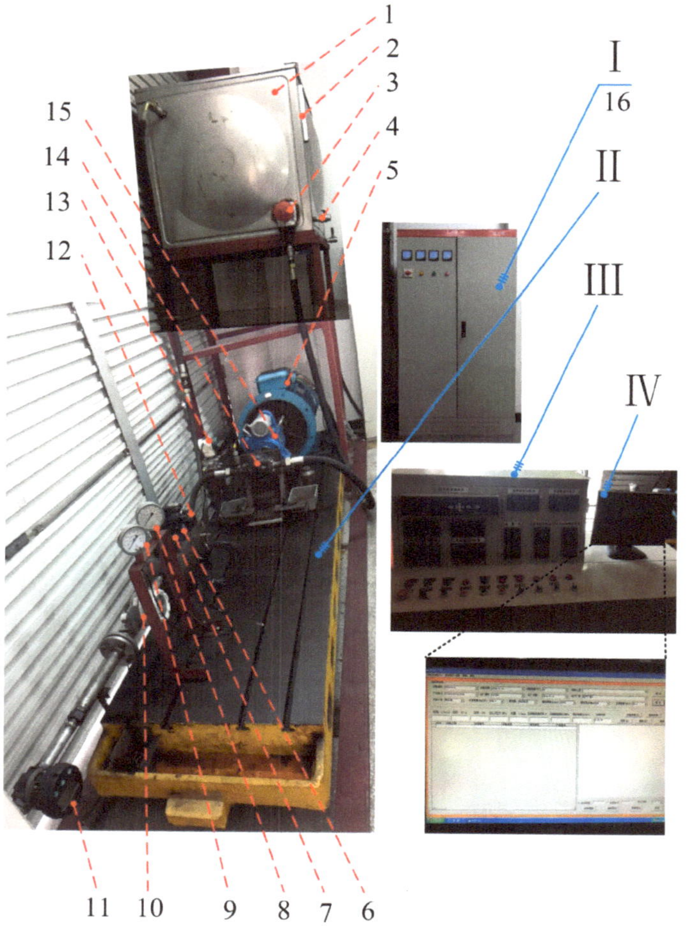

Figure 18. Overall layout of test system:I–test bench; II–power cabinet; III–control cabinet; IV–industrial computer; 1–tank; 2–level gauge; 3–suction filter; 4–shut-off valve; 5–variable frequency motor; 6–electromagnetic spill valve; 7–proportional relief valve; 8–pressure gauge 1; 9–pressure gauge 2; 10–digital flow meter; 11–flow meter; 12–check valve; 13–pressure sensor; 14–test pump; 15–torque speed sensor; 16–frequency converter.

It can be known from Figure 18 that the test system is composed of four parts: the test bench, the power cabinet, the control cabinet, and the industrial control computer. Among them, the frequency converter is installed in the power cabinet. After the test system is started normally, the current or voltage signal is keyed in the secondary instrument on the control cabinet, and then converted them into the corresponding driving signal through the proportional amplifier to control the action of the corresponding electrical components. The internally-developed data acquisition software is installed in the industrial computer to automatically save all data detected by the instruments.

4.2. Performance Evaluations and Result Analysis

With reference to the national test standard for the hydraulic gear pumps, the performance experiment content of the straight line conjugate internal meshing gear pump prototype is established, which mainly includes the running-in test. That is, the rotation speed of the frequency conversion motor is adjusted within the working speed range under the condition of no load. In this way, the assembled new pump can achieve the best working condition to avoid the problems such as over-tight fit or inflexible rotation. To be consistent with the conditions of simulation calculation, the test speed is also set to 2000 r/min, the working medium is also selected 46# hydraulic oil, and the performance evaluations are conducted at room temperature.

4.2.1. Analysis of the Outlet Average Flow Rate

In the course of the experiment, the prototype runs stably and no abnormal vibration and noise are found. Because the flow pulsation frequency ($z_1 n_1/60$) is very high, the current digital flowmeter cannot directly detect the instantaneous flow rate, and can only measure the average flow rate through the main oil circuit. Obtain the flowmeter's detection data within one minute and fit them into curve, as shown in Figure 19.

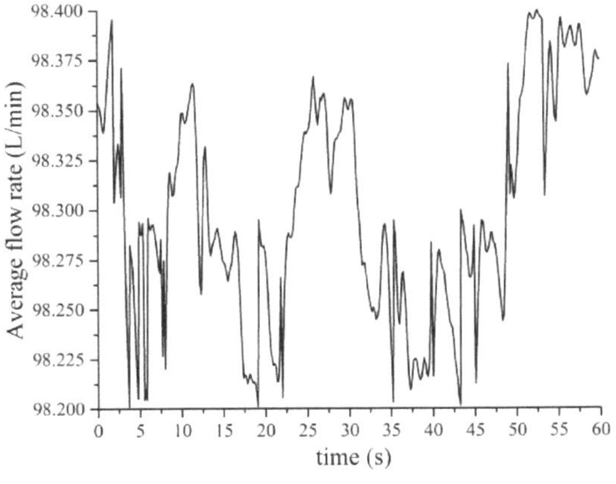

Figure 19. Average flow rate curve detected by the flowmeter.

It can be known from Figure 19, when the rotation speed of variable frequency motor is set to 2000 r/min, the average flow curve detected by the flowmeter is not a horizontal straight line, but fluctuates up and down with the passage of time, and the change amplitude is less than 0.2 L/min. The reason may be that the power supply of the frequency converter is unstable, the speed feedback is unstable, the program problem itself, and the harmonic content of external power supply is too high, etc. [45], which causes the speed of frequency conversion motor to fluctuate up and down within a certain range. Therefore, to obtain the average flow rate at a speed of 2000 r/min, average the flow rate within one minute, which is 98.30 L/min, and the corresponding volumetric efficiency is 97.71%. In this way, the flow characteristics under no-load conditions obtained by theoretical analysis, numerical simulations, and experimental stufies can be listed, as shown in Table 6.

Table 6. Flow characteristics under no-load conditions.

Theoretical Calculation		Simulated Results		Experimental Results
Flow rate (L/min)	Flow pulsation rate (%)	Flow rate (L/min)	Flow pulsation rate (%)	Flow rate (L/min)
103.71	1.76	100.12	8.86	98.30

It can be known from Table 6, the experimental average flow rate decreases by 1.82 L/min when compared with the simulation average flow rate, which is due to factors such as the cumulative error of the measuring instruments and the large actual leakage. The decline rate is 1.8%, less than 2%, which can be ignored. The consistency between the simulation and experimental results verifies the applicability of the test system and the correctness of the test method.

4.2.2. Analysis of the Flow Rate-Speed Characteristic and Volume Efficiency–speed Characteristic

When the rotational speed of hydraulic pump is too low, the original small flow rate is almost all lost to the leakage. At this time, the hydraulic pump can hardly discharge the oil, the higher the pressure, the larger the leakage. Therefore, according to the minimum speed requirement of continuous output pressure in the gear pumps, and the table of the technical performance and the application range of hydraulic pumps [41,44], the speed range of the test pump is 750 r/min to 2600 r/min.

According to the running-in test method, the flow rate-speed and volume efficiency–speed characteristic curves of the gear pump at the main speeds of 750 r/min, 1000 r/min, 1500 r/min, 2000 r/min, and 2600 r/min are obtained, respectively, as shown in Figures 20 and 21.

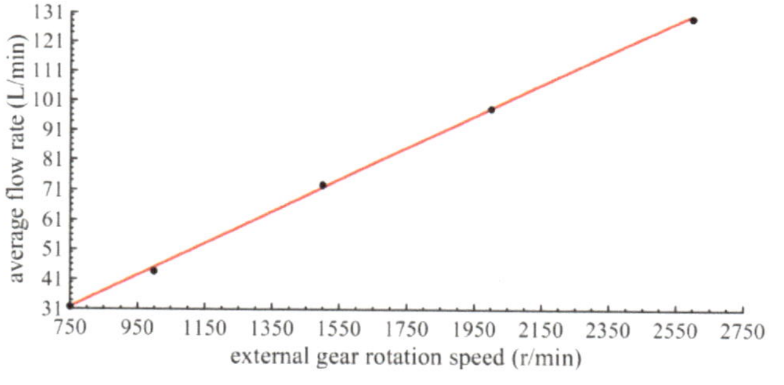

Figure 20. Flow rate-speed characteristic curve.

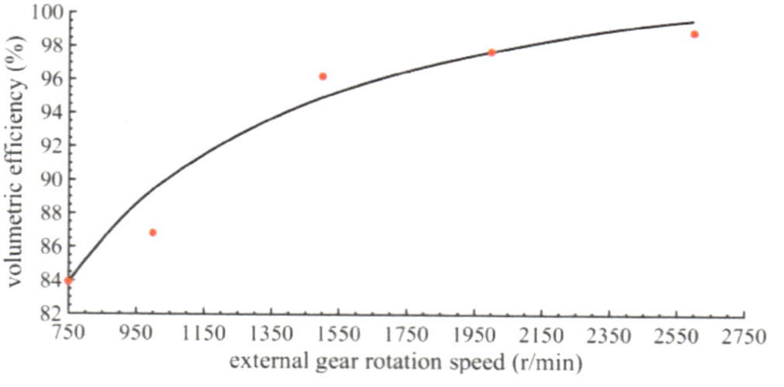

Figure 21. Volumetric efficiency–speed characteristic curve.

It can be known from Figure 20, under no-load conditions, the average flow rate at the outlet of the gear pump linearly increases with the increase in the rotation speed of the external gear, which is completely in line with the objective fact that the average flow rate of the fixed displacement pumps is proportional to the rotation speed. Since the internal leakage is different at different rotation speeds, further, the corresponding volumetric efficiency is also different. The specific volumetric efficiency–speed characteristic curve is shown Figure 21.

It can be known from Figure 21, under no-load conditions, with the increase in the rotation speed of the external gear, the volume efficiency increases as a logarithmic function with a base number greater than 1. When the rotation speed is greater than 1000 r/min, the corresponding volumetric efficiency is greater than 89%. When the rotation speed is greater than 1500 r/min, the corresponding volumetric efficiency is greater than 94%. When the rotational speed reaches the rated value of 2600 r/min, the corresponding volumetric efficiency is 98.87%, which is very close to 1. The reason is that the radial clearance leakage inside gear pump decreases slowly with the increase in the rotation speed, so the corresponding volumetric efficiency increases slowly.

Therefore, to increase the average flow rate at the outlet of the gear pump and improve the volumetric efficiency, the rotation speed of the external gear should be increased as much as possible within the allowable speed range.

5. Conclusions

In this paper, a general mathematical model of the theoretical instantaneous flow rate is established using the volume change method. By using the number-shape combination method, the theoretical instantaneous flow rate equations based on changes in the distance from the node to the meshing point and the rotation angle of the external gear, the initial and ending meshing angle equation sets, the theoretical coincidence degree equation, the theoretical oil trap time equation, the maximum and minimum instantaneous flow rate equations, the theoretical flow rate equation, and the geometric flow pulsation rate equation are analyzed and determined. Taking the straight line conjugate internal meshing gear pump as the research object, the theoretical flow rate curves that correspond to the theoretical and actual coincidence degrees are compared. Then, the specific values of the theoretical flow rate and the geometric flow pulsation rate are calculated and analyzed. The design variables affecting the theoretical flow rate and geometric flow pulsation rate and the corresponding value ranges are determined, and the change laws of the gear pair's geometric parameters on theoretical flow rate and geometric flow pulsation rate are studied. By using the dynamic grid technology in CFD and considering the clearance leakage of the axial and radial, the simulation model of the internal flow channel is developed. The numerical simulation results are analyzed and compared. The influence factors and the influence degree of the flow pulsation and the average flow rate at the outlet of the gear pump are investigated. The high-pressure positive displacement pump test system is also designed and built. The performance evaluations for the test pump are implemented, and the experimental results are analysed and compared with the simulation results. The specific conclusions are as follows:

1. The periodic change of the meshing point position is the root cause of the geometric flow pulsation. Gear pumps with different tooth profiles exhibit different flow pulsation characteristics. When the meshing point coincides with the node, the instantaneous flow rate reaches the maximum. When the rotation angle of the external gear is the same as the ending meshing angle, the instantaneous flow rate is at the minimum.
2. The ending meshing angle affects the instantaneous flow rate curves corresponding to the theoretical and actual coincidence degrees. On one hand, the theoretical instantaneous flow rate curve is smooth, and the geometric flow pulsation rate is low. On the other hand, the theoretical average flow rate curve is a straight line along the horizontal distribution. The theoretical coincidence degree is 1.1212, the theoretical flow rate is 103.71 L/min, and the geometric flow pulsation rate is 1.76%.

3. The tip circle radius of the external gear exerts the greatest influence on theoretical flow rate and geometric flow pulsation rate. As this radius increases, the theoretical flow rate sharply increases, and the geometric flow pulsation rate sharply decreases. The tip circle radius of the internal gear ring only influences the theoretical flow rate, which constantly decreases with the increase in this radius. The remaining variables exert no significant effect on theoretical flow rate and geometric flow pulsation rate. To increase the theoretical flow rate whilst decreasing the geometric flow pulsation rate, the tip circle radius of the external gear should be increased as much as possible within the allowable range of the design calculation.
4. The oil film between the surfaces of the friction pair separates the oil absorption cavity from the oil discharge cavity. The pressure distribution in the flow channel is evenly transitioned. The movement of fluid particles in the rotor area is in a complex and irregular state, and the oil velocity in the gap reaches the maximum value at 11 m/s. To prevent the erosion of high-speed oil, the surface material modification should be applied to the surfaces of the gear and the crescent diaphragm. The upper wall surface of the outlet channel is affected by the oil at a relatively high speed (approximately 7 m/s). Hence, increasing the wall thickness of the upper wall surface is necessary.
5. The factors that affect the flow pulsation include geometric flow pulsation, oil compressibility and uneven internal leakage. Under the working conditions applied in this study, the oil compressibility will not affect flow pulsation. The uneven internal leakage is the main influencing factor, accounting for 80.1% of the total flow pulsation rate. By contrast, the geometric flow pulsation only accounts for a small proportion (19.9%).
6. The factors that affect average flow rate are the oil compressibility and the internal leakage. Under the working conditions applied in this study, oil compressibility does not affect average flow rate. The internal leakage reduces the simulated average flow rate by 3.59 L/min compared with the theoretical flow rate. The simulated average flow rate curve demonstrates a non-linear periodic change with a small amplitude (less than 0.4 L/min).
7. The difference between the experimental and simulated average flow rates is 1.82 L/min. The difference rate is less than 2%. The consistency between the simulation and test results verifies the applicability of the test system and the correctness of the test method. Under no-load conditions, with the increase in the rotation speed of the external gear, the average flow rate linearly increases, and the volume efficiency increases as a logarithmic function with a base number greater than 1. To increase the average flow rate at the outlet of the gear pump and improve the volumetric efficiency, the rotation speed of the external gear should be increased as much as possible within the allowable speed range.

Future research on straight line conjugate internal meshing gear pumps will be conducted based on three aspects. Firstly, on the basis of the accurate equations are obtained in this study, taking the maximum theoretical flow rate and the minimum geometric flow pulsation rate as the objectives, the value range of the design variables is determined according to the constraint conditions. The design variables are sorted in the primary and secondary order based on orthogonal experiments. The multi-objective genetic algorithm is used to optimize the basic parameters of the gear pair. Secondly, given that the axial and radial clearances of the straight line conjugate internal meshing gear pump are fixed, the moving parts are sealed by a small fixed clearance. However, the internal leakage from the high-pressure cavity to the low-pressure cavity through the fixed gap is inevitable. When the clearance is small, the clearance leakage decreases, the velocity gradient of the liquid in the gap increases, and the viscous friction loss of the liquid increases. In contrast, when the clearance is large, the viscous friction loss of the liquid decreases, the leakage loss in the gap increases, and the volume efficiency decreases. The oil film thickness between the friction pair's surfaces becomes the main influencing factor of the dynamic characteristics and the volumetric efficiency of the gear pump. Lastly, the load capacity is one of important indicators in measuring the performance of gear pumps. The working pressure is an essential factor that affects the form and the content of air in hydraulic oil and is also the main factor that changes the leakage flow rate. Moreover, this factor is a key source of pressure shock noises. The rotation speed change of the external gear under the loading conditions not only affects the average flow rate and the volume efficiency, but also influences the oil absorption characteristics and the oil properties. Consequently, the flow pulsation and the cavitation noise of gear pump are affected. In conclusion, the changes in the working parameters are essential influencing factors of the flow characteristics and working noises of gear pumps.

Author Contributions: Writing—original draft, writing—review and editing: H.C.; supervision, validation: G.Y.; formal analysis: G.W.; investigation, data curation: G.B.; conceptualization: W.L. All authors have read and agreed to the published version of the manuscript.

Funding: This research was funded by Open Foundation of the State Key Laboratory of Fluid Power and Mechatronic Systems, grant number GZKF-201823; Gansu Province Science and Technology Major Special Project-Research and Promotion of New Energy Comprehensive Utilization Technology, grant number 17ZD2GA010; Science and Technology Research Program of Chongqing Municipal Education Commission, grant number KJQN201801329; and The APC was funded by 17ZD2GA010.

Conflicts of Interest: The authors declare that they have no conflicts of interest.

Nomenclature

l	Vertical distance from external gear center to the tooth profile of the external gear	Z_2	Number of the internal gear ring
θ	Center angle corresponding to the gear thickness of the external gear reference circle	ω_1	Angular velocity of the external gear
β	Profile half angle of the external gear	ω_2	Angular velocity of the internal gear ring
r_1	Reference circle radius of the external gear	i_{12}	Gear pair transmission ratio
r_2	Reference circle radius of the internal gear ring	a	Center distance
B	Gear width	φ	Rotation angle of the external gear
r_{a1}	Tip circle radius of the external gear	Φ_0	Initial meshing angle
r_{f1}	Root circle radius of the external gear	Φ_1	Ending meshing angle
r_{a2}	Tip circle radius of the internal gear ring	φ_{lim}	Rotation angle of the external gear when the instantaneous flow rate reaches the limit
r_{f2}	Root circle radius of the internal gear ring	φ'	Angle between the line between the center of the external gear and the meshing point and the vertical axis
$\Delta\varphi_1$	Rotation angle of the external gear	n_1	Rotation speed of the external gear
$\Delta\varphi_2$	Rotation angle of the internal gear ring	ε_α	Coincidence degree
ΔV_1	Reduced volume of the oil discharge cavity	T_α	Theoretical oil trap time
ΔV_2	Enlarged volume of the oil discharge cavity	$\Delta\Sigma$	Oil discharge volume of each pair of gears
ΔV	Volume change of the oil discharge cavity	Σ	Theoretical displacement
q_{sh}	Theoretical instantaneous flow rate	q_t	Theoretical flow rate
ρ_1	Distance from the external gear center to the meshing point	q_{shmin}	Minimum instantaneous flow rate
ρ_2	Distance from the internal gear ring center to the meshing point	δ_q	Geometric flow pulsation rate
Z_1	Number of the external gear		

References

1. Castilla, R.; Gamez-Montero, P.J.; Del Campo, D.; Raush, G.; Garcia-Vilchez, M.; Codina, E. Three-dimensional numerical simulation of an external gear pump with decompression slot and meshing contact point. *Trans. ASME* **2015**, *137*, 539–550. [CrossRef]
2. Nishimoto, T. Noise, vibration and pulsation problems in oil hydraulic components and systems. *Thought-Evoking Approaches Eng. Probl. Toyama Jpn. Ohta-Kouyoudai* **2014**, *7*, 153–180.
3. Zhao, X.; Vacca, A. Numerical analysis of theoretical flow in external gear machines. *Mech. Mach. Theory* **2017**, *108*, 41–56. [CrossRef]

4. Zhao, X.; Vacca, A. Theoretical investigation into the ripple source of external gear pumps. *Energies* **2019**, *12*, 535. [CrossRef]
5. Chen, Z.; Lv, Z.; Xu, R.; Liao, J. Simulation and test of gear pump flow pulsation. *Mach. Syst.* **2018**, *11*, 265–272.
6. Truninger, P. Gear Pump. U.S. Patent 3,491,698, 27 January 1970.
7. Wang, W.; Yin, Y.-M.; He, S.-H.; Liu, G.-M. Study on flow characteristic of gear pumps by gear tooth shapes. *Appl. Sci. Eng.* **2017**, *20*, 367–372.
8. Huang, L.-Z. Tooth profile analysis of straight line-conjugate linetype internal meshing gear pump. *Mech. Transm.* **2004**, *28*, 16–22.
9. Dong, Y.-C.; Cui, J.-K.; Li, K.; Wei, Z. Study on sliding ratio of the straight conjugate internal gear pair. *Mach. Des. Manuf.* **2006**, *7*, 9–12.
10. Cong, X.-Q.; Liu, M.-X.; Wu, J. Design method of straight conjugate Internal gear pump. *Drain. Irrig. Mach.* **2008**, *26*, 33–36.
11. Zhang, Y.-S.; Hu, X.-H.; Liu, Y.-B.; Chen, L. The study of straight conjugate internal meshing gear pump. *Mach. Des. Manuf.* **2010**, *6*, 138–140.
12. Liu, L.-C. Gear's Research and Analysis of the Linear Conjugated Internal Gear Pump. Master's Thesis, Tianjin University of Technology, Tianjin, China, 2014.
13. Xu, X.-Z.; Zhang, K.; Tan, H.-M. Research on mesh characteristics of straight conjugated internal gear pump. *Mach. Tool Hydraul.* **2011**, *39*, 48–51.
14. Yang, G.-Y.; Bai, G.-X. The tooth profile equation of straight conjugate internal gear pump based on pressure angle function. *Hydraul. Pneum.* **2012**, *7*, 39–42.
15. Wei, W.-F.; Zhang, G.-P.; Du, Z.-Y.; Kang, N. Design method of internal rotary gear pump by parameterized line conjugated tooth profile. *J. Mech. Eng.* **2014**, *50*, 49–55. [CrossRef]
16. Gu, K.-L.; Wang, Z.-H.; Li, G.; Liu, X.-R. Optimization of geometric parameters of the straight conjugate internal gear pump based on GA. *Electron. Sci. Technol.* **2017**, *30*, 39–44.
17. Hu, H.-L.; Cui, J.-K.; Xu, J.-B. Optimal design of tooth profile parameter of straight-conjugate internal gear pair. *Mech. Transm.* **2015**, *39*, 77–80.
18. Li, X.; Cui, J.-K. Analysis of the interference of straight conjugate internal gear pair based on matlab. *Mech. Transm.* **2014**, *38*, 128–131.
19. Xu, X.-X.; Song, T.-L. Optimal design of basic parameters of straight line conjugate gear pump. *Mech. Transm.* **2007**, *31*, 69–72.
20. Rundo, M. Models for flow rate simulation in gear pumps: A review. *Energies* **2017**, *10*, 1261. [CrossRef]
21. Zhao, X.; Vacca, A. Analysis of continuous-contact helical gear pumps through numerical modeling and experimental validation. *Mech. Syst. Signal Process.* **2018**, *109*, 352–378. [CrossRef]
22. Pellegri, M.; Vacca, A. Numerical simulation of gerotor pumps considering rotor micro-motions. *Meccanica* **2016**, *52*, 1851–1870. [CrossRef]
23. Li, S.-D.; Cui, J.-K. The finite element analysis of pump body line. *Manuf. Autom.* **2011**, *33*, 33–37.
24. Wang, Z.-W.; Zhang, Z.-S.; Liang, W.-G.; Zhang, F. Analysis on meshing efficiency of straight conjugate internal gear pair. *Torpedo Technol.* **2013**, *21*, 369–374.
25. Xu, X.-Z.; Song, T.-L. Study on sliding coefficients of straight line conjugate tooth profile gear pairs. *J. Huaiyin Teach. Coll.* **2007**, *3*, 225–228.
26. Yuan, D.-Q.; Kang, S.-B.; Cong, X.-Q.; Liu, M. Finite element analysis on gear tangency stress of straight line conjugate internal gear pump. *Drain. Irrig. Mach.* **2009**, *27*, 111–114.
27. Gao, M.-L.; Cui, J.-K.; Huang, D.-P.; Liu, S.; Zhang, L. The calculation analysis of contact ratio for straight conjugate internal gear transmission. *Fluid Mach.* **2017**, *45*, 28–32.
28. Duan, G.; Cui, J.-K.; Gao, M.-L. Contact ratio of an internal meshing gear pair with a straight line tooth profile out gear. *J. Univ. Shanghai Sci. Technol.* **2018**, *40*, 335–341.
29. Rundo, M. Theoretical flow rate in crescent pumps. *Simul. Model. Pract. Theory* **2017**, *71*, 1–14. [CrossRef]
30. Rundo, M.; Corvaglia, A. Lumped parameters model of a crescent pump. *Energies* **2016**, *9*, 876. [CrossRef]
31. Rydlewicz, W.; Rydlewicz, W.; Pałczyński, T. Experimental investigation of the influence of an orifice plate on the pressure pulsation amplitude in the pulsating flow in a straight pipe. *Mech. Syst. Signal Process.* **2019**, *117*, 634–652. [CrossRef]

32. Liu, D.-W.; Ba, Y.-B.; Ren, T.-Z. Flow fluctuation abatement of high-order elliptical gear pump by external noncircular gear drive. *Mech. Mach. Theory* **2019**, *134*, 338–348. [CrossRef]
33. Wang, Z.-L.; Zhang, Z.-S.; Zhang, F.-F.; Liu, J. Analysis of the effect of parameter on trapping oil of straight conjugate internal gear pump. *Mech. Transm.* **2014**, *38*, 91–93.
34. Xu, X.-Z.; Song, T.-L. Analysis of flow characteristics of gear pump with straight line profile. *Coal Min. Mach.* **2008**, *29*, 47–50.
35. Wang, Z.-W.; Cui, J.-K.; Zhang, L.-H. Instantaneous flow analysis of straight line conjugate internal meshing gear pump. *Fluid Transm. Control* **2013**, *4*, 23–27.
36. Zhang, H.; Wang, Z.-H. Analysis of the flow pulsation of straight conjugated internal gear pump. *Mech. Transm.* **2015**, *39*, 98–102.
37. Kilani, M.I. Computer aided design tools in the development of surface micromachined mechanisms. *Jordan J. Mech. Ind. Eng.* **2011**, *5*, 167–176.
38. Zhang, Y. The Research of New Processing Method in Large Module Internal Engage Gear Based on Conjugate Principle. Master's Thesis, Dalian University of Technology, Dalian, China, 2007.
39. Wei, B.; Xiong, H.-G. *Gear Mechanism and Its Design. Mechanical Principle*, 1st ed.; Huazhong University of Science & Technology Press: Wuhan, China, 2007; Volume 7, pp. 137–180.
40. Chen, Y.-Q.; Xue, D.-Y. *Basic Programming of MATLAB Language. System Simulation Technology and Application Based on MATLAB/Simulink*, 2nd ed.; Tsinghua University Press: Beijing, China, 2002; Volume 2, pp. 17–93.
41. Buch Hydraulics. *Technical Data & Performance Graphs. High Pressure Internal Gear Pumps*; Buch Hydraulics Co. Ltd.: Bern, Switzerland, 2014; Volume 2 & 3, pp. 5–10.
42. Xu, Y.-M. *Oil Film Design in Gears and Vane Pumps (Motors). Oil Film Theory and Friction Pair Design of Hydraulic Pumps and Motors*, 1st ed.; Machinery Industry Press: Beijing, China, 1987; Volume 13, pp. 354–381.
43. Jiang, Y.; Furmanczyk, M.; Lowry, S.; Zhang, D.; Perng, C.-Y. *A Three-Dimensional Design Tool for Crescent Oil Pumps*; SAE World Congress & Exhibition: Detroit, MI, USA, 2008; pp. 4–14.
44. Wang, Y.-Q.; Gao, D.-R. *Hydraulic Medium. Technical Manual for Hydraulic Engineers*, 1st ed.; Chemical Industry Press: Beijing, China, 2009; Volume 2, pp. 66–71.
45. Rao, S.-H. *Asynchronous Motor. Electrical and Electronic Technology Foundation*, 1st ed.; Beijing Institute of Technology Press: Beijing, China, 2008; Volume 6, pp. 103–119.

processes

MDPI

Article

A Numerical Study on the Flow Mechanism of Performance Improvement of a Wide-Angle Diffuser by Inserting a Short Splitter Vane

Xu Meng [1], Zhigang Zuo [1,*], Michihiro Nishi [2] and Shuhong Liu [1,*]

[1] State Key Laboratory of Hydro Science and Engineering, Department of Energy and Power Engineering, Tsinghua University, Beijing 100084, China; zgbh.mx@163.com

[2] Senior Academy, Kyushu Institute of Technology, Tobata, Kitakyushu 804-8550, Japan; nishi.michihiro@gmail.com

* Correspondence: zhigang200@mail.tsinghua.edu.cn (Z.Z.); liushuhong@mail.tsinghua.edu.cn (S.L.); Tel.:+86-10-62773947 (Z.Z. & S.L.)

Received: 25 December 2019; Accepted: 19 January 2020; Published: 22 January 2020

Abstract: Usage of a wide-angle diffuser may result in unfavorable separated flow and a significant diffuser loss. To improve the performance of the diffusers, inserting short splitter vanes is known as a useful method that has been demonstrated experimentally. Regarding the role of the vane in the diffuser flow, Senoo & Nishi (1977) qualitatively explained that the lift force acting on the vane should be a key factor. However, its quantitative verification remains since then. To challenge this issue, numerical simulations of incompressible flow in a wide angle of 28° two-dimensional diffuser with and without a short splitter vane were conducted in the present study. An improvement of pressure-recovery by the vane and oscillatory flows in the diffuser are reasonably reproduced from comparison with the experimental results made by Cochran & Kline (1958). It is also found that the lift force acting on the vane varies periodically in an opposite phase with the detachment point moved back and forth on a diverging wall, since one vane is not sufficient to fully suppress the flow separation that occurred on the wall and the incoming main-flow shifts toward the other diverging wall in the diffuser. Thus, as a role of splitter vane in the diffuser, "the lift force of the vane is a key factor" may be quantitatively verified from the present numerical simulation. Further, it is confirmed by the local loss analysis that the turbulent kinetic energy production observed in mixing layers contributes most of the loss in the diffuser. Consequently, the present numerical technique may be usable to investigate the flow character in a diffuser with splitter vanes at a design stage.

Keywords: wide-angle diffuser; splitter vane; flow simulation; loss analysis; flow mechanism

---

1. Introduction

Diffusers are widely used in fluid machines and model test facilities (e.g., draft tubes of turbines [1–3], subsonic wind tunnels [4,5], etc. [6]) as a component to convert the dynamic pressure to the static pressure by decelerating the main flow velocity. If the diffuser loss Δp_{Loss} is required to design the above-mentioned systems, the following relationships derived from the one-dimensional theory for the incompressible steady flow have been used [7–9]:

$$\zeta_d\left(=\frac{\Delta p_{Loss}}{q_1}\right)=C_{p_i}-C_p=C_{p_i}\left(1-\frac{C_p}{C_{pi}}\right)=C_{p_i}(1-\eta_d) \quad (1)$$

$$C_{p_i}=1-\left(\frac{A_1}{A_2}\right)^2=1-\left(\frac{1}{A_R}\right)^2 \quad (2)$$

$$C_p = \frac{p_2 - p_1}{q_1} \tag{3}$$

$$q_1 = \frac{1}{2}\rho V_1{}^2 \tag{4}$$

where ζ_d, Cp_i, Cp, and η_d are the diffuser loss coefficient, the ideal and the actual static-pressure recovery coefficients, and the diffuser effectiveness, respectively. And A_R denotes the area-ratio of the diffuser exit area A_2 to the inlet A_1. p_1, p_2, q_1, V_1, and ρ are the inlet and the exit static pressures [Pa], the dynamic pressure at the diffuser inlet [Pa], inlet velocity [m/s] and fluid density [kg/m^3], respectively.

It is known that the diffuser performances given by ζ_d and Cp primarily depend on Reynolds number, inflow conditions (i.e., velocity distribution, main-flow turbulence), and diffuser geometries (i.e., divergence angle, area-ratio, and diffuser length) [8]. Taking into account those parameters mentioned above, extensive studies have been conducted experimentally to prepare selection charts of diffuser geometries associated with pressure recovery and flow characteristics as a convenient tool to design two-dimensional [9], conical [10] and annular diffusers [11].

According to the selection chart (or performance map), diffusers with the divergence angle around 7° may have favorable geometry of duct to provide a low loss, a high pressure-recovery and a stable flow. In some cases of application, however, much wider divergence-angles have to be selected due to the limitation of the axial diffuser length for a given area-ratio. Owing to this selection, the adverse pressure gradient becomes steeper along the diffuser wall, and separation of flow may occur in the diffuser. Once the flow-separation (or stall) takes place in a diffuser, unsteadiness, and non-uniformity of flow are apt to appear at the diffuser exit, and the diffuser loss usually increases together with the decrease of pressure recovery. Thus, many investigations have been carried out to develop separation-control methods including boundary layer suction [12], jet injection [13], high-level turbulence at the inlet [14], passive and active vortex generators [15–17], installation of splitter vanes [7,8,18–20], etc. [21]. Note that Sun et al. [22] performed a numerical investigation of the flow rectification of a vaned micro diffuser.

Historically, flow in a diffuser near the stall condition had been treated almost experimentally because the prediction of separation point was hard to achieve from Prandtl's boundary layer analysis, where the interaction between the boundary layer and the adjacent free-stream wasn't considered [23,24]. Furthermore, a fundamental issue from the viewpoint of fluid mechanics was presented by the results from a wide-angle diffuser with splitter vanes [25]. That is, why is the separating flow suppressed by inserting short splitter vanes into the inlet region of a wide-angle diffuser? Though Senoo and Nishi [26] qualitatively explained that the lift force acting on the vane should be a key factor, its quantitative verification still remains to be done.

In order to challenge the unresolved issue mentioned above, numerical simulation of incompressible flow in a wide-angle two-dimensional diffuser with and without a short splitter vane is conducted in this study.

2. Methodology

2.1. Test Diffuser and Pressure Recovery

In the present study, a specific wide-angle diffuser with and without a splitter vane was investigated. As a test straight diffuser for numerical simulation, we reproduced the geometry used by Cochran and Kline [7]. Referring to their experimental results obtained at the throat Reynolds number of 2.4×10^5, the divergence angle of 28° and the area-ratio of 4.9 were chosen. Thus, the mean velocity at the diffuser throat is 47.85 m/s in the present cases.

The following pressure recovery coefficient C_{PR} is used to evaluate the diffuser performance in the present simulation for direct comparison with the experimental values:

$$C_{PR} = \frac{\frac{1}{A_2}\int_{A_2} p_2 dA_2 - \frac{1}{A_1}\int_{A_1} p_1 dA_1}{\frac{1}{A_1}\int_{A_1} q_1 dA_1} \tag{5}$$

2.2. Numerical Technique

Figure 1 shows the top view of the computational domain, and its depth is 611 mm. As shown in the figure, upstream and downstream channels are added to the domain to take account for the boundary conditions. Specifically, the upstream channel is 1.5 times the length of its width (width of the upstream channel = $8W_1$), and the downstream-channel [5] is 3 times the length of its width (width of the downstream channel = $16W_1$), where W_1 denotes the throat width.

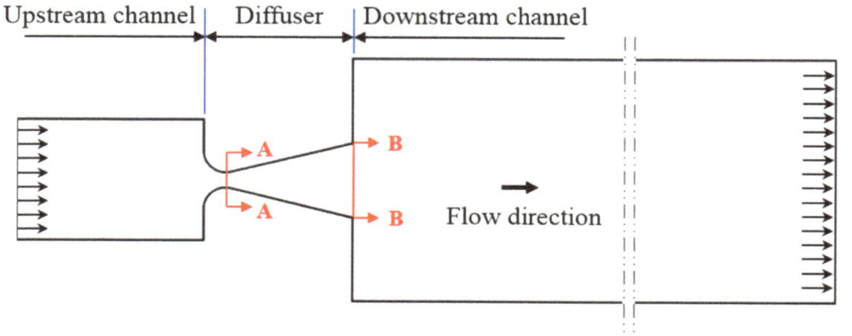

Figure 1. Top view of the computational domain.

Ideal dry air (20 °C) was used as the working fluid. Its uniform velocity at the inlet of the upstream channel and atmospheric pressure at the outlet of the downstream channel were chosen as the boundary conditions. According to the Reynolds number at the diffuser throat mentioned before, the mean velocity was set as 5.96 m/s at the inlet of the upstream channel. The specification method of "Intensity and Viscosity Ratio" was chosen. The turbulence intensity is 5% and the turbulent viscosity ratio is 10. "No-slip" conditions were imposed on all stationary walls. Figure 2 shows the geometrical parameters of the diffuser with and without a short splitter vane (or flat vane), whose major dimensions are listed in Table 1, where it is seen that the numerical test has only one vane length, which is nearly a quarter of the diffuser length. Note that all wall surfaces were smooth following the experiment [7].

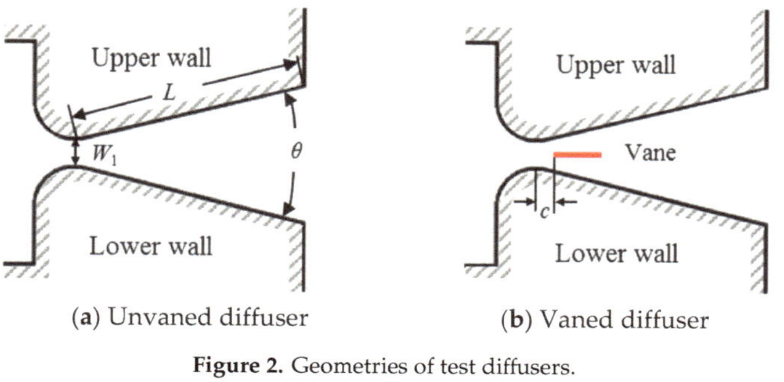

Figure 2. Geometries of test diffusers.

Table 1. Major dimensions of test diffuser and vane.

	Symbol	Nomenclature	Value
Diffuser	θ (°)	divergence angle	28
	W_1 (mm)	throat width	76
	L/W_1	dimensionless diffuser wall length	8
Vane	L_v (mm)	length	152.4
	δ (mm)	thickness	2.8
	c (mm)	location	44

To study the boundary layer and the large separated flow in the diffuser before and after inserting the splitter vane, a low-Reynolds-number RANS (Reynolds Averaged Navier-Stokes) turbulence model, i.e., SST k-ω turbulence model was applied to this double precision simulation, where wall function was not used. It has been shown that SST k-ω has good compatibility in simulating flow separations [27,28]. SIMPLEC (Semi-Implicit Method for Pressure-Linked Equations Consistent) algorithm with a second-order spatial discretization and a first-order implicit transient formulation were chosen as the pressure-velocity coupling method.

Structured grid-systems for the whole calculation domain were developed by using commercial software ICEM-CFD (Integrated Computer Engineering and Manufacturing code for Computational Fluid Dynamics) in this study. Figure 3 shows the calculation grids. Figure 3a is the local grid refinements near the edge of the diffuser throat and Figure 3b is the top view of the topology of the mesh. To ensure that the first boundary grid layer is located in a viscous sublayer, required by the SST k-ω turbulence model, local grid refinements to the boundary layers were applied at the vane and diffuser walls. Figure 4 shows y^+ distributions on the diverging walls obtained at steady flow simulation of the diffuser without a vane (or unvaned diffuser). "Ideally, while using enhanced wall treatment, the wall y^+ should be in the order of 1 (at least less than 5) to resolve the viscous sublayer" [29]. In our simulations, it is controlled less than 5.

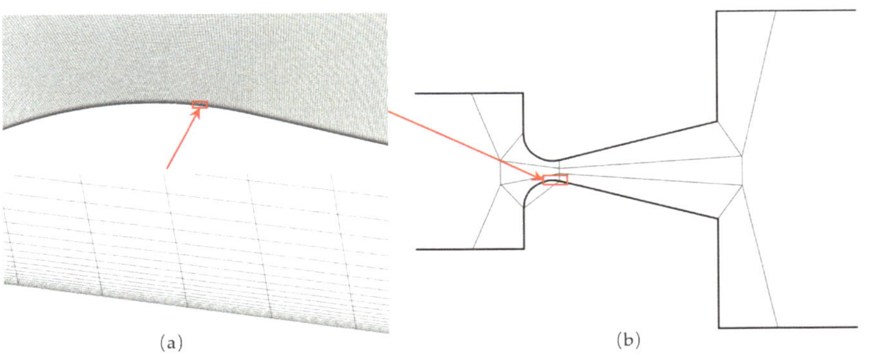

Figure 3. Calculation grids. (**a**) Local grid refinements near the edge of the diffuser throat; (**b**) Topology of the mesh.

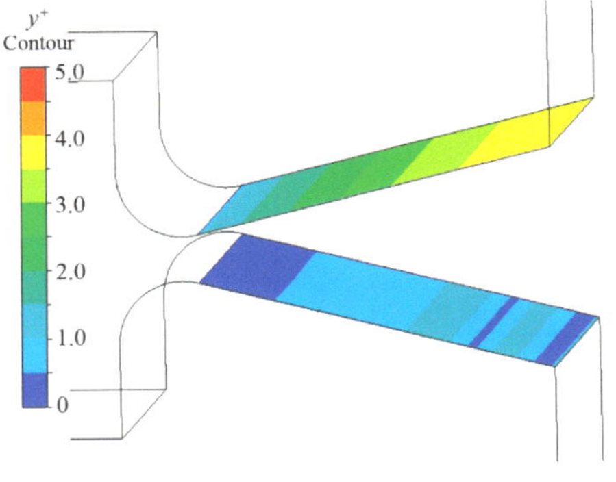

Figure 4. Distributions of y^+ on the diffuser walls.

Unsteady numerical simulations also based on RANS method were applied in the present study. Considering the grid-scale and the Courant number requirement, i.e., $C = u\Delta t/\Delta s < 10$ (where u, Δt, and Δs are the characteristic flow velocity, the time step size and typical cell size, respectively), Δt was set as 0.001 s. The maximum number of iterations was set as 20. Simulation convergence was achieved when the residual error was lower than 10^{-6}, or the variation of the operational parameter C_{PR} was below 0.01%. For the purpose of measuring the operational parameter C_{PR}, static pressure monitors were set to record the average static pressure of the diffuser inlet and outlet.

As preliminary tests, predicted C_{PR} of steady flow simulations by using five mesh systems for both unvaned and vaned diffusers were carried out to find the plausible number of grids for the computational domain. The corresponding results are shown in Figure 5 and the mesh sensitivity analysis is summarized in Table 2, where the following features are observed: C_{PR} becomes almost constant in the region of the number of grids greater than 4.2×10^5 for the unvaned diffuser and 5.9×10^5 for the vaned diffuser. In Table 2, relative change in C_{PR} given by the following expression is used to evaluate the mesh sensitivity. Note that i (2~5) is the serial number of the grid.

$$\frac{\left|C_{PR}(i) - C_{PR}(i-1)\right|}{C_{PR}(i-1)} \times 100\%$$

Therefore, the mesh system with an 8.3×10^5 grid for the unvaned diffuser and that with a 9.7×10^5 grid for the vaned diffuser were chosen for the present investigations.

Table 2. Mesh sensitivity analysis.

Serial Number	Number of Grids (Unvaned)	Relative Change in C_{PR}	Number of Grids (Vaned)	Relative Change in C_{PR}
1	103,096	-	143,550	-
2	207,480	5.92%	282,122	6.78%
3	417,606	12.00%	587,028	2.01%
4	832,944	0.32%	972,532	0.95%
5	1,680,826	1.06%	1,992,446	0.15%

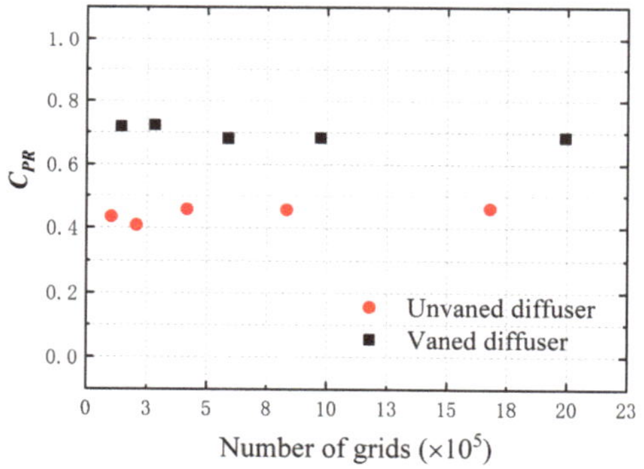

Figure 5. Grid independence test.

3. Results and Discussions

3.1. Validation of the Simulation from Diffuser Performance

Table 3 shows the comparison of pressure recovery between the steady flow simulation and Cochran & Kline measurement [7], where Equation (5) was applied to the pressure and velocity data at the mid-depth of diffuser for calculation of C_{PR}. If the uncertainty of C_{PR} being estimated between 0.012 and 0.092 in the reference [7] is considered, it may be said that C_{PR} is reasonably predicted by the present simulation. The numerical results also show that an increase of about 43% in C_{PR} is achieved after inserting a short splitter vane in the diffuser.

Table 3. Comparison of C_{PR} between simulation and experiment.

	C_{PR}: Simulation	C_{PR}: Experiment	Error (%)
Unvaned diffuser	0.456	0.437	4.35
Vaned diffuser	0.653	0.623	4.82

Distributions of pressure in the cross-sections of diffuser inlet (or throat) and the exit are displayed as pressure contours in Figure 6, where the dash-dot line shows the mid-depth. It is seen that the flow characteristics in the mid-depth section may be treatable as the flow in a two-dimensional diffuser since the pressure patterns in the transverse direction (i.e., the horizontal direction in the figure) are not much different in the vicinity of mid-depth.

Figure 7 shows the velocity field in the unvaned diffuser. It is seen that flow detachment occurs about 27.5 mm downstream of the throat on one diverging wall (i.e., the lower wall in the present case), and the main through-flow shifts toward the other diverging wall (i.e., upper wall), where no separation is observed. Note that the location of the detachment point [30] was determined by using the limiting streamline method. According to the reference [7], flow characteristics of a wide-angle of 28° of the two-dimensional diffuser are explained as follows: The flow was relatively steady, and the overall separation that was nearly two-dimensional in form extended to within a small distance from the throat on one diverging wall. In fact, these are observed in Figure 7.

Consequently, the mid-depth section of the diffuser was chosen for further analyses of computational data.

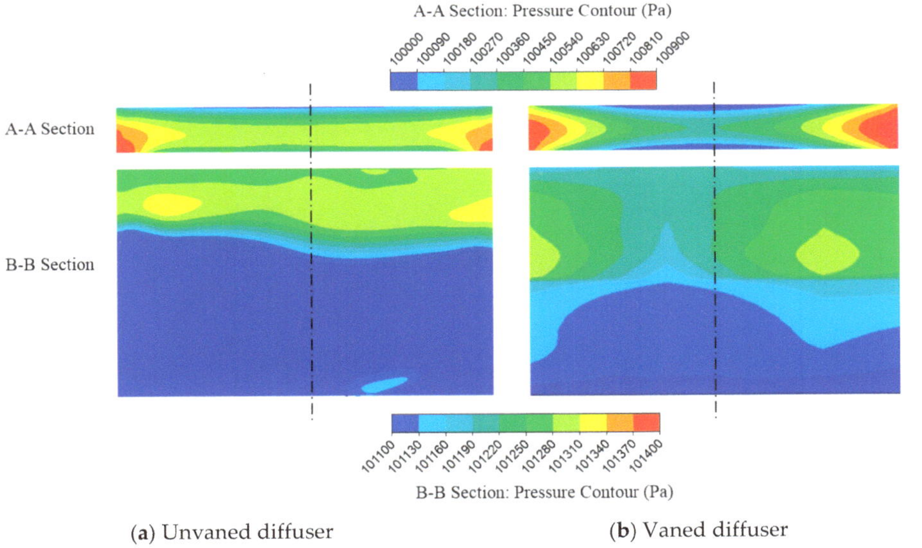

Figure 6. Pressure contours in the cross-section viewed in the downstream direction. (A-A section: diffuser throat, B-B section: diffuser exit).

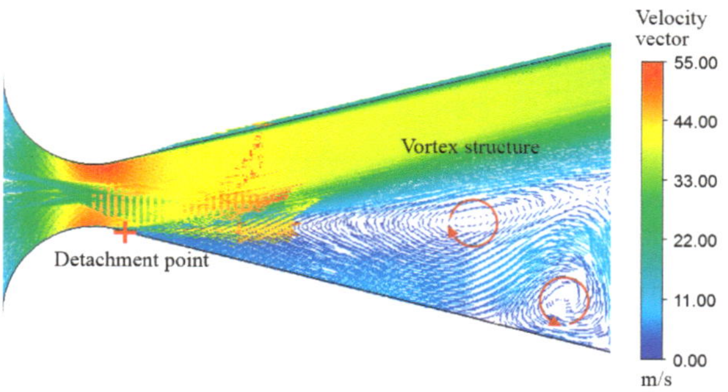

Figure 7. Velocity field in the unvaned diffuser (steady flow simulation).

3.2. Flow Characteristics

A red curve in Figure 8 shows the variation of instantaneous pressure-recovery C_{PR} with time in the case of the vaned diffuser. Though the pressure-recovery of the diffuser is really improved, flow steadiness is deteriorated. This may be because one splitter vane cannot fully eliminate the stall zone in the present diffuser. As shown in Figure 9, the fundamental frequency of C_{PR} variation is around 6.1 Hz from FFT (Fast Fourier Transform) analysis. In reference [7], it is described that pulsating flow was observed due to the installation of a short vane, but there was no reliable information on its frequency and amplitude. Keeping this in mind, we examined the periodic variation of detachment-point location, the distance of which is X_S measured from the diffuser throat along the wall, and added it in Figure 8. Understandable are those features that oscillations of C_{PR} and X_S are almost in-phase, and the minimum C_{PR} appears a little bit later than the minimum X_S appears. The occurrence of the time lag may be related to the main flow inertia.

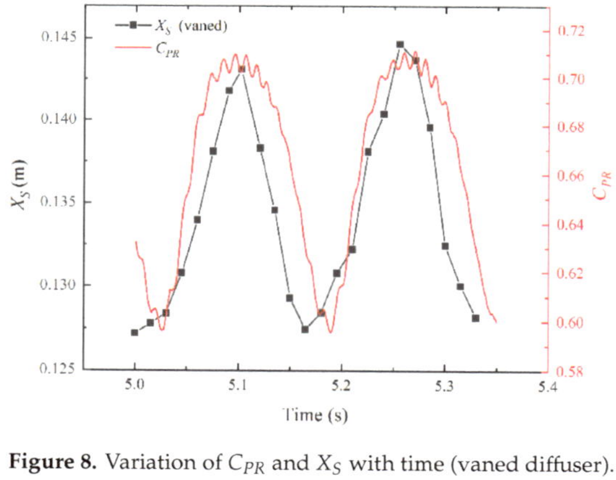

Figure 8. Variation of C_{PR} and X_S with time (vaned diffuser).

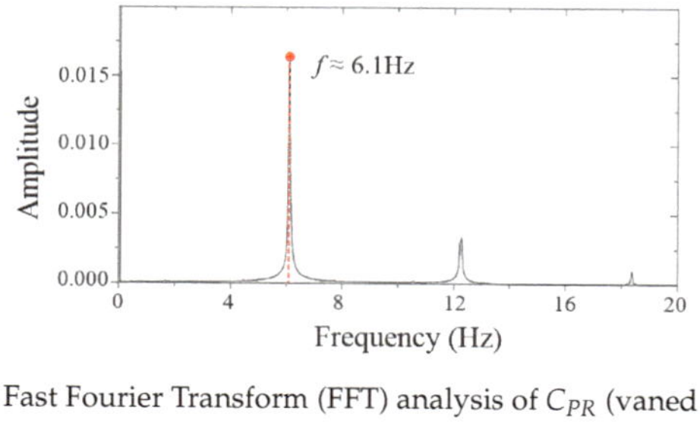

Figure 9. Fast Fourier Transform (FFT) analysis of C_{PR} (vaned diffuser).

Figure 10 shows six snapshots of the velocity field in the vaned diffuser during one time period T (=0.164 s) of flow oscillation. It is seen that the oscillatory flow is primarily related to the behaviors of the detachment point, and the stall zone extended downstream of the diffuser exit. This phenomenon occurs because one short vane cannot fully suppress the flow separation in the diffuser so that incoming main-flow tends toward the upper wall all the time. Due to this flow condition, the angle of incidence may be larger than the stall angle of a flat vane, and large-scale wake flow from the vane is always observed in the main through-flow, but it doesn't directly contact the boundary layer on the upper diverging wall.

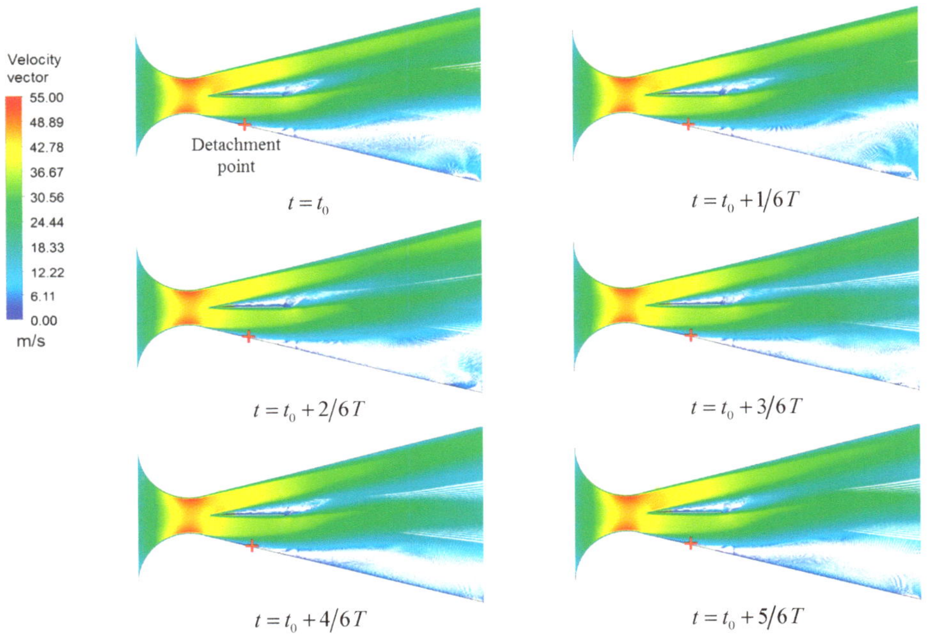

Figure 10. Snapshots of the velocity field in the vaned diffuser (t_0 = 5 s).

To see the pressure rise along the diverging wall, the following dimensionless wall-pressure $p_w{}^*$ is introduced:

$$p_w^* = \frac{p_w - p_{atm}}{\frac{\rho}{2}\overline{V}_1{}^2} \tag{6}$$

where p_w and p_{atm} denote local wall-pressure and atmospheric pressure. And $\overline{V}_1$ is the mean velocity at the throat.

By using Equation (6), six wall-pressure distributions along both upper and lower walls are obtained corresponding to the velocity fields given in Figure 11. From the results shown in Figure 11, where the right end of each $p_w{}^*$ curve corresponds to the diffuser exit, the following features are observed:

- As atmospheric pressure is given at the domain outlet, dimensionless wall-pressure generally decreases toward diffuser inlet from the diffuser exit.
- Since the separation of flow is observed on the lower wall and no separation on the upper wall, recovery of wall-pressure along the upper wall is greater than that along the lower wall.
- The steep adverse pressure gradient is observed in the region between the throat and location of the vane's leading-edge.
- Depending on the location of the detachment point, the shape of $p_w{}^*$ curve varies to some extent.

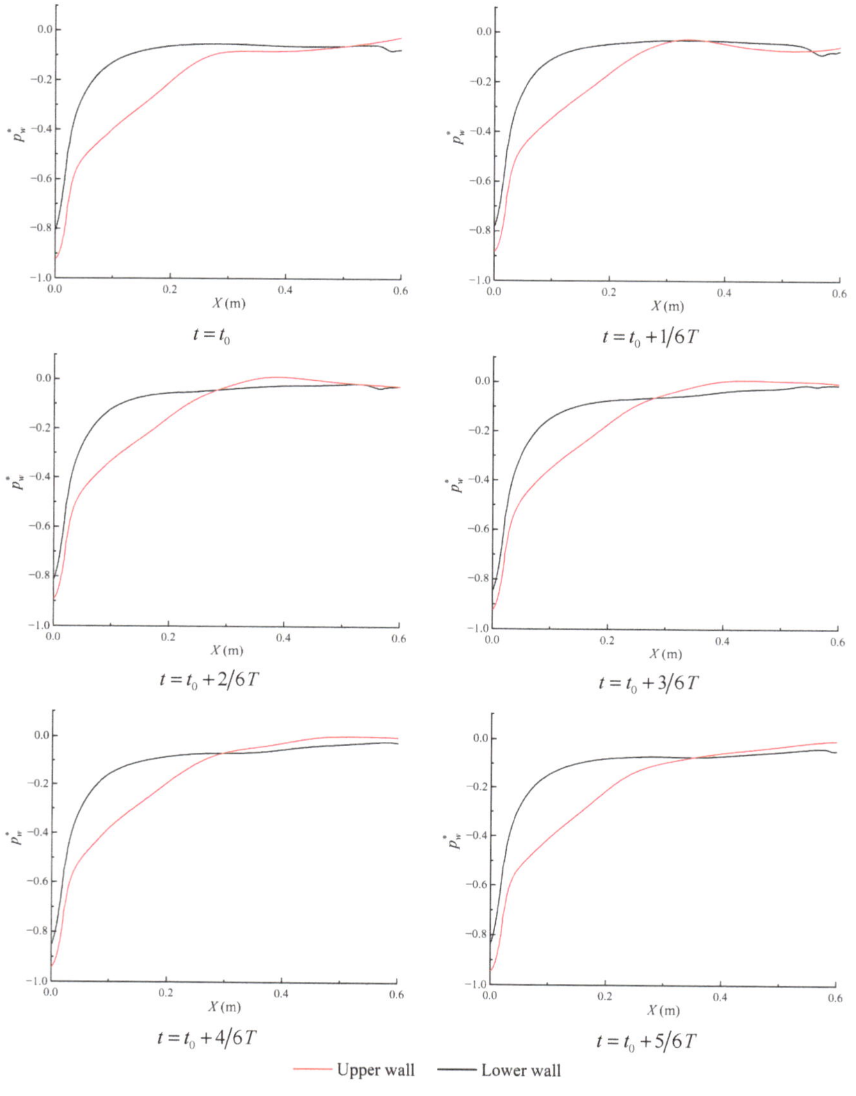

Figure 11. Distributions of wall pressure along diverging walls of the vaned diffuser.

Time-averaged pressure distributions being derived during a cycle of flow oscillation are shown in Figure 12a for the vaned diffuser and Figure 12b for the unvaned diffuser. In Figure 12a, $p_w{}^*$ curves calculated from the steady flow simulation are drawn for comparison. The following features are found from the figures:

- In the vaned diffuser case, time-averaged $p_w{}^*$ curves are almost reproduced by the curves calculated from the steady flow simulation.
- In the case of the unvaned diffuser, the detachment point occurs much closer to the throat than that of the vaned case, and amounts of wall-pressure recovery reduce greatly.
- As large stall zone occupies in the unvaned diffuser, pressure rise downstream of the detachment point is almost suppressed along the lower wall.

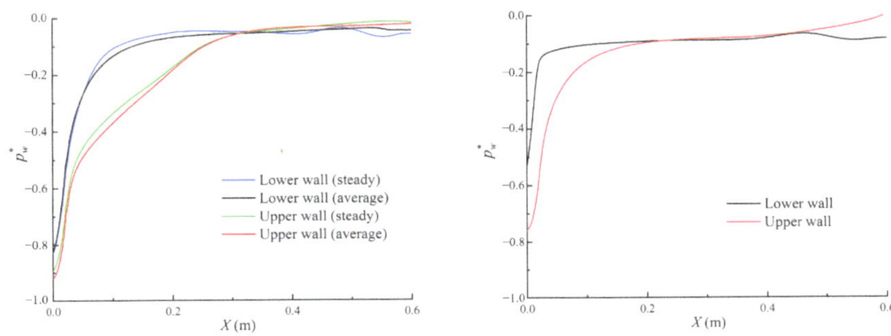

(**a**) Vaned diffuser (**b**) Unvaned diffuser

Figure 12. Time-averaged wall-pressure distributions.

It is noted that p_w* curves for unvaned diffuser calculated from the steady flow simulation almost coincide with those of unsteady flow simulation, as the internal flow is fairly steady in the 28° two-dimensional diffuser having area-ratio of 4.9 [7].

3.3. Force on the Vane

The near-wall velocity-distributions on upper and lower walls at the throat section are shown in Figure 13, where Y is the distance from the center axis of the diffuser. The effects of a splitter vane are clearly observed in the boundary layer of the lower wall. That is, the boundary layer thickness decreases, and the main–flow velocity increases by inserting the vane.

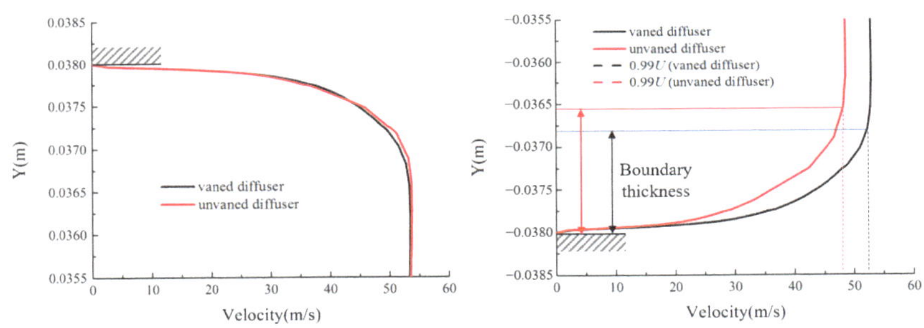

(**a**) Upper wall (**b**) Lower wall

Figure 13. Velocity distributions in the boundary layer at the throat.

To investigate the force acting on the vane in a two-dimensional diffuser, the following equation is used to calculate the dimensionless lift F_L* assuming that the lift force is approximated by the pressure-force normal to the vane.

$$F_L^* = \frac{F_L}{\frac{\rho}{2}\overline{V}_1{}^2 L_v} \tag{7}$$

where L_v denotes the length of splitter vane. Lift force F_L directing toward the upper wall is set as positive.

In fact, the positive value of lift force was always calculated from instantaneous pressures on the vane. After processing those data by using Equation (7), the variation of dimensionless lift with time is correlated in Figure 14, where fluctuation of detachment-point location X_S is also plotted for comparison. It is seen that the dimensionless lift oscillates between F_L* = 0.15 and F_L* = 0.25, having the same fundamental frequency as the pressure recovery and the location of the detachment point. Unlike the pressure-recovery (see Figure 8), the anti-phase character is observed between the location of detachment point X_S and the dimensionless lift F_L*. The smaller X_S is, the larger F_L* becomes, and the larger X_S is, the smaller F_L* becomes. If the lift force acts on a vane in the flow, the reaction force of the vane pushes the flow. Thus, a so-called self-controlling mechanism is recognized in Figure 14, though the flow separation on the lower wall cannot fully be suppressed in the test case. This numerical result may quantitatively verify the qualitative explanation for the role of a splitter vane made by Senoo & Nishi [26].

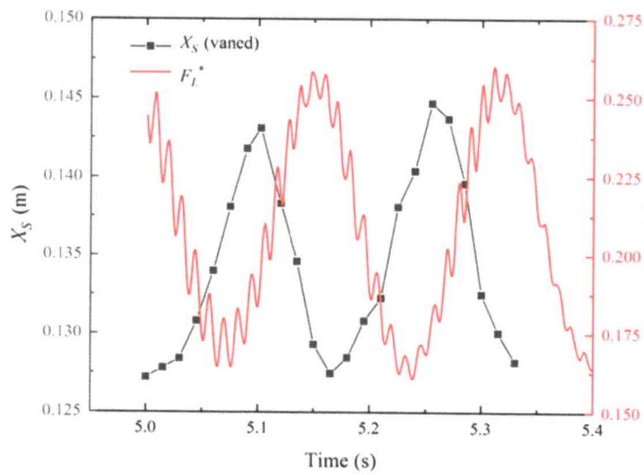

Figure 14. Lift force acting on a splitter vane.

3.4. Analysis of Unsteady Local Loss

A local energy analysis method from reference [31] was used to investigate the relationship between the flow patterns and the loss in diffusers [32]. The loss in the diffuser P_L is expressed as follows:

$$P_L = \underbrace{\iiint_V \frac{\partial}{\partial t}\left(\frac{1}{2}\rho\overline{u}^2\right)dV}_{Term\ I} - \underbrace{\iiint_V \frac{\partial(-\overline{u_i}\overline{\rho u'_i u'_j})}{\partial x_j}dV}_{Term\ II} - \underbrace{\iiint_V \mu\frac{\partial(\overline{u_i}\overline{D_{ij}})}{\partial x_j}dV}_{Term\ III} + \underbrace{\iiint_V (-\overline{\rho u'_i u'_j})\frac{\partial\overline{u_i}}{\partial x_j}dV}_{Term\ IV} + \underbrace{\iiint_V \mu\overline{D_{ij}}\frac{\partial\overline{u_i}}{\partial x_j}dV}_{Term\ V} \quad (8)$$

where μ is the kinematic viscosity and $D_{ij} = \frac{\partial u_i}{\partial x_j} + \frac{\partial u_j}{\partial x_i}$. If a denotes a physical quantity, $\overline{a}$ and a' show the time-averaged value and the fluctuation respectively. Among the five terms, *Term I* represents the variation of the kinetic energy of the mean flow. *Term II* and *Term III* correspond to the diffusion of mean kinetic energy acting by Reynolds stress and viscous stress. *Term IV* is the turbulent kinetic energy production. And *Term V* contributes to the loss by viscous dissipation of the mean kinetic energy.

As SST *k-ϖ* turbulence model (RANS) was used in this study, Reynolds stress $\rho\overline{u'_i u'_j}$ was calculated as follows:

$$\rho\overline{u'_i u'_j} = -\frac{1}{2}\mu_t\left(\frac{\partial\overline{u_i}}{\partial x_j} + \frac{\partial\overline{u_j}}{\partial x_i}\right) + \frac{2}{3}\rho k\delta_{ij} \quad (9)$$

$$\delta_{ij} = \begin{bmatrix} 1 & 0 & 0 \\ 0 & 1 & 0 \\ 0 & 0 & 1 \end{bmatrix} \quad (10)$$

where μ_t, k and δ_{ij} are the eddy viscosity, the turbulent kinetic energy, and the Kronecker delta respectively.

The local loss analysis method was applied to see the contribution of each term to the loss in the diffusers. Figure 15 illustrates the terms of loss in unvaned and vaned diffusers separately (Note: λ is the loss factor that is the ratio of each term to the total loss of the mid-depth section in the unvaned diffuser). As the total loss in the vaned diffuser is around 65% of the vaned diffuser loss, the internal flow may be improved by inserting the splitter vane. The results also show that the turbulent kinetic energy production is predominant and it is the primary component of the loss for both diffusers.

Considering that the total loss in the present diffuser is mostly caused by the turbulent kinetic energy production, visualization of its distribution in the vaned diffuser was conducted to see the effect of vane installation. Figure 16 shows such six pictures corresponding to the time used in Figure 10, where the velocity fields are displayed. There are two apparent zones that correspond to the shear layer, where the amount of turbulent kinetic energy production is large. One is observed at the boundary between the main flow and the detached boundary-layer flow (or stall zone) on the lower wall of the diffuser. The other appears in the mixing layer between the main flow and the separating flow from the vane surface (i.e., suction side) faced the upper wall. And it is seen that the highest production occurs in the boundary caused by the leading-edge separation of the vane.

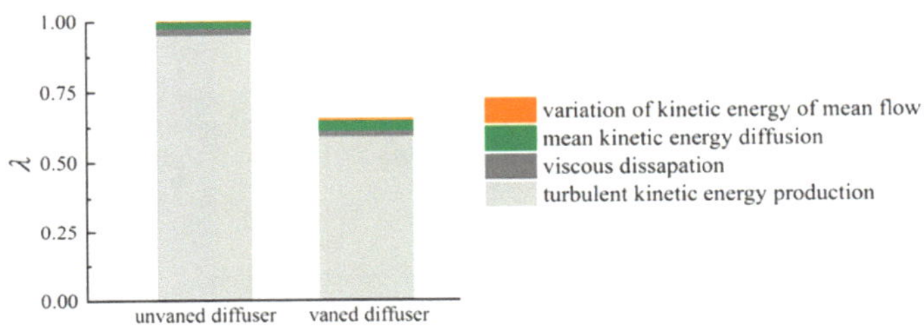

Figure 15. Composition of diffuser loss predicted in unvaned and vaned diffusers.

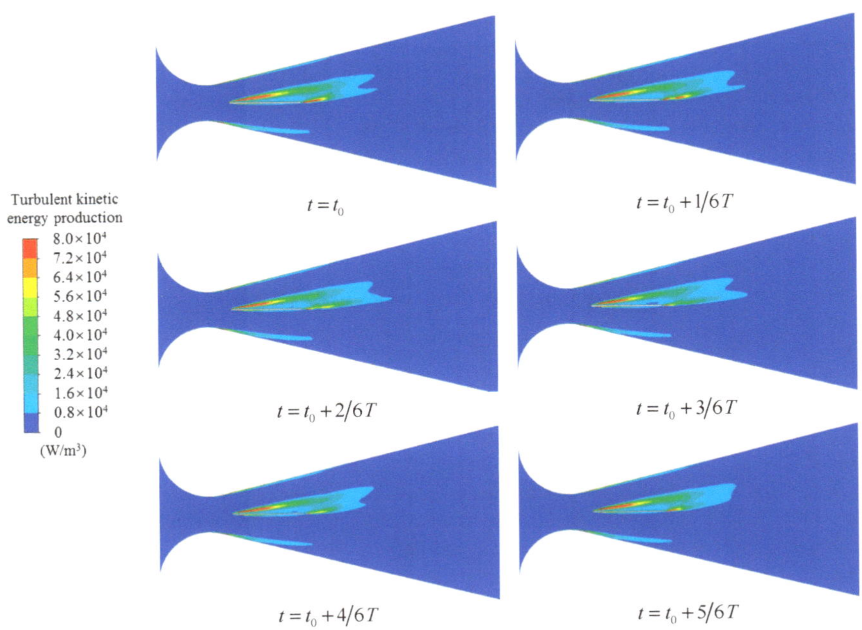

Figure 16. Distributions of turbulent kinetic energy production in vaned diffuser.

4. Concluding Remarks

From the numerical simulation of incompressible flow in a wide-angle of 28° two-dimensional diffuser with and without a short splitter vane, the following concluding remarks are drawn:

1. Improvement of pressure-recovery by the vane and oscillatory diffuser-flow reported in the experimental study done by Cochran & Kline (1958) are reasonably simulated.

2. The unsteady flow simulation shows that the lift force acting on the vane varies periodically in the opposite phase with the detachment point moved back and forth on a diverging wall since one vane is not sufficient to fully suppress the flow separation and the incoming main-flow shifts toward the other diverging wall in the diffuser.

3. As a role of splitter vane in the diffuser, "the lift force of the vane is a key factor" may be verified quantitatively from the present numerical simulation.

4. The local loss analysis shows that the turbulent kinetic energy production observed in mixing layers contributes most of the loss in the diffuser.

5. Distributions of time-averaged velocity and wall-pressure calculated from unsteady flow simulation are reproduced from steady flow simulation reasonably.

6. Consequently, the present numerical technique may be usable to investigate the flow character in a wide-angle diffuser with splitter vanes at a design stage.

Author Contributions: Investigation, X.M.; data curation, X.M.; writing—original draft preparation, X.M.; writing—review and editing, Z.Z., M.N., and S.L.; visualization, M.X., M.N., and S.L.; supervision, Z.Z., M.N., and S.L. All authors have read and agreed to the published version of the manuscript.

Funding: This research was funded by National Key R&D Program of China (2018YFB0606101) and National Natural Science Foundation of China (No. 51876100).

Conflicts of Interest: The authors declare that they have no conflict of interest.

References

1. Liu, S.; Wu, Y.; Chen, T.; Nishi, M. Development of Numerical Performance Test Stand for a Kaplan Turbine. *Int. J. Turbo Jet-Engines* **2009**, *24*, 253–262. [CrossRef]
2. Guénette, V.; Houde, S.; Ciocan, G.D.; Dumas, G.; Huang, J.; Deschênes, C. Numerical Prediction of a Bulb Turbine Performance Hill Chart through RANS simulations. In Proceedings of the 26th IAHR Symposium on Hydraulic Machinery and Systems, Beijing, China, 19–23 September 2012.
3. Ruopp, A.; Ruprecht, A.; Riedelbauch, S.; Arnaud, G.; Hamad, I. Development of a Hydro Kinetic River Turbine with Simulation and Operational Measurement Results in Comparison. In Proceedings of the 27th IAHR Symposium on Hydraulic Machinery and Systems, Montreal, QC, Canada, 22–26 September 2014.
4. Eckert, W.T.; Mort, K.W.; Jope, J. *Aerodynamic Design Guidelines and Computer Program for Estimation of Subsonic Wind Tunnel Performance*; NASA-TN-D-8243; NASA Ames Research Center: Moffett Field, CA, USA, 1976.
5. King, C.D.; Ölçmen, S.M.; Sharif, M.A.R.; Presdorf, T. Computational Analysis of Diffuser Performance for Subsonic Aerodynamic Research Laboratory Wind Tunnel. *Eng. Appl. Comput. Fluid Mech.* **2013**, *7*, 419–432. [CrossRef]
6. Vassiliev, V.; Irmisch, S.; Florjancic, S. CFD Analysis of Industrial Gas Turbine Exhaust Diffusers. In Proceedings of the ASME TURBO EXPO2002, Amsterdam, The Netherlands, 3–6 June 2002.
7. Cochran, D.L.; Kline, S.J. *The Use of Short Flat Vanes for Producing Efficient Wide-Angle Two-Dimensional Subsonic Diffusers*; NACA-TN-4309; National Advisory Committee for Aeronautics: Washington, DC, USA, 1958.
8. Moore, C.A., Jr.; Kline, S.J. *Some Effects of Vanes and of Turbulence in Two-Dimensional Wide-Angle Subsonic Diffusers*; NACA-TN-4080; National Advisory Committee for Aeronautics: Washington, DC, USA, 1958.
9. Reneau, L.R.; Johnston, J.P.; Kline, S.J. Performance and Design of Straight, Two-dimensional Diffusers. *ASME Trans. J. Basic Eng.* **1967**, *89*, 141–150. [CrossRef]
10. Cockrell, D.J.; Markland, E. A Review of Incompressible Diffuser Flow. *Aircr. Eng. Aerosp. Technol.* **1963**, *35*, 286–292. [CrossRef]
11. Sovran, G.; Klomp, E.D. Experimentally Determined Optimum Geometries for Rectilinear Diffusers with Rectangular, Conical or Annular Cross-Section. In *Fluid Mechanics of Internal Flow, Proceedings of the symposium on the fluid mechanics of internal flow, Warren, MI, USA, 1965*; Elsevier: New York, NY, USA, 1967; pp. 270–312.
12. Furuya, Y.; Fujimoto, T.; Yamazato, E.; Tsuzuki, I.; Nishiura, I. Performance of the Two-Dimensional Diffusers with Suction at the Entrance. *Bull. JSME* **1970**, *13*, 264–271. [CrossRef]

13. Nicoll, W.B.; Ramaprian, B.R. Performance of Conical Diffusers with Annular Injection at Inlet. *ASME Trans. J. Basic Eng.* **1970**, *92*, 827–835. [CrossRef]
14. Hoffmann, J.A. Effects of Free-stream Turbulence on Diffuser Performance. *ASME Trans. J. Fluids Eng.* **1981**, *103*, 385–390. [CrossRef]
15. Senoo, Y.; Nishi, M. Improvement of the Performance of Conical Diffusers by Vortex Generators. *ASME Trans. J. Fluids Eng.* **1974**, *96*, 4–10. [CrossRef]
16. Reichert, B.A.; Wendt, B. J Improving Curved Subsonic Diffuser Performance with Vortex Generators. *AIAA J.* **1996**, *34*, 65–72. [CrossRef]
17. Nishi, M.; Yoshida, K.; Morimitsu, K. Control of Separation in a Conical Diffuser by Vortex Generator Jets. *JSME Int. J. Ser. B Fluids Therm. Eng.* **1998**, *41*, 233–238. [CrossRef]
18. Feil, O.G. Vane Systems for Very-wide-angle Subsonic Diffusers. *ASME Trans. J. Basic Eng.* **1964**, *86*, 759–764. [CrossRef]
19. Majumdar, B.; Singh, S.N.; Agrawal, D.P. Flow Characteristics in a Large Area Ratio Curved Diffuser. *Proc. Inst. Mech. Eng. Part G J. Aerosp. Eng.* **1996**, *210*, 65–75. [CrossRef]
20. Chong, T.P.; Joseph, P.F.; Davies, P.O.A.L. Design and Performance of an Open Jet Wind Tunnel for Aero-Acoustic Measurement. *Appl. Acoust.* **2009**, *70*, 605–614. [CrossRef]
21. Chiekh, M.B.; Béra, J.-C.; Sunyach, M. Synthetic Jet Control for Flows in a Diffuser: Vectoring, Spreading and Mixing Enhancement. *J. Turbul.* **2003**, *4*, 37–41. [CrossRef]
22. Sun, C.L.; Lin, G.C. The Flow Rectification of a Vaned Micro diffuser: A Numerical Investigation. *J. Chin. Inst. Eng.* **2010**, *33*, 1–13. [CrossRef]
23. Ghose, S.; Kline, S.J. The Computation of Optimum Pressure Recovery in Two-Dimensional Diffusers. *ASME J. Fluids Eng.* **1978**, *100*, 419–426. [CrossRef]
24. Bardina, J.; Lyrio, A.; Kline, S.J.; Ferziger, J.H.; Johnston, J.P. A Prediction Method for Planar Diffuser Flows. *ASME J. Fluids Eng.* **1981**, *103*, 315–321. [CrossRef]
25. Kline, S.J. On the Nature of Stall. *ASME Trans. J. Basic Eng.* **1959**, *81*, 305–320. [CrossRef]
26. Senoo, Y.; Nishi, M. Prediction of Flow Separation in a Diffuser by a Boundary Layer Calculation. *ASME Trans. J. Fluids Eng.* **1977**, *99*, 379–389. [CrossRef]
27. Menter, F.R.; Kuntz, M.; Langtry, R. Ten Years of Industrial Experience with the SST Turbulence Model. *Turbul. Heat Mass Transf.* **2003**, *4*, 625–632.
28. El-Behery, S.M.; Hamed, M.H. A Comparative Study of Turbulence Models Performance for Separating Flow in a Planar Asymmetric Diffuser. *Comput. Fluids* **2011**, *44*, 248–257. [CrossRef]
29. ANSYS Help. *ANSYS Help Viewer 14.5*; SAS IP, Inc.: Cary, NC, USA, 2012.
30. Simpson, R.L. Review—A Review of Some Phenomena in Turbulent Flow Separation. *ASME Trans. J. Fluids Eng.* **1981**, *103*, 520–533. [CrossRef]
31. Lu, G.; Zuo, Z.; Liu, D.; Liu, S. Energy Balance and Local Unsteady Loss Analysis of Flows in a Low Specific Speed Model Pump-Turbine in the Positive Slope Region on the Pump Performance Curve. *Energies* **2019**, *12*, 1829. [CrossRef]
32. Wilhelm, S.; Balarac, G.; Métais, O.; Ségoufin, C. Analysis of Head Losses in a Turbine Draft Tube by Means of 3D Unsteady Simulations. *Flow Turbul. Combust.* **2016**, *97*, 1255–1280. [CrossRef]

processes

MDPI

Article

Study of the Affinity Law of Energy and Cavitation Characteristics in Emergency Drainage Pumps at Different Rotating Speeds

Weidong Cao [1,2,*] and Jiayu Mao [1,2]

1 Research Institute of Fluid Engineering Equipment Technology, Jiangsu University, Zhenjiang 212013, China; mjy1006a@163.com

2 China National Research Center of Pumps, Jiangsu University, Zhenjiang 212013, China

* Correspondence: cwd@ujs.edu.cn; Tel.: +86-1395-281-6468

Received: 2 November 2019; Accepted: 2 December 2019; Published: 6 December 2019

Abstract: The affinity law is widely used in pump design and experiments. The applicability of the affinity law in an emergency drainage pump at different rotating speeds was studied. Experiments and numerical simulation through ANSYS CFX (Computational Fluid Dynamics X) 15.0 software were used to research the affinity law characteristics. Results show that the simulation of characteristics is basically consistent with the experimental curves. In small flow rate conditions, due to the existence of obvious differential pressure between the pressure side and the suction side in the impeller blade tip area, the leakage flow occurs at the tip clearance, which collides with the main stream at the inlet and generates vortices at the leading edge of the impeller. The tip leakage flows of the pump at four different rotating speeds were compared, and it was found that the tip leakage increased with increasing rotation speed, and at the same rotation speed, the tip leakage flow was large in the small flow rate condition, which led to the simulation value of the characteristics being greater than the scaling value. As the flow rate increased, the anti-cavitation performance of the pump became worse and the hydraulic loss was larger, so the pump's performance curve deviated from the scaling curve.

Keywords: affinity; emergency drainage pump; rotating speed; cavitation

1. Introduction

High-speed emergency drainage pumps have been widely applied in rapid responses to solve urgent water-logging problems because they are of light weight and have powerful water transmission capacity [1]. Similar to ordinary centrifugal pumps, the main design methods of drainage pumps are the affinity law and the velocity coefficient method, but with the in-depth study of the traditional basic theory and design methods, some new design methods have been created, such as the increasing flow rate design method, clogging flow channel design method, non-clogging pump design method, two-phase flow design method, and so on [2]. Some pumps are difficult to experiment on because of their large size and higher rotating speed, the traditional affinity law is still a dominant method that is widely used in the process of designing pump products. This method can shorten the product development cycle; however, the applicability of the affinity law for high-speed, axial-flow emergency drainage pumps is challenged. Under the conditions where the pumps runs at small flow rates and where high lift or cavitation occurs at large flow rates, the similarity of the internal flow in high-speed, axial-flow emergency drainage pumps seems to be quite different from that of ordinary centrifugal pumps, making their internal flow characteristics worth studying.

Shi verified the similarity of the scaling model of a typical submersible well pump with a scaling factor of 0.66, and it was found that the designed pump and the model pump basically showed consistent variation tendency for their head, efficiency, and shaft power under 0.4–1.6-times rated flow conditions, and the highest efficiency points were both achieved under 1.2-times rated flow rate conditions, which means the predicted performance met similar conversion rules [3]. Si studied the applicability of the similarity law under the condition of gas–liquid two-phase flow in centrifugal pumps, and the comparisons between the experimental and simulated results showed that the greater the inlet gas volume fractions was, the worse the applicability to the similarity law [4]. Naoki studied the air–water two-phase flow performance in a conventional centrifugal pump and found the similarity law of the impeller diameter was experimentally confirmed, even in two-phase flow condition; influences of blade height on air–water two-phase flow performances indicated little difference from the similarity law [5]. Zhu studied the hydraulic characteristics between prototype and model pumps in pumping stations, because the local hydraulic loss was the main part in the inlet and outlet channels of the pumping station, but during the efficiency correction, the local loss had not been separated; in fact, it did not need to be converted and had nothing to do with the scale effect [6]. Chen analyzed the approximation of hydraulic simulation, mechanical constraints, and geometric similarity between the model and the prototype for pumps and pump sets. The hydraulic efficiency expression was deduced in terms of its original definition, where pump and pump-set efficiency conversion formulas were derived, which concerned hydraulic friction loss only or all hydraulic losses, including shape resistance and shock vortex resistance [7].

Zhen studied the rotating stall characteristics of an axial-flow pump. The results showed that there were two reversed-flow field areas in the leading edge close to the shroud and in the trailing edge close to the hub of the blade suction surface under rotating stall conditions, and the dominant frequency under the designed conditions in the impeller inlet and the middle of the guide vane was blade passing frequency. However, the dominant frequency under the designed conditions in the impeller outlet was guide passing frequency, because of rotor–stator interaction [8]. Li studied the flows in an axial-flow pump at the design flow rate for five different tip gap sizes by using large eddy simulation. The results indicated that the head, shaft power, and efficiency of the pump decreased as the tip gap size increased, and the velocity of the tip leakage flow in the tip gap increased gradually along the radial direction; when the tip gap size was larger than 1.0‰ of the impeller diameter, the dominant tip leakage vortex extended to the pressure side of the neighbor blade, and the tip separation vortex and several secondary vortices were also found [9]. Zhang simulated and analyzed the formations of a three-dimensional tip leakage vortex cavitation cloud and the periodic collapse of vortex-induced suction side perpendicular cavitation vortices. The improved turbulence model and the homogeneous cavitation model were validated by comparing the simulation results with an experiment by measuring unsteady cavitation shedding flow around a hydrofoil. The unsteady cavitation cloud occurred near the blade trailing edge and the shapes of the sheet cavitation and normal cavitation fluctuated [10]. Zhang also studied the tip leakage vortex in an axial flow pump under small flow rate conditions. Owing to the large pressure difference between the pressure side and suction side at the tip, the axial velocity of flow in the gap was negative, and the absolute value of the axial velocity increased gradually from the pressure side to the suction side. High-speed photography experimental results showed that as the cavitation number decreased gradually, the leakage vortex cavitation was initiated in the tip clearance under small flow rate conditions [11]. Wang studied the internal flow field in a pump installation, where an unstable pattern behind the pump and significant numbers of vortices occurred in the outlet passage, the pressure distribution was irregular in the middle of the blade suction side, and a high-pressure zone existed on the blade pressure side near the leading edge and on the blade suction side near the trailing edge [12]. Zhou studied the stability of the reverse-power generation of a pumping station; the frequency of the pressure pulsation was influenced by the frequency of the runner and was concentrated at low frequencies. The flow pressure fluctuation in the middle and edge of the inlet of the runner was obvious, and the maximum pressure pulsation occurred in the middle of the runner outlet. The amplitude of the pressure pulsation was nearly 3 times that of the outlet edge of the runner, and it was nearly 2 times the amplitudes of the middle and edge of the inlet of the runner [13].

When the pumps work in off-design conditions, it is easy to observe stall, leakage, cavitation, and other phenomena, which lead to a certain deviation between the scaling value and the simulation value after the affinity law is used. The above scholars have mainly focused on the study of the scaling methods of hydraulic characteristics of the actual pump and the model pump. The affinity law was modified according to improving various losses, such as hydraulic loss and volume loss, and various scaling formulas have been proposed to improve the affinity law of pumps. However, there are few studies on the reasons for the differences between the scaling results and the experimental results after the affinity law is used, particularly the study of the affinity law characteristics in emergency drainage pumps at different rotating speeds. In order to study the influencing factors on the affinity law and to figure out the applicability of the similarity theory at variable operating conditions, commercial ANSYS CFX 15.0 software was adopted to simulate the whole flow field of an emergency drainage pump at different flow rates and rotating speeds. The internal flow characteristics were analyzed, and the simulation results were compared with those of experiments. The distributions of cavitation volume fraction in the impeller under different flow rates and different cavitation numbers, the variations of cavitation flow fields in the blade tip region, and the relationship between leakage loss and affinity law in the pump under different rotating speeds were all analyzed. Through analyzing the leakage flow with small flow rates and cavitation flow with large flow rates, affinity law mechanisms related to energy and cavitation characteristics will be discussed.

2. Numerical Model and Method

2.1. Flow Control Equations

The basic equations of fluid mechanics include the continuity equation, momentum equations, and energy equation. The fluid flow in pumps changes with time and space, and it is a complex, three-dimensional, and unsteady process. The water transported by the emergency drainage pump can be considered an incompressible medium. The equations adopted are as follows.

The continuity equation is the mathematical expression of the law of mass conservation:

$$\frac{\partial \rho}{\partial t} + \frac{\partial(\rho v_x)}{\partial x} + \frac{\partial(\rho v_y)}{\partial y} + \frac{\partial(\rho v_z)}{\partial z} = 0 \tag{1}$$

where ρ is the fluid density; t is the time; v_x, v_y and v_z represent the velocity components in x, y, and z directions.

The momentum equations, named N-S equations, are the mathematical expression of the law of momentum conservation. Although the N-S equations can describe the motion of turbulence, due to the huge difference between the time and space characteristic scales in turbulent flow fields, it is difficult to solve N-S equations in actual flows. The Reynolds time-averaged equations are generally used, and the tensor expression is as follows,

$$\frac{\partial}{\partial x_j}\left(\rho u_i u_j\right) = -\frac{\partial P}{\partial x_i} + \frac{\partial}{\partial x_j}\left(\mu \frac{\partial u_i}{\partial x_j} - \rho \overline{u_i' u_j'}\right) + S_i \tag{2}$$

where u_i is the Reynolds average velocity; u_i' is the velocity fluctuation; μ is the kinematic viscosity.

The energy equation, meaning the amount of energy change in the controlling body, is equal to the sum of the heat entering the controlling body minus the heat escaping from the controlling body per unit time, and the equation is expressed as,

$$\frac{\partial(\rho T)}{\partial t} + \frac{\partial(\rho v_x T)}{\partial x} + \frac{\partial(\rho v_y T)}{\partial y} + \frac{\partial(\rho v_z T)}{\partial z} = div\left(\frac{k}{c_{\mathrm{p}}} \mathrm{grad} T\right) + S_{\mathrm{T}} \tag{3}$$

where T is the temperature; k is the heat transfer coefficient of the medium; c_{p} is the specific heat capacity; S_{T} is the heat source inside the fluid.

2.2. Turbulence Mode

According to the number of differential equations used, the commonly used turbulence models can be divided into the zero-equation model, one-equation model, two-equation model, and multiple-equation model. The two-equation turbulence model is widely used at present. Strong shear stress and a reverse pressure gradient exist in the emergency drainage pump. Therefore, the SST (Shear Stress Transport) k-ω model is selected to close the equations, with its turbulent energy k and turbulent fluctuation frequency ω equation.

Turbulent kinetic energy k equation:

$$\frac{\partial(\rho k)}{\partial t} + \frac{\partial(\rho k u_i)}{\partial x_i} = \frac{\partial}{\partial x_j}[(\mu_m + \frac{\mu_T}{\sigma_k})\frac{\partial k}{\partial x_j}] + P_k - \beta' \rho k \omega \tag{4}$$

Turbulent pulsation frequency ω equation:

$$\frac{\partial(\rho \omega)}{\partial t} + \frac{\partial(\rho \omega u_i)}{\partial x_i} = \frac{\partial}{\partial x_j}[(\mu_m + \frac{\mu_T}{\sigma_\omega})\frac{\partial \omega}{\partial x_j}] + \alpha \frac{\omega}{k} P_k - \beta \rho \omega^2 \tag{5}$$

The relationship between turbulent viscosity μ_T, turbulent kinetic energy k and turbulent pulsation frequency ω is,

$$\mu_T = \rho \frac{k}{\omega} \tag{6}$$

where P_k is the turbulence generation rate, and also the pressure generation term produced by the velocity gradient; other variables are obtained using the N-S formula. In addition, the constant terms in the model are $\beta' = 0.09$, $\alpha = 5/9$, $\beta = 0.075$, and $\sigma_k = 1$, $\sigma_\omega = 2$.

2.3. Hydraulic Model

The emergency drainage pump studied in this paper is mainly used in urban emergency drainage projects. Its compact structure makes it convenient to be transported and used. As shown in Figure 1, the pump mainly includes the motor, inlet section, impeller, and guide vane. The design parameters are as follows: rated flow Q_{opt} is 220 m^3/h, head H is 7.5 m, rated rotating speed is 3800 r/min, power is 15 kW, the specific speed is 756, the impeller blade number is 4, the diameter of the impeller is 131.6 mm, the hub diameter of the impeller is 61.8 mm, the clearance between impeller and drum wall is 0.7 mm, the guide vane blade number is 6, the distance between inlet edge of the guide vane and outlet of the impeller is 5 mm, and the diameter at the outlet of the guide blade is 137 mm.

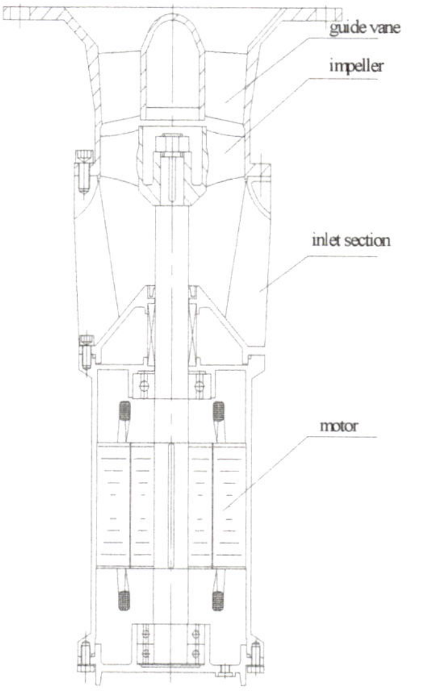

Figure 1. Emergency drainage pump.

2.4. Mesh Division

Commercial software Icem-cfd(The Integrated Computer Engineering and Manufacturing code for Computational Fluid Dynamics) 15.0 was used to divide the flow field into structured meshes, which are shown in Figure 2. In order to control the mesh distribution in the boundary layer near the wall, the O-type topology is selected. The mesh in the blade boundary layer and its adjacent zone is controlled by the O-grid, and the mesh in the impeller is encrypted. A ten-layer grid is placed in the tip clearance to better capture the flow. As shown in Table 1, in order to obtain the appropriate number of meshes, five grid sizes with different precision of division were compared, and the simulated head and efficiency values were chosen as reference indexes, as shown in Figure 3. When the mesh number reaches over 2.75 million, the simulated results of efficiency (η) and head (H) do not fluctuate obviously with the changing of numbers of meshes, and the total simulation mesh number is 2,756,481.

Figure 2. Structured mesh in the computational domain.

Table 1. Comparison of mesh numbers.

Scheme	Node Number	Mesh Number
1	1983484	2021348
2	2302738	2527547
3	2624871	2756481
4	2995718	3122487
5	3327496	3463789

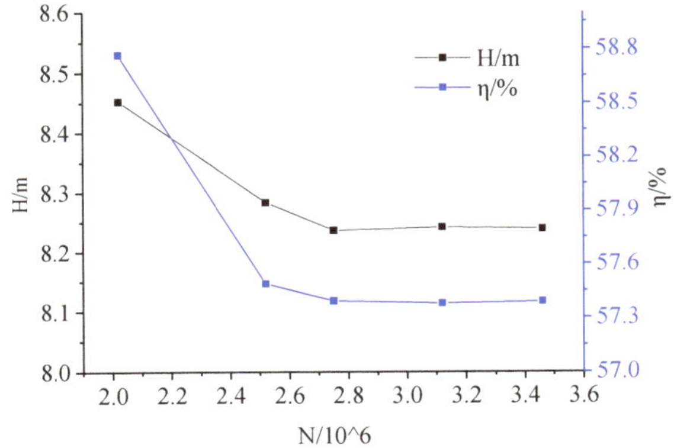

Figure 3. Mesh independence analysis.

2.5. *Simulation Methods and Boundary Conditions*

The SST *k-ω* turbulent model and CFX 15.0 software were adopted in this paper. A steady numerical simulation was adopted. For the boundary conditions, the inlet was set as the total pressure, while the outlet was set as mass flow rate. The solid wall was set as a non-slip and adiabatic wall surface. The roughness of the main flow passage components, such as the impeller and guide vane, was set at 0.025 mm. The standard wall function was adopted for the near wall surface. The water in the impeller was set as the rotation area, while the rest of the water body was set as the static area, and the outer wall surface of the impeller was set as the counter-rotating wall. The coupling interfaces between the rotating areas and the static areas were set in transient frozen rotor mode, and the coupling interfaces between the static regions were not set in any mode. The convergence accuracy was set at 10^{-5} with the higher-order solution format.

The default ZGB(Zwart-Gerber-Belamri) cavitation model was chosen to simulate the cavitation flow, total pressure was selected as the inlet boundary condition, the volume fraction of the liquid phase was set to 1, the volume fraction of the gas phase was set to 0, and mass flow rate was selected as the outer boundary condition. The medium for computational fluid was 25 °C water. The critical cavitation pressure was set to 3574 pa according to the temperature of the saturated steam pressure, the average diameter of cavity was set to 10^{-6} mm, and the convergence residual was set to 10^{-5}.

3. Characteristics and Analysis

3.1. *Characteristics Comparison*

The experimental scheme and the tested pump are shown in Figure 4. The main instruments included the LWGY-150A0A3T turbine flow meter, the precision level for which was 0.5%; the ϕ150 flow rate control valve; and the WT2000 pressure transmitter whose precision level was 0.2%. The precision level for the current and voltage sensors was 0.5%, and the HSJ-2010 hydraulic mechanical comprehensive test instrument was used, the precision level for which was 0.2%.

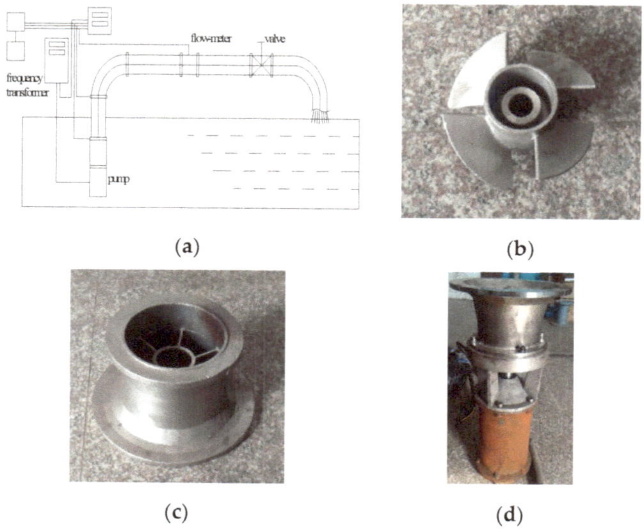

Figure 4. Experimental scheme and the tested pump: (**a**) experimental scheme; (**b**) impeller; (**c**) guide vane; (**d**) pump.

As shown in Figure 5, the simulation results are basically consistent with the variation trend of the experiment curves according to the performance comparison of the emergency drainage pump at the design rotating speed of 3800 r/min. The simulation results are almost same as those of the experiments at low flow rates. When the flow rate changes to larger than 300 m^3/h, the head and efficiency of the simulation results are gradually higher than those of the experiments. At the design flow rate (1.0Q_{opt}), the relative head and efficiency errors between the simulation and experiment are 2.2% and 1.5%, respectively. The head increases when the flow rate decreases from 1.0Q_{opt} to 0.8Q_{opt}, but the change is small. As the flow rate continues to decrease from 0.8Q_{opt} to 0.6Q_{opt}, the head shows a significant increasing trend. However, at large flow rates of 1.0Q_{opt} and 1.2Q_{opt}, the experimental results and the simulation results deviate greatly, and the simulation values of head and efficiency are greater than the experimental values.

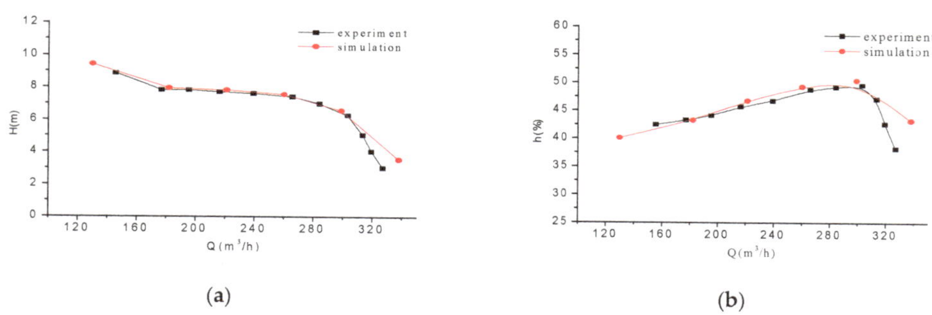

Figure 5. Comparison of performance curves: (**a**) *Q-H*(Flow rate-Head) curve; (**b**) *Q-η* (Flow rate-Efficency) curve.

3.2. Analysis of the Applicability of the Similarity Theory

The similarity theory is widely used in the pump design process and in experiments. Because the similarity theory is based on a hypothesis, it is empirical; the similarity theory can only be used conditionally in practical applications. The results show that if the size and speed of the model pump and real pump are similar, when two similar geometric pumps work in similar conditions, their efficiency can be thought of as equal. If the corresponding size of the pumps is equal, or for the same pump and the same transmission medium, the affinity law can be simplified as,

$$Q/Q' = n/n',\ H/H' = (n/n')^2, P/P' = (n/n')^3 \tag{7}$$

where Q is the flow rate in m^3/h, n is the pump speed in r/min, H is the pump head in m, and P is the pump shaft power in kW. The above formulas, known as scaling laws, represent the relationship between the performance parameters when the pump speed changes. During the experiments, the pump speed changes under different working conditions; in general, the data at each experimental speed must be converted to the rated speed. The scaling method relies on the above scaling laws.

The high-speed emergency drainage pump studied in this paper will be applied in different conditions by adjusting the rotating speeds. Based on the experimental data and a rotating speed of 3800 r/min, similar flow rates and heads at rotating speeds of 3600 r/min, 3400 r/min, and 3200 r/min are obtained through Equations (7). Additionally, the true experiment results at rotating speeds of 3600 r/min, 3400 r/min, and 3200 r/min are given and compared, as shown in Figure 6a. The experiment results at rotating speeds of 3600 r/min and 3400 r/min are consistent with the corresponding scaling results under design flow rates, and the experimental value is slightly larger than the scaling value for small flow rates. However, when the rotating speed drops to 3200 r/min, there is a large deviation between the experimental result and the scaling result, the experimental values are far greater than the scaling values, indicating that the scaling of the affinity law has a certain scope of application. Beyond this scope, the affinity law will not be applicable.

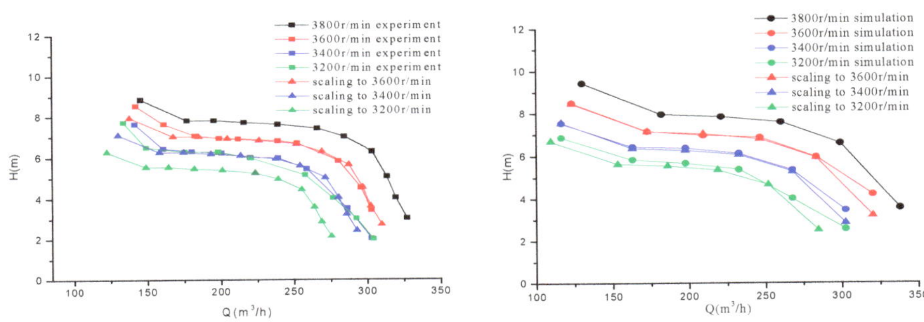

(a) (b)

Figure 6. Comparison of head curves: (**a**) experimental scaling results and comparison; (**b**) simulation scaling results and comparison.

The simulation heads at rotating speeds of 3800 r/min, 3600 r/min, 3400 r/min, and 3200 r/min are shown in Figure 6b. Based on the simulation head at the rotating speed of 3800 r/min, the heads at rotating speeds of 3600 r/min, 3400 r/min, and 3200 r/min are also obtained through Equation (7). The trends in Figure 6b are generally consistent with those in Figure 6a. When the rotating speed drops to 3200 r/min, the simulation result and the scaling result also show great deviation. This indicates that the affinity law can only be applied to the emergency drainage pump within the range of a 15% speed reduction.

3.3. *Internal Flow Field Analysis*

3.3.1. Surface Pressure Distribution of the Impeller

Figure 7 shows the static pressure on the impeller blades one the pressure side at different flow rates and a rotating speed of 3800 r/min. In general, the pressure on the pressure side of the blades increases gradually from the inlet to the outlet and from the hub to the rim. However, the pressure distribution under the flow rate of $0.8Q_{opt}$ does not apply to this rule. As shown in Figure 7a, the pressure on the pressure side gradually decreases from the inlet side to the outlet side, and a relatively high-pressure area forms in the inlet area. As the flow rate increases, the maximum pressure and the high-pressure area gradually decrease, and the pressure gradient on the blades becomes more obvious. At the flow rate of $1.2Q_{opt}$, the low pressure area begins to appear in the inlet area on the pressure side, and the low pressure area gradually expands with the further increase of the flow rate.

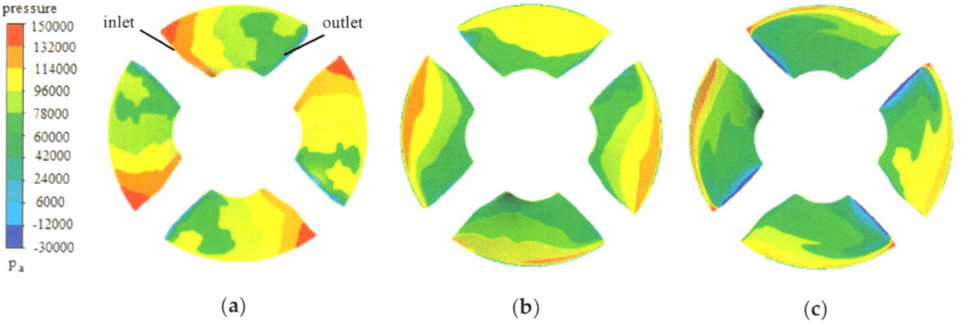

Figure 7. Static pressure on impeller blades on the pressure side: (**a**) $0.8Q_{opt}$; (**b**) $1.0Q_{opt}$; (**c**) $1.2Q_{opt}$.

Figure 8 shows the static pressure on the impeller blades on the suction side at different flow rates and a rotating speed of 3800 r/min. The static pressure on the suction side increases gradually from the inlet to the outlet. At the flow rate of $0.8Q_{opt}$, an obvious low pressure zone forms in the area near the inlet edge and on the rim near the inlet. This verifies that the cavitation phenomenon in the impeller of the emergency drainage pump usually happens at the inlet suction side and in tip clearance area. With the increase of the flow rate, the static pressure of the inlet area gradually increases. When the flow rate is $1.2Q_{opt}$, almost the entire outer edge of the blade becomes a relatively high-pressure area.

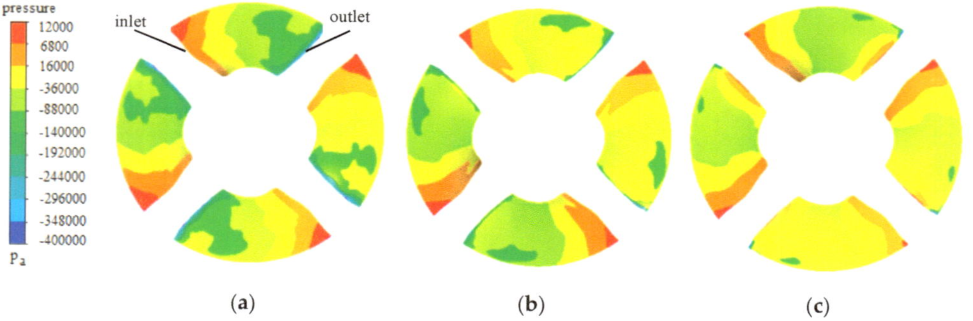

Figure 8. Static pressure on impeller blades on the suction side: (**a**) $0.8Q_{opt}$; (**b**) $1.0Q_{opt}$; (**c**) $1.2Q_{opt}$.

Comparing the pressure distribution, it can be seen that the static pressure on the pressure side and the suction side near blade inlet will change with the alternations of flow rate. When the pump works at the design flow rate, the flow path conforms to the blade shape, and the shock loss can be neglected. When the pump performs under low flow rate, a negative angle exists between the flows and the blade, flow separation on the suction side happens, and a low pressure area appears at the inlet area on the suction side. When the pump performs under a higher flow rate, a positive angle exits between the flows and the blade, flow separation occurs on the blade pressure side, and a low pressure area appears at the inlet area on the pressure side.

3.3.2. Streamline between Impeller Blades

Because there is a clearance between the impeller and the cylinder wall, due to the pressure difference between the blades on the pressure and suction side, leakage flow occurs at the tip region and interacts with the incoming flow, then vortices form, which result in a very complex flow in the tip regions.

As shown in Figure 9, the radial coefficient *span* of the blade is defined as the dimensionless distance of the blade from the hub along the radial direction to the edge of the wheel rim.

$$span = (r - r_h)/(r_t - r_h) \tag{8}$$

where r is the radius in mm, r_t the impeller rim radius in mm, and r_h the impeller hub radius in mm.

Figure 9. Impeller blades with different radial coefficients.

Figure 10 shows the streamline on the expansion plane as *span* = 0.5 in the impeller and guide vane, the rotating speed is 3800 r/min, and the flow rate is $0.8Q_{opt}$. Flow separation happens at the inlet edge of the impeller blade on the suction side and forms a low pressure area. There are strong vortices near the guide vane blades on the suction side. With the development of these new vortices, the vortices gradually fall off from the guide vane blades and spread to whole outlet. At the flow rate of $1.0Q_{opt}$, there is no flow separation in the impeller and the flow is stable. There are a few vortices between the guide vane blades on the suction side near the outlet. At the flow rate of $1.2Q_{opt}$, the flow separation is relatively weak and occurs on the guide vane blades on the pressure side in the inlet area.

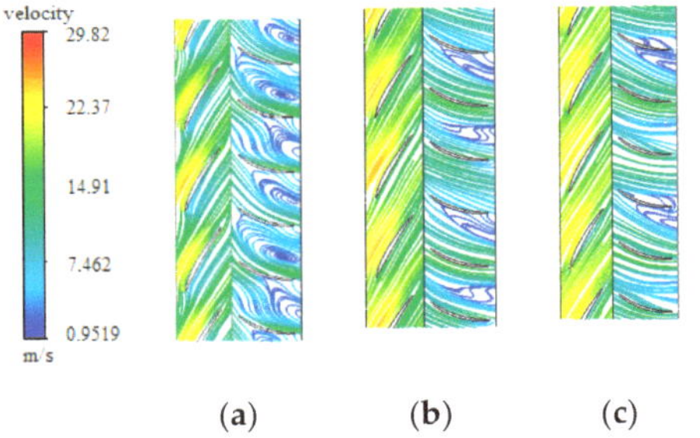

Figure 10. Streamline in the pump: (**a**) $0.8Q_{opt}$; (**b**) $1.0Q_{opt}$; (**c**) $1.2Q_{opt}$.

3.3.3. Turbulent Kinetic Energy

The turbulent kinetic energy represents the mechanical energy loss of the fluid. Figure 11 shows the distribution of turbulent kinetic energy at the axial section of the pump at the flow rates of $0.8Q_{opt}$, $1.0Q_{opt}$, and $1.2Q_{opt}$ respectively, at a rotating speed of 3800 r/min. The turbulent kinetic energy in the pump is mainly distributed in the impeller, and it gradually decreases as the flow rate increases from $0.8Q_{opt}$ to $1.2Q_{opt}$.Therefore, the flow in the impeller is the most disordered, and the hydraulic loss is the largest with small flow rates.

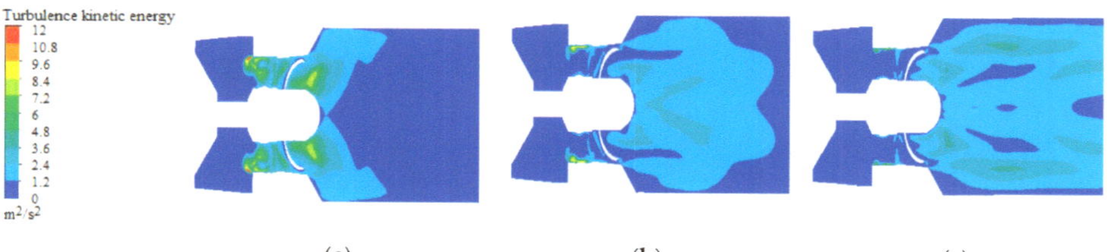

(a) (b) (c)

Figure 11. Distribution of turbulent kinetic energy at different flow rates: (**a**) $0.8Q_{opt}$; (**b**) $1.0Q_{opt}$; (**c**) $01.2Q_{opt}$.

Figure 12 shows the distribution of turbulent kinetic energy in the impeller at different rotating speeds, with a flow rate of $1.0Q_{opt}$. It can be seen from the figure that the turbulent kinetic energy gradually increases with increasing rotating speed at the same flow rate. The maximum turbulent kinetic energy appears in the rim region, because the pressure difference between the impeller pressure and the suction surfaces causes leakage and secondary flow at the clearance, and the interaction with the incoming flow forms vortices, which result in large hydraulic loss.

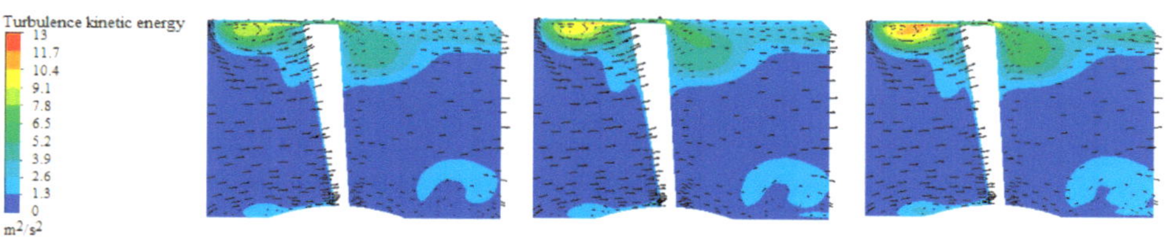

(a) (b) (c)

Figure 12. Turbulent kinetic energy distributions at different rotating speed, at a flow rate of $1.0Q_{opt}$: (**a**) 3400 r/min; (**b**) 3600 r/min; (**c**) 3800 r/min.

3.3.4. Clearance Leakage

Leakage flow exists in the clearance between the impeller rim and the cylinder wall. This is the volume of fluid passing through the tip gap section in unit time, which can be obtained by integrating the velocity component of the normal direction with the square value of the section. The leakage amount in the unit area of the feature section is defined as follows:

$$Q_1 = \int_A v_i ds \tag{9}$$

where v_i is the normal velocity through the characteristic section in m/s; A is the area of the characteristic section in m^2. The characteristic section is defined as the plane formed by two axial sections of the impeller and the extension surfaces of the pressure side and suction side, the impeller rim face, and the wheel chamber wall surface. The area containing the cross shaft and blade center is taken as the reference section. Based on this reference section, eight other characteristic sections are recorded by turning left or right by 10°, 20°, 30°, and 35°. A total nine characteristic sections are named as I, II, ... IX, as shown in Figure 13.

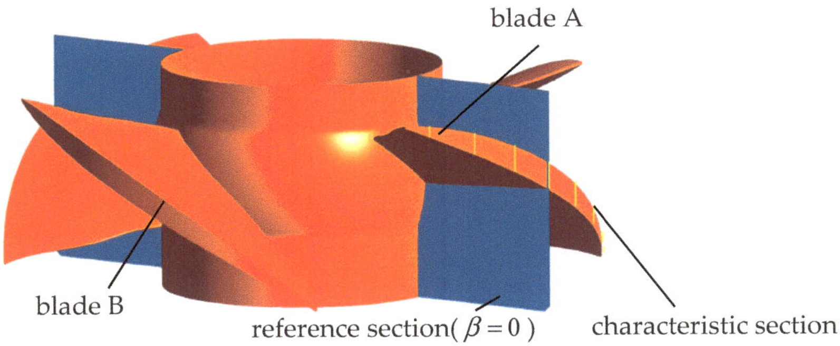

Figure 13. Characteristic sections.

Figure 14 shows the variation trends of tip clearance leakage at different flow rates, with a rotating speed of 3800 r/min. Since the area of the characteristic section increases first and then decreases with the change of the chord length coefficient, the leakage increases first and then decreases, forming a parabolic-shaped line. Because of the large pressure difference, the maximum leakage occurs near the reference section. The total amount of leakage under the flow rate of $0.7Q_{opt}$ is the largest. The flows in the tip clearance area are more complex and it is easy for back-flow to occur, leading to large hydraulic loss. As the flow rate increases, the total amount of tip leakage decreases, the tip leakage is more uniform with the change of chord length coefficient, and the energy loss is relatively small.

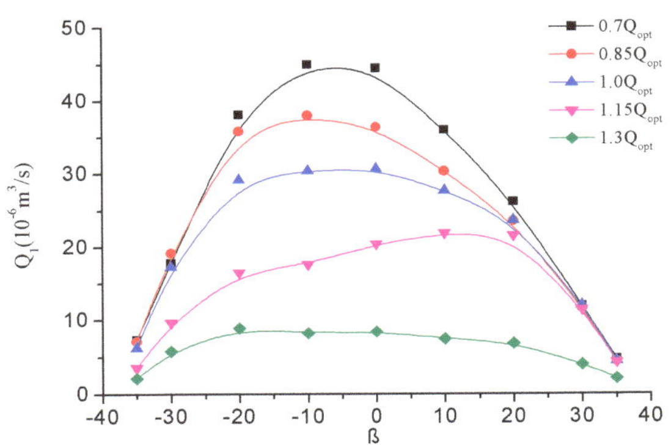

Figure 14. Leakage at different flow rates.

Figure 15 shows the variation trends of tip clearance leakage at different rotating speeds, with a flow rate of $1.0Q_{opt}$. The leakage increases first and then decreases with the change of chord length coefficient at various rotational speeds, and it gradually increases with the increase of rotating speed. The leakage reaches the maximum at the rotating speed of 3800 r/min. The blade tip leakage at different rotating speeds is disproportionate. As a result, the simulation value of the head at each speed is slightly higher than the scaling value; that is to say, the leakage does not meet the affinity law.

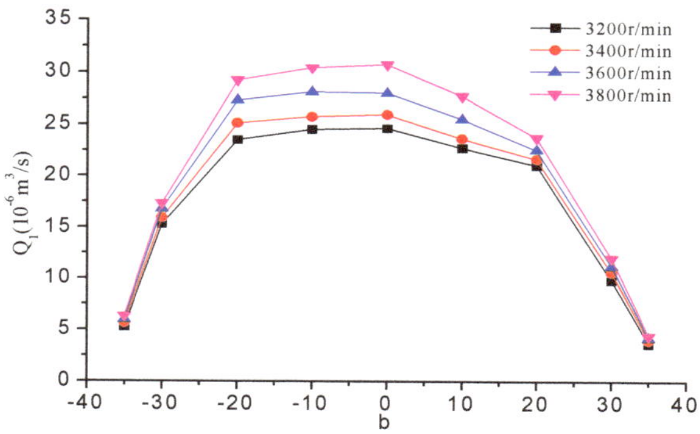

Figure 15. Leakage at different rotating speeds.

4. Analysis of Cavitation

4.1. Cavitation Characteristic Curve

The ZGB model is used for the cavitation model, and its expression is as follows,

$$m = \begin{cases} F_{\mathrm{v}} \frac{3\alpha_{\mathrm{ruc}}(1-\alpha_v)\rho_v}{R_B}\sqrt{\frac{2}{3}\frac{|p_v-p|}{\rho_l}} & (\ if\ p < p_v) \\ F_c \frac{3\alpha_v\rho_v}{R_B}\sqrt{\frac{2}{3}\frac{|p-p_v|}{\rho_l}} & (\ if\ p > p_v) \end{cases} \tag{10}$$

where p_v is the vaporization pressure of the liquid; R_B is the radius of a single bubble; a_v is the volume fraction of the gas; ρ_l, ρ_{v} are the density values of fluid and steam, respectively; the empirical condensation coefficient is $F_c = 0.01$, and the evaporation coefficient is $F_v = 50$.

The cavitation characteristic of the emergency drainage pump was numerically simulated through altering the inlet pressure. The cavitation characteristic curve was obtained at the flow rate of $1.0Q_{opt}$ and rotating speed of 3800 r/min, as shown in Figure 16. $\mathrm{NPSH_a}$ is defined as,

$$NPSH_a = \frac{p_{in} - p_v}{\rho g} + \frac{{v_{in}}^2}{2g} \tag{11}$$

where p_{in} is the static pressure at the pump inlet in $\mathrm{p_a}$; p_v is the saturated vapor pressure, 3574 $\mathrm{p_a}$; ρ is the density in kg/m^3; g is the acceleration of gravity in m/s^2; v_{in} is the velocity at the pump inlet in m/s. When the $\mathrm{NPSH_a}$ is higher than 12 m, the head remains almost the same and no cavitation occurs. With the decreasing value of $\mathrm{NPSH_a}$, the head begins to change. If $10 < \mathrm{NPSH_a} < 12$, due to a slight development of cavitation on the blades, the head slightly increases. The cavitation in the impeller starts but only a few bubbles exist, and the influence on the flow is small. If $\mathrm{NPSH_a} < 10$, the head drops obviously. When $\mathrm{NPSH_a} = 9.75$ m, the drop in head reaches about 3%. When $\mathrm{NPSH_a}$ decreases further, the head drops sharply. In general, the critical $\mathrm{NPSH_a}$ is defined as the head dropping by 3%. The emergency drainage pump's critical $\mathrm{NPSH_a}$ is value 9.75 m. Cavitation will not occur under submersible conditions at the design flow rate and rated rotating speed, however, this does not mean the pump is of good cavitation character. The compact structure design method and high rotating speed bring about a common cavitation character.

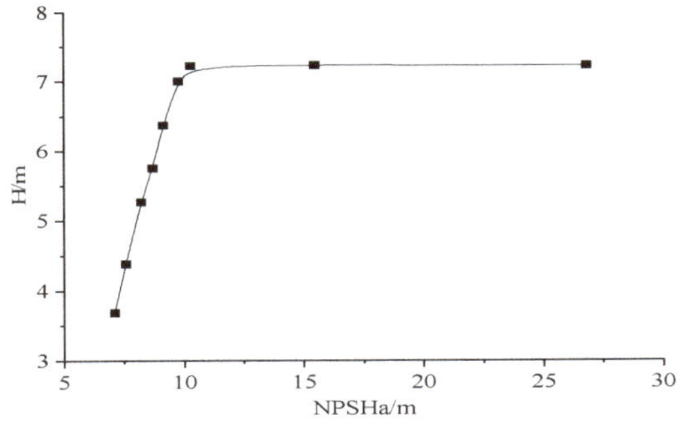

Figure 16. Cavitation performance at 3800 r/min and a flow rate of $1.0Q_{opt}$.

4.2. Cavitation Bubbles in the Impeller

Figure 17 shows the distribution of cavitation volume fraction on the surface of the impeller at a design flow rate of $1.0Q_{opt}$, at a rotating speed of 3800 r/min. The white part represents the contour with a cavitation vapor volume fraction of 0.1. If u_2 is the impeller rim rotating velocity in m/s, the cavitation number σ is defined as,

$$\sigma = \frac{p_{in} - p_v}{\rho {u_2}^2/2} \tag{12}$$

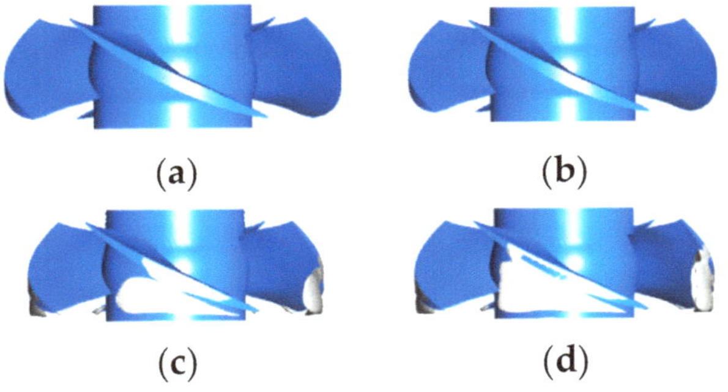

Figure 17. Vapor volume distribution on the impeller with different cavitation numbers: (**a**) σ = 1.16; (**b**) σ = 0.58; (**c**) σ = 0.43; (**d**) σ = 0.37.

When σ equals 1.16 or 0.58, the inlet pressure is relatively high and there is almost no distribution of cavitation bubbles on the surface of the impeller. When the inlet pressure decreases to σ = 0.43, a small amount of bubbles begin to appear near the rim of the impeller blades on the suction side close to the inlet. As the pressure further decreases, bubbles diffuse towards the outlet edges of the blades and the hub side. As σ = 0.37, the middle areas between blades are of low pressure, but there are no obvious cavitation bubbles. These phenomena indicate that cavitation of this emergency drainage pump mainly occurs in the tip clearance area of the impeller under the design flow conditions, manifesting as angular vortex cavitation, entrainment area cavitation, and tip leakage vortex cavitation. Among them, angular vortex cavitation and tip leakage vortex cavitation occur firstly in the tip clearance area. With the reduction of the cavitation number, the tip leakage vortex cavitation appears in the entrainment area near the suction side, and the cavitation area and cavitation bubbles increase rapidly.

Three cylindrical cross-sections of *span* are built at values of 0.75, 0.8, and 0.95 near the impeller blade rim. The cavitation vapor volume fractions are shown in Figure 18. As σ = 0.4, no cavitation vapor appears when the cross-section of *span* = 0.75, however, a small amount of cavitation vapor appears when the cross-section of *span* = 0.8, and a large amount of cavitation vapor appears when the cross-section of *span* = 0.95. This indicates that the cavitation forming in the impeller is tip leakage vortex cavitation.

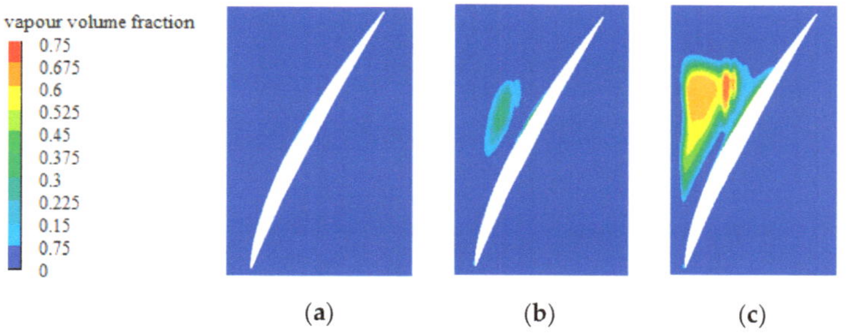

Figure 18. Vapor volume distribution on the impeller with different cross-section values: (**a**) *span* = 0.7; (**b**) *span* = 0.8; (**c**) *span* = 0.95.

4.3. Cavitation Performance at Different Flow Rates

Figure 19 shows the simulation cavitation characteristic curves of the emergency drainage pump at three different flow rates with a rotating speed of 3800 r/min. The critical $NPSH_a$ values are 6.42 m, 9.75 m, and 14.76 m, respectively, in the order of flow rate from low to high. The minimum critical $NPSH_a$ value for the three flow rates occurs at $0.7Q_{opt}$, which indicates that anti-cavitation performance is relative better with low flow rates. As the flow rate increases, the corresponding critical $NPSH_a$ value also increases, which means that the anti-cavitation performance of the pump decreases. It can be seen that the cavitation performances of the pump cannot meet the affinity law at different flow rates.

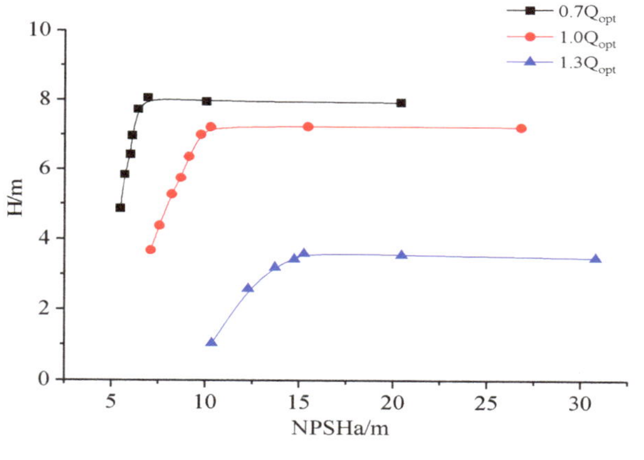

Figure 19. Cavitation performance at different flow rates.

Figure 20 shows the cavitation volume fraction of 0.1 in the impeller at the critical $NPSH_a$ value. The flow rates are $0.8Q_{opt}$, $1.0Q_{opt}$, and $1.2Q_{opt}$, and the rotating speed is 3800 r/min. Under a small flow rate of $0.8Q_{opt}$, not only there is leakage vortex cavitation and entrainment zone cavitation in the pump, but there is also attached cavitation on the suction side. At the design flow rate of $1.0Q_{opt}$, the cavitation vortex rope has a slender shape compared with those at small flow rate, and the vane tip leakage vortex cavitation and entrainment zone increase. At a large flow rate of $1.2Q_{opt}$, the cavitation gradually favors the suction side, and the attached cavitation occurs on the suction surfaces of the leading edge, which connects with leakage vortex cavitation and entrainment area cavitation. This is because the main flow forms the separation area on the blades' suction surfaces, leading to the weak cavitation resistance of the pump, especially with large flow rates. The head decreases with the deterioration of cavitation performance. Cavitation is prone to occur at high flow rates, which results in small deviations between the simulation value and the scaling value. As a result, the cavitation performances of the pump cannot meet the affinity law under different flow rates.

(a) (b) (c)

Figure 20. Cavitation distributions on the impeller at different flow rates: (**a**) $0.8Q_{opt}$; (**b**) $1.0Q_{opt}$; (**c**) $1.2Q_{opt}$

The Q-H characteristics of $1.15Q_{opt}$ and $1.3Q_{opt}$ at rotating speeds of 3600 r/min and 3400 r/min were simulated by adding the cavitation model. Based on the simulation result at 3800 r/min, similar flow rates and heads at rotating speeds of 3600 r/min and 3400 r/min are obtained again through Equation (7). Scaling and simulation results are all shown in Figure 21. Under large flow rates, the head simulation results at rotating speeds of 3600 r/min and 3400 r/min decrease significantly compared with Figure 6b, and the simulation results are closer to the scaling results. However, there are still some deviations. It is proven that the affinity law is slightly affected by cavitation at large flow rates.

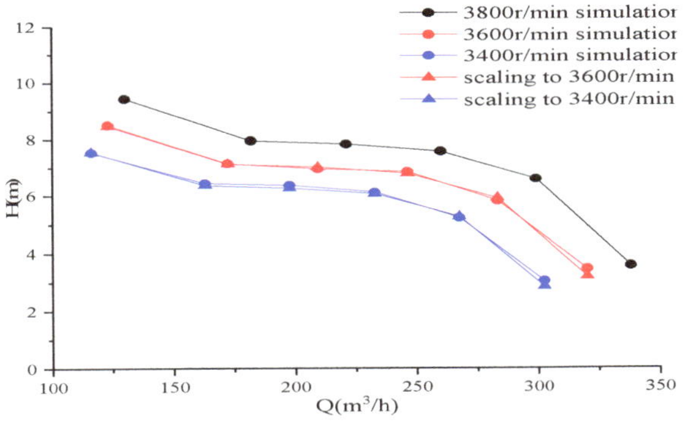

Figure 21. Q-H curves of the simulation with the cavitation model.

5. Conclusions

Through the comparison of experiments and numerical simulations, the flow details for the emergency drainage pump operating at high speed under the conditions of small and large flow rates are shown, along with cavitation. The reason for the deviation in similar theories for this type of pump is analyzed.

(1) For small flow rates, the static pressure on the pressure side of the blade is significantly higher than that on the suction side, and the static pressure from the inlet to the outlet of the impeller blade shows an upward trend. Because of the obvious pressure difference between the pressure side and the suction side in the tip area, there is tip leakage flow in the tip clearance area, which collides with the main flow at the inlet, generating an eddy current at the leading edge of the impeller and generating greater turbulent kinetic energy in the tip area. Leakage has a great influence on the external characteristic curve, especially at small flow rates.

(2) For large flow rates, with the decrease of $NPSH_a$, the main impeller cavitation is the tip leakage cavitation. The hydraulic performance of the emergency drainage pump is affected by cavitation.

(3) If the tip leakage is considered and the cavitation model is used for simulation, the affinity law can be improved a lot. The affinity law can only be applied to the emergency drainage pump within the range of up to 15% speed reduction

Author Contributions: W.C. designed and tested the scheme, performed simulations, and organized the paper. J.M. helped to analyze the data.

Funding: This work was supported by the National Key R&D Program of China (2018YFC0810506) and the Key R&D Program of Zhenjiang (SH2017049).

Conflicts of Interest: The author(s) declared no potential conflicts of interest with respect to the research, authorship, and/or publication of this article.

References

1. Li, B.L.; Wang, H.Y.; Tang, X.F. Impacts of solid particles on drainage pump performances of emergency drainage vehicles. *Chin. J. Constr. Mach.* **2016**, *14*, 26–31.
2. Wang, F.J. A variable-condition similar-design method in pump CAD system. *Trans. Chin. Soc. Agric. Eng.* **1996**, *12*, 35–40.
3. Shi, W.D.; Yang, Y.; Zhou, L.; Lu, W.; Pan, B. Verification of comparability and analysis of inner flow fields on scaling models of submersible well pump. *Trans. Chin. Soc. Agric. Eng.* **2017**, *33*, 50–57.
4. Si, Q.R.; Cui, Q.L.; Yuan, S.Q.; Zhang, K.Y.; Cao, R.; Tang, Y.F. Influence of inlet gas volume fraction on similarity law in centrifugal pumps under gas-liquid two-phase condition. *Trans. Chin. Soc. Agric. Mach.* **2018**, *49*, 107–112, 268.
5. Matsushita, N.; Furukawa, A.; Watanabe, S.; Okuma, K. Study on design of air-water two-phase flow centrifugal pump based on similarity law. *Int. J. Fluid Mach. Syst.* **2009**, *2*, 499–506. [CrossRef]
6. Zhu, H.G.; Yuan, S.Q.; Liu, H.L.; Shi, W. Research on conversion methods of hydraulic characteristics between prototype and model pumps and pumping stations. *Trans. Chin. Soc. Agric. Mach.* **2006**, *37*, 91–95.
7. Chen, S.S.; Ma, X.Z.; Chen, J.Q.; Zhou, Z.F.; He, Z.N. The conversion method of characteristic parameters of pump and pump set between prototype and model. *J. Yangzhou Univ. Nat. Sci. Ed.* **2015**, *18*, 45–48.
8. Zhen, Y.; Chen, Y.J.; Zhang, R.; Ge, X.F.; Sun, A. Analysis on unsteady stall flow characteristics of axial-flow Pump. *Trans. Chin. Soc. Agric. Mach.* **2017**, *48*, 127–135.
9. Li, Y.-J.; Shen, J.-F.; Yan, H.-J.; Zhang, Z.-M. Investigation of the effects of tip-gap size on the tip-leakage flow in an axial-flow pump using LES. *J. Hydraul. Eng.* **2014**, *45*, 235–242.
10. Zhang, D.S.; Shi, L.; Shi, W.D.; Zhao, R.J.; Wang, H.Y.; van Esch, B.B. Numerical analysis of unsteady tip leakage vortex cavitation cloud and unstable suction-side-perpendicular cavitating vortices in an axial flow pump. *Int. J. Multiph. Flow* **2015**, *77*, 244–259. [CrossRef]
11. Zhang, D.S.; Wu, S.Q.; Shi, W.D.; Pan, D.Z.; Wang, H.Y.; Li, T.T. Characteristics of tip leakage vortex cavitation in axial flow pump at small flow rate condition. *Trans. Chin. Soc. Agric. Eng.* **2013**, *29*, 68–75.
12. Wang, M.Q.; Li, Y.J.; Yuan, S.Q.; Meng, F. Runaway characteristics of two-way passage axial-flow pump installation. *J. Drain. Irrig. Mach. Eng.* **2018**, *36*, 384–390.
13. Zhou, Y.; Zhen, Y.; He, Z.W.; Sun, A.R.; Zhang, F.L.; Wang, H.L. Pressure fluctuation and fluid-solid coupling in reverse generation of large axial flow pump. *J. Drain. Irrig. Mach. Eng.* **2019**, *37*, 480–485.

processes

MDPI

Article

Numerical Study of the Unsteady Flow Characteristics of a Jet Centrifugal Pump under Multiple Conditions

Rong Guo [1,2,*], Rennian Li [1,2,*], Renhui Zhang [1,2] and Wei Han [1,2]

1 School of Energy and Power Engineering, Lanzhou University of Technology, Lanzhou 730050, China; zhangrh@lut.cn (R.Z.); hanwei@lut.cn (W.H.)
2 Key Laboratory of Fluid Machinery and Systems, Lanzhou 730050, China
* Correspondence: xizhilang88666@163.com (R.G.); lirn@lut.cn (R.L.); Tel.: +86-093-1297-3750 (R.G.)

Received: 28 September 2019; Accepted: 28 October 2019; Published: 1 November 2019

Abstract: To study the reasons for the low efficiency of jet centrifugal pumps (JCPs) and the mechanism of unsteady flow characteristics under multiple conditions, taking a JET750G1 JCP as the object, three-dimensional steady and unsteady numerical calculations of the model pump were carried out using the k–ω turbulence model. The transient fluctuation characteristics of the flow field in the major flow passage components and the spatial and temporal evolution laws of vortices in the rotor–stator cascades were analyzed. The accuracy of the numerical method was verified by experiments. The results show that there are various scales of flow distortion phenomena in the chamber of the JCP, such as eddies, blockage of the flow passage, recirculation, secondary flow, and circulation, which not only cause great hydraulic loss, but also destroy the flow stability, symmetry, and balance in the other flow passage components. This is an important reason for the obviously lower efficiency of a JCP compared to a general centrifugal pump. The spatial and temporal evolution laws of vortices in the rotor–stator cascades are mainly related to the relative positions of the impeller blades and guide vane blades. The formation mechanism of the unsteady flow field fluctuation characteristics of JCPs is mainly related to the number of blades in the rotor–stator cascades and the operation parameters of the pump. The fluctuation intensity of the flow field inside the impeller and guide vane is obviously greater than that in the other flow areas, reflecting that the rotor–stator interaction is the decisive factor affecting the unsteady flow characteristics of a JCP under multiple conditions.

Keywords: jet centrifugal pump (JCP); unsteady flow; numerical calculation; pressure fluctuation; rotor–stator interaction

1. Introduction

The flow passage components of a jet centrifugal pump (JCP) have relatively more complex internal flow, which leads them to have shortcomings such as low efficiency, easy cavitation, and high noise. However, JCPs have been widely used in many industries and fields because of their good low-level water-lifting capacity and self-priming performance [1,2]. Compared with a general centrifugal pump, the JCP has an additional self-circulating ejector device at the impeller inlet [3]; therefore, its working principle is rather special. As a self-priming pump, the analysis, optimization, and improvement of the self-priming performance of the JCP using numerical or experimental methods has been the research focus of scholars. Wang et al. [4] performed unsteady calculations of a typical self-priming centrifugal pump and simulated the gas–fluid two-phase flow in the self-priming process. It was found that the whole self-priming process of a self-priming pump can be divided into three stages: the initial self-priming stage, the middle self-priming stage, and the final self-priming stage. Wang et al. [5] numerically studied the influence of adding anti-circulation ribs at the back of the guide vane on gas–liquid mixing and separation in the pump chamber using the Euler–Euler multiphase model. The results showed that adding ribs could reduce the gas-phase volume fraction at the inlet of the ejector and the outlet of the nozzle, improving the gas-phase volume fraction at the outlet of the pump chamber and the self-priming performance. In addition, the authors of [6] optimized the structural parameters of the jet and impeller by an orthogonal experimental method, which improved the pump efficiency by about 5% under the design conditions. Liu et al. [7,8] studied the effects of different combinations of geometric parameters such as the total length, angle, outlet length, and outlet diameter of the ejector on the gas–liquid flow of the pump and obtained the optimum combination of parameters of the ejector. They added a diversion device at the outlet of the impeller of the JCP and found that the gap between the separating flow board and impeller could not only prevent the formation of circulation of the fluid, reduce hydraulic loss, and improve the efficiency of the pump, but also quicken the mixing of the gas and liquid near the gap and shorten the self-priming time of the pump. Wang et al. [9] studied the self-priming process of a JCP by numerical and experimental methods and analyzed the gas–liquid mixing, separation, and escape, as well as the variation of the velocity, pressure, and volume fraction of gas in the pump during the self-priming process. Zhang et al. [10] experimentally studied the cavitation and energy characteristics of a JCP at different installation heights and numerically analyzed flow characteristics at 0-m heights based on the k–ω turbulence model and Zwart–Gerber–Belamri cavitation model. They reported that high-speed reflux and strong shear flow in the ejector are important reasons for the low efficiency and easy cavitation of the pump.

With the rapid development of computer technology and numerical theory, computational simulation has been widely used in the field of hydraulic machinery. Scholars have performed many studies on the steady or unsteady flow field characteristics of JCPs [11–15], but they have not yet given a reasonable and full account of the low efficiency and the formation mechanism of unsteady fluctuation characteristics. In this paper, a numerical method was employed to study the unsteady flow characteristics of a JET750G1 JCP under multiple conditions. The reasons for the low efficiency and the formation mechanism of the unsteady flow characteristics of the pump are analyzed with respect to the flow pattern, distribution characteristics of the time-averaged flow field and fluctuation intensity, spatial and temporal evolution process of vortices in the rotor–stator cascades, and pressure fluctuation characteristics in the major flow passage components. This provides a theoretical basis for the design and optimization of JCPs with high efficiency and low energy consumption.

2. Research Method and Model

2.1. The Model Pump

The parameters of the model JCP are as follows: rated flow Q_d = 2.5 m^3/h, rated water head H_d = 23 m, rated efficiency η_d = 20%, rotational speed n = 2850 r/min, shaft-passing frequency SPF = 47.5 Hz, blade-passing frequency of impeller BPF_I = 285 Hz, and blade-passing frequency of guide vane BPF_G = 237.5 Hz. The main geometric parameters of the impeller and guide vane are listed in Table 1. Two- and three-dimensional structure diagrams of the model pump are shown in Figure 1.

Table 1. Parameters of the model jet centrifugal pump.

	Parameter	Value
Impeller	Inlet Diameter D_j (mm)	40
	Outlet Diameter D_2 (mm)	123
	Blade Number Z_1	6
	Blade Wrap Angle φ (°)	76
	Blade Outlet Width b_2 (mm)	5.3
Guide Vane	Base Diameter (mm)	125
	Outlet Diameter D_3 (mm)	64
	Blade Number Z_2	5

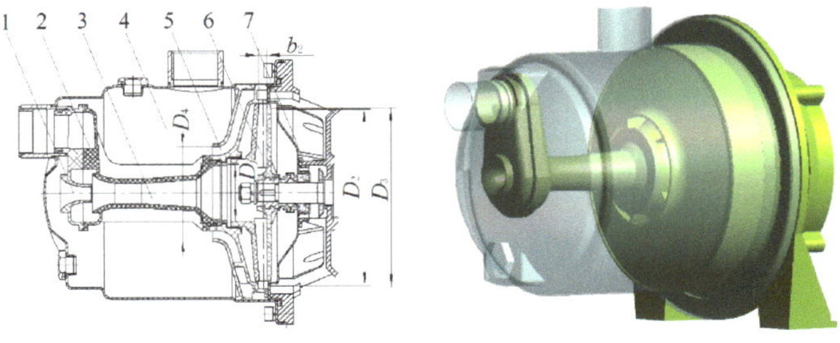

(a) (b)

Figure 1. Structural diagrams of the model pump. (**a**) Two-dimensional structure diagram; (**b**) Three-dimensional structure diagram. 1: Ejector nozzle; 2: Ejector; 3: Pump chamber; 4: Ejector tube; 5: Guide vane; 6: Impeller; 7: Pump cover.

2.2. Numerical Method

The whole computational domain included the impeller, guide vane, ejector, front pump cavity, back pump cavity, inlet pipe, outlet pipe, and pump chamber. ANSYS-ICEM (16.0 ANSYS, Inc., Canonsburg, PA, USA) was employed to construct the discrete computational domain; hexahedral structured meshes were employed to form the computational domain of the impeller, guide vane, and ejector nozzle and tube, and tetrahedral unstructured meshes were employed to form the other flow passage components. The grid independence verification is shown in Table 2, which is divided into six mesh schemes according to the density of the grid. While the accuracy and efficiency of the calculation were taken into account, comparing the head prediction value of different schemes, among schemes 4, 5 and 6, increasing the amount of mesh does not have too much of an influence on the result, which means that scheme 4 has enough accuracy. Therefore, scheme 4 was chosen as the final computational mesh to conduct the simulation. The meshes of the impeller, guide vane, and jet of the model pump are shown in Figure 2.

Table 2. The grid independence analyzation.

Schemes	Scheme 1	Scheme 2	Scheme 3	Scheme 4	Scheme 5	Scheme 6
Number of the grid	1,483,718	1,975,598	2,526,764	3,030,389	3,494,674	4,046,652
Head (m)	23.97	24.22	25.43	26.15	26.17	26.16

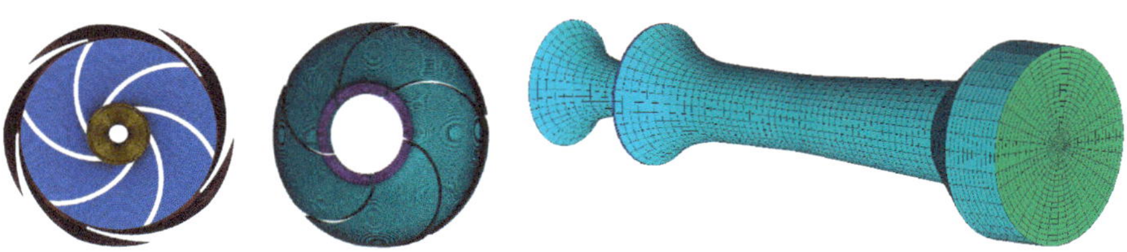

(a) (b) (c)

Figure 2. Grid diagrams of the impeller, guide vane, and jet. (**a**) Impeller; (**b**) Guide vane; (**c**) Ejector nozzle and tube.

CFX 16.0 (16.0, ANSYS Inc., Pittsburgh, PA, USA) was employed for the flow field calculations. The impeller flow fields were calculated in the rotating coordinate system in the multi-coordinate system, and the flow fields of the other components were calculated in the stationary coordinate system. The k–ω turbulence model was adopted, as the performance of this model is better than that of the k–ε model in near-wall flow analysis [16]. GGI (general grid interface) technology was employed to exchange data between the rotor and stator. Pressure inlet and velocity outlet boundary conditions were set, and the inlet pressure values were set according to the values obtained from experiments. All the solid walls were provided with non-slip wall conditions, and the roughness was set to 25 μm according to actual processing. In the steady simulation process, the interface models between rotating–stationary and stationary–stationary domains were selected as a frozen-rotor and general connection, while the governing equations were solved by a second-order upwind formula. For transient simulation, the transient rotor–stator interface model was used for the rotating–stationary interface, while the time-term was discreted by a second-order implicit scheme. The time-step size determines the precision of post-processing; therefore, the setting of the time-step requires an adequate response resolution to the flow field fluctuation. Finally, the time-step is 0.000117 s, which means that the impeller rotates about 2 degrees per time-step and 180 steps in a circle. All residuals were less than 10^{-5}.The simulation was completed on a workstation with 20 cores and 128 GB of memory. For a model, it takes 90 min to complete 2000 steps of steady calculation, and about 1000 min to complete the transient calculation of 13 impeller cycles.

2.3. Numerical Validation

Validation of the numerical calculation of the flow fields was performed at the JCP test bed for hydraulic performance and hydrodynamic noise in the Key Laboratory of Fluid Machinery and Systems, Gansu Province. The test device consists of two parts: an operation system and a data acquisition system. The operation system meets the requirements of the operation of the pump under various conditions, including pump, circulation pipeline, motor, valves, frequency converter, and water box; the data acquisition system collects the relevant transient and steady physical quantities generated during the operation of the pump by various sensors, and then converts it into identifiable data signals through a series of processing. The related instruments include the computer, pressure sensors, flowmeter, measuring instrument of electric power, and tachometer. The flow of the pump can be changed by adjusting the inlet valve. The test system is shown in Figure 3a.

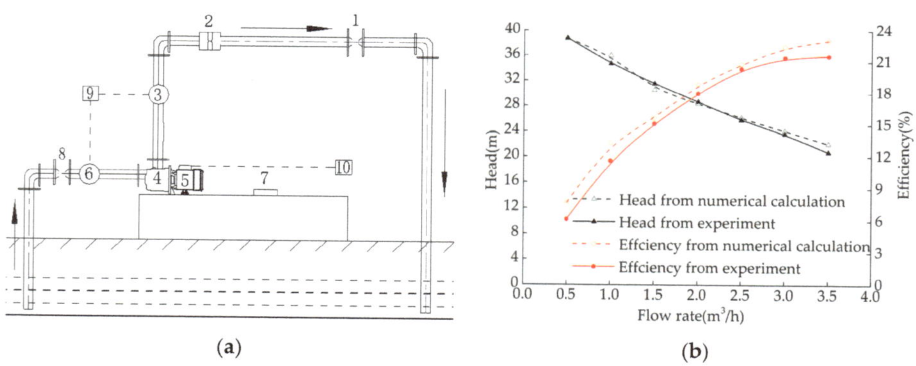

Figure 3. Experimental system and result. (**a**) Experimental system; (**b**) A comparison of hydraulic performance between the experiment and the numerical simulation. 1: Valve at pump outlet; 2: Flowmeter; 3: Pressure sensor at pump outlet; 4: The model pump; 5: Electric motor; 6: Pressure sensor at pump inlet; 7: Tachometer; 8: Valve at pump inlet; 9: Computer; 10: Measuring instrument of electric power.

Comparison curves of the hydraulic performance between the experiment and numerical simulation are shown in Figure 3b. It can be seen that the variation trends of the performance curves are consistent between the numerical calculation and experiment, and the values at different working points are close. The maximum error in the water head was 4.6%, and that in efficiency was 4.1%, which proves the accuracy of the calculation results.

3. Analysis of the Velocity Vector and Streamline

The velocity vector and streamline diagrams through the inlet and outlet tangent planes of the JCP under four conditions are shown in Figure 4; 0.2 Q_d, 0.6 Q_d, 1.4 Q_d represent conditions at which the pump provides 20%, 60%, and 140% of the rated flow, respectively. It can be seen that there are obvious eddies in the lower half of the pump chamber near the inlet side under each condition; the flow velocity in this area is slow, and the flow passage is seriously blocked, which affects the normal flow in the upstream and downstream flow passage components and leads to obvious recirculation at the border of the impeller and ejector. Under the conditions of small flow rates, there are eddies near the outlet in the upper half of the pump chamber, and the smaller the flow rate, the more obvious the eddies. The eddies in the upper and lower sides of the pump chamber cause a blockage of the outlet channels in both sides of the ejector tube, where the flow is relatively more symmetrical. With the increasing flow rate, the eddies in the upper side of the pump chamber gradually weaken, while the eddies in the lower side gradually strengthen, and the flow in the ejector tube shows a tendency of upward deviation. There is obvious circulation movement in the pump chamber, and the flow patterns of the left and right sides are also not balanced; this is related to the asymmetry of the guide vane blades, of which there are five. There is a counter-clockwise eddy along the circumferential direction near the right-hand side of the outlet pipeline in the pump chamber, and the greater the flow rate, the greater the eddy. The area with the fastest velocity in the pump chamber is located in the circle near the ejector tube, and the flow in this area is very complicated.

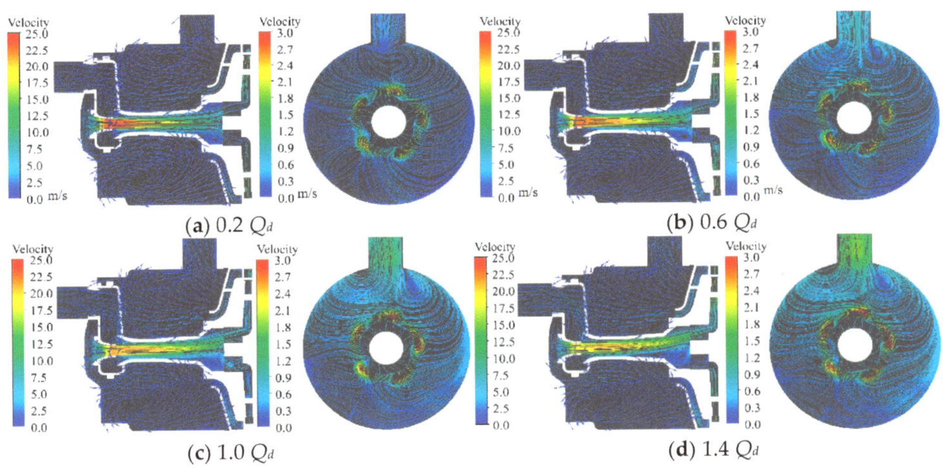

Figure 4. Velocity vector and streamline diagrams under different working conditions.

The above analysis shows that there exist various sizes and scales of eddies, channel blockage, back flow, secondary flow, circulation, and other phenomena of flow field distortion in the pump chamber under multiple conditions. These not only result in great hydraulic losses but also destroy the stability, symmetry, and balance of other flow passage components, and are an important reason for the JCP's efficiency being significantly lower than that of a general centrifugal pump.

4. Analysis of Unsteady Flow Field Characteristics Based on a Statistical Method

The unsteady flow characteristics of the model pump were analyzed by a statistical method. The unsteady physical quantities include two parts on each grid node (x, y, z) of the flow domain, the time-averaged component $\overline{\phi}$ and the periodic component $\widetilde{\phi}$, where the periodic component represents the variation of the physical quantities in a rotation period of the impeller [17]. The two components are respectively expressed as follows:

$$\overline{\phi}(n) = \frac{1}{N}\sum_{j=0}^{N-1} \phi(n, t_0 + j\Delta t), \tag{1}$$

$$\widetilde{\phi}(n) = \phi(n,t) - \overline{\phi}(n) \tag{2}$$

where n, N, and t_0 are grid nodes, the number of samples in an impeller rotating period, and the starting time of the rotation cycle, respectively.

The non-dimensional standard deviation of the periodic component $\widetilde{\phi}$ is defined as the flow field fluctuation intensity, including the pressure fluctuation intensity, velocity fluctuation intensity, turbulence energy fluctuation intensity, etc.

The pressure fluctuation intensity is expressed as:

$$C_P^* = \frac{\sqrt{\frac{1}{N}\sum_{j=0}^{N-1} \widetilde{P}(n, t_0 + j\Delta t)^2}}{0.5\rho U_2^2}. \tag{3}$$

The velocity fluctuation intensity is expressed as:

$$C_U^* = \sqrt{\frac{1}{N}\sum_{j=0}^{N-1} \widetilde{U}(n, t_0 + j\Delta t)^2 / U_2}. \tag{4}$$

The turbulence energy fluctuation intensity is expressed as:

$$C_T^* = \sqrt{\frac{1}{N}\sum_{j=0}^{N-1} \widetilde{T}(n, t_0 + j\Delta t)^2 / 0.5U_2^2}. \tag{5}$$

where U_2 means the circumferential speed at the impeller outlet.

4.1. Time-Averaged Distribution Characteristics of the Flow Field

Figure 5 shows the time-averaged distribution contours of pressure on two planes under four conditions. It can be seen that the pressure in each flow passage component decreases gradually with the increasing flow rate. At small flow rates, the lowest area of pressure is located at the ejector nozzle outlet. As the flow rate increases, the lowest area gradually shifts and expands to the middle of the ejector tube, which shows that unlike the cavitation of a general centrifugal pump, which first occurs in the inlet area of the impeller, the cavitation of a JCP first occurs in the ejector.

Figure 6 shows the time-averaged distribution contours of absolute velocity on two planes under four conditions. It can be seen that the difference in velocity is small in the pump chamber, guide vane, and impeller among the different working conditions, but it is large in the ejector nozzle and tube, and the largest velocity appears in the area of the nozzle exit. As the flow rate increases, the velocity in the ejector nozzle and tube decreases gradually and the distribution symmetry becomes worse and worse, showing characteristics of upward inclination. By comparing with Figure 4, we can see that this is because with the increasing flow rate, the blockage in the upper side of the ejector exit gradually weakens while it gradually increases in the lower side, and the flow then moves upward.

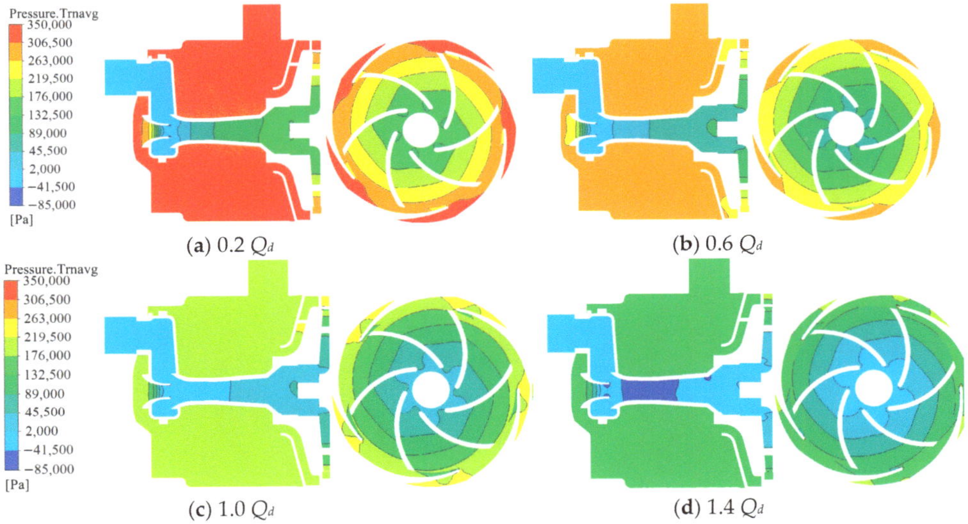

Figure 5. Time-averaged distribution contours of pressure on two planes under four conditions.

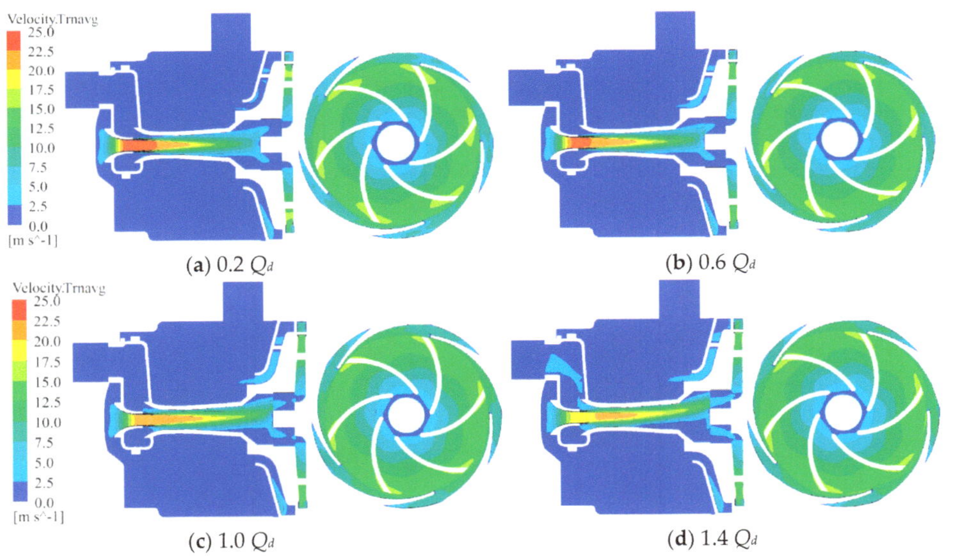

Figure 6. Time-averaged distribution contours of absolute velocity on two planes under four conditions.

Figure 7 shows the time-averaged distributions of turbulent kinetic energy contours on two planes under four conditions. As can be seen in the figure, the turbulent kinetic energy is greater in the ejector tube and nozzle under every condition. At small flow rates, the maximum value is located around the nozzle outlet. By comparing with Figure 6, it can be seen that this is due to the mixing of high-speed flow at the nozzle outlet and low-speed flow from the pump inlet in this area, which leads to some larger vortices around the nozzle area and very violent turbulent movement. As the flow rate increases, the turbulent kinetic energy at the inlet of the impeller increases gradually, while it decreases gradually at the junction area of the rotor–stator cascades. Similar to the distribution of the velocity field, the turbulent kinetic energy field in the ejector tube shifts upward with the increasing flow rate.

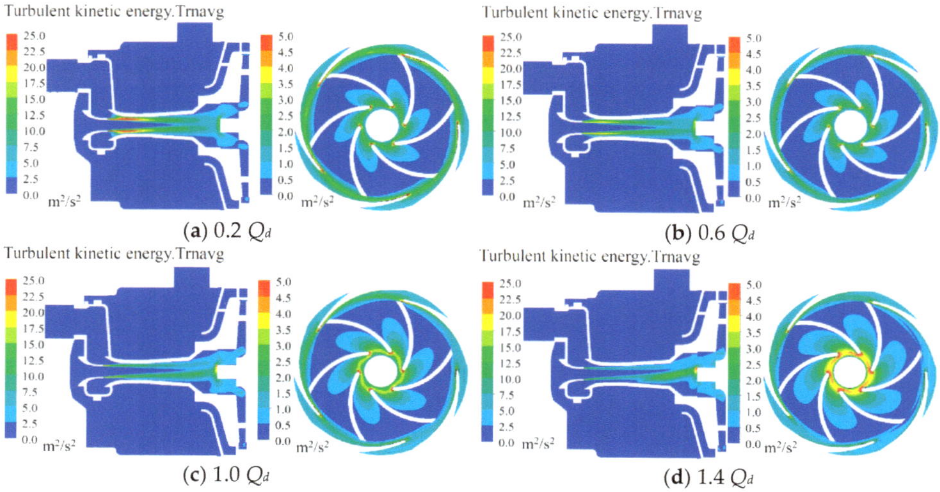

Figure 7. Time-averaged distribution contours of turbulent kinetic energy on two planes under four conditions.

4.2. Distribution Characteristics of the Fluctuation Intensity of the Flow Field

Figure 8 shows the distributions of pressure fluctuation intensity contours on two planes under four conditions. It can be seen that the pressure fluctuation intensity in the impeller and guide vane is obviously greater than that in other flow passage components, which is influenced by the rotor–stator interaction. The maximum value of the intensity appears near the pressure surface of the impeller outlet. As the flow rate increases, the fluctuation intensity decreases gradually near the junction area of the impeller and guide vane, but it increases gradually at the impeller inlet.

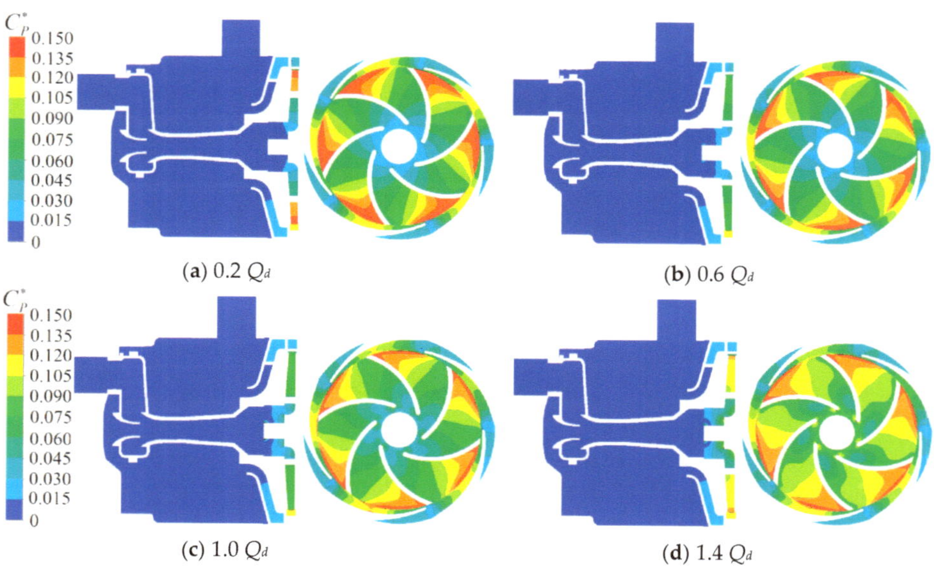

Figure 8. Fluctuation intensity distribution contours of pressure on two planes under four conditions.

Figure 9 shows the distributions of velocity fluctuation intensity contours on two planes under four conditions. It can be seen that, similar to the law of the pressure fluctuation intensity distribution, the velocity fluctuation intensity in the impeller and guide vane is obviously greater than that in other flow passage components due to the rotor–stator interaction, and the maximum intensity appears in the impeller inlet area; with the increasing flow rate, the velocity fluctuation intensity increases gradually at all positions.

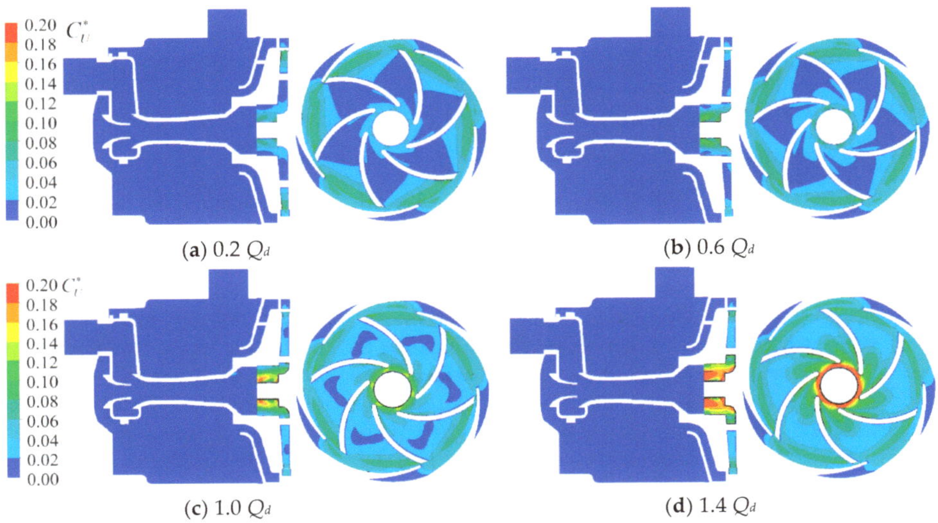

Figure 9. Fluctuation intensity distribution contours of velocity on two planes under four conditions.

Figure 10 shows the distribution contours of the turbulent kinetic energy fluctuation intensity on two planes under four conditions. It can be seen that the region of the largest intensity is at the impeller inlet, which is similar to the distribution law of the velocity fluctuation intensity. As the flow rate increases, the fluctuation intensity in the impeller inlet increases gradually, while it decreases gradually in the junction area of the rotor–stator cascades.

The analysis of Figures 8–10 shows that the fluctuation intensity of the flow field in the impeller and guide vane is obviously greater than that in other flow passage areas, reflecting that the rotor–stator interaction plays a decisive role in the unsteady fluctuation intensity of the flow field of a JCP.

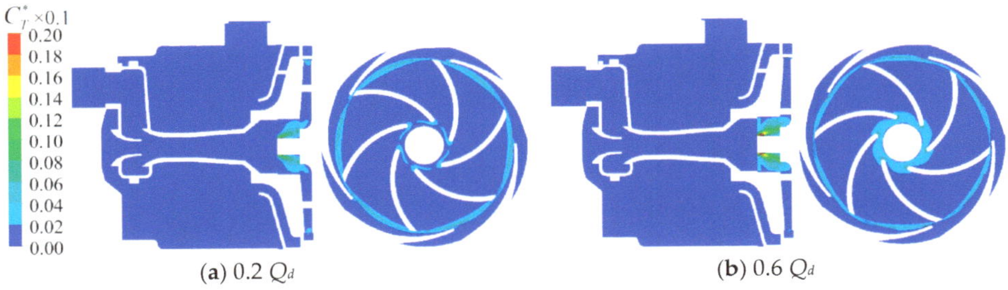

Figure 10. *Cont.*

(**c**) 1.0 Q_d (**d**) 1.4 Q_d

Figure 10. Fluctuation intensity distribution contours of velocity on two planes under four conditions.

5. Analysis of Spatial–Temporal Evolution of Vortices in the Rotor–Stator Cascades

To understand the formation mechanism of the unsteady characteristics of the flow field in the JCP, in this section, the formation and evolution process of vortices in the impeller and guide vane of the JCP under four conditions is studied by a transient analysis method. Figures 11–14 show the spatial and temporal evolution processes of the vortices in the stator and rotor cascades during a 1/6 cycle (60 degrees) of impeller rotation under four conditions. Since the impeller of the model pump contains six blades evenly arranged in the circumferential direction, a 1/6 cycle of the flow process actually reflects the transient flow characteristics of the pump during the whole operating period.

It can be seen that the vortices in every impeller passage underwent a process from generation to disappearance in each operating condition within 1/6 of a cycle, but the intensity and pattern of the vorticity were unique in different passages and time. This is because, on the one hand, the number of impeller and guide vane blades was different, and on the other hand, from the velocity vector and streamline analysis of the pump chamber in the previous section, there is flow asymmetry in the upper, lower, left, and right parts of the pump chamber. The process from generation to disappearance of the vortices is related to the constant change of the relative positions of the impeller blades and guide vane blades. When the outlet flow in a channel of the impeller is divided by a guide vane and flows into two adjacent guide vane channels, backflow and vorticity will arise in the impeller outlet near the pressure surface. When a vortex exists in the impeller channel, blockage of the flow path will occur in the inlet of the guide vane. The more obvious the vortex is, the more serious the blockage, which will affect the normal and steady flow in the guide vane. By comparing the vortex evolution processes under different conditions, it can be seen that the smaller the flow rate of the pump, the more obvious the vortices and the more serious the blockage of the flow passage, but their evolution law is basically the same. The above analysis results show that the temporal and spatial evolution processes and law of vortices in the rotor and stator cascades are closely related to the relative positions of the impeller blades and the guide vane blades. Therefore, the design of impeller or guide vane geometry and the optimization of the internal flow field of JCPs should focus on the overall matching of rotor and stator cascades.

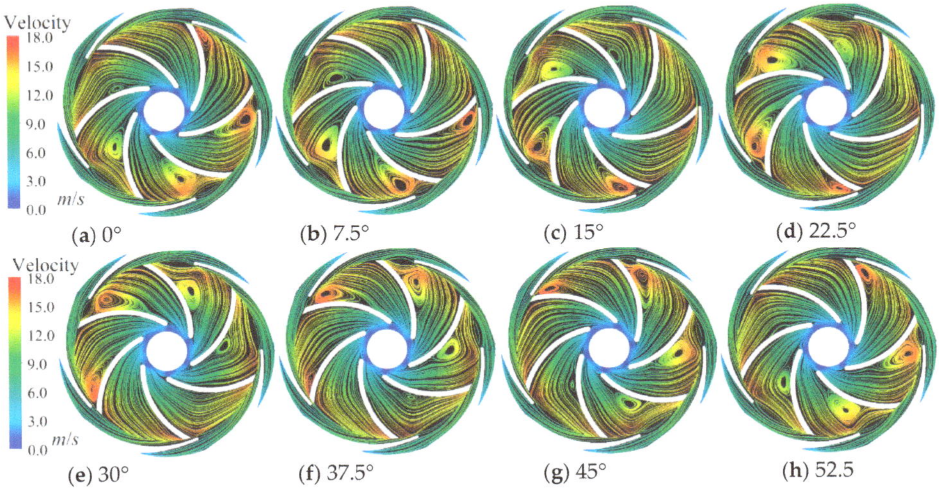

Figure 11. Evolution processes of the vortices in the stator and rotor cascade during a 1/6 period of impeller rotation under condition 0.2 Q_d.

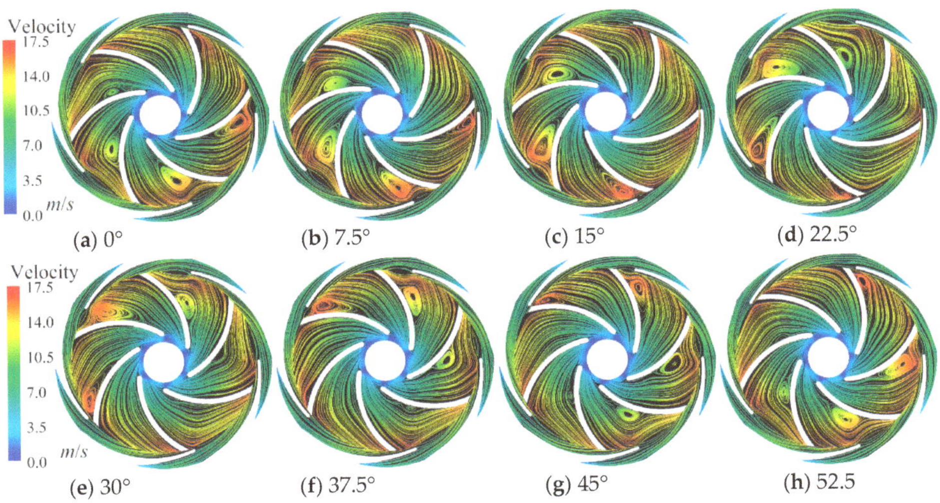

Figure 12. Evolution processes of the vortices in the stator and rotor cascade during a 1/6 period of impeller rotation under condition 0.6 Q_d.

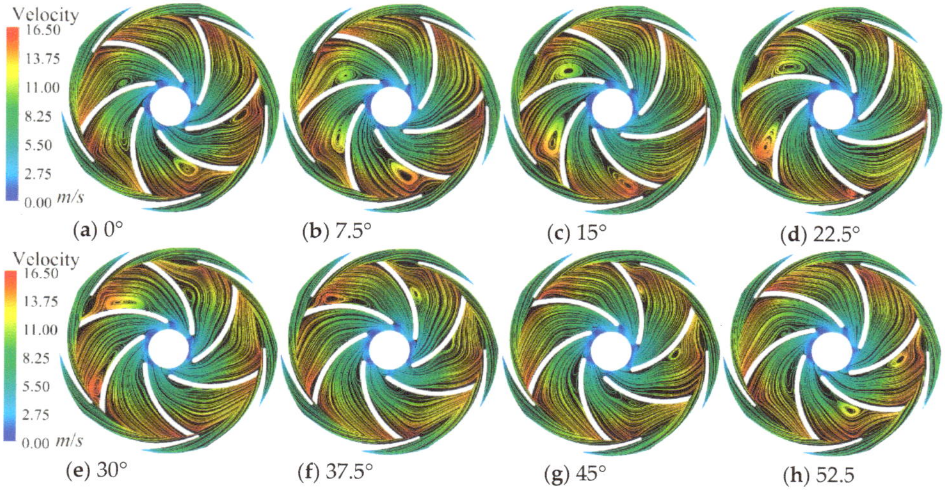

Figure 13. Evolution processes of the vortices in the stator and rotor cascade during a 1/6 period of impeller rotation under condition 1.0 Q_d.

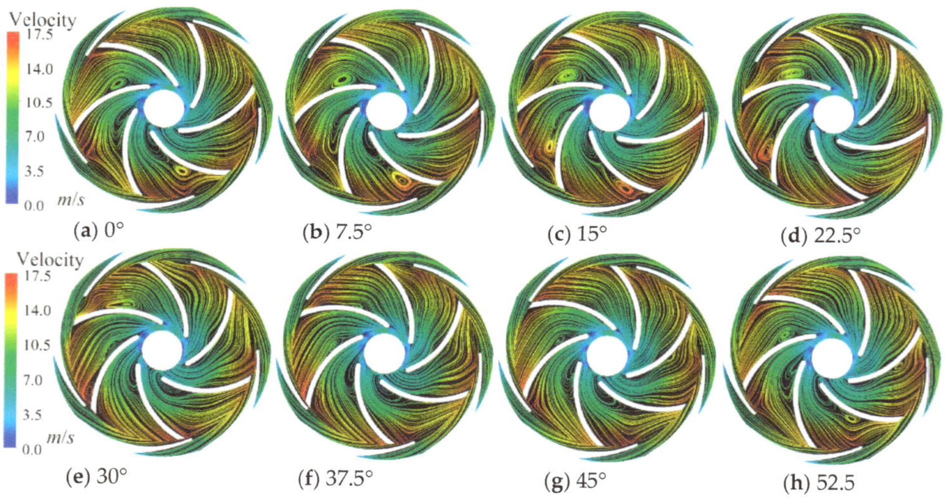

Figure 14. Evolution processes of the vortices in the stator and rotor cascade during a 1/6 period of impeller rotation under condition 1.4 Q_d.

6. Pressure Fluctuation Characteristics of the Main Flow Passage Components

In this section, we analyze in the time and frequency domains the pressure fluctuation characteristics of the flow field inside the main flow passage components of the model pump under the rated operating conditions. Figure 15 shows a schematic diagram of the positions of the monitoring points in each component.

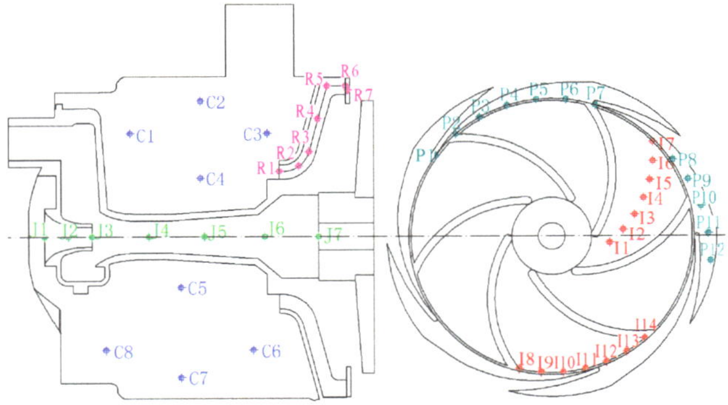

Figure 15. A schematic diagram of the positions of the monitoring points in each component.

6.1. Impeller

6.1.1. Monitoring Points Stationary Relative to the Fixed Coordinate System

Figure 16 shows the time and frequency domain diagrams of pressure fluctuation for the monitoring points at the middle streamline of a single impeller channel when the monitoring points are stationary relative to the fixed coordinate system. The time domain diagram is the variation curve over two impeller rotation cycles (the same is true below). It can be seen that when the impeller rotates one full circle, the number of pressure fluctuation waves at each monitoring point is consistent with the number of impeller blades—six similar periodic waveforms. From Figure 16b, we can see that the dominant frequency of pressure fluctuation at each monitoring point is BPF_I, and it has obvious amplitudes at other low-order frequency multiplications of BPF_I. Along the direction of radius increase, the pressure fluctuation curves rise overall, but the fluctuation amplitude does not change much.

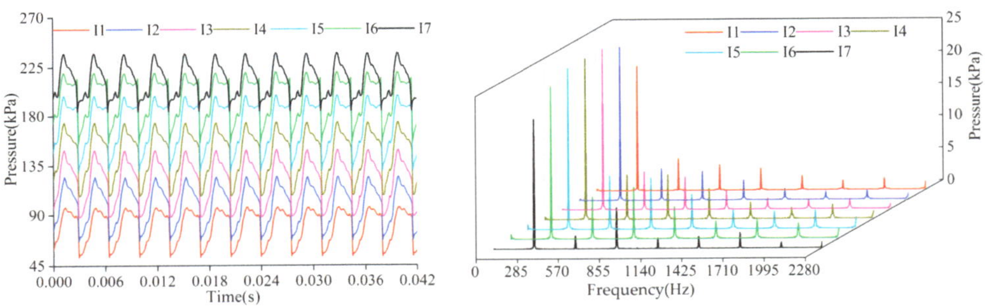

(**a**) Time domain (**b**) Frequency domain

Figure 16. The time and frequency domain diagrams of pressure fluctuation for the monitoring points at the middle streamline of a single impeller channel when the monitoring points are stationary relative to the fixed coordinate system.

6.1.2. Monitoring Points Stationary Relative to the Rotating Coordinate System

Figure 17 shows the time and frequency domain diagrams of pressure fluctuation for the monitoring points at the middle streamline of a single impeller passage when the monitoring points are stationary relative to the rotating coordinate system. From Figure 17a, it can be seen that there are five obvious similar waveforms at each monitoring point when the impeller rotates one full cycle, which is consistent with the number of guide vanes (of which there are five). The pressure fluctuation curve at each monitoring point rises overall with the radius, but the law of variation with time is basically the same. Figure 17b shows that the dominant frequency of pressure fluctuation at all the monitoring points is

237.5 Hz, which is the BPF_G, and obvious amplitudes appear at low-order frequency multiplications of the dominant frequency. As the monitoring point gets closer to the impeller outlet, the rotor and stator interference becomes more and more violent, so the fluctuation amplitude at each characteristic frequency becomes larger and larger. In addition, relatively small fluctuation amplitudes are also shown at the shaft-passing frequency and its multiplication frequency, which reflects the influence of the impeller operating parameters (pump rotation speed) on the unsteady flow field of the JCP.

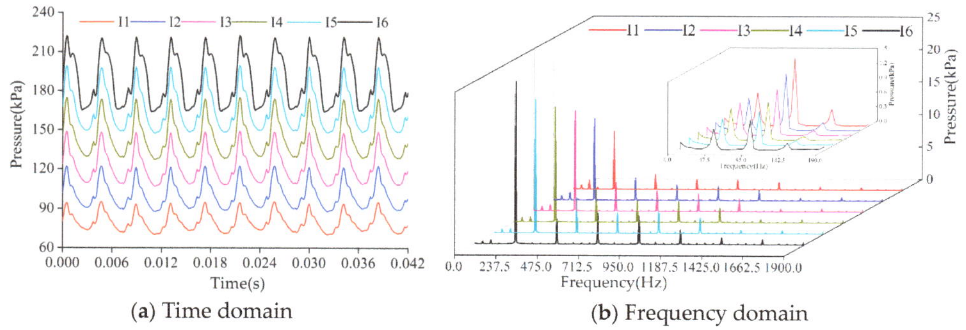

Figure 17. The time and frequency domain diagrams of pressure fluctuation at monitoring points on the middle streamline of a single impeller channel when the monitoring points are stationary relative to the rotational coordinate system.

Figure 18 shows the time and frequency domain diagrams of pressure fluctuation for monitoring points at the outlet edge of a single impeller channel. It can be seen from Figure 18a that when the impeller rotates one full circle, the pressure fluctuates five times, which is consistent with the number of guide vanes. However, the variation law with time is different: peaks and troughs occur at different times because the pressure fluctuation in the impeller is closely related to the relative positions of the guide vanes, and the guide vanes pass the monitoring point one by one, so the wave peaks or troughs appear in turn. Figure 18b shows that the dominant frequency of pressure fluctuation for all of the points is BPF_G, and there are obvious amplitudes at low-order frequency multiplications of BPF_G. At the dominant frequency, the fluctuation amplitude in the middle region of the flow channel is larger, while that at the points near the pressure or suction surface is relatively small.

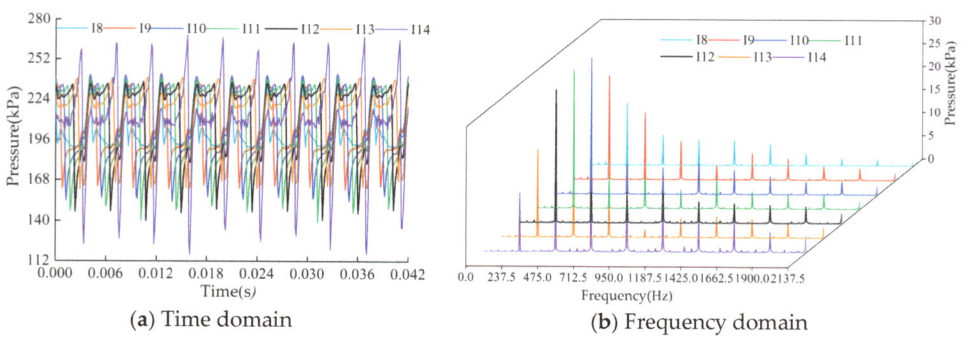

Figure 18. The time and frequency domain diagrams of pressure fluctuation for the monitoring points at the outlet edge of a single impeller channel.

The analysis results from Figures 16–18 show that the pressure fluctuation characteristics of the impeller flow field are mainly related to the number of impeller blades and the speed of the impeller when the monitoring points are stationary relative to the fixed coordinate system. The pressure fluctuation characteristics of the impeller flow field are determined by the number of guide vane blades and the speed of the impeller when the monitoring points are stationary relative to the rotating coordinate system, and the farther the monitoring point is from the interface of the rotor and stator cascades, the weaker the fluctuation amplitude at the characteristic frequencies.

6.2. Guide Vane

Figure 19 shows the time and frequency domain diagrams of the pressure fluctuation characteristics for monitoring points at the inlet edge of a guide vane channel. Figure 19a shows that the pressure fluctuation at each monitoring point exhibits six periods in an impeller cycle, which is consistent with the number of impeller blades, but the variation laws with time are not identical. The times when peaks and troughs appear are different because the pressure fluctuation characteristics in the guide vane channel are closely related to the relative position of the impeller blades. The impeller blades pass the monitoring point one by one, so the wave peaks and troughs at that point appear in turn. Figure 19b shows that the dominant frequency of pressure fluctuation at all the monitoring points is BPF_I, and there are obvious amplitudes at low-order frequency multiplications of BPF_I.

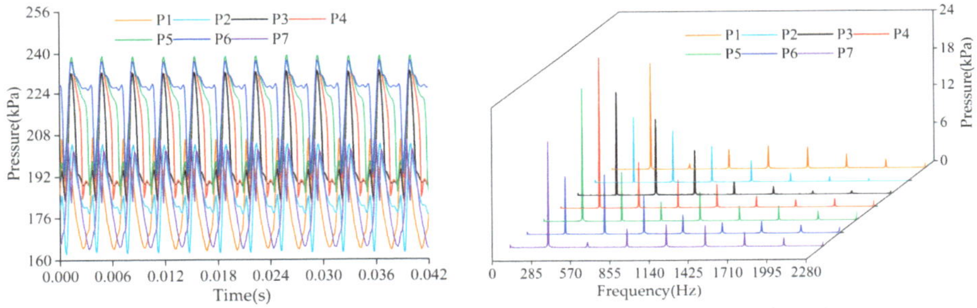

(**a**) Time domain (**b**) Frequency domain

Figure 19. The time and frequency domain diagrams of pressure fluctuation for monitoring points at the inlet edge of a guide vane channel.

Figure 20 shows the time and frequency domain diagrams of the pressure fluctuation characteristics for monitoring points at the single-channel middle streamline of the guide vane. Figure 20a shows that the time domain characteristic curve of pressure fluctuation is related to the number of impeller blades. As the radius increases, the average pressure at each monitoring point increases as a whole, but it decreases at the outlet position of the positive guide vane; this is due to the gradual transition of the flow direction from the radial to the axial direction. Figure 20b shows that the dominant frequency of the pressure fluctuation is BPF_I, and there are amplitudes at other frequency multiplications of BPF_I. As the radius increases, the influence of the impeller on the guide vane gradually weakens, so the amplitude, in turn, decreases at characteristic frequencies, but it rises slightly at the outlet of the positive guide vane, where the flow is complex.

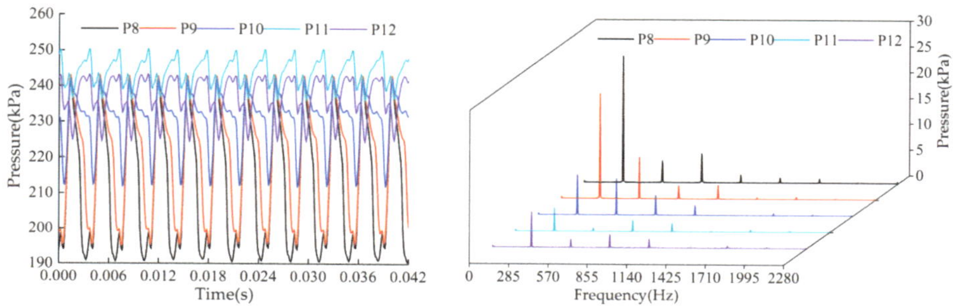

(**a**) Time domain (**b**) Frequency domain

Figure 20. The time and frequency domain diagrams of pressure fluctuation for monitoring points at the middle streamline of a guide vane channel.

Figure 21 shows the time and frequency domain diagrams of the pressure fluctuation characteristics for monitoring points at the middle streamline of a return guide vane channel. Figure 21a shows that the fluctuation characteristics of the pressure inside the return guide vane in the time domain are basically the same as those of the positive guide vane. After the fluid passes through the monitoring point R5, the average pressure increases gradually along the outlet direction. Figure 21b shows that the characteristic frequencies of pressure fluctuation are BPF_I and its low-order frequency multiplications, among which the dominant frequency is BPF_I along the outlet direction. The amplitude of pressure fluctuation at the dominant frequency and other characteristic frequencies decreases monotonically, which is due to the influence of rotor–stator interaction becoming weaker and weaker as the monitoring point is farther away from the impeller.

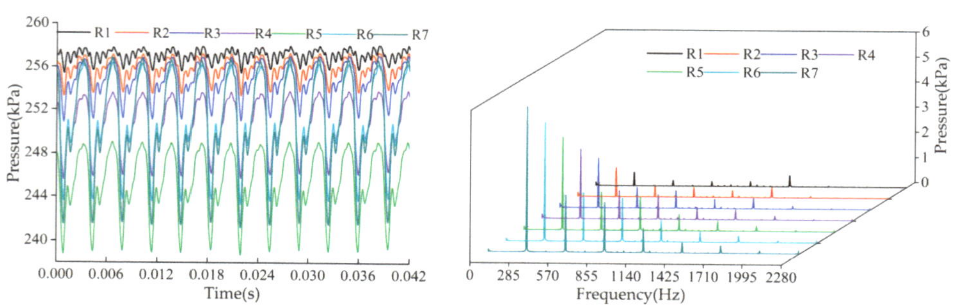

(**a**) Time domain (**b**) Frequency domain

Figure 21. The time and frequency domain diagrams of pressure fluctuation for monitoring points at the middle streamline of a return guide vane channel.

The analysis results of Figures 19–21 show that the pressure fluctuation characteristics of the fluid in the guide vane are mainly related to the number of impeller blades and impeller speed. The farther the monitoring point is from the interface of the rotor and stator cascades, the weaker the fluctuation amplitude at the characteristic frequency.

6.3. Pump Chamber and Ejector

Figure 22 shows the time and frequency domain diagrams of the pressure fluctuation characteristics for monitoring points in the pump chamber. Figure 22a shows that the pressure at the monitoring points in the pump chamber presents obvious periodic fluctuation characteristics in the time domain, and their fluctuation ranges and variation laws are very consistent. In a rotation period of the impeller, the pressure fluctuation has 30 peaks and troughs, which is the product of the number of impeller blades and the number of guide vane blades. Figure 22b shows that the amplitude of pressure fluctuation inside the pump chamber is obviously smaller than that at the impeller and guide vane, which is consistent with the analysis results of the pressure fluctuation intensity in Section 4.2. The dominant frequency of pressure fluctuation at each monitoring point is 1425 Hz, which is the frequency of the rotor and stator cascade encounter (impeller speed × impeller blade number × guide vane blade number). The difference in fluctuation amplitude is small among the different monitoring points at the dominant frequency and at the other characteristic frequencies. In addition, there are obvious fluctuation amplitudes when the frequency of the *SPF* doubles, which have an important influence on the unsteady fluctuation characteristics of the pump chamber.

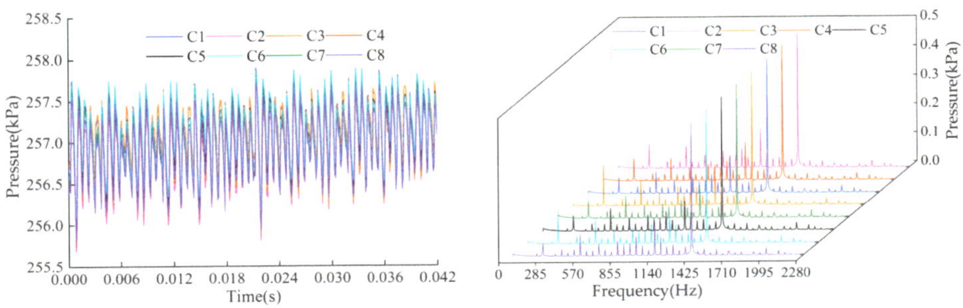

(**a**) Time domain (**b**) Frequency domain

Figure 22. The time and frequency domain diagrams of pressure fluctuation for monitoring points in the pump chamber.

Figure 23 is the frequency domain diagram of the pressure fluctuation characteristics for the monitoring points in the ejector nozzle and tube. It can be seen that the pressure fluctuation characteristics of the flow field in the ejector are basically the same as those in the pump chamber. That is, the dominant frequency is the encounter frequency of the rotor and stator cascades. Within the ejector nozzle, the fluctuation amplitude at the dominant frequency decreases gradually along the flow direction. The closer the monitoring point is to the impeller, the greater the pressure fluctuation amplitude. It can be seen from Figure 23b that there are non-negligible fluctuation amplitudes at frequency multiplications of the *SPF*.

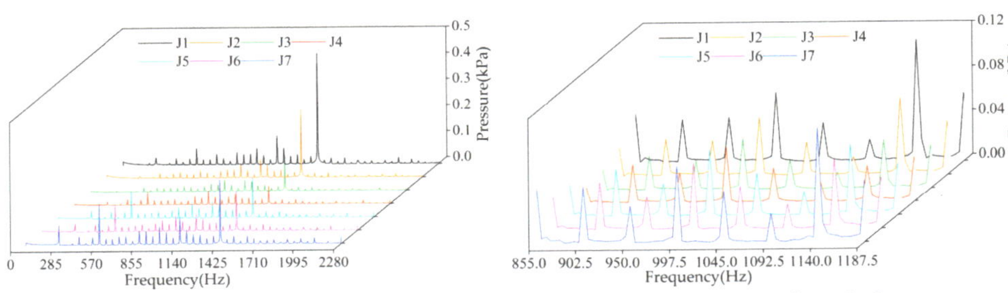

(**a**) Frequency domain 1 (**b**) Frequency domain 2

Figure 23. Frequency domain diagrams of pressure fluctuation for monitoring points in the ejector.

The analysis results of Figures 22 and 23 show that the pressure fluctuation characteristics of the flow field in the pump chamber and ejector are the result of combined action, including the interaction of the number of impeller blades, the number of guide vane blades, and the speed of the impeller. The rotor–stator interaction affects not only the unsteady flow characteristics in the impeller and guide vanes, but also the pump chamber and ejector.

7. Conclusions

In this paper, the transient fluctuation characteristics of a flow field in the major flow passage components of a jet centrifugal pump, and the spatial and temporal evolution laws of vortices in rotor–stator cascades were analyzed. The main results are as follows:

(1) Under multiple operating conditions, various scales of flow distortion phenomena occur in the JCP chamber, such as eddies, blockage of the flow passage, recirculation, secondary flow, and circulation, which not only cause great hydraulic loss, but also destroy the flow stability, symmetry, and balance of the other flow passage components' flow fields. This is an important reason for the obviously lower efficiency of JCPs than that of general centrifugal pumps.

(2) The spatial and temporal evolution laws of vortices in rotor–stator cascades are mainly related to the relative positions of impeller blades and guide vane blades; the scale of vortices is not only related to the structures of the impeller blade and guide vane blade, but also to their matching relationship. Therefore, the design of impeller or guide vane geometry and the optimization of the internal flow field of a JCP should focus on the overall matching of rotor and stator cascades.

(3) The pressure fluctuation characteristics in the impeller are mainly determined by the number of guide vane blades and the pump rotation speed; those in the guide vane by the number of impeller blades and the pump rotation speed; and those in the pump chamber and ejector by the numbers of impeller blades and guide vane blades and the pump rotation speed. All of this reflects that the formation mechanism of the unsteady flow field fluctuation characteristics of a JCP is mainly related to the number of blades in the rotor–stator cascades and the operation parameters of the pump.

(4) The fluctuation intensity of the flow field inside the impeller and guide vane is obviously greater than that in other flow areas, reflecting that rotor–stator interaction is the decisive factor affecting the unsteady flow characteristics of a JCP under multiple conditions.

Author Contributions: Formal analysis, R.G.; Funding acquisition, R.L.; Methodology, R.G.; Resources, R.Z.; Validation, W.H.; Writing—original draft, R.G.; Writing—review and editing, R.L.

Funding: This research was funded by Natural Science Foundation of China, grant number 51579125; National Key Research and Development Program of China, grant number 2016YFB0200901.

Acknowledgments: The authors gratefully acknowledge the financial support from the National Natural Foundation of China (51579125) and the National Key Research and Development Program of China (2016YFB0200901).

Conflicts of Interest: The authors declare no conflict of interest.

Nomenclature

Q_d	Rated volume flow m^3/h
H_d	Rated water head m
η_d	Rated efficiency %
n	Rotational speed r/min
SPF	Shaft passing frequency Hz
BPF_I	Blade passing frequency of impeller Hz
BPF_G	Blade passing frequency of guide vane Hz
D_j	Inlet diameter of impeller mm
D_2	Outlet diameter of impeller mm
Z_1	Blade number of impeller
φ_1	Blade wrap angle of impeller °
b_2	Blade outlet width of impeller mm
D_3	Base diameter of guide vane mm
D_4	Outlet diameter of guide vane mm
Z_2	Blade number of guide vane
GGI	General grid interface
$\overline{\phi}$	Time-averaged components
$\widetilde{\phi}$	Periodic components

n Grid nodes
N Number of samples in an impeller rotating period
t Total time
t_0 Starting time of the rotation cycle
C_P^* Non-dimensional pressure fluctuation intensity
C_U^* Non-dimensional velocity fluctuation intensity
C_T^* Non-dimensional turbulence energy fluctuation intensity
$\widetilde{P}$ Pressure fluctuation intensity
$\widetilde{U}$ Velocity fluctuation intensity
$\widetilde{T}$ Turbulence energy fluctuation intensity

References

1. Liu, J.R.; Zhou, G.P. Research and prospects for the self-priming irrigation pumps. *Trans. Chin. Soc. Agric. Mach.* **2007**, *8*, 177–180. [CrossRef]
2. Wang, W.J.; Tang, Y.; Ma, H.; Yang, Z. Design and research of jet self-priming centrifugal pump. *Equip. Manuf. Technol.* **2017**, *5*, 32–34.
3. Li, W.; Chang, H.; Liu, J.R.; Zhou, L.; Wang, C. Research situation and prospect of jet type self-priming centrifugal pump. *Drain. Irrig. Mach. Eng.* **2016**, *34*, 947–952. [CrossRef]
4. Wang, C.; Hu, B.; Zhu, Y.; Wang, X.L.; Luo, C.; Cheng, L. Numerical study on the gas-water two-phase flow in the self-priming process of self-priming centrifugal pump. *Processes* **2019**, *7*, 330. [CrossRef]
5. Wang, Y.; Li, G.D.; Cao, P.Y.; Yin, G.; Cui, Y.R.; Li, Y.C. Effects of internal circulation flow on self-priming performance of flow-ejecting self-priming pump. *Trans. Chin. Soc. Agric. Mach.* **2014**, *45*, 129–133. [CrossRef]
6. Wang, Y.; Han, Y.W.; Zhu, X.X.; Sun, W.; Cao, P.Y.; Wu, W. Optimization and experiment on performance of flow-ejecting self-priming pump based on CFD. *Trans. Chin. Soc. Agric. Eng.* **2016**, *32*, 16–21. [CrossRef]
7. Liu, J.R.; Wen, H.G.; Gao, Z.J.; Guo, C.X.; Wang, P. Effects of geometric parameters for jet nozzle on self-priming performance of spray pump. *Trans. Chin. Soc. Agric. Eng.* **2012**, *28*, 47–54. [CrossRef]
8. Liu, J.R.; Wen, H.G.; Xiang, H.J.; Guo, C.X.; Gao, Z.J. Effects of diffuser on self-priming performance of flow-ejecting self-priming centrifugal pump. *Trans. Chin. Soc. Agric. Mach.* **2013**, *44*, 43–47. [CrossRef]
9. Wang, C.; Shi, W.D.; Li, W.; Zhang, D.S.; Jiang, X.P. Unsteady numerical calculation and validation on self-priming process of self-priming spray irrigation pump. *Trans. Chin. Soc. Agric. Eng.* **2016**, *32*, 65–72. [CrossRef]
10. Zhang, R.H.; Zhang, S.D.; Tian, L.; Chen, X.B. Research on performance and inner flow characteristics of jet centrifugal pump. *J. Drain. Irrig. Mach. Eng.* **2019**, *37*, 763–768+775. [CrossRef]
11. Chen, N.; Liu, W.H.; Lai, H.Q. Numerical analysis of flow field inside injection pump by CFD. *Ship Sci. Technol.* **2016**, *38*, 53–56. [CrossRef]
12. Li, G.D.; Wang, Y.; Yang, X.M.; Zhao, L.F.; Wu, W.; Hu, R.X. Features of internal flow in jet nozzle of self-priming centrifugal pump based on large eddy simulation. *J. Drain. Irrig. Mach. Eng.* **2017**, *35*, 369–374. [CrossRef]
13. Wang, Y.; Cui, Y.R.; Wang, K.; Wang, W.J.; Li, G.D.; Yin, G. Analysis of unsteady flow in flow-ejecting self-priming centrifugal pump based on LES. *J. Drain. Irrig. Mach. Eng.* **2015**, *33*, 481–487. [CrossRef]
14. Wang, W.J.; Wang, Y.; Li, G.D.; Yin, G.; Cao, P.Y. Numerical simulation and testing analysis of three dimensional turbulence flow in flow-ejecting self-priming centrifugal pump. *Trans. Chin. Soc. Agric. Mach.* **2014**, *45*, 54–60. [CrossRef]
15. Li, G.D.; Wang, Y.; Cao, P.Y.; Zhou, G.H.; Wu, W.; Han, Y.W. Internal flow of flow-ejecting centrifugal pump under off-design conditions. *Trans. Chin. Soc. Agric. Mach.* **2015**, *46*, 48–53. [CrossRef]
16. Ansys Inc. *Ansys Fluent Theory Guide*; Ansys Inc.: Canonsburg, PA, USA, 2015.
17. Pei, J.; Yuan, S.Q.; Yuan, J.P.; Wang, W.J. Comparative study of pressure fluctuation intensity for a single blade pump under multiple operating conditions. *J. Huazhong Univ. Sci. Technol.* **2013**, *41*, 29–33. [CrossRef]

processes

MDPI

Article

Design and Verification of a Single-Channel Pump Model based on a Hybrid Optimization Technique

Jin-Hyuk Kim [1,2,*], Sang-Bum Ma [3], Sung Kim [1,2], Young-Seok Choi [1,2] and Kwang-Yong Kim [3]

1 Thermal & Fluid System R&D Group, Korea Institute of Industrial Technology 89 Yangdaegiro-gil, Ipjang-myeon, Seobuk-gu, Cheonan, Chungcheongnam-do 31056, Korea; ks2928@kitech.re.kr (S.K.); yschoi@kitech.re.kr (Y.-S.C.)

2 Industrial Technology (Green Process and Energy System Engineering), Korea University of Science & Technology, 217 Gajeong-ro, Yuseong-gu, Daejeon 34113, Korea

3 Department of Mechanical Engineering, Inha University, 100 Inha-ro, Michuhol-Gu, Incheon 22212, Korea; msb927@inha.edu (S.-B.M.); kykim@inha.ac.kr (K.-Y.K.)

* Correspondence: jinhyuk@kitech.re.kr

Received: 16 September 2019; Accepted: 12 October 2019; Published: 15 October 2019

Abstract: This paper handles a hybrid multiple optimization technique to concurrently enhance hydraulic efficiency and decrease unsteady radial forces resulting from fluid-induced vibration of a single-channel pump for wastewater treatment. A single-channel impeller and volute was optimized systematically by using a hybrid particle swarm optimization and genetic algorithm coupled with surrogate modeling. Steady and unsteady Reynolds-averaged Navier–Stokes analyses were conducted to optimize the internal flow path in the single-channel pump. Design variables for controlling the internal flow cross-sectional area of the single-channel impeller and volute in the single-channel pump were chosen to concurrently optimize objective functions with hydraulic efficiency and the unsteady radial forces resulting from impeller–volute interaction. The optimization results clearly showed that the arbitrary cluster optimum design considerably enhanced hydraulic efficiency and decreased the unsteady radial forces concurrently, compared to the reference design. Finally, the hydraulic performance of the optimized prototype model was verified experimentally. Then, it was proved that the proposed technique is a practical tool for designing a single-channel pump.

Keywords: single-channel impeller; computational fluid dynamics (CFD); unsteady RANS; optimization; hybrid PSO-GA; radial force

1. Introduction

Recently, the water industry has been receiving global attention as an industry with potential growth in the future. Current trends indicate that the wastewater treatment industry is a rapidly growing industry globally. Thus, the demand for special pumps for transferring sewage containing solid wastes is increasing rapidly. However, grinder and vortex pumps that are mostly used for wastewater treatment are inefficient and cannot transfer sewage with solid wastes because of low efficiency and the need for frequent maintenance, among other factors. To overcome these problems, many research studies have been conducted on treating sewage containing solid wastes [1–4]. One such study proposed a single-channel impeller with a single free annulus passage, which can smoothly transfer sewage containing solid wastes. Moreover, it has a high efficiency and wide flow range compared with those of grinder and vortex pumps. However, the single-channel impeller has an unsymmetrical flow passage, and there is a limit to stabilizing unstable vibrations with the generated radial forces due to unsteady features between the stationary volute and rotating impeller. Moreover, a single-channel pump impeller produces unstable radial force sources at a general frequency specified by the impeller rotating speed [4].

Researches on the interaction between the impeller and stator have been conducted over the past several years. Qin and Tsukamoto [5] analyzed the unsteady flow characteristics caused by impeller–diffuser interaction in a diffuser pump using a singularity method. The pressure fluctuation characteristics generated by rotor–stator interaction in a diffuser pump were investigated numerically by Shi and Tsukamoto [6]. As the result, it was confirmed that the flow characteristics generated by the impeller–diffuser interaction can be analyzed through an unsteady flow analysis. The unsteady flow characteristics between the radial impeller and diffuser in a radial pump were investigated by Feng et al. [7]. The flow between the impeller and diffuser was analyzed effectively by using the unsteady Reynold-averaged Navier–Stokes (RANS) analysis and laser Doppler velocimetry (LDV), and it was reported that there are two types of interaction effects between a rotor and stator. One was the downstream effect induced by the impeller, which has an unsteady flow characteristic because of the highly distorted flow field and the wake of the impeller. The other was the upstream effect induced by the stator, which induces unsteady pressure fluctuations. Experimental and computational numerical studies for investigating unsteady hydrodynamic forces on the diffuser pump impeller were performed by Zhang and Tsukamoto [8]. The hydrodynamic unsteady forces acting on the pump impeller were analyzed by varying the number of impeller blades. It was found that the hydrodynamic force was smaller when the blade number of impellers was equal to the diffuser vane number. However, pressure fluctuation was the largest compared with that in other cases. Benra [9] measured the hydrodynamic forces acting on the blade of a single-channel pump. Fluid-structure interaction (FSI) was used for numerical analysis of hydrodynamic forces, and the deflection of the blade was measured using proximity sensors. The vibration induced by the blade was transmitted to the volute and pipes. This phenomenon occurred periodically according to the rotation of the blade, and it was confirmed that the hydrodynamic force increases as the flow rate of the pump increases. Nishi et al. [10] proposed a single-channel pump design to increase the amount of solid waste passing through the impeller and to improve the pump performance. The results of the study showed that the pump efficiency exceeded 62%. Pei et al. [11] analyzed the flow characteristics of a single-blade centrifugal pump through numerical analysis and confirmed that unsteady flow characteristics are distributed according to the position of the blade. In addition, it has been reported that the stability of a single-channel pump can be improved by optimizing the flow characteristic or flow path between the blade and volute. These unsteady radial force sources were generally induced by the impeller–volute interaction, and consequentially, these sources adversely affected the entire hydraulic and mechanical performances (durability, life expectancy, etc.) of a single-channel pump. Therefore, an advanced design technique that can concurrently enhance hydraulic efficiency and decrease these unsteady sources is required.

To achieve multiple goals simultaneously, optimization methods using numerical analysis are widely used. Classical methods such as decomposition, Lagrangian techniques and cutting plane methods were previously applied to find optimal points. However, these classic optimization methods are ineffective for practical engineering problems. Hence, many studies were focused on exploratory optimization algorithms, such as a particle swarm optimization (PSO) [12] and genetic algorithm (GA) [13] to identify global optima, because the classic methods can be trapped in local optima according to the initially set value. These exploratory optimization methods combined with three-dimensional (3-D) RANS analysis have been generally used to find the best design solutions in the last decades—especially for turbomachinery designs [14,15]. Recently, a hybrid optimization algorithm was developed that combines the advantages of several optimization algorithms to improve computational efficiency. Robinson et al. [16] sought to combine GA and PSO and introduced an initial model of hybrid PSO-GA. Shi et al. [17] introduced an improved hybrid PSO-GA algorithm. The hybrid PSO-GA and GA were used to calculate benchmark test problems, and the performances of the algorithms were compared. The hybrid PSO-GA algorithm showed an improved performance with respect to the ability to identify optimal solutions and the speed of the convergence. Gandelli et al. [18] established another hybrid PSO-GA algorithm by applying a novel combinations strategy and tested its performance using various benchmark test problems. The optimization using this hybrid PSO-GA algorithm led to consistent and effective results for the benchmark problems.

In fact, as shown in the above literatures, even though many computational optimization methodologies on various pump designs have been proposed, no systematic design works have been attempted for single-channel pumps. In the present work, the geometric shapes of the volute and impeller flow paths of a single-channel pump were systematically optimized to concurrently enhance hydraulic efficiency and decrease the unsteady radial forces by using hybrid PSO-GA based on a surrogate model. Here, for multi-objective optimization, the hydraulic efficiency and unsteady radial forces resulting from impeller–volute interaction were used as objective functions. These were then approximated using a radial-bias neural network (RBNN) model [19]. Furthermore, as one of the design-of-experiment (DOE) methods, Latin hypercube sampling (LHS) [20] was employed to efficiently choose the design points within the design space. This work is the follow-up study of the authors' previous works [21–23]. In the previous works, the authors verified that the design technique combined with the numerical analysis and optimization method is a practical tool for realizing the high-efficiency and low-fluid-induced vibration of the single-channel pump. Based on these results, the importance for objective functions was understood. Then, two objective functions were selected to design the high-efficiency and low-fluid-induced vibration single-channel pump. In addition to the hybrid PSO-GA, the advanced technique is used to improve the prediction accuracy and performance of the optimal design. The hybrid PSO-GA proved to be superior through some benchmark function tests, but rarely applied to the engineering problems [16–18]. Therefore, in this study, it was verified whether the hybrid PSO-GA technique could be practically used in the actual pump design step.

2. Numerical Methods

2.1. Single-Channel Pump Model

In this study, from the Stepanoff theory applied in the authors' preceding works [21–24], the initial geometry of the volute and impeller of the single-channel pump was designed, as shown in Figure 1. The Stepanoff theory is a method of controlling the cross-sectional area to minimize the flow losses caused by the flow rate differences. The Stepanoff theory is normally used in a volute design, but in this study it was used for designing both the volute and impeller because they each consist of a single flow path. Therefore, the initial area distributions of the volute and impeller are continuously increased as the theta angle between the volute and impeller increases from the inlet to outlet, as shown in Figure 2.

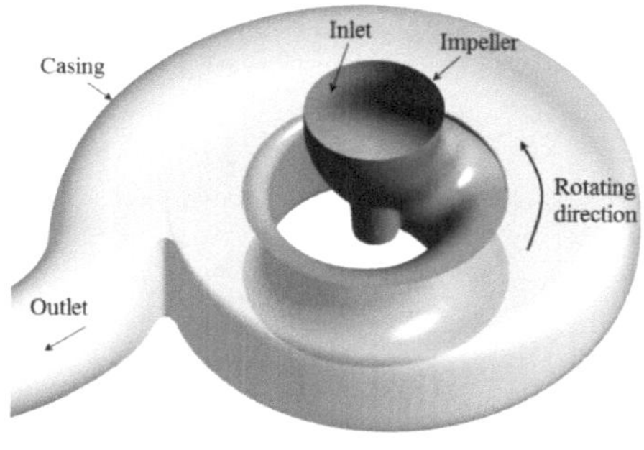

Figure 1. Geometry of single-channel pump.

As shown in Figure 2, the cross-sectional area of the impeller of the reference model was designed considering the minimum thickness in the range 0–70° of the impeller. The impeller was designed to have the maximum cross-sectional area at the 360°. In the case of the volute, as shown in Figure 2, the cross-sectional area was designed to constantly increase from 0 to 360°. A more detailed description can be referred to in the authors' previous work.

The head and flow coefficients, defined as below, for the reference pump are 0.074 and 0.019, respectively.

$$\psi = \frac{gH}{N^2D^2}, \tag{1}$$

$$\phi = \frac{Q}{ND^3}, \tag{2}$$

where, N, D, g, H and Q are the rotational speed, impeller diameter, gravity acceleration, total head and volume flow rate, respectively. The additional design information is listed in Table 1 [23].

Table 1. Design major specifications of the single-channel pump [23].

Flow coefficient (ϕ)	0.019
Head coefficient (ψ)	0.074
Rotational speed (RPM)	1760
Impeller inlet-outlet diameter ratio	1.9

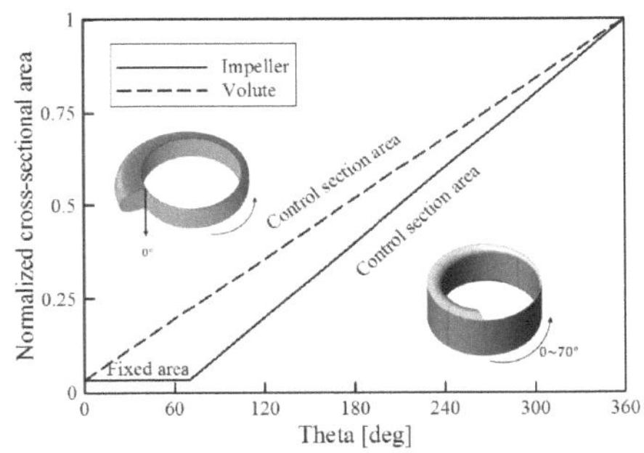

Figure 2. Cross-sectional area distributions along the theta angle of impeller and volute.

2.2. Numerical Analysis

Commercial code ANSYS CFX-15.0 was employed for the hydraulic analysis of the single-channel pump. For turbulence closure, the k-ω based shear stress transport (SST) model [25] was applied and three-dimensional steady and unsteady incompressible RANS equations were solved. Generally, the SST model combines a k-ω model of the near-wall flow and k-ε model of the bulk flow. A blending function was used to connect these two models. The SST model is well known to predict precisely flow separations under adverse pressure gradients.

The 3-D model for the volute and impeller was generated by Solidworks 2015 and the computational grids system for numerical analysis was created by ICEM-CFD. Since the computational domain includes both stationary and rotating domains, the specific processing technique provided by ANSYS CFX-15.0 was used at the interface. The stage (or mixing plane) and the transient Rotor-Stator interface methods were used at the interface between the rotating and stationary domains in the RANS and unsteady RANS analyses, respectively.

Freshwater was employed for working fluid and the inlet boundary condition of the pump was set to atmospheric pressure. The surfaces of the volute and impeller are considered as adiabatic and no-slip conditions. The outlet boundary condition was set to the mass flow rate, and the numerical analysis was performed while changing the mass flow rate.

The grid system consists of a tetrahedral grid in the volute and impeller domains. In this study, y^+ was kept below 2 to use the low Reynolds SST model, so that the flow near the wall was accurately analyzed. To eliminate the grid dependency of the numerical solutions, grid dependency test was performed at the design mass flow rate for the hydraulic performance of the pump. Based on the results in the previous work [23], the optimal grids system (a total of 2,500,000 nodes; 1,300,000 nodes in the impeller domain and 1,200,000 nodes in the volute domain) was selected, as shown in Figure 3.

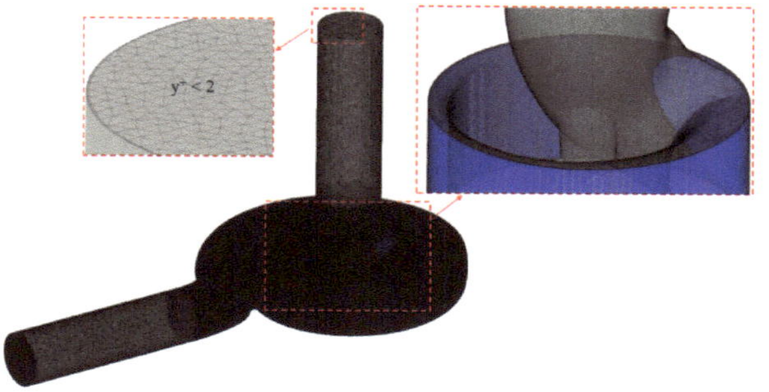

Figure 3. Grid systems of computational domain.

To determine the convergence of the numerical calculations, the root-mean-square (RMS) values of the residuals of the governing equations were set to be less than 10^{-5}. The calculation for the steady RANS analysis were performed using an Intel Xeon CPU (2.70 GHz) processor and the CPU running time for each analysis was about 4 h.

Steady RANS results were employed as initial values for unsteady RANS analysis, and coefficient loop with time scale control and time step were set to three times and 0.000947 s, respectively. Unsteady RANS analysis results were obtained after five revolutions (180 iterations with a total time of 0.01704775 s), and the CPU running time for each analysis was approximately 8 h. Through unsteady RANS analysis, characteristics of radial force sources due to flow interaction at the interface between the rotating and stationary domains were obtained and analyzed.

3. Optimization Techniques

The flowchart for the multi-objective optimization process using surrogate modeling is shown in Figure 4. First, the objective functions and constraints are defined according to the design goal. Subsequently, the design variables considering constraints are chosen. The LHS [20] is used as DOE to select the design points and the objective functions are evaluated by (U)RANS analysis at these design points. The next step is the construction of surrogate model to approximate the objective functions and a searching algorithm, that is, hybrid PSO-GA, is used to combine and formulate these surrogate models. Finally, Pareto-optimal solutions—that is, a collection of non-dominated solutions—are derived and the multi-objective optimization procedure is terminated [26].

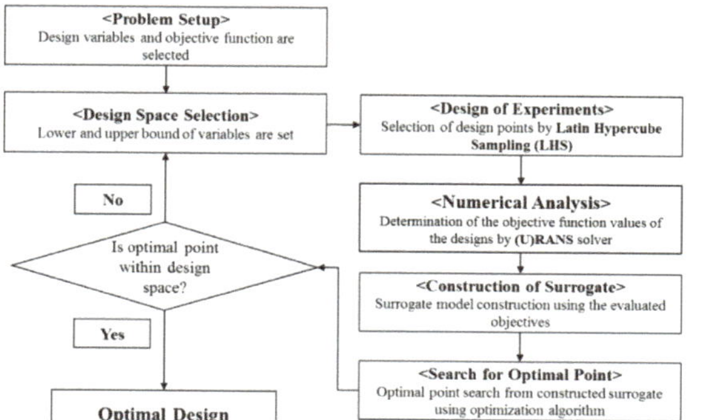

Figure 4. Optimization procedure.

3.1. Optimization Goal

The design goals of the current multiple optimization were to concurrently reduce the radial force sources and improve the hydraulic efficiency. The hydraulic efficiency, one of the objective functions, is defined as follows:

$$F_\eta = -\frac{\rho g H Q}{P}, \tag{3}$$

where, H, Q, ρ, g, and P represent total head, volume flow rate, the density, gravity acceleration, and power, respectively.

Other objective functions for the radial force sources are the distance of the mass center of the sweep area from the origin and the sweep area of the radial force during one revolution, as follows:

$$A_s = \frac{1}{2}\sum_{i=0}^{n-1}(x_i y_{i+1} - x_{i+1} y_i), \tag{4}$$

where, A_s represents the signed area of the polygon, considered as the sweep area of the radial force during one revolution. The centroid of a non-self-intersecting closed polygon defined by n vertices (x_0, y_0), (x_1, y_1), …, ($x_{(n-1)}$, $y_{(n-1)}$) is defined as the point (C_x, C_y), as follows:

$$C_x = \frac{1}{6A_s}\sum_{i=0}^{n-1}(x_i + x_{i+1})(x_i y_{i+1} - x_{i+1} y_i), \tag{5}$$

$$C_y = \frac{1}{6A_s}\sum_{i=0}^{n-1}(y_i + y_{i+1})(x_i y_{i+1} - x_{i+1} y_i). \tag{6}$$

The vertices in these equations are defined as being numbered in sequence along the perimeter of the polygon. Hence, the distance of the mass center of the sweep area from the origin can be finally defined as follows:

$$F_{radial} = \sqrt{C_x^2 + C_y^2}. \tag{7}$$

Thus, in order to fulfil the design goals described above, multi-objective optimization was performed using F_η and F_{radial} as objective functions.

In this study, five geometric parameters associated to the internal flow characteristics of the pump were chosen as design variables to control the cross-sectional area of the volute and impeller. The distribution of the cross-sectional area of the internal flow of impeller and volute can be controlled smoothly by adjusting the control points represented by a third and fourth order Bezier-curve, respectively, as shown in Figure 5. Hence, the variations in the y-axes for two control points (CP_1 and CP_2) for the impeller and three control points (CP_3, CP_4 and CP_5) for the volute were selected. Figure 5 shows the defined design variables and ranges are listed in Table 2.

Table 2. Ranges of design variables.

	LB	Ref.	UB
CP 1	0.03	0.32	0.61
CP 2	0.42	0.71	1.00
CP 3	0.00	0.25	0.50
CP 4	0.00	0.50	1.00
CP 5	0.50	0.75	1.00

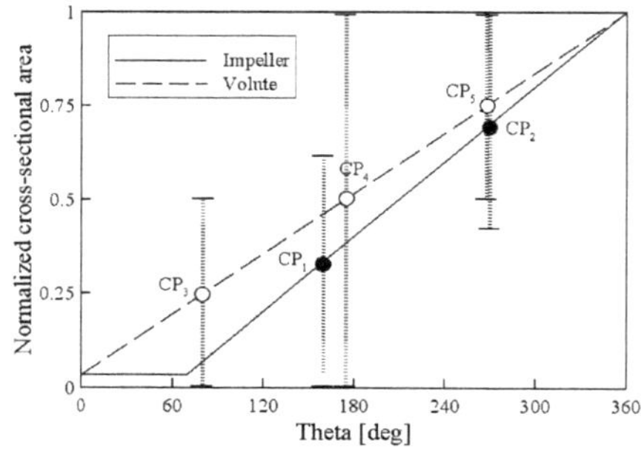

Figure 5. Defined design variables.

3.2. Surrogate Modeling

In the present work, 54 design points were generated for the five design variables by LHS using the MATLAB function 'lhsdesign' [27]. LHS is an effective sampling technique that uses an $m \times n$ simulation matrix where m is the number of levels (sampling points) to be examined and n is the number of design parameters. Each of the n columns of the matrix containing the levels, 1, 2, ... , m, is randomly paired to form a Latin hypercube. This approach produces random sample points, ensuring that all portions of the design space are represented. The objective function values at these design points were calculated by (U)RANS analysis, as shown in Table A1 (Appendix A).

The RBNN model was applied for the surrogate modeling from the objective functions. The RBNN as a neural network model, presented in Figure 6, consists of a hierarchical structure that starts at the input stage and ends at the output stage, where each stage is known as a layer. There are intermediate layers between the input and output stages, known as hidden layers. By controlling the number and size of these hidden layers, arbitrary complex functions can be learned, and the surrogate models can be constructed. The RBNN includes a two-layered hidden network containing a layer of radial neurons and an output of linear neurons. The hidden layer uses a series of radial primitives to nonlinearly modify the input space to the intermediate space. The output layer then executes a linear combiner to produce the desired target. In this study, MATLAB built-in code 'radbas' and 'newrb' were used to perform radial basis function-based machine learning [27]. 'Newrb' makes a two-layer network. As mentioned above, the first layer constructed by 'radbas' is radial basis neuron network and the second layer is linear neuron network. This combination model $f(\mathrm{x})$ for the formulation can be represented as a linear combination of a set of N radially-symmetric functions as follow:

$$f(\mathrm{x}) = \sum_{j=1}^{N} w_j \varnothing_j, \tag{8}$$

where w_j is the weight and Φ_j is radial basis function. Machine learning finds the input vector with the largest error in the network and adjusts the weight or neurons of this vector to minimize the error in the network. This process is repeated until the network's mean squared error falls below the set goal. The construction process can be accessed through graphical user interface (GUI) environment using the application 'neural net fitting', which is supported within MATLAB [27].

In order to construct the RBNN model, an initial value must be assigned to the spread constant (SC), that is, a constant for the first hidden layer (nonlinear function) of the neural network. Because machine learning is based on this initial SC value, it should be chosen carefully.

The network training was implemented by changing the SC values to identify the minimum cross-validation error. The parameters SC_1 and SC_2 are constants that are utilized to configure the surrogate model with respect to the efficiency and radial force, respectively. The SC values were chosen using a k-fold cross-validation test [28], which is a validation method that estimates how the results of statistical analysis are generalized to independent datasets. The process of the k-fold cross-validation test used in the present work is described as follows:

Step 1 Construct a surrogate model using the 53 experimental points, except for one point of the 54 experimental points.

Step 2 Compare the value of the objective function at the location of the experimental point excluded from Step 1 (between CFD simulation value and predicted value by the surrogate model).

Step 3 This process is carried out at all experimental points. Then, evaluate the sum of the errors between predicted and CFD simulation values.

As shown in Figure 7, the error was minimized when the values of SC_1 and SC_2 were 0.2 and 1.0, respectively. Neural networks with various numbers of neurons were trained 10 times and the averaged adjusted R^2 values are compared (Figure 8). The same quantity of neurons was used in each layer; when using 12 neurons, the statistically most accurate models with adjusted R^2 equals to 1.00 and 0.78 could be achieved.

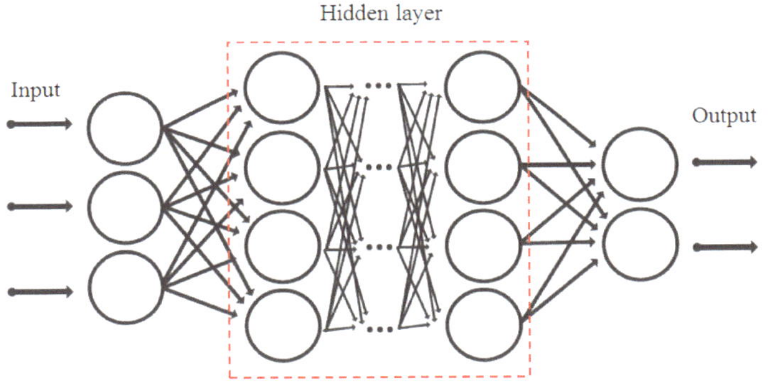

Figure 6. Schematic of neural network.

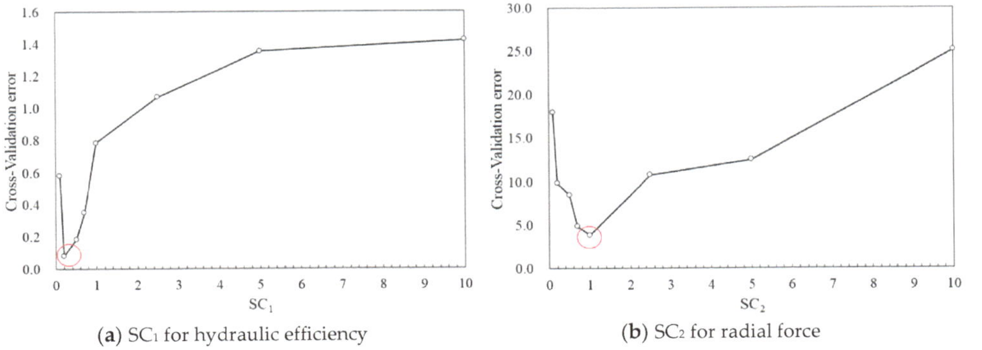

(**a**) SC_1 for hydraulic efficiency (**b**) SC_2 for radial force

Figure 7. Cross-validation error vs. Spread constant (SC) value.

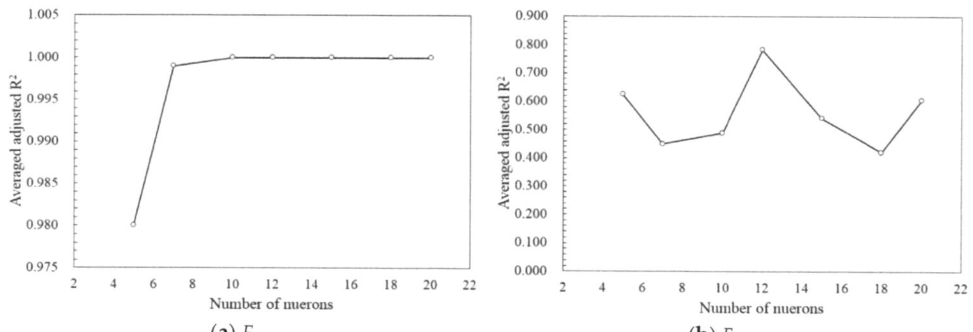

Figure 8. Comparison of averaged adjusted R^2 of RBNN.

3.3. Searching Algorithm

Once the surrogate model is constructed, it is necessary to find the optimal point on the fitness assignment, which is the output in the form of a function of the surrogate model. In the present work, a hybrid PSO-GA algorithm that was most recently developed [18] was used as searching algorithm to identify a global optimal solution within the fitness assignment. It is known to have a high reliability. Hybrid strategies combine two or more algorithms to maximize the benefits of each algorithm. One benefit of the PSO algorithm is its simplicity and another one is that the PSO algorithm controls the convergence criteria to identify the optimal solution in a short time. The GA, on the other hand, has the advantage of finding the global optimal solution by controlling the genetic search operators—that is, the crossover and mutation rates—which affect the convergence. The locations of the global best particles are updated by the genetic search operators to prevent premature convergence. By using the genetic search operators, the locations of the particles can be exchanged among the particles and thus have the ability to escape the local optima.

The procedure of the hybrid PSO-GA algorithm is represented in Figure 9 [15]. This algorithm shares the initial number of particles of the PSO with the number of populations of the GA. The first step of this procedure is to identify an initial optimal solution using the particles of the PSO. The next step is to modify the locations of the particles using the genetic search operators and identify the improved optimal solution. The procedure is competed when the termination criteria are met by referring to the updated location; otherwise, it is repeated. In the present work, the following parameters and termination criteria were used: population size (i.e., number of particles) = 400, crossover fraction = 0.35, mutation fraction = 0.70, generations = 1000, function tolerance = 10^{-6}, and initial velocity of particles = 2.5. The pseudo-code of the hybrid PSO-GA used in this study is shown in Table A2 (Appendix B).

3.4. Optimization Results

For the optimization, hybrid PSO-GA with RBNN was used to search the Pareto-optimal front in a feasible solution space bounded by the lower and upper bounds of the design variables. Pareto-optimal solutions with the optimum trade-off between the two conflicting objective functions are plotted in Figure 10.

The arbitrary optimum design (AOD) was obtained as the final design near the center of the Pareto-optimal front, to improve the prediction of both the objective functions compared to the reference. The values of the design variables and objective functions of the AOD, along with those for the reference design, are listed in Table 3. The AOD showed an improvement of 3.07% and 44.80% for the hydraulic efficiency and radial force, respectively, compared with the reference design. In contrast, the relative errors between (U)RANS and the predicted values for these objective functions were 0.16% and 8.77%, respectively.

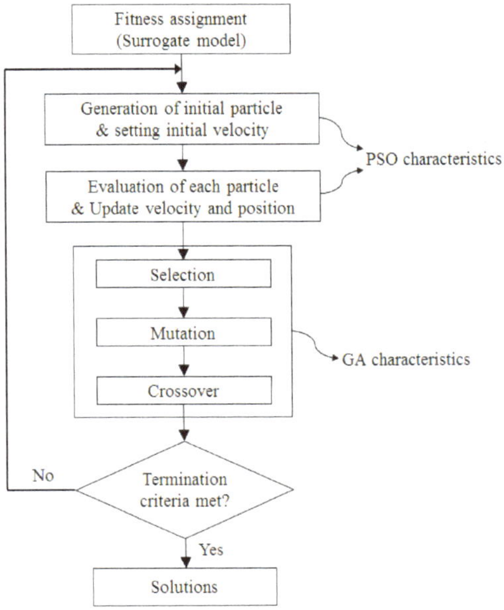

Figure 9. Algorithm of hybrid PSO-GA.

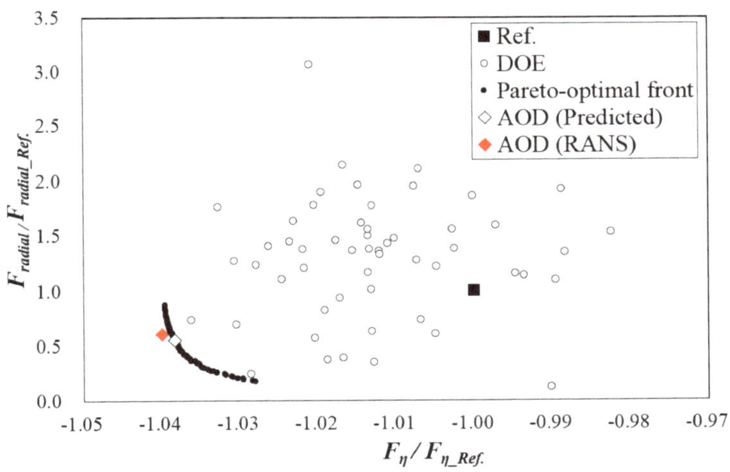

Figure 10. Pareto-optimal solutions.

Table 3. Results of optimization.

	Design Variables					Predicted Values		(U)RANS		Relative Error (%)	
AOD	CP 1	CP 2	CP 3	CP 4	CP 5	$F_{\eta}/F_{\eta\_Ref.}$	$F_{radial}/F_{radial\_Ref}$	$F_{\eta}/F_{\eta\_Ref.}$	$F_{radial}/F_{radial\_Ref}$	$F_{\eta}/F_{\eta\_Ref.}$	$F_{radial}/F_{radial\_Ref}$
	0.600	0.004	0.003	0.214	0.837	−1.038	0.5520	−1.040	0.6050	0.16	8.77

4. Results and Discussion

4.1. Unsteady Analyses of Internal Flow Field

In a previous work of the authors [29], the accuracy of the numerical analysis used in this work was proved in comparison with the experimental result and then the results of the numerical analysis were consequentially considered valid.

To quantitatively verify the improvement in hydraulic performance of the AOD owing to the optimization, the internal flow paths for the reference model and AOD were compared systematically with their cross-sectional area distributions, as shown in Figures 11–17. Figure 11 shows the cross-sectional area distributions in the single-channel impeller and volute for the reference model and AOD. The cross-sectional area distribution of the flow path of the impeller in AOD most closely resembles a concave shape, as shown in Figure 11a. This implies that the flow path in the impeller of AOD was narrowed slightly compared with the reference model. The cross-sectional area distribution in the region of the volute upstream (from 0 to 240°) in AOD also decreased considerably, while the cross-sectional area distribution in the downstream region after 240° increased slightly compared with the reference model, as shown in Figure 11b. The concave flow path in the region of the impeller downstream and volute upstream was found to be considerably sensitive for the high efficiency and low radial force design of the single-channel pump. As a result, it can be concluded that the single-channel impeller and volute shapes should be optimized concurrently to design a single-channel pump with high hydraulic efficiency and low radial forces.

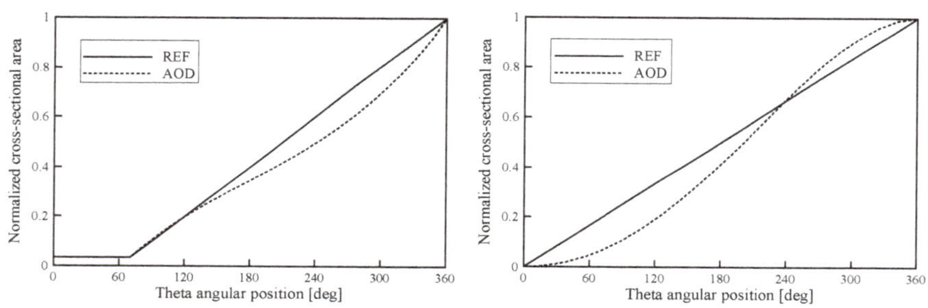

(**a**) Impeller (**b**) Volute

Figure 11. Cross-sectional area distributions of the flow path.

(**a**) Reference model (**b**) Arbitrary optimum design (AOD)

Figure 12. Time-averaged iso-surfaced contours at a low velocity of 1.5 m/s.

Figure 12 shows time-averaged iso-surfaced contours having a low velocity component (1.5 m/s), which is the flow separation zone, for the reference model and AOD. For the reference impeller, the large flow separation zone formed along with the internal annulus wall in the single-channel impeller, as shown in Figure 12a. On the contrary, the flow separation zone was reduced remarkably in the downstream region of the optimized impeller, as shown in Figure 12b.

Figure 13 shows the time-averaged streamline distributions in the impeller surface wall of the reference and optimized models. Extreme flow separation was occurred in the hub wall by abruptly changing the flow path for both models. However, the limited streamline in the AOD was moved from the impeller outside wall to the inside wall by the optimized flow path. It can be seen that these changing flows contributed to the decrease in flow separation and increase in the hydraulic efficiency owing to optimization. Thus, the decrease in the flow separation from the impeller can be made by the smooth pressure increase by matching with the optimized volute shape, as shown in the time-averaged pressure contours on the mid-span in Figure 14. These results illustrate the improvement in the entire hydraulic efficiency of the pump.

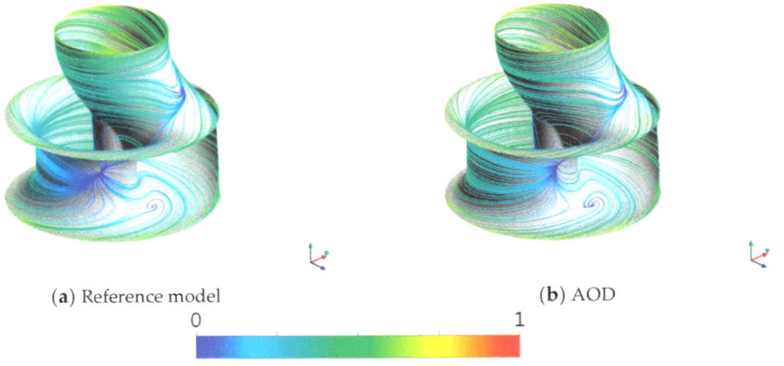

Figure 13. Time-averaged pressure contours on impeller surface wall.

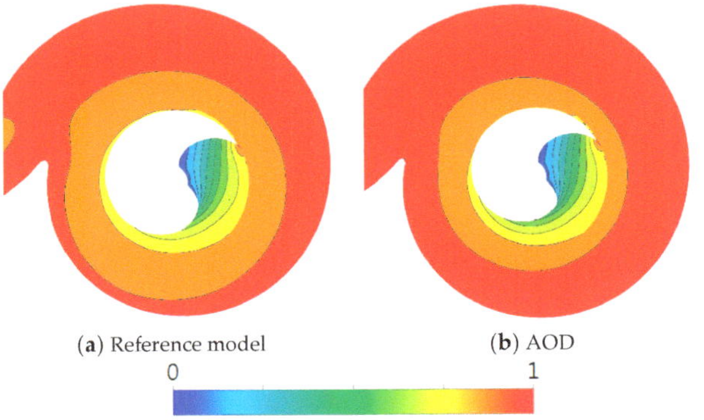

Figure 14. Time-averaged pressure contours at mid-span.

The sweep area distributions generated by the unsteady radial force components during the impeller one rotation for the reference model and AOD are plotted in Figure 15; they are averaged at the cylindrical surface near exit of the single-channel impeller. Here, all F_x and F_y values of the radial force were normalized based on the maximum radial force of the reference design. As shown in Figure 15, the sweep area distribution in the reference design leaned a little toward the fourth quadrant, whereas it was closer to the origin for AOD. Moreover, the sweep area in AOD reduced significantly compared with the reference model. As discussed previously in Section 3.4, the radial force in AOD decreased by 14.73, compared with the reference design.

Figure 16 shows the unsteady fluctuation distributions in the net radial forces during the impeller one rotation in the reference model and AOD. Here, the net radial force values were normalized using the maximum net radial force value of the reference design. The fluctuation amplitude in AOD generally decreased during one revolution of the impeller, especially for flat formed angles from 0 to 240°. In addition, the net radial force distribution in AOD was lower and mostly flat compared with the normalized values of the reference model. The remarkable reduction in the radial force sources can be explained clearly by the optimization of this work.

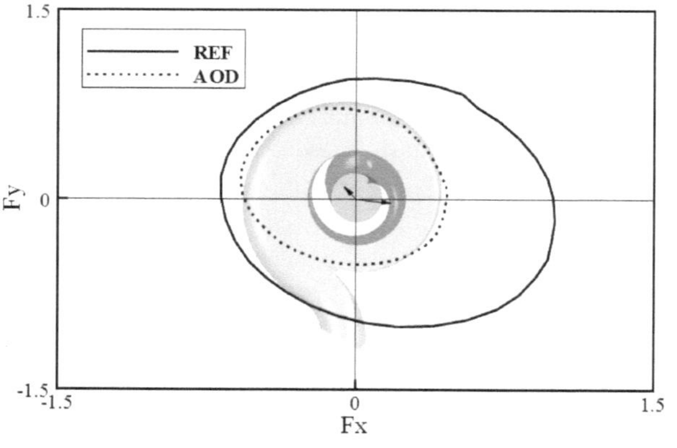

Figure 15. Sweep area distributions during one revolution of impeller.

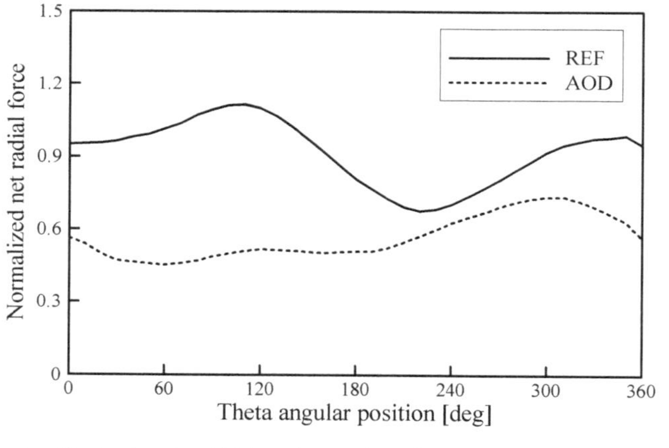

Figure 16. Unsteady net radial force fluctuations during one rotation of impeller.

The instantaneous change in the unsteady pressure values on the cylindrical surface near the exit of the impeller for the reference model and AOD are shown in Figure 17. Here, the impeller one rotation is divided into five steps to clearly explain the variation in the flow phenomenon with time during impeller one rotation. In the reference model, the pressure variation caused by the impeller–volute interaction is prominent and then increases gradually, especially in Figure 17b. As discussed previously in Figure 15, the high-pressure zone is the fourth quadrant where the sweep area distribution leans slightly. Owing to this repetitive rotation, fluid-induced vibration caused by unsteady radial forces are observed. These unbalancing phenomena have an adverse effect on pump performance. In contrast, unsteady pressure distribution is generally uniform during the impeller one rotation; especially, at the same time of the reference model. With these trends, the high-pressure variation induced by the impeller–volute interaction clearly disappeared in Figure 17. Therefore, the AOD resulted in mostly stable flows in the pump with high efficiency and low radial forces. There is a remarkable decrement in the fluid-induced vibration induced by the impeller–volute interaction from the current multiple optimization result.

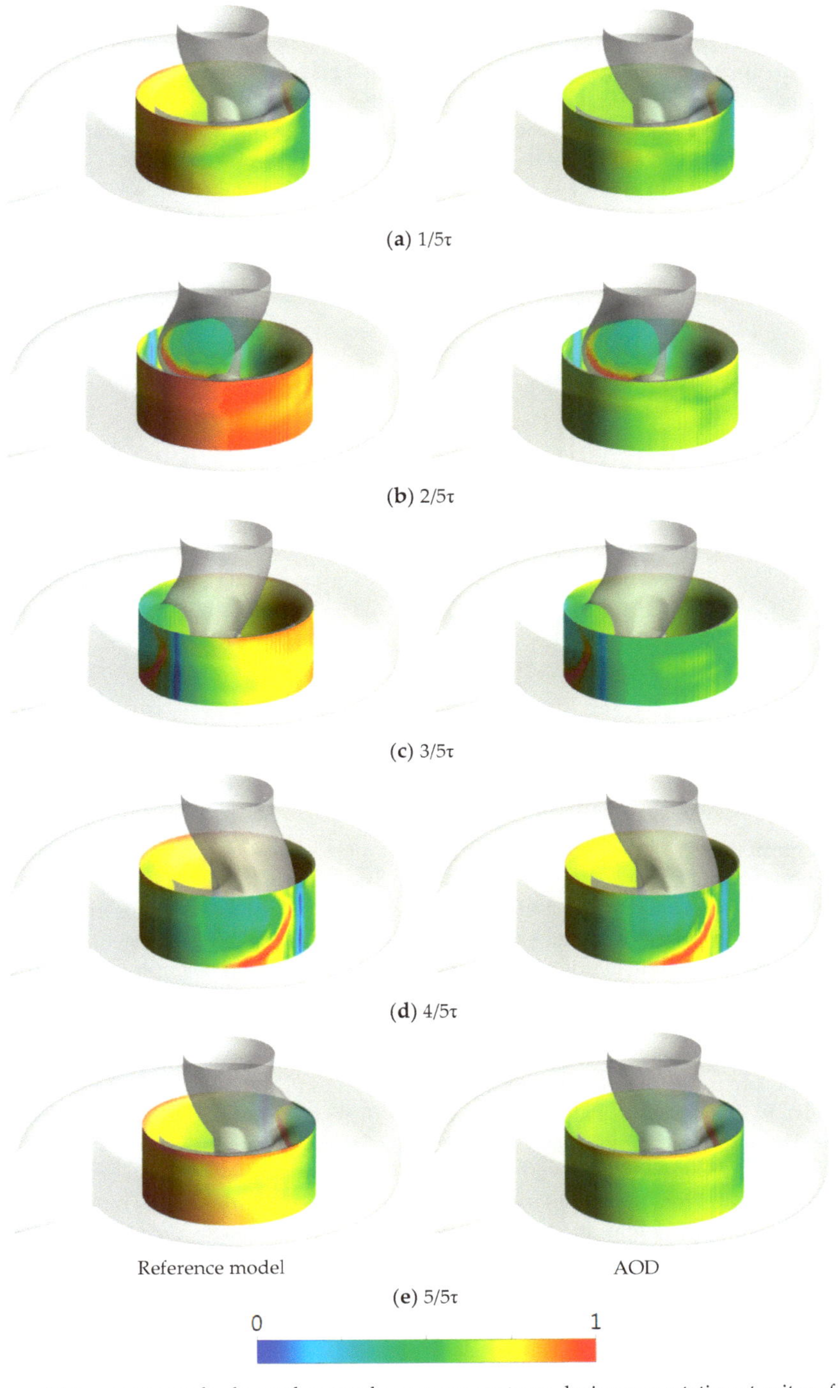

Figure 17. Instantaneously changed unsteady pressure contours during one rotation at exit surface of impeller.

Figure 18 shows the spectra of the normalized amplitude values based on the wall pressure fluctuations measured at the observation point on the casing wall near the tongue of the volute located in the fourth quadrant in Figure 15. In Figure 18, the first blade passing frequency (BPF) is measured at approximate 30 Hz, and this is represented by the fluid-induced vibration characteristics from the impeller–volute interaction. These peak amplitude values are clarified at every harmonic BPF in 30 Hz steps due to the periodic motion of impeller rotation. In the AOD, a remarkable decrement in the amplitude values at the first BPF was occurred, and the decrement in fluid-induced vibration induced by the impeller–volute interaction was confirmed.

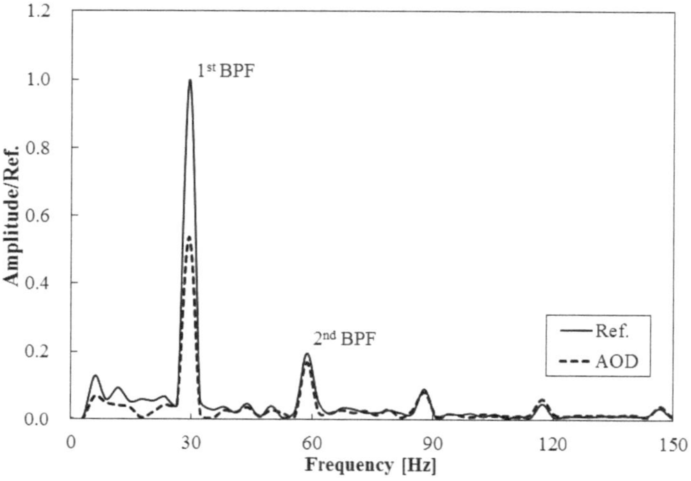

Figure 18. Spectra of amplitude values on casing wall near the tongue of volute.

4.2. *Performance Verification of the Optimized Prototype Model*

To confirm the hydraulic performance of the optimized model obtained through the numerical optimization technique, the prototype model was manufactured in Figure 19. Figure 20 shows the schematic of the experimental apparatus constructed in Korea Institute of Industrial Technology to measure the hydraulic head and efficiency of the prototype model. The experimental apparatus was constructed based on the standards of KS B 6301 and 6321. In addition, the measuring devices for the pressure sensor, temperature sensor, magnetic flow meter, and digital power meter were calibrated from the qualified office of Korea. Figure 21 shows the experimental results for the head and efficiency of the prototype model, with the numerical analysis results. As shown in Figure 21, the head coefficient curves of the experimental and numerical results are similar in the entire range, the head value, in particular, is almost the same at the design flow coefficient. In contrast, the efficiency curves of the numerical analysis including the motor efficiency are relatively higher than those of the experimental results, because the numerical results do not include the mechanical losses. However, the general curve trend matched well. From these results, the required single-channel pump could be designed through the proposed numerical optimization technique; hence, the technique was demonstrated to be feasible.

Figure 19. Prototype model for experimental test.

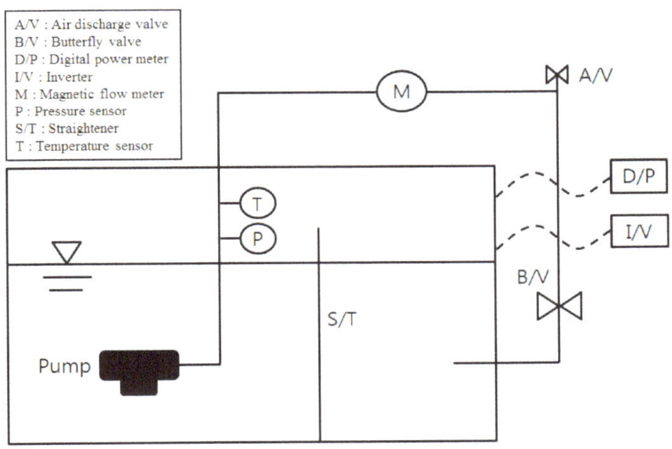

Figure 20. Experimental apparatus for pump test.

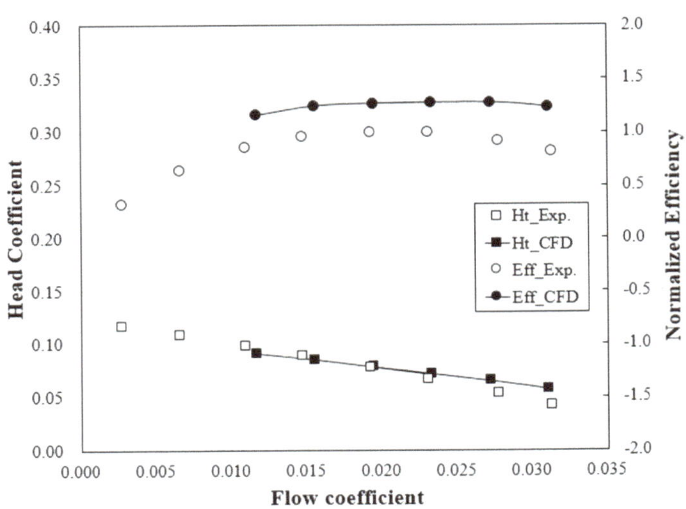

Figure 21. Verification results of prototype model.

5. Conclusions

In this study, a multiple optimization technique using hybrid PSO-GA was introduced to effectively enhance the efficiency and radial force sources with fluid-induced vibration of the single-channel pump. 3-D steady and unsteady RANS analyses were performed to investigate the hydraulic characteristics of the single-channel pump using ANSYS CFX-15.0. As the design variables, design parameters for controlling the internal cross-sectional areas of the single-channel impeller and volute were chosen in the authors' previous study. Two performance functions, i.e., hydraulic efficiency and radial force, were defined as the objective functions. The optimization based on the surrogate model was performed considering reasonable calculation time. The RBNN method was used as a surrogate model, and it was repeatedly constructed 10 times with a different number of neurons constituting the layer for statistical analysis. As a result, when 12 neurons were used, the most accurate surrogate model was constructed. Through optimization with RBNN consisting of 12 neurons, the hydraulic efficiency of the optimum design increased by 3.07%, and the radial force caused by impeller–volute interaction reduced by 44.80%, compared to the reference design.

In the optimum design, the flow path in the impeller was narrow in comparison with the reference design. The flow path in the volute upstream of the optimum design was narrowed, while that in the volute downstream was slightly increased, in comparison with the reference design. These structural changes balanced the pressure distribution along the impeller–volute interface and delayed the flow separation on the impeller surface. As a result, it was confirmed through (U)RANS analysis that the fluid-induced vibration of the optimum design reduced in comparison with the reference design. The optimum design obtained in the present study was used to manufacture a prototype model, and the hydraulic performances, i.e., efficiency and head, were verified experimentally. Thus, reasonable results were obtained with the required design specification. However, the precise vibration test was not conducted, and only the vibration values were measured briefly using the simple vibration meter. This is because it was very difficult to simultaneously measure detailed vibration with the hydraulic performances in the experimental apparatus, and there was no submersible three-axis vibration sensor. Therefore, in a future work, the vibration characteristics of the single-channel pump will be investigated in more detail using a three-axis vibration meter, and the flow characteristics including solid wastes will be studied.

Author Contributions: Conceptualization, J.-H.K., S.K. and Y.-S.C.; methodology, S.-B.M.; software, J.-H.K., S.K. and S.-B.M.; validation, J.-H.K., S.K. and Y.-S.C.; formal analysis, J.-H.K. and S.-B.M.; investigation, J.-H.K., S.K. and S.-B.M.; writing original draft preparation, J.-H.K. and S.-B.M.; supervision, J.-H.K. and K.-Y.K.; project administration, J.-H.K; funding acquisition, J.-H.K;

Funding: This work was supported by the Demand-based-Manufacturing Technique Commercialization R&D Project of the Korea Institute of Industrial Technology (KITECH) (No. JB190004), which was funded by the Ministry of Science and ICT (MSIT). The authors gratefully acknowledge this support.

Conflicts of Interest: The author(s) declared no potential conflicts of interest with respect to the research, authorship, and/or publication of this article.

Nomenclature

AOD	Arbitrary optimum design
BPF	Blade passing frequency
CFD	Computational fluid dynamics
CP	Control point
D	Diameter of impeller
DOE	Design of experiments
FSI	Fluid-structure interaction
g	Gravity acceleration
GA	Genetic algorithm
H	Total head

LDV	Laser Doppler velocimetry
LHS	Latin hypercube sampling
N	Rotational speed
P	Power
PSO	Particle swarm optimization
Q	Volume flow rate
RANS	Reynolds-averaged Navier-Stokes
RBNN	Radial basis neural network
SC	Spread constant
SST	Shear stress transport
(U)RANS	Unsteady Reynolds-averaged Navier-Stokes
ρ	Density
ϕ	Flow coefficient
Ψ	Head coefficient

Appendix A

Table A1. Design variables and objective functions values at LHS design points.

	CP 1	CP 2	CP 3	CP 4	CP 5	$F_{\eta}/F_{\eta\_Ref.}$	$F_{radial}/F_{radial\_Ref}$
1	0.000	0.000	1.000	0.000	0.000	−1.021	3.064
2	0.000	0.000	0.000	0.000	1.000	−1.033	1.759
3	0.000	0.000	1.000	1.000	0.000	−1.017	2.144
4	1.000	1.000	0.000	0.000	1.000	−0.982	1.527
5	0.265	0.306	0.429	0.980	0.837	−1.026	1.403
6	1.000	0.469	0.286	0.592	0.571	−0.994	1.133
7	0.551	0.735	0.327	0.490	0.000	−1.012	1.355
8	0.959	0.184	0.857	0.531	0.327	−1.007	2.109
9	0.020	0.857	0.224	0.449	0.265	−1.014	1.161
10	0.939	0.653	0.796	0.673	0.878	−0.988	1.344
11	0.449	0.510	0.735	0.000	0.469	−1.016	1.362
12	0.347	0.408	0.776	0.918	0.163	−1.015	1.958
13	0.694	0.041	0.510	0.694	0.694	−1.024	1.441
14	0.633	0.878	0.122	0.735	0.633	−1.003	1.377
15	0.796	0.714	0.592	0.061	0.082	−1.000	1.861
16	0.571	0.265	1.000	0.388	0.714	−1.020	1.778
17	0.857	0.245	0.061	0.184	0.898	−1.013	0.628
18	0.918	0.694	0.449	0.122	0.776	−1.013	0.347
19	0.388	0.959	0.551	0.265	0.408	−1.011	1.429
20	0.245	0.918	0.980	0.755	0.245	−1.014	1.495
21	0.184	0.000	0.184	0.551	0.755	−1.036	0.738
22	0.306	0.020	0.878	0.857	0.551	−1.023	1.630
23	0.837	0.061	0.633	0.143	0.673	−1.013	1.011
24	0.327	0.429	0.020	0.510	0.306	−1.020	0.569
25	0.878	0.980	0.245	0.327	0.367	−0.990	0.122
26	0.469	0.939	0.673	0.837	0.816	−1.003	1.557
27	0.429	0.796	0.000	0.224	0.612	−1.029	0.244
28	0.898	0.571	0.694	0.939	0.347	−0.997	1.591
29	0.816	0.837	0.918	0.306	0.490	−0.989	1.920
30	0.714	0.143	0.367	0.653	0.143	−1.013	1.559
31	0.143	0.082	0.653	0.082	0.592	−1.031	1.271
32	0.735	0.592	0.898	0.612	0.020	−1.008	1.950
33	0.286	0.776	0.959	0.286	0.796	−1.007	1.275
34	0.224	0.327	0.265	0.041	0.102	−1.025	1.103
35	0.061	0.286	0.306	0.878	0.204	−1.022	1.207
36	0.755	0.490	0.041	0.163	0.184	−1.007	0.738

Table A1. *Cont.*

	CP 1	CP 2	CP 3	CP 4	CP 5	$F_\eta/F_{\eta\_Ref.}$	$F_{radial}/F_{radial\_Ref}$
37	0.102	0.633	0.143	0.633	0.857	−1.019	0.373
38	0.673	1.000	0.612	0.776	0.286	−0.995	1.152
39	0.980	0.755	0.102	0.816	0.122	−0.989	1.095
40	0.776	0.122	0.490	0.102	0.061	−1.018	1.457
41	0.612	0.347	0.082	0.714	0.918	−1.019	0.821
42	0.000	0.388	0.469	0.429	0.449	−1.022	1.374
43	0.592	0.449	0.163	1.000	0.388	−1.012	1.328
44	0.490	0.367	0.531	0.245	0.939	−1.017	0.387
45	0.367	0.163	0.755	0.408	0.224	−1.019	1.892
46	0.510	0.531	0.571	0.571	0.510	−1.013	1.373
47	0.408	0.898	0.408	0.347	1.000	−1.005	0.605
48	0.204	0.816	0.347	0.959	0.429	−1.005	1.218
49	0.653	0.224	0.939	0.898	0.980	−1.013	1.769
50	0.041	0.551	0.837	0.796	0.653	−1.010	1.476
51	0.122	0.612	0.388	0.020	0.735	−1.017	0.936
52	0.163	0.673	0.714	0.367	0.041	−1.014	1.613
53	0.082	0.204	0.816	0.469	0.959	−1.028	1.234
54	0.531	0.102	0.204	0.204	0.531	−1.030	0.696

Appendix B

Table A2. Pseudo-code of the hybrid PSO-GA.

```
g = 0          * g: generation number
for i = 1 to M do   * M: population (particles) size
      Initialize particles of PSO x_i to random values
      x_i^b = x_i   * x^b: initial information of particle
      F_i = f(x_i)   * f: fitness assignment
end for
x_g^b = best{x_i^b; i = 1, … , M}   * x_g: initial global best particle
Pop = {x_1, x_2, … , x_M}
F = {F_1, F_2, … , F_M}
<Main Loop>
while do
     {Evaluation Loop 1}
     for i = 1 to M do
     if f(x_i) is better than f(x_i^b) then
     x_i^b = x_i
     end if
if f(x_i^b) is better than f(x_g^b) then
     x_g^b = x_i^b
     end if
     end for
     {Genetic Operators – Update particles' position}
     Pop ← Selection(Pop, F)
     Pop ← Crossover(Pop, C)      * C: crossover rate
     Pop ← Mutation(Pop, M)   * M: mutation rate
     {Evaluation Loop 2}
     for i = 1 to M do
     F_i = f(x_i)
    end for
F = {F_1, F_2, … , F_M}
g = g+1
end while
```

References

1. Steinmann, A.; Wurm, H.; Otto, A. Numerical and Experimental Investigations of the Unsteady Cavitating Flow in a Vortex Pump. *J. Hydrodyn.* **2010**, *22*, 319–324. [CrossRef]
2. Nishi, Y.; Matsuo, N.; Fukutomi, J. A Study on Internal Flow in a New Type of Sewage Pump. *J. Fluid Sci. Technol.* **2009**, *4*, 661–671. [CrossRef]
3. Liu, J.; Wang, R. Simulation and Research on Flow of Whirl Cavity in a Vortex Pump. In Proceedings of the International Conference on Computer Design and Applications, Qinhuangdao, China, 25–27 June 2010.
4. Hansen, B.F.; Henning, P.J. Waste Water Pump. U.S. Patent Application US9500204B2, 11 November 2016.
5. Qin, W.; Tsukamoto, H. Theoretical Study of Pressure Fluctuations Downstream of a Diffuser Pump Impeller-Part 1: Fundamental Analysis on Rotor-Stator Interaction. *J. Fluid Eng.* **1997**, *199*, 647–652. [CrossRef]
6. Shi, F.; Tsukamoto, H. Numerical Study of Pressure Fluctuations Caused by Impeller-Diffuser Interaction in a Diffuser Pump Stage. *J. Fluid Eng.* **2001**, *123*, 466–474. [CrossRef]
7. Feng, J.; Benra, F.K.; Dohmen, H.J. Investigation of Periodically Unsteady Flow in a Radial Pump by CFD Simulations and LDV Measurements. *J. Turbomach.* **2010**, *133*, 011004. [CrossRef]
8. Zhang, M.; Tsukamoto, H. Unsteady Hydrodynamic Forces due to Rotor-Stator Interaction on a Diffuser Pump with Identical Number of Vanes on the Impeller and Diffuser. *J. Fluid Eng.* **2005**, *127*, 743–751. [CrossRef]
9. Benra, F.K. Numerical and Experimental Investigation on the Flow Induced Oscillations of a Single-Blade Pump Impeller. *J. Fluid Eng.* **2006**, *128*, 783–793. [CrossRef]
10. Nishi, Y.; Fujiwara, R.; Fukutomi, J. Design Method for Single-Blade Centrifugal Pump Impeller. *J. Fluid Sci. Technol.* **2009**, *4*, 786–800. [CrossRef]
11. Pei, J.; Yuan, S.; Yuan, J. Numerical analysis of periodic flow unsteadiness in a single-blade centrifugal pump. *Sci. China Technol. Sci.* **2013**, *56*, 212–221. [CrossRef]
12. Eberhart, R.C.; Kennedy, J. A New Optimizer Using Particle Swarm Theory. In Proceedings of the Sixth International Symposium on Micro Machine and Human Science, Nagoya, Japan, 4–6 October 1995; pp. 39–43.
13. Deb, K.; Agrawal, S.; Pratap, A.; Meyarivan, T. A Fast Elitist Non-Dominated Sorting Genetic Algorithm for Multi-Objective Optimization: NSGA-II. In Proceedings of the International Conference on Parallel Problem Solving form Nature, Paris, France, 18–20 September 2000; pp. 849–858.
14. Ma, S.B.; Kim, K.Y. Optimization of Discrete Cavities in a Centrifugal Compressor to Enhance Operating Stability. *Aerosp. Sci. Technol.* **2017**, *68*, 308–319. [CrossRef]
15. Dinh, C.T.; Ma, S.B.; Kim, K.Y. Aerodynamic optimization of a single-stage axial compressor with stator shroud air injection. *AIAA J.* **2017**, *55*, 2739–2754. [CrossRef]
16. Robinson, J.; Sinton, S.; Rahmat-Samii, Y. Particle swarm, genetic algorithm, and their hybrids: Optimization of a profiled corrugated horn antenna. In Proceedings of the IEEE Antennas and Propagation Society International Symposium, San Antonio, TX, USA, 16–21 June 2002; pp. 314–317.
17. Shi, X.H.; Wan, L.M.; Lee, H.P.; Yang, X.W.; Wang, L.M.; Liang, Y.C. An improved genetic algorithm with variable population-size and a PSO-GA based hybrid evolutionary algorithm. In Proceedings of the 2003 International Conference on Machine Learning and Cybernetics, Xi'an, China, 5 November 2003; pp. 1735–1740.
18. Gandelli, A.; Grimaccia, F.; Mussetta, M.; Pirinoli, P.; Zich, R.E. Development and validation of different hybridization strategies between GA and PSO. In Proceedings of the 2007 IEEE Congress on Evolutionary Computation, Singapore, 25–28 September 2007; pp. 2782–2787.
19. Orr, M.J.L. Introduction to radial basis function networks. *Time*, April 1996; 1–67.
20. McKay, M.D.; Beckman, R.J.; Conover, W.J. A comparison of three methods for selecting values of input variables in the analysis of output from a computer code. *Technometrics* **2000**, *42*, 55–61. [CrossRef]
21. Cho, B.M.; Kim, J.H.; Choi, Y.S.; Lee, K.Y.; Oh, H.Y.; Kim, K.Y.; Kim, Y.S.; Kim, J.H. High-efficiency design technique of a single-channel submersible pump impeller for waste treatment. In Proceedings of the ASME/JSME/KSME 2015 Joint Fluids Engineering Conference, Seoul, Korea, 26–31 July 2015; Volume 1A.
22. Kim, J.-H.; Cho, B.-M.; Choi, Y.-S.; Lee, K.-Y. Multi-objective Optimization Based on Unsteady Analysis Considering the Efficiency and Radial Force of a Single-Channel Pump for Wastewater Treatment. *J. Mech. Eng. Autom.* **2016**, *6*, 234–245.

23. Song, W.-G.; Ma, S.-B.; Choi, Y.-S.; Lee, K.-Y.; Kim, Y.-S.; Kim, K.-Y.; Kim, J.-H. Multi-Objective Optimization for Designing a High-Efficiency and Low-Fluid-Induced-Vibration Single-Channel Pump. *KSFM J. Fluid Mach.* **2018**, *21*, 30–38. [CrossRef]
24. Kim, J.-H.; Choi, Y.S. *State-of-the-Art Design Technique of a Single-Channel Pump for Wastewater Treatment*; Wastewater and Water Quality: London, UK, 2018.
25. Menter, F.R. Two-equation eddy-viscosity turbulence models for engineering applications. *AIAA J.* **1994**, *32*, 1598–1605. [CrossRef]
26. Afzal, A.; Kim, K.Y. Optimization of pulsatile flow and geometry of a convergent-divergent micromixer. *Chem. Eng. J.* **2015**, *281*, 134–143. [CrossRef]
27. MATLAB®. *The Language of Technical Computing*; MathWork Inc.: Natick, MA, USA, 2014.
28. Society, R.S. Cross-Validatory Choice and Assessment of Statistical Predictions. *J. R. Stat. Soc. Ser. B (Methodol.)* **2009**, *36*, 111–147.
29. Kim, J.H.; Song, W.G.; Choi, Y.S.; Lee, K.Y.; Ma, S.B.; Kim, K.Y. Three-objective optimization of a single-channel pump for wastewater treatment. *IOP Conf. Ser. Earth Environ. Sci.* **2019**, *240*, 370–381. [CrossRef]

processes

MDPI

Article

Performance Optimization of High Specific Speed Centrifugal Pump Based on Orthogonal Experiment Design Method

Zikang Li [1], Hongchang Ding [1,*], Xiao Shen [1] and Yongming Jiang [2]

[1] College of Mechanical and Electronic Engineering, Shandong University of Science and Technology, Qingdao 266590, China; powerlzk@163.com (Z.L.); shenxiao555@163.com (X.S.)
[2] Group of Shanghai Liancheng, Shanghai 201812, China; m13256922556@163.com
* Correspondence: dhchang@sdust.edu.cn; Tel.: +86-13853292764

Received: 16 September 2019; Accepted: 3 October 2019; Published: 11 October 2019

Abstract: A high specific speed centrifugal pump is used in the situation of large flow and low head. Centrifugal pump parameters need to be optimized in order to raise its head and efficiency under off-design conditions. In this study, the orthogonal experiment design method is adopted to optimize the performance of centrifugal pump basing on three parameters, namely, blade outlet width b_2, blade outlet angle β_2 and blade wrap angle φ. First, the three-dimensional model of the centrifugal pump is established by CFturbo and SolidWorks. Then nine different schemes are designed by using orthogonal table, and numerical simulation is carried out in CFX15.0. The final optimized combination of parameters is b_2 = 24 mm, β_2 = 24°, φ = 112°. Under the design condition, the head and efficiency of the optimized centrifugal pump are appropriately improved, the increments of which are 0.74 m and 0.48%, respectively. However, the efficiency considerably increases at high flow rates, with an increase of 6.9% at 1.5 Q_d. The anti-cavitation performance of the optimized centrifugal pump is also better than the original pump. The results in this paper can provide references for parameter selection (b_2, β_2, φ) in the centrifugal pump design.

Keywords: high specific speed centrifugal pump; performance optimization; orthogonal design method

1. Introduction

High specific speed centrifugal pumps with low head and large flow are generally used in rural irrigation, urban drainage, fish pond aquaculture, aerospace engineering, etc. As the internal flow during the operation of centrifugal pump is extremely complicated, especially under off-design working conditions, it is often accompanied by unstable phenomena, such as overload operation, cavitation, and vibration, which will affect the normal operation. Therefore, it is particularly important to optimize the performance to improve their stability and reliability (Zhang et al. [1]). Numerical simulation has become a common research method with the development of computational fluid dynamics (CFD) (Wei et al. [2]; Zhou et al. [3]; Han et al. [4]).

Generally, there are two ways to optimize the performance of centrifugal pump. One is to study the effect of a single parameter or structure on the performance of centrifugal pump. Skrzypacz and Bieganowski [5] studied the influence of micro grooves on centrifugal pump performance by numerical simulation and experiment. The results illustrated that micro-blades can make the velocity distribution in the impeller passage more uniform, which thus improves the head and efficiency of the pump. Chen, He and Liu [6] put forward the concept of twisted vice blade, and verified that the use of twisted vice blades can effectively improve the comprehensive performance of centrifugal pumps. Nishi, Fukutomi, and Fujiwara [7] found the influence of blade outlet angle on radial thrust and modeled components is obvious by experiments and CFD analysis. Fu, Zhu, Jiang, and Li [8] uncovered how the diffuser vane height affects the pump performance. They concluded that reducing the diffuser vane height could improve the output work of impeller. Cui, Wang, Zhu, and Jin [9] found that a larger blade outlet angle could improve the internal flow of low-specific-speed centrifugal pump and improve its working efficiency.

The other common optimization method is to use some mathematical models or algorithms to study the comprehensive effects of several parameters on the performance of centrifugal pumps, such as DOE (Design of Experiments), approximation models, genetic algorithm, etc. (Zhou et al. [10]; Lyn et al. [11]; Nasruddin et al. [12]). Wang, Feng, Ye, and Luo, Liu [13] selected blade inlet angle, outlet angle, and blade wrap angle as the optimization variables. The multi-objective optimization was carried out based on NSGA-II genetic algorithm. The optimization results show that the algorithm can effectively improve the performance of centrifugal slurry pump. Derakhshan, Pourmahdavi, Abdolahnejad, Reihani, and Ojaghi [14] adopted the ABC (artificial bee colony) and ANNs (artificial neural networks) algorithm to design a new flow passage shape of centrifugal pump. Zhang, Hu, Wu, Zhang, and Chen [15] used Kriging metamodels to optimize double suction centrifugal pump based on four different parameters. Lomakin, Chaburk, and Kuleshova [16] chose six parameters of impeller and guide vane as optimization parameters, and used LP-tau algorithm to optimize the centrifugal pump to improve its comprehensive performance, including increasing head, reducing cavitation, and vibration phenomenon. Koor, Vassiljev and Koppel [17] used the LMA (Levenberg-Marquardt algorithm) to maximize the total efficiency of the pump system and thereby minimize energy consumption.

Among these optimization methods, the orthogonal design experiment in DOE is a more efficient and economical method. Pei, Yin, Yuan, and Wang [18] used the orthogonal design experiment to study the influence of impeller geometric parameters on the cavitation of the pump, and found that the optimum combination of parameters had a better anti-cavitation and hydraulic performance than the original one. Huang and Liu [19] studied the influence of four parameters on centrifugal pump by orthogonal experiment, which provided reference for the design of rotor and volute. Wang and Huo [20] concluded that the optimized pump had no obvious "jet-wake" phenomenon by an orthogonal test. The blade outlet angle and blade leading edge position of impeller are critical factors affecting efficiency and anti-cavitation performance. Although orthogonal design experiments have been used to optimize centrifugal pumps in some references, few references have been made to optimize cavitation performance, and there is basically no research regarding centrifugal pumps with high specific speed. In this paper, the orthogonal design method is adopted to optimize the performance of high specific speed centrifugal pump basing on three parameters, namely, the blade outlet width b_2, the blade outlet angle β_2, and the blade wrap angle φ. First, the hydraulic experiment of the prototype pump is carried out to verify that the three-dimensional (3D) model can be simulated. Subsequently, nine groups of representative parameter combinations are obtained according to the orthogonal table. The number of simulations are greatly reduced. Finally, the best combination of parameters is obtained through numerical simulation results. For the high specific speed centrifugal pump that was studied in this paper, the hydraulic performance and anti-cavitation performance can be improved by properly increasing the blade outlet width, reducing the blade outlet angel, and blade wrap angle.

2. Calculation Model and Method

2.1. Calculation Model

The calculation model in this paper is a single stage single suction centrifugal pump with a high specific speed of 51 ($n_s = \frac{n\sqrt{Q}}{H^{3/4}}$). The design parameters are as follows: flow rate Q = 100 m^3/h, head H = 20 m, rotational speed n = 2900 r/min. Table 1 shows other main geometric parameters of the model.

Table 1. Main parameters of centrifugal pump impeller.

Parameter	Value
Impeller	
Suction diameter (D_j)	96 mm
Impeller diameter (D_2)	135 mm
Outlet width (b_2)	23 mm
Blade inlet angle (β_1)	42°
Blade outlet angle (β_2)	27°
Blade wrap angle (φ)	120°
Blade number (Z)	5
Blade thickness (δ)	3 mm
Blade shape	Twist blade
Volute	
Base diameter (D_3)	140 mm
Inlet width (b_3)	46 mm
Tongue angle (φ_0)	34°
Diffuser outlet diameter (D_d)	125 mm
Diffuser length (L)	265 mm

Figure 1 shows the 3D model of pump established by software CFturbo (CFturbo 10, CFturbo GmbH, Dresden, Germany) and SolidWorks (SolidWorks 2017, Dassault Systèmes SolidWorks Corporation, Waltham, MA, US), which include four parts, namely inlet, impeller, volute, and outlet. The impeller inlet and the volute outlet are properly extended to reduce the influence of water backflow.

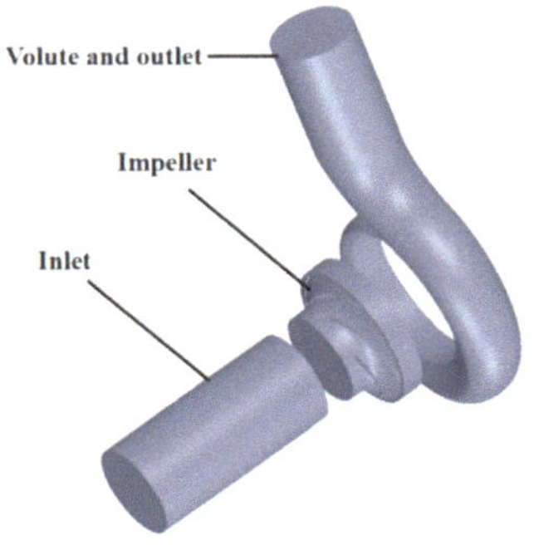

Figure 1. The 3D model of centrifugal pump.

2.2. Governing Equation and Turbulence Model

The Navier–Stokes governing equations consisting of momentum conservation equation and energy conservation equation are used in the numerical simulation (Cohen and Eyal [21]). They can be expressed as:

$$\frac{\partial u}{\partial t} = -(u \cdot \nabla)u + v\nabla^2 u - \frac{1}{\rho}\nabla P + F \tag{1}$$

$$\nabla \cdot u = 0 \tag{2}$$

where u is velocity, t is time; $\frac{\partial u}{\partial t}$ is the rate of velocity vs time; ∇ is gradient operator; v is viscosity coefficient; P is pressure; ρ is density; and, F is external force.

The turbulence model chosen in this paper is the Renormalization-group (RNG) k-ε model. Table 2 shows that the numerical prediction of the RNG k-ε turbulence model is in good agreement with the experimental data at the design flow rate. Accordingly, the RNG k-ε model is adopted in this paper.

Table 2. Selection of turbulence model.

Turbulence Model	Head (m)	Efficiency (%)	Head Error (%)	Efficiency Error (%)
standard k-ε	20.0726	83.09	2.65	2.44
RNG k-ε	19.9939	82.26	2.28	1.41
realizable k-ε	20.1325	83.41	2.96	2.83
Experiment	19.5538	81.11		

The RNG k-ε model is a mathematical model that is derived from instantaneous N-S equation by using the "renormalization group method" (Coutier et al. [22]; Smith [23]). It can capture turbulent diffusion at multiple scales.

2.3. Cavitation Model

The cavitation model is based on the Rayleigh–Plesset equation, which is provided by ANSYS CFX software (ANSYS 15.0, ANSYS Inc., Canonsburg, PA, USA). The cavitation flow is usually regarded as a single-phase flow with average fluid characteristics. It is used to calculate the mass transfer process between the gas phase and liquid phase (Singhal et al. [24]).

$$R_b \frac{d^2 R_B}{dt^2} + \frac{3}{2}\left(\frac{dR_B}{dt}\right)^2 + \frac{2S}{\rho_1 R_B} = \frac{P_{sat} - P}{\rho_1} \quad (3)$$

where R_B is bubble radius, P_{sat} is saturated vapor pressure, and ρ_1 is liquid density.

Ignoring the surface tension and the second-order term of the bubble during vaporization and condensation, the growth rate of the bubble can be obtained from Equation (9).

$$\frac{dR_B}{dt} = \sqrt{\frac{2(P_{sat} - P)}{3\rho_1}} \quad (4)$$

2.4. Grid Independence Analysis

The computational grids are generated by using ICEM CFD (ICEM CFD 15.0, ANSYS Inc., Canonsburg, PA, USA). Aiming at the complex flow passage structure of centrifugal pump, unstructured tetrahedral mesh with strong adaptability is selected for mesh generation (Ahn and Kwon [25]). The local refinement is carried out at the surface of blades and the tongue of volute to deal with large flow gradient change. Figure 2 shows the computational domain grids of model pump.

The grid independence verification is required in order to reduce the influence of grid number on simulation results. Table 3 shows the simulation head and efficiency under different grid numbers. From the table, we can find that the error of head and efficiency is less than 2% between scheme II and III, so the influence of grid number on calculation can be neglected. Scheme II is finally selected for numerical simulation after considering the computational cost and grid independence.

Table 3. Grid Independence Verification.

Scheme	Grid Number	Head/m	Efficiency/%
I	869,171	20.48	83.18
II	1,330,820	19.99	82.26
III	2,102,155	19.72	81.79

The global element scale factor of scheme II is set to 1 and the max element of global element seed size is set to 4. The minimum grid quality is bigger than 0.3 and the minimum angle is bigger than 14°. The grid quality is satisfied. The y^+ values of each part ranged from 30 to 100 (Zhou, Shi, Lu, Hu and Wu [26]).

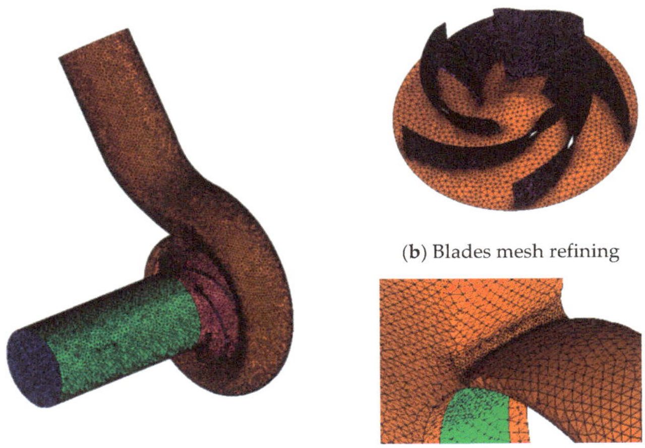

(a) Global grid (b) Blades mesh refining (c) Tongue mesh refining

Figure 2. Computational domain grid.

2.5. Calculation Method

The simulation in this paper is carried out in the software ANSYS-CFX. The RNG k-ε turbulence model is used to close the N-S equations and the standard wall function method is used to deal with the near-wall region. The SIMPLE algorithm is used for the coupling of velocity and pressure. The second-order upwind method is used for the discretization of momentum, turbulent kinetic energy, and dissipation rate equations. The impeller region is set as the rotating domain and the rest region are set as stationary domain. Frozen rotor method is used to connect the dynamic-static interface. Turbulence intensity is set to medium (intensity = 5%). Total pressure (1 atm) is used as the inlet boundary condition and mass flow rate (27.78 kg/s) is used as the outlet boundary condition. The boundary conditions at the wall of hub, shroud, outlet, inlet, and volute are set to no-slip condition. In solver control, the maximum iteration step is set to 1000 and the residual target of convergence criteria is set to 10^{-5} (Babayigit et al. [27]). Table 4 shows all of the boundary conditions settings.

Table 4. Boundary condition setting.

Position	Boundary Condition
Inlet	Total pressure (1 atm)
Outlet	Mass flow rate (27.78 kg/s)
Walls	No-slip wall
Impeller	Rotating domain
Volute	Stationary domain
Interface	General connection
Near-wall region	Standard wall function

The cavitation model is based on the Rayleigh–Plesset equation. The single-phase calculation results without cavitation are taken as the initial values of the two-phase flow of cavitation in order to improve the convergence rate of the calculation. The internal cavitation of the pump is realized by gradually reducing the inlet total pressure. The volume of the vapour at the inlet is set to be 0 and the volume fraction of water is set as 1. For fluid pair models, the mean diameter of bubbles is set to 2×10^{-6}, the saturation pressure is set to 3326 Pa, the cavitation condensation coefficient is set to 0.01, and the cavitation vaporization coefficient is set to 50.

2.6. Prediction Algorithm

ANSYS-CFX simulation cannot directly obtain parameters, such as head and efficiency of centrifugal pump, so it needs to input formulas for calculation.

The calculation formula of centrifugal pump head can be expressed as (Ding et al. [28]):

$$H = \frac{p_{out} - p_{in}}{\rho g} \quad (5)$$

where H is the centrifugal pump head, p_{out} is the volute outlet pressure, p_{in} is the impeller inlet pressure, ρ is the fluid density, and g is the gravity acceleration.

The total efficiency (η) is calculated, as follows:

$$\eta = \left(\frac{1}{\eta_v \eta_h} + \frac{\Delta p_d}{p_e} + 0.03 \right)^{-1} \quad (6)$$

where η_v is the volume efficiency, η_h is the flow efficiency, Δp_d is the friction loss of impeller disk, and p_e is the effective power of fluid.

2.7. Model Verification

The characteristic curve of the model pump is measured on the experimental platform in order to verify that the model can simulate correctly, as shown in Figure 3. Pressure meters are installed in the inlet and outlet pipe of centrifugal pump. Before starting the centrifugal pump, the pump and the inlet pipe should be filled with water to discharge the air in the pump.

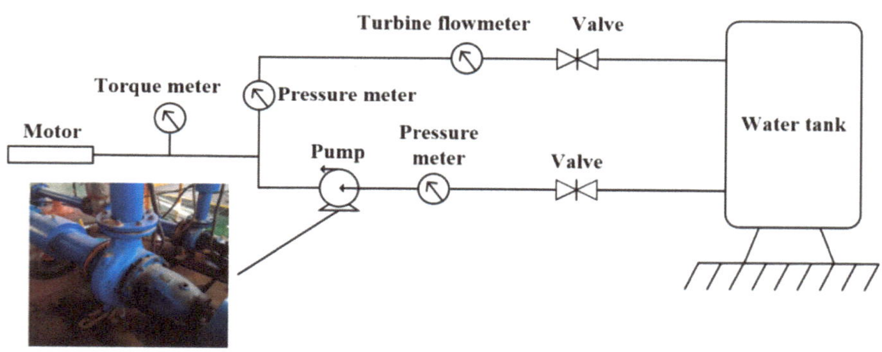

Figure 3. Hydraulic performance test system.

Table 5 shows all of the equipment and parameters of the experiment. The centrifugal pump is driven by Y160M1-2 motor, its rated voltage is 380 V, and rated power is 11 kw.

Table 5. Instruments for experiment.

Measured Parameter	Equipment Name	Model	Minimum Scale	Precision
Pressure	Pressure meter	YM3	100 Pa	±0.1
Flow rate	Turbine meter	LWGY-MIK	0.01 Pa	±0.5
Test speed			0.1 r/min	
Shaft power	Torque meter	YXS-NJ	0.1 kw	±0.5
Torque			0.1 N m	

In the process of experiment, uncertainty will arise due to the accuracy of measuring equipment, artificial operation errors, and experimental environment. In each experiment, the uncertainty of hydraulic efficiency that is caused by different variables can be expressed by the following formula (Babayigit et al. [29]).

$$\omega_{\eta_h} = \pm 100 \left[\left(\frac{\partial \eta_h}{\partial p} \omega_p \right)^2 + \left(\frac{\partial \eta_h}{\partial \dot{Q}} \omega_{\dot{Q}} \right)^2 + \left(\frac{\partial \eta_h}{\partial N} \omega_N \right)^2 \right]^{0.5} / \eta_h \quad (7)$$

where η_h is flow efficiency. (Ding et al. [28]). ω_p, $\omega_{\dot{Q}}$, and ω_N represent the precision of turbine meter, pressure meters, and torque meter, respectively. Precision values of the measurement parameters for pressure (Pa), flow rate (m^3/h), and torque are shown in Table 5. The variation of uncertainty value again flow rate is shown in Figure 4. During the experiment, the uncertainty value is less than ±1%. The uncertainty value continues to decline, as the flow rate of change affects the calculation of hydraulic efficiency more than other parameters.

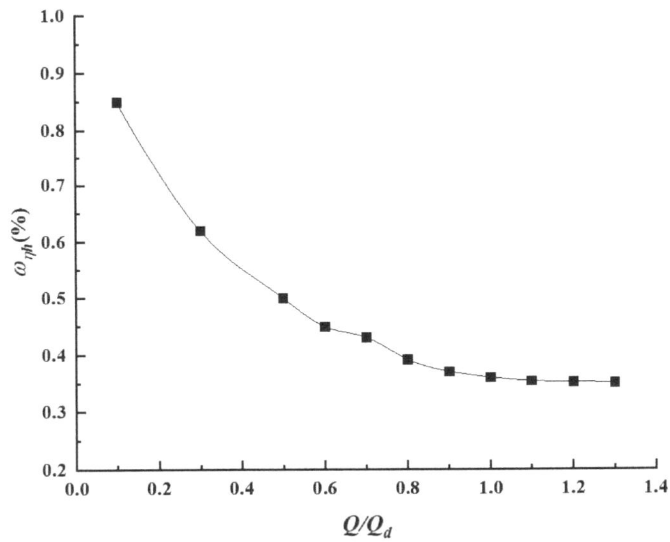

Figure 4. Variation of uncertainty value with flow rate.

Figure 5 shows the characteristic curves of simulation and experiment. It can be seen that the simulated head and efficiency characteristic curves are consistent with the experimental curves. As the simulation does not take into account many losses, the values of head and efficiency are always slightly higher than the experimental values. The maximum difference is less than 5% for the head and efficiency values of simulation and experiment under different flow rate, so the model used in this paper can be correctly simulated.

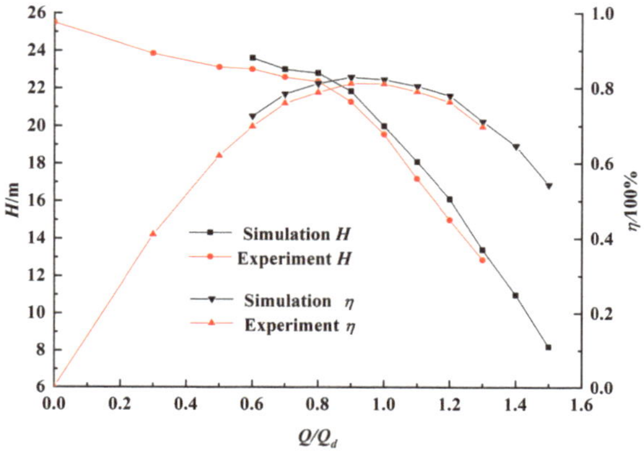

Figure 5. Characteristic curves of simulation and experiment.

3. Orthogonal Experimental Scheme

The main purpose of this paper is to optimize the centrifugal pump by orthogonal experiment design. Orthogonal experiment design is a method to analyze the influence of many factors on the whole situation by using the orthogonal Table. The principle of this method is to select some representative horizontal combinations from all the combinations of experimental factors through orthogonal relations and simulate the experimental results of these combinations. Ultimately, we can obtain the influence of each parameter on the overall performance and find the optimal horizontal combination. It is an efficient and convenient optimization design method.

3.1. Factors Selection

High specific speed centrifugal pump has wide channel and the blade wrap angle size will directly affect its hydraulic performance (Tan et al. [30]). The blade wrap angle represents the diffusion degree of blade passage. Figure 6 shows the schematic diagram of blade wrap angle; it is defined as the included angle of two lines, one is the connecting line of the blade inlet edge and the centre, another is the connecting line of blade outlet edge and the center.

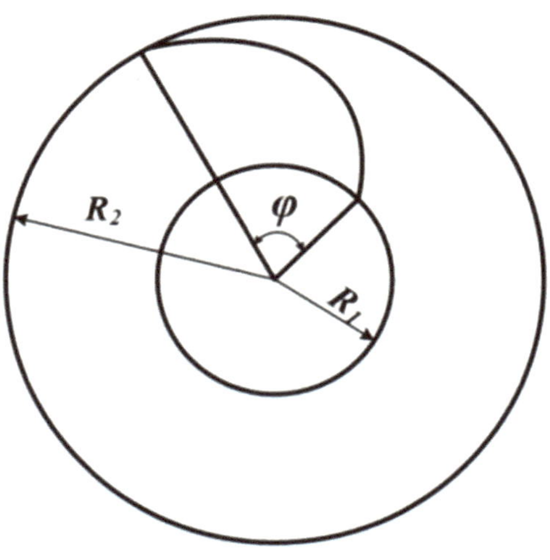

Figure 6. Schematic diagram of blade wrap angle.

Table 6 shows the optimal value of $Z\varphi/360$ corresponding to specific speed n_s. According to the number of blades (Z = 5) and the specific speed of centrifugal pump (n_s = 51), the warp angle range is calculated to be 108–129.6. Therefore, three different blade wrap angles of 112°, 120°, and 128° are selected in this study.

Table 6. Selection of wrap angle.

n_s	10–14	14–19	22–33	36–60	63–77
$Z\varphi/360$	2.1–2.3	1.9–2.1	1.7–1.9	1.5–1.8	1.46–1.65

The blade outlet angles are selected as 24°, 27°, and 30° according to previous studies (Ding et al. [28]). The blade outlet widths are selected to 22 mm, 23 mm, and 24 mm.

3.2. Orthogonal Table Design

According to the principle of orthogonal design experiment method, the orthogonal tables of the above three parameters are established. Three different levels are selected for each factor, as shown in Table 7.

Table 7. Level of factors in orthogonal experiments.

Level	Factor			
	A	B	C	D
	b_2/mm	β_2/(°)	φ/(°)	—
1	23	27	120	1
2	22	24	112	2
3	24	30	128	3

According to L_9 (3^4) orthogonal table, nine groups of orthogonal test schemes are set up, as shown in Table 8. An experiment-independent factor is added to supplement the orthogonal table in the last column, as there are only 3^4 orthogonal table in the orthogonal experimental and no 3^3 table.

Table 8. The orthogonal table design.

Scheme	Serial Number				Corresponding Parameters			
	A	B	C	D	b_2/mm	β_2/(°)	φ/(°)	—
1	A_1	B_1	C_1	D_1	23	27	120	1
2	A_1	B_2	C_2	D_2	23	24	112	2
3	A_1	B_3	C_3	D_3	23	30	128	3
4	A_2	B_1	C_2	D_3	22	27	112	3
5	A_2	B_2	C_3	D_1	22	24	128	1
6	A_2	B_3	C_1	D_2	22	30	120	2
7	A_3	B_1	C_3	D_2	24	27	128	2
8	A_3	B_2	C_1	D_3	24	24	120	3
9	A_3	B_3	C_2	D_1	24	30	112	1

3.3. Orthogonal Analysis

Through numerical simulation in ANSYS CFX, the head (Equation (5)) and efficiency (Equation (6)) of centrifugal pump for nine orthogonal design schemes are obtained, as shown in Table 9. The results can be used to find suitable parameters combination.

Table 9. Simulation results in orthogonal design.

Scheme	*H*/m	*η*/100%
1	19.99	82.26
2	20.29	82.77
3	19.58	81.03
4	20.07	82.36
5	19.83	82.67
6	19.70	81.46
7	19.83	81.90
8	20.74	82.81
9	20.44	81.60

For the orthogonal design, range analysis is used to seek the optimal parameters combination based on the simulation results in Table 9, and the algorithm of range analysis can be expressed, as follows:

$$k_i = \frac{1}{3}K_i = \frac{1}{3}\sum_{j=1}^{N_i} y_{i,j} \tag{8}$$

$$R = \max(k_1, k_2, \ldots, k_i) - \min(k_1, k_2, \ldots, k_i) \tag{9}$$

where K_i is the sum of i levels of each factor, k_i is the average value, and R is the range, which reflects the influence degree of factors.

K_i, k_i, and R are calculated by the above formula. The larger the K_i, the higher the head and efficiency at this level. The larger the R, the greater the influence of this factor on pump performance. Table 10 shows the results of range analysis.

Table 10. Range analysis of head.

Index	Factor		
	A	B	C
K_1	59.86	59.89	60.43
K_2	59.6	60.86	60.8
K_3	61.01	59.72	59.24
k_1	19.95	19.96	20.14
k_2	19.87	20.29	20.27
k_3	20.34	19.91	19.75
R	0.47	0.38	0.52

In Table 10, the influence order of factors on head is as follows: φ, b_2, and β_2, so the optimal combination for head is $A_3B_2C_2$. In Table 11, the influence order of factors on efficiency is: β_2, φ, and b_2, so the optimal combination of efficiency is $A_2B_2C_2$. Here, the highest efficiency of centrifugal pump is selected as the evaluation criterion, and the optimal combination is $A_3B_2C_2$. Therefore, the geometric parameters of the optimal impeller model are outlet width b_2 = 24 mm, blade outlet angle β_2 = 24°, and blade wrap angle φ = 112°, respectively.

Table 11. Variation in vapor volume fraction.

Inlet Pressure (kPa)	Vapor Volume Fraction	
	Origin	Optimal
30.397	0.094769443	0.059625714
40.530	0.062375526	0.036388726
50.662	0.050064852	0.028880441
60.794	0.022115802	0.014879935
70.927	0.016335408	0.013927614
81.060	0.006443528	0.0005504
121.589	0	0

4. Results and Discussion

4.1. Numerical Simulation of Optimal Model

When compared with the original model, the blade outlet width of the optimized model increases by 1 mm, the blade outlet angle decreases by three degrees and the blade wrap angle decreases by eight degrees. Figure 7 gives a comparison between the optimal centrifugal pump blade and the original pump blade.

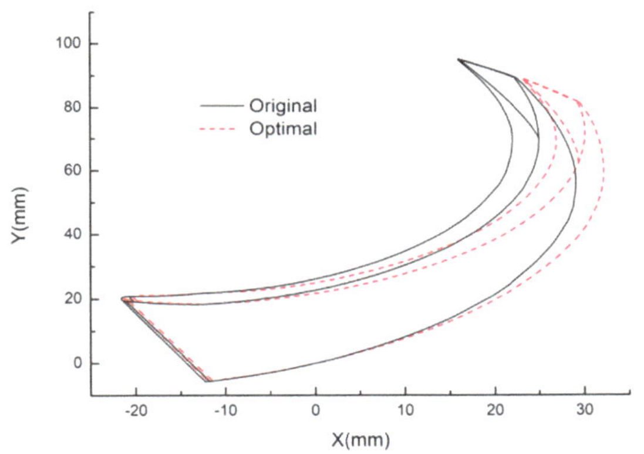

Figure 7. Optimal blade shape.

A new three-dimensional model of centrifugal pump is built and simulated according to the optimized parameters. Figure 8 shows the numerical simulation results of the external characteristics of centrifugal pumps before and after optimization.

From Figure 8a, it can be seen that the head of the optimal pump is higher than the original pump under all working conditions. At low flow rate (0.6 Q_d–1.0 Q_d), the head increment of optimal pump is small, but it becomes large with the increase of flow rate. At the design flow rate, the head of original and optimal pump are 19.99 m, 20.74 m, respectively, and the head after optimization has been increased by 3.75%. Under 1.5 times design flow rate, the head of the original and optimal pump are 8.19 m and 9.17 m, respectively, the head increment reaches to 12%. As can be seen from Figure 8b, the trend of efficiency change is different from that of head change. In the low flow rate region (0.6 Q_d–1.0 Q_d), the efficiency of the original pump is higher than the optimal pump. At 0.6 Q_d, the efficiency of the original pump is 3.7% higher than that of the optimal pump. However, in the high flow rate region (1.0 Q_d–1.5 Q_d), the efficiency of the optimal pump is obviously higher than the original pump and the value increases by 6.9% at 1.5 Q_d.

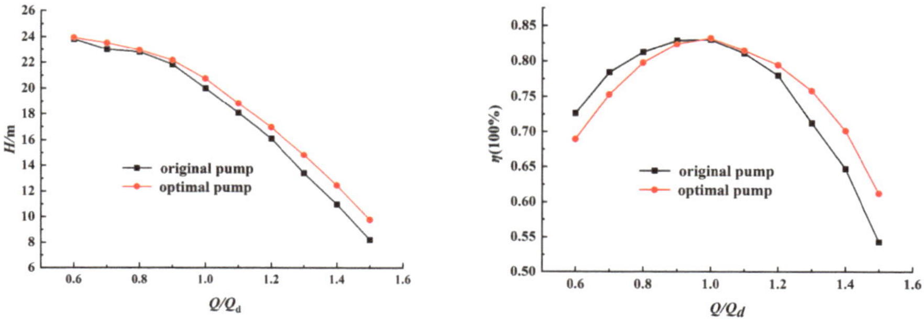

(**a**) (**b**)

Figure 8. Head and efficiency of original pump and optimal pump. (**a**) Q-H; (**b**) Q-η.

As shown in Figure 9, with the decrease of $NPSH_a$ (available net positive suction head), the change of head is very small. When $NPSH_a$ decreases to a certain value, the head begins to decline faster. $NPSH_r$ (require net positive suction head) corresponds to that when head drops by 3%. The larger $NPSH_r$ means the greater the pressure drops, and the worse the anti-cavitation performance of the pump. The $NPSH_r$ for the optimal pump is 5.38 m, which is smaller than that of 5.92 m for the original pump, achieving a appropriate improvement on the anti-cavitation performance.

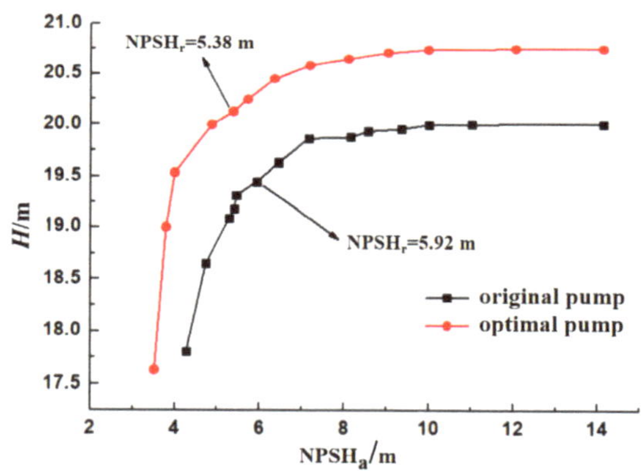

Figure 9. Cavitation performance of original pump and optimal pump.

4.2. Internal Flow Analysis

By comparing the internal flow characteristics of the original pump and the optimal pump, the reasons for the influence of three factors on the performance of high specific speed centrifugal pump are found. Figure 10 shows the total pressure distributions under 0.6 Q_d, 1.0 Q_d, and 1.5 Q_d, respectively. It can be seen that the pressure reaches its lowest value at the suction side near the leading edge, where cavitation might begin to appear. The pressure increases along with the passage and reaches the maximum at the trailing edge. In addition, the pressure distribution at the trailing edge of the original impeller is less uniform than that of the optimal one. The optimal pump has a smaller minimum pressure area, which means less possibility of cavitation, especially in high flow rate. As shown in Figure 10e,f, optimization, the low pressure area (represented by the numbers I and II) of the pump is obviously reduced after optimization.

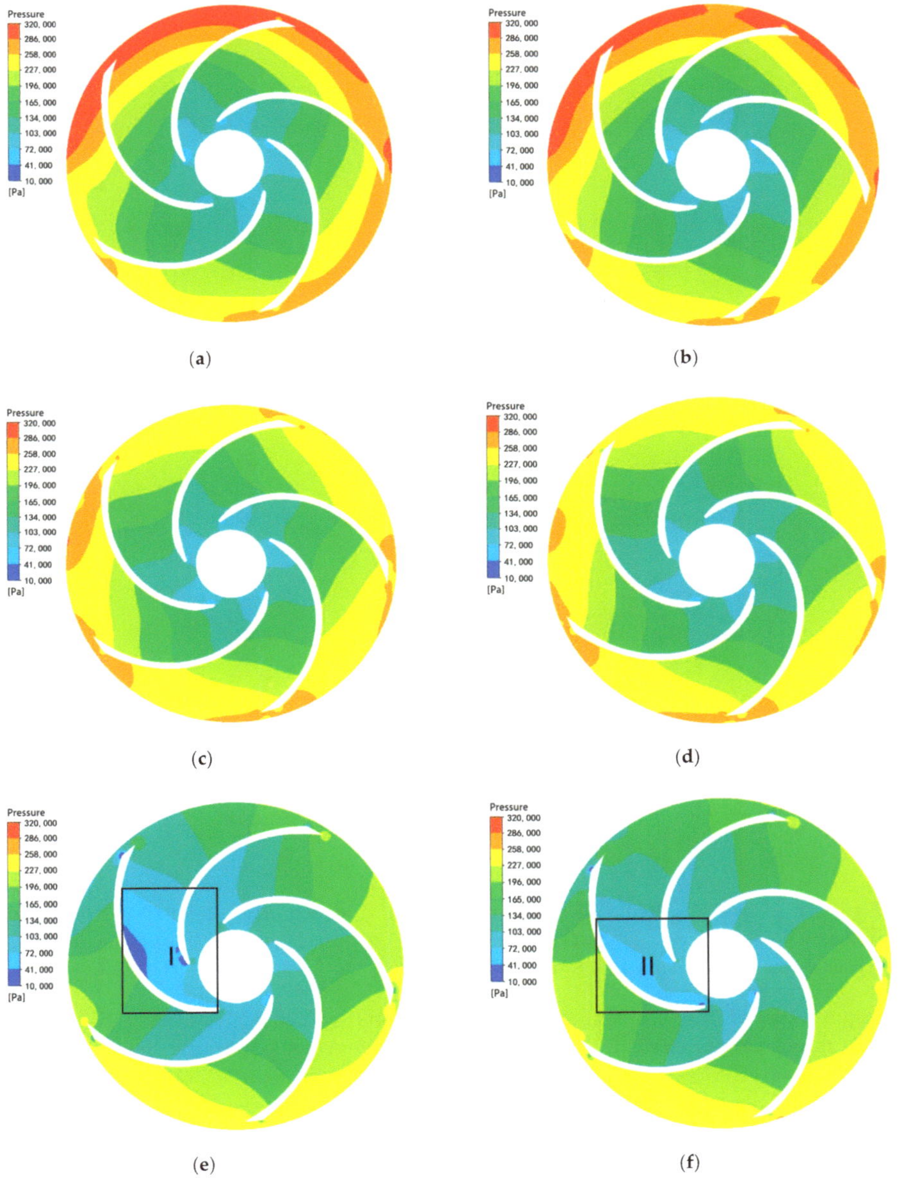

Figure 10. Pressure distributions in impeller of original and optimal pump. (**a**) 0.6 Q_d (original); (**b**) 0.6 Q_d (optimal); (**c**) 0.8 Q_d (original); (**d**) 0.8 Q_d (optimal); (**e**) 1.0 Q_d (original); (**f**) 1.0 Q_d (optimal).

Figure 11 shows the relative velocity distributions of the original and optimal pumps under three different flow rates. It can be seen that the velocity vector distribution in each channel is smoother after optimization, the internal vortex of impeller is reduced, and the flow state is obviously improved. The low velocity area in the optimal impeller is smaller than that in the original one, and this will reduce hydraulic loss and improve efficiency.

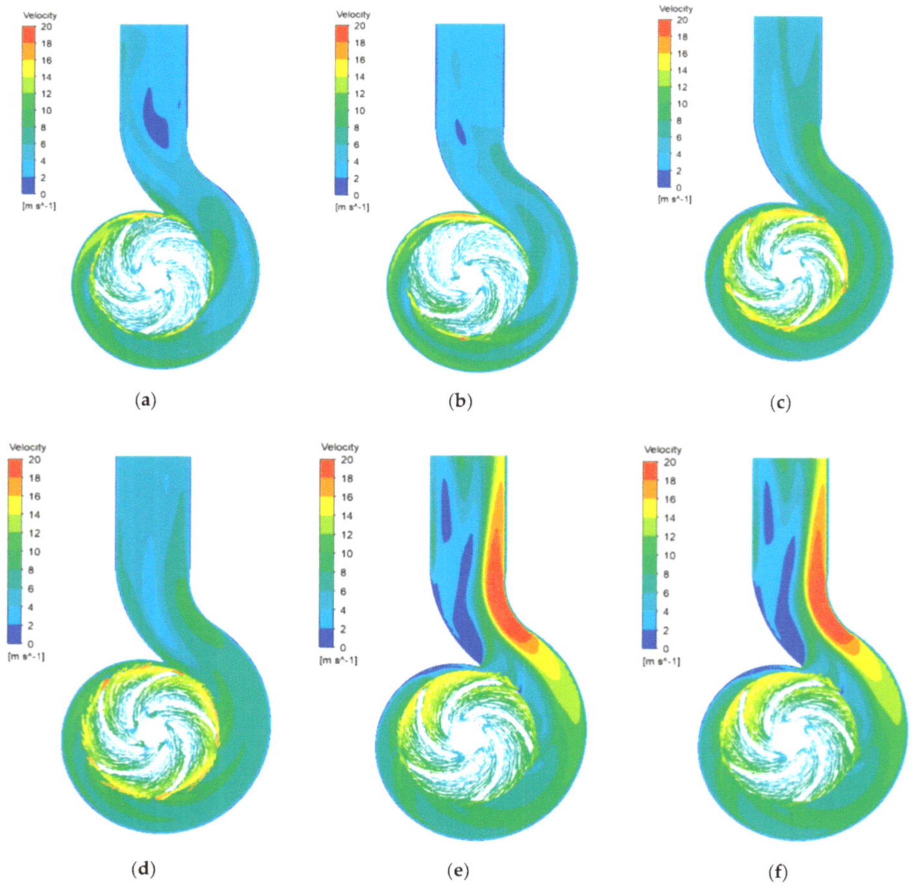

Figure 11. Velocity distributions in volute and vector distribution in impeller midspan of original pump and optimal pump. (**a**) 0.6 Q_d (original); (**b**) 0.6 Q_d (optimal); (**c**) 0.8 Q_d (original); (**d**) 0.8 Q_d (optimal); (**e**) 1.0 Q_d (original); (**f**) 1.0 Q_d (optimal).

Figure 12 shows the turbulent kinetic energy distribution in blade-to-blade view at 0.8 span of original pump and optimal pump. Turbulent kinetic energy refers to the part of energy that converts mechanical energy into heat energy in the fluid movement. The magnitude of turbulent kinetic energy is proportional to the intensity of turbulent vortices. l_{span} is defined as the dimensionless distance from the hub to shroud of impeller, which ranges from 0 to 1. At low flow rate, the turbulent kinetic energy is not significantly reduced after optimization, and even some larger turbulent kinetic energy appears at the inlet. With the increase of the flow rate, the more violent the turbulent flow energy changes in the impeller passage, as shown in Figure 12b,c. Under the condition of large flow rate, the strong turbulent energy region in the impeller mainly concentrates at the inlet of the impeller and the turbulent energy fluctuation at the outlet of the impeller is relatively weak, as shown in Figure 12e,f. This is because, as the flow rate increases, the pressure difference between the impeller outlet and inlet decreases. The turbulent energy of the optimized impeller changes more smoothly, which indicates that the turbulent fluctuation of the optimal impeller is weaker and the hydraulic performance is better.

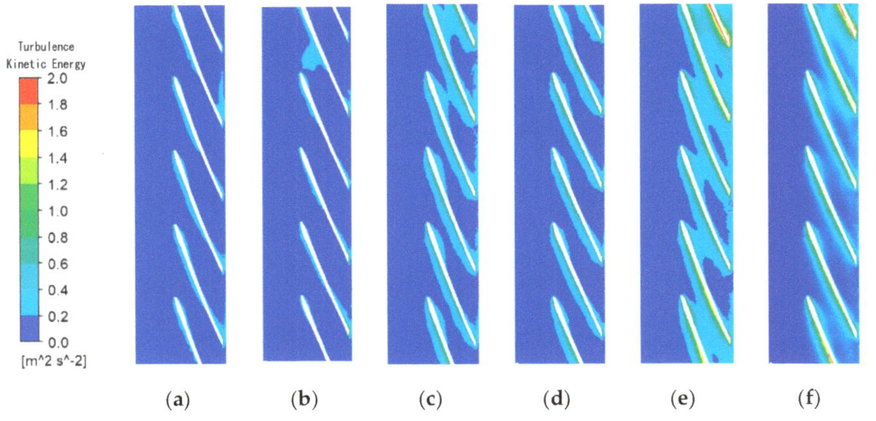

Figure 12. Turbulence kinetic energy distributions in blade-to-blade view at 0.8 l_{span} of original pump and optimal pump. (**a**) original impeller at 0.6 Q_d, (**b**) optimal impeller at 0.6 Q_d, (**c**) original impeller at 1.0 Q_d, (**d**) optimal impeller at 1.0 Q_d, (**e**) optimal impeller at 1.5 Q_d, and (**f**) original impeller at 1.5 Q_d.

Figure 13 shows the vapor volume fraction distribution in impeller of original pump and optimal pump under different inlet pressure. As the inlet pressure gradually decreases to the saturated vapor pressure of water at 25 °C, vapor bubbles begin to appear in the channel of the centrifugal pump. It means that the cavitation phenomenon appears. When the inlet pressure of the pump dropped to 60.8 kPa, critical cavitation occurred in the original impeller. However, as can be seen from Figure 13a,b, the vapor volume fraction of optimal impeller is lower than the original impeller at the same condition, which means that the optimal impeller provided better anti-cavitation performance. When the inlet pressure is further reduced to 50.7 kPa, the vapor volume fraction is larger, which means that the cavitation becomes more serious. However, the vapor volume fraction (represented by numbers I and II) of the optimal pump is still significantly less than the original pump, as shown in Figure 13c,d. Figure 13e,f also show the same situation. The optimized centrifugal pump reduces the possibility of cavitation occurring.

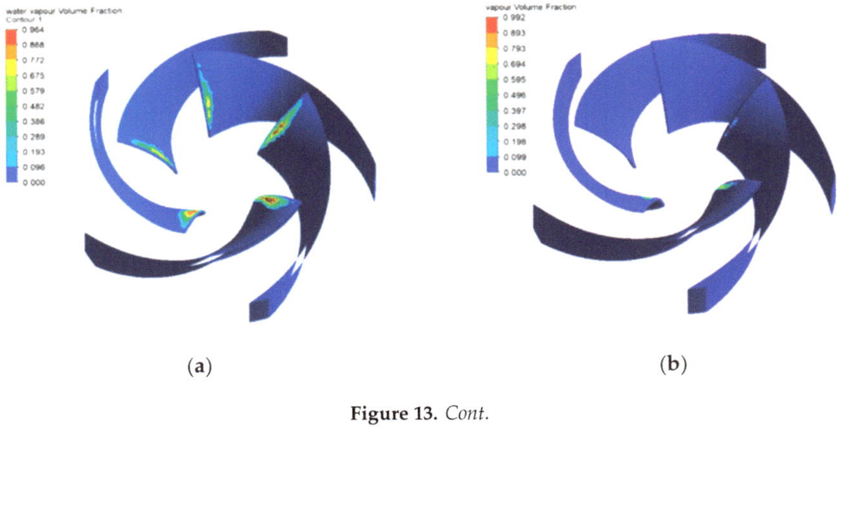

Figure 13. *Cont.*

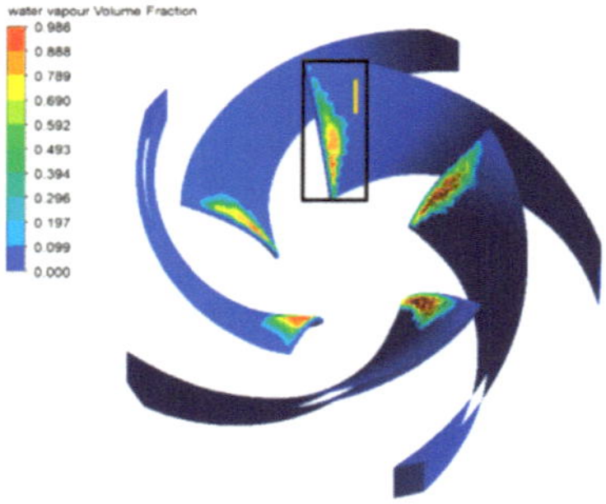

(c)

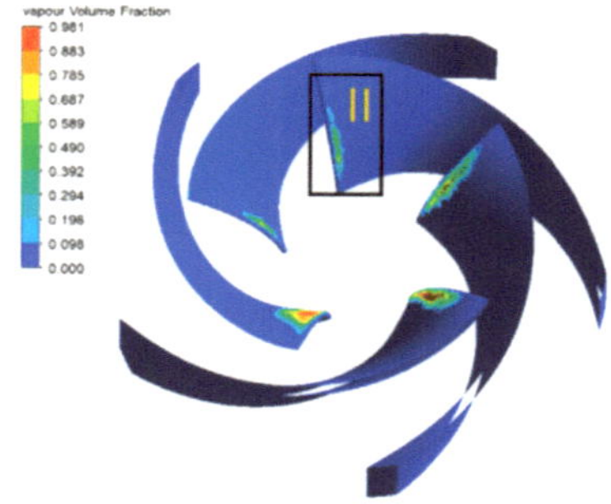

(d)

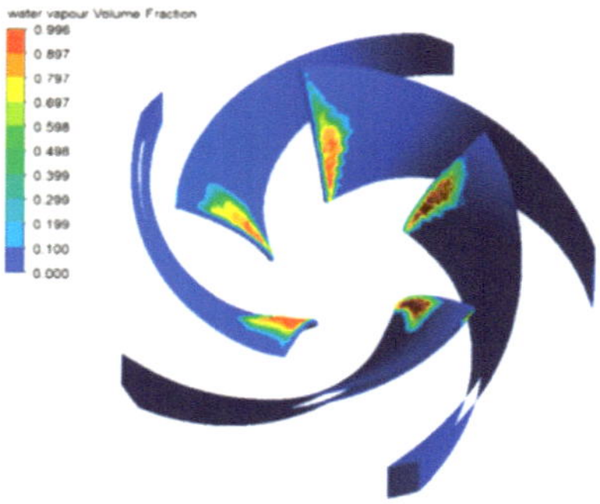

(e)

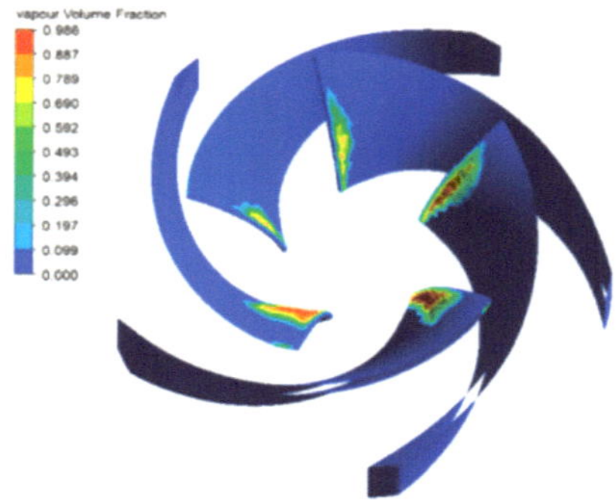

(f)

Figure 13. Vapor volume fraction distribution in impeller of original pump and optimal pump at different inlet pressure. (**a**) 60.8 kpa (original); (**b**) 60.8 kpa (optimal); (**c**) 50.7 kpa (original); (**d**) 50.7 kpa (optimal); (**e**) 40.5 kpa (original); (**f**) 40.5 kpa (optimal).

Table 11 shows the specific vapor volume fraction under different inlet pressure. When combining with Figure 13, it can be seen more intuitively that the optimized centrifugal pump has less vapor volume fraction and better anti-cavitation performance under the same condition.

5. Conclusions

In this paper, three impeller parameters are selected for orthogonal design experiments, and the optimal combination of parameters is obtained according to the results of numerical simulation. The main conclusions are as follows:

(1) Orthogonal design experiments can be used to study the influence of impeller parameters on the head and efficiency of centrifugal pumps. Among the three parameters studied, the wrap angle φ has the greatest impact on the head, while the blade outlet angle β_2 has the greatest impact on the efficiency.

(2) The head of optimal pump is obviously higher than that of the original pump, which increases by 0.74 m under the design flow rate and 1.58 m under 1.5 Q_d. However, the optimal pump's efficiency is lower than the original pump when the flow rate is less than 1.0 Q_d. When the flow rate is greater than 1.0 Q_d, the efficiency significantly increases; its value increases by 6.9% at 1.5 Q_d.

(3) The total pressure distribution of optimal pump is improved under the three operation points, especially at 1.5 Q_d. At the same time, the low velocity area of optimal pump is reduced, which will

cause less hydraulic loss and increase efficiency. Additionally, the optimized pump has less turbulent kinetic energy loss and better flow characteristics.

(4) The $NPSH_r$ value of optimized centrifugal pump is 0.54 m lower than the original model and its anti-cavitation performance is better.

The major optimizing objective in this paper is to improve efficiency. On this basis, the cavitation performance is also compared. However, the selected parameters are not enough to improve the cavitation performance, and some inlet parameters should be selected. In future work, the influence of volute parameters on the unsteady performance of centrifugal pump can also be studied. The optimization method can also adopt more advanced methods, such as Optimal Latin hypercube design.

Author Contributions: H.D. and Z.L. conceived and designed the content structure and experiments; Y.J. performed the experiments; Z.L. analyzed the data; Z.L. and X.S. did the simulation; H.D. and Z.L. wrote the paper.

Funding: This research was funded by National Natural Science Foundation of China (61803235), China Postdoctoral Science Foundation (2015M582112), key research and development project of Shandong province (2017GGX203005). The supports are gratefully acknowledged.

Conflicts of Interest: The authors declare no conflict of interest, financial or otherwise.

Symbols

b_2	Blade outlet width
b_3	Volute inlet width
D_2	Impeller diameter
D_d	Diffuser outlet diameter
D_3	Volute base diameter
D_j	Suction diameter
F	External force
g	gravity acceleration
H	Head
i	Factors
k	Turbulent energy
k_i	The average of K_i
K_i	The sum of i levels of each factor
L	Diffuser length
n	Rotational speed
n_s	Specific speed
p	Pressure
p_e	Effective power of fluid
p_{in}	Impeller inlet pressure
p_{out}	Volute outlet pressure
Q	Flow rate
Q_d	Design flow rate
R	Range
R_B	Bubble radius
t	Time
u	velocity
v	viscosity coefficient
Z	Blade number
β_2	Blade outlet angle
δ	Blade thickness
ρ	Density
η_h	Flow efficiency

η_v	Volume efficiency
Δ_{pd}	The friction loss of impeller disk
Δ	Radient operator
φ	Blade wrap angle
φ_0	Tongue angle
ω_N	Precision of torque meter
ω_p	Precision of pressure meters
ω_{η_h}	Uncertainty of hydraulic efficiency
$\omega_{\dot{Q}}$	Precision of turbine meter

References

1. Zhang, N.; Yang, M.; Gao, B.; Li, Z.; Ni, D. Experimental and numerical analysis of unsteady pressure pulsation in a centrifugal pump with slope volute. *J. Mech. Sci. Technol.* **2015**, *29*, 4231–4238. [CrossRef]
2. Wei, Z.; Yang, W.; Xiao, R. Pressure fluctuation and flow characteristics in a two-stage double-suction centrifugal pump. *Symmetry* **2019**, *11*, 65. [CrossRef]
3. Zhou, P.; Dai, J.; Yan, C.; Zheng, S.; Ye, C.; Zhang, X. Effect of StallCells on Pressure Fluctuations Characteristic in a Centrifugal Pump. *Symmetry* **2019**, *11*, 1116. [CrossRef]
4. Xiangdong, H.; Yong, K.; Deng, L.; Weiguo, Z. Impeller optimized design of the centrifugal pump: A numerical and experimental investigation. *Energies* **2018**, *11*, 1444. [CrossRef]
5. Skrzypacz, J.; Bieganowski, M. The influence of micro grooves on the parameters of the centrifugal pump impeller. *Int. J. Mech. Sci.* **2018**, *144*, 827–835. [CrossRef]
6. Chen, H.; He, J.; Liu, C. Design and experiment of the centrifugal pump im-pellers with twisted inlet vice blades. *J. Hydrodyn.* **2017**, *29*, 1085–1088. [CrossRef]
7. Nishi, Y.; Fukutomi, J.; Fujiwara, R. Effect of blade outlet angle on radial t-hrust of single-blade centrifugal pump. *Earth Environ. Sci.* **2012**, *15*, 47–54. [CrossRef]
8. Fu, L.; Zhu, W.; Jiang, W.; Guo, J.; Li, J. Numerical investigation on influence of diffuser vane height of centrifugal pump. *Int. Commun. Heat Mass Transf.* **2017**, *82*, 114–124. [CrossRef]
9. Cui, B.; Wang, C.; Zhu, Z.; Jin, Y. Influence of Blade Outlet Angle on Perf-ormance of Low-specific-speed Centrifugal Pump. *J. Therm. Sci.* **2013**, *22*, 117–122. [CrossRef]
10. Zhou, L.; Wang, Z.X.; Liu, Z.W. Investigation on influence of design parameters for tandem cascades diffuser using DOE method. *Eng. Appl. Comput. Fluid Mech.* **2014**, *8*, 240–251. [CrossRef]
11. Lyu, Q.; Xiao, Z.; Zeng, Q.; Xiao, L.; Long, X. Implementation of design of experiment for structural optimization of annular jet pumps. *J. Mech. Sci. Technol.* **2016**, *30*, 585–592. [CrossRef]
12. Alhamid, N.M.I.; Aisyah, N. Exergetic-economic analysis and optimization of solar assisted heat pump using multi-objective genetic algorithm. *IOP Conf. Ser. Earth Environ. Sci.* **2018**, *105*. [CrossRef]
13. Wang, C.; Feng, Y.; Ye, J.; Luo, B.; Liu, K. Multi-objective parameter optimi-zation of slurry pump based on RBF neural network and NSGA-II algorithm. *Trans.-Ns Chin. Soc. Agric. Eng.* **2017**, *33*, 109–155. (In Chinese) [CrossRef]
14. Derakhshan, S.; Pourmahdavi, M.; Abdolahnejad, E.; Reihani, A.; Ojaghi, A. Numerical shape optimization of a centrifugal pump impeller using artificial bee colony algorithm. *Comput. Fluids* **2013**, *8*, 145–151. [CrossRef]
15. Zhang, Y.; Hu, S.; Wu, J.; Zhang, Y.; Chen, L. Multi-objective optimization of double suction centrifugal pump using Kriging metamodels. *Adv. Eng. Softw.* **2014**, *74*, 16–26. [CrossRef]
16. Lomakin, V.O.; Chaburko, P.S.; Kuleshova, M.S. Multi-criteria optimization of the flow of a centrifugal pump on energy and vibroacoustic characteristic. *Procedia Eng.* **2017**, *176*, 476–482. [CrossRef]
17. Koor, M.; Vassiljev, A.; Koppel, K. Optimization of pump efficiencies with different pumps characteristics working in parallel model. *Adv. Eng. Softw.* **2016**, *101*, 69–75. [CrossRef]
18. Pei, J.; Yin, T.; Yuan, S.; Wang, W.; Wang, J. Cavitation Optimization for a Centrifugal Pump Impeller by Using Orthogonal Design of Experiment. *Chin. J. Mech. Eng.* **2017**, *30*, 103–109. [CrossRef]
19. Huang, X.B.; Liu, Z.Q. Research of factors influencing centrifugal pump external characteristics based on orthogonal test. *IOP Conf. Ser. Mater. Sci. Eng.* **2013**, *52*, 022008. [CrossRef]
20. Wang, Y.; Huo, X. Multiobjective Optimization Design and Performance Prediction of Centrifugal Pump Based on Orthogonal Test. *Adv. Mater. Sci. Eng.* **2018**, *2018*, 1–10. [CrossRef]

21. Bouvier, E.; Cohen, E.; Najman, L. From crowd simulation to airbag deployment: Particle systems, a new paradigm of simulation. *J. Electron. Imaging* **1997**, *6*, 94–107. [CrossRef]
22. Coutier-Delgosha, O.; Patella, R.F.; Reboud, J.L. Evaluation of the Turbulence Model Influence on the Numerical Simulations of Unsteady Cavitation. In Proceedings of the 2001 ASME FEDSM, New Orleans, LA, USA, 1 June 2001. [CrossRef]
23. Smith, L.M. Development of renormalization group analysis of turbulence. *J. Sci. Comput.* **1990**, *1*, 3–51. [CrossRef]
24. Singhal, A.K.; Athavale, M.M.; Li, H.; Jiang, Y. Mathematical basis and validation of the full cavitation model. *J. Fluids Eng.* **2002**, *124*, 617–624. [CrossRef]
25. Ahn, S.J.; Kwon, O.J. Numerical investigation of a pump-jet with ring rotor using an unstructured mesh technique. *J. Mech. Sci. Technol.* **2015**, *29*, 2897–2904. [CrossRef]
26. Zhou, L.; Shi, W.; Lu, W.; Hu, B.; Wu, S. Numerical investigations and performance experiments of a deep-well centrifugal pump with different diffusers. *J. Fluids Eng.* **2012**, *134*, 071102. [CrossRef]
27. Babayigit, O.; Kocaaslan, O.; Aksoy, M.H.; Guleren, K.M.; Ozgoren, M. Numerical identification of blade exit angle effect on the performance for a multistage centrifugal pump impeller. *EPJ Web Conf.* **2015**, *92*, 02003. [CrossRef]
28. Ding, H.C.; Li, Z.K.; Gong, X.B.; Li, M.S. The influence of blade outlet angle on the performance of centrifugal pump with high specific speed. *Vacuum* **2019**, *159*, 239–246. [CrossRef]
29. Babayigit, O.; Kocaaslan, O.; Aksoy, M.H.; Guleren, K.M.; Kocaaslan, O. Experimental and CFD investigation of a multistage centrifugal pump including leakages and balance holes. *Desalin. Water Treat.* **2017**, *67*, 28–40. [CrossRef]
30. Tan, M.; Ji, Y.; Liu, H.; Wu, X.; Zhu, Z. Effect of Blade Wrap Angle on Performance of a Single-Channel Pump. *Exp. Tech.* **2018**, *42*, 481–490. [CrossRef]

processes

Article

Unsteady Characteristics of Forward Multi-Wing Centrifugal Fan at Low Flow Rate

Yuxin Lun [1], Xinxue Ye [2], Limin Lin [2], Cunlie Ying [2] and Yikun Wei [1,*]

1 National-Provincial Joint Engineering Laboratory for Fluid Transmission System Technology, Zhejiang Sci-Tech University, Hangzhou 310018, China; yuxin_lun@sina.com
2 Zhejiang Yilida Ventilator Co., Ltd., Taizhou 318056, China; yexinxue@yilida.com (X.Y.); linlimin@yilida.com (L.L.); yingcunlie@yilida.com (C.Y.)
* Correspondence: yikunwei@zstu.edu.cn

Received: 8 August 2019; Accepted: 26 September 2019; Published: 2 October 2019

Abstract: The unsteady flow characteristics of a forward multi-wing centrifugal fan under a low flow rate are studied using the computational fluid dynamics (CFD) method. This paper emphasizes the eddy current distribution in terms of the Q criterion method, as well as pressure fluctuation, frequency spectrum, and kinetic energy spectrum analysis of internal monitoring points in a forward multi-wing centrifugal fan. The numerical results show that abnormal eddies mainly appear at the volute outlet and near the volute tongue, boundary layer separation occurs near the suction surface of the blade, and shedding eddies appear at the trailing edge of the blade with the time evolution. The unsteady flow characteristics of a forward multi-wing centrifugal fan at a small flow rate provide significant physical insight into understanding the internal flow law.

Keywords: centrifugal fan; large eddy simulation; unsteady flow; pressure fluctuation; frequency spectrum

1. Introduction

The rotation stall phenomenon of a centrifugal fan under small flow conditions destructively strikes centrifugal fans. The unsteady flows for a forward multi-wing centrifugal fan are essential to study the rotation stall flow field characteristics at small flow conditions. Lee et al. [1] studied the unsteady flows in turbomachinery by means of the large eddy simulation. Conway et al. [2] analyzed the flow through the blades in a swirl generator by the large eddy simulation. Tajadura et al. [3] reported the complete unsteady flow of a centrifugal fan impeller-volute system by implementing three-dimensional numerical simulations. They accessed the information of the pressure fluctuations in some critical positions over the volute wall. Younsi et al. [4,5] studied the unsteady flow in a centrifugal fan and the overall and local behavior of a centrifugal fan. Recently, large eddy simulation has been continually considered as an effective tool to study the internal flow field of centrifugal fans [6–9].

In general, the unstable flow phenomena will appear when the fan operates at a small flow rate. When the small flow rate decreases, the flow loss of the fan extremely increases, and the efficiency of the fan greatly reduces. Even more, the fan operation is destroyed. When the flow rate of the fan is lower than the designed flow rate, the separated eddy will occur in the impeller passage. When the flow rate further decreases, the outlet of the centrifugal fan and the pipeline of the centrifugal fan periodically produce and relieve the backflow [10–15], which brings the surge of the whole fan system with a decreasing flow rate. As is known to all, the rotating stall and surge are two common phenomena that cause the failure of the centrifugal fan. What are the preconditions for controlling these two instabilities? The most important thing is to reveal the unsteady flow characteristics of a centrifugal fan in low flow conditions, so as to provide significant physical insight into the unstable flow in a centrifugal fan.

Based on the above discussion, we mainly studied the flow field characteristics of a centrifugal fan in low flow conditions. The unsteady flow characteristics of centrifugal fan at a small flow rate are investigated to understand the unsteady flow characteristics by numerical simulations. Through the analysis of the unsteady flow characteristics of the centrifugal fan at a small flow rate, it is revealed that the boundary layer separation near the suction surface—the shedding vortices at the end of the blade—will cause a pulsating reaction in the volute port. This provides a precondition for avoiding the unstable phenomenon of a centrifugal fan in low flow conditions and increasing the safety. This paper is organized as follows. In Section 2, the governing equation and the gird will be briefly introduced at first. Afterwards, the physical parameters under steady and unsteady calculation are analyzed, and some conclusions are drawn.

2. The Gird and the Governing Equation

2.1. Centrifugal Fan Model

In general, numerical simulations are an effective tool to study the internal flow laws and unsteady flow characteristics of centrifugal fans. The unsteady numerical analysis is executed on a centrifugal fan with 40 forward wing blades at 1300 RPM. The centrifugal fan is composed by one impeller, a volute, an exit, and two entrances. The volume flow rate of the design point (Q_n) is 321.6 m^3/h. The geometric parameters of the model are shown in Table 1, and the structure of the model is shown in Figure 1. In our numerical simulations, the aerodynamic performance and efficiency characteristic curves are obtained by implementing the Renormalization Group RNG k-ε turbulence of steady state, and the unsteady flows are used to study the pressure fluctuation and energy spectrum in a centrifugal fan by implementing the large eddy simulation.

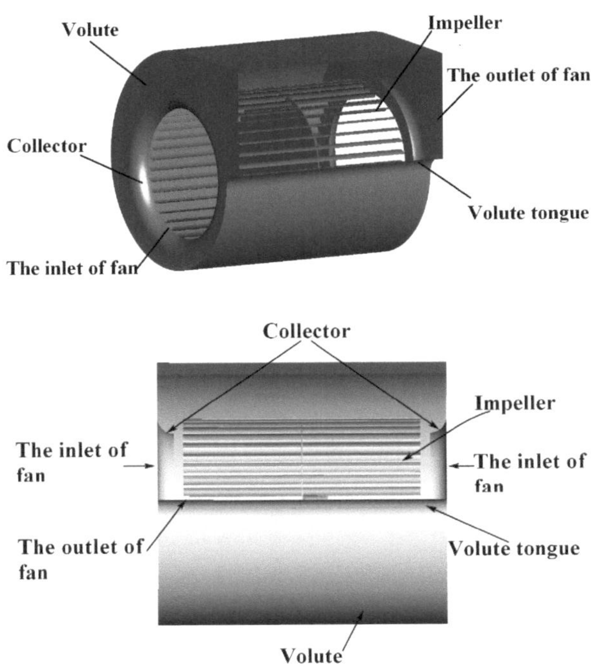

Figure 1. Three-dimensional model of a centrifugal fan.

Table 1. Design dimensions for the centrifugal fan.

Parameter	Dimension
Impeller inlet diameter (D_1)	131.6 mm
Impeller outlet diameter (D_2)	150 mm
Impeller width (b)	186 mm
Blade arc radius (R_k)	8.95 mm
Blade thickness (δ)	0.45 mm
Volute width (B)	227.5 mm
Impeller-tongue distance (Δt)	11.26 mm
Blade inlet angle (β_{1A})	84.83°
Blade outlet angle (β_{2A})	152.04°
Number of blades (Z)	40

2.2. Laboratory Testing

This performance experiment of the centrifugal fan was tested in the Zhejiang Yilida aerodynamics and acoustic laboratory. Figure 2 shows the exterior and interior of the semi-anechoic room in the laboratory. The performance and noise data of a centrifugal fan in a semi-anechoic room can be tested synchronously. The installation method of a centrifugal fan and instrument are also shown in Figure 2.

(**a**) Laboratory Appearance and Internal Scene

Figure 2. *Cont.*

(b) Installation Method of Centrifugal Fan

(c) Speed Sensor and Instrument Cabinet

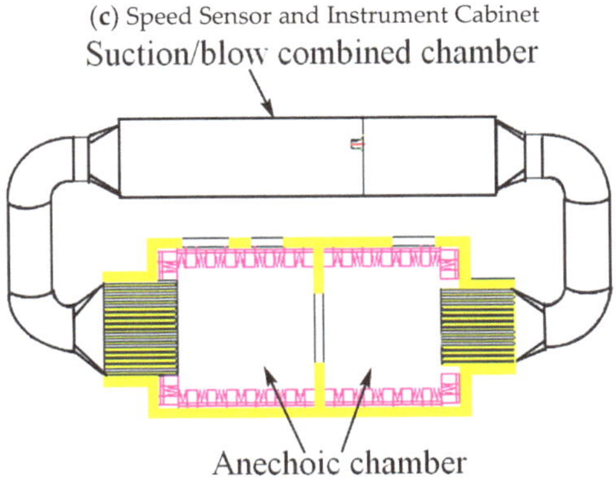

(d) Two-Dimensional Map of Semi-Anechoic Room

Figure 2. Laboratory test system.

2.3. Grid System

In our numerical simulations, the centrifugal fan model is supposed to be generated mesh at first. Common grids are divided into unstructured grids and structured grids in Figure 3. Generally, compared with unstructured grids, a structured grid has a fewer number of and more uniform distribution of grid nodes. As is known to all, a large eddy simulation requires more accurate grids in the numerical simulations of centrifugal fans. In order to obtain more precise calculation data, structural grids are adopted in all fluid regions of centrifugal fans. Figure 3b,c show the distribution of structural grids in the impeller domain, respectively. The grid details of the volute are shown in Figure 3d. The flow of fluid on a solid wall will produce a velocity gradient that is perpendicular to the wall direction; accordingly, the grid near the wall needs to be refined, and the details of the mesh near the blade are shown in Figure 3e.

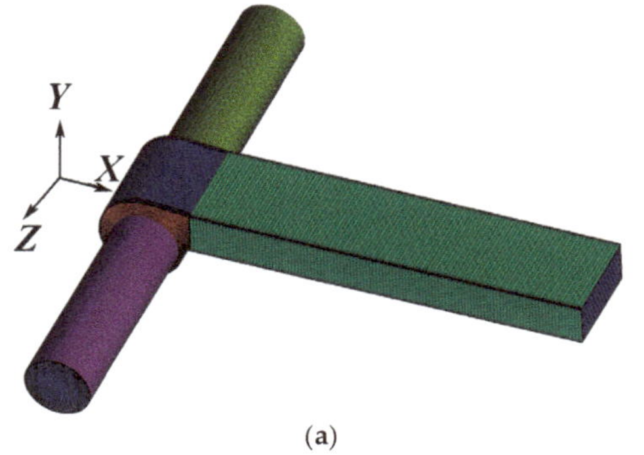

(a)

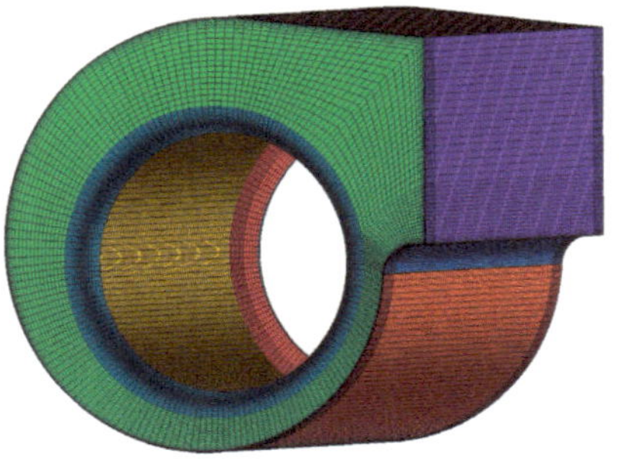

(b)

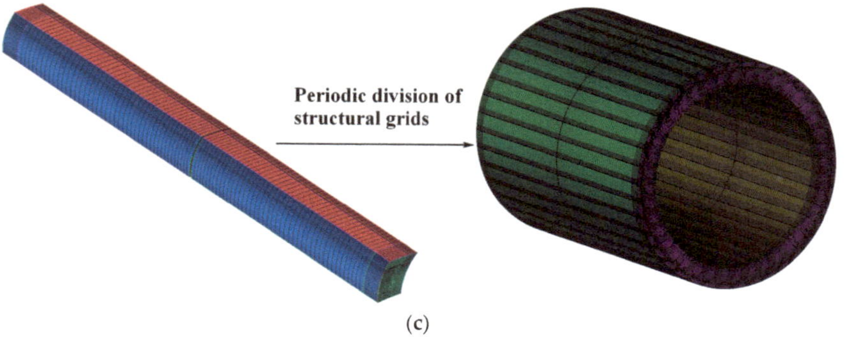

(c)

Figure 3. *Cont.*

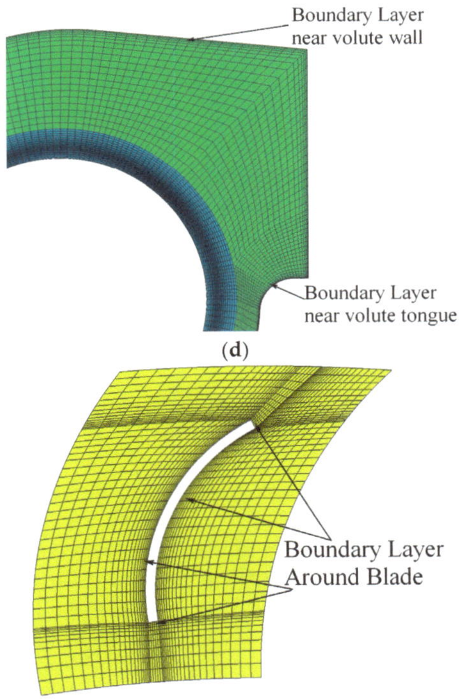

(e)

Figure 3. Mesh of compute domain: (**a**) the overall structural mesh of the centrifugal fan; (**b**) the mesh of the volute; (**c**) the mesh of the impeller; (**d**) the boundary layer of the volute; (**e**) detail of the mesh near the blade.

A numerical simulation based on the finite-volume numerical method using FLUENT 16.0 software was carried out to analyze the internal flow of the centrifugal fan at steady and unsteady computations, respectively. All numerical calculations were carried out using a DELL computer including two CPU (2.30 GHz). In order to achieve the verification of grid independence in numerical simulations, four different grid number models, 4.1[M], 4.7[M], 5.4[M], and 6.2[M] were implemented at standard working condition, respectively. Figure 4 shows the static pressure for different numbers of grids. As shown in Figure 4, we can clearly see that the change rate of static pressure is less than 1% at different mesh numbers. Thus, we can ignore the static pressure change. The calculation results are not affected by the number of grids. In the following numerical simulations of the forward multi-wing centrifugal fan, the number of girds in each fluid region and the total grid number are shown in Table 2.

Table 2. Number of grids in each fluid region of the centrifugal fan.

Fluid Region	Number of Grids	Grid Quality (Determinant 2*2*2)
Inlet (left)	265,933	A range from 0.735 to 0.996
Inlet (right)	293,210	A range from 0.655 to 0.997
Impeller	4,168,320	A range from 0.875 to 0.997
Volute	417,540	A range from 0.612 to 0.999
Outlet	302,621	A range from 0.95 to 1
Total	5,447,624	

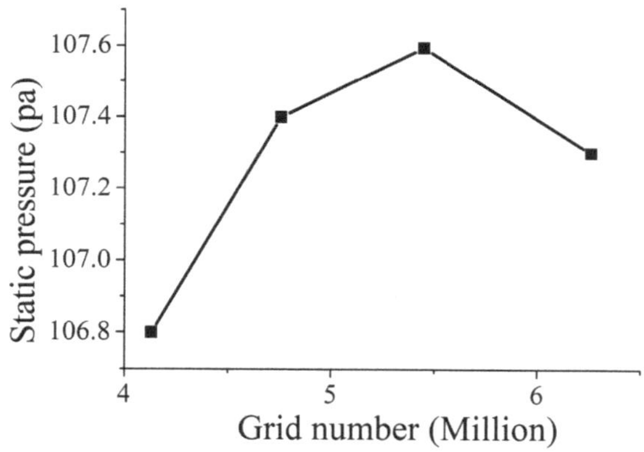

Figure 4. Static pressure for different numbers of grids.

2.4. Steady State Simulation Method

The steady internal flow in a centrifugal fan is studied by implementing the Reynolds-averaged Navier–Stokes Equations method (RANS). As a general rule, the RANS model solves the average information of the flow field. The unsteady turbulence problem is transformed into a steady one, which is sufficient to study the aerodynamic performance of centrifugal fans.

The RNG k-ε was mainly used for the steady calculations in this paper. The *e* equation in the RNG turbulence model improves the ability to simulate high strain flow. The RNG turbulence model takes turbulent vortices into account, improves the accuracy in this respect, and provides an analytical formula considering the viscous flow of Landmine Nobel Number. This makes the RNG turbulence model more accurate and reliable, so that a swirling flow with moderate intensity and low Reynolds number flow will be well predicted.

The governing equation for a steady flow calculation of the internal flow field of a centrifugal fan is as follows [16,17]:

$$\frac{\partial(\rho u_i)}{\partial x_i} = 0 \tag{1}$$

$$\frac{\partial\left(\rho u_j u_i\right)}{\partial x_j} = f_i - \frac{\partial P^*}{\partial x_i} + \frac{\partial\left[\mu_e\left(\frac{\partial u_i}{\partial x_j} + \frac{\partial u_j}{\partial x_i}\right)\right]}{\partial x_j} \tag{2}$$

where ρ denotes the fluid density; x_i and x_j represent the components x, y, z, u_i, and u_j in the three-dimensional coordinate system; u, v, w, and P^* denote the average relative velocity components; *fi* denotes the component of volume force; μ_ε denotes the coefficient of viscosity; $\mu_\varepsilon = \mu + \mu_t$; μ denotes the molecular viscous coefficient, and μ_t is the viscous coefficient of the turbulent eddy.

$$\mu_t = \rho C_\mu \frac{k^2}{\varepsilon}\ (C_\mu = 0.0845) \tag{3}$$

The equation of turbulent energy and the RNG k-ε turbulence model diffusion equation is given as:

$$\frac{\partial}{\partial t}(\rho k) + \frac{\partial}{\partial x_i}(\rho k u_i)_k = \frac{\partial}{\partial x_j}\left(\alpha_k \mu_{eff} \frac{\partial k}{\partial x_j}\right) + G_k + G_b - \rho\varepsilon - Y_M + S_k \tag{4}$$

$$\frac{\partial}{\partial t}(\rho\varepsilon) + \frac{\partial}{\partial x_i}(\rho\varepsilon u_i) = \frac{\partial}{\partial x_j}\left(\alpha_\varepsilon \mu_{eff} \frac{\partial \varepsilon}{\partial x_j}\right) + C_{1\varepsilon}\frac{\varepsilon}{k}(G_k + G_{3\varepsilon}G_b) - C_{2\varepsilon}\rho\frac{\varepsilon^2}{k} - R_\varepsilon + S_\varepsilon \tag{5}$$

where G_k is the turbulent energy due to the laminar velocity gradient, and G_b denotes the turbulent energy g due to buoyancy. $C_{1\varepsilon}$, $C_{2\varepsilon}$, and $C_{3\varepsilon}$ denote constants. α_k and α_ε are the Prandtl numbers of the turbulent equation, respectively.

2.5. Transient Simulation Method

The RANS model is usually used to calculate the time-averaged flow field. The RANS model essentially changes the actual problem, abandons the simulation of unsteady turbulence information, and only seeks the flow results in the average time [18–22]. Hence, it is impossible to simulate unsteady turbulent flow accurately. Nevertheless, the large eddy simulation is a turbulent model to simulate unsteady flow. It solves the instantaneous information of the flow field. The unsteady numerical calculations of the centrifugal fan are also carried out using Large Eddy Simulation (LES), RNG k-ε, realizable k-ε, and standard k-ε models in view of same mesh system at a flow rate of $Q/Q_n = 0.031$. The same time step was adopted for all the unsteady simulations, and the results at the same time were also analyzed. The difference of Q value distribution is revealed at the Z = 0.015 m plane in Figure 5. The LES model clearly shows more details of the internal vortex structure. On the contrary, the numerical results calculated by the RANS models have limitations on small-scale vortices. The LES model is very useful for most studies of unsteady flow phenomena. Since the small vortices are easily dissipated in the calculation process, the accuracy of the simulation can be improved by using the model instead of the small vortices. Therefore, the following explanations of unsteady flow are all based on numerical results using the LES model.

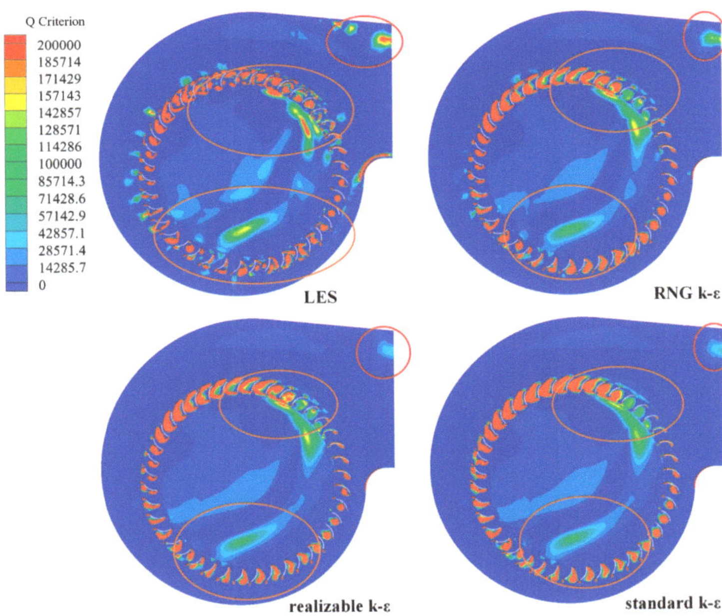

Figure 5. Q value distribution between Large Eddy Simulation and Reynolds-averaged Navier–Stokes Equations method (RANS) at the same computing time t = 0.0125 s.

The basic aim of LES is that turbulence is composed of many eddies of various scales. A large-size vortex has a great influence on the average flow, while a small-size vortex plays a major role in dissipation [11]. Therefore, the mainstream energy in LES is divided into a large-scale vortex and small-scale vortex by some filtering method. A large-scale vortex is directly solved by the Navier-Stokes (N–S) equation, while a small-scale vortex is calculated by establishing the relationship with a large-scale vortex through a sub-grid scale model [23–28]. The LES equation can be obtained by filtering the time term from the N–S equation in the Fourier equation or the spatial domain, which can effectively filter out the smaller eddies in the filter mesh, and thus obtain the momentum equation of the larger eddies.

The filtered variables are defined as:

$$\overline{\phi}(X) = \int_D \phi(X')G(X,X')dX' \tag{6}$$

in which D represents the flow field area, and G denotes the function to determine the size of the filter. The box filtration is frequently employed:

$$\overline{\phi}(X) = \frac{1}{V}\int_v \phi(X')dX' \tag{7}$$

in which V denotes the computational element volume. The filter function, G(X,X′), is defined as

$$G(X,X') = \begin{cases} \frac{1}{V}, & X' \in v \\ 0, & otherwise \end{cases} \tag{8}$$

By filtering the incompressible N–S equation, the following equations are given as [29]:

$$\frac{\partial \rho}{\partial t} + \frac{\partial}{\partial x_i}(\rho\overline{u_i}) = 0 \tag{9}$$

and:

$$\frac{\partial}{\partial t}(\rho\overline{u_i}) + \frac{\partial}{\partial x_j}\left(\rho\overline{u_i u_j}\right) = \frac{\partial}{\partial x_j}\left(\mu\frac{\partial\overline{u_i}}{\partial x_j}\right) - \frac{\partial\overline{p}}{\partial x_i} - \frac{\partial\tau_{ij}}{\partial x_j} \tag{10}$$

Among them, τ_{ij} denotes the tension of a sub-grid, which is defined by:

$$\tau_{ij} = \rho\overline{u_i u_j} - \rho\overline{u_i}\overline{u_j} \tag{11}$$

If the sub-grid tension is obtained after filtration, the eddy viscous model equation is used by the following equation [30,31]:

$$\tau_{ij} - \frac{1}{3}\tau_{kk}\delta_{ij} = -2\mu_t\overline{S}_{ij} \tag{12}$$

in which u_t is the sub-grid turbulent viscous force and $\overline{S}_{ij}$ denotes the tensor spin, which is defined as:

$$\overline{S}_{ij} = \frac{1}{2}\left(\frac{\partial\overline{u}_i}{\partial x_j} + \frac{\partial\overline{u}_j}{\partial x_i}\right) \tag{13}$$

where u_t is obtained using the Samagorin–Lilly model, which is the basis of the sub-grid model. The equation is:

$$\mu_t = \rho L_s^2\left|\overline{S}\right| \tag{14}$$

where:

$$\left|\overline{S}\right| \equiv \sqrt{2\overline{S}_{ij}\overline{S}_{ij}} \tag{15}$$

where L_s represents the grid hybrid length; its expression is:

$$L_s = \min\left(\kappa d, C_s V^{1/3}\right) \tag{16}$$

in which k is the von Kármán constant, d is the distance to the nearest wall, and V is the volume of the computational element. $C_s = 0.1$ is currently implemented.

2.6. Q Criterion

In general, the criteria that is used to judge vorticity includes the Q criterion and vorticity criterion. The application of vorticity criterion requires a given threshold, but different flow thresholds may be different, and different thresholds will result in different results. The Q criterion reflects a balance between the rotation and deformation of a fluid cluster in the flow field. Q >0 reflects the dominance of rotation in flow. At the same time, the Q criterion has no threshold selection, which is more objective than the vorticity criterion.

Different vortex structures reflect the flow characteristics of fluids in various computational domains. The Q criterion is regarded as providing significant physical insight into understanding the internal flow law and the local flow loss in the flow field [19–22]. To better analyze the internal flow of a centrifugal fan, the Q criterion is considered as a rule of discriminating eddy current in this paper.

The formula is given as:

$$\Omega_{ij} = \frac{1}{2}\left(\frac{\partial u_i}{\partial x_j} - \frac{\partial u_j}{\partial x_i}\right) \tag{17}$$

$$S_{ij} = \frac{1}{2}\left(\frac{\partial u_i}{\partial x_j} + \frac{\partial u_j}{\partial x_i}\right) \tag{18}$$

$$Q = \frac{1}{2}\left(||\Omega^2|| - ||S^2||\right) \tag{19}$$

3. Results and Discussion

In the following sections, the calculation method and numerical verification are presented firstly. Then, the distribution of physical parameters on different planes is analyzed to understand the internal flow of a centrifugal fan. Finally, the unsteady flow characteristics of the centrifugal fan are comprehended by analyzing the variation of the internal flow field with time evolution.

3.1. Numerical Verification

Through steady and unsteady simulations of the centrifugal fan based on computational fluid dynamics (CFD), the internal flow field of the centrifugal fan has attracted a large number of researchers [32–42]. The discrete format at steady calculations is SIMPLE; SIMPLE is also adopted at the unsteady simulations, and the second-order format is carried out in the steady and unsteady simulations. The second-order upwind form is mainly based on a coupled solution of the nonlinear finite difference equations according to the multigrid technique. It is necessary to set a frozen rotor in the steady numerical stimulation, and sliding mesh is applied for unsteady numerical simulation between dynamic and static domains. The time requirement for an impeller to rotate 1 degree 1.25×10^{-4} s) is considered as the time-step size of unsteady calculation for the sake of more precise detail data. Ten cycles are implemented at the unsteady simulations of the centrifugal fan, so the time steps are 3700.

In order to ensure the correctness and rationality of the calculation results, the forward multi-wing centrifugal fan is tested to acquire the performance data, and is also simulated to obtain calculation data at multiple operating conditions. The static pressure curves of both are shown in Figure 6. As shown in Figure 6, it can be clearly observed that the static pressure obtained by numerical simulations is well consistent with that of the experimental results, which indicates that the numerical simulation method is reasonable in this paper [25–27].

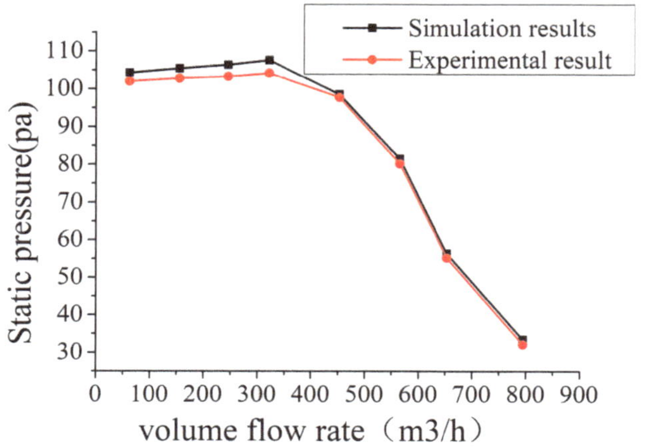

Figure 6. Comparison of experimental and numerical results under different volume flow rates.

3.2. Numerical Results and Discussions of Steady Flow

In this subsection, the static pressure, streamline, and Q value distribution on planes at different positions will be introduced in the internal flow of the centrifugal fan, respectively.

Figure 7 shows the planes of different positions. As shown in Figure 7, the Z = 0 mm plane represents the middle disk, and there are three planes along the positive direction the Z-axis; their locations are Z = 0.015 m, Z = 0.030 m, and Z = 0.045 m, respectively. The parameters on the three planes will be presented by the analysis of the steady results.

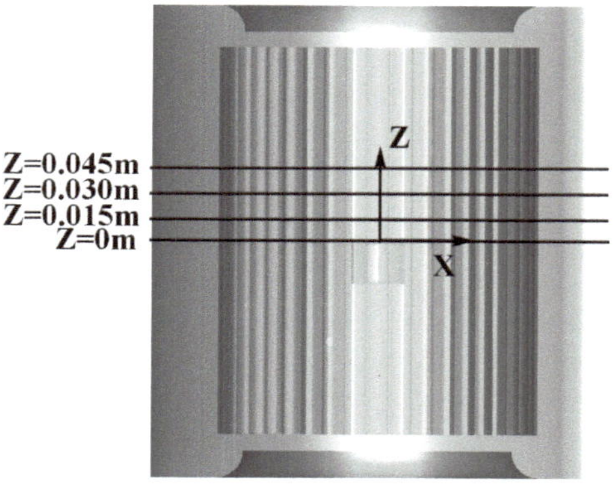

Figure 7. Different section positions of the centrifugal fan.

Figure 8 exhibits the static pressure distribution of different planes. From Figure 8, it can be seen that the static pressure along the positive direction of the Z-axis presents different static pressure distribution at three locations. The large area of low pressure appears near the blade channel, and the plane low-pressure area at Z = 0.015 is smaller than that of the other two planes. In the volute tongue position and volute outlet position, high hydrostatic pressure distribution and gradient change are observed in all three planes, and the greatest pressure gradient occurs at the Z = 0.030 m plane. It is further indicated that the fluid reflux phenomenon occurs at the high-pressure gradient location of fluid flows from the high static pressure region to the low static pressure region.

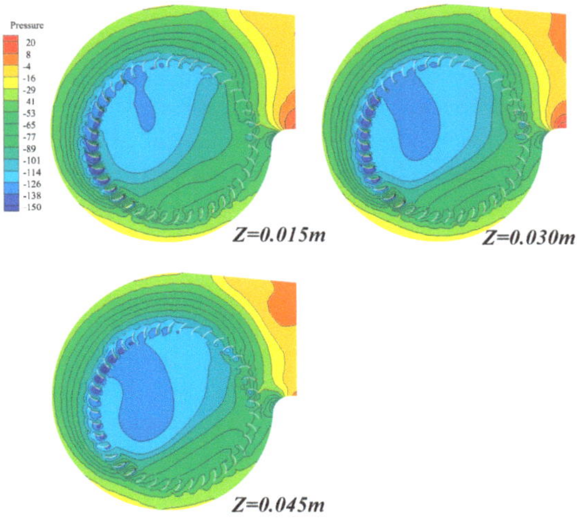

Figure 8. Static pressure of cross-sections at different positions (Q/Qn = 0.031).

Figure 9 expresses the streamline distribution at different planes. As shown in Figure 9, the difference of streamlines in the internal flow of centrifugal fan with the increase of Z-axis distance is clearly observed. The comparison of streamline distribution on different planes indicates that near the impeller entrance, no vortex streamlines occur on the two planes z = 0.015 m and z = 0.030 m, but some small vortex streamlines appear on the plane z = 0.045 m. Most importantly, a large number of small-scale eddies are caused by boundary layer separation on the wall above the volute outlet. It is further found that fluid refluxes obviously occur on the Z = 0.045 m plane and the direction of fluid flow has changed greatly, followed by the Z = 0.015 m plane in the vicinity of the volute outlet, which is well consistent with the distribution of the static pressure in Figure 7.

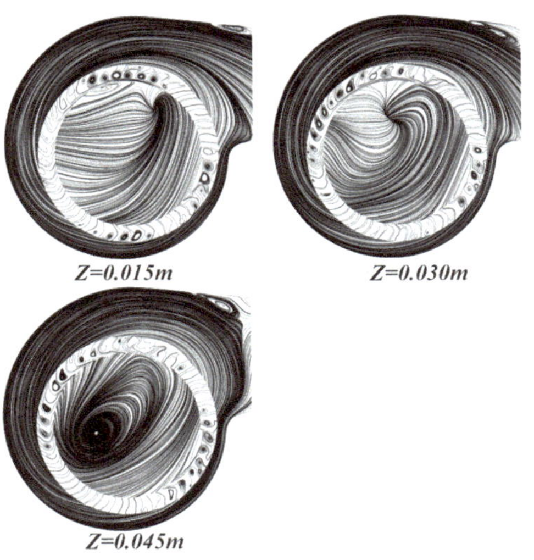

Figure 9. Streamline of cross-sections at different positions (Q/Qn = 0.031).

Figure 10 gives the distribution of Q values at different planes in the internal flow of centrifugal fans. The Q criterion is implemented to analyze the internal flow of centrifugal fans, which is regarded as one of the criteria of judging vortices. The region with a larger Q value represents the phenomenon of vortices. We can see clearly that the distribution of Q values assumes differences at three planes in Figure 6. It is clearly observed that the significant distribution differences of the Q values occur at the marked position of the impeller entrance. The high Q value appears on the Z = 0.045 m plane, the significant distribution differences of the Q values occur at the marked position of the impeller, a slight reduction occurs on the Z = 0.030 m plane at the same position of Z = 0.045 m, and a significant reduction occurs on the Z = 0.015 m plane compared with that of Z = 0.03 m and Z = 0.045 m. In the marked position of the impeller exit, all three planes have higher Q value distribution; however, the area with a high Q value on the Z = 0.045 m plane is obviously larger than that of the other two planes. A small part district of a high Q value appears in the upper part of the volute outlet. At the cochlear volute tongue position, the Q value of the Z = 0.045 m plane is significantly higher than that of the other planes. The different distributions of Q values on the planes represent the flow conditions of different fluids. The Q value shown in Figure 10 coincides with the static pressure in Figure 8 and the streamline in Figure 9.

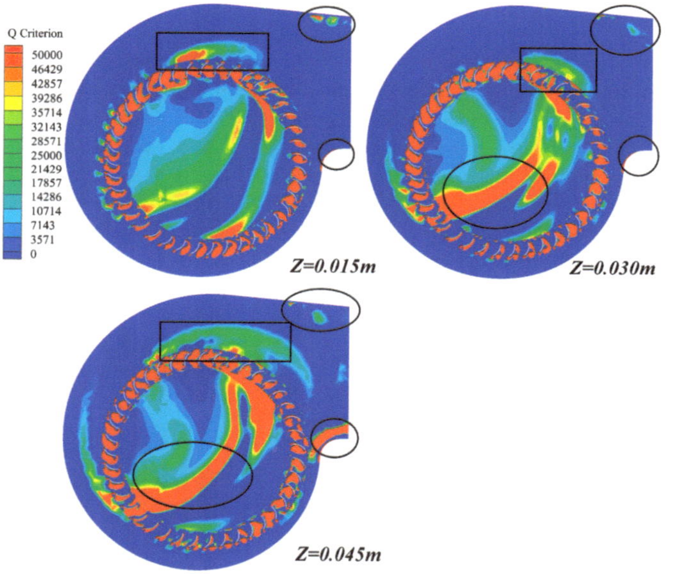

Figure 10. Q value distribution of cross-sections at different positions (Q/Qn = 0.031).

Figure 11 describes static pressure of cross-sections at different positions at the flow rate Q/Qn = 1. As we can see from Figure 11, the static pressure distribution is slightly different, mainly in the vicinity of the volute tongue. The static pressure distribution is similar on the Z = 0.015 m and Z = 0.030 m planes, but it is very uniform on the z = 0.045 m plane.

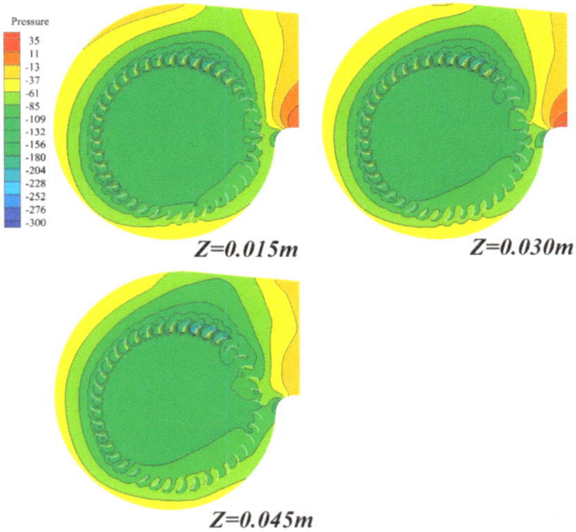

Figure 11. Static pressure of cross-sections at different positions (Q/Qn = 1).

Figure 12 exhibits the streamline distribution of different planes. As shown in Figure 11, one can see that the airflow can flow smoothly along the volute outlet, but it is different near the volute tongue. This also reflects that the air flow at the volute tongue is more complex.

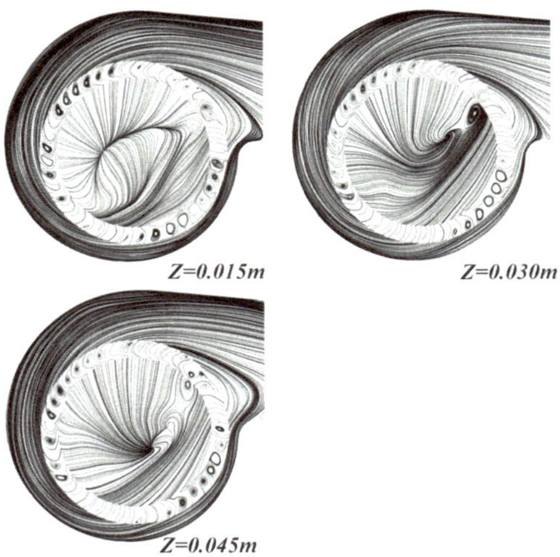

Figure 12. Streamline of cross-sections at different positions (Q/Qn = 1).

Figure 13 shows the Q value distribution of different planes at a flow rate of Q/Qn = 1. It is clearly observed that the Q value distribution in the impeller is not uniform, and the Q value is higher in the upper region of the volute and volute tongue. This corresponds to the streamline chart.

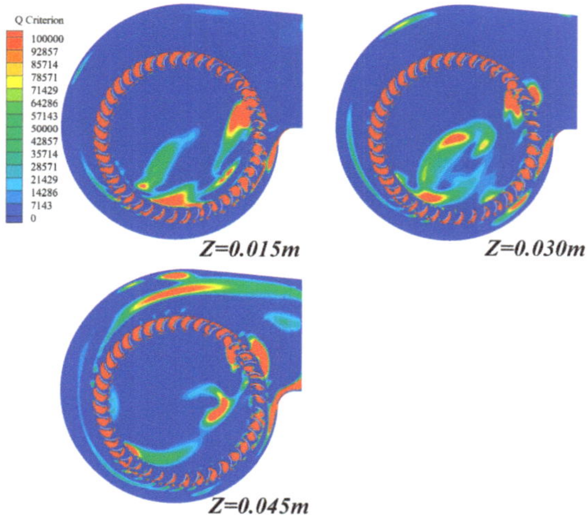

Figure 13. Q value distribution of cross-sections at different positions (Q/Qn = 1).

Figure 14 represents that the static pressure of different positions at the flow rate Q/Qn = 1.758. Figure 14 clearly shows that the static pressure distributions of different sections are similar. The static pressure gradient varies greatly, and a higher static pressure appears near the cochlear tongue, which makes the airflow in the cochlear tongue unstable.

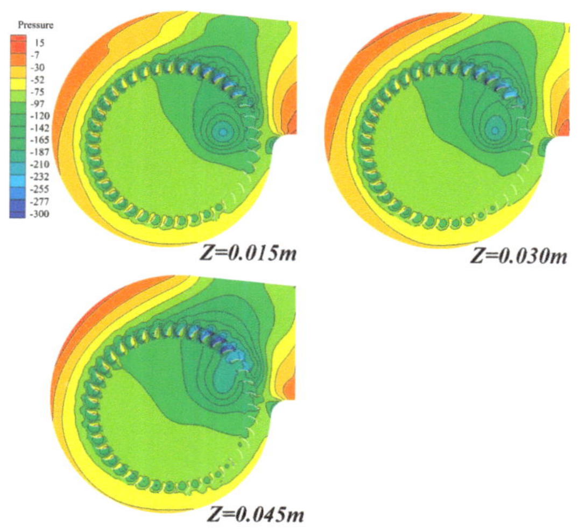

Figure 14. Static pressure of cross-sections at different positions (Q/Qn = 1.758).

Figure 15 shows that streamline of cross-sections at different planes at a flow rate of Q/Qn = 1.758. From Figure 15, it can be seen that there is an eccentric eddy current at the impeller inlet at three positions. Due to the complex flow near the volute tongue, the airflow here is unstable.

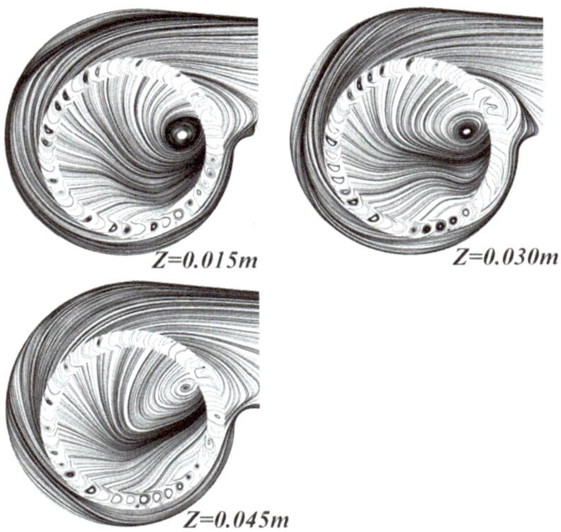

Figure 15. Streamline of cross-sections at different positions (Q/Qn = 1.758).

Figure 16 shows the Q value distribution of cross-sections at different positions at a flow rate of Q/Qn = 1.758. From Figure 16, it can be seen that there is a large Q value distribution at the impeller entrance, which corresponds to the eccentric vortices in the streamline diagram. A large Q value distribution also appears in the upper part of the volute on the Z = 0.015 m plane.

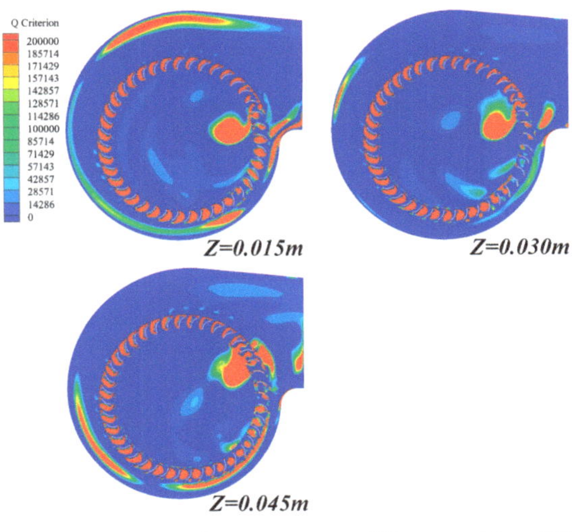

Figure 16. Q value distribution of cross-sections at different positions (Q/Qn = 1.758).

3.3. Unsteady Internal Flow Structure

In this subsection, the static pressure, streamline, and Q value distribution on the Z = 0.015 m plane with time evolution will be introduced in the internal flow of the centrifugal fan. Then, the three-dimensional eddy phenomena of the impeller and volute domains are analyzed at different Q values.

Figure 17 proves the static pressure distribution with time evolution on the Z = 0.015 m plane. Plotted in Figure 17, one can see that the static pressure distribution constantly transforms with time evolution near the impeller inlet and volute outlet. There is a low-pressure area near the impeller entrance, and a little gradient change of the low-pressure area occurs near the impeller entrance with time evolution. The most important part is that the static pressure gradient of the volute tongue and the volute outlet changes greatly; there are two high-pressure areas at the time of T_1, a wide range of high-pressure areas at the time of T_2, and a high-pressure area at the time of T_3 only appears at the cochlear volute tongue position. The most obvious static pressure gradient of the volute tongue appears at the time of T_2; however, the static pressure gradient of the volute at the time of T_3 is evidently decreased.

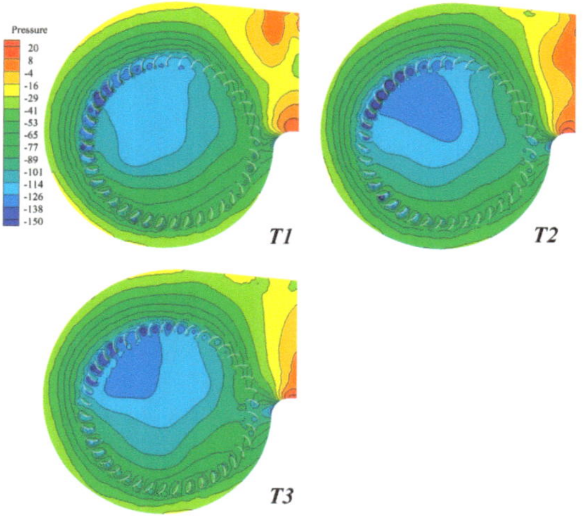

Figure 17. Static pressure at different times of Z = 0.015 m section (Q/Qn = 0.031).

Figure 18 exhibits the streamline distribution in a centrifugal fan with time evolution at the Z = 0.015 m plane. From Figure 18, we can distinctly observe that streamlines in the middle of a plane do not change much in the course of time lapse, but the streamlines at the volute outlet gradually become smooth, and reflux occurs with time evolution. The phenomenon of boundary layer separation above the volute outlet gradually weakens, which coincides with the static pressure distribution in Figure 16.

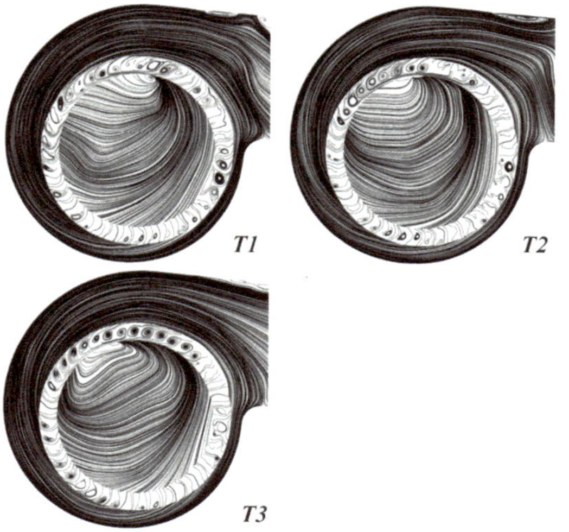

Figure 18. Streamline with time evolution at Z = 0.015 m section (Q/Qn = 0.031).

Figure 19 displays the distribution of Q values with time evolution on the Z = 0.015 m plane. As shown in Figure 19, it is clearly observed that the distribution of the Q value changes with time evolution appear in the upper part of the volute outlet and near the volute tongue. The Q value above the volute outlet decreases gradually from a high value, and the area diminishes over time evolution. Near the cochlear volute tongue, a higher Q value appears at the time of T_3. This corresponds to the flow phenomena described in Figures 16 and 17.

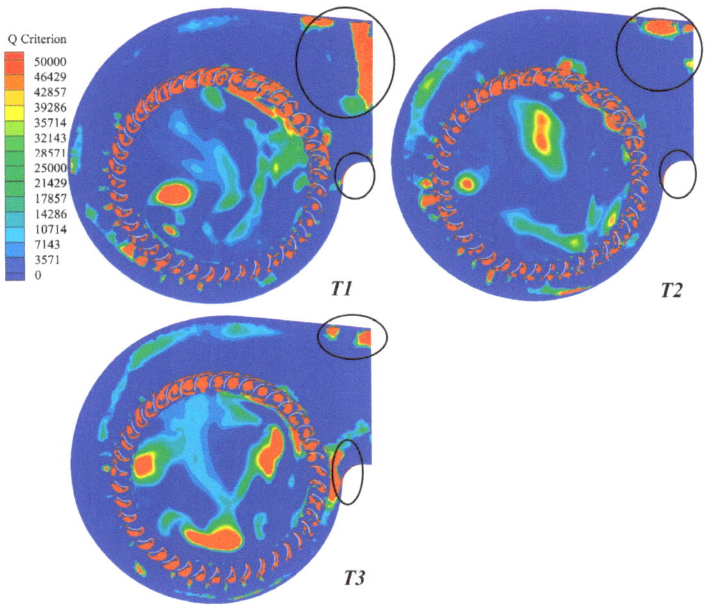

Figure 19. Q value distribution with time evolution at the Z = 0.015 m section (Q/Qn = 0.031).

In order to further explain the internal flow physical essence of a centrifugal fan, the Q value distribution of the blade passage is presented with evolution at different times on the Z = 0.015 m section in Figure 20. As shown in Figure 20, it is clearly seen that the Q value is higher near the suction surface and at the end of the blade. When airflow flows through the blade passage, the flow pattern around the blade is complex. The separate flow appears near the suction surface, and an obvious eddy current exists at the tail of some blades, whereas the flow pattern around the blade wake is disordered. For the local enlarged image, the boundary layer separation occurs near the suction surface in the blade passage, and the Q value is obviously becomes high at T_3. With the impeller rotating, the suction surface airflow separation cluster will fall off at the end of the blade. At the time of T1, the eddy current is not obvious at the end of the blade, and with the fan running, the eddy at the end of the blade will gradually become obvious, and the area will increase. The boundary layer separation on the suction surface and the shedding vortices at the end of the blade will lead to the pulsating reaction in the flow passage of volute.

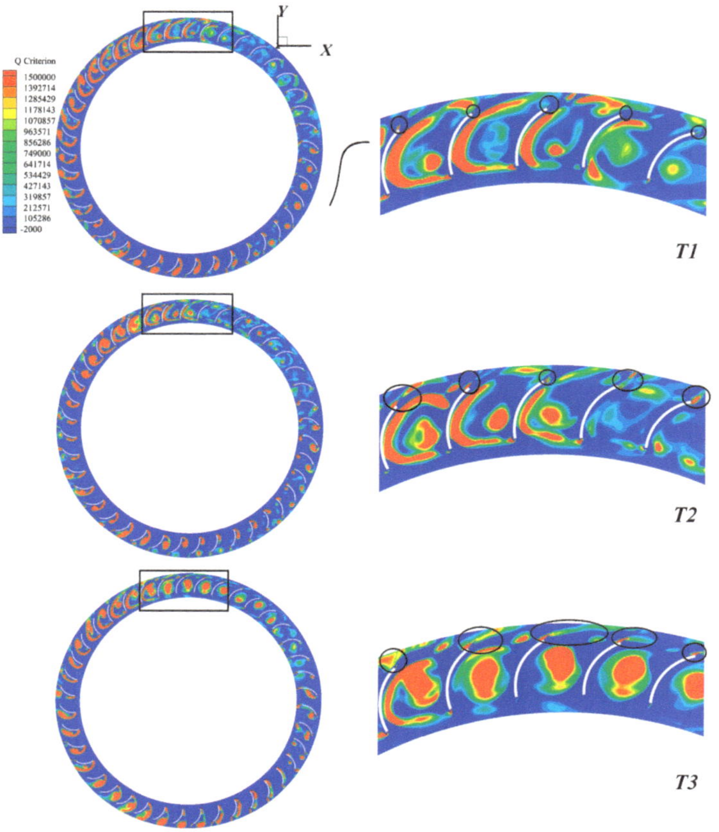

Figure 20. Q value distribution at different times of blade passage on the Z = 0.015 m section (Q/Qn = 0.031).

Figure 21 shows the skin friction coefficients of the pressure surface and suction surface of the blade. The magnitude of the friction coefficient on the blade surface indicates the velocity near the blade surface. If the friction coefficient is positive or negative in a region, which indicates that the direction of local flow is inconsistent, then flow separation or reattachment occurs in the region. As we can see from Figure 21, there is no negative value of friction coefficient on the suction surface or pressure surface of the blade, expressing that no obvious flow separation has occurred. At Mark 1, the friction coefficient is higher, because the rotation of the impeller makes the air accelerate. However, the friction coefficients of the pressure surface and suction surface are similar. At Mark 2, the friction coefficient on the suction surface of the blade changes slightly with time, and a high friction coefficient appears on both ends of the blade at T_2. The variation of the friction coefficient gradient at both ends of the blade makes the local flow unstable. The friction coefficient of the blade pressure surface is smaller than that of the suction surface over time. This indicates that the flow becomes unstable due to the large velocity variation in the blade passage.

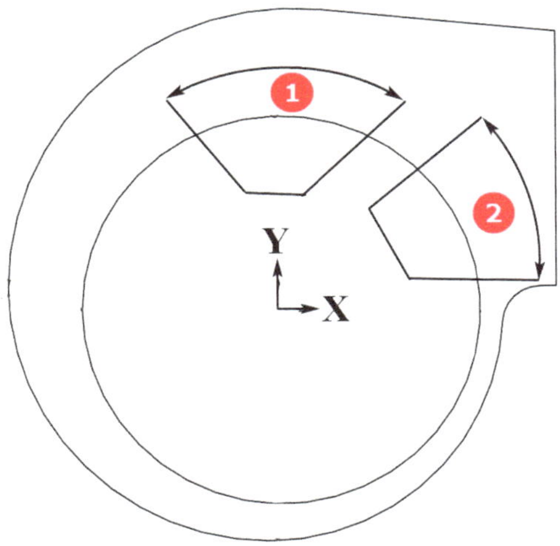

(a)

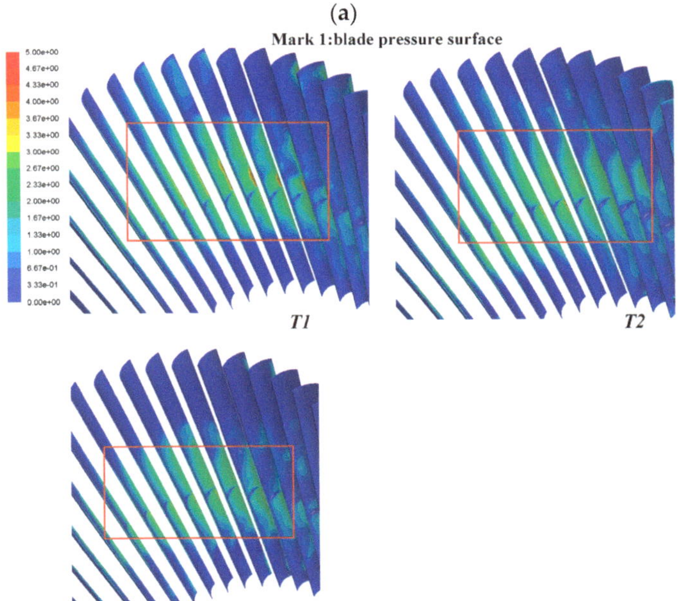

(b)

Figure 21. *Cont.*

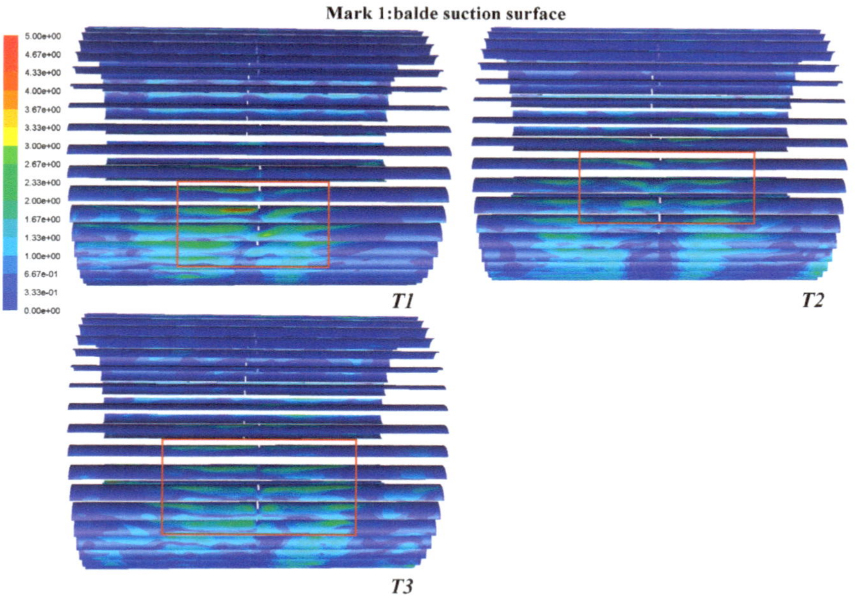

(c)

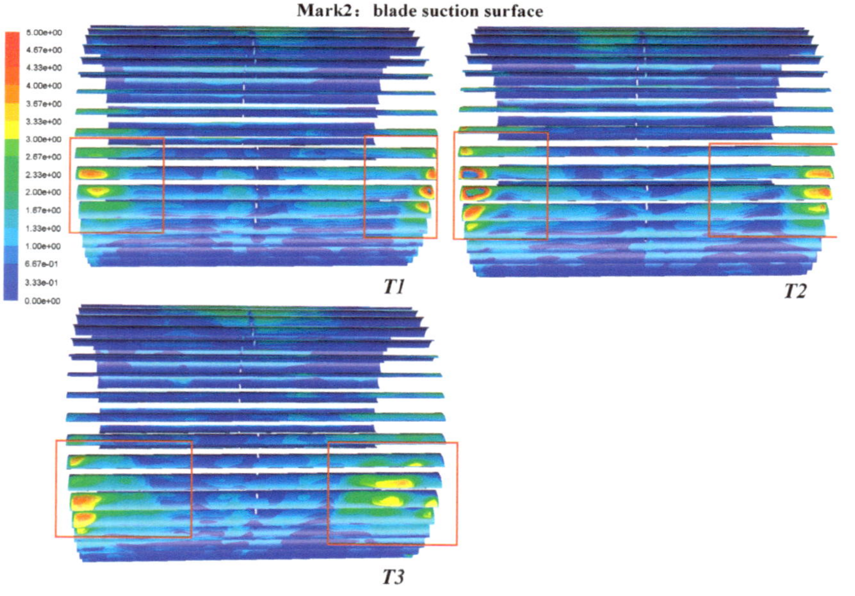

(d)

Figure 21. *Cont.*

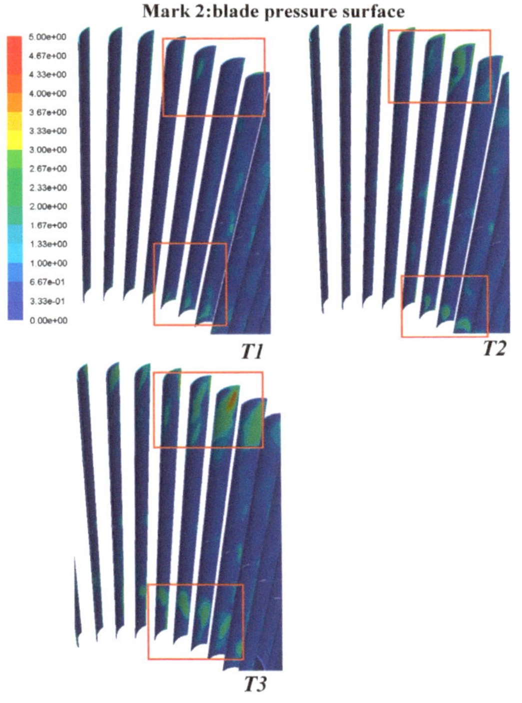

(e)

Figure 21. Skin friction coefficient of the pressure surface and suction surface of the blade.

3.4. Pressure Fluctuation of Internal Flow in Centrifugal Fan

In this subsection, the pressure fluctuation and frequency spectrum at key points are mainly introduced in order to further analyze the internal flow of the centrifugal fan with time at a non-dimensional flow rate of Q/Qn = 0.031.

Figure 22 shows the location of monitoring points on the Z = 0 m plane in the process of unsteady calculation. According to Figure 22, one can obviously see that nine material points are monitored. According to the coordinate system in the drawing, four monitoring points are set at the outlet of the impeller, marked as a, b, c, and d, respectively. Three monitoring points (b1, c1, and d1) are located on the wall of the volute. b0 is the middle of b and b_1, and c_0 is the central position between c and c1. The static pressure monitoring of these nine points is carried out separately; in addition, the velocity of points a, b, c, d, b_0, and c_0 is monitored. The static pressure and velocity data are extracted once at each time-step size.

Figure 23 shows a schematic diagram of the pressure fluctuation at each monitoring point. As we can see from Figure 23, the pressure fluctuation at each monitoring point fluctuates in a certain range. The monitoring points (a, b, c, and d) at the impeller outlet all include the pressure fluctuation caused by the rotating of the blade. Due to the close distance from the volute tongue, the airflow near point A is intricate, so the pressure distribution at point A is unstable, and the regularity is weak. The monitoring points (b_0, c_0, b_1, c_1, and d_1) in the volute channel and on the wall of the volute are far away from the impeller, and the pressure fluctuation at each point is mainly caused by boundary layer separation on the suction surface of the blade, the shedding vortex at the end of the blade and the pulsation of the boundary layer on the volute wall; therefore, no obvious regularity is represented. Figure 22 illustrates that the acceleration of the blade passage makes the gas fluctuate greatly when the airflow flows out of the blade passage. The gas velocity near the volute wall decreases due to the diffusive action of the volute, which leads to the separation of the boundary layer. The velocity of gas is larger when the gas flows out of the blade passage. Since the existence of the volute tongue clearance results in the velocity of gas decreasing rapidly, so the fluctuation of gas near the volute tongue is also larger.

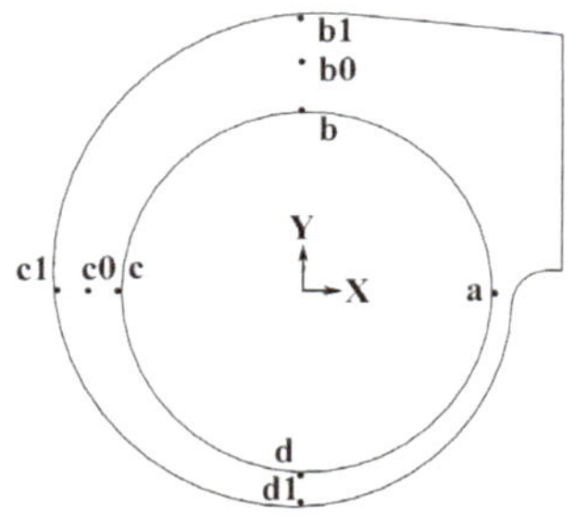

Figure 22. Location of monitoring points on the Z = 0 m plane.

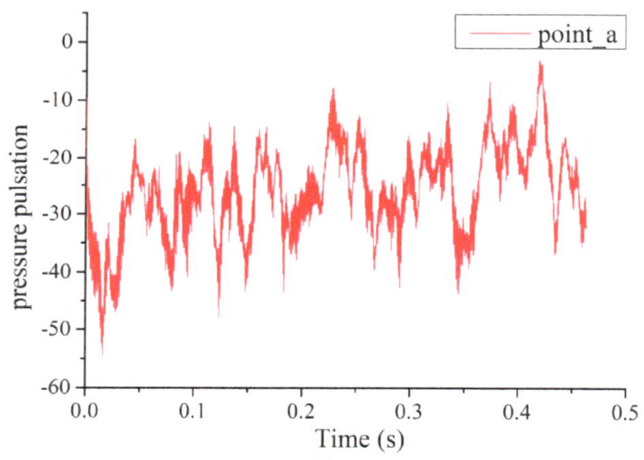

(**a**)

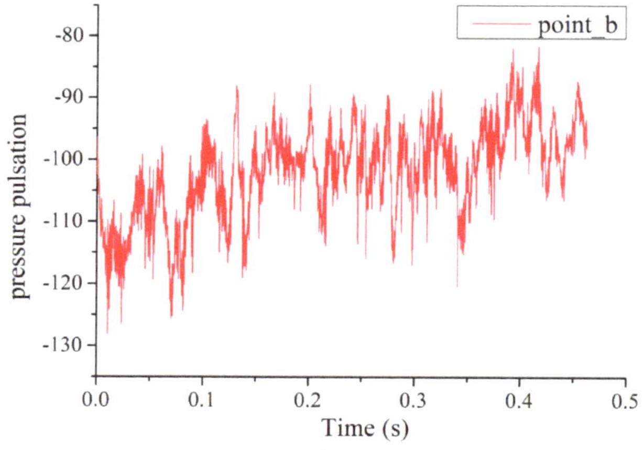

(**b**)

Figure 23. *Cont.*

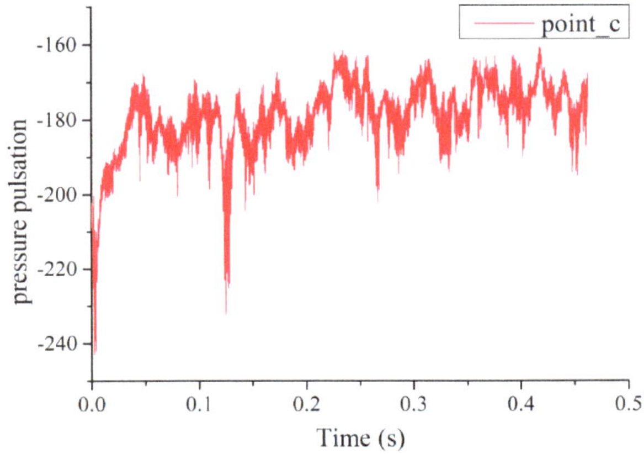

(**c**)

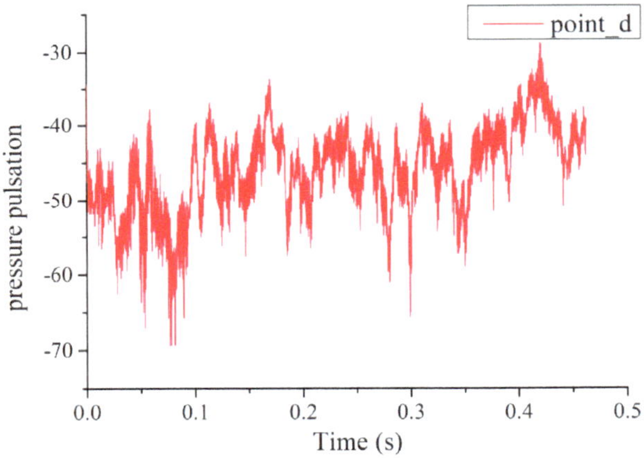

(**d**)

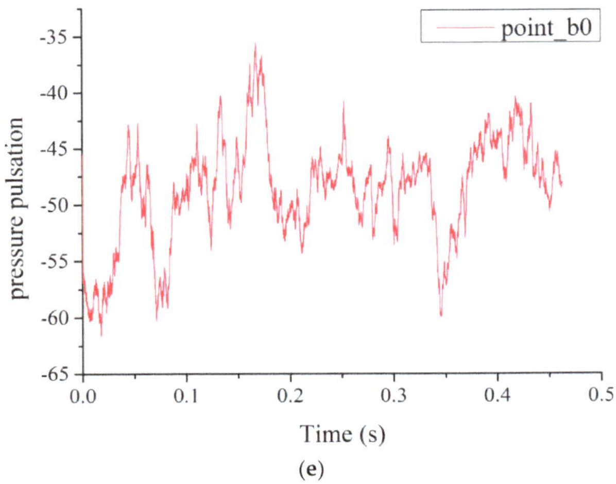

(**e**)

Figure 23. *Cont.*

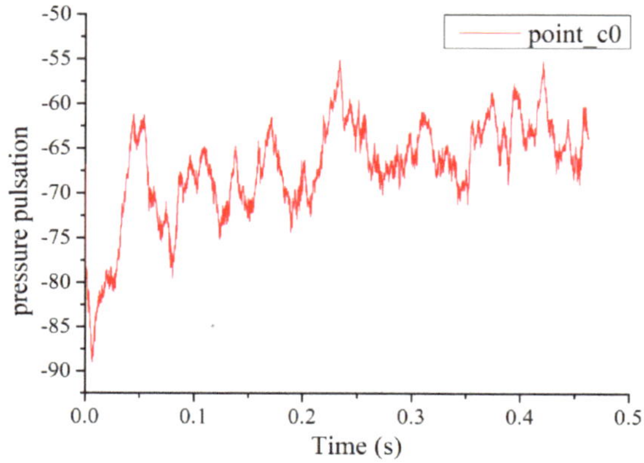

(f)

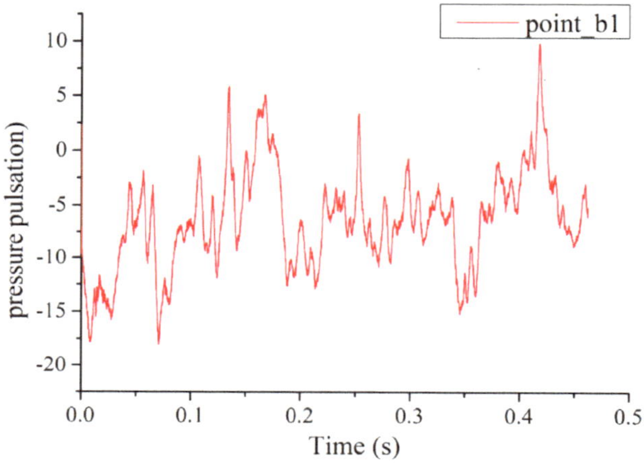

(g)

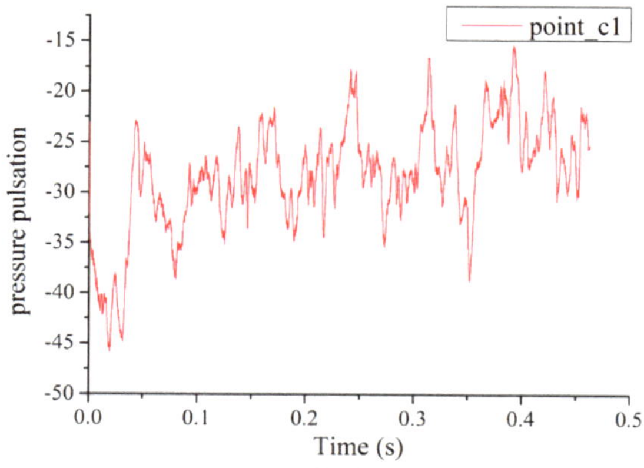

(h)

Figure 23. *Cont.*

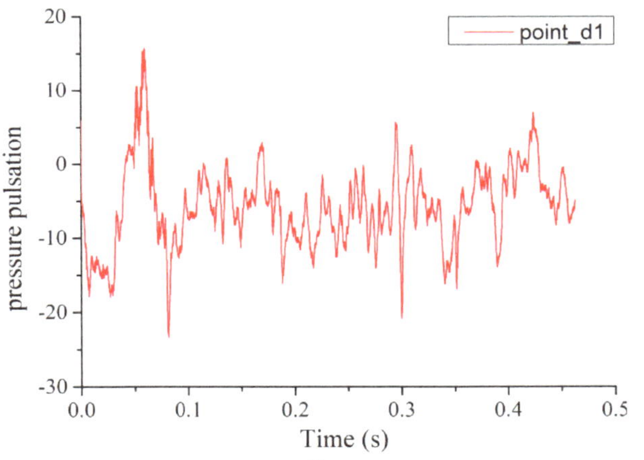

(i)

Figure 23. Static pressure fluctuation at different monitoring points. (**a**) at point a, (**b**) at point b, (**c**) at point c, (**d**) at point d, (**e**) at point b0, (**f**) at point c0, (**g**) at point b1, (**h**) at point c1, (**i**) at point d1.

Figure 24 shows the variation of static pressure coefficients at different monitoring points over time variation. Generally, the distribution of static pressure coefficients can be analyzed from the wall pressure gradient. When *dCp/dt* is greater than 0, the boundary layer flow decelerates until separation occurs. On the contrary, when *dCp/dt* is less than 0, the boundary layer flow is accelerated, which effectively suppresses the separation. From Figure 24, we can see that the gradient of the static pressure coefficient at the monitoring point is less than 0. Therefore, the flow near the monitoring points is accelerated, and no flow separation occurs inside the volute.

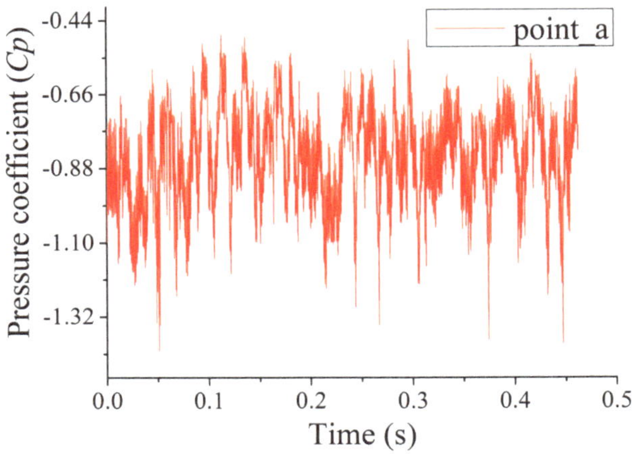

(a)

Figure 24. *Cont.*

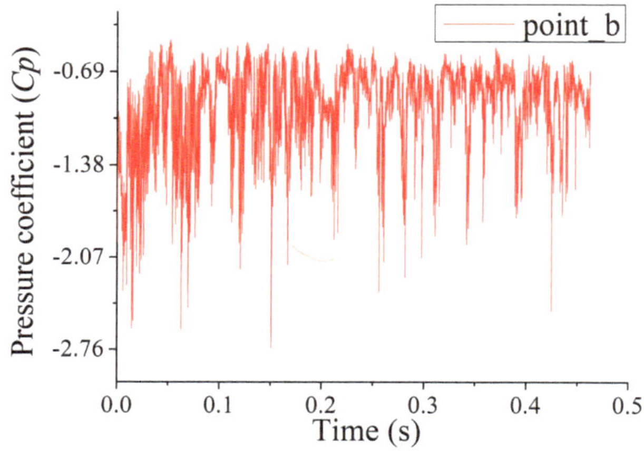

(**b**)

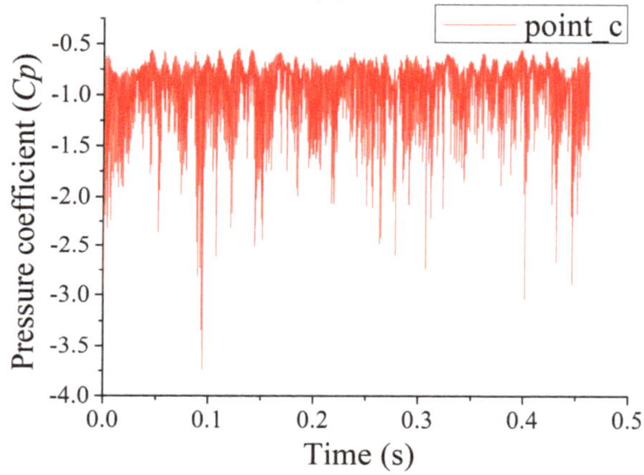

(**c**)

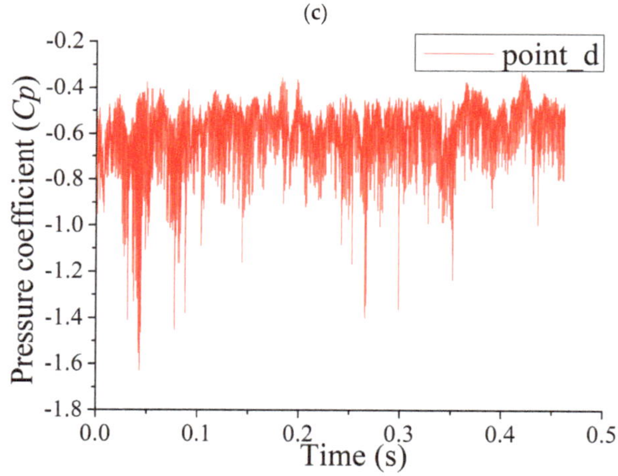

(**d**)

Figure 24. *Cont.*

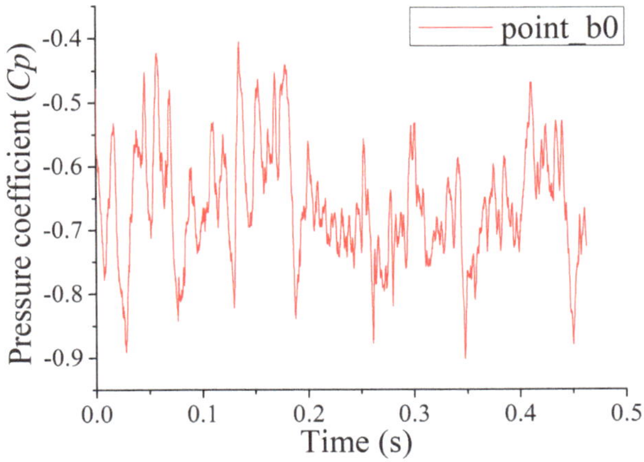

(e)

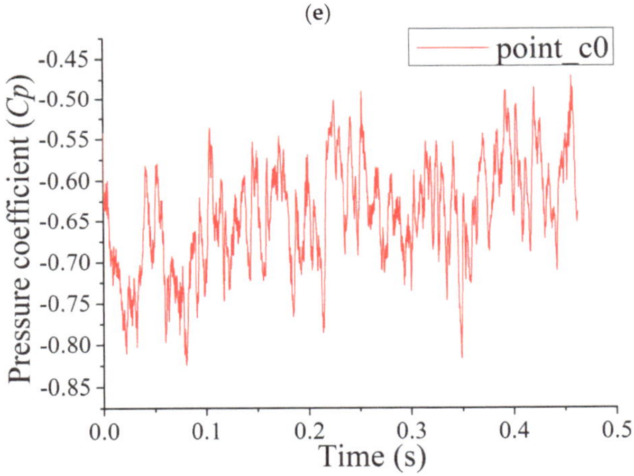

(f)

Figure 24. Pressure coefficient at monitoring points under a flow rate of Q/Qn = 0.031. (**a**) at point a, (**b**) at point b, (**c**) at point c, (**d**) at point d, (**e**) at point b0, (**f**) at point c0.

Figure 25 presents the static pressure fluctuation spectrum of each monitoring point. The abscissa is the frequency value, and the ordinate is the pressure fluctuation energy amplitude corresponding to each frequency value. As shown in Figure 25, it is observed that the static pressure fluctuations at each monitoring point show distinct discrete spectrum characteristics on the spectrum. Figure 25a, at the impeller outlet position (a, b, c, and d), the obvious peak amplitudes appear at the low-frequency band (2.16 Hz to 160 Hz) and at 866 Hz, and the amplitudes at 15 Hz and 866 Hz are larger than the others. It is mainly indicated that the shedding vortices occur at the blade tip and the rotating action of the impeller is obvious. Since point c is close to the volute tongue, the amplitude of point c at 866 Hz is obviously higher than that of the other three points due to the periodic impact of flow on the volute tongue. In Figure 25b, we can see that the frequency spectrum of monitoring points b_1, c_1, and d_1 in the volute wall position fluctuates within 2.16 Hz to 160 Hz, but it is not obvious at 866 Hz. Due to the diffusive effect of the volute, the velocity of the airflow near the volute wall is small, and a vortex shedding exists on the volute wall, which leads to the fluctuation of the boundary layer. However, the peak value is not obvious at the basic frequency, but there is a larger peak density at the low-frequency band. In Figure 25c, it is seen that the frequency spectrum of the b_0 and c_0 monitoring points in the volute passage fluctuates from 2.16 Hz to 160 Hz, and the amplitude of monitoring point c_0 occurs at 866 Hz, while the amplitude of b_0 is not obvious. It is shown that the flow separation of the blade and the shedding vortices at the end of the blade also produce pressure fluctuation in the volute passage. In Figure 25d, it is obtained that only monitoring point b has a significant amplitude at 866 Hz. In Figure 25e, it is also noted that the amplitude of point c is larger than that of point c_0, but the amplitude of point c1 is not obvious. It is clearly stated that the closer the impeller outlet, the fluctuation frequency is well consistent with the passage frequency of the blade due to the rotating effect of the impeller.

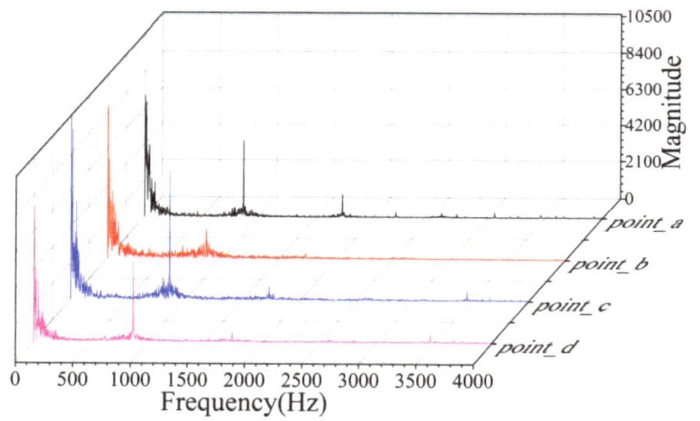

(a)

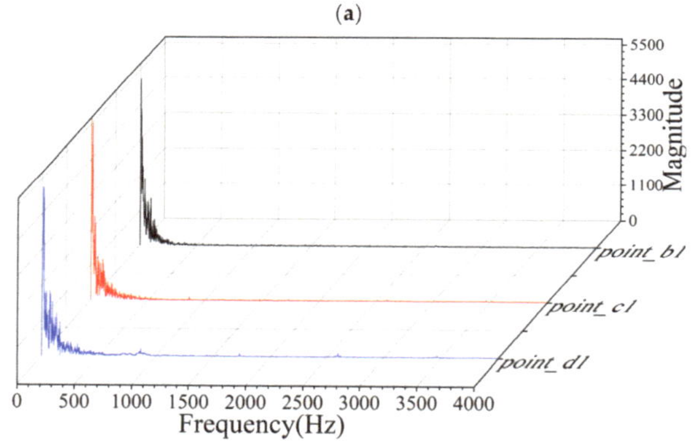

(b)

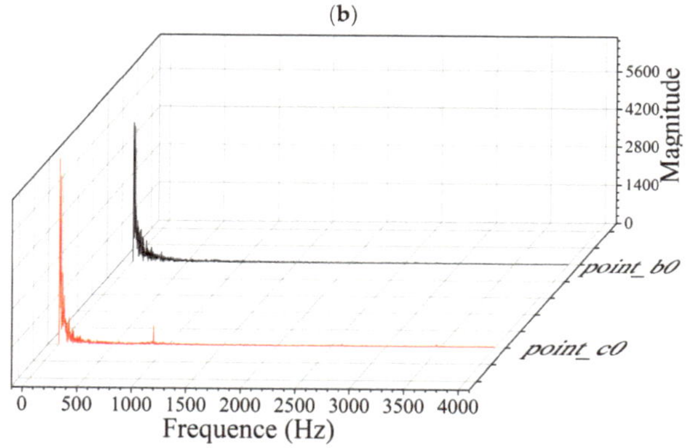

(c)

Figure 25. *Cont.*

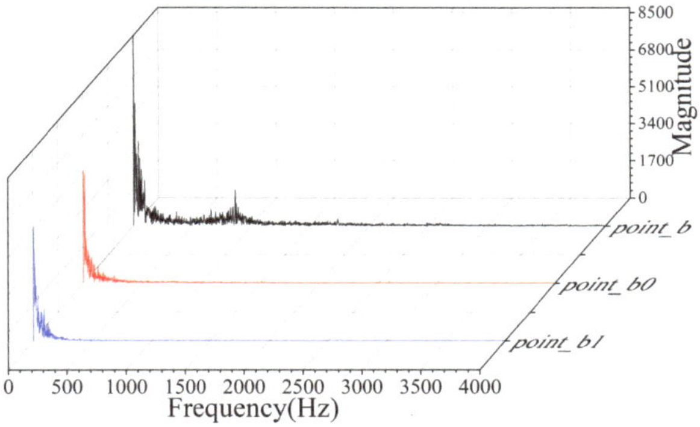

(d)

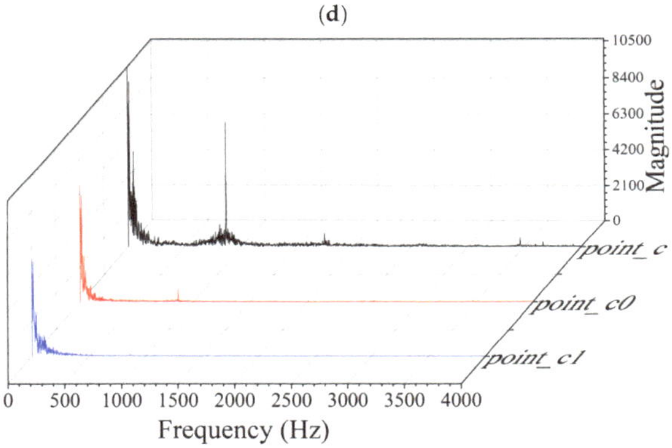

(e)

Figure 25. Spectrum characteristics of monitoring points on the Z = 0 m plane. (**a**) at point a, b, c, and d, (**b**) at point b1, c1, and d1, (**c**) at point c0, and b0, (**d**) at point b, b0, and b1, (**e**) at point c, c0, and c1.

3.5. Kinetic Energy Spectrum of Internal Flow

In this subsection, the internal flow kinetic energy spectrum will be introduced in some critical points of the centrifugal fan. Kolmogorov proposed a statistical theory of locally isotropic turbulence. This theory engenders a famous energy spectrum −5/3 rule of shear turbulence [28].

The drawing of a turbulent flow energy spectrum is divided into three steps. The first step is to obtain the velocity fluctuation (u) at the monitoring point and get the square (u^2). The second step is to make a Fourier transform of the square(u^2). Finally, the logarithm of the transverse and longitudinal coordinates is used to obtain the kinetic energy spectrum.

Figure 26 exhibits the kinetic energy spectrum of monitoring points on the Z = 0 m plane (logarithmic coordinates). Figure 26a shows the kinetic energy spectrum of the monitoring points (a, b, c, and d) of the outer diameter of the impeller. As illustrated in Figure 26a, we can see that the energy of each frequency at point b is obviously higher than that at the other three points, because point b is close to the volute outlet, and the tangential direction is consistent with the main flow direction. Four monitoring points have large energy peaks at the wave number of 3. It is obtained that a turbulent flow in a centrifugal fan gradually develops at a range of frequency from $10^{0.2}$ to $10^{1.7}$. A full turbulent flow in a centrifugal fan occurs at a range of frequency from $10^{1.7}$ to 10^3, which is well consistent with the rake ratio(−5/3) rule of shear turbulence. A turbulent dissipation flow in a centrifugal fan occurs at a range of frequency from 10^3 to $10^{3.7}$. Figure 25b shows the kinetic energy spectrum of the monitoring points (b_0,c_0) inside the volute. The wave number range from $10^{0.2}$ to $10^{1.7}$ is a turbulent region; between the $10^{1.7}$ and 10^3 wave numbers, it is a complete turbulent region that presents obvious turbulent characteristics. A large energy peak appears at the wave number of 10^3. Compared with the monitoring points at the outlet of the impeller and the internal monitoring points of the volute, the internal monitoring points of the volute are more satisfied with the rake ratio (−5/3) rule of shear turbulence.

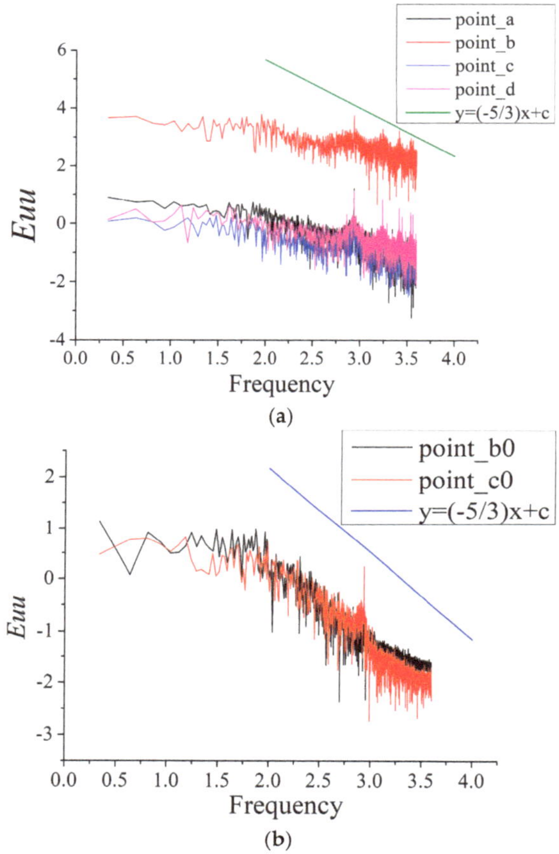

Figure 26. Kinetic energy spectrum of monitoring points on the Z = 0 m plane. (**a**) at point a, b, c, and d, (**b**) at point b0 and c0.

4. Conclusions

The unsteady internal flow characteristic was researched under a low flow rate by means of large eddy simulation based on computational fluid dynamics. Through the analysis of its internal flow, it is found that the flow instability does exist under the condition of a small flow rate.

First of all, it is can be clearly observed that the internal Q value distribution of a centrifugal fan at different times on the Z = 0.015 m surface is significantly higher near the volute exit and volute tongue. This indicates that the flow instability appears at the volute exit and volute tongue.

In addition, the Q value is higher near the suction surface of the blade and at the end of the blade. This also shows that when airflow flows through the blade passage, flow separation appears around the suction surface, and an evident vortex exists at the blade trailing edge. As the impeller rotates, the airflow from the blade passage falls off at the blade tail, and it gradually becomes apparent. The boundary layer separation near the suction surface and the shedding vortices at the end of the blade will cause a pulsating reaction in the volute port.

What's more, the pressure fluctuation at each monitoring point shows periodic variation, and its frequency spectrum exhibits obvious discrete characteristics. The frequency spectrum of each monitoring point indicates that the fluctuation frequency is consistent with the fundamental frequency of impeller rotation near the impeller position. The farther away from the impeller position, the higher the density peak appears at the lower frequency band instead of at the fundamental frequency band due to the influence of the blade-shedding vortices and the boundary layer fluctuation on the volute wall.

Finally, for the impeller outlet monitoring points (a, b, c, and d) and the volute channel internal monitoring points (b_0, c_0) similarly, a full turbulent flow in a centrifugal fan occurs at a range of frequencies from $10^{1.7}$ to 10^3, which is well consistent with the $-5/3$ rule of shear turbulence.

Author Contributions: The following statements could be used Y.W. and Y.L. conceived and designed the experiments; X.Y. performed the experiments; L.L. and C.Y. analyzed the data; Y.L. contributed reagents/materials/analysis tools; Y.W. wrote the paper.

Funding: This work was supported by the National Natural Science Foundation of China (11872337, 11902291 and 51706205), the Natural Science Foundation of Zhejiang Province (LY18A020010), and the Fundamental Research Funds of Zhejiang Sci-Tech University (2019Y004).

Conflicts of Interest: The authors declare no conflict of interest.

References

1. Lee, S.; Kim, H.J.; Runchal, A. Large eddy simulation of unsteady flows in turbomachinery. *Proc. Inst. Mech. Eng. Part A J. Power Energy* **2004**, *218*, 463–475. [CrossRef]
2. Conway, S.; Caraeni, D.; Fuchs, L. Large eddy simulation of the flow through the blades of a swirl generator. *Int. J. Heat Fluid Flow* **2001**, *21*, 664–673. [CrossRef]
3. Tajadura, R.-B.; Suarez, V.-S.; Cruz, J.-P.-H.; Morros, C.-S. Numerical calculation of pressure fluctuations in the volute of a centrifugal fan. *J. Fluids Eng.* **2006**, *128*, 359–369. [CrossRef]
4. Younsi, M.; Bakir, F.; Kouidri, S.; Rey, R. Numerical and experimental study of unsteady flow in centrifugal fan. In Proceedings of the 7th European Turbomachinery Conference, Athens, Greece, 5–9 March 2007; pp. 175–189.
5. Younsi, M.; Bakir, F.; Kouidri, S.; Rey, R. 2D and 3D unsteady flow in squirrel-cage centrifugal fan and aeroacoustic behaviour. In Proceedings of the FEDSM2006-98457, ASME Joint US–European Fluids Engineering Summer Meeting, Miami, FL, USA, 17–20 July 2006; pp. 805–813.
6. Kato, C.; Kaiho, M.; Manabe, A. An overset finite-element large eddy simulation method with application to turbomachinery and aeroacoustics. *J. Appl. Mech.* **2003**, *70*, 32–43. [CrossRef]
7. Moon, Y.J.; Cho, Y.; Nam, H.S. Computation of unsteady viscous flow and aeroacoustic noise of cross flow fans. *Comput. Fluids* **2002**, *32*, 995–1015. [CrossRef]
8. Velarde, S.S.; Ballesteros, T.R.; Santolaria, M.C. Unsteady Flow Pattern Characteristics Downstream of a Forward-Curved Blades Centrifugal Fan. *J. Fluids Eng.* **2001**, *123*, 265–272. [CrossRef]
9. Yang, J.; Meng, L.; Zhou, L.J.; Luo, Y.Y.; Wang, Z.W. Unsteady internal flow field simulations in a double suction centrifugal fan. *Eng. Comput.* **2013**, *30*, 345–356. [CrossRef]
10. Wei, Y.K.; Yang, H.; Lin, Z.; Wang, Z.D.; Qian, Y.H. A novel two-dimensional coupled lattice Boltzmann model for thermal incompressible flows. *Appl. Math. Comput.* **2018**, *339*, 556–567. [CrossRef]
11. Deardorff, J.W. A Numerical Study of Three-Dimensional Turbulent Channel Flow at Large Reynolds Number. *J. Fluid Mech.* **1970**, *41*, 453–480. [CrossRef]
12. Germano, M.; Piomelli, U.; Moin, P. A Dynamic Subgrid Scale Eddy Viscosity Model. *Phys. Fluids A* **1991**, *3*, 1760–1765. [CrossRef]
13. Liang, H.; Li, Y.; Chen, J.X.; Xu, J.R. Axisymmetric lattice Boltzmann model for multiphase flows with large density ratio. *Int. J. Heat Mass Transf.* **2019**, *130*, 1189–1205. [CrossRef]
14. Zhang, W.; Chen, X.P.; Yang, H.; Liang, H.; Wei, Y.K. Forced convection for flow across two tandem cylinders with rounded corners in a channel. *Int. J. Heat Mass Transf.* **2019**, *130*, 1053–1069. [CrossRef]
15. Lilly, D.K. A Proposed Modification of The Germano sub-grid scaleclosure method. *Phys. Fluids A* **1992**, *4*, 633–635. [CrossRef]

16. Meneveau, C.; Lund, T.S.; Cabot, W.H. A Lagrangian Dynamic Subgrid Scale Model of Turbulence. *J. Fluid Mech.* **1996**, *319*, 353–385. [CrossRef]
17. Piomelli, U. High Reynolds Number Calculations Using The Dynamic Subgrid Scale Stress Model. *Phys. Fluids* **1993**, *5*, 1484–1490. [CrossRef]
18. Bardina, J.; Ferziger, J.H.; Reynolds, W.C. Improved subgrid scale models for large eddy simulation. *Am. Inst. Aeronaut. Astronaut.* **1980**, *80*, 13–57.
19. Zhou, H.; Cai, G.B. Research of wall roughness effects based on Q criterion. *Microfluid. Nanofluid.* **2017**, *21*, 1–14. [CrossRef]
20. Li, X.J.; Gao, P.L.; Zhu, Z.C.; Li, Y. Effect of the blade loading distribution on hydrodynamic performance of a centrifugal pump with cylindrical blades. *J. Mech. Sci. Technol.* **2018**, *32*, 1161–1170. [CrossRef]
21. Zhang, S.F.; Li, X.J.; Hu, B.; Liu, Y.; Zhu, Z.C. Numerical investigation of attached cavitating flow in thermosensitive fluid with special emphasis on thermal effect and shedding dynamics. *Int. J. Hydrogen Energy* **2019**, *44*, 3170–3184. [CrossRef]
22. Kolář, V.; Šístek, J. Corotational and Compressibility Aspects Leading to a Modification of the Vortex-Identification Q-Criterion. *AIAA J.* **2015**, *53*, 2406–2410. [CrossRef]
23. Sipp, D.; Jacquin, L. A Criterion of Centrifugal Instabilities in Rotating Systems. *Lect. Notes Phys.-N. Y. Berl.* **2000**, *55*, 293–302.
24. Billant, P.; Gallaire, F. Generalized Rayleigh criterion for non-axisymmetric centrifugal instabilities. *J. Fluid Mech.* **2005**, *542*, 365–377. [CrossRef]
25. Lun, Y.X.; Lin, L.M.; He, H.J.; Zhu, Z.C.; Wei, Y.K. Effects of Vortex Structure on Performance Characteristics of a Multiblade Fan with Inclined tongue. *Proc. Inst. Mech. Eng. Part A J. Power Energy* **2019**, *233*, 1007–1021. [CrossRef]
26. Wang, C.; Hu, B.; Zhu, Y.; Wang, X.; Luo, C.; Cheng, L. Numerical study on the gas-water two-phase flow in the self-priming process of self-priming centrifugal pump. *Processes* **2019**, *7*, 330. [CrossRef]
27. Wang, C.; He, X.; Zhang, D. Numerical and experimental study of the self-priming process of a multistage self-priming centrifugal pump. *Int. J. Energy Res.* **2019**, *43*, 1–19. [CrossRef]
28. Kolmogorov, A.N. The Local Structure of Turbulence in Incompressible Viscous Fluid for Very Large Reynolds Numbers. *C.R. Acad. Sci. U.R.S.S.* **1941**, *30*, 301–312. [CrossRef]
29. Li, X.J.; Jiang, Z.W.; Zhu, Z.C.; Si, Q.R.; Li, Y. Entropy generation analysis for the cavitating head-drop characteristic of a centrifugal pump. *Proc. Inst. Mech. Eng. Part C J. Mech. Eng. Sci.* **2018**, *232*, 4637–4646. [CrossRef]
30. Liu, Q.; Ye, J.H.; Zhang, G.; Lin, Z.; Xu, H.; Jin, H.; Zhu, Z. Study on the metrological performance of a swirlmeter affected by flow regulation with a sleeve valve. *Flow Meas. Instrum.* **2019**, *67*, 83–94. [CrossRef]
31. Liu, Q.; Ye, J.H.; Zhang, G.; Lin, Z.; Xu, H.; Zhu, Z. Metrological performance investigation of swirl flowmeter affected by vortex inflow. *J. Mech. Sci. Technol.* **2019**, *33*, 1–10. [CrossRef]
32. Xu, H.; Cantwell, C.D.; Monteserin, C.; Eskilsson, C.; Engsig-Karupet, A.P.; Sherwin, S.J. Spectral/hp element methods: Recent developments, applications, and perspectives. *J. Hydrodyn.* **2018**, *30*, 1–22. [CrossRef]
33. Xu, H.; Mughal, S.M.; Gowree, E.; Atkin, C.J.; Sherwin, S.J. Destabilisation and modification of Tollmien–Schlichting disturbances by a three-dimensional surface indentation. *J. Fluid Mech.* **2017**, *819*, 592–620. [CrossRef]
34. Zhang, W.; Li, X.J.; Zhu, Z.C. Quantification of wake unsteadiness for low-Re flow across two staggered cylinders, Proceedings of the Institution of Mechanical Engineers, Part C. *J. Mech. Eng. Sci.* **2019**. [CrossRef]
35. Xu, H.; Sherwin, S.J.; Hall, P.; Wu, X. The behaviour of Tollmien–Schlichting waves undergoing small-scale localised distortions. *J. Fluid Mech.* **2016**, *792*, 499–525. [CrossRef]
36. Xu, H.; Lombard, J.E.; Sherwin, S.J. Influence of localised smooth steps on the instability of a boundary layer. *J. Fluid Mech.* **2017**, *817*, 138–170. [CrossRef]
37. Zheng, X.; Lin, Z.; Xu, B.Y. Thermal conductivity and sorption performance of nano-silver powder/FAPO-34 composite fin. *Appl. Therm. Eng.* **2019**, *5*. [CrossRef]
38. Zhu, Y.J.; Yang, Z.W.; Luo, K.H.; Pan, J.F.; Pan, Z.H. Numerical investigation of planar shock wave impinging on spherical gas bubble with different densities. *Phys. Fluids* **2019**, *31*, 056101–056109.
39. Zhu, Y.J.; Yang, Z.W.; Pan, Z.H.; Zhang, P.G.; Pan, J.F. Numerical investigation of shock-SF6 bubble interaction with different mach numbers. *Comput. Fluids* **2018**, *177*, 78–86. [CrossRef]

40. Yang, H.; Zhang, W.; Zhu, Z.C. Unsteady mixed convection in a square enclosure with an inner cylinder rotating in a bi-directional and time-periodic mode. *Int. J. Heat Mass Transf.* **2019**, *136*, 563–580. [CrossRef]
41. Wei, Y.K.; Wang, Z.D.; Dou, H.S.; Qian, Y.H.; Yan, W.W. Simulations of natural convection heat transfer in an enclosure at different Rayleigh number using lattice Boltzmann method. *Comput. Fluids* **2016**, *124*, 30–38. [CrossRef]
42. Yang, H.; Yu, P.Q.; Xu, J.; Ying, C.L.; Wen, B.C.; Wei, Y.K. Experimental investigations on the performance and noise characteristics of a forward-curved fan with the stepped tongue. *Measu. Contr.* **2019**. [CrossRef]

processes

MDPI

Article

Effects of Single-arc Blade Profile Length on the Performance of a Forward Multiblade Fan

Yikun Wei [1], Cunlie Ying [2], Jun Xu [2], Wenbin Cao [2], Zhengdao Wang [1,*,†] and Zuchao Zhu [1,*,†]

1 National-Provincial Joint Engineering Laboratory for Fluid Transmission System Technology, Faculty of Mechanical Engineering and Automation, Zhejiang Sci-Tech University, Hangzhou 310018, China; yikunwei@zstu.edu.cn

2 Zhejiang Yilida Ventilator Co., Ltd., Taizhou 318056, China; yingcunlie@yilida.com (C.Y.); xujun@yilida.com (J.X.); caowenbin@yilida.com (W.C.)

* Correspondence: dao@zstu.edu.cn (Z.W.); zhuzuchao@zstu.edu.cn (Z.Z.); Tel.: +86-13906507577 (Z.Z.)

† These authors contributed equally to this work.

Received: 30 July 2019; Accepted: 12 September 2019; Published: 17 September 2019

Abstract: The effects of single-arc blade profile length on the performance of a forward multiblade fan are investigated in this paper by computational fluid dynamics and experimental measurement. The present work emphasizes that the use of a properly reduced blade inlet angle (β_{1A}) and properly improved blade outlet angle (β_{2A}) is to increase the length blade profile, which suggests a good physical understanding of internal complex flow characteristics and the aerodynamic performance of the fan. Numerical results indicate that the gradient of the absolute velocity among the blades in model-L (reducing the blade inlet angle and improving blade outlet angle) is clearly lower than that of the baseline model and model-S (improving the blade inlet angle and reducing blade outlet angle), where a number of secondary flows arise on the exit surface of baseline model and model-S. However, no secondary flow occurs in model-L, and the flow loss at the exit surface of the volute (scroll-shaped flow patterns) for model-L is obviously lower than that of the baseline model at the design point. The comparison of the test results further shows that to improve the blade profile length is to increase the static pressure and the efficiency of the static pressure, since the improved static pressure of the model-L rises as much as 22.5 Pa and 26.2%, and the improved static pressure efficiency of the model-L rises as much as 5 % at the design flow rates. It is further indicated that increasing the blade working area provides significant physical insight into increasing the static pressure, total pressure, the efficiency of the static pressure and the total pressure efficiency.

Keywords: performance; blade profile length; forward multiblade fan; numerical simulation; experimental test

1. Introduction

It is well known that centrifugal fans have been widely adopted to industrial application, such as cooling units in air-conditioning, heating, ventilating and air conditioning systems, and home appliance machines [1–5]. In general, it is essential for the centrifugal fan to improve its aerodynamic performance to achieve the purpose of energy conservation [5–9]. Many centrifugal fans with moderate efficiency and low aero-acoustic performances still need to be improved by making use of today's technology. Velarde and Ballesteros [10] reported that to improve the blade number, tongue length and scroll contour is to improve the aerodynamic performance of a centrifugal blower. Younsi et al. [11] investigated that the internal flow of the blower for various rotational frequencies effects the performance characteristics of a turbo blower in a refuse collecting system. Lin and Huang [12] investigated that the performance of a centrifugal fan is affected by changing the blade inlet angle and the housing outlet geometry through experimental and computational methods.

Lun et al. [13] evaluated the total pressure and efficiency of a centrifugal fan by changing the impeller gap and blade inlet angle. Although varying the blade's outlet angle does not appear to be very advantageous herein, it certainly shows the potential benefits of such a modification, especially in cases where the fan design is not optimal and demands vigorous enhancements [14]. The transient surface of the models with different outlet angles can be found [15–17]. Computational fluid dynamics (CFD) results were implemented to provide insight into revealing flow loss in the internal flow of fan. Thus, the prediction method of CFD is represented to study the internal flow mechanisms in the fan. The section of computational methodology includes solver descriptions of the Navier–Stokes and the used turbulence models associated, computational procedures for non-rotating and rotating components of the fan, gridding strategy and a study of grid density, and flow parameters for the performance of predicting the fan [18,19]. To validate the predictions of CFD for the performance of fan data, comparisons with one-fifth-scale model fan test data are provided [20]. The trend is quite straightforward: Increasing the outlet angle leads to a higher pressure rise and torque across the whole operating range, and vice versa.

Complex turbulent structures due to rotating Coriolis force and narrow blade passage in the internal flow of a forward multiblade fan play a dominant role in directly affecting the performance of the fan. Once these complex turbulent internal flows of the fan have been described, and the underlying internal flow mechanisms have been understood, new open questions will emerge in the internal flow of the fan. How do we restrain turbulent flow to reduce flow loss in the internal flow of our fan? In the low to mid-range flow rates, changing the outlet angle does not considerably affect the static efficiency of the fan, whereas the ensuing effects in the higher range are significant, and the impellers with larger outlet angles show a better performance.

Based on the above discussions, our study mainly focuses on the effects of the outlet angle of the impeller blades on the aerodynamic performance of the centrifugal fan for potential energy conservation. The origins and effects of a single-arc blade profile length on the performance of a forward multiblade fan are discussed with the emphasis on the understandings of the physical mechanisms of internal complex flow characteristics. Our work indicates that by reducing the blade inlet angle (β_{1A}, $\beta_{1A} > 60°$), improving the blade outlet angle (β_{2A}, $\beta_{2A} < 175°$) and increasing the blade's working area may provide significant physical insight into increasing the static pressure and the efficiency of this static pressure. It gives insight into improving the aerodynamic load of fans to achieve the purpose of energy conservation. The performance of numerical results is quantitatively and qualitatively compared with that of the experimental results. This paper is organized as follows. In Section 2, governing equations and the numerical method will be briefly described at first. After that, the detailed numerical and experimental results are discussed. In final, some conclusions are addressed.

2. Governing Equations and Numerical Method

In the following section, the three-dimensional equation of fluid dynamics is implemented in fluid flow analysis of the computational domain [15,16]. The incompressible Reynolds Averaged Navier–Stokes (RANS) Turbulence Models are discretized by the finite volume method [16]. The equation of fluid dynamics is:

$$\frac{\partial(\rho u_i)}{\partial x_i} = 0 \tag{1}$$

$$\frac{\partial(\rho u_j u_i)}{\partial x_j} = f_i - \frac{\partial P^*}{\partial x_i} + \frac{\partial\left[\mu_e\left(\frac{\partial u_i}{\partial x_j} + \frac{\partial u_j}{\partial x_i}\right)\right]}{\partial x_j} \tag{2}$$

in which ρ is fluid density, x_i and x_j denote the components x, y and z in the rectangular coordinate system in three directions, u_i and u_j are the average relative velocity components u, v and w, P^* denotes the reduced pressure, f_i is the volume force component, and μ_e is the efficient viscosity coefficient [17,21].

The Re-Normalization Group (RNG) k-epsilon model is implemented in steady flow. The empirical formula of turbulence model RNG k-epsilon is as follows [22,23].

$$\frac{\partial}{\partial t}(\rho k) + \frac{\partial}{\partial x_i}(\rho k u_i)_k = \frac{\partial}{\partial x_j}\left(\alpha_k \mu_{eff} \frac{\partial k}{\partial x_j}\right) + G_k + G_b - \rho\varepsilon - Y_M + S_k \tag{3}$$

$$\frac{\partial}{\partial t}(\rho\varepsilon) + \frac{\partial}{\partial x_i}(\rho\varepsilon u_i) = \frac{\partial}{\partial x_j}\left(\alpha_\varepsilon \mu_{eff} \frac{\partial \varepsilon}{\partial x_j}\right) + C_{1\varepsilon}\frac{\varepsilon}{k}(G_k + G_{3\varepsilon}G_b) - C_{2\varepsilon}\rho\frac{\varepsilon^2}{k} - R_\varepsilon + S_\varepsilon \tag{4}$$

in which G_k is the turbulent kinetic energy of laminar velocity gradient, G_b denotes the turbulent kinetic energy, Y_M is the effect of compressible turbulent fluctuation expansion, $C_{1\varepsilon}$, $C_{2\varepsilon}$ and $G_{3\varepsilon}$ are constants, α_k and α_ε are the turbulent Prandtl numbers of the k equation and ε equation, respectively. The above equations are solved by implementing the finite volume method. In first, the time discretization of the Navier–Stokes equations coupled with some governing equation for a pollutant is implemented in this paper. The spatial schemes are used to account for convective terms, viscous terms, mean pressure gradient effects and source terms associated with momentum or scalar equations, focusing on triangular meshes [20,24].

3. Computational Model and Grid of Fan

The forward multiblade centrifugal fan with a double-width double-inlet impeller is implemented in the current investigation. Figure 1 shows the double inlet radial fan with a forward curved blade. It consists of the impeller hub plate, volute, the mid-disc of impeller and the symmetry plane. The blade shape of test cases and some key parameters of the baseline model are presented in Figure 2. The complex vortex structure of the baseline model on the internal flow and performance is referred to in the previous work by Lun [13]. As shown in Figure 2, we can obtain that the blade profile of the baseline model is a single circular arc with the length of 10.904 mm; model-L refers to the model with a longer blade, and model-S with a shorter blade. Plotted in Figure 2, it is seen that the blade profile of model-L (reducing the blade inlet angle and improving blade outlet angle) is 150% that of the baseline model, and model-S (improving the blade inlet angle and reducing blade outlet angle) is 80%. The blade profiles of both modified models are still single circular arcs which are shown in Figure 2. The blade length of model-S, the baseline model and model-L increase successively in Figure 2.

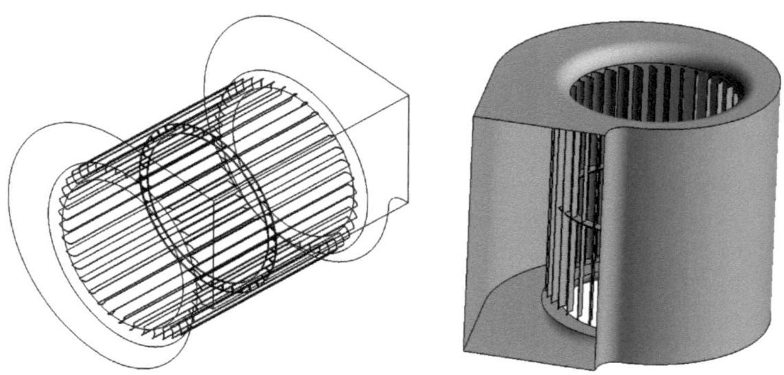

Figure 1. Example of double inlet radial fan with forward curved blade.

Table 1 shows the blade inlet angle and blade outlet angle of three models. As shown in Table 1, one can see that the blade inlet angles of the baseline model, model-S and model-L, are 84.83°, 91.27° and 71.34°, respectively, and the blade outlet angles of the baseline model, model-S and model-L are 152.36°, 146.69° and 170.13°, respectively. The numerical model of the radial fan is divided into five independent domains, as shown in Figure 3. Plotted in Figure 3, the impeller domain (domain 3) is set as the rotor and others are set as stators, while each domain is connected by an interface. The structured mesh is implemented for the whole fan model, and details of the numerical meshes of the volute and static pressure at a different number of grids are given in Figure 4. We can obtain that the change rate of static pressure is less than 1% at different mesh numbers. Thus, we can ignore the static pressure change. The numerical results are not affected by the number of grids. In all simulations, the ANSYS CFX software (CFX 16.0, Ansys, Canonsburg, PA 15317, USA) is implemented to study the internal complex flow of the forward multiblade centrifugal fan. The Multi Frames of Reference (MFR) technique is one such interface model. The MFR approach obtained by ANSYS CFX 16.0 is implemented to calculate the part of the impeller in the reference frame and the other parts in the stationary reference frame [12,13]. An interface is used at each junction where the frame change of reference takes place. There are two kinds of interface techniques available to exchange the information between the different frames of reference [12,13].

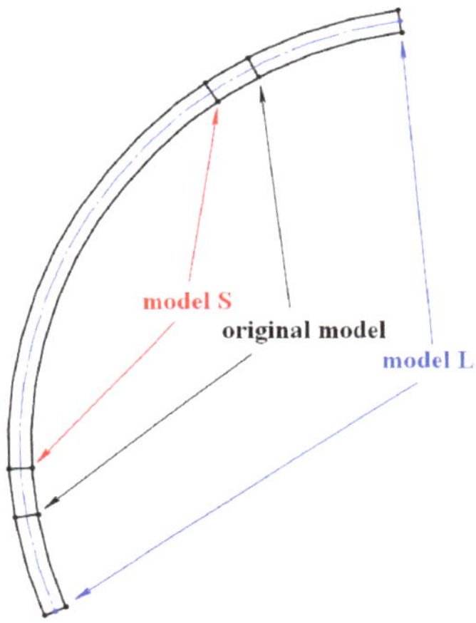

Number of blades	z=40
Volume flow rate	Qv=513.87 m^3/h
Revolving speed	n=1200 rpm
Inlet diameter	D1=131.6 mm
Outlet diameter	D2=150 mm
Rotor width	b=200 mm
Volute width	B=202.5 mm

Figure 2. Blade shape of test cases and geometrical parameters of the baseline model.

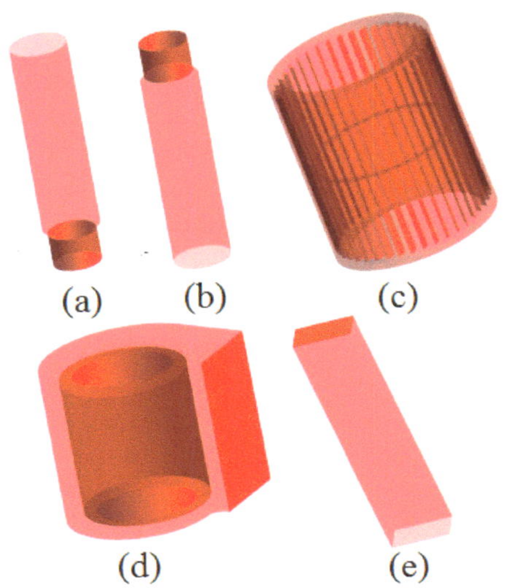

Figure 3. Domain composition: Inlet left (**a**), inlet right (**b**), impeller (**c**), volute (**d**) and outlet (**e**). The domain interface is colored in deep red.

Table 1. Blade inlet angle and blade outlet angle of the three models.

	Baseline Model	Model-S	Model-L
Blade inlet angle (β_{2A})	84.83°	91.27°	71.34°
Blade outlet angle (β_{1A})	152.36°	146.69°	170.13°

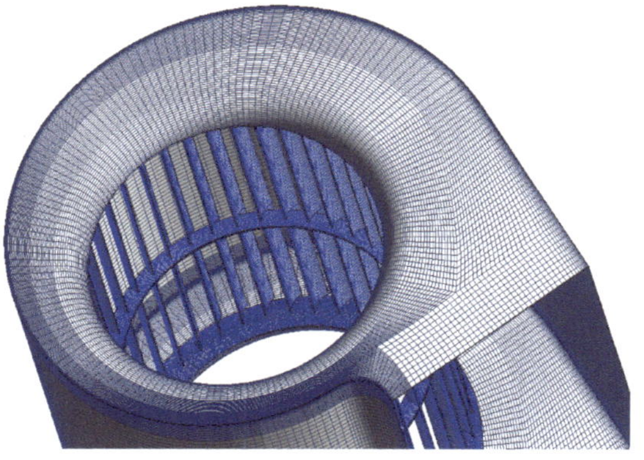

(**a**)

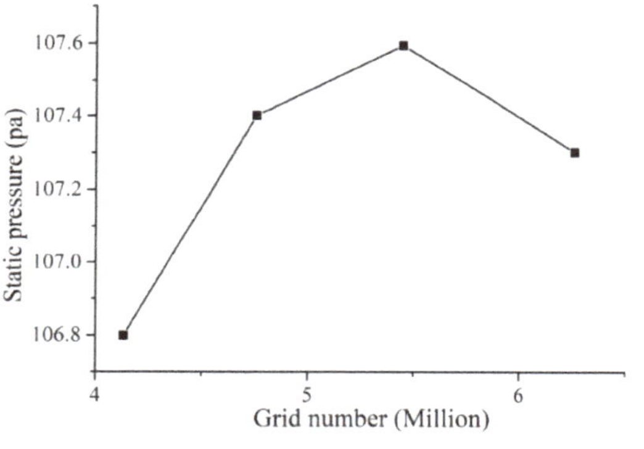

(**b**)

Figure 4. Detail of the numerical mesh and Static pressure at different number of grids. (**a**) numerical meshes of the volute; (**b**) static pressure at different number of grids.

4. Results and Discussions

Finite element methods [24–27] and the finite volume method [28,29] are very effective tools to solve some partial differential equations (PDEs) on complex geometries, which is applied in a wide range of engineering and biomedical disciplines [30–34]. In this paper, the finite volume method is implemented to calculate the NS equation. Numerical simulations are carried out using steady RANS based on the RNG k-ε turbulence model in our studies. The empirical formula of the turbulence model RNG k-epsilon is introduced in detail in Section 2. The mass flow rate is implemented as the inlet condition, and static pressure is implemented as the outlet condition. The wall condition is no slip boundary. It is well known that a moving fluid in contact with a solid body will not have any velocity relative to the body at the contact surface. This condition of not slipping over a solid surface has to be satisfied by a moving fluid. The second order upwind discretized form is used for all cases in incompressible flows [21,22]. The second order upwind discretized form is mainly based upon a coupled solution of the nonlinear finite difference equations according to the multigrid technique. Calculations have been made of the driven turbulent flow for several Reynolds numbers and finite difference grids. In comparison with the hybrid discretized form, the second order upwind discretized form is somewhat more accurate, but it is not monotonically accurate with mesh refinement.

4.1. Numerical Verification

In order to confirm the accuracy of numerical results, the baseline model is implemented to study its static pressure and the efficiency of this static pressure at various flow rates. Figure 5 shows the comparison of static pressure as well as the static pressure efficiency between numerical results and experimental results. As shown in Figure 5, one can clearly see that the static pressure (P, Pa) decreases with the increase of flow rate (Q, m^3/h) in both the experiment and numerical simulation. It is also obtained that the static pressure curve and the static pressure efficiency curve of the baseline model is well consistent with the experimental results at various flow rates, which demonstrates that the numerical accuracy is reliable.

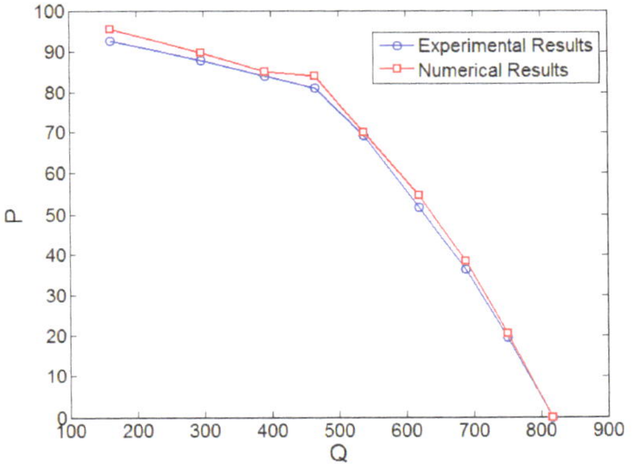

(a)

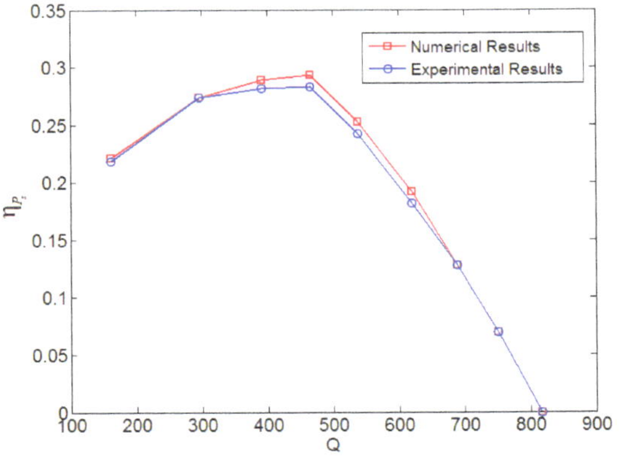

(b)

Figure 5. Comparison of static pressure and the static pressure efficiency between numerical results and experimental results. (**a**) curve of static pressure -flow (**b**) curve of static pressure efficiency-flow.

4.2. Complex Internal Flow in Different Forward Multiblade Fans

In the following section, complex internal flow in different forward multiblade fans will be introduced. The analysis refers to the centrifugal fan operating at three mass flow rate conditions of 0.0403 kg/s ($0.235Q_n$), 0.1713 kg/s (Q_n) and 0.2809 kg/s ($1.640Q_n$). Figure 6 shows the overview of the flow behavior within the impeller with the absolute velocity contours. Plotted in Figure 6, S_1, S_2 and S_3 represent the cutting planes at the hub, at the mid-span and at the tip of the impeller, respectively.

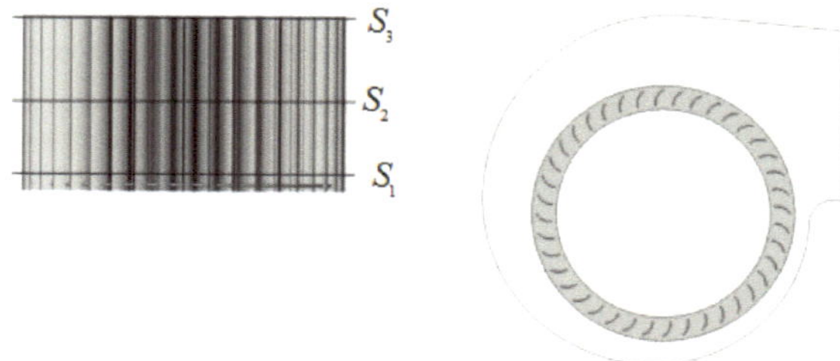

Figure 6. Different heights within the impeller.

Figure 7 shows the absolute velocity at different heights within the impeller. As shown in Figure 7, it is obviously seen that the gradient of the absolute velocity among the blades decreases with the successive increase of the blade length of model-S, the baseline model and model-L at the cross section of S_1, S_2 and S_3, which mainly indicates that the size of the flow separation among the blades becomes smaller with increasing the blade outlet angle. We can see that on the hub surface (S_1), the velocity difference between inlet and outlet of the impeller increases, and the velocity gradient in the impeller passage of model-L is more uniform than that of others. The area of the low velocity region at the inlet of the impeller decreases because of lengthening the blade in model-L. The average velocity decreases with the increase of axial height in the impeller passage. The main reason is that the flow in the passage of the tip cutting plane (S_3) is greatly affected by the inflow near the inlet of the volute.

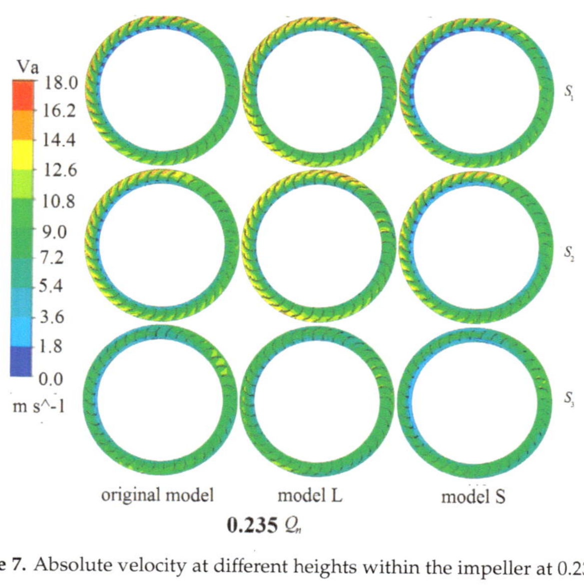

Figure 7. Absolute velocity at different heights within the impeller at $0.235Q_n$.

Figure 8 shows the absolute velocity at different heights within the impeller at design point Q_n. It is clearly observed that at the design point, the high velocity region exists at the impeller outlet of model-L at the hub plane (S_1) and midspan (S_2). However, model-L possesses more uniform velocity distribution in the impeller passage than that of the baseline model and model-S. Figure 9 shows the absolute velocity at different heights within the impeller at $1.64Q_n$. One can see that the high absolute velocity region in model-L is obviously larger than that of the baseline model and model-S. A large number of low absolute velocity regions occur at the inlet of blades in the baseline model and model-S at sections of S_1 and S_2. Nevertheless, the gradient of the absolute velocity among the blades in model-L is clearly lower than that of the baseline model and model-S.

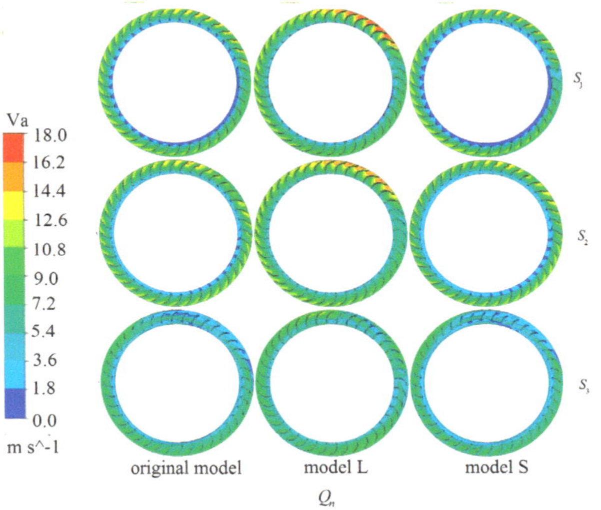

Figure 8. Absolute velocity at different heights within the impeller at Q_n.

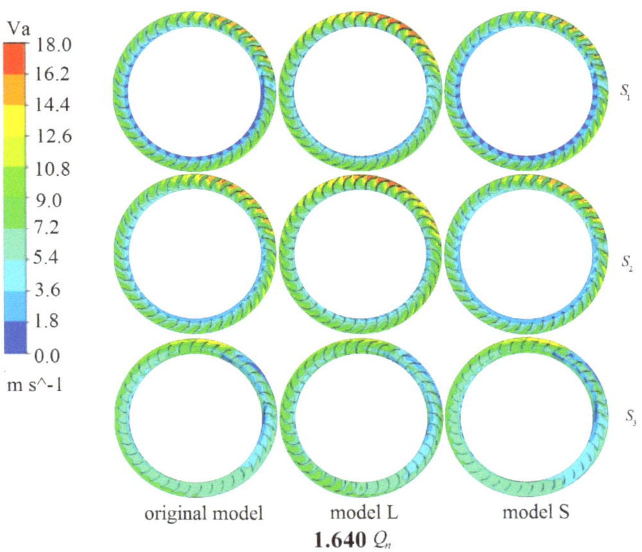

Figure 9. Absolute velocity at different heights within the impeller at $1.640Q_n$.

From Figure 7 to Figure 9, it is also obtained that the high velocity region at the impeller outlet moves toward the outlet of volute, and the difference of velocity in the blade passage increases accordingly with the increase of the mass flow rate. It is also noted that the high absolute velocity region in model-L is obviously higher than that of the baseline model and model-S. However, the gradient of the absolute velocity among the blades in model-L is clearly lower than that of the baseline model and model-S. It is further found that the gradient of the absolute velocity among the blades decreases in model-L, which brings about the decreasing size of the flow separation among the blades.

Figure 10 delineates the turbulence kinetic energy (TKE) of impeller outlet at 0.235Q_n. It is clearly observed that the high turbulence kinetic energy area in the baseline model occurs at the inlet of the volute. Meanwhile, the high turbulence kinetic energy area in model-L and model-S also occurs at the inlet of the volute. It is noted that for the turbulence kinetic energy, no significant difference occurs near the inlet of the volute in the baseline model, model-L and model-S.

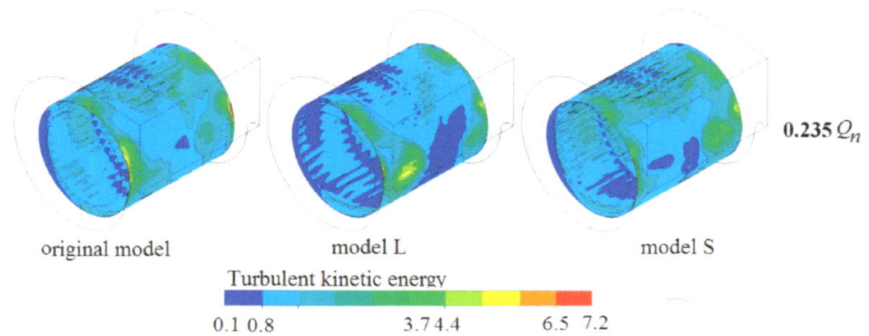

Figure 10. Distribution of turbulence kinetic energy of the impeller at 0.235Q_n.

Figure 11 shows the TKE of the impeller outlet at the design point Q_n. One can obviously see that at design point Q_n, the turbulence kinetic energy area near the inlet of the volute for model-L is lower than that of baseline model and model-S. It is also noted that the TKE at the impeller outlet of model-L and model-S is obviously reduced as compared with that of the baseline model and model-S.

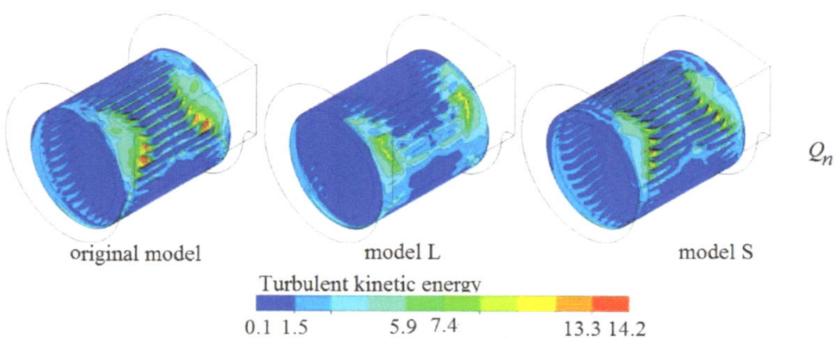

Figure 11. Distribution of turbulence kinetic energy of the impeller at Q_n.

Figure 12 demonstrates the TKE of the impeller outlet at 1.64Q_n. The variation tendency at 1.64Q_n for the turbulence kinetic energy area near the inlet of the volute is same as that of the design point Q_n. From Figure 10 to Figure 12, it is obtained that model-L has a better status as a whole, in not only reducing the peak of TKE at the impeller outlet, but also making the TKE change more uniformly along the circumferential direction of impeller outlet, especially in the area far from the volute outlet.

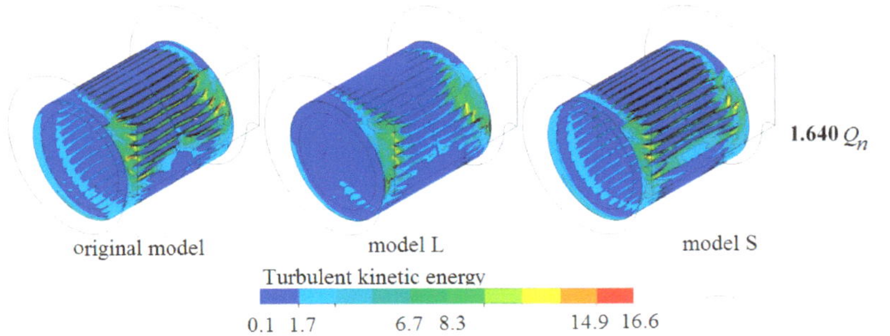

Figure 12. Distribution of turbulence kinetic energy of the impeller at 1.640Q_n.

Figure 13 shows the static pressure distribution on the surface of the volute tongue at 0.235Q_n. One can see that a low-pressure region and a high-pressure region appear on the surface of the volute tongue along the flow direction in each model at the condition of the lower flow rate, and the distribution range is large along the axial direction. For the model-L, the high-pressure region mainly locates at the middle part of tongue surface; the pressure value near middle part of tongue surface is obviously higher than that of the model-S and baseline model at 0.235Q_n.

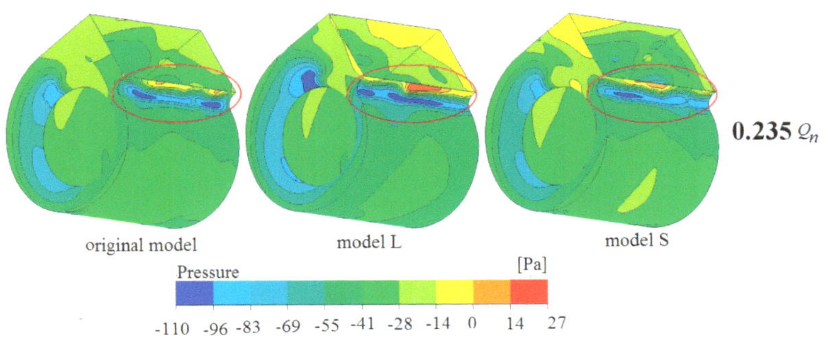

Figure 13. Static pressure distribution on the surface of the volute tongue at 0.235Q_n.

Figures 14 and 15 show the static pressure distribution on the surface of the volute tongue at design points Q_n and 1.64Q_n, respectively. Plotted in Figure 14, it is observed that the low-pressure region disappears at design point Q_n, and is replaced by a high-pressure region mainly in the middle of the volute tongue, which occupies about one third of the area. As shown in Figure 15, it is noted that the area of the high-pressure region on the middle of volute tongue surface increases at a high flow rate condition while the low-pressure region appears on both sides of the tongue.

Figure 16 shows the velocity profile (along the circumferential direction) near the blade trailing edge of impeller exit for the baseline model, model-S and model-L at different flow rates. As shown in this figure, it is observed that at 0.235Q_n, the velocity distribution at the outlet of each model impeller varies slightly and seemed uniform along the circumferential direction. Little difference appears for the peak velocity of each model; the velocity slowly decreases near the volute outlet and the volute tongue area. It is also obtained that at design rate (Q_n) and high flow rate (1.640Q_n), the velocity of all models greatly increases near the outlet of the volute and then rapidly decreases near the volute tongue, and the difference between the highest and the lowest velocity magnitude on the profile increases along the circumferential direction. The velocity of model-L is the highest near the volute outlet, but the amplitude of the velocity fluctuation is the smallest over the whole circumferential distance. In other words, the velocity profile (along the circumferential direction) near the blade trailing edge of the impeller exit obviously increases with the increase of the blade length at various flow rates. Nevertheless, the velocity amplitude of fluctuation decreases with the increase of the blade length.

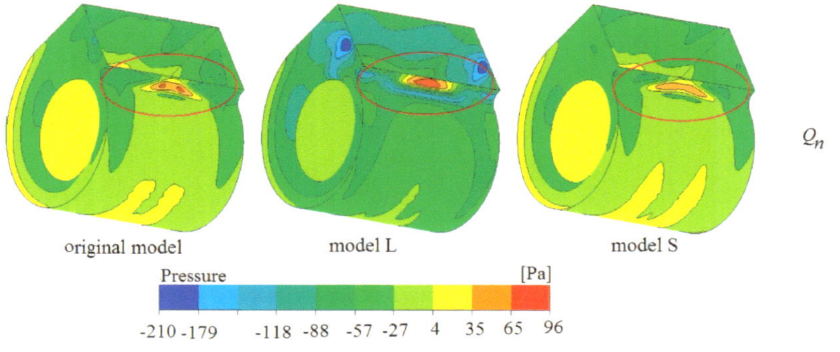

Figure 14. Static pressure distribution on the surface of the volute tongue at Q_n.

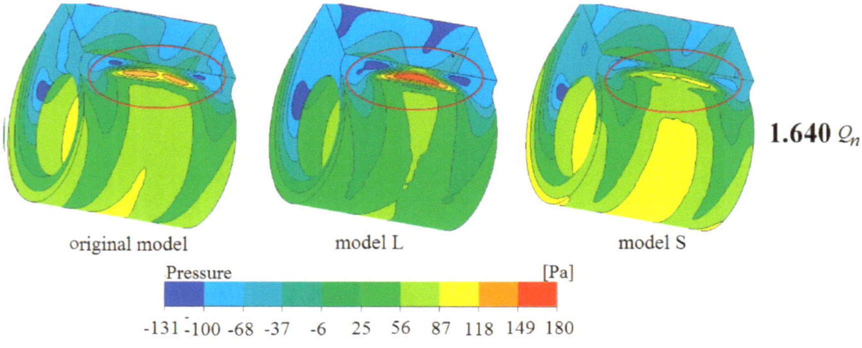

Figure 15. Static pressure distribution on the surface of the volute tongue at 1.640Q_n.

The streamline at the exit surface of the volute is shown in Figure 17. One can see that at a lower flow rate condition, a number of small vortices form on both sides of the volute exit in the baseline model, where one small vortex forms on the bottom of model-S and no small vortex forms in model-L, which indicates that the flow loss at the exit surface of volute for model-L is obviously lower than that of the baseline model. At design point, a number of secondary flows arise at the exit surface of the baseline model and model-S. However, no secondary flow occurs in model-L, which also indicates that the flow loss at the exit surface of the volute for model-L is obviously lower than that of the baseline model and model-S.

It is also obtained that the symmetry of streamline on the exit surface has been improved with the increase of the flow rate, and this also greatly promotes the development of secondary flow with the influence on the flow in the volute discharge.

Figure 18 shows the nine sections of volute outlet from bottom to top. Figure 19 shows the velocity distribution along the height (from 5% H to 80% H) of the volute outlet at the design condition. Plotted in Figure 19, it is observed that at the design point, the backflow area of model-L is the largest in the middle and lower part of the volute outlet. The backflow area of the model-L is reduced to a minimum at 40% H and disappears completely at 50% H, while a small range of backflow occurs in the baseline model and the model-S. The backflow completely disappears in the middle and upper part of the volute outlet for all models. At a high flow rate condition, the backflow range of all models is close to the lower part of the volute outlet. With the increase of the outlet height, the backflow range of the baseline model and model-S decreases rapidly and disappears at 50% H, while a small range of backflow still occurs in the model-L, and the velocity gradient increases in the axial direction.

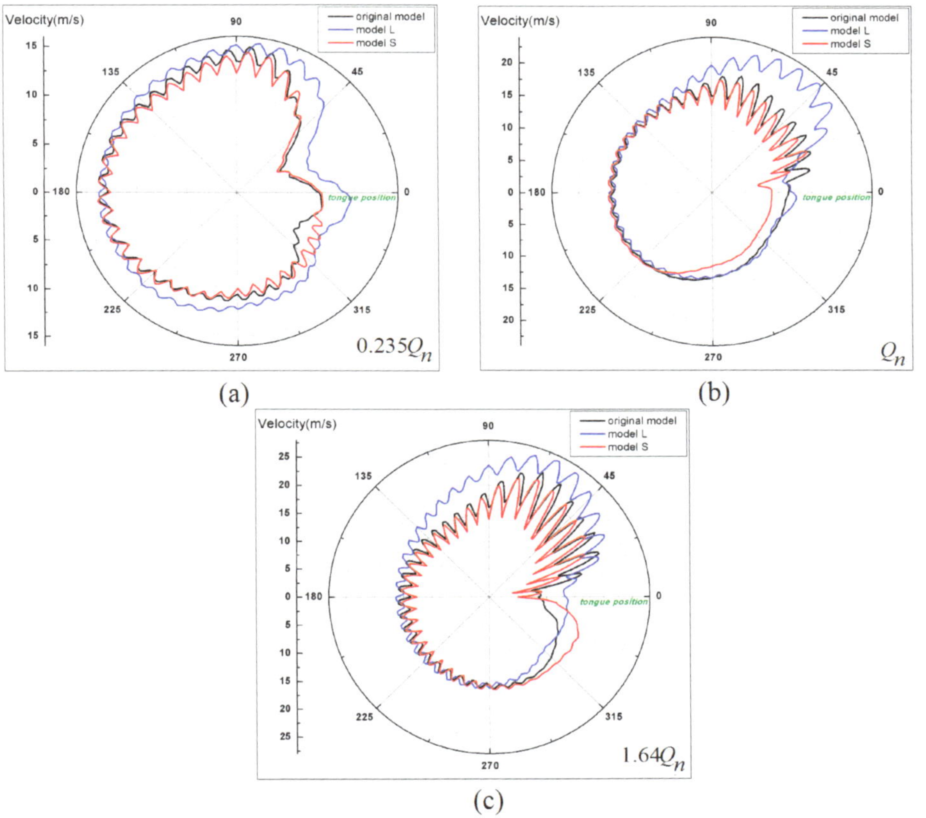

Figure 16. Velocity profile above the impeller. (**a**) velocity profile at 0.235 Q_n; (**b**) velocity profile at Q_n; (**c**) velocity profile at 1.64 Q_n.

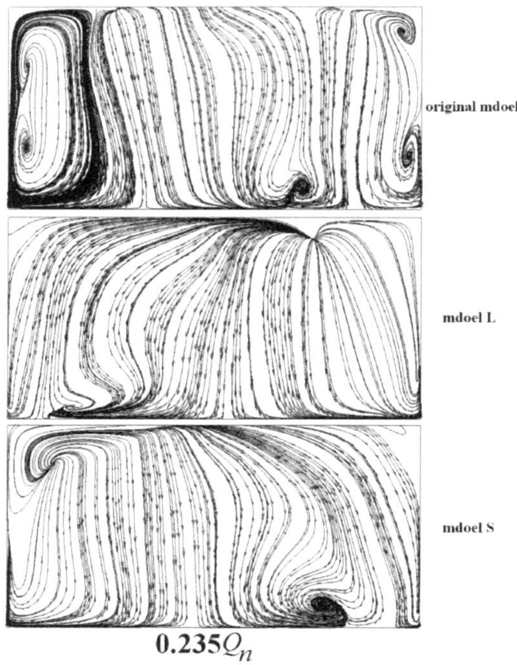

(**a**)

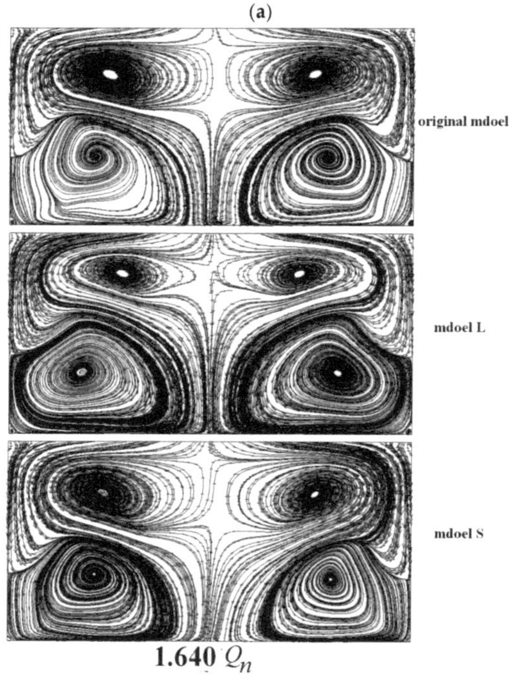

(**b**)

Figure 17. *Cont.*

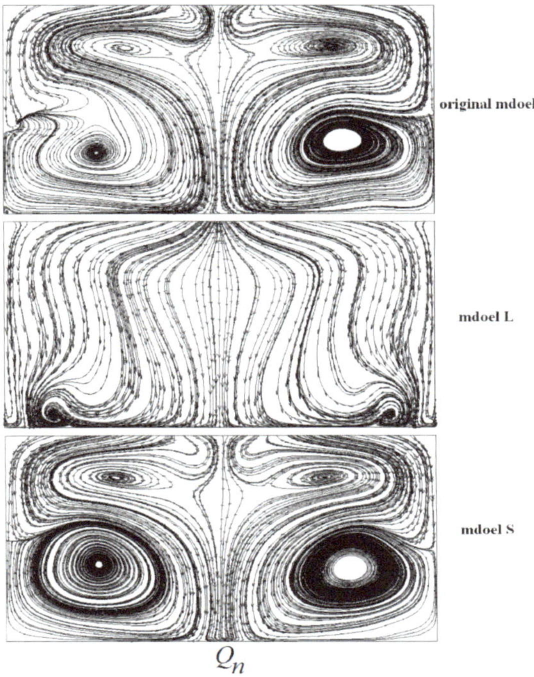

(c)

Figure 17. The streamline on the exit surface of volute. (**a**) at 0.235 Q_n; (**b**) at 1.64 Q_n; (**c**) at Q_n.

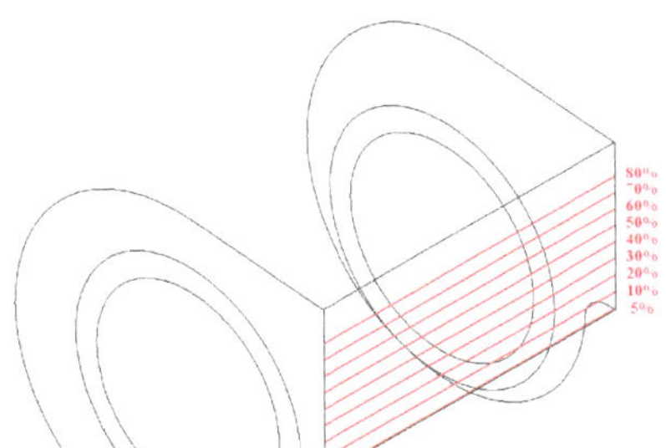

Figure 18. Section of volute outlet from bottom to top.

Figure 20 illustrates the velocity distribution along the height (from 5% H to 80% H) of volute outlet at the high flow rate condition (1.64Q_n). The x-axis refers to the axial position and the y-axis refers to the velocity perpendicular to the volute outlet. The negative value of velocity (below the 0 scale line of the ordinate as shown in Figure 19) indicates that the backflow appears at this height of the volute outlet. The backflow at the volute outlet mainly converges at the low height region of the outlet, and the range of backflow gradually narrows and finally disappears with the increase of the height.

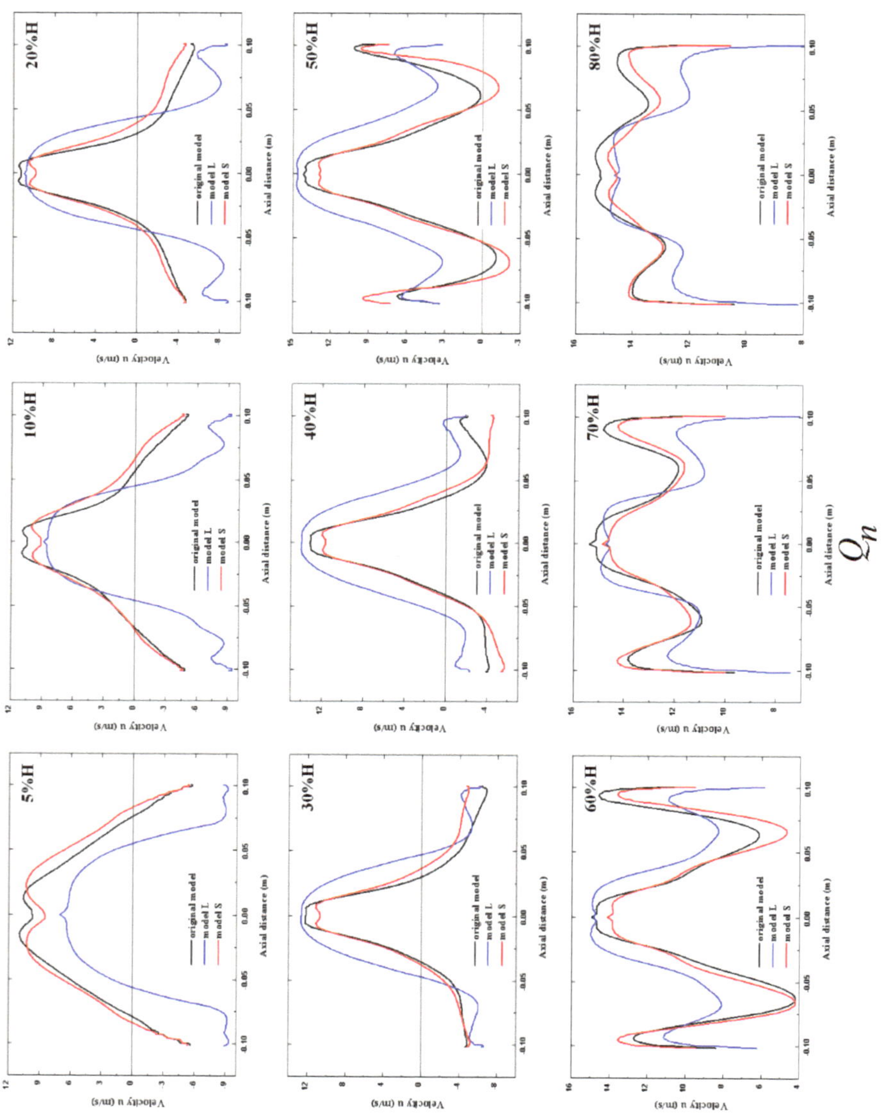

Figure 19. Velocity distribution along the height of volute outlet at the design condition.

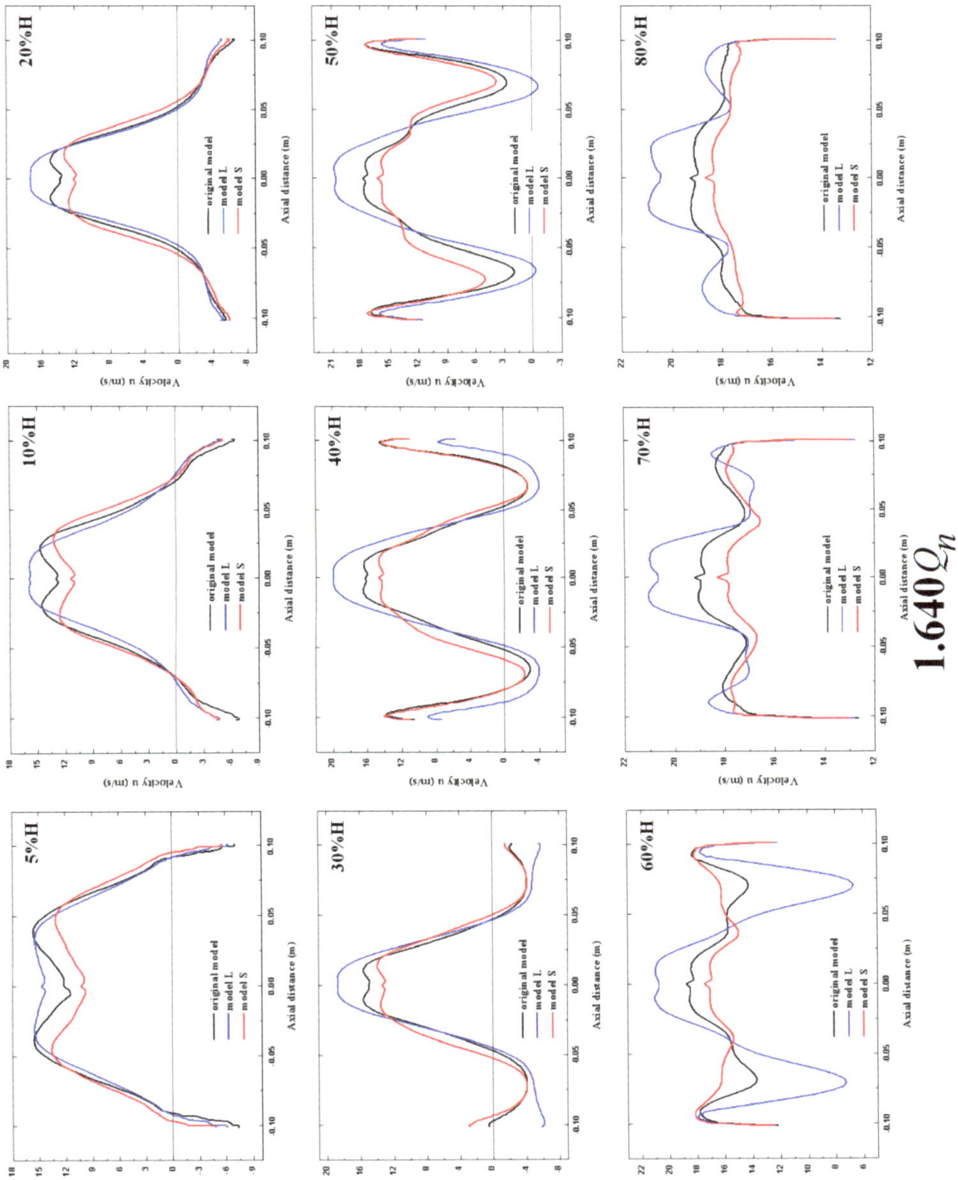

Figure 20. Velocity distribution along the height of volute outlet at $1.64Q_n$.

4.3. Performance Results of Numerical Simulations

The relevant parameters for the performance of the fan, i.e., output power, shaft power and total-to-total efficiency for the impeller flow field, are defined by

$$SPWR = T_{imp}\omega \tag{5}$$

$$impPWR_{out} = (\Delta P_t)T_{imp}Q \tag{6}$$

$$\eta_{imp} = \frac{impPWR_{out}}{SPWR} \tag{7}$$

in which T_{imp}, Q, and P_{imp} denote the impeller torque, rate of flow, total pressure increases across the bell mouth and impeller, and the fan flow rate. The performance of the entire fan is calculated differently from calculation of the impeller. The $SPWR$ is again evaluated by Equation (5), but T_{imp} is evaluated by integrating the torque from all the impeller blades. The lift-side total and static efficiency are defined by:

$$(\eta_t)_{lift} = \frac{(\Delta P_t)_{lift}Q_{lift}}{T_{imp}\omega}t \tag{8}$$

$$(\eta_S)_{lift} = \frac{(\Delta P_S)_{lift}Q_{lift}}{T_{imp}\omega}t \tag{9}$$

Figure 21 shows the static pressure and static pressure efficiency for various models obtained by steady numerical simulations. As shown in Figure 22, it is clearly observed that the static pressure and static pressure efficiency of model-L are obviously higher than that of model-S and the baseline model at the design flow rate, and the static pressure and static pressure efficiency of model-S are obviously lower than that of the baseline model at the design flow rate. The static pressure and static pressure efficiency obviously increases with the increase of the blade length of model-S, the baseline model and model-L at the design flow rate. It is also obtained that the static pressure value of the model-S is about 78 Pa, the static pressure value of the baseline model is about 82 Pa, and the static pressure value of the model-L is about 104 Pa. We can obtain that at the designed flow rate, the improved static pressure of the model-L rises as much as 23 Pa. It is further found that the static pressure efficiency of the baseline model is about 27.5%, and the static pressure efficiency of the model-L is about 33.5%, and also the improved static pressure efficiency of the model-L rises as much as 6% at design flow rates. Our work suggests that reducing the blade inlet angle ($\beta_{1A} > 60°$) and improving the blade outlet angle ($\beta_{2A} < 175°$) can provide a significant increase on the static pressure and the efficiency of static pressure, which improves the aerodynamic load of the forward multiblade fan to achieve the aim of energy conservation.

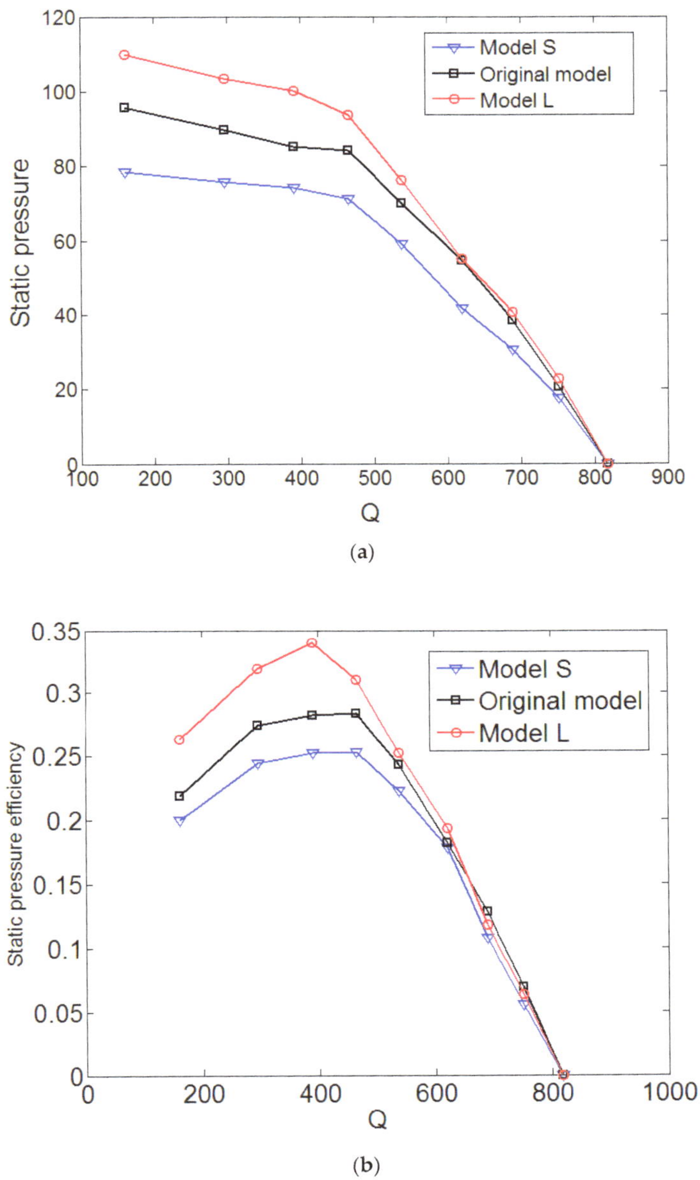

Figure 21. Static pressure and static pressure efficiency of different model (Q, m^3/h). (**a**) static pressure-flow; (**b**) static pressure efficiency-flow.

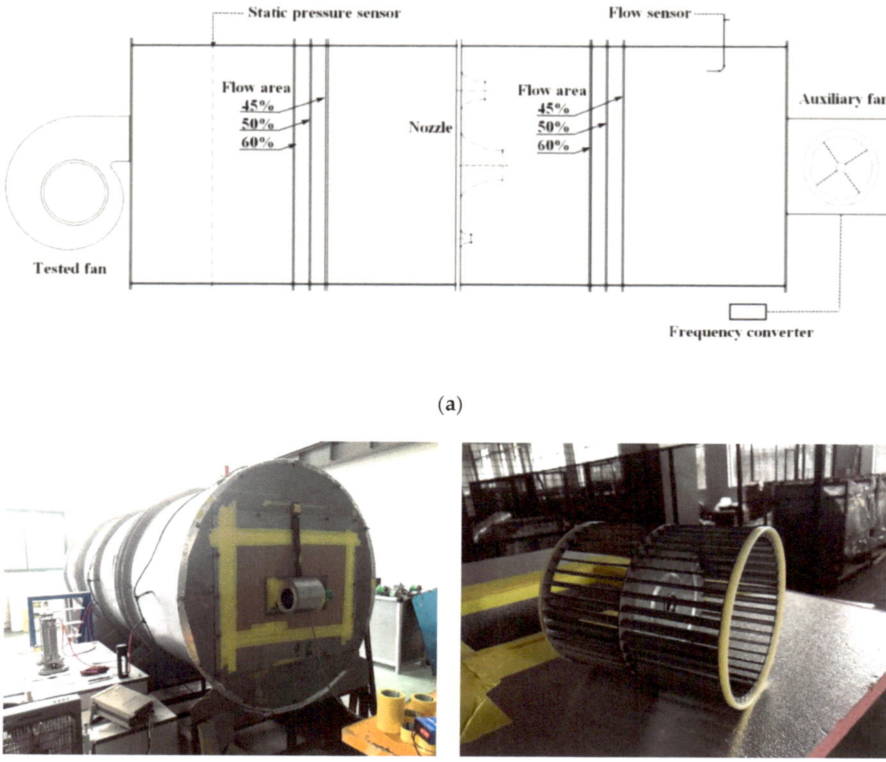

(a)

(b) (c)

Figure 22. Experimental device and impeller model. (**a**) two-dimensional cross-section of performance measurement device; (**b**) on-site testing equipment; (**c**)impeller of model-L in experimental performance measurement.

4.4. Performance Results of Experimental Test

Figure 22 shows the performance measurement device of testing the centrifugal fan. Plotted in Figure 22, the centrifugal fan is mounted on the inlet of Yilida's performance measurement device. At the outlet of the wind tunnel an auxiliary fan is mounted. The network of steady flow is mounted in the chamber to measure homogeneous flow. Multiple nozzles are implemented in the chamber. The details of the test rig are shown in Figure 22b. Figure 22c presents the impeller of model-L in experimental performance measurement.

The performance characteristics of the baseline model and model-L are performed on a chamber test rig. The performance data, such as static pressure, total pressure, static pressure efficiency and total pressure efficiency, are obtained at a range of flow rates. The measuring device for testing the centrifugal fan is manufactured by Air Movement and Control Association International (AMCA) standard. The centrifugal fan is placed at the entrance of the performance measuring device, and an auxiliary fan is placed at the outlet as shown in Figure 22. A steady flow network in the chamber can provide stable flow patterns for measurement. Pairs of nozzles are installed in the chamber to obtain various flow rates. A throttling device with an auxiliary fan to control the operating point of the testing centrifugal fan is used at the outlet of the test chamber.

Figure 23 shows the static pressure and static pressure efficiency for the baseline model and model-L by experimental test. One can see that the static pressure and static pressure efficiency of model-L are obviously higher than that of the baseline model near the design flow rate. The static pressure value of the baseline model is about 82.5 Pa, and the static pressure value of the model-L is about 105 Pa. It is obvious that compared to the baseline model, the static pressure of the model-L rises as much as 22.5 Pa at the design flow rate. It is also obtained that the static pressure efficiency of the baseline model is about 26.4% (including efficiency of the motor), and the static pressure efficiency of the model-L is about 31.5% (including efficiency of the motor), thus the static pressure efficiency of the model-L rises as much as 5% at the design flow rate.

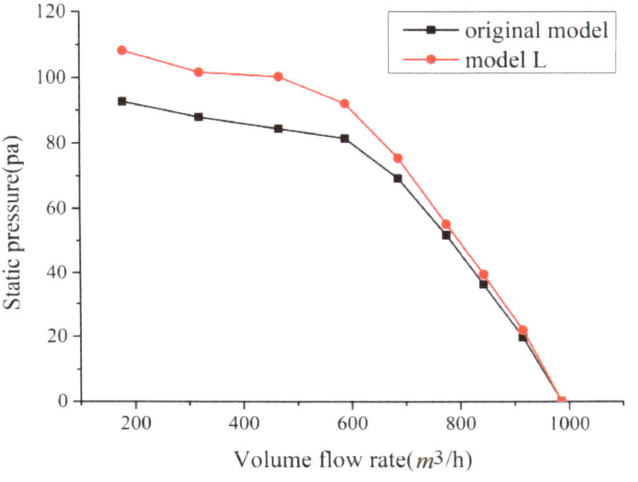

(**a**)

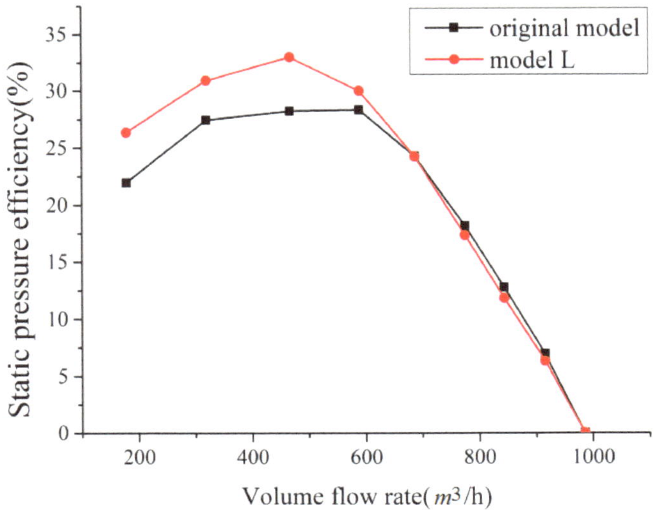

(**b**)

Figure 23. Experimental static pressure and static pressure efficiency for the baseline model and model-L. (**a**) static pressure-flow; (**b**) static pressure efficiency-flow.

Figure 24 demonstrates the total pressure and total pressure efficiency for the baseline model and model-L by experimental test. The total pressure and total pressure efficiency of model-L are obviously higher than that of the baseline model at all flow rates. The total pressure value of the baseline model is about 101.5 Pa, and the total pressure value of the model-L is about 125.4 Pa, thus the static pressure of the model-L rises as much as 25 Pa compared with that of the baseline model.

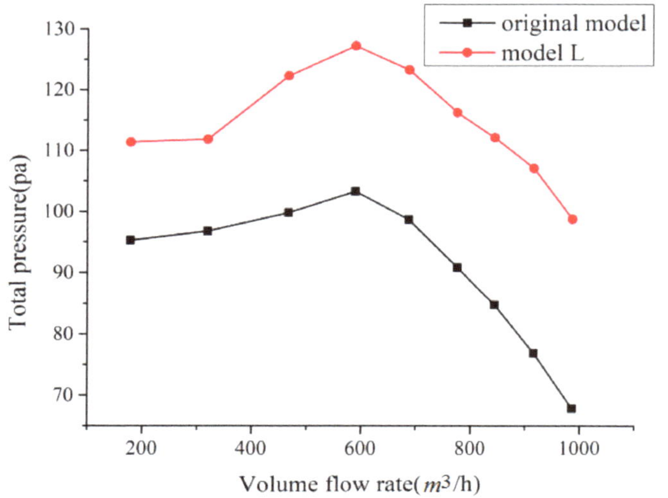

(**a**)

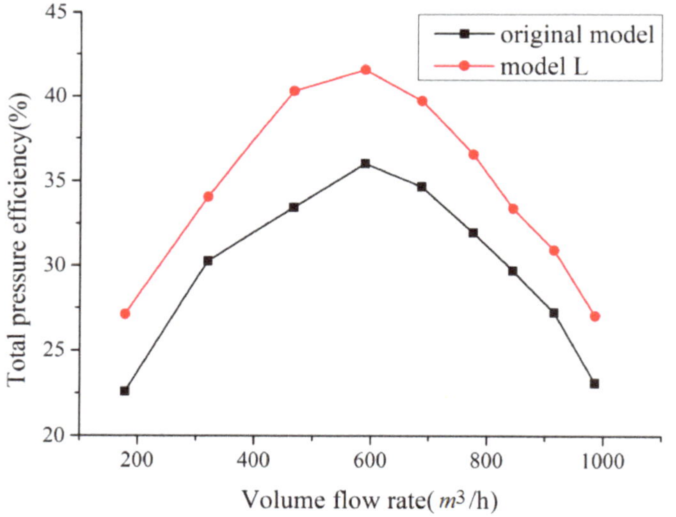

(**b**)

Figure 24. Experimental total pressure and total pressure efficiency for baseline model and model-L. (**a**) total pressure-flow; (**b**) total pressure efficiency-flow.

It is also found that the total pressure efficiency of the baseline model is about 35.5% (including the efficiency of the motor), and the total pressure efficiency of the model-L is about 41.5% (including efficiency of the motor), thus the total pressure efficiency of the model-L rises as much as 6% at the design flow rate.

5. Conclusions

In this paper, the effects of single-arc blade profile length on the performance of a forward multiblade fan are presented and demonstrated. The internal complex flow characteristics of fans are investigated by numerical simulations by implementing the RANS turbulence model, and the performance of fans is studied by numerical simulations and experimental test. Several conclusions can be summarized as follows:

The gradient of the absolute velocity among the blades in model-L is clearly lower than that of the baseline model and model-S, and the gradient of the absolute velocity among the blades decreases in model-L, which brings about the size of the flow separation among the blades.

At the design point Q_n, the area of turbulent kinetic energy at the inlet of the volute for model-L is smaller than that of the baseline model and model-S. The flow loss at the exit surface of the volute for model-L is obviously lower than that of the baseline model and model-S

Experimental results demonstrate that the static pressure of model-L rises as much as 22.5 Pa and 26.2%, while the static pressure efficiency of the model-L rises as much as 5% at the design flow rate. Meanwhile, the total pressure of the model-L rises as much as 25 Pa and 23.6%, and the total pressure efficiency of the model-L rises as much as 6% at the design flow rates.

It is found that a properly increasing blade working area provides a significant increasing on the static pressure, total pressure, the efficiency of static pressure and total pressure efficiency. The increase of the blade working area obviously improves the aerodynamic load of fans to achieve the purpose of energy conservation.

Author Contributions: The following statements could be used Y.W. and C.Y. conceived and designed the experiments; J.X. performed the experiments; W.C. and Z.W. analyzed the data; Z.Z. contributed reagents/materials/analysis tools; Y.W. wrote the paper.

Funding: This work was supported by the National Natural Science Foundation of China (11872337, 11902291 and U1709209), Natural Science Foundation of Zhejiang Province (LY18A020010), public welfare technology application research project of Zhejiang province (2017C31075), and Fundamental Research Funds of Zhejiang Sci-Tech University (2019Y004).

Acknowledgments: The authors appreciate sincerely the referees' valuable comments and suggestions on our work.

Conflicts of Interest: The authors declare no conflict of interest.

References

1. Cau, G.; Mandas, N.; Manfrida, N.; Nurzia, F. Measurements of primary and secondary flows in an industrial forward-curved centrifugal fan. *ASME J. Turbomach.* **1990**, *112*, 84–90. [CrossRef]
2. Cao, S.; Chu, L. Experimental study on the matching between centrifugal impeller and volute. *Chin. Fluid Mach.* **1991**, *10*, 2–4.
3. Wei, Y.K.; Yang, H.; Lin, Z.; Wang, Z.D.; Qian, Y.H. A novel two-dimensional coupled lattice Boltzmann model for thermal incompressible flows. *Appl. Math. Comput.* **2018**, *339*, 556–567. [CrossRef]
4. Liang, H.; Li, Y.; Chen, J.X.; Xu, J.R. Axisymmetric lattice Boltzmann model for multiphase flows with large density ratio. *Int. J. Heat Mass Transf.* **2019**, *130*, 1189–1205. [CrossRef]
5. Zhang, W.; Chen, X.P.; Yang, H.; Liang, H.; Wei, Y.K. Forced convection for flow across two tandem cylinders with rounded corners in a channel. *Int. J. Heat Mass Transf.* **2019**, *130*, 1053–1069. [CrossRef]
6. Zhou, H.; Ca, G.B. Research of wall roughness effects based on Q criterion. *Microfluid. Nanofluid.* **2017**, *21*, 114–121. [CrossRef]

7. Šístek, V. Corotational and Compressibility Aspects Leading to a Modification of the Vortex-Identification Q-Criterion. *AIAA J.* **2015**, *53*, 2406–2410.
8. Gong, W.; Lv, W.; Xi, G.; Zhang, W. Experimental study on the velocity field near the tongue of a forward curved multi-blades centrifugal fan. *Chin. J. Mech. Eng.* **2005**, *41*, 166–171. [CrossRef]
9. Velarde-Suárez, S.; Santolaria-Morros, C.; Ballesteros-Tajadura, R. Experimental study on the aeroacoustic behavior of a forward-curved blades centrifugal fan. *ASME J. Fluid Eng.* **1999**, *121*, 276–281. [CrossRef]
10. Velarde-Suárez, S.; Ballesteros-Tajadura, R.; Santolaria-Morros, C.; Gonzalez-Perez, J. Unsteady flow pattern characteristics downstream of a forward-curved blades centrifugal fan. *ASME J. Fluid Eng.* **2001**, *123*, 265–270. [CrossRef]
11. Younsi, M.; Bakir, F.; Kouidri, S.; Rey, R. Numerical and experimental study of unsteady flow in centrifugal fan. *Proc. Inst. Mech. Eng. Part A J. Power Energy* **2007**, *221*, 1025–1036. [CrossRef]
12. Lin, S.; Huang, C. An integral experimental and numerical study of forward curved centrifugal fan. *Exp. Fluid Sci.* **2002**, *26*, 421–434. [CrossRef]
13. Lun, Y.X.; Lin, L.M.; Zhu, Z.C.; Wei, Y.K. Effects of Vortex Structure on Performance Characteristics of a Multiblade Fan with Inclined tongue. *Proc. Inst. Mech. Eng. Part A J. Power Energy* **2019**, *233*, 1007–1021. [CrossRef]
14. Montazerin, N.; Damangir, A.; Mirian, S. A new concept for squirrel cage fan inlet. *J. Power Energy* **1998**, *212*, 343–349. [CrossRef]
15. Montazerin, N.; Damangir, A.; Mirzaie, H. Inlet induced flow in squirrel-cage fans. *J. Power Energy* **2000**, *214*, 243–253. [CrossRef]
16. Bayomi, N.N.; Osman, A.M. Effect of inlet straighteners on centrifugal fan performance. *Energy Convers. Manag.* **2006**, *47*, 3307–3318. [CrossRef]
17. Zheng, X.; Lin, Z.; Xu, B.Y. Thermal conductivity and sorption performance of nano-silver powder/FAPO-34 composite fin. *Appl. Therm. Eng.* **2019**, *160*, 114055. [CrossRef]
18. Hosangadi, A.; Lee, R.A.; York, B.J.; Sinha, N.; Dash, S.M. Upwind Unstructured Scheme for Three-Dimensional Combusting Flows. *J. Propul. Power* **1996**, *12*, 494–502. [CrossRef]
19. Hosangadi, A.; Lee, R.A.; Cavallo, P.A.; Sinha, N.; York, B.J. Hybrid, Viscous, Unstructured Mesh Solver for Propulsive Applications. In Proceedings of the AIAA 34th JPC, Cleveland, OH, USA, 13–15 July 1998; Volume 98, p. 3153.
20. Lee, Y.T.; Mulvihill, L.; Coleman, R.; Ahuja, V.; Hosangadi, A.; Birkbeck, R.; Becnel, A.; Slipper, M. LCAC Lift Fan Redesign and CFD Evaluation. *Nav. Surf. Warf. Cent. Rep.* **2007**, *23*, 112–119.
21. Wang, C.; Hu, B.; Zhu, Y.; Wang, X.; Luo, C.; Cheng, L. Numerical study on the gas-water two-phase flow in the self-priming process of self-priming centrifugal pump. *Processes* **2019**, *7*, 330. [CrossRef]
22. Zhang, S.F.; Li, X.J.; Hu, B.; Liu, Y.; Zhu, Z.C. Numerical investigation of attached cavitating flow in thermo-sensitive fluid with special emphasis on thermal effect and shedding dynamics. *Int. J. Hydrog. Energy* **2019**, *44*, 3170–3184. [CrossRef]
23. Li, X.J.; Jiang, Z.W.; Zhu, Z.C.; Si, Q.R.; Li, Y. Entropy generation analysis for the cavitating head-drop characteristic of a centrifugal pump. *Proc. Inst. Mech. Eng. Part C J. Mech. Eng. Sci.* **2018**, *232*, 4637–4646. [CrossRef]
24. Liu, Q.; Ye, J.H.; Zhang, G.; Lin, Z.; Xu, H.; Jin, H.; Zhu, Z. Study on the metrological performance of a swirlmeter affected by flow regulation with a sleeve valve. *Flow Meas. Instrum.* **2019**, *67*, 83–94. [CrossRef]
25. Xu, H.; Cantwell, C.D.; Monteserin, C.; Eskilsson, C.; Engsig-Karupet, A.P.; Sherwin, S.J. Spectral/hp element methods: Recent developments, applications, and perspectives. *J. Hydrodyn.* **2018**, *30*, 1–22. [CrossRef]
26. Xu, H.; Mughal, S.M.; Gowree, E.R.; Atkin, C.J.; Sherwin, S.J. Destabilisation and modification of Tollmien–Schlichting disturbances by a three-dimensional surface indentation. *J. Fluid Mech.* **2017**, *819*, 592–620. [CrossRef]
27. Zhang, W.; Li, X.J.; Zhu, Z.C. Quantification of wake unsteadiness for low-Re flow across two staggered cylinders. *Proc. Inst. Mech. Eng. Part C J. Mech. Eng. Sci.* **2019**. [CrossRef]
28. Xu, H.; Sherwin, S.J.; Hall, P.; Wu, X. The behaviour of Tollmien–Schlichting waves undergoing small-scale localised distortions. *J. Fluid Mech.* **2016**, *792*, 499–525. [CrossRef]
29. Xu, H.; Lombard, J.E.; Sherwin, S.J. Influence of localised smooth steps on the instability of a boundary layer. *J. Fluid Mech.* **2017**, *817*, 138–170. [CrossRef]

30. Wang, C.; Shi, W.; Wang, X.; Jiang, X.; Yang, Y.; Li, W.; Zhou, L. Optimal design of multistage centrifugal pump based on the combined energy loss model and computational fluid dynamics. *Appl. Energy* **2017**, *187*, 10–26. [CrossRef]
31. Liu, Q.; Ye, J.; Zhang, G.; Lin, Z.; XU, H.; Zhu, Z. Metrological performance investigation of swirl flow meter affected by vortex inflow. *J. Mech. Sci. Technol.* **2019**, *33*, 2671–2680. [CrossRef]
32. Chen, X.P.; Li, X.P.; Dou, H.S.; Zhu, Z.C. Effects of variable specific heat on energy transfer in a high-temperature supersonic channel flow. *J. Turbul.* **2018**, *19*, 365–389. [CrossRef]
33. Chen, X.P.; Li, X.L.; Zhu, Z.C. Effects of dimensional wall temperature on velocity-temperature correlations in supersonic turbulent channel flow of thermally perfect gas. *Sci. China Phys. Mech. Astron.* **2019**, *62*, 064711–064719. [CrossRef]
34. Yang, H.; Zhang, W.; Zhu, Z.C. Unsteady mixed convection in a square enclosure with an inner cylinder rotating in a bi-directional and time-periodic mode. *Int. J. Heat Mass Transf.* **2019**, *136*, 563–580. [CrossRef]

processes

MDPI

Article

The Role of Blade Sinusoidal Tubercle Trailing Edge in a Centrifugal Pump with Low Specific Speed

Bowen Li [1], Xiaojun Li [1,*], Xiaoqi Jia [1], Feng Chen [2] and Hua Fang [1]

[1] National-Provincial Joint Engineering Laboratory for Fluid Transmission System Technology, Zhejiang Sci-Tech University, Hangzhou 310018, China; libowen669@163.com (B.L.); jiaxq01@163.com (X.J.); fh15957175818@163.com (H.F.)

[2] Department of Mechanical Technology, Zhejiang Institute of Mechanical and Electrical Engineering, Hangzhou 310053, China; wzh2811@sina.com

* Correspondence: lixj@zstu.edu.cn; Tel.: +86-571-86843348

Received: 8 August 2019; Accepted: 11 September 2019; Published: 17 September 2019

Abstract: Pressure pulsations may cause high-amplitude vibrations during the process of a centrifugal pump. The trailing edge shape of the blade has a critical influence on the pump's pressure fluctuation and hydraulic characterization. In this paper, inspired by the humpback whale flipper, the authors research the impact of applying the sinusoidal tubercles to the blade suction side of the trailing edge. Numerical calculation and experiments are carried out to investigate the impact of the trailing edge shape on the pressure pulsations and performance of a centrifugal pump with low specific speed. Two designed impellers are tested, one is a sinusoidal tubercle trailing edge (STTE) impeller and the other is the original trailing edge (OTE) prototype. The detailed study indicates that the sinusoidal tubercle trailing edge (STTE) reduces pressure pulsation and enhances hydraulic performance. In the volute tongue region, the pressure pulsation amplitudes of STTE at f_{BPF} decrease significantly. The STTE impeller also effectively changes the vortex structure and intensity in the blade trailing edge area. This investigation will be of great benefit to the optimal design of pumps.

Keywords: pressure pulsations; centrifugal pump; blade trailing edge; sinusoidal tubercle

1. Introduction

The centrifugal pump is an important energy conversion and fluid transfer tool and is widely used in various fields, such as the agriculture, industry, urban water supply, aerospace, and fire engines. According to an annual report made by the European Commission, pump systems account for about 22% of the electric energy supply in the world [1]. However, due to the low operating efficiency of pumps, most of the energy is not used efficiently. Therefore, there is an urgent need to improve the energy conversion utilization of pump systems.

In fact, when a centrifugal pump operates under partial-flow conditions, fluid transport is quite disordered. The operation efficiency is very low, especially for a pump with low specific speed. Some unsteady phenomena, such as backflow at the impeller inlet and outlet, the local cavitation, etc., affect the performance and induce large pressure pulsations and noise [2–4]. Pressure pulsation may cause high-amplitude vibrations [5,6], which may cause cracks in pump components and shorten the life of the pump. As for the centrifugal pump, many investigations were implemented to explore the performance and pressure pulsations through experiments or numerical simulations [7–9]. Fu et al. [10] analyzed the flow phenomena in a low-specific-speed centrifugal pump, and found the cavitation occurs under low flow rate, the inlet low-frequency pressure fluctuations were closely correlated to cavitation phenomena. Pei and Wang [11] studied the pressure fluctuations using statistical analysis. The highest pressure fluctuation intensity occurs at the trailing edge of the pressure side and near the volute tongue, and modifying the geometry of the leading edge in the impeller and the tongue in the volute can reduce the intensity. Jafarzadeh [12] simulated the flow of a low-specific-speed high-speed pump, and highlighted the effect of the number of blades on efficiency and head. Chu and Dong [13,14] found that the unsteady flow pattern and the rotor–stator interaction are two main excitation components of the pressure pulsations and noise. Keller et al. [15] used particle image velocimetry technology to measure the rotor–stator interaction, and found that the vortex sheds from trailing edge and affects the flow field in the volute, their collision generates strong pressure pulsation. Jia [16] explored the pressure distribution in a low-specific-speed centrifugal pump, and found that the asymmetry of pressure distribution is caused by the tongue. It may induce vibration and noise. Heskestad and Olberts [17] explored the effect of different blade trailing edge geometries on pressure pulsation and vibration, and revealed that the vortex intensity and the vortex shedding frequency are regulated by the trailing edge.

Many researchers have investigated the methods to improve the performance and weaken pressure pulsation amplitude in a low-specific-speed centrifugal pump. Gao and Zhang [18] numerically investigated the impeller with five especial blade trailing edge shapes. The final conclusions show that well-designed trailing edges can significantly improve pump efficiency and effectively reduce pressure pulsations. Spence [19] analyzed the effect of the geometric parameters of the impeller and volute on pressure pulsation, and found that increasing gap between the impeller and volute will reduce the pressure pulsation, but the efficiency may decline. Solis [20] used numerical simulation to study the influence of radial gap and splitter blades on pressure pulsation and hydraulic performance. Adding splitter blades can effectively reduce pressure pulsation, and improve performance. Amir [21] designed an original impeller and volute with CFturbo, and carried out Computational Fluid Dynamics (CFD) simulation. Then, he modified the design parameters to get new impellers and volutes, and conducted simulation calculation again until the desired results were achieved. Zhu and Chen [22] designed the gap structure blade to suppress cavitation and pressure pulsation induced by cavitation. Khalifa [23] tested cutting the impeller blade outlet to increase the effective clearance with volute, and the results show that it is useful for reducing pressure pulsations. For the pump as a turbine, the position of proper blade trailing edges also produces an effective impact on enhancing the performance and reducing pressure pulsations [24].

More recently, a large number of researchers have focused attention on the application of bionics [25–27]. The humpback is one of them. Despite humpback whales being bulky, they can move flexibly underwater and even jump out of the water. This is due to the fact that some sinusoidal tubercles grow in their pectoral fins. Shi et al. [28] investigated two types of 3D hydrofoil turbine blades, including one original hydrofoil and one hydrofoil with leading-edge tubercles. By comparison, the latter's hydraulic performance is superior, and the strength of the tip vortex is also significantly reduced. Therefore, inspired by bionics, we designed a new impeller with sinusoidal tubercles on the blade trailing edge. The purpose of this case is to study the effect of the sinusoidal tubercle on the pressure pulsations and the performance of a centrifugal pump with low specific speed, and one original impeller and one new tubercle impeller are investigated by numerical calculation and experiment. The sinusoidal tubercles can divide the vortex shedding from blade trailing edge into several small vortexes. The vortex core region in the pump is reduced, and the rotor–stator interference of the impeller and volute is weakened, so as to reduce the pressure pulsation and energy loss. The pressure signals are captured by monitoring points arranged in the pump. Finally, the effect of the sinusoidal tubercle on pressure pulsations and performance are obtained and analyzed successfully.

2. Test Model Pump and Setup

2.1. Test Model Pump

In this paper, a single-stage test model pump with low specific speed is designed to investigate the performance and pressure pulsations. The specific speed *Ns* ($N_s = 3.65 * n * \sqrt{Q}/H^{0.75}$) of the model pump is 40.1. Q and H represent the nominal flow rate and nominal head, respectively. n is the speed of rotation of the pump in rpm. The test impeller has five cylindrical blades. The inlet and outlet diameters are 50 mm and 160 mm, respectively. Table 1 lists the essential geometric dimensions and design parameters. The geometry of the test model pump impeller and volute is shown in Figure 1.

Table 1. Geometric dimensions and design parameters.

Main Parameters	Signs	Values
Nominal rotating speed (rpm)	n	3000
Nominal flow rate (m^3/h)	Q_N	10
Nominal head (m)	H	35
Specific speed	Ns	40.1
Impeller blades number	Z	5
Impeller inlet diameter (mm)	D_1	50
Impeller exit diameter (mm)	D_2	160
Impeller exit width (mm)	B_2	10
Impeller inlet width (mm)	B_1	19
volute base circle diameter(mm)	D_3	165
volute exit diameter (mm)	D_4	40
Blade inlet angle	β_1	23
Blade outlet angle	β_2	23
Nominal flow rate coefficient	φ_N	0.017
Nominal head coefficient	ψ_N	0.622
Roughness in impeller passage(μm)	Ra_1	0.4
Roughness in volute(μm)	Ra_2	1.6
Inlet Reynolds number	Re	70396

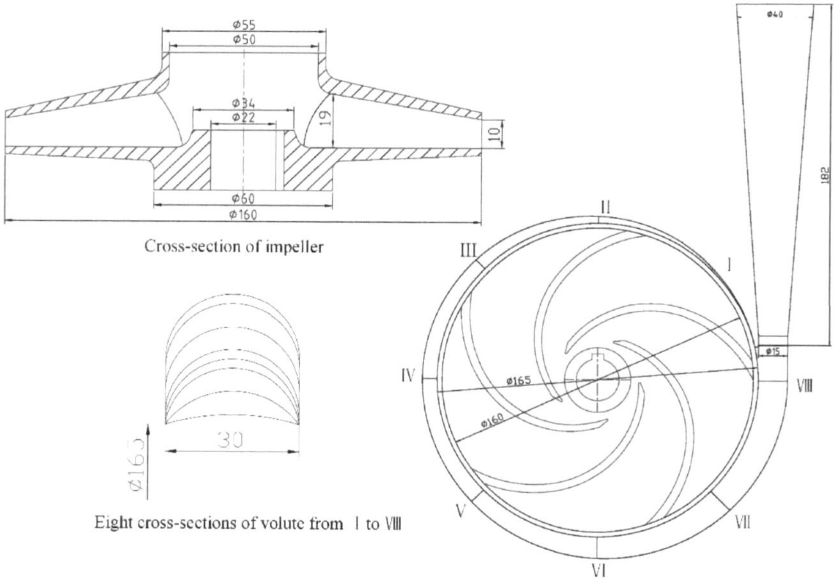

Figure 1. The geometry of the test model pump impeller and volute.

2.2. Impeller Blade Trailing Edge

Two impellers are designed and manufactured by 3DP technology. Figure 2 shows the differences between the two trailing edges, including the original trailing edge (OTE) prototype and the new designed sinusoidal tubercle trailing edge (STTE). OTE is the original trailing edge, which is a sharp corner. The vortexes shedding from the OTE are intensive. The STTE only exists on the blade suction surface, the pressure side remains constant. There are also sinusoidal grooves on the suction surface at the blade outlet, and the length of the longitudinal extension is L. Groove depth gradually decreases from A to zero along the blade profile line. As shown in Figure 2, λ is the sinusoidal wavelength, indicating the spacing of sinusoidal tubercle. A is the amplitude of the sinusoid, which is half the height of the sinusoidal tubercle. According to the width of the impeller outlet, the values of λ, A, and L are 2.5 mm, 1 mm, and 10mm, respectively. In the design process, the essential parameters of two impellers are the same.

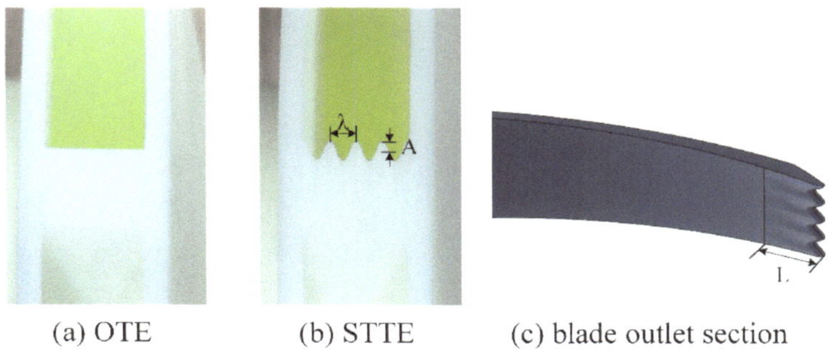

Figure 2. Differences between two trailing edges: (**a**) original trailing edge (OTE), (**b**) sinusoidal tubercle trailing edge (STTE), (**c**) blade outlet section

2.3. Experimental Setup

In order to obtain accurate test model pump data, the experiment was tested in a closed test rig [29], as shown in Figure 3. Pressure gauges with uncertainties of ± 0.2% are installed at the pump inlet and outlet for recording the pressure values. The flow rates of the experimental process are recorded by an electromagnetic flowmeter, the uncertainty is 0.2% class. Various flow rates ($0.2–1.6Q_N$) are regulated by valves located at the model pump outlet pipe. The test model pump is driven by a variable frequency motor with a speed of 3000 rpm. The uncertainty of the speed of rotation is 10 rpm. In order to ensure a constant speed during the experiment, a frequency inverter was used. In the experiment, the shaft torque is measured by a torque meter with uncertainties of ±0.5%. Sample signals of monitoring points are recorded by dynamic pressure sensors arranged around the volute.

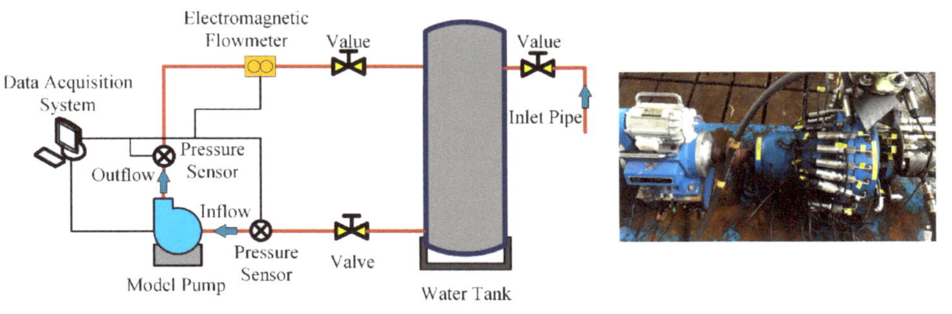

Figure 3. Closed test platform and model pump.

3. Numerical Investigation

3.1. Numerical Method

Considering the influence of leakage flow on the flow field, the computational domain of the full flow field was used, which has seven components, including, inlet extension, front chamber, impeller, volute, back chamber, outlet extension, and wear ring, as shown in Figure 4. The test STTE impeller is also shown in Figure 4. A pair of interfaces connects each adjacent computational domain. The working fluid is water at 298 K. To avoid a reverse flow appearing in the inlet boundary, the extension length of the inlet pipe is three times that of the impeller inlet diameter in the calculation process.

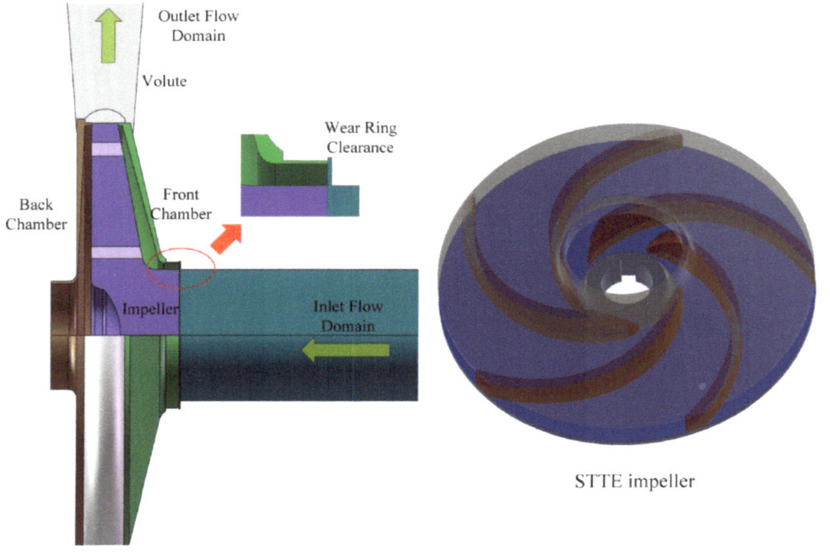

Figure 4. Computational domain and the test STTE impeller.

More recently, CFX software has been widely used in the study of fluid simulation, which has greatly improved work efficiency [30–35]. The ANSYS CFX-18.0 is performed in the numerical simulation. Reasonable boundary conditions are conducive to convergence. The inlet boundary condition is set as uniform velocity, which is obtained by various flow rates and inlet pipe areas. Constant static pressure $p = 1.013 \times 10^5$ Pa is set as the outlet boundary condition. All solid walls are set as no-slip walls, and the roughness is neglected. The turbulent intensity of 5% is set in the inlet boundary. The high-resolution scheme is selected for the advection scheme, and the second order is used for the turbulence numeric. In the steady-state simulation process, the impeller is a rotating computational domain, and the others are stationary domains. The interface models between rotating–stationary and stationary–stationary domains are selected frozen-rotor and general connection, respectively. The shear stress transport (SST) k-ω model is applied to solve the turbulent flow and obtain the head and efficiency of the test pump, which has been used in many studies with reliable accuracy [36–38]. For transient simulation, the transient rotor–stator interface model is used for the rotating–stationary interface. The transient scheme option selects the second order backward Euler for transient simulation. The time step size determines the acquisition and post-processing of pressure signals [39]. Therefore, the setting of the time step requires an adequate response resolution to the turbulence pulsation. Finally, 240 steps are calculated in one revolution, and the time step is $\Delta t = 8.3 \times 10^{-5}$ s, corresponding to impeller rotation 1.5 degrees, as suggested by Li et al. [40]. The root mean square (RMS) is set to 1×10^{-5} to converge the result.

During the transient-state simulation, we choose the detached-eddy simulation (DES) model based on the SST k-ω model to capture the unsteady transient pressure signals in the exit region of the impeller. Compared with the large eddy simulation (LES), DES uses Reynolds Averaged Navier-Stokes (RANS) on the near wall and LES in the turbulent core. DES model can not only accurately capture the flows closed to the wall [41,42], but also save the computing resource. In the transient-state simulation, to obtain a quick convergence result, the results obtained from the steady-state simulation are set as the incipient flow condition of transient DES model. To obtain the steady operation results, we simulated the transient state for 15 revolutions, the last five revolutions data are only retained for analysis of pressure fluctuations.

3.2. Computational Mesh

Mesh has a very significant impact on the accuracy of calculations in numerical simulation [43,44]. To ensure computational accuracy and computational efficiency, the entire computational domains adopt the structured hexahedral mesh generated by ANSYS-ICEM software. It is well known that the near-wall grid distribution directly affects calculation results [45]. Therefore, a boundary layer is generated near the wall to refine the mesh to meet the y+ requirements of the turbulence model, especially on the impeller blade wall and the volute tongue wall, as shown in Figure 5. After grid independence check [46–48], the total number of computational domains grids is determined to be 9.94 million. The average y+ value of the impeller is about 4.1. Detailed components of the grid distribution are displayed in Table 2.

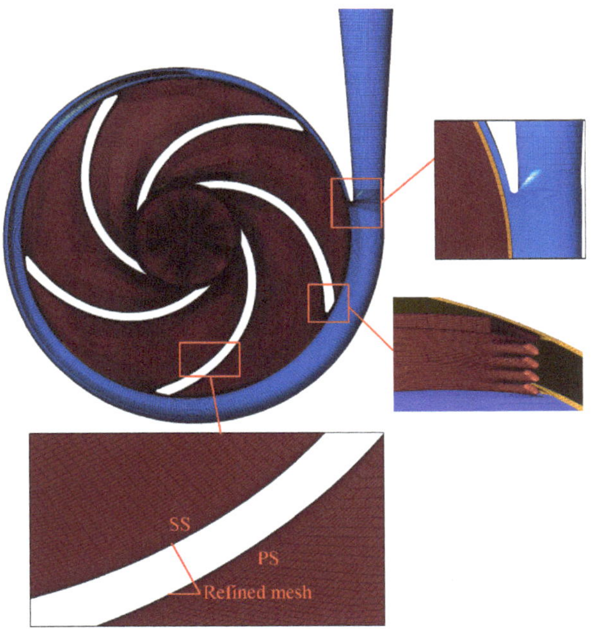

Figure 5. Structured hexahedral mesh of the impeller and volute.

Table 2. Detailed components of the grid distribution.

Components	Impeller	Volute	Back Chamber	Front Chamber	Inlet Extension	Outlet Extension
Number	7,636,570	932,106	391,136	390,150	365,792	228,096

3.3. Monitoring Points Arrangement

To get fluctuating pressure signals in the volute flow channel and the tongue region, 26 monitoring points are placed at the tongue region and the middle cross-section of the volute passage. However, in order to capture exhaustive pressure signals in the rotor–stator interaction region, five monitoring points T1–T5 are set near the tongue. And the angle is only 6 degrees between adjacent points. Detailed location distributions are shown in Figure 6.

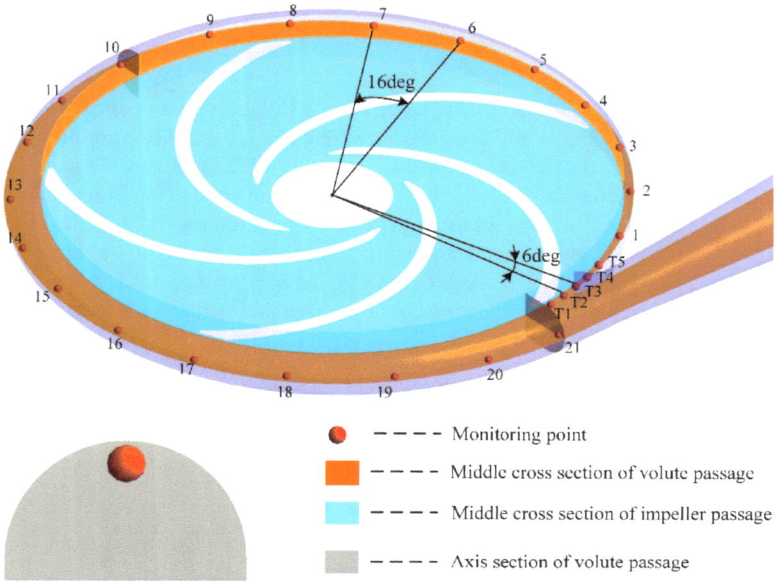

Figure 6. Location of monitoring points.

4. Results and Analysis

4.1. Performance Analysis of Two Test Pumps

Figure 7 shows the comparison of the numerical and tested head performance curve (φ-ψ) of the OTE pump to verify the accuracy of the calculation. The numerical results in Figure 7 are obtained by steady-state calculation using the SST k-ω model. The flow rates are tested from $0.2\varphi_N$ to $1.6\varphi_N$. Due to the uncertainties of the pressure gauges, the experimental error was 0.2%. The amplitude of the error bar represents the difference of the OTE model pump between the simulated value and the experimental value under different operating conditions. In general, the trends are well consistent. At the nominal flow rate, the experimental and calculated head coefficient difference is 2.15%. Under the condition of large flow rates, the head coefficients of numerical simulation and the experiment are almost identical. The maximum calculation difference is less than 3%. The difference may be mainly due to the grid resolution and leakage problems. Meanwhile, the spectrum analysis of pressure is carried out and compared between the numerical simulation and experiment of the OTE pump. The experiment and CFD simulation are implemented at the nominal flow rate Q_N, and rotating speed of 3000 r/min. The CFD fluctuating pressure signals are obtained by transient simulation, and recorded by monitoring points set in the volute. The DES model is applied in the transient-state simulation. The pressure values from the last five revolutions are used for the Fast Fourier transform to get Figure 8. The number of pressure signals used for analysis is 1200. Obviously, the blade passing frequency (f_{BPF}) and other harmonic frequencies can be clearly detected and distinguished. The theoretical blade passing frequency (f_{BPF}) is 250 Hz. The calculated blade passing frequency (f_{BPF}) is 249.02 Hz, the difference is 0.39%. The experimental blade passing frequency (f_{BPF}) is 249.50 Hz, the difference is 0.2%. The cause for these difference is that the model pump has a certain degree of deviation during operation due to the influence of the motor, voltage, etc. It is believed that the current calculation model and method are reliable to obtain precise simulation results.

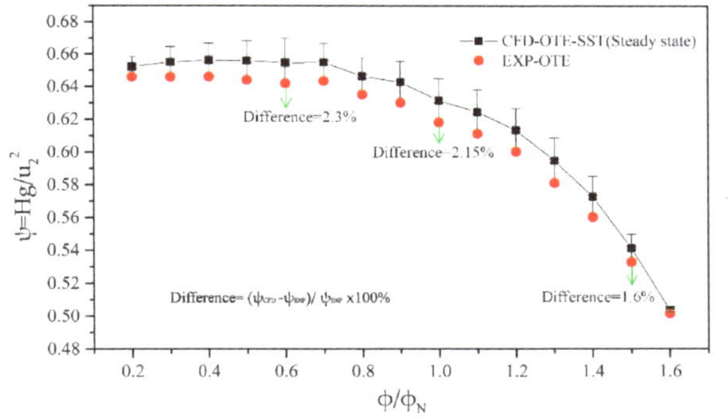

Figure 7. Numerical and experimental results comparison of the OTE model pump.

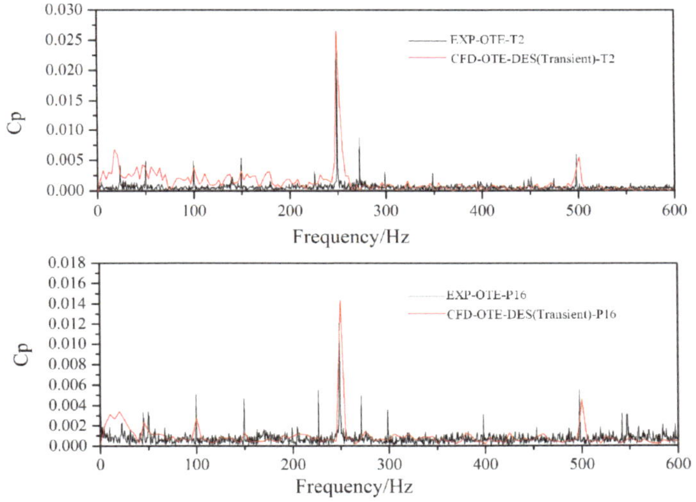

Figure 8. Comparison of experimental and numerical pressure spectra of OTE pump at the nominal flow rate at monitoring point16 and T2.

During the analysis of pressure spectra, the pressure value is dimensionless and expressed as C_p.

$$C_p = \frac{2(p - \overline{p})}{\rho u_2^2} \quad (1)$$

where p and $\overline{p}$ represent the pressure sample values at each time step and time-average pressure, respectively. u_2 is the circumferential speed at the impeller exit, ρ is the fluid density.

The performance of pumps are implemented under different flow rates from $0.2\varphi_N$ to $1.6\varphi_N$. The CFD calculated results are obtained by steady-state simulation with the SST k-ω model. Figure 9 shows the head coefficient and efficiency curves of two blade trailing edge (BTE) profiles model pumps. The amplitude of the error bar represents the performance difference between the two model pumps under different operating conditions. For all relevant flow rates, the STTE profile pump is observed to have a better head hydraulic performance. Under the small flow conditions, there is little difference in the head performance, and the efficiency is almost constant. The difference in the head coefficient and efficiency between the two model pumps with the OTE and STTE profiles is about 2.3% and 0.93% at the nominal flow rate, respectively. The increase of the head and efficiency is due to the gentler flow of fluid from the STTE trailing edge. A typical wake-jet structure exists in the outlet area of the impeller, which causes the main energy loss of the centrifugal pump [49]. By modifying the trailing edge shape, the structure and frequency of the shedding vortex at the impeller exit are changed, and the development of the wake-jet structure is suppressed. It reduces the energy loss of the pump.

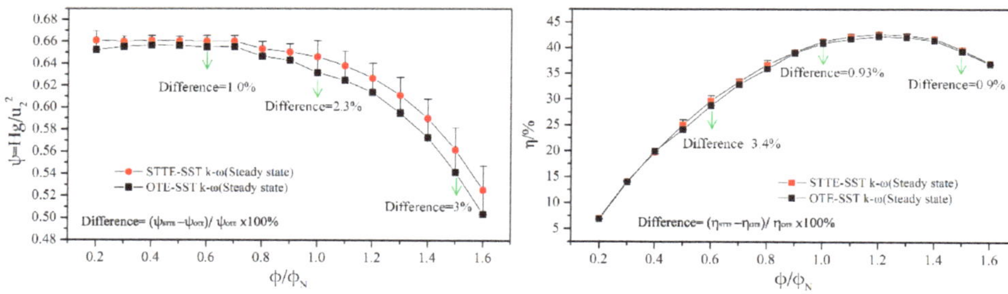

Figure 9. Numerical calculation results comparison of two model pumps (steady-state, SST k-ω).

4.2. Pressure Pulsations for the Two Pumps

The pressure values are dimensionless to represent the pressure distribution in the pump cavity of the two models. The average pressure coefficient is expressed as $\overline{Cp}$.

$$\overline{Cp} = \frac{1}{N}\sum_{i=1}^{N}\frac{p_i}{0.5\rho u_2{}^2} \tag{2}$$

where p_i represents the pressure sample values at each time step, N represents the total number. We choose the pressure values of the last five revolutions to calculate. The value of N is 1200.

Figure 10 shows the pressure distribution of two model pumps in the volute at the nominal flow rate Q_N. No. Centrifugal pumps with two trailing edge profiles have a very similar variation in the circumferential pressure of the volute. However, the STTE profile has a smaller value at each monitoring point. The pressure value gradually increases from No.1, and the maximum value occurs at the region of the volute outlet. The maximum value of pressure is on monitoring point T2 and the minimum value of pressure is on monitoring point T3. This is because the shedding vortex on the blade hits the volute tongue and causes large pressure fluctuations [50–53].

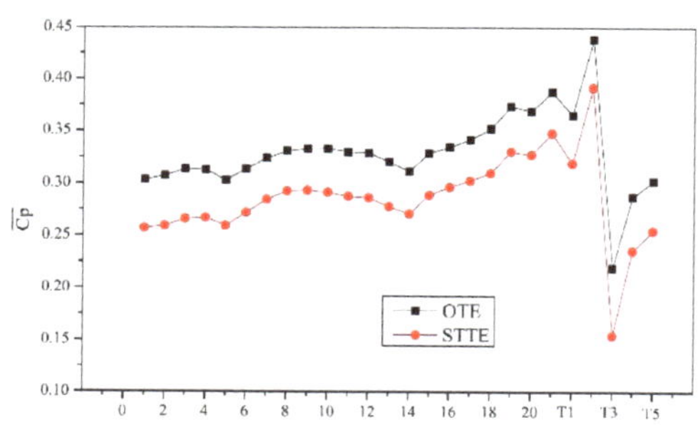

Figure 10. Pressure distribution of two model pumps at the nominal flow rate coefficient φ_N.

To clearly explain how the STTE profile affects unsteady pulsations, the DES model is used in a transient-state simulation to capture the fluctuating pressure signals and the flow patterns. Figure 11 reveals the comparison of two impellers time domain pressure signals at the monitoring point T2 at the nominal flow rate coefficient φ_N. The pressure fluctuation curve of the three periods is shown in

Figure 11. As observed, periodic pressure fluctuations are evident, and five distinct peak and valley values occur alternately in a cycle. It depends on the number of blades in the impeller. At the same time, the STTE profile has lower pressure signals than the OTE profile. Obviously, the model pump with the STTE profile has better pressure stability. It is worth noting that the STTE profile does have a more positive effect on the pressure pulsations.

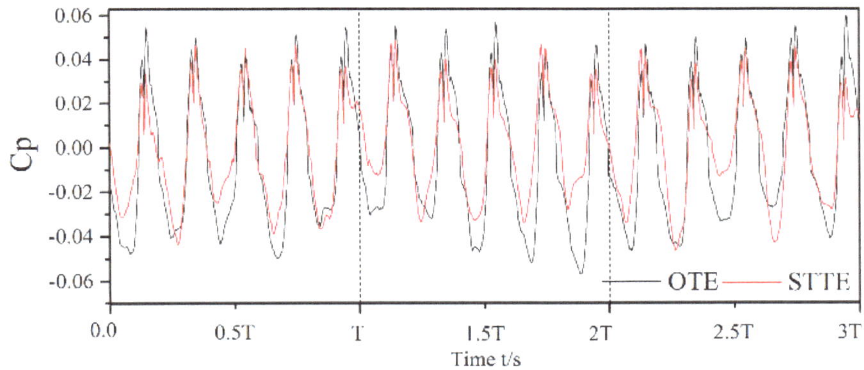

Figure 11. Time-domain pressure pulsations at point T2 at the nominal flow rate coefficient φ_N.

For a better analysis of the effect of the blade BTE shapes on pressure fluctuation, the collected fluctuating pressure signals in the tongue region are converted into a spectrum using fast Fourier transform (FFT). The pressure values of the last five revolutions at the nominal flow rate are selected. The violent rotor–stator interaction occurs in the tongue area. We select five monitoring points (T1–T5) in the tongue area for frequency domain analysis. As can be seen from Figure 12, the primary excitation frequency is f_{BPF} and its harmonic frequency $2f_{BPF}$ and $3f_{BPF}$. The peak at f_{BPF} always governs spectra analysis. Meanwhile, pressure coefficient amplitudes of T3, T4, and T5 are much higher than points T1 and T2, which indicates that the rotor–stator interaction is more intense in this region. This results from the small radial clearance between impeller and volute. The points T1 and T2 are far away from the volute tongue, so the rotor–stator interaction is less intense, which results in smaller pressure amplitudes. By comparing the pressure pulsation amplitudes between the two model pumps, the C_p of STTE at points T1–T5 are reduced by varying degrees at f_{BPF} and its harmonic frequencies. In the low-frequency zone below f_{BPF}, the model pump with the OTE shape has several more complex frequencies compared with STTE shape at the volute area. There are some more complicated flow structures in the OTE model pump, such as flow separation. Table 3 lists the contrast of the pressure coefficient C_p at blade passing frequency between two pumps.

Table 3. Comparison of C_p at blade passing frequency.

Monitoring Points	C_p of OTE Profile	C_p of STTE Profile	Reduction (%)
T1	0.0316	0.028	10.4
T2	0.0281	0.0266	5.3
T3	0.081	0.0745	8.2
T4	0.0577	0.0558	3.3
T5	0.065	0.0623	5.5

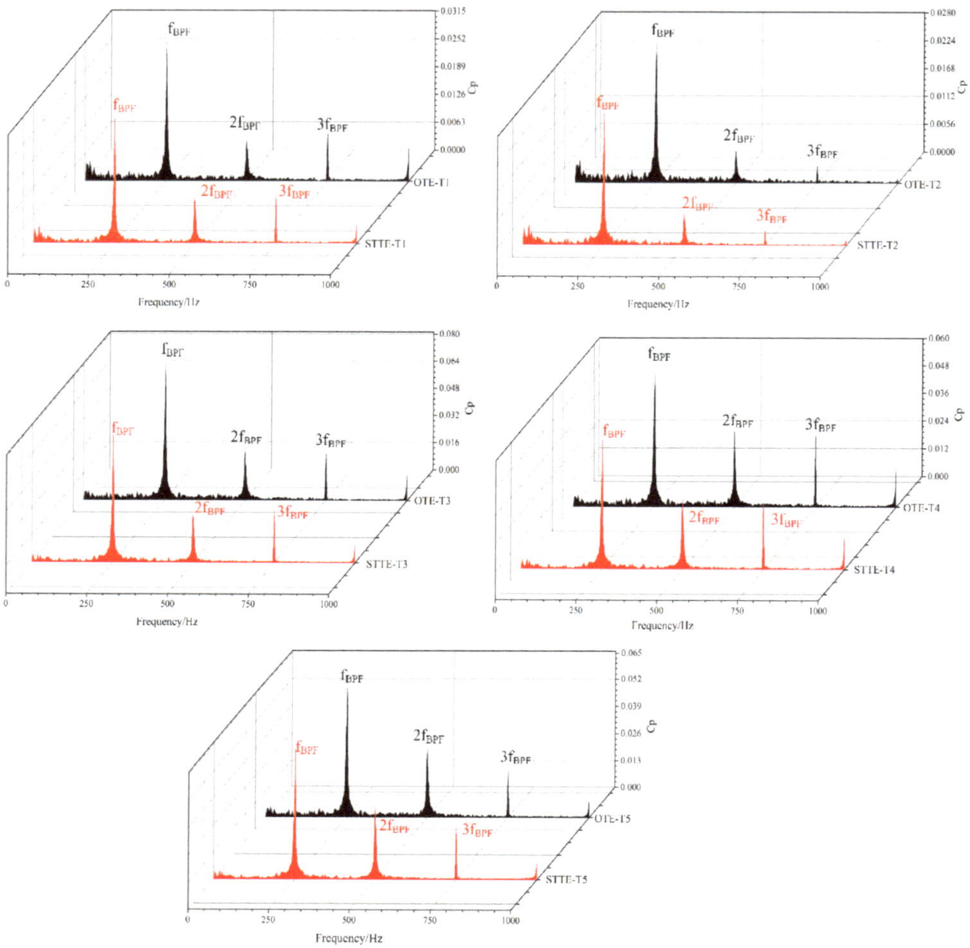

Figure 12. Frequency-domain pressure pulsations at points T1–T5 at the nominal flow rate coefficient φ_N.

4.3. Flow Structures of Two Model Pumps

The internal flow field of the model pumps is analyzed to reveal the reasons for the pressure pulsation amplitude reduction at the nominal rate. In Figure 13, the vorticity contours of two model pumps at 50% span are shown. There are high vorticity areas at the inlet and outlet of the blade. One is associated with flow separation and the flow strike at the inlet region. The other one is produced by the wake-jet structure and shedding vortex. The rotor–stator interaction is very intense due to the hitting of the vortex shedding from the blade trailing edge with the volute tongue. Through comparing the vorticity contours of two BTE shapes, the vorticity of STTE is obviously more diminished than OTE in the blade inlet region. The shedding vortex is also weakened in the blade outlet region.

Figure 14 shows the partially enlarged view of vorticity contours at the blade trailing edge. The high vorticity area of the STTE impeller is shortened and the vorticity magnitude is also decreased. To further understand the internal flow pattern in the impeller, velocity streamlines are presented in Figure 15. It can be clearly distinguished that the STTE profile makes the flow field more uniform. For the OTE shape, flow field is quite disordered, and some vortices even occur on the pressure side of the blade. The STTE shape decreases flow separation in the blade inlet region and the middle flow channels.

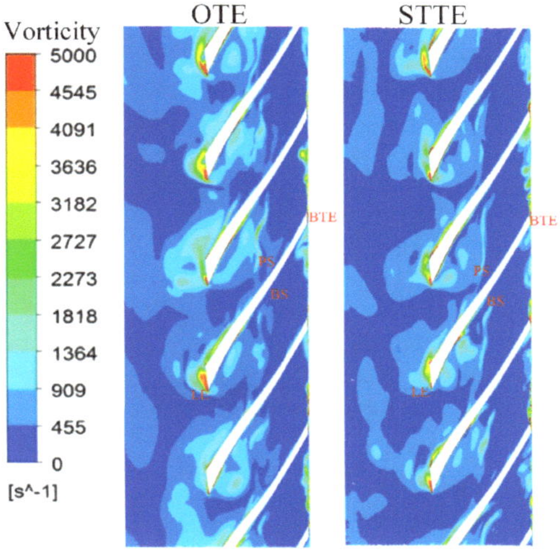

Figure 13. Vorticity distributions on the blade-to-blade surface at 0.5 span.

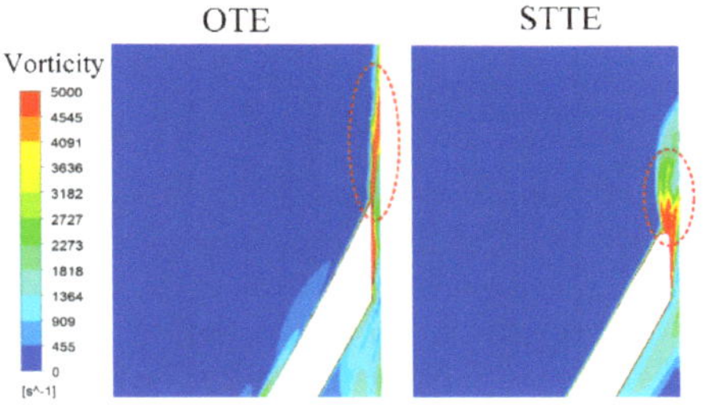

Figure 14. Partially enlarged view of vorticity contours at 0.5 span.

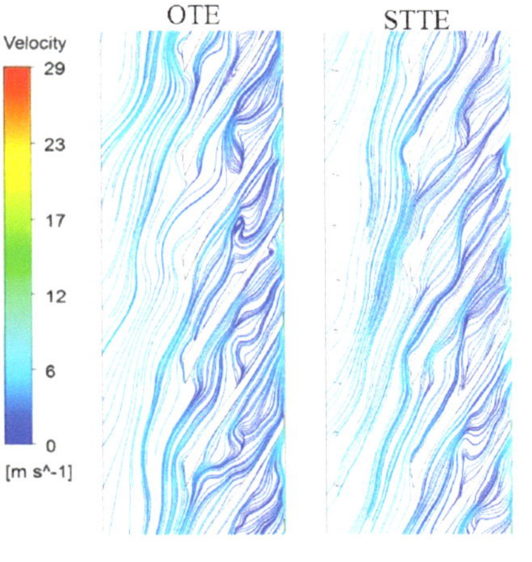

Figure 15. Velocity streamline distributions at 0.5 span.

In order to understand the mechanism of the STTE trailing edge improving the pump performance and pressure pulsation, and better identify the shedding vortex pattern at the impeller outlet, the vortex core distribution in the blade trailing edge (BTE) area is presented in Figure 16. As shown in Figure 16. Comparing the two model pumps, it is found that the structure of shedding vortex changed greatly in the impeller BTE region. The large intact vortexes shedding from the blade trailing edge are divided into several small vortexes by the STTE profile. The total vortex core region of the STTE model pump is significantly smaller than OTE model pump. The interaction of small vortexes with the volute is also weaker than with large vortexes. It is acceptable to control vortex shedding by modifying the trailing edge profile, which is very important for pressure stability.

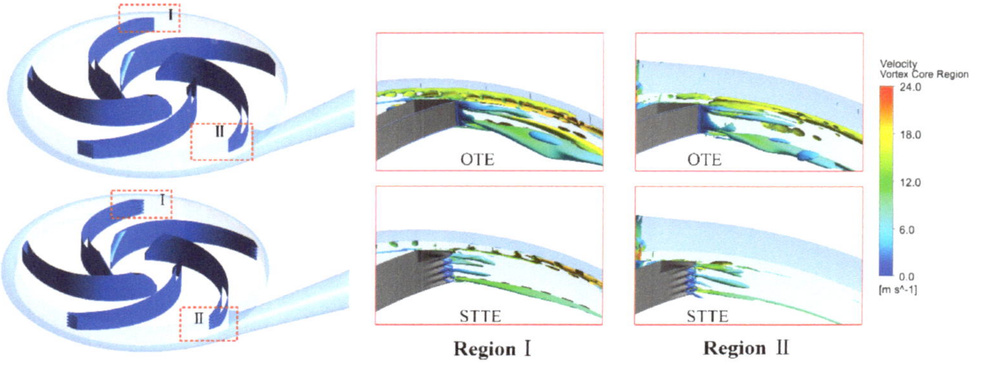

Figure 16. Vortex core distributions in the blade trailing edge (BTE) region at the nominal flow rate.

5. Conclusions

In this paper, a five-blade centrifugal pump is tested and calculated to explore the effect of two BTE shapes on pressure fluctuation and performance. Time domains and frequency domains of the pressure pulsations are analyzed. Some relevant conclusions are drawn.

The impeller with the STTE profile shows better performance. The head and efficiency are increased by 2.3% and 0.93% at the nominal flow rate, respectively.

The model pump with the STTE shape has lower pressure pulsation amplitudes than the OTE shape at various monitoring points. The peak at f_{BPF} always governs the pressure spectra. By comparison of two BTE profiles, the STTE profile makes the pressure pulsation amplitudes at f_{BPF} reduces by varying degrees.

The vortex core distributions in the BTE region at the nominal flow rate show that the STTE shape divides the large vortex shedding from the BTE into several small vortices, which reduces the rotor–stator interaction. The internal flow of the STTE model pump becomes uniform. So, the pressure pulsation of the STTE pump is decreased. The impeller with STTE would provide a new design method to reduce pressure pulsation and improve the performance of pump.

Because this case only studied two models with specific design parameters, which may not be optimal designs, there may be little effect on efficiency. Different design parameters will be studied to find the optimal design in the future. The analysis in this paper is based on the application of STTE in a low-specific-speed centrifugal pump with low efficiency, and the conclusions might only be applicable to such a pump. The conclusions should be verified when applied to other pumps.

Author Contributions: Conceptualization, X.L.; formal analysis, B.L. and F.C.; experiment investigation, B.L., X.J., and H.F..; writing-original draft preparation, B.L.; writing-review and editing, X.L.

Funding: This research was funded by the National Natural Science Foundation of China, grant No.51776189 and U1709209, Zhejiang Province Science Foundation, grant No.LGG18E060005, and the Talent Project of Zhejiang Association for Science and Technology, grant No.2018YCGC013. The supports are gratefully acknowledged.

Conflicts of Interest: The authors declare no conflict of interest.

Nomenclature

D_1	Impeller entrance diameter [mm]
D_2	Impeller exit diameter [mm]
D_3	volute base circle diameter [mm]
Q	Nominal flow rate [m^3/h]
H	Nominal head [m]
B_1	Impeller inlet width [mm]
B_2	peller exit width [mm]
N	Nominal rotating speed [rpm]
Z	Impeller blades number
Cp	Pressure coefficient [= $(pi-\overline{p})/(0.5\rho\ u_2^2)$]
u_2	circumferential speed at the impeller exit [$= \pi * D_2 * n/60$].
Ns	Specific speed
ρ	Fluid density [kg/m^3]
$\overline{Cp}$	Mean pressure coefficient
φ	Flow coefficient [$Q/(u_2R_2^2)$]
ψ	Head coefficient [Hg/u_2^2]
η	Efficiency [%]
p_i	Monitoring point pressure value [Hg/u_2^2]
BTE	blade trailing edge
STTE	sinusoidal tubercle trailing edge
β_1	Blade inlet angle [°]
β_2	Blade outlet angle [°]
φ_N	Nominal flow rate coefficient
ψ_N	Nominal head coefficient

References

1. Shankar, A.; Subramaniam, U.; Shanmugam, P.; Hanigovszki, N. A comprehensive review on energy efficiency enhancement initiatives in centrifugal pumping system. *Appl. Energy* **2016**, *181*, 495–513. [CrossRef]
2. Majidi, K. Numerical Study of Unsteady Flow in a Centrifugal Pump. *J. Turbomach.* **2004**, *127*, 805–814.
3. Li, X.; Jiang, Z.; Zhu, Z.; Si, Q.; Li, Y. Entropy generation analysis for the cavitating head-drop characteristic of a centrifugal pump. *Proc. Inst. Mech. Eng. Part C* **2018**, *232*, 4637–4646. [CrossRef]
4. Bai, L.; Zhou, L.; Han, C.; Zhu, Y.; Shi, W. Numerical Study of Pressure Fluctuation and Unsteady Flow in a Centrifugal Pump. *Processes* **2019**, *7*, 354. [CrossRef]
5. Fu, D.; Wang, F.; Zhou, P.; Xiao, R.; Yao, Z. Impact of Impeller Stagger Angles on Pressure Fluctuation for a Double-Suction Centrifugal Pump. *Chin. J. Mech. Eng.* **2018**, *31*, 10. [CrossRef]
6. Wang, C.; Chen, X.; Qiu, N.; Yong, Z.; Shi, W. Numerical and experimental study on the pressure fluctuation, vibration, and noise of multistage pump with radial diffuser. *J. Braz. Soc. Mech. Sci. Eng.* **2018**, *40*, 481. [CrossRef]
7. Fu, Q.; Yuan, S.; Zhu, R. Pressure Fluctuation of the Low Specific Speed Centrifugal Pump. *Appl. Mech. Mater.* **2012**, *152–154*, 935–939. [CrossRef]
8. Cui, B.; Chen, D.; Xu, W.; Jin, Y.; Zhu, Z. Unsteady Flow Characteristic of Low-Specific-Speed Centrifugal Pump under Different Flow-Rate Conditions. *J. Therm. Sci.* **2015**, *24*, 17–23. [CrossRef]
9. Zhang, W.; Yu, Y.; Zhu, S. Characteristic analysis on the pressure fluctuation in the impeller of a low specific speed mixed flow pump. *IOP Mater. Sci. Eng.* **2016**, *129*, 012035. [CrossRef]
10. Fu, Y.X.; Yuan, J.P.; Yuan, S.Q.; Pace, G. Numerical and Experimental Analysis of Flow Phenomena in a Centrifugal Pump Operating Under Low Flow Rates. *J. Fluids Eng.* **2014**, *137*, 011102. [CrossRef]
11. Pei, J.; Wang, W.J.; Yuan, S.Q. Statistical analysis of pressure fluctuations during unsteady flow for low-specific-speed centrifugal pumps. *J. Cent South Univ.* **2014**, *21*, 1017–1024. [CrossRef]
12. Jafarzadeh, B.; Hajari, A.; Alishahi, M.M.; Akbari, M.H. The flow simulation of a low-specific-speed high-speed centrifugal pump. *Appl. Math. Model.* **2010**, *35*, 242–249. [CrossRef]

13. Chu, S.; Dong, R.; Katz, J. Relation between Unsteady Flow, Pressure Fluctuations, and Noise in a Centrifugal Pump-Part B: Effects of Blade-tongue Interactions. *J. Fluids Eng.-Trans. ASME* **1997**, *117*, 30–35. [CrossRef]
14. Chu, S.; Dong, R.; Katz, J. Relation between Unsteady Flow, Pressure Fluctuations, and Noise in a Centrifugal Pump-Part A: Use of PDV Date to Compute the Pressure Field. *J. Fluids Eng.-Trans. ASME* **1997**, *117*, 24–29. [CrossRef]
15. Keller, J.; Blanco, E.; Barrio, R.; Parrondo, J. PIV Measurements of the Unsteady Flow Structures in a Volute Centrifugal Pump at a High Flow Rate. *Exp. Fluids* **2014**, *55*, 1820. [CrossRef]
16. Jia, X.; Cui, B.; Zhu, Z.; Yu, X. Numerical Investigation of Pressure Distribution in a Low Specific Speed Centrifugal Pump. *J. Therm. Sci.* **2018**, *27*, 25–33. [CrossRef]
17. Heskestad, G.; Olberts, D.R. Influence of Trailing-Edge Geometry on Hydraulic-Turbine-Blade Vibration Resulting From Vortex Excitation. *J. Eng. Power* **1960**, *82*, 103. [CrossRef]
18. Gao, B.; Zhang, N.; Li, Z.; Ni, D.; Yang, M. Influence of the Blade Trailing Edge Profile on the Performance and Unsteady Pressure Pulsations in a Low Specific Speed Centrifugal Pump. *J. Fluids Eng.-Trans. ASME* **2016**, *138*, 051106. [CrossRef]
19. Spence, H.; Aaral-Teixeira, I. Investigation into pressure pulsations in a centrifugal pump using numerical methods supported by industrial tests. *Comput. Fluids* **2008**, *37*, 690–704. [CrossRef]
20. Solis, M.; Bakir, F.; Khelladi, S.; Noguera, R. Numerical Study on Pressure Fluctuations Reduction in Centrifugal Pumps: Influence of Radial Gap and Splitter Blades. *ISRN Mech. Eng.* **2011**, *2011*, 479594. [CrossRef]
21. Amir, J.; Hossein, A. Impeller and volute design and optimization of the centrifugal pump with low specific speed in order to extract performance curves. *JCAMECH* **2018**, *49*, 359–366.
22. Zhu, B.; Chen, H. Cavitation suppression of low specific speed centrifugal pump with gap drainage blades. *J. Hydrodyn.* **2012**, *24*, 729–736. [CrossRef]
23. Khalifa, A.E. Effect of Blade Exit Shape on Performance and Vibration of a Double Volute Centrifugal Pump. *Int. J. Mater. Mech. Manuf.* **2014**, *2*, 261–264. [CrossRef]
24. Binama, M.; Su, W.; Cai, W.; Li, B.; Alexis, M.; Li, B.; Emmanuel, B. Blade trailing edge position influencing pump as turbine (PAT) pressure field under part-load conditions. *Renew. Energy* **2019**, *136*, 33–47. [CrossRef]
25. Xue, G.; Liu, Y.; Zhang, M.; Ding, H. Numerical Analysis of Hydrodynamics for Bionic Oscillating Hydrofoil Based on Panel Method. *Appl. Bionics Biomech.* **2016**, *2016*, 6909745. [CrossRef] [PubMed]
26. Xin, H.; Zhang, C.; Wei, J.; Hu, X.; Wei, H. Wind turbine bionic blade design and performance analysis. *J. Vis. Commun. Image Represent.* **2019**, *60*, 258–265.
27. Yang, Z.; Li, G.; Song, Y.; Bai, Y. Effects of setting angle and chord length on performance of four blades bionic wind turbine. *IOP Earth Environ. Sci.* **2017**, *93*, 012041. [CrossRef]
28. Shi, W.; Atlar, M.; Norman, R.; Aktas, B.; Turkmen, S. Numerical optimization and experimental validation for a tidal turbine blade with leading-edge tubercles. *Renew. Energy* **2016**, *96*, 42–55. [CrossRef]
29. Gu, Y.; Pei, J.; Yuan, S. Clocking effect of vaned diffuser on hydraulic performance of high-power pump by using the numerical flow loss visualization method. *Energy* **2019**, *170*, 986–997. [CrossRef]
30. Cheng, H.Y.; Bai, X.R.; Long, X.P.; Ji, B.; Peng, X.X.; Farhat, M. Large eddy simulation of the tip-leakage cavitating flow with an insight on how cavitation influences vorticity and turbulence. *Appl. Math. Model.* **2019**, *77*, 788–809. [CrossRef]
31. Wang, C.; Shi, W.; Wang, X. Optimal design of multistage centrifugal pump based on the combined energy loss model and computational fluid dynamics. *Appl. Energy* **2017**, *187*, 10–26. [CrossRef]
32. Li, D.; Fu, X.; Zuo, Z.; Wang, H.; Li, Z.; Liu, S.; Wei, X. Investigation methods for analysis of transient phenomena concerning design and operation of hydraulic-machine systems-A review. *Renew. Sustain. Energ Rev.* **2019**, *101*, 26–46. [CrossRef]
33. Zhang, Y.; Chen, T.; Li, J.; Yu, J. Experimental study of load variations on pressure fluctuations in a prototype reversible pump turbine in generating mode. *ASME J. Fluids Eng.* **2017**, *139*, 074501. [CrossRef]
34. Liu, M.; Tan, L.; Cao, S. Cavitation–vortex–turbulence interaction and one-dimensional model prediction of pressure for hydrofoil ALE15 by large eddy simulation. *ASME J. Fluids Eng.* **2019**, *141*, 021103. [CrossRef]
35. Huang, B.; Qiu, S.; Li, X.; Wu, Q.; Wang, G. A review of transient flow structure and unsteady mechanism of cavitating flow. *J. Hydrodyn.* **2019**, *31*, 429–444. [CrossRef]

36. Yousefi, H.; Noorollahi, Y.; Tahani, M.; Fahimi, R.; Saremian, S. Numerical simulation for obtaining optimal impeller's blade parameters of a centrifugal pump for high-viscosity fluid pumping. *Sustain. Energy Technol. Assess.* **2019**, *34*, 16–26. [CrossRef]
37. Yi, S.; Zhu, H.; Zhang, J. Experiment and numerical study of a new generation three-stage multiphase pump. *J. Pet. Sci. Eng.* **2018**, *169*, 471–484.
38. Pei, J.; Zhang, F.; Appiah, D.; Hu, B.; Yuan, S.; Chen, K.; Asomani, S.N. Performance Prediction Based on Effects of Wrapping Angle of a Side Channel Pump. *Energies* **2019**, *12*, 139. [CrossRef]
39. Wang, C.; Hu, B.; Zhu, Y.; Wang, X.; Luo, C.; Cheng, L. Numerical Study on the Gas-Water Two-Phase Flow in the Self-Priming Process of Self-Priming Centrifugal Pump. *Processes* **2019**, *7*, 330. [CrossRef]
40. Li, X.; Gao, P.; Zhu, Z.; Li, Y. Effect of the blade loading distribution on hydrodynamic performance of a centrifugal pump with cylindrical blades. *J. Mech. Sci. Technol.* **2018**, *32*, 1161–1170. [CrossRef]
41. Xu, C.; Chen, L.; Lu, X. Large-eddy and detached-eddy simulations of the separated flow around a circular cylinder. *J. Hydrodyn.* **2007**, *B19*, 559–563. [CrossRef]
42. Lin, D.; Su, X.; Yuan, X. DDES Analysis of the Wake Vortex Related Unsteadiness and Losses in the Environment of a High-Pressure Turbine Stage. *ASME J. Turbomach.* **2018**, *140*, 041001. [CrossRef]
43. Wei, Y.; Yang, H.; Lin, Z.; Wang, Z.; Qian, Y. A novel two-dimensional coupled lattice Boltzmann model for thermal incompressible flows. *Appl. Math. Comput.* **2018**, *339*, 556–567. [CrossRef]
44. Liu, Q.; Ye, J.; Zhang, G.; Lin, Z.; Xu, H.; Zhu, Z. Metrological performance investigation of swirl flowmeter affected by vortex inflow. *J. Mech. Sci. Technol.* **2019**, *33*, 1–10. [CrossRef]
45. Zhang, W.; Li, X.; Zhu, Z. Quantification of wake unsteadiness for low-Re flow across two staggered cylinders. *Proc. Inst. Mech. Eng. Part C JMES* **2019**, *233*, 0954406219866478. [CrossRef]
46. Zhang, S.; Li, X.; Hu, B.; Liu, Y.; Zhu, Z. Numerical investigation of attached cavitating flow in thermo-sensitive fluid with special emphasis on thermal effect and shedding dynamics. *Int. J. Hydrog. Energy* **2019**, *44*, 3170–3184. [CrossRef]
47. Liu, Q.; Ye, J.; Zhang, G.; Lin, Z. Study on the metrological performance of a swirlmeter affected by flow regulation with a sleeve valve. *Flow Meas. Instrum.* **2019**, *67*, 83–94. [CrossRef]
48. Wang, W.; Osman, M.K.; Pei, J.; Gan, X.; Yin, T. Artificial Neural Networks Approach for a Multi-Objective Cavitation Optimization Design in a Double-Suction Centrifugal Pump. *Processes* **2019**, *7*, 246. [CrossRef]
49. Zhang, N.; Liu, X.; Gao, B.; Xia, B. DDES analysis of the unsteady wake flow and its evolution of a centrifugal pump. *Renew. Energy* **2019**, *141*, 570–582. [CrossRef]
50. Wang, C.; He, X.; Shi, W.; Wang, X.; Wang, X.L.; Qiu, N. Numerical study on pressure fluctuation of a multistage centrifugal pump based on whole flow field. *AIP Adv.* **2019**, *9*, 035118. [CrossRef]
51. Li, D.; Wang, H.; Qin, Y. Mechanism of High Amplitude Low Frequency Fluctuations in a Pump-Turbine in Pump Mode. *Renew. Energy* **2018**, *126*, 668–680. [CrossRef]
52. Liu, Y.; Tan, L. Tip clearance on pressure fluctuation intensity and vortex characteristic of a mixed flow pump as turbine at pump mode. *Renew. Energy* **2018**, *129*, 606–615. [CrossRef]
53. Lun, Y.; Lin, L.; He, H.; Zhu, Z.; Wei, Y. Effects of Vortex Structure on Performance Characteristics of a Multiblade Fan with Inclined tongue. *Proc. Inst. Mech. Eng. Part A J. Power Energy* **2019**, *233*, 0957650919840964. [CrossRef]

processes

MDPI

Article

Numerical Simulation and Performance Prediction of Centrifugal Pump's Full Flow Field Based on OpenFOAM

Si Huang, Yifeng Wei *, Chenguang Guo and Wenming Kang

School of Mechanical and Automotive Engineering, South China University of Technology, Guangzhou 510640, China

* Correspondence: weiyifeng95@foxmail.com

Received: 30 July 2019; Accepted: 4 September 2019; Published: 7 September 2019

Abstract: The open-source software OpenFOAM 5.0 was used as a platform to perform steady-state and transient numerical simulation for full flow field of a pipeline centrifugal pump (specific speed n_s = 65) in a wide operating capacity range of $0.3Q_d$~$1.4Q_d$. The standard k-ε and k-ω SST (Shear-Stress Transport) turbulence models were selected in the flow governing equations. The simpleFoam and pimpleDyMFoam solvers were used for the steady-state and transient calculations, respectively. ParaView, the postprocessor in OpenFOAM, was used to display the calculated flow velocity, pressure and streamline distributions, and to analyze the relationship between the vortex and the hydraulic loss in the pump. The external performance parameters of the pump such as head, input power and efficiency were also calculated based on the simulated flow fields. The predicted pump performances by OpenFOAM and Ansys-Fluent are compared with the test results under the same calculation model, grids and boundary conditions. The comparison indicates that OpenFOAM had high accuracy in the prediction of pump performance in the current case.

Keywords: OpenFOAM; centrifugal pump; numerical simulation; performance prediction

1. Introduction

Centrifugal pumps are widely applied in the processing industry, agricultural, civil and various other fields [1]. With the development of computational technology, the computational fluid dynamics (CFD) technique becomes very powerful and spans a wide range of industrial and non-industrial areas [2]. Commercial CFD software is convenient to use but the programming is usually unknown, so it is hard for users to modify or add models and algorithms according to their own ideas on certain issues. OpenFOAM, an object-oriented CFD open-source software, not only provides a wide range of standard solvers but also allows users to edit them in terms of the source code [3]. Due to the expandability and openness of source codes, recently some researches have been carried out on the flow field of pump machines by means of OpenFOAM. Petit et al. [4,5] and Nilsson [6], respectively, performed steady-state (frozen rotor method) and transient (sliding grid method) simulations on the flow field of centrifugal pumps and hydro turbines, and compared the results with the actual measurement. Xie [7] used 2D and 3D steady-state and transient methods to simulate the flow fields of a low specific speed centrifugal pump. Zhang [8] performed numerical calculations of turbulent flow for a high head turbine with k-ω SST (shear-stress transport) turbulence model and obtained the distribution of cavitating vapor in the turbine. Nuernberg and Tao [9] simulated the 3D tidal turbine array using a moving mesh with a rotor-stator interface to account for the effects of rotating tidal turbine blades and the flow field characteristics in tidal turbine arrays. Liu et al. [10] conducted a steady-state numerical simulation of the flow field in a centrifugal pump with specific speed n_s = 117.8 under three operating conditions ($0.8Q_d$, $1.0Q_d$ and $1.2Q_d$) and found that the pump under the off-design operating conditions was well predicted by OpenFOAM 1.5-dev. However, due to the limitations of the solvers and post-processing functions of the previous OpenFOAM versions, the numerical results of the above-mentioned studies were far from satisfactory on the prediction accuracy of pump performance. Also, the researches based on OpenFOAM lacked in a full flow field of centrifugal pumps in a wide range of flow rates.

With the upgrade of the open source software OpenFOAM, the potential for computing pumps' flow field and performance has significantly improved. In this work, therefore, a common pipeline centrifugal pump was selected by using OpenFOAM 5.0 to study the steady-state and transient full flow fields in a wide range ($0.3Q_d$~$1.4Q_d$) of operating conditions and to analyze the relationship between flow field and hydraulic loss in the pump with the updated post-processors. By comparing the measured results, the accuracy of OpenFOAM 5.0 software in predicting the performance of the centrifugal pump was analyzed.

2. Numerical Method and Model

2.1. Governing Equations

The governing equations of the fluid flow in this work included the continuity equation and the momentum equation. For turbulent flow, the k-ε model is used widely for internal flows, which provides an efficient method of calculating engineering flows [11]. Also, Liu et al. [10] found that the standard k-ε turbulence model had higher accuracy in computation of centrifugal pump performance after comparing with RNG (ReNormalization Group) k-ε and k-ω SST models. However, Shojaeefard et al. [12] found that the near-wall flow could be evaluated with high precision using the k-ω SST model, and the results had better accuracy than those of the k-ε model in a centrifugal pump simulation. Therefore, both the standard k-ε model and the k-ω SST model were chosen for the simulation in this study. All the governing equations can be expressed in a general differential formulation:

$$\frac{\partial}{\partial t}(\rho\phi) + \frac{\partial}{\partial x_j}(\rho u_i \phi) = \frac{\partial}{\partial x_j}\left(\Gamma\frac{\partial\phi}{\partial x_j}\right) + S \tag{1}$$

where ρ is the density of the fluid, u_i is the i-directional component of the velocity vector, ϕ is the arbitrary scalar, Γ is the diffusion coefficient, and S is the source term. The items in Equation (1) from left to right are transient items, convection items, diffusion items and source items. For different equations, ϕ, Γ and S have different forms. Table 1 shows the detailed forms of the parameters ϕ, Γ and S for different equations, in which the turbulence model is the standard k-ε model.

Table 1. Specific forms of the symbols in Equation (1).

Equation	ϕ	Γ	S
Continuity equation	1	0	0
Momentum equation	u_i	μ	$-\frac{\partial p}{\partial x_i} + S_i$
Turbulent kinetic energy equation	k	$\mu + \frac{\mu_t}{\sigma_k}$	$G_k + S_k - \rho\varepsilon$
Turbulence eddy dissipation equation	ε	$\mu + \frac{\mu_t}{\sigma_\varepsilon}$	$C_{1\varepsilon}\frac{\varepsilon}{k}G_k - C_{2\varepsilon}\rho\frac{\varepsilon^2}{k} + S_\varepsilon$

In these equations, μ is dynamic viscosity. The turbulent viscosity μ_t is computed by combining k and ε as follows:

$$\mu_t = \rho C_\mu \frac{k^2}{\varepsilon} \tag{2}$$

C_μ, $C_{1\varepsilon}$, $C_{2\varepsilon}$, σ_k and σ_ε are model constants. G_k represents the generation of turbulence kinetic energy due to the mean velocity gradients. S_k and S_ε are source terms. S_i is a generalized source term of the momentum conservation equation.

2.2. Computational Domain and Meshing

A common pipeline centrifugal pump of type IS125-100 was chosen for the research as shown in Figure 1. The design parameters for the centrifugal pump were: capacity (Q_d = 54 L/s), rotational speed (n = 2900 r/min), specific speed (n_s = 65), impeller inlet diameter (D_0 = 125 mm), impeller outlet diameter (D_2 = 216 mm), impeller outlet width (b_2 = 26.5 mm), impeller blade outlet angle (β_2 = 28°), number of impeller blades (Z = 6), discharge diameter (D = 100 mm). The computational domain was composed of an inlet pipe, impeller and volute. A 3D model was created by Pro-E software and imported into the ICEM software to generate the computational grid, which was converted into a format that OpenFOAM recognized. Since the domain of a centrifugal pump with such specific speed is complex, the grid is generally generated in an unstructured type for applications.

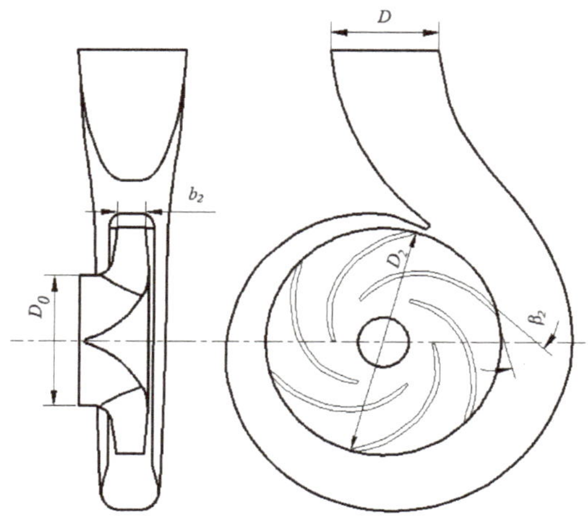

Figure 1. Geometry of type IS125-100.

To choose a proper computational grid for simulation, the grid independency was examined. Figure 2 shows the relation between element number and calculated pump head with the standard k-ε turbulence model by Fluent and OpenFOAM. When the number of grids exceeded 1,000,000, the effect of the grids increase on the head was negligible, so the total number of elements in the simulation was 1,024,853 as shown in Figure 3.

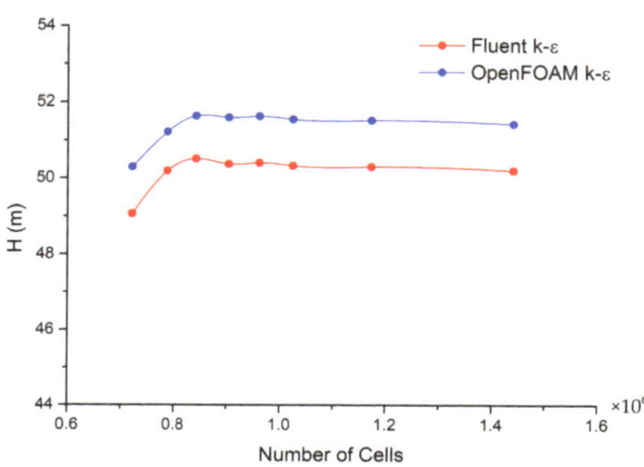

Figure 2. Curves of grids independency.

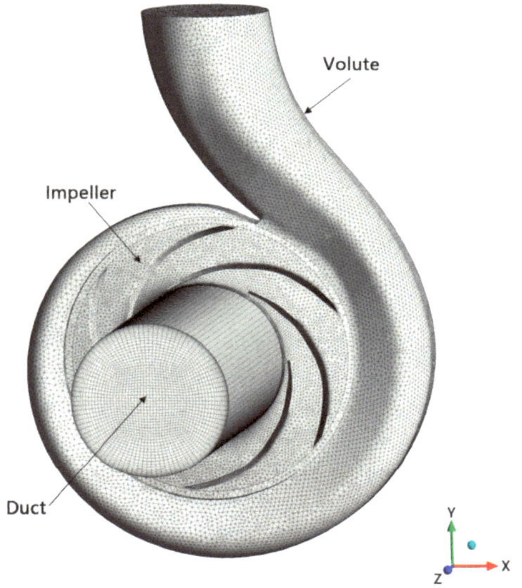

Figure 3. Unstructured grids in the computational domain.

2.3. Solvers and Boundary Conditions

The solver is the key tool for solving the discretization forms of the flow governing equations. The solvers, simpleFoam and pimpleDyMFoam, were used for the steady-state and transient simulations in this work, respectively. It was assumed that the pressure at the inlet of the pipeline pump was high enough and no cavitation occurred in the pump.

2.3.1. The SimpleFoam Solver

The simpleFoam is a steady-state incompressible flow solver in OpenFOAM, based on the SIMPLE (Semi-Implicit Method for Pressure Linked Equations) algorithm for pressure-velocity coupling [13]. In this algorithm, for a given pressure field (initial value or the result of the last iteration calculation), the discrete form of the flow governing equation is solved to obtain the velocity field for the next iteration. Under-relaxation factors α_u and α_p ($\alpha_u + \alpha_p = 1$) were used to maintain iteration stability for velocity and pressure correction [14]. Also, the optimum convergence results with the combination of under-relaxation factors were found to be $\alpha_u = (0.7 \sim 0.8)$ and $\alpha_p = (0.3 \sim 0.2)$ [15]. Finally, the under-relaxation factors of velocity v, pressure p, turbulence kinetic energy k and turbulence dissipation rate ε or specific dissipation rate ω were 0.7, 0.3, 0.3, 0.3, respectively.

The "frozen rotor method" [16] was used in the simulations based on the multiple reference frame (MRF). The MRF approach implies that there is no relative mesh motion of the rotating and stationary parts. The solution of the steady-state, turbulent and incompressible flow of the impeller is governed by the Reynolds-averaged Navier-Stokes equations (RANS) in the rotating reference frame [4].

Arbitrary mesh interface (AMI) [17] was adopted to transfer the flux between the rotational impeller and stationary domain grids by a set of weighting factors both for steady-state and transient cases. The boundary conditions for the steady state computation were: (1) The type of velocity v, turbulent kinetic energy k and turbulent eddy dissipation ε or ω at the inlet were "fixedValue". The value of v was given by the pump operation capacity, the initial values of k, ε and ω were calculated and the pressure type at the inlet was "zeroGradient"; (2) The pressure type at the outlet was "fixedValue"; (3) The no-slip solid wall condition was adopted in wall surfaces, and the standard wall function method was used to determine the flow around the solid wall. The turbulent parameters of wall functions, μ_t, k, ε and ω were specified through the "nutkWallFunction", "kqRWallFunction", "epsilonWallFunction" and "omegaWallFunction" in terms of OpenFOAM code, respectively. Yplus (y+) is the dimensionless distance from the wall and is used to check the location of the first node. When standard wall functions are used, the value of y+ in the range 30~500 satisfies the log-law [18]. In the present study, thus, the values of y+ were set with the locations of the first nodes in the range of 30~300; (4) The type of the impeller wall was "movingwallVelocity"; (5) The impeller domain was defined by the "MRFProperties" file according to the "frozen rotor method". For the discretizations, the second-order Gauss linear upwind convection schemes were adopted for a higher accuracy. For the gradient terms, the Gauss linear scheme was selected, and the residual criterion was set to 10^{-3}.

2.3.2. The PimpleDyMFoam Solver

The pimpleDyMFoam solver, based on the PIMPLE (merged PISO (Pressure Implicit with Splitting of Operators)-SIMPLE) algorithm [19], is a reasonable time step transient for incompressible turbulent flow with dynamic meshes. The basic method in the PIMPLE algorithm is to use the PISO algorithm to step the time, and at each step the pressure field of a non-final iteration step and velocity field are solved. The solution of the time-dependent, turbulent and incompressible flow system of the pump is governed by the unsteady Reynolds-averaged Navier-Stokes equations (URANS) that can be in the formulations written for a moving control volume [20].

All boundary conditions for the transient case were consistent with those for the steady-state case except condition (5). The computational domain of the impeller was defined by the "dynamicMeshDict" file according to the "sliding mesh method" [21,22]. The first order implicit Euler scheme was adopted for the time discretization. The time step was set to be 5×10^{-5} s equivalent to the time for the impeller to rotate 1 degree, and the maximum Courant number was set to 5. The residual criterion was set to 10^{-6} and the maximum iterative loop was 100 in each time step.

3. Results and Analysis

3.1. *Pump Flow Field Distribution*

The calculated results of flow fields can be displayed by ParaView (OpenFOAM's postprocessor) or converted to other post-processing data formats. Figure 4 shows the flow velocity vectors in the pump by steady-state calculation under three operating conditions ($0.3Q_d$, $1.0Q_d$ and $1.4Q_d$). It can be seen from the figure that due to the rotation of the impeller, the flow velocity in the impeller gradually increases along the radius, and the flow velocity at the impeller outlet reaches a maximum value. After entering the volute, the flow velocity decreases. Since the volute passage near the outlet is designed as a bend, the flow velocity outside the volute is higher than that inside near the outlet due to centrifugal force. Apparently, the velocity vectors in the volute appear too small or too large under the condition of $0.3Q_d$ or $1.4Q_d$, indicating that the size of the volute is not suitable for these operating conditions.

Figures 5 and 6 respectively show the distributions of the streamlines in the centrifugal pump projected on the central xy and zx planes (corresponding to Figure 3) by the steady-state calculation under three operating conditions. On the xy plane, under $0.3Q_d$ operating condition, the fluid passes the volute diffuser region in a lower velocity (Figure 5a). When the flow rate is $1.0Q_d$, the main flow tends to follow the curvature of the volute outer wall (Figure 5b). Under the $1.4Q_d$ operating condition, a recirculation region is obviously observed on the inner wall near the volute exit due to a higher centrifugal force (Figure 5c). It is noted that flow velocity departing from the outlet of the impeller is more uniform (magnitude is about 20 m/s) under $1.0Q_d$ than at $0.3Q_d$ and $1.4Q_d$ operating conditions.

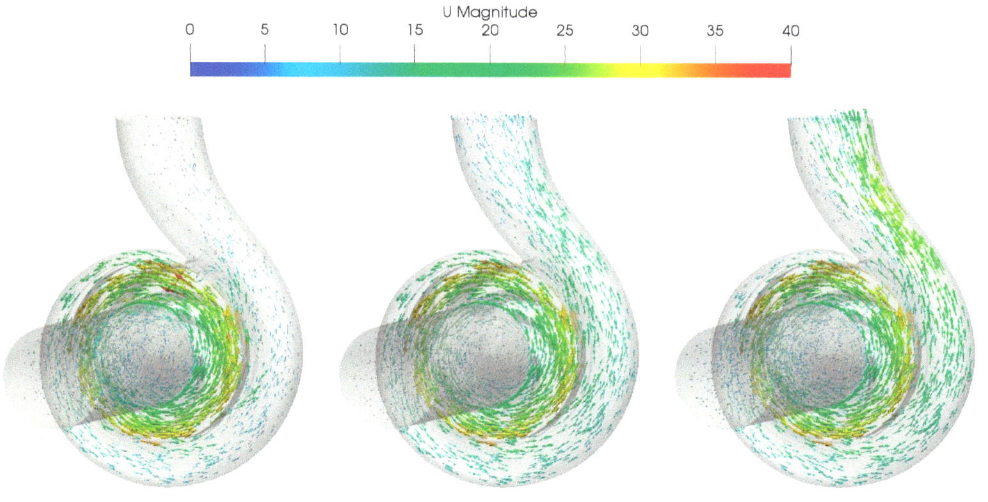

(**a**) 0.3Q$_d$ (**b**) 1.0Q$_d$ (**c**) 1.4Q$_d$

Figure 4. Velocity vectors in the pump (steady-state, OpenFOAM k-ε).

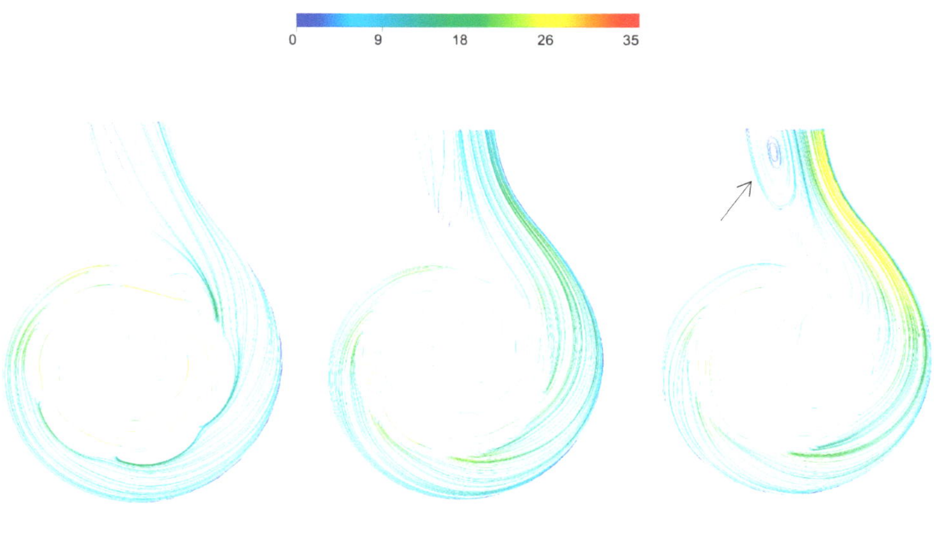

(**a**) 0.3Q$_d$ (**b**) 1.0Q$_d$ (**c**) 1.4Q$_d$

Figure 5. Distribution of streamlines in the pump on the xy plane (steady-state, OpenFOAM k-ε).

On the zx plane, there are some vortexes (see arrows in Figure 6) with different sizes in the inlet, impeller and volute under three operating conditions. Under the 0.3Q$_d$ operating condition (Figure 6a), the inlet velocity is insufficient to overcome adverse pressure gradient and the separation flow phenomenon occurs at the duct, generating a couple of strong swirling vortexes in the separation point. In Figure 6, some vortexes are clearly observed in the volute cross sections, which belong to the secondary flow vortex which usually occurs in a bent pipe. In contrast, the vortexes under 1.0Q$_d$ are less comparable to those under off-design conditions. These phenomena explain the reason that the hydraulic loss under off-design conditions is significant compared to that under the design condition.

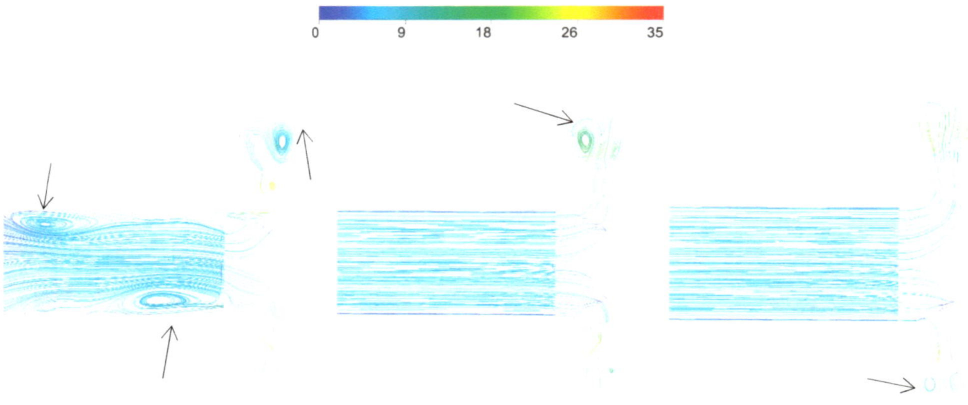

(**a**) $0.3Q_d$ (**b**) $1.0Q_d$ (**c**) $1.4Q_d$

Figure 6. Distribution of streamlines in the pump on the zx plane (steady-state, OpenFOAM k-ε).

Figure 7 shows the static pressure distribution per density of the fluid in the centrifugal pump by the steady-state calculations under the three operating conditions. It can be seen from the figure that the lowest static pressure zone is located around the leading edges of the impeller blades, while the highest pressure value appears at a certain position in the volute passage. From the point of view of fluid mechanics, hydraulic loss mainly includes friction loss and diffusion loss along the flow passage. Under the $0.3Q_d$ flow condition (Figure 7a), lower velocity in a larger size of volute leads to serious diffusion loss. Due to the increase of flow velocity under the high flow condition of $1.4Q_d$ (Figure 7c), the frictional loss increases and the static pressure also reduces at the outlet of the volute.

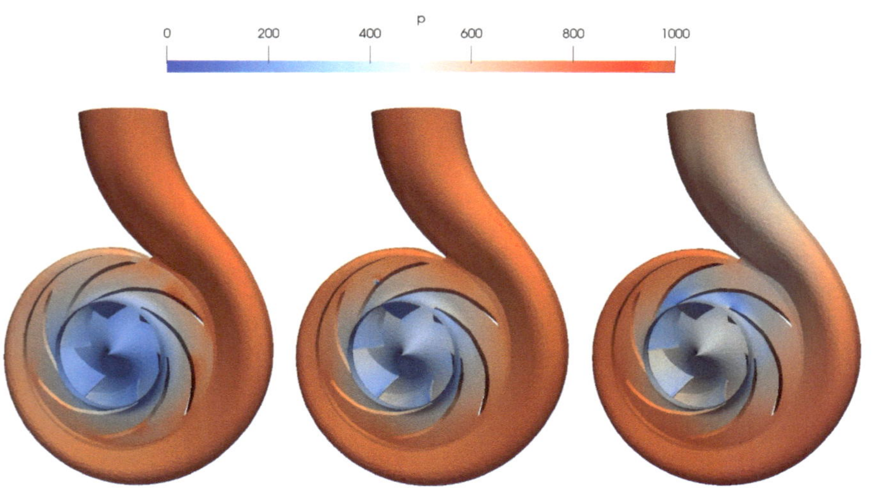

(**a**)$0.3Q_d$ (**b**)$1.0Q_d$ (**c**)$1.4Q_d$

Figure 7. Static pressure distribution in the pump (steady-state, OpenFOAM k-ε).

3.2. Pump Characteristics Prediction

Based on the numerical results of flow field in the pump, the external characteristic parameters such as head H, input power N and efficiency η can be calculated by Equations (3)~(5):

$$H = \frac{P_{out} - P_{in}}{\rho g} + h \quad (3)$$

$$N = M\omega \quad (4)$$

$$\eta = \frac{\rho g Q H}{N} \quad (5)$$

where, P_{out}, P_{in} is the average total pressure of the pump at outlet and inlet, respectively; h is the distance in gravity direction from the pump outlet to the inlet; M is the torque acting on the inner flow surface of the impeller; $\omega = 2\pi n/60$.

Figure 8 shows the steady-state and transient calculated results of the head H over time t. From the H transient curve, it can be seen that after about $2T$ time (T = 0.0207 s, impeller rotational period), the H transient value has a stable harmonic variation with t. The time-average value of transient H is approximately equal to the steady-state value of H (i.e., the pump is in a stable operating stage). Within the period T, the H transient value has six pulses corresponding to the number of impeller blades. The pulsation amplitude increases significantly with operating capacity.

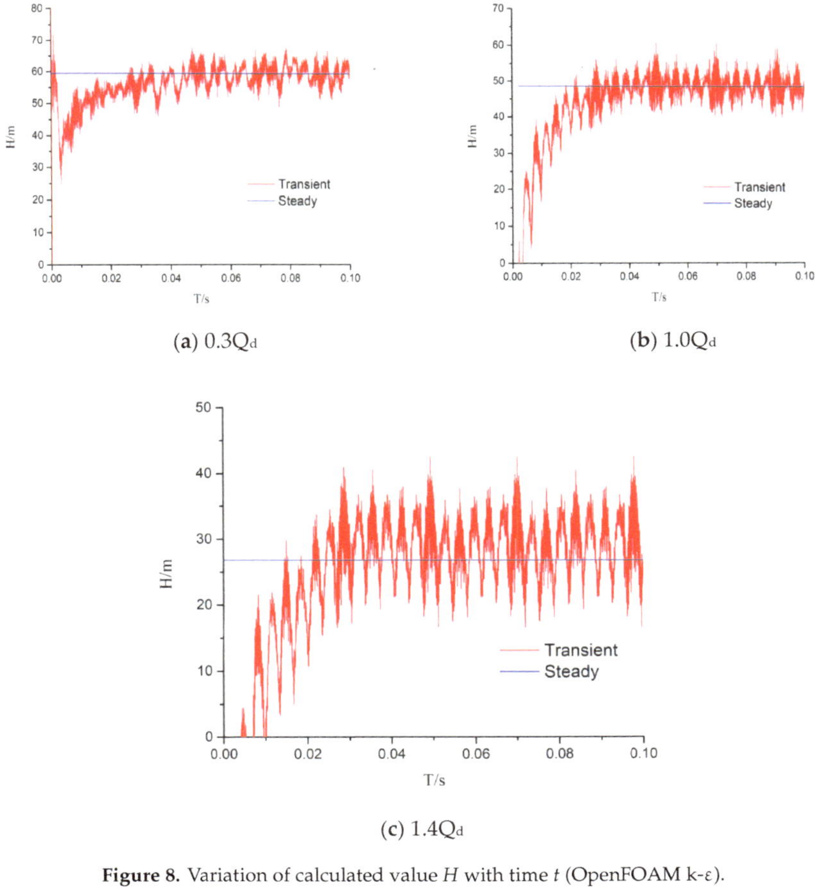

Figure 8. Variation of calculated value H with time t (OpenFOAM k-ε).

3.3. CFD Model Validation

In this study, flow field simulation was also performed with the standard k-ε turbulence model by Ansys Fluent in the same geometry model, computational grids and boundary conditions. In order to validate the computational method and results, the performance tests of the type IS125-100 centrifugal pump were performed at Guangyi Pump Co. Ltd. in China [23].

Figure 9 shows the comparisons between the predicted performance results and the test data. The predicted values were calculated by Equations (3)~(5) with steady-state OpenFOAM and Fluent computation, respectively. It is noted that the measured values of shaft power and efficiency are related to the volumetric and mechanical loss of the pump. These two kinds of losses could not be considered in CFD software, so the predicted results of power and efficiency were only consumed by flow and the hydraulic efficiency, respectively.

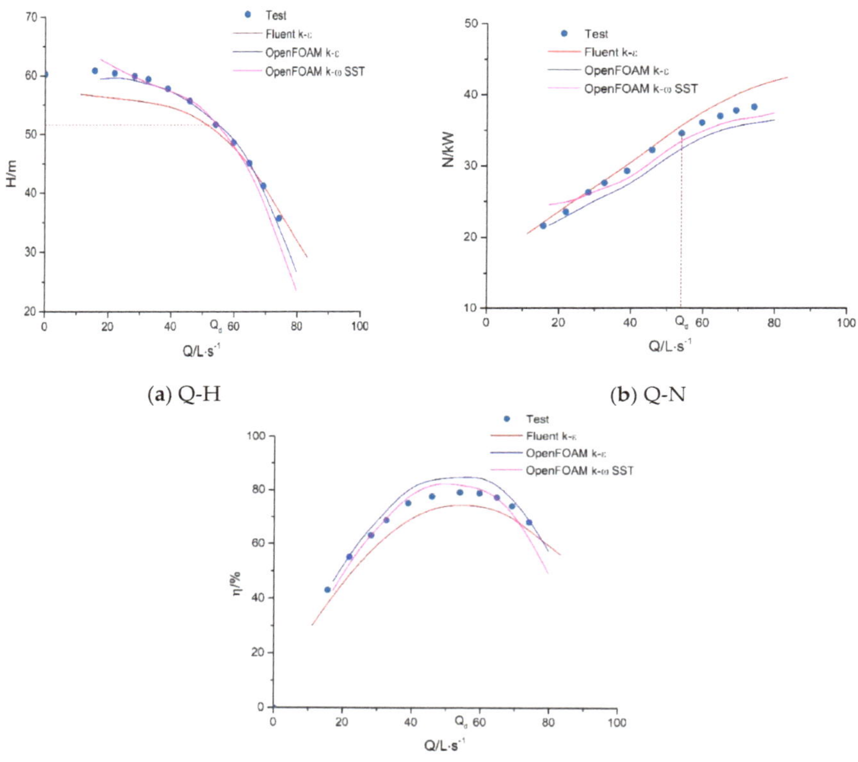

(a) Q-H (b) Q-N

(c) Q-η

Figure 9. Comparisons between predicted and tested performances.

Compared to the results by Fluent as seen in Figure 9a, the head curves by OpenFOAM are closer to the test values. The error analysis was conducted, and the results are shown in Table 2. The mean relative error of the head curve computed with k-ε model by Fluent is 4.9% relative to the test values, while the corresponding values with k-ε and k-ω SST models by OpenFOAM are 1.2% and 2.3%, respectively. In terms of head maximum relative error, the value with the k-ε model by Fluent reaches 7.5%, but the values with k-ε and k-ω SST by OpenFOAM are 4.7% and 6.5%. This indicates that OpenFOAM is able to obtain more acceptable results than Fluent under the flow conditions of $0.3Q_d$~$1.4Q_d$ in the current case. Meanwhile, in Figure 9b,c, the measured values of the shaft power and efficiency lie between the predicted results by Fluent and OpenFOAM. If volumetric and mechanical losses are considered, the predicted values by Fluent will deviate further from the measured results, and the predicted results by OpenFOAM will be close to the measured results.

Table 2. Error analysis of head.

	Fluent k-ε	OpenFOAM k-ε	OpenFOAM k-ω SST (shear-stress transport)-
Mean relative error	4.9%	1.2%	2.3%
Maximum relative error	7.5%	4.7%	6.5%

4. Conclusions

With the steady state and transient simulations of the full flow field of the centrifugal pump in a wide range of operating conditions by OpenFOAM, the following conclusions are summarized:

(1) When the pump runs stably, the transient value of the head changes with time in a stable harmonic way, and the time-average value of the head is approximately equal to the steady state value. In one impeller rotation period, the number of head pulses corresponds to that of the impeller blades. The pulsation amplitude increases significantly with capacity.
(2) The flow simulation by OpenFOAM could well capture the vortexes in the pump. There are more vortexes under off-design conditions than those under design operating conditions. The flow velocity departing from the outlet of the impeller is more uniform under design operating conditions than those under off-design conditions. Diffusion loss and friction loss are the main hydraulic losses under lower and higher flow conditions, respectively.
(3) The predicted results by OpenFOAM with the standard k-ε and the k-ω SST turbulence models agree well with the measured results of the pump characteristics in the current case.
(4) The present study assumed that the inlet pressure was high enough to ensure no cavitation. In fact, cavitation always occurs and seriously affects pump performance. Thus, the effect of cavitation will be considered by OpenFOAM in further pump performance simulations.

Author Contributions: S.H. contributed significantly to the conception of the study and drafted the paper. S.H. and Y.W. contributed to numerical simulation. C.G. and Y.W. performed the data analysis; W.K. helped perform the analysis with constructive discussions.

Funding: This research received no external funding.

Conflicts of Interest: The authors declare no conflict of interest.

References

1. Karassik, I.J.; Messina, J.P.; Cooper, P.; Heald, C.C. *Pump Handbook*; McGraw-Hill: New York, NY, USA, 2001; Volume 3.
2. Sakran, H.K. Numerical analysis of the effect of the numbers of blades on the centrifugal pump performance at constant parameters. *Technology* **2015**, *6*, 105–117.
3. Li, W.; Yang, C.; Yu, L.; Wang, G.; You, Y. Prediction on hydrodynamic performances of propeller by OpenFOAM. In Proceedings of the ASME 2009 28th International Conference on Ocean, Offshore and Arctic Engineering, Honolulu, HI, USA, 31 May–5 June 2009; pp. 633–638.
4. Petit, O.; Page, M.; Beaudoin, M.; Nilsson, H. The ERCOFTAC centrifugal pump OpenFOAM case-study. In Proceedings of the 3rd IAHR International Meeting of the Workgroup on Cavitation and Dynamic Problem in Hydraulic Machinery and Systems, Brno, Czech Republic, 14–16 October 2009.
5. Petit, O.; Nilsson, H. Numerical investigations of unsteady flow in a centrifugal pump with a vaned diffuser. *Int. J. Rotating Mach.* **2013**, *2013*. [CrossRef]
6. Nilsson, H. Evaluation of OpenFOAM for CFD of turbulent flow in water turbines. In Proceedings of the 23nd IAHR symposium on Hydraulic Machinery and Systems, Yokohama, Japan, 17–21 October 2006.

7. Xie, S. Studies of the ERCOFTAC Centrifugal Pump with OpenFOAM. Master's Thesis, Chalmers University of Technology, Gothenburg, Sweden, 2010.
8. Zhang, H.; Zhang, L. Numerical simulation of cavitating turbulent flow in a high head Francis turbine at part load operation with OpenFOAM. *Procedia Eng.* **2012**, *31*, 156–165. [CrossRef]
9. Nuernberg, M.; Tao, L. Three dimensional tidal turbine array simulations using OpenFOAM with dynamic mesh. *Ocean Eng.* **2018**, *147*, 629–646. [CrossRef]
10. Liu, H.L.; Yun, R.E.N.; Kai, W.A.N.G.; Wu, D.H.; Ru, W.M.; Tan, M.G. Research of inner flow in a double blades pump based on OpenFOAM. *J. Hydrodyn. Ser. B* **2012**, *24*, 226–234. [CrossRef]
11. Nallasamy, M. Turbulence models and their applications to the prediction of internal flows: A review. *Comput. Fluids* **1987**, *15*, 151–194. [CrossRef]
12. Shojaeefard, M.H.; Tahani, M.; Ehghaghi, M.B.; Fallahian, M.A.; Beglari, M. Numerical study of the effects of some geometric characteristics of a centrifugal pump impeller that pumps a viscous fluid. *Comput. Fluids* **2012**, *60*, 61–70. [CrossRef]
13. Patankar, S.V.; Spalding, D.B. A calculation procedure for heat, mass and momentum transfer in three-dimensional parabolic flows. *Int. J. Heat Mass Transf.* **1972**, *15*, 1787–1806. [CrossRef]
14. Demirdzic, I.; Gosman, A.D.; Issa, R.I.; Peric, M. A calculation procedure for turbulent flow in complex geometries. *Comput. Fluids* **1987**, *15*, 251–273. [CrossRef]
15. Patankar, S. *Numerical Heat Transfer and Fluid Flow*; CRC Press: Boca Raton, FL, USA, 1980.
16. Luo, J.Y.; Gosman, A.D. Prediction of impeller-induced flow in mixing vessels using multiple frames of reference. *Inst. Chem. Eng. Symp. Ser.* **1994**, *136*, 549–556.
17. Farrell, P.E.; Maddison, J.R. Conservative interpolation between volume meshes by local Galerkin projection. *Comput. Methods Appl. Mech. Eng.* **2011**, *200*, 89–100. [CrossRef]
18. Versteeg, H.K.; Malalasekera, W. *An Introduction to Computational Fluid Dynamics: The Finite Volume Method*; Pearson Education: London, UK, 2007.
19. Jasak, H. Error Analysis and Estimation for the Finite Volume Method with Applications to Fluid Flows. Bachelor's Thesis, Department of Mechanical Engineering Imperial College of Science, London, UK, 1996.
20. Auvinen, M.; Ala-Juusela, J.; Pedersen, N.; Siikonen, T. Time-accurate turbomachinery simulations with Open-Source® CFD: Flow analysis of a single-channel pump with OpenFOAM®. In Proceedings of the ECCOMAS CFD 2010, Lisbon, Portugal, 14–17 June 2010.
21. Arnone, A.; Pacciani, R. Rotor-stator interaction analysis using the Navier-Stokes equations and a multigrid method. *J. Turbomach.* **1996**, *118*, 679–689. [CrossRef]
22. Dawes, W.N. A simulation of the unsteady interaction of a centrifugal impeller with its vaned diffuser: Flow analysis. *J. Turbomach.* **1995**, *117*, 213–222. [CrossRef]
23. Huang, S.; Ruan, Z.; Liu, L. Performance predication of piping centrifugal pumps using computational fluid dynamics (CFD) technique. *Chem. Eng. Mach.* **2009**, *2*, 128–130.
24. Cheah, K.W.; Lee, T.S.; Winoto, S.H.; Zhao, Z.M. Numerical flow simulation in a centrifugal pump at design and off-design conditions. *Int. J. Rotating Mach.* **2007**, *2007*, 1–8. [CrossRef]
25. Limbach, P.; Kimoto, M.; Deimel, C.; Skoda, R. Numerical 3D simulation of the cavitating flow in a centrifugal pump with low specific speed and evaluation of the suction head. In Proceedings of the ASME Turbo Expo 2014: Turbine Technical Conference and Exposition, Dusseldorf, Germany, 16–20 June 2014.
26. Shim, H.S.; Kim, K.Y. Evaluation of Rotor-Stator interface models for the prediction of the hydraulic and suction performance of a centrifugal pump. *J. Fluids Eng.* **2019**, *141*, 111106. [CrossRef]

processes

MDPI

Article

Numerical Investigation of the Effect of Incorporated Guide Vane Length with SCC Piston for High-Viscosity Fuel Applications

Mohd Fadzli Hamid [1], Mohamad Yusof Idroas [1], Mazlan Mohamed [2], Shukriwani Sa'ad [3], Teoh Yew Heng [1], Sharzali Che Mat [4], Muhamad Azman Miskam [1], Zainal Alimuddin Zainal Alauddin [1] and Muhammad Khalil Abdullah [5,*]

1 School of Mechanical Engineering, Universiti Sains Malaysia, Engineering Campus, Seri Ampangan, Nibong Tebal 14300, Pulau Pinang, Malaysia; Mohd_Fadzli@outlook.com (M.F.H.); meyusof@usm.my (M.Y.I.); yewhengteoh@usm.my (T.Y.H.); azman@usm.my (M.A.M.); mezainal@usm.my (Z.A.Z.A.)

2 Advanced Material Research Cluster, Faculty of Bioengineering and Technology, Universiti Malaysia Kelantan, Jeli 17600, Kelantan, Malaysia; mazlan.m@umk.edu.my

3 Faculty of Computer, Media and Technology Management, TATI University College, Kijal 24000, Kemaman, Terengganu, Malaysia; shukriwani@tatiuc.edu.my

4 Faculty of Mechanical Engineering, Universiti Teknologi Mara, Pulau Pinang, Permatang Pauh 13500, Pulau Pinang, Malaysia; sharzali.chemat@uitm.edu.my

5 School of Material and Mineral Resources Engineering, Universiti Sains Malaysia, Engineering Campus, Seri Ampangan, Nibong Tebal 14300, Pulau Pinang, Malaysia

* Correspondence: mkhalil@usm.my

Received: 7 September 2020; Accepted: 1 October 2020; Published: 22 October 2020

Abstract: Compression ignition (CI) engines that run on high-viscosity fuels (HVF) like emulsified biofuels generally demonstrate poor engine performance. An engine with a consistently low performance, in the long run, will have a negative effect on its lifespan. Poor combustion in engines occurs mainly due to the production of less volatile, less flammable, denser, and heavier molecules of HVF during injection. This paper proposes a guide vane design (GVD) to be installed at the intake manifold, which is incorporated with a shallow depth re-entrance combustion chamber (SCC) piston. This minor modification will be advantageous in improving the evaporation, diffusion, and combustion processes in the engine to further enhance its performance. The CAD models of the GVD and SCC piston were designed using SolidWorks 2018 while the flow run analysis of the cold flow CI engine was conducted using ANSYS Fluent Version 15. In this study, five designs of the GVD with varying lengths of the vanes from 0.6D (L) to 3.0D (L) were numerically evaluated. The GVD design with 0.6D (L) demonstrated improved turbulence kinetic energy (TKE) as well as swirl (R_s), tumble (R_T), and cross tumble (R_{CT}) ratios in the fuel-injected zone compared to other designs. The suggested improvements in the design would enhance the in-cylinder airflow characteristics and would be able to break up the penetration length of injection to mix in the wider area of the piston-bowl.

Keywords: alternative fuels; engine modelling; piston; guide vane; biofuels

1. Introduction

Diesel engines, otherwise known as the compression ignition (CI) engines are widely used in industrial, commercial transportation, agricultural, watercraft propulsion power, and non-road construction applications, particularly in off-grid situations [1]. To date, petro-diesel fuels have dominated as the preferred fuel in CI engines [2]. However, the periodic shortage of petroleum, fluctuations in petroleum price in the global market, and strict regulations on vehicular emissions in several countries have encouraged automotive manufacturers, scientists, and researchers to look for an alternative source that could replace existing fossil fuels [3]. Modern lifestyle and rapid development in urban areas have surged the global demand for fossil fuels to approximately 53%, according to the report of the International Energy Agency [4]. For this reason, the challenges in the new era of the twenty-first century are to resolve the dependency on fossil fuels and to minimize the emission of exhaust pollutants for the sake of human civilization.

Biofuel is a potential fuel alternative since it can be easily extracted from various feedstocks such as plants, vegetable oil, edibles/non-edibles, and animal fats. The similarity in the properties of biofuel and petro-diesel make the former an attractive source to replace the conventional petro-diesel fuel [5]. The commercial production of biofuel has been reported as an economically feasible process, with an added advantage of improved emission characteristics [6]. Although the burning of biofuel could reduce the emissions of pollutants such as soot, particulate matter (PM), carbon monoxide (CO), and total hydrocarbon (THC); nevertheless, the emission of nitrogen oxide (NO_x) is still high due to the higher auto-ignition temperature of biofuel than that of petro-diesel [7]. The problem can be alleviated by enforcing biofuel emulsification, wherein the biofuel burning and thermal inefficiency in the combustion chamber are improved.

Numerous reports have shown that emulsified biofuel emits lower emissions of CO_2, NO_x, and CO compared to neat biofuels [8–10]. Prakash et al. investigated the engine performance and emissions from burning emulsified Jatropha methyl ester (JME), wherein about 35% of the reduction was observed in the HCs emissions compared to that of neat JME [11]. In another study, it was found that CO_2 emission was lower in higher load operated engines using emulsified Jatropa blends as fuel [12]. In general, the emission level of emulsified biofuels may vary based on the level of water concentration.

Generally, the direct use of biofuels without modification either in their properties or the engine compartment is not recommended as it may lead to inferior engine performance, increased build-up of carbon, injector choking, filter clogging, excessive fuel consumption, and reduced lifespan [13]. Similar outcomes can also be ascribed to the biofuel properties, such as higher viscosity, lower volatility, higher density, and presence of heavier molecules during injection. High viscosity fuels are less prone to evaporation, thus detrimental to the air-fuel mixture in the cylinder. Furthermore, on using neat biofuel without modifications, a build-up of carbon deposits in the piston-bowl and cylinder wall were found on disassembling the engine [14]. However, when the biofuel properties were modified closer to the physiochemical properties of petro-diesel, for instance, biodiesel, a higher thermal efficiency compared to that of petro-diesel, was observed. It is due to the presence of oxygen molecules and different auto-ignition constituents in the biodiesel itself [15].

In order to improve the performance and reduce the emission of CI engines to run on high viscosity fuel, particularly biofuels, researchers introduced various techniques to improve and resolve the above-mentioned problems. Some of the popular techniques to reduce the viscosity of biofuels are preheating of the biofuels at a specific temperature [16], blending the biofuel with petro-diesel [17], and double-blending of biofuel with different cetane numbers [18]. In the engine configuration, modifications of piston-bowl, intake manifold, and injection pressure and timing to suit the applications of biofuel were investigated [3,19]. Meanwhile, in the modification of the physicochemical properties of the biofuel, techniques such as emulsification, double-blending with different viscosity fluid, and transesterification were introduced [20]. All of these techniques were proven to improve the performance and emission of CI engines, yet their performance when using biofuel was still lower compared to that of petro-diesel [21]. So far, very few studies related to the modification of in-cylinder airflow characteristics to enhance the engine performance and reduce the emissions when running on high viscosity fuels like emulsified biofuel have been conducted.

Studies to improve the in-cylinder airflow characteristics inside the engine combustion chamber are beneficial as their capabilities to stimulate and break down the highly viscous fuel during the injection stage could be investigated [22]. Furthermore, the penetration length of spraying can be reduced, and the spreading angle along the piston-bowl can be enhanced. Several designs to improve the engine in-cylinder airflow were proposed by researchers, such as baffles or butterfly valves, intake runners, guide vane design (GVD), combustion chamber designs, and piston bowl geometry designs. Among the proposed designs, GVD signifies the simplest design with a promising improvement of in-cylinder airflow compared to other designs [23]. GVD also escapes the need for an extensive physical modification on CI engines. Therefore, this study suggests that the GVD will improve CI engines in terms of swirl, turbulence, kinetic energy, tumble, and cross tumble ratios. However, none of the researchers had explored the idea of combining GVD with piston geometry design to enhance the in-cylinder airflow characteristics and regulate fluid flow inside the combustion chamber.

In this paper, the identified research gap has been considered by investigating the in-cylinder airflow characteristics using the combination of guide vane design (GVD) and piston bowl design of the shallow depth combustion chamber (SCC) type to enhance and regulate the in-cylinder airflow characteristics of CI engines. Concerning GVD, the four main parameters of the study are vane height, vane angle, vane number, and vane length. However, this research is limited to optimizing the length of the vane, while the other parameters were kept constant at fixed values based on the findings from previous studies [24–26]. As for the piston bowl design, the SCC type was selected since it is capable of generating uniform fluid flow throughout the piston-bowl design, as suggested by several researchers [27–29].

2. Computer Simulation

SolidWorks 2018, a computer-aided drawing software, was used in this research to draw the GVD and SCC piston-bowl design. ANSYS FLUENT 15, a computational fluid dynamics (CFD) analysis software, was employed to simulate the CI engine. The simulation was conducted based on three stages: the SCC piston-bowl design, base model, and GVD with various lengths of the vane. The simulation of the CI engine is comprised of four main steps, namely, drawing the model, meshing the elements, setting up the boundary conditions, and computing the CI engine airflow characteristics. The details of the constructed computer simulation models are discussed in the following sections.

2.1. Engine Specification Model

The numerical investigation was conducted on the CI engine, which was readily available in the automotive laboratory of the School of Mechanical Engineering, Universiti Sains Malaysia. The engine was a Yanmar L70AE, which can constantly run at 2000 rpm. Table 1 shows the specifications of the Yanmar L70AE engine.

Table 1. Technical specifications of the Yanmar L70AE engine.

Engine Parameter	Detail
Model	Yanmar L70AE
Bore	78 mm
Stroke	62 mm
Compression ratio	19.1
Number of cylinders	1
Weight	36 kg
Type of injection	Direct injection
Fuel injection pressure	19.6 Mpa
Displacement	0.296 L
High idle speed	3600 rpm
Injection timing	14 + 1 BTDC
Intake	Naturally aspirated
Direct of rotation	Counterclockwise

2.2. Design Geometry of Piston-Bowl

The selection of the piston-bowl geometry design is crucial because a very close relationship exists between the air–fuel mixture and emission formation in a diesel engine. The piston–bowl design employed in this study was the shallow depth re-entrance combustion chamber (SCC). The SCC piston-bowl design was chosen as it was capable of producing homogenous and well-organized in-cylinder airflow characteristics applicable for high viscosity fuels such as emulsified biofuel [27]. Figure 1 illustrates the SCC piston–bowl design.

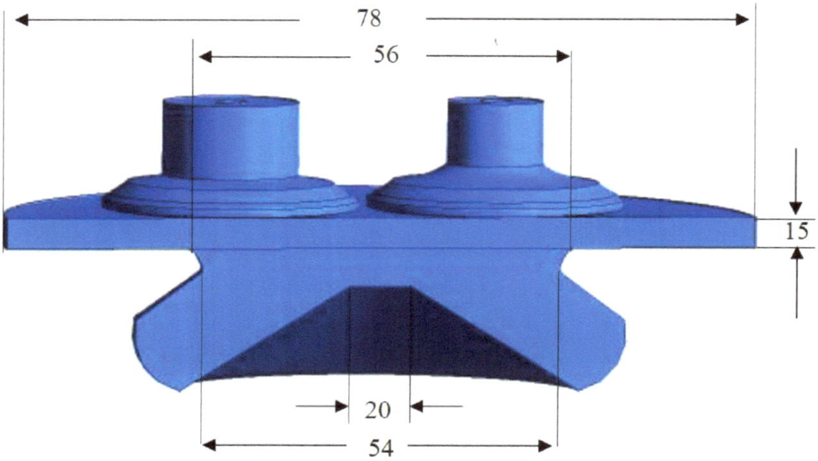

Figure 1. The SCC piston–bowl design (all unit in mm).

2.3. Guide Vane Design (GVD)

The primary purpose of a guide vane in the intake runner is to guide the airflow before it enters into the cylinder. It is a static vane used according to its shape to deflect the airflow [25]. It is usually placed within the intake system between the air filter and the inlet valve. In the existing engine, using alternative fuel of higher fuel viscosity will result in inferior engine performance due to the higher viscosity fuel. The pattern of spraying is also different when using diesel and alternative fuels like emulsified biofuel. The injection pattern when using diesel produced a wide cone angle and short-length penetration. However, using emulsified biofuel shows an inverse pattern of injection, which produces a low cone angle and long-length of penetration, promoting problems like carbon deposition in the piston bowl. In order to mitigate this limitation, the in-cylinder airflow characteristics need to be improved. The engine performance is enhanced by using alternative fuels or fuels with a higher viscosity, and it is necessary to improve the in-cylinder air flow swirl and tumble in the CI engine, i.e., at the start of injection (SOI), start of combustion (SOC) or in the ignition delay region. Attention is paid to the ignition delay because of the in-cylinder air flow strength, which is capable of breaking down the higher viscosity of the alternative fuel or emulsified biofuel molecules allowing them to spread outward and facilitating better mixing of air and fuel. Guide vane design considers the arrangement of vanes in the intake runner; the vanes serve to guide the airflow before the airflow is introduced into the combustion chamber during the intake stroke [26]. Since guide vanes have fins to guide the airflow, the developed swirl is predicted to be better than that produced by a typical engine configuration, as it will combine with the existing swirl created by the piston-bowl and swirl chamber.

The computer-aided design software, SolidWorks 2018, was utilized for drawing the guide vane model and for describing the study parameters. Limited literature is available concerning the effects of parameters such as the number of vanes, vane height and length, and their corresponding angles. The effect of guide vane length was investigated in this research, and the specifications of the GVD are given in Table 2. Based on studies found in the literature, the length of the guide vane is a key factor for determining the momentum and structure of the inlet airflow during a combustion process. However, an over-designed GVD length could result in hindering the airflow, which negatively affects the volumetric efficiency. In this sense, the current research varied this key parameter, i.e., GVD length, based on the diameter of the intake runner (*R*). The lengths were varied from 0.6 to 3.0 times the diameter (D) in steps of 0.6 and were labelled as 0.6D (L), 1.2D (L), 1.8D (L), 2.4D (L), and 3.0D (L), respectively. Altogether, a total of 5 GVD models with varying vane lengths were evaluated in this study. L represents the vane length, with variations, while the other parameters such as the angle, width, and number were correspondingly fixed at 35°, 0.2 times R, and four vanes were retained perpendicular to one another. Figure 2 shows the GVD design employed in this study.

Table 2. Specification parameters of the GVD.

GVD Parameter	Parameter Detail
Number of vanes	4
Width of vane (W)	0.5 mm
Height of vane (H)	0.60R
Length of vane (L)	0.6D (L), 1.2D (L), 1.8D (L), 2.4D (L) and 3.0D (L)
Vane twist angle (α)	35°

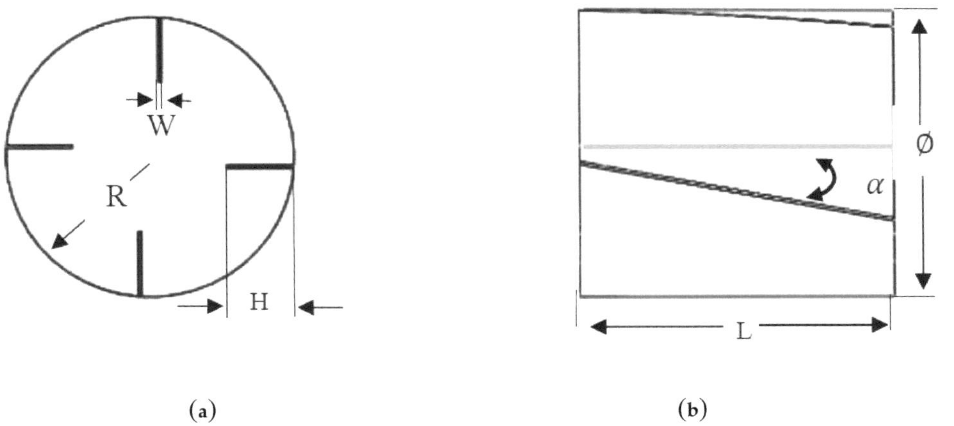

Figure 2. The GVD design: (**a**) front view and (**b**) side view.

3. Numerical Setting

SolidWorks 2018 software was used to design the geometry of the YANMAR L70AE CI generator diesel engine, based on its specifications. The intake runner, intake valve, exhaust runner, and GVD were drawn separately as individual parts and were assembled as a single model, as shown in Figure 3. The complete assembled YANMAR L70AE CI generator diesel engine model was then exported to the ANSYS-FLUENT v15 CFD software to numerically investigate and solve the complex CFD model in a cost- and time-effective manner. ANSYS-FLUENT v15 software equipped with adaptive mesh refinery (AMR) was even capable of refining the mesh during piston motion from the bottom dead center (BDC) to the top dead center (TDC), as shown in Figure 4. In order to compute and analyze the parameters representing fluid flow, the overall engine body compartments were set as a fluid domain, except for the valves, which were set as a solid domain. However, the main challenge in the 3D simulation of the piston-bowl, valves, and cylinder of the CI engine was the correct handling of the moving boundaries, set for dynamic simulation. The strategies of mesh remapping and grid moving are the conventional techniques used by various researchers [30,31]. The mesh generation technique employed in this research was based on the level assembly meshing technique. The CFD simulation setting was based on the cold flow CI engine specifications published by the company ANSYS Inc. [32].

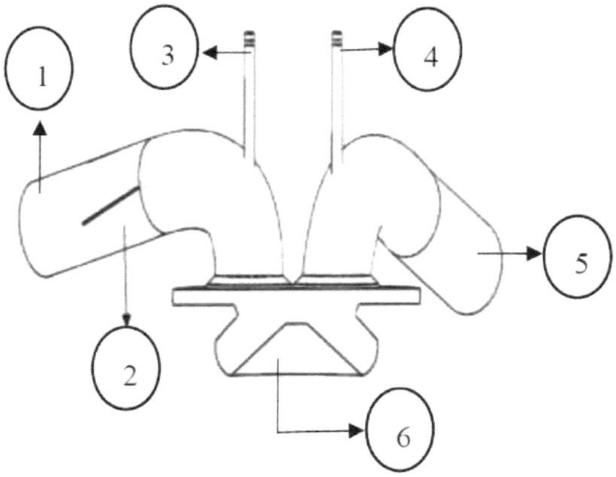

No.	Name
1	Intake runner
2	GVD
3	Intake valve
4	Exhaust valve
5	Exhaust runner
6	SCC piston

Figure 3. YANMAR L70AE CI generator diesel engine SolidWorks model.

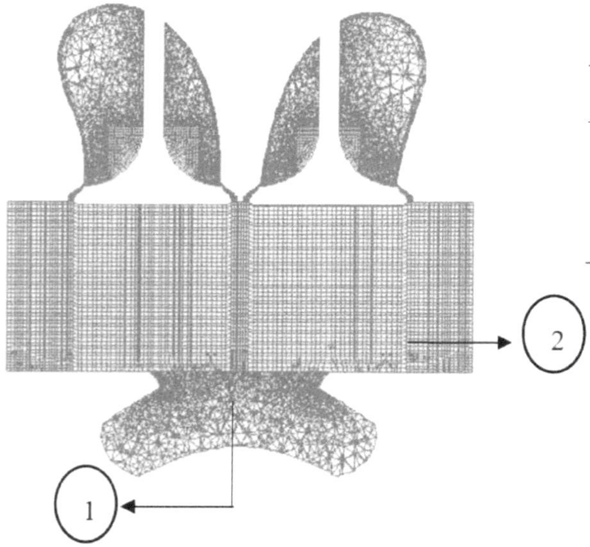

No.	Name
1	Tetrahedral Mesh
2	Hexahedral Mesh

Figure 4. The separation mesh in a different area.

The numerical simulation presented in this work focused on the intake and compression strokes, using the algorithms of mesh simplification and dynamic mesh moving boundaries [32]. An arbitrary Lagrangian-Eulerian (ALE) approach was taken to treat the moving mesh for the typical events that occurred in the CI engine, encompassing the valve and piston motions. Every event represents different mesh types and boundary geometries for every single different step of crank angle. Therefore, to perform a proper numerical simulation for the CI engine, several analyses and calculations should be considered, including the unsteady flow (transient) condition, high compressible Reynolds number (Re), moving mesh, momentum, heat, and mass transfer computations.

The conservation of mass, momentum, energy (energy equations) and k-ϵ turbulence models are governed by the Navier-Stokes equations [33], given below:

$$\frac{\partial \rho}{\partial t} + \nabla.(\rho U) = 0 \quad (1)$$

where ρ is the density of the fluid and U is the three-dimensional speed in the direction of the x, y and z axes. The momentum equation is based on Newton's second law of motion, in which the forces considered are the body and the surface forces on the volume of control, written as follows:

$$\frac{\partial(\rho U)}{\partial t} + \nabla.(\rho U \otimes U) = -\nabla p + \nabla.\tau + S_M \quad (2)$$

where p is the pressure of the fluid, τ is the strain rate and S_M is a source of momentum. All of these equations are referred to as Navier-Stoke equations [33]. The energy equation is defined as the rate of energy change within the fluid element, given in the conservation form by

$$\frac{\partial(\rho h_{tot})}{\partial t} - \frac{\partial \rho}{\partial t} + \nabla.(\rho U h_{tot}) = \nabla.(\lambda \nabla T) + \nabla.(U.\tau) + U.S_M \quad (3)$$

where h_{tot} is total enthalpy and λ is thermal conductivity, respectively.

Shear stress transport (SST) is a two-equation eddy–viscosity model that is employed in this analysis. The SST model is a combination of k-ω and k-ϵ turbulence models which can overcome the limitation of the shear stress until a 5% turbulence intensity is reached in the adverse gradient region. The pressure and temperature conditions considered in this study are 101.3 kPa and 300 K, respectively. Detailed information on the settings and boundary conditions of the models can be referred to in the ANSYS FLUENT v15-Solver Theory Guide [32].

On the theoretical basis of physical boundary conditions of the engine, two distinct phases of simulation studies were carried out: intake port analysis and intake analysis. Intake port analysis is associated with the clearance volume before airflow is inducted into the cylinder along the y-direction, while the intake analysis is related to the intake runner. The components that are not applicable in this work were suppressed as there was no contribution to the results. This is to minimize the computation time and increase the accuracy of analysis during the calculations.

4. Results and Discussion

The use of high viscosity fuels like emulsified biofuel aimed to improve the engine's performance, reduce the emissions, and increase the engine lifespan. Therefore, this research focuses on improving the characteristics of in-cylinder airflow to stimulate diffusion, evaporation, air-fuel mixing, and combustion processes in the CI engine, within the injection delay period. Based on the specifications of the Yanmar L70 engine as shown in Table 1, the injection timing was set 14° before approaching the top dead center (BTDC). Thus, this section also presents the in-cylinder airflow characteristics concentrating on the start of injection (SOI) between the crank angle (CA) of 346° and the start of combustion (SOC) between 352° CA. The in-cylinder airflow characteristics, such as swirl (R_s), turbulence kinetic energy (TKE), tumble (R_T), and cross tumble ratio (R_{CT}) are the determining components responsible for stimulating diffusion, evaporation and mixing, thereby improving the combustion efficiency. Hence, the simulation results were compared for various vane lengths with the base model of the engine to examine the effects on the in-cylinder airflow characteristics and eventually determine the optimum vane length.

4.1. Model Validation

This section discusses the experimental setup adopted in the study for validating the results of numerical simulation. The experiment was carried out in a test rig of the CI generator diesel engine YANMAR L70AE model, as shown in Figure 5. The existing data were used for numerical validation, considering some important technical aspects. During the experiment, all the engine boundary conditions were fixed, and a Kistler 7061B sensor with a highly sensitive water-cooled precision was screwed directly into the standard M14 hole. An optical sensor at the TDC position with a magnetic pickup shaft encoder was used to measure the in-cylinder pressure and record the data of crank angle positions. For better accuracy, the sensor was mounted near the valve. K-type thermocouples were installed at the intake and exhaust boundaries as close as possible to the cylinder head to measure the temperature. In this experiment, the SCC piston was fabricated and installed first-hand in the YANMAR L70AE CI generator diesel engine before running the engine test rig to collect experimental data.

No.	Name
1	GVD
2	Kistler 7061B
3	Transaxle shaft
4	Water Cooling
5	Dynamometer

Figure 5. The YANMAR L70AE CI generator diesel engine test rig.

In order to characterize the large scale in-cylinder air-fuel mixing during the intake and compression strokes, three main components that influence the in-cylinder air flow motion, namely Rs, R_T and R_{CT} were calculated based on the crank angle engine stroke. Figure 6 shows the swirl, tumble and cross tumble ratio orientation diagrams acting on the cylinder. The dimensionless parameter used to quantify the rotational and angular motion inside the cylinder commonly characterized the swirl, tumble and cross tumble ratio. These values were calculated by dividing the engine speed by the effective angular velocity of the in-cylinder air motion. The three variables investigated in this paper (swirl, tumble and cross tumble ratio) are presented in non-dimensionless equations [32], as follows:

Swirl ratio is defined as:

$$R_s = \frac{60H_z}{2\pi I_z\omega} \tag{4}$$

Tumble ratio is defined as:

$$R_T = \frac{60H_x}{2\pi I_x\omega} \tag{5}$$

Cross tumble is defined as:

$$R_{CT} = \frac{60H_y}{2\pi I_y\omega} \tag{6}$$

where H_x, H_y, and H_z represent the angular momentum of the cylinder gas about the x-axis, y-axis and z-axis, respectively. The symbol Ω represents the crankshaft rotation or engine speed in the unit of rotation/minutes.

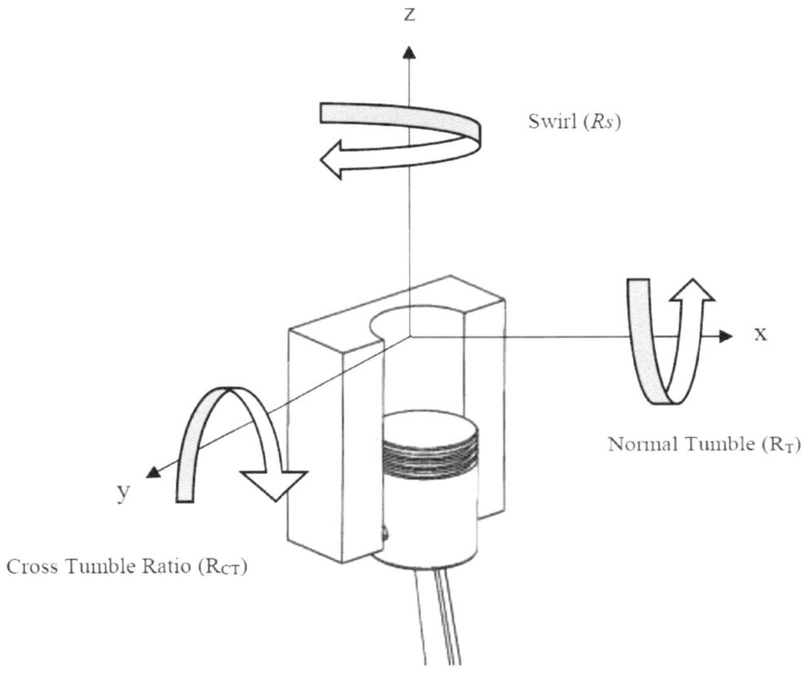

Figure 6. Swirl, tumble and cross tumble ratio orientation diagram.

Figure 7 illustrates the in-cylinder pressure (Pa) against the crank angle (θ) for both the experimental and numerical results of the engine operation without combustion at a rotational speed of 2000 rpm for the intake temperature and pressure of 302 K and 102 kPa, respectively. The numerical data collected with the GVD consist of intake, compression, and expansion stroke between 0° to 540°. Based on the results, a good agreement between numerical and experimental results with a slight difference of about 7% at the peak pressure was observed. This might be due to the friction loss that occurred inside the cylinder while running the experiment, a factor which is not possible to include in simulation through numerical analysis.

4.2. Grid Independence Test (GIT)

A grid independence test (GIT) is a term used to describe the process of improving the results by successfully using smaller cell sizes for the calculations. A calculation should approach the correct answer as the mesh gets finer. The components involved in the numerical calculations are the valves, intake port, cylinder head and piston-bowl. GIT is essentially performed as it will affect the time and cost of the computational analysis. Typically, the cell size ranges from coarse to ultrafine, and the number of elements was set between 100 and 400 k. For the sake of grid sensitivity and reasonable computation time, half of the cells were used to generate meshes in the cylinder head and piston-bowl. The hexahedral mesh was applied in the mesh generation step at the cylinder motion from the bottom dead center (BDC) to top dead center (TDC), owing to their stability and better accuracy compared to other types of mesh. Table 3 shows the comparison of elements average cylinder of mesh and the pressure generated into the cylinder. From Table 3, Case 3 exhibits the optimum pressure generated;

when the elements of the mesh were increased, the pressure indicated similar values to Case 3. Therefore, Case 3 was selected for further analysis.

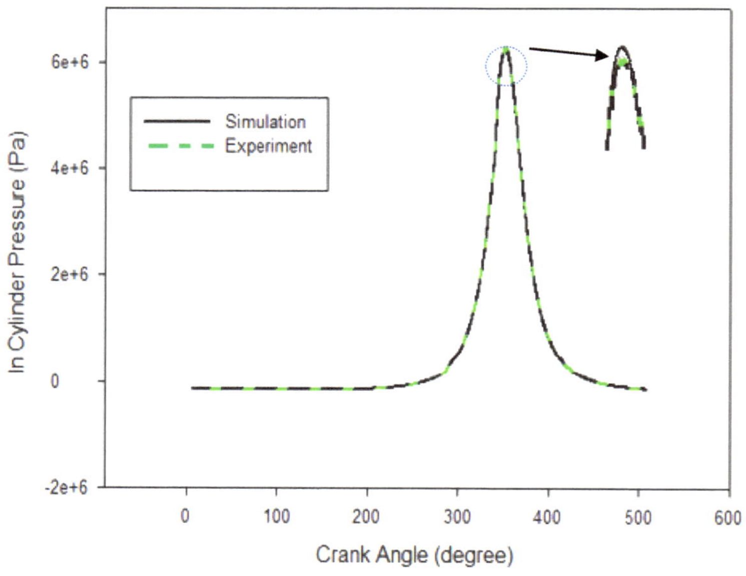

Figure 7. The plot of the in-cylinder pressure (Pa) against the crank angle (degree).

Table 3. Grid independence test.

Case	1	2	3	4	5
Elements average cylinder Pressure (Pa)	101,191 2.013×10^6	249,543 3.109×10^6	301,142 3.160×10^6	369,178 3.160×10^6	446,769 3.160×10^6

4.3. Turbulence Kinetic Energy (TKE)

Theoretically, TKE is the mean kinetic energy per unit mass associated with eddies in a turbulent flow [3,34]. In other terms, it is the fluctuation of velocity in the in-cylinder airflow and is calculated based on the root-mean-square. During the movement of the piston in the compression stroke, the piston moves from the bottom dead centre (BDC) towards the top dead centre (TDC) and reduces the volume of the combustion chamber, hence limiting the fluctuation of the cylinder airflow motion. When the piston approaches the TDC, the velocity of air starts to decrease and becomes zero after reaching the TDC. Therefore, TKE is expected to decrease during this time. Several researchers studied and confirmed that the trend of turbulence pattern diminished linearly while the piston was heading towards the TDC [35,36]. Furthermore, Payri et al. investigated and found that in the early stage of the intake stroke, the piston geometry had little effect on the in-cylinder airflow [35]. However, the piston-bowl geometry played a significant role during the movement of the piston approaching the TDC in the compression stroke and the early stage of the expansion stroke. Figure 8 demonstrates that the base engine without GVD showed the lowest magnitude of TKE compared to the other modifications of GVD. The TKE of the vane modified engine compared with the baseline engine showed a 30% difference with 1.8D (L), 35% with 2.4D (L), 40% with 1.2D (L), and 45% with 0.6D (L).

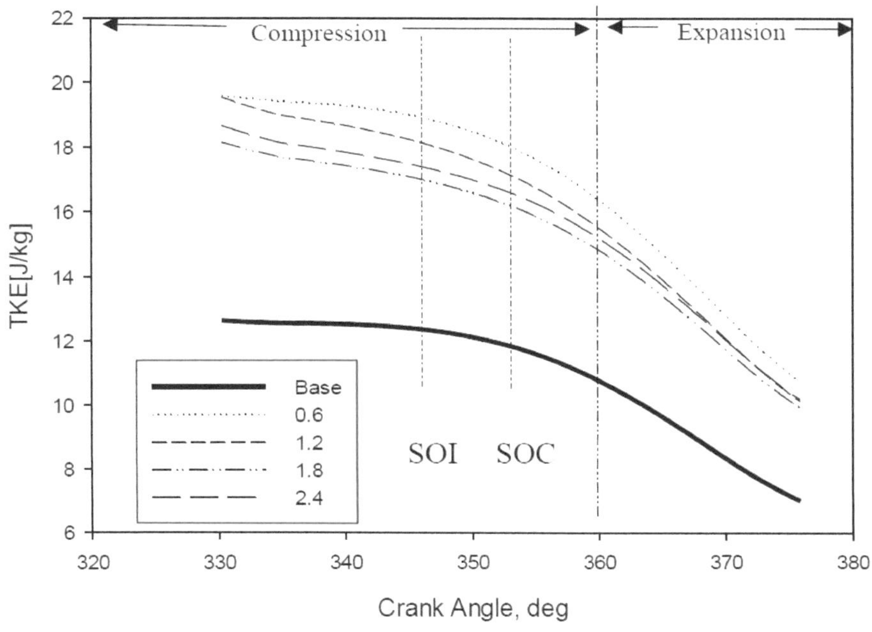

Figure 8. The plot of TKE against the crank angle.

The GVD model with 0.6D (L) displayed the highest TKE produced, followed by the other designs. The graph also implies that further increase in the length of the GVD does not considerably increase the TKE. However, it could be an indication of the minimal length of GVD required for rapid guidance of airflow, less air resistance, and swift initial turbulence in the intake manifold. A minimal length of vanes would be able to guide the airflow while the SCC piston produced a higher magnitude of TKE. The swirl, which already existed in the SCC piston-bowl and the extra swirl created via the 0.6D (L) GVD, could match the wind and create a larger magnitude of total swirl. Based on the result, the GVD successfully improved and increased the average magnitude of TKE.

4.4. In-Cylinder Swirl (R_s), Tumble (R_T), and Cross Tumble Ratio (R_{CT})

The in-cylinder airflow motion in the combustion chamber is an important factor that will determine the scale of air and fuel mixing during the intake and compression stroke, particularly in the velocity streamline. The three main components of air structure that will influence the combustion efficiency are R_S, R_T, and R_{CT}. Figure 9 shows the in-cylinder air flow swirling strength (R_s) of the local swirling motion. R_s is defined by the organized movement of fresh air around the cylinder axis. Swirl is created by an initial angular momentum of the intake airflow into the cylinder. In a diesel engine, the swirl ratio is important for the preparation of the air-fuel mixture. The swirl effect can be induced by a special intake port(s) shapes created by the rotational flow in the intake valve area transferred to the cylinder during the intake process [37]. The simulation results for various vane lengths are illustrated in Figure 9, which shows the magnitude of R_s. Based on Figure 9, it is found that all the variations of the GVD model witness an improvement in the R_s compared to the base model. In combustion theory, the generation of swirl inside the combustion chamber during the injection period assists in breaking down the molecular chain and practically improves the mixing process of air and fuel, thereby ultimately improving the combustion and engine efficiency [38]. The GVD design of 0.6D (L) represents the optimum guide vane length to enhance R_s produced during the start of injection (SOI) and the start of combustion (SOC).

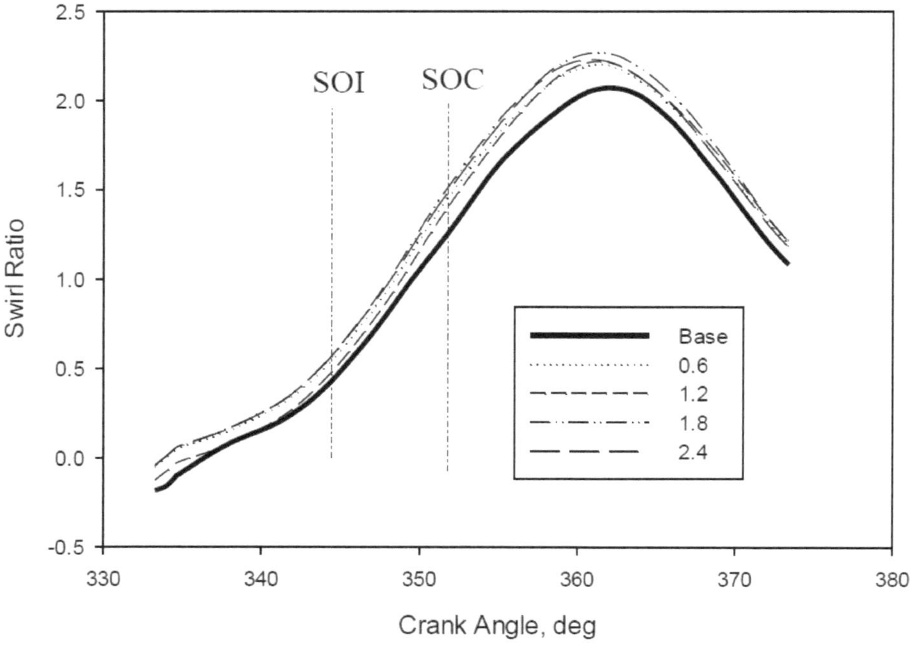

Figure 9. The plot of the swirl strength against the crank angle.

Figure 10 illustrates the behavior of R_T for different lengths of GVD against the crank angle before approaching the TDC. R_T is defined as the ratio between the equivalent angular velocity of the instant solid body and the angular velocity of the engine. The angular velocity (ω) is the angular velocity of a rotating solid angular momentum body equal to that of the instantaneous in-cylinder mass, evaluated perpendicular to the cylinder axis and centered between the head and the instantaneous piston position [39]. The negative or positive sign of R_T is neglected as it is arbitrary and dependent on the magnitude obtained. From the graph, it can be seen that the design of the GVD type 0.6D (L) produced a value closer to that of the base model. However, the GVD model of 0.6D (L) showed a slightly higher R_T value magnitude, about 7% different from that of the baseline engine in the area of the start of injection (SOI). The other GVD designs showed a lower magnitude of R_T values compared to the baseline engine. This implies that GVD is capable of generating a strong lateral airflow within the cylinder. Strong lateral airflow generated in the cylinder is imperative and will aid in enhancing the speed and spread of the atomization of the fuel during the injection. As a result, an improvement in the flame propagation and combustion efficiency is observed while eliminating the carbon deposition that will reduce the lifespan of the engine.

Figure 11 shows the R_{CT} for the different lengths of GVD against the crank angle before approaching the TDC. R_{CT} is defined as the rotational ratio of airflow on the cross tumble axes [40]. In combustion theory, the enhancement of R_{CT} facilitates the increment of turbulent flow and assists in homogenous mixing of the fuel and air along the piston-bowl. Nevertheless, R_{CT} has a close correlation with R_s and R_T. Bari et al. found that in order to improve R_{CT} in a CI engine, R_s and R_T values need to be improved [41,42]. Prasad et al. reported that enhancing the R_{CT} air motion is a crucial issue, as it has an impact on increasing the air-fuel mixing process, achieving faster burning rates, increasing indicated efficiency, and improving the air management [43]. As previously mentioned, the use of higher viscosity fuels such as emulsified biofuel will deteriorate the atomization process, subsequently decreasing the cone angle and increasing the length of penetration. Hence, with the correct length of GVD, R_{CT} would be able to significantly break down the length of penetration, which is expected to widen the cone angle during the injection period. Based on the figure, the 0.6D (L) GVD shows the highest magnitude of values compared to the base engine and other models of GVD. It is noticeable that the 0.6D (L) GVD can produce a homogeneous mixture during compression and consequently improves the lateral flow of air within the cylinder.

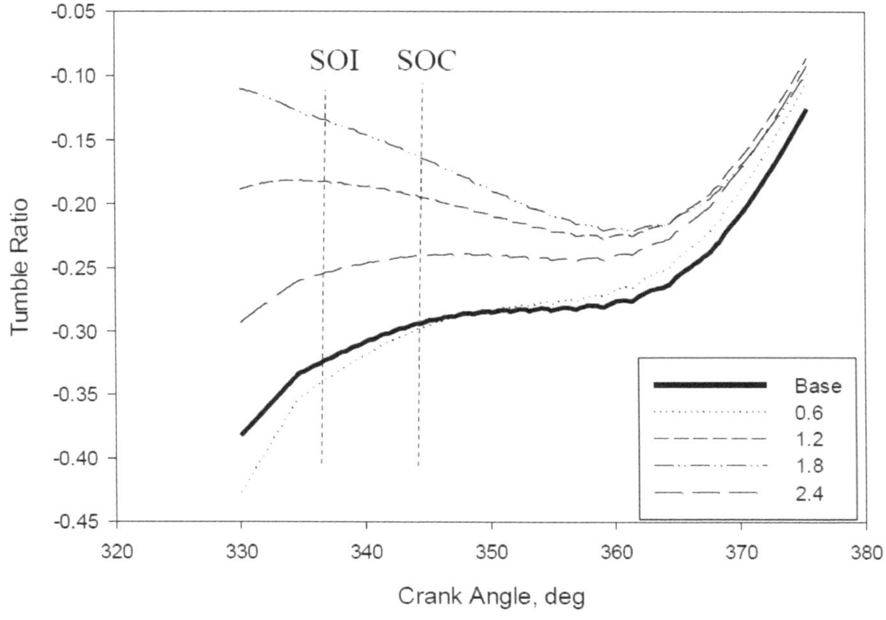

Figure 10. The plot of the tumble ratio against the crank angle.

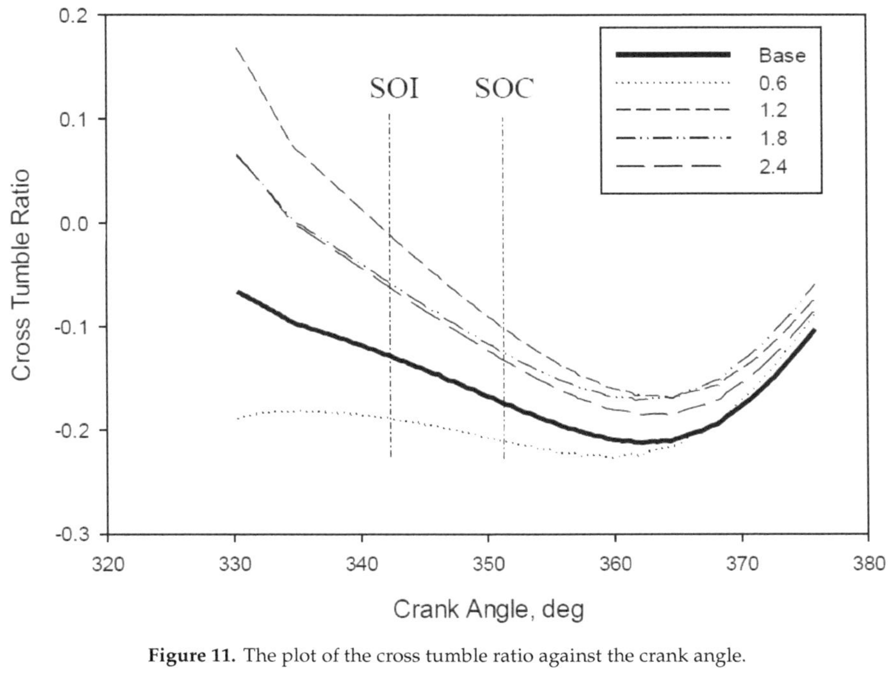

Figure 11. The plot of the cross tumble ratio against the crank angle.

4.5. Characteristics of Cylinder Airflow During Intake and Compression Stroke

Figure 12 illustrates the intake stroke of the streamline of 10° and 100° crank angle (CA). During the intake stroke at 10° CA, the airflow was initially induced through the SCC piston. As a result of the lack of guidance in the intake stroke, tumbled flow occurred when the intake valve was opened. However, when considering the 0.6D (L) GVD, the in-cylinder air flow phenomena behaved differently. When the intake valve was opened, the airflow initially produced swirl. From the graph, it can also be seen that in-cylinder airflow velocity improved when using the GVD compared to that of the standard engine. This outcome is crucial to enhance the in-cylinder airflow by promoting the atomization of heavy molecules, like emulsified biofuels. The engine that does not use vanes showed lower air velocity in the combustion chamber during the intake stroke. These results were supported by studies of Heywood [44], where it was recommended that turbulence intensity in the combustion chamber could be increased by improving the vortex flow during the intake stroke.

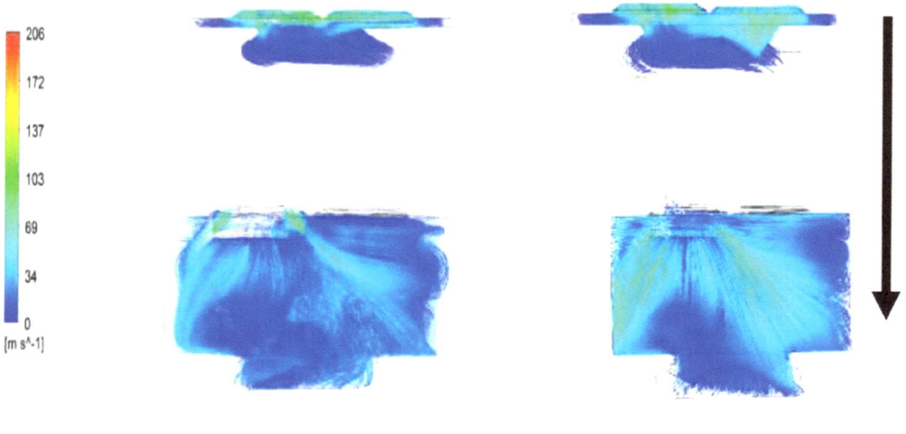

Figure 12. The in-cylinder process of intake stroke with CA at 10° (top) and 100° (bottom).

Figure 13 illustrates the compression stroke of the streamline of 350° and 310° CA. In the compression stroke, the piston starts to compress while moving from the BDC to the TDC, and the volume in the combustion chamber begins to decrease due to the motion of the piston. When both the intake valve and the exhaust valves are closed, air density, temperature and pressure witness an increase. Figure 13 shows that the velocity of air is higher during the compression stroke of 310° and begins to decrease when the piston approaches the TDC. The combination of the SCC piston and the GVD successfully improves the uniformity of airflow on both the sides and also increases the velocity of air throughout the SCC piston compared to the engine standards. It implies that the engine which uses vanes in its operation produces higher velocity airflow, high turbulent flow and is able to transport heavy fuel molecules.

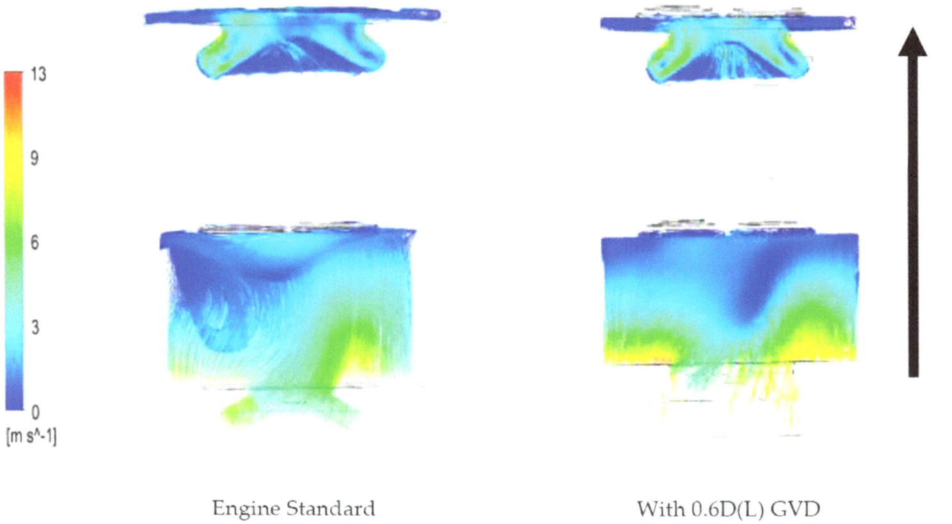

Figure 13. The in-cylinder process of compression stroke CA at 350° (top) and 310° (bottom).

5. Conclusions

Alternative fuels, particularly emulsified biofuel, possess high potential to replace conventional diesel fuels in the near future. The renewable nature and the similarities in the chemical properties of the emulsified fuel show that the existing YANMAR L70AE CI generator diesel engine can use these alternative fuels with minor modifications. However, high viscosity fuels are generally characterized by low volatility, comprised of heavier molecules of fuel which may result in a longer length of penetration and shorter cone angles during injection in the combustion chamber. These patterns of injection result in incomplete combustion and carbon deposition in the cylinder wall and piston head. The combination of the GVD and SCC piston-bowl design has successfully enhanced the in-cylinder airflow characteristics as well as those of R_s, R_T, R_{CT}, and TKE in order to assist in evaporating higher viscosity fuel to improve the combustion and reduce the carbon deposition. ANSYS FLUENT V15 was used to run the simulation, and it was found that most of the results showed that the 0.6D (L) GVD design incorporated with SCC piston was capable of improving the in-cylinder airflow characteristics with better-organized airflow during SOI to SOC period (injection delay) compared to the baseline engine (without GVD). This combination of improved characteristics of the in-cylinder airflow is expected to effectively break the penetration length of highly viscous fuels and increase the volatility of the alternative fuels, resulting in better combustion and reduced carbon deposits. Therefore, it can be concluded that installing the 0.6D (L) GVD incorporated with the SCC piston will enhance the in-cylinder airflow characteristics and subsequently improve the performance of engine running on high viscosity fuels. In the future, the study will be extended to investigate the effect on the injection profile, structure of spraying, and the combustion characteristics based on the combination of the GVD incorporated with SCC piston in diesel engines.

Author Contributions: Conceptualization, M.F.H. and M.Y.I.; Methodology, M.F.H., M.K.A., M.Y.I. and S.S.; Experimentation, S.C.M., M.F.H. and T.Y.H.; Data analysis, M.F.H., M.M. and M.A.M.; Resources, M.Y.I., M.F.H. and M.A.M.; Writing—original draft preparation, M.F.H. and M.K.A.; Writing, review and editing: Z.A.Z.A. All authors have read and agreed to the published version of the manuscript.

Funding: The authors would like to thank Universiti Sains Malaysia (USM) for providing financial support through the Research University Grant Scheme under Grant No. 1001/PBAHAN/8014006 and USM Fellowship RU (1001/CIPS/AUPE001).

Conflicts of Interest: The authors declare no conflict of interest.

Nomenclature

English Symbol	Description	Unit
U	Three-dimensional flow velocities in the x, y, and z directions	m/s
P	Pressure	Pa
t	Time	s
h	Total enthalpy	J/kg
S_M	Momentum source	N/m^3
R	Radius	mm
l	Length	mm
x,y,z	Cartesian coordinates	mm
Greek Symbol	**Description**	**Unit**
ρ	Fluid density	kg/m^3
ω	Angular acceleration	rad/s
λ	Thermal conductivity	W/m.K
θ	Crank angle degree	–
τ	Strain rate	1/s
α	Vane twist angle	–
∇	Gradient operator	–
Abbreviation	**Description**	
GVD	Guide vane design	
HVF	Higher viscosity fuel	
SCC	Shallow depth re-entrance combustion chamber	
S_M	Momentum source	
R_T	Tumble ratio	
SOI	Start of Injection	
SOC	Start of Combustion	
R_S	Swirl ratio	
CI	Compression Ignition	
JME	Jatropha methyl ester	
HC	Hydrocarbons	
CA	Crank Angle	
R_{CT}	Cross tumble ratio	
TKE	Turbulence kinetic energy	

References

1. Mahmudul, H.M.; Hagos, F.Y.; Mamat, R.; Adam, A.A.; Ishak, W.F.W.; Alenezi, R. Production, characterization and performance of biodiesel as an alternative fuel in diesel engines—A review. *Renew. Sustain. Energy Rev.* **2017**, *72*, 497–509. [CrossRef]
2. Sivaganesan, S.; Chandrasekaran, M. Impact of various compression ratio on the compression ignition engine with diesel and mahua biodiesel. *Int. J. ChemTech Res.* **2016**, *9*, 63–70.
3. Hamid, M.F.; Idroas, M.Y.; Basha, M.H.; Sa'ad, S.; Che Mat, S.; Abdullah, M.K.; Zainal Alauddin, Z.A. Numerical study on dissimilar guide vane design with SCC piston for air and emulsified biofuel mixing improvement. *MATEC Web Conf.* **2016**, *90*, 01065. [CrossRef]
4. International Energy Agency. *World Energy Outlook Special Report: Energy and Climate Change*; International Energy Agency: Paris, France, 15 June 2015.
5. Popp, J.; Lakner, Z.; Harangi-Rákos, M.; Fári, M. The effect of bioenergy expansion: Food, energy, and environment. *Renew. Sustain. Energy Rev.* **2014**, *32*, 559–578. [CrossRef]
6. Azad, A.K.; Rasul, M.G.; Khan, M.M.K.; Sharma, S.C.; Hazrat, M.A. Prospect of biofuels as an alternative transport fuel in Australia. *Renew. Sustain. Energy Rev.* **2015**, *43*, 331–351. [CrossRef]
7. Nalgundwar, A.; Paul, B.; Sharma, S.K. Comparison of performance and emissions characteristics of di CI engine fueled with dual biodiesel blends of palm and jatropha. *Fuel* **2016**, *173*, 172–179. [CrossRef]

8. Reham, S.S.; Masjuki, H.H.; Kalam, M.A.; Shancita, I.; Rizwanul Fattah, I.M.; Ruhul, A.M. Study on stability, fuel properties, engine combustion, performance and emission characteristics of biofuel emulsion. *Renew. Sustain. Energy Rev.* **2015**, *52*, 1566–1579. [CrossRef]
9. Annamalai, M.; Dhinesh, B.; Nanthagopal, K.; SivaramaKrishnan, P.; Isaac JoshuaRamesh Lalvani, J.; Parthasarathy, M.; Annamalai, K. An assessment on performance, combustion and emission behavior of a diesel engine powered by ceria nanoparticle blended emulsified biofuel. *Energy Convers. Manag.* **2016**, *123*, 372–380. [CrossRef]
10. Ayhan, V.; Tunca, S. Experimental investigation on using emulsified fuels with different biofuel additives in a DI diesel engine for performance and emissions. *Appl. Therm. Eng.* **2018**, *129*, 841–854. [CrossRef]
11. Prakash, R.; Singh, R.K.; Murugan, S. Experimental studies on combustion, performance and emission characteristics of diesel engine using different biodiesel bio oil emulsions. *J. Energy Inst.* **2015**, *88*, 64–75. [CrossRef]
12. Raheman, H.; Kumari, S. Combustion characteristics and emissions of a compression ignition engine using emulsified jatropha biodiesel blend. *Biosyst. Eng.* **2014**, *123*, 29–39. [CrossRef]
13. Dhinesh, B.; Annamalai, M. A study on performance, combustion and emission behaviour of diesel engine powered by novel nano nerium oleander biofuel. *J. Clean. Prod.* **2018**, *196*, 74–83. [CrossRef]
14. Che Mat, S.; Idroas, M.Y.; Hamid, M.F.; Zainal, Z.A. Performance and emissions of straight vegetable oils and its blends as a fuel in diesel engine: A review. *Renew. Sustain. Energy Rev.* **2018**, *82*, 808–823. [CrossRef]
15. Ramesh, A.; Ashok, B.; Nanthagopal, K.; Ramesh Pathy, M.; Tambare, A.; Mali, P.; Phuke, P.; Patil, S.; Subbarao, R. Influence of hexanol as additive with Calophyllum Inophyllum biodiesel for CI engine applications. *Fuel* **2019**, *249*, 472–485. [CrossRef]
16. Ashok, B.; Nantha Gopal, K.; Rajagopal, T.K.R.; Alagiasingam, S.; Appu, S.; Murugan, A. Design and analysis of a fuel preheating device for evaluation of ethanol based biofuel blends in a diesel engine application. *SAE Int. J. Engines* **2017**, *10*, 39–45. [CrossRef]
17. Mofijur, M.; Rasul, M.G.; Hyde, J.; Azad, A.K.; Mamat, R.; Bhuiya, M.M.K. Role of biofuel and their binary (diesel-biodiesel) and ternary (ethanol-biodiesel-diesel) blends on internal combustion engines emission reduction. *Renew. Sustain. Energy Rev.* **2016**, *53*, 265–278. [CrossRef]
18. Vallinayagam, R.; Vedharaj, S.; Yang, W.M.; Lee, P.S.; Chua, K.J.E.; Chou, S.K. Pine oil-biodiesel blends: A double biofuel strategy to completely eliminate the use of diesel in a diesel engine. *Appl. Energy* **2014**, *130*, 466–473. [CrossRef]
19. Mat, S.C.; Idroas, M.Y.; Teoh, Y.H.; Hamid, M.F. Physicochemical, performance, combustion and emission characteristics of melaleuca cajuputi oil-refined palm oil hybrid biofuel blend. *Energies* **2018**, *11*, 3146. [CrossRef]
20. Koc, A.B.; Abdullah, M. Performance and NOx emissions of a diesel engine fueled with biodiesel-diesel-water nanoemulsions. *Fuel Process. Technol.* **2013**, *109*, 70–77. [CrossRef]
21. Tan, Y.H.; Abdullah, M.O.; Nolasco-Hipolito, C. The potential of waste cooking oil-based biodiesel using heterogeneous catalyst derived from various calcined eggshells coupled with an emulsification technique: A review on the emission reduction and engine performance. *Renew. Sustain. Energy Rev.* **2015**, *47*, 589–603. [CrossRef]
22. Agarwal, A.K. Biofuels (alcohols and biodiesel) applications as fuels for internal combustion engines. *Prog. Energy Combust. Sci.* **2007**, *33*, 233–271. [CrossRef]
23. Bari, S.; Saad, I. Performance and emissions of a compression ignition (CI) engine run with biodiesel using guide vanes at varied vane angles. *Fuel* **2015**, *143*, 217–228. [CrossRef]
24. Bari, S.; Saad, I. Optimization of vane numbers through simulation and experiment, and investigation of the effect on the performance and emissions of a CI (compression ignition) engine run with biodiesel. *Energy* **2015**, *79*, 248–263. [CrossRef]
25. Bari, S.; Saad, I. Effect of guide vane height on the performance and emissions of a compression ignition (CI) engine run with biodiesel through simulation and experiment. *Appl. Energy* **2014**, *136*, 431–444. [CrossRef]
26. Saad, I.; Bari, S. *Improving Air-Fuel Mixing in Diesel Engine Fuelled by Higher Viscous Fuel Using Guide Vane Swirl and Tumble Device (GVSTD)*; SAE Technical Papers; SAE: Warrendale, PA, USA, 2013.
27. Hamid, M.F.; Idroas, M.Y.; Sa'ad, S.; Saiful Bahri, A.J.; Sharzali, C.M.; Abdullah, M.K.; Zainal, Z.A. Numerical investigation of in-cylinder air flow characteristic improvement for Emulsified biofuel (EB) application. *Renew. Energy* **2018**, *127*, 84–93. [CrossRef]

28. Lalvani, J.I.J.R.; Prakash, E.; Parthasarathy, M.; Jayaraj, S.; Annamalai, K. Structural Analysis on Swirling Grooved SCC Piston. In *Advanced Materials Research*; Trans Tech Publications Ltd.: Stafa-Zurich, Switzerland, 2014.
29. Yadav, P.; Saravanan, C.G.; Edward, J.G.; Perumal, R. *Experimental and Numerical Investigation of Flow and Combustion in a di Diesel Engine with Different Piston Geometries*; SAE Technical Papers; SAE: Warrendale, PA, USA, 2015.
30. Gaburro, E.; Dumbser, M.; Castro, M.J. Direct Arbitrary-Lagrangian-Eulerian finite volume schemes on moving nonconforming unstructured meshes. *Comput. Fluids* **2017**, *159*, 254–275. [CrossRef]
31. Bianchi, G.; Rane, S.; Kovacevic, A.; Cipollone, R. Deforming grid generation for numerical simulations of fluid dynamics in sliding vane rotary machines. *Adv. Eng. Softw.* **2017**, *112*, 180–191. [CrossRef]
32. ANSYS Inc. *ANSYS Fluent, Release 19.1, Help System, Theory Guide*; ANSYS Inc.: Washington, PA, USA, 2018.
33. Gao, R.; Liu, K.; Li, A.; Fang, Z.; Yang, Z.; Cong, B. Study of the shape optimization of a tee guide vane in a ventilation and air-conditioning duct. *Build. Environ.* **2018**, *132*, 345–356. [CrossRef]
34. Hamid, M.F.; Abdullah, M.K.; Idroas, M.Y.; Alauddin, Z.A.Z.; Sharzali, C.M.; Khimi, S.R.; Sa'Ad, S. Effect of vane numbers on the in-cylinder air flow characteristic in compression ignition (CI) engine run with emulsified biofuel. *Mater. Today Proc.* **2019**, *17*, 989–994. [CrossRef]
35. Payri, F.; Benajes, J.; Margot, X.; Gil, A. CFD modeling of the in-cylinder flow in direct-injection Diesel engines. *Comput. Fluids* **2004**, *33*, 995–1021. [CrossRef]
36. Payri, R.; Salvador, F.J.; Gimeno, J.; Zapata, L.D. Diesel nozzle geometry influence on spray liquid-phase fuel penetration in evaporative conditions. *Fuel* **2008**, *87*, 1165–1176. [CrossRef]
37. Broatch, A.; Olmeda, P.; García, A.; Salvador-Iborra, J.; Warey, A. Impact of swirl on in-cylinder heat transfer in a light-duty diesel engine. *Energy* **2017**, *119*, 1010–1023. [CrossRef]
38. Benajes, J.; Olmeda, P.; Martín, J.; Blanco-Cavero, D.; Warey, A. Evaluation of swirl effect on the Global Energy Balance of a HSDI Diesel engine. *Energy* **2017**, *122*, 168–181. [CrossRef]
39. Wang, T.; Liu, D.; Tan, B.; Wang, G.; Peng, Z. An investigation into in-cylinder tumble flow characteristics with variable valve lift in a gasoline engine. *Flow Turbul. Combust.* **2015**, *94*, 285–304. [CrossRef]
40. Buhl, S.; Gleiss, F.; Köhler, M.; Hartmann, F.; Messig, D.; Brücker, C.; Hasse, C. A Combined Numerical and Experimental Study of the 3D Tumble Structure and Piston Boundary Layer Development During the Intake Stroke of a Gasoline Engine. *Flow Turbul. Combust.* **2017**, *98*, 579–600. [CrossRef]
41. Bari, S.; Saad, I. CFD modelling of the effect of guide vane swirl and tumble device to generate better in-cylinder air flow in a CI engine fuelled by biodiesel. *Comput. Fluids* **2013**, *84*, 262–269. [CrossRef]
42. Hamid, M.F.; Idroas, M.Y.; Sa'ad, S.; Yew Heng, T.; Che Mat, S.; Zainal Alauddin, Z.A.; Shamsuddin, K.A.; Shuib, R.K.; Abdullah, M.K. Numerical Investigation of Fluid Flow and In-Cylinder Air Flow Characteristics for Higher Viscosity Fuel Applications. *Processes* **2020**, *8*, 439. [CrossRef]
43. Prasad, B.V.V.S.U.; Sharma, C.S.; Anand, T.N.C.; Ravikrishna, R.V. High swirl-inducing piston bowls in small diesel engines for emission reduction. *Appl. Energy* **2011**, *88*, 2355–2367. [CrossRef]
44. Heywood, J.B. *Internal Combustion Engine Fundamentals*; McGraw Hill International: New York, NY, USA, 1988.

Publisher's Note: MDPI stays neutral with regard to jurisdictional claims in published maps and institutional affiliations.

processes

Article

Effects of Bending of Fluidic Oscillators on Aerodynamic Performance of an Airfoil with a Flap

Nam-Hun Kim and Kwang-Yong Kim *

Department of Mechanical Engineering, Inha University, Incheon 22212, Korea; knh308@naver.com or ahnsy9028@inha.edu
* Correspondence: kykim@inha.ac.kr; Tel.: +82-32-860-7317

Abstract: The present work investigated the effects of bending the outlet nozzles of fluidic oscillators installed on the NACA0015 airfoil with a flap on the flow control performance and, thus, the aerodynamic performance of the airfoil. The effects of bending on fluidic oscillators have not been reported so far in previous works. The aerodynamic analysis was performed numerically using unsteady Reynolds-averaged Navier-Stokes equations. Three different cases were considered: Case 1 changes only the bending angle with a fixed pitch angle, Case 2 changes only the pitch angle without bending, and Case 3 changes both the bending and pitch angles. Although the bending of the oscillators was introduced inevitably due to a geometrical limitation in the installation, the results indicated that the bending rather improved the lift coefficient and lift-to-drag ratio of the airfoil by improving the characteristics of the fluidic oscillators, such as the jetting angle and peak velocity ratio.

Keywords: fluidic oscillator; airfoil; flow separation control; aerodynamic performance; bending angle; unsteady Reynolds-averaged Navier-Stokes equations

Citation: Kim, N.-H.; Kim, K.-Y. Effects of Bending of Fluidic Oscillators on Aerodynamic Performance of an Airfoil with a Flap. *Processes* **2021**, *9*, 1429. https://doi.org/10.3390/pr9081429

Academic Editor: Krzysztof Rogowski

Received: 1 July 2021
Accepted: 16 August 2021
Published: 18 August 2021

1. Introduction

Flow separation that occurs on an aircraft wing negatively affects aerodynamic performance. Therefore, many studies on the control of flow separation have been conducted [1–5]. Flow control technologies are classified into passive flow control (PFC) [1], which attaches various structures to the surface, and active flow control (AFC) [2], which uses actuators such as plasma actuators [3], backward traveling waves [4], and periodic surface morphing [5]. AFC has an advantage of removing the drag-increase problem, which is a disadvantage of PFC, and enables effective separation control [6]. In particular, among the AFC technologies, the method using fluidic oscillators showed a superior control effect than other methods [7].

Fluidic oscillator has a simple shape and generates a vibrating jet from a steady inlet flow. Figure 1 shows the working principle of a fluidic oscillator with two feedback channels. When a constant pressure is applied to the inlet, the main stream flows along one wall of the mixing chamber by the Coanda effect. Most of the main stream proceeds to the outlet, but a part of the stream is introduced into a feedback channel to change the direction of the main stream at the inlet throat. This phenomenon is repeated to generate a vibrating jet at the outlet.

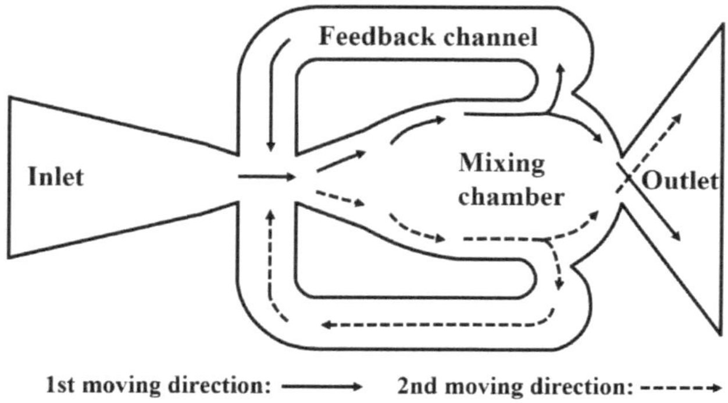

Figure 1. The mechanism of operation of a fluidic oscillator.

Fluidic oscillators are being investigated in recognition of their excellent performance of flow excitation in various fields such as acoustic control [8], film cooling [9], heat transfer [10], wind turbine [11], flow separation control [12,13], etc., [14–17]. Melton [14] evaluated the separation control performance under various conditions by installing fluidic oscillators on the NACA0015 airfoil equipped with a simple-hinged flap. In addition, Melton et al. [15] analyzed comparatively the flow control performance of steady jet actuators and fluidic oscillators on the airfoil and reported that the fluidic oscillators showed a similar lift increase as the steady jet actuators with a smaller flow rate. Seele et al. [16] improved the efficiency of an aircraft rudder by installing fluidic oscillators on the trailing edge of a vertical stabilizer of a commercial aircraft. Cerretelli et al. [11] applied fluidic oscillators to wind turbine blades and confirmed 10–60% lift increase. Meng et al. [17] showed that the total pressure loss coefficient decreased by 6.8% by installing fluidic oscillators on the blade suction side of a compressor cascade.

For more effective separation control using fluidic oscillators, various studies have been conducted on the installation conditions of oscillators, such as arrangement, mounting angles, mounting position, etc. Kim and Kim [18,19] conducted a study by adjusting the arrangement and mounting angles of fluidic oscillators on a hump surface and an airfoil. As a result, it was reported that the pitch angle of the oscillators had the most sensitive effect on separation control. The effect of the mounting position of the oscillators on the separation control was also investigated according to the angle of attack of the airfoil. The lift was effectively increased when the oscillators were installed downstream of the separation point, and larger drag reduction was obtained closer to the leading edge at all angles of attack. Kim and Kim [20] conducted a study by adjusting the pitch angle of fluidic oscillators to control the flow separation occurring on the flap of an airfoil using a numerical analysis, referring to the experimental model of Melton [14]. It was reported that the pitch angle of the fluidic oscillators had a greater effect on lift increase as the oscillator exits became more parallel to the flap surface.

The effects of fluidic oscillator geometry on the flow control performance have also been investigated [21–24]. Melton et al. [21] evaluated the performance of three types of fluidic oscillators with the same orifice size and reported that the fluidic oscillators with the largest jet deflection angle were the most effective for separation control. Ostermann et al. [22] investigated the pressure required to inject the same mass flow in the cases where the edge of the fluidic oscillator is straight and curved. They reported that the curved oscillator was superior in terms of energy requirement, requiring 20% less supply pressure. Jung and Kim [23] conducted a multi-objective optimization of a fluidic oscillator using unsteady Reynolds-averaged Navier-Stokes (URANS) analysis and investigated the effects of inlet nozzle width and distance between splitters on the fluidic oscillator performance.

Kim and Kim [20] applied fluidic oscillators to the flow control on an airfoil with a flap, but they adjusted the pitch angle by bending the outlets of fluidic oscillators inevitably due to the geometrical limitations of the installation for some pitch angles closer to the flap angle, where the best aerodynamic performance of the airfoil was obtained. However, the effects of the bent outlets on the performance improvement were not resolved in their work. In a following work, Kim and Kim [25] investigated how the bending angle of a fluidic oscillator outlet affects the characteristics of the fluidic oscillator with and without external flow. But, in their work, the effects of the bent outlet was evaluated only on the performance parameters of the oscillator, such as peak velocity ratio at the exit and pressure drop (not on the aerodynamic performance of an airfoil).

Therefore, the contributions of the bending angle and the pitch angle to the enhancement of aerodynamic performance confirmed in the previous study [20], have not yet been identified separately. The main purpose of the present study is to find the effects of the bending angle of fluidic oscillator outlets on the aerodynamic performance of the airfoil isolated from the effects of the pitch angle of the jet injection in the flow control of an airfoil using fluidic oscillators. This is needed to find how the bending of fluidic oscillators affects the flow control performance of the oscillators and what the best installation condition is, when they are installed on an airfoil. In this study, the fluidic oscillators were installed on a NACA0015 airfoil equipped with a simple-hinged flap, and the effects of the bending angle of the oscillator outlet and injection pitch angle on the aerodynamic performance of the airfoil were identified separately using URANS analysis.

2. Fluidic Oscillator Model, Installation Conditions and Computational Domain

The fluidic oscillator model used in this study (Figure 1) is the model with two feedback channels proposed by Melton et al. [21], which was also used by Kim and Kim [20]. An array of fluidic oscillators are installed on a NACA0015 airfoil equipped with a simple-hinged flap of 30% chord length.

Only three oscillators are included in the computational domain using periodic conditions at the sided boundaries, as shown in Figure 2. The total computational domain consists of the external flow domain of the airfoil (Figure 2a) and the internal domain of the fluidic oscillator (Figure 2b). The chord length of the airfoil was set to 305 mm, and the span length was set to 99 mm including three fluidic oscillators. The angle of attack (α) was fixed at 8°, and the flap angle (δ_f) was fixed at 40°. The detailed dimensions of the fluidic oscillator (Figure 2b) are shown in Table 1.

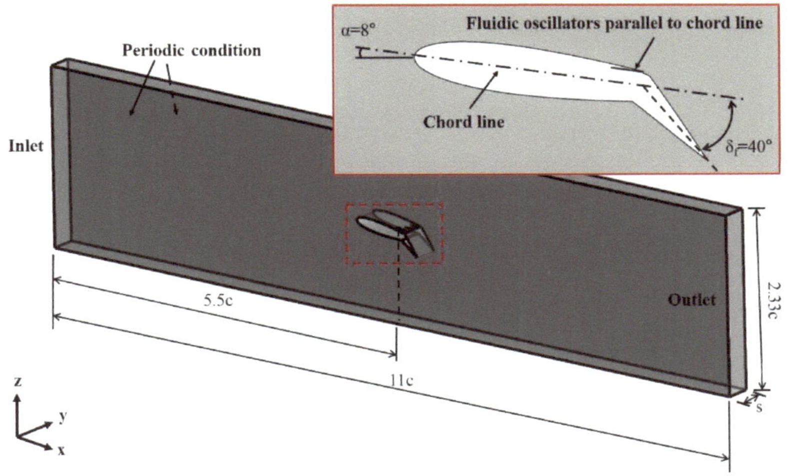

(a)

Figure 2. *Cont.*

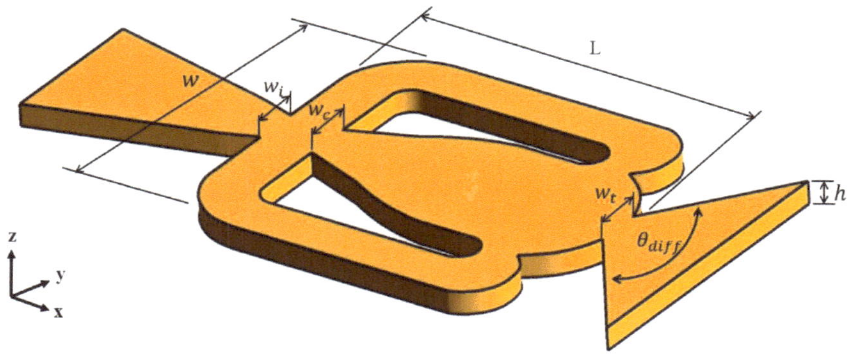

(b)

Figure 2. Computational domain: (**a**) external domain and (**b**) internal domain of the oscillator.

Table 1. Geometrical parameters of the tested fluidic oscillator [20].

Parameter	Value
Fluidic oscillator width, w (mm)	14.3
Inlet nozzle width, w_i (mm)	2.0
Inlet chamber width, w_c (mm)	2.03
Outlet throttle width, w_t (mm)	2.0
Fluidic oscillator height, h (mm)	1.0
Inlet nozzle feedback channel width, h_{ft} (mm)	2.41
Diffuser angle of outlet nozzle, θ_{diff} (°)	107
Distance between the inlet of mixing chamber and the throat, L (mm)	15.6

Figures 3 and 4 show the installation conditions of the fluidic oscillators. The fluidic oscillator is placed at B-B′ plane as shown in Figure 3, which is located 7 mm downstream of 70% chord position from the leading edge of the airfoil. The B-B′ plane is always normal to the centerline of the oscillator outlet and, the zero pitch angle ($\beta = 0$) is defined as the condition where the oscillator outlet is mounted parallel to the chord line of the airfoil. Figure 4 shows the installation conditions of the fluidic oscillators. In this figure, the symbol γ indicates the bending angle of the outlet parts of the oscillators, and β indicates the injection pitch angle of the oscillators.

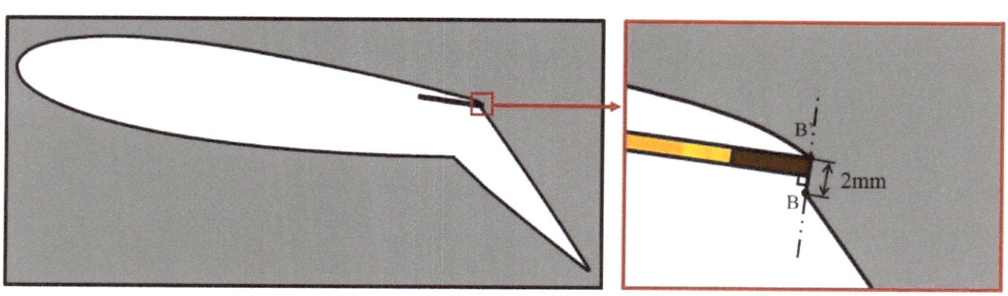

Figure 3. Installation location of fluidic oscillators.

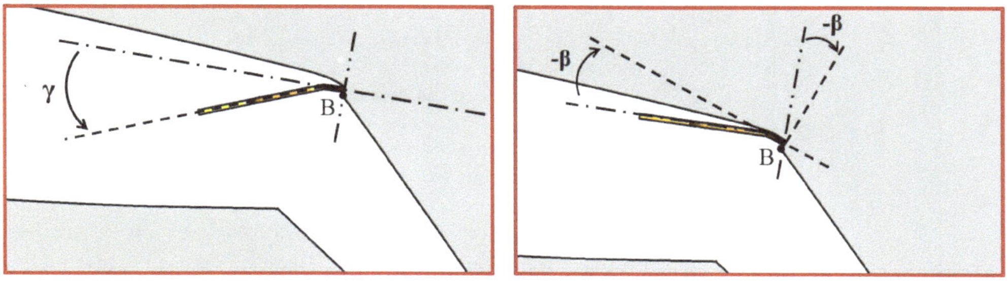

Figure 4. Installation conditions of fluidic oscillators.

First, to investigate the effect of bending angle on the aerodynamic performance, only the bending angle of the fluidic oscillator was adjusted with the pitch angle fixed at $\beta = 0°$. The bending angle test was carried out in a range of $\gamma = 0{\sim}40°$. Next, the negative pitch angle was tested in the same range of $-\beta = 0{\sim}40°$ without bending the oscillators ($\gamma = 0°$). The present study considered only the negative pitch angles, which showed good flow control performance in the previous work [20]. Actually, this test using negative pitch angles cannot be realized without bending the oscillators. However, to isolate the effect of the bending angle on the aerodynamic performance from that of the pitch angle when both the bending and pitch angles are changed with $-\beta = \gamma$, this unrealistic condition was considered numerically only for comparison.

3. Performance Parameters

To evaluate the aerodynamic performance of the airfoil, the lift, drag, and pressure coefficients were defined as follows:

$$C_L = \frac{L}{\frac{1}{2}\,\rho_\infty\,U_\infty^2\,c\,s} \tag{1}$$

$$C_D = \frac{D}{\frac{1}{2}\,\rho_\infty\,U_\infty^2\,c\,s} \tag{2}$$

where L, D, ρ, U, c and s indicate the lift force, drag force, fluid density, velocity, airfoil chord length and width of the computational domain, respectively. The subscript ∞ indicates free-stream value.

To evaluate the exit velocity change caused by the bending of the fluidic oscillators, peak velocity ratio of the oscillating jet at the exit of the fluidic oscillator (F_{VR}) was defined as follows:

$$F_{VR} = \frac{U_{peak}}{U_{ref}} \tag{3}$$

where the reference velocity U_{ref} is the velocity at the throat (width w_t in Figure 2b).

$$U_{ref} = \frac{\dot{m}_{inlet}}{A_{ref}\rho} \tag{4}$$

U_{peak} is the peak value of the time-averaged jet velocity at the oscillator outlet, $\dot{m}_{inlet}$ is the mass flow rate at the inlet, A_{ref} is the area of the throat, and ρ is the air density at 25 °C.

4. Numerical Analysis

In this study, three-dimensional URANS and continuity equations were solved numerically for the flow analysis using the commercial software ANSYS CFX 15.0® [26].

The shear stress transport (SST) model was used to analyze the turbulence. Pendey and Kim [27] reported that the SST model predicted the internal flow of a fluidic oscillator accurately. The SST model is known to predict flow separation in adverse pressure gradients well [28,29]. The governing differential equations were discretized using a finite-volume method. Multiple inner iterations were employed using the physical time scale for transient analysis of the flow. A high-resolution scheme of Barth and Jesperson [30] was used to discretize the advection terms and a second-order backward Euler scheme were used for the transient terms in time.

Air at 25 °C is used as the working fluid, which is assumed to be an ideal gas. For the external domain shown in Figure 2a, a uniform velocity of 25 m/s is assigned to the inlet, which corresponds to a Reynolds number of 5.0×10^5 based on the chord length. A constant pressure condition of 1 atm is applied to the outlet, and no-slip boundary conditions are used at the walls. In addition, to reduce the computational time, periodic conditions are assigned to both sides of the external domain considering the periodicity of the flow, and the domain includes three fluidic oscillators following the work of Kim and Kim [20], where the optimum number of oscillators included in the computational domain was determined through a preliminary test. In the case of the internal domain shown in Figure 2b, Outlet of the domain is connected to the external domain using an interface condition. The uniform velocity condition is also used at the inlet of the oscillators. The inlet velocity can be calculated from the momentum coefficient (C_μ).

In this work, the C_μ value is fixed at C_μ = 2.64%. The momentum coefficient is defined as a ratio of total momentum flux supplied by the oscillators to the momentum flux of the external free stream as follows:

$$C_\mu = \frac{n\, \rho_{jet} U_{ref}\, A_{ref}\, U_{ref}}{\frac{1}{2}\rho_\infty\, U_\infty^2\, c\, n\, l} = 2\frac{A_{ref}}{c\, l}\left(\frac{U_{ref}}{U_\infty}\right)^2 \quad (5)$$

where l and n are the space between the two adjacent oscillators and the number of oscillators in the domain. The fluid density in the fluidic oscillator (ρ_{jet}) is assumed to be the same as that of the external flow (ρ_∞).

A typical grid system is shown in Figure 5. Unstructured tetrahedral meshes were created using ICEM CFD 15.0® in all domains and, to resolve the high-velocity gradients in the boundary layers, 19 and 12 prism layers were placed on the airfoil surface and the oscillator walls, respectively, with an expansion ratio of 1.2. In all cases, the first grid points on a wall surface were located at y+ < 2.

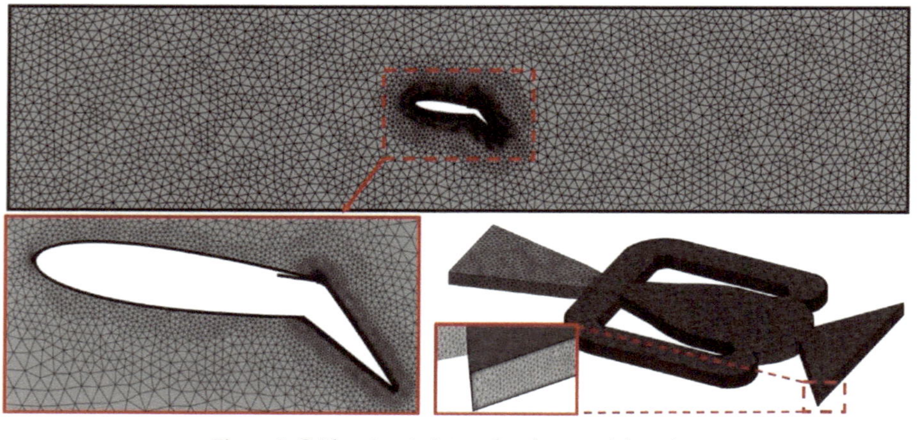

Figure 5. Grid system in internal and external domains.

The computational results for the steady RANS analysis were used as initial values for the URANS analysis. Using the option of adaptive time step selection in ANSYS CFX 15.0®, 5×10^{-6} s was found as a time step by 2–4 repeating calculations, where the root-mean-square values of the relative residuals of the governing equations were less than 1.0×10^{-4}. The calculations were performed using an Intel Xeon Phi 7250/1.4 GHz processor with 68 CPU cores. The analysis of the entire computational domain took about 48 h.

5. Results and Discussion

Because the same computational domain as in the present work was used in the previous work [20], the grid system used in the previous work was also employed in the present work. The grid system was determined in the previous work based on a grid-convergence index (GCI) analysis [31]. For an internal domain, the analysis was performed for the jet frequency of the fluidic oscillator, and the relative discretization error of the finally selected grid with 4.7×10^5 cells was 0.957%. For the external domain, a grid system with 2.2×10^6 cells was selected with a relative discretization error of 0.072%.

Numerical results were also validated in the previous work [20] using the experimental data of Melton et al. [15]. The results for the frequency of the fluidic oscillator in the calculation of the internal domain showed a relative error of about 2%, compared to the experimental measurement at a mass flow rate of 0.7 g/s, and the lift coefficient showed a relative error of 9% at $\alpha = 8°$ in the external flow calculation. The distributions of the pressure coefficient on the airfoil surfaces also showed reasonable agreement with the measurements.

The present study considered the three cases shown in Table 2. Case 1 is a case where only the bending angle is changed with a fixed pitch angle at $\beta = 0$, and Case 2 is a case where only the pitch angle is changed with $\gamma = 0$, and Case 3 changes both the bending and pitch angles with $-\beta = \gamma$, which was used to adjust the injection pitch angle in the study of Kim and Kim [22]. Thus, in Case 3, their results are introduced here without new calculations.

Table 2. Tested Cases.

	Change
Case 1	γ changed with $\beta = 0$
Case 2	β changed with fixed $\gamma = 0$
Case 3 (Kim and Kim [22].)	Both β and γ changed with $\gamma = -\beta$

Figures 6–8 show the variations of lift and drag coefficients and lift-to-drag ratio, respectively, with the bending and/or pitch angles. It is noticeable that an increase in the bending angle enhances lift coefficient without changing the pitch angle in a range of $\gamma > 20°$ compared to the case with zero bending angle (Case 1). The highest C_L is achieved at $\gamma = 35°$ in Case 1. Kim and Kim [25] reported that, at $\gamma = 40°$, the oscillation was suppressed at low oscillator flow rates and entirely disappeared for the mass flow rates larger than 0.3 g/s in the case of a single isolated fluidic oscillator which is the same as used in the present work. In Figure 6, a sudden decrease in C_L is also found at $\gamma = 40°$ in Case 1 and Case 3. An increase in $-\beta$ enhances C_L largely for $-\beta > 10°$ in Case 2 and Case 3. When γ and $-\beta$ are changed together in Case 3, the interaction appears effective in improving the lift coefficient, especially at $\gamma = -\beta = 35°$. The differences in C_L between Case 2 and Case 3 at $\gamma = -\beta = 35°$ and 40° are obviously due to the bending angle effect shown in the results of Case 1.

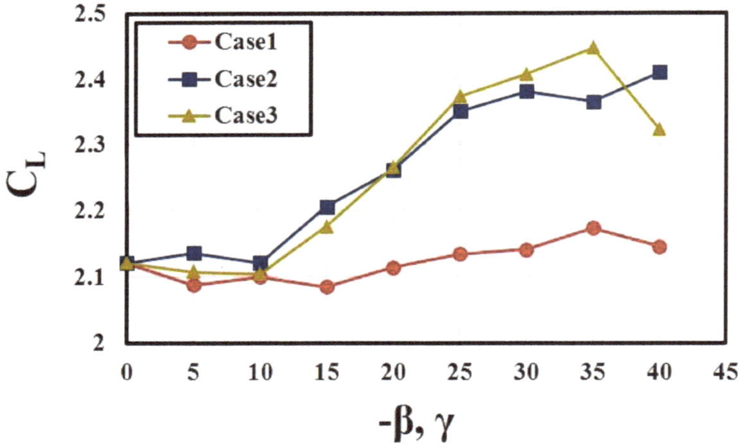

Figure 6. Variations of lift coefficient with pitch and/or bending angles of fluidic oscillators ($\alpha = 8°$, $\delta_f = 40°$, $C_\mu = 2.64\%$).

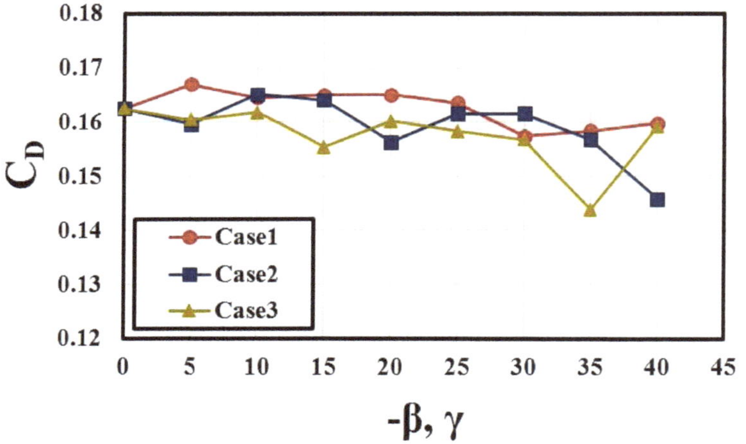

Figure 7. Variations of drag coefficient with pitch and/or bending angles of fluidic oscillators ($\alpha = 8°$, $\delta_f = 40°$, $C_\mu = 2.64\%$).

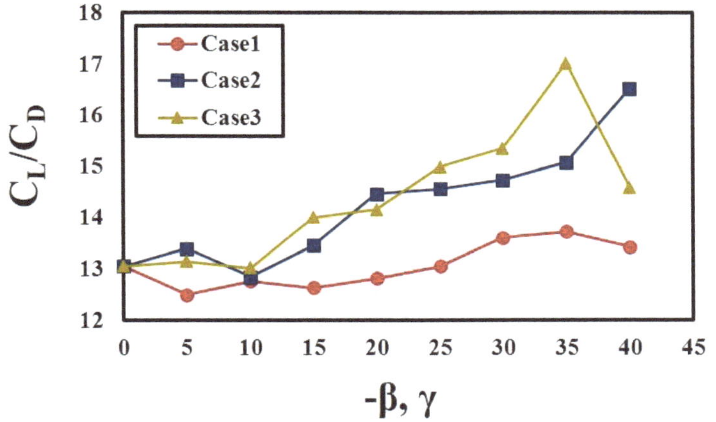

Figure 8. Variations of lift-to-drag ratio with pitch and/or bending angles of fluidic oscillators ($\alpha = 8°$, $\delta_f = 40°$, $C_\mu = 2.64\%$).

Figure 7 shows that the drag coefficient does not vary largely in all the cases except at the angles larger than 30°. The differences in C_D among the cases are not large, either. It is interesting that an increase in the bending angle with $\beta = 0°$ also slightly reduces C_D at $\gamma = 30$–$40°$ compared to the case without bending (Case 1). A sudden decrease in C_D is found at $\gamma = -\beta = 35°$ in Case 3, but there is no noticeable change at this angle in Case 1.

The lift-to-drag ratio is also improved by the bending of outlet nozzle in a range of $25° < \gamma < 40°$ as can be seen in Figure 8. Case 2 and Case 3 show similar variations of C_L/C_D with β for $\gamma < 25°$, but Case 3 shows higher values thereafter, except at $-\beta = 40°$, which reflects the coupling effect of bending and pitching angles same as in Figure 6 for C_L.

To further analyze the causes of the above results for the aerodynamic performance, changes in the characteristics of the fluidic oscillators depending on the bending and/or pitching angles are presented in Figures 9–11. Figure 9 shows variations of the peak velocity ratio (F_{VR}) with the angles in the three cases. The results show that F_{VR} is far more affected by the bending angle (Case 1 and Case 3) than by the pitch angle (Case 2). The effect of pitch angle on F_{VR} appears beyond $-\beta = 15°$ (Case 2). However, the effect of bending angle on F_{VR} is found over the whole tested range of γ and $-\beta$ (Case 1 and Case 3). Case 3 shows higher F_{VR} than that of Case 1 for $-\beta > 15°$, where contribution of the pitch angle to F_{VR} is found in Case 2.

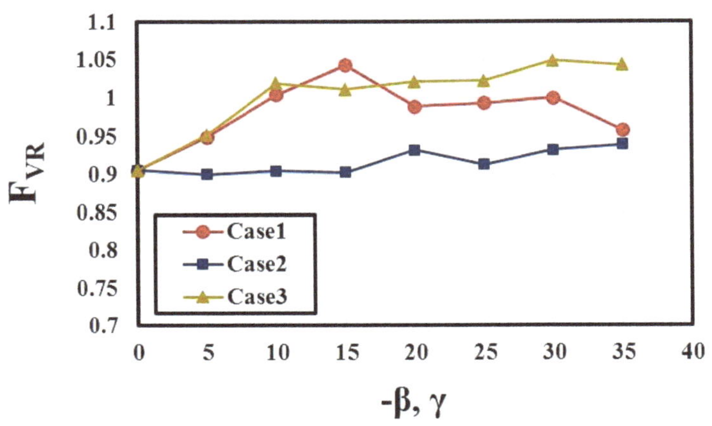

Figure 9. Variations of peak velocity ratio with pitch and/or bending angles of fluidic oscillators.

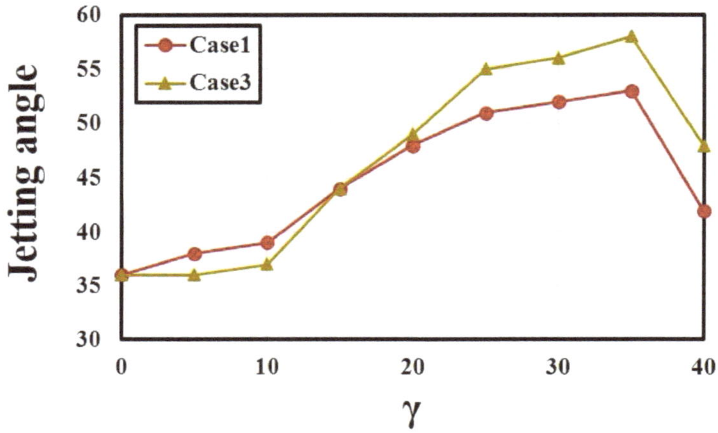

Figure 10. Variations of jetting angle with various pitch and/or bending angles of fluidic oscillators.

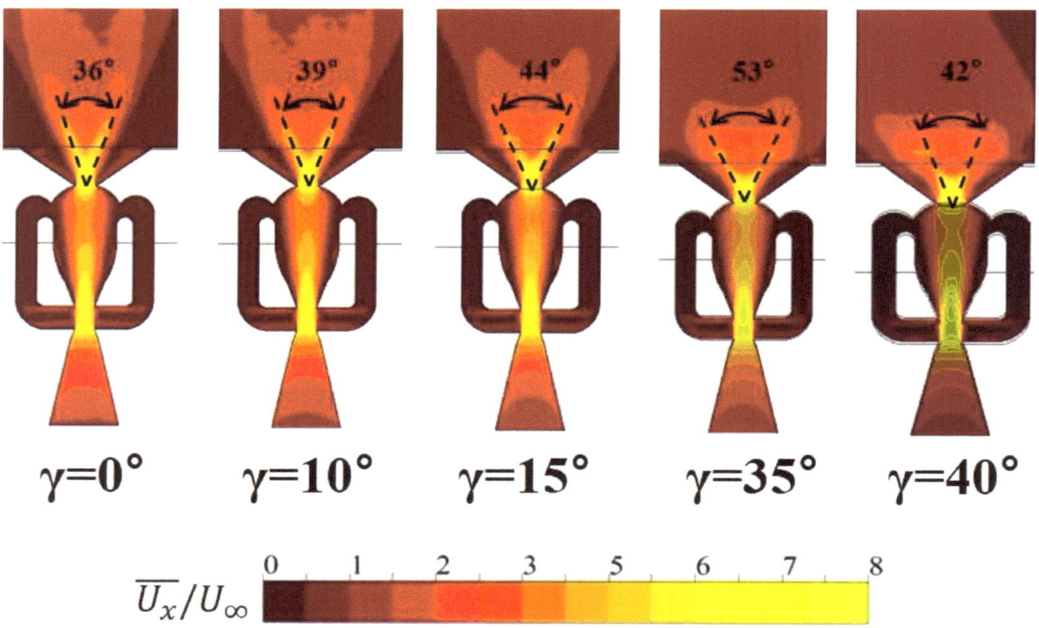

Figure 11. Time-averaged velocity contours and exit jetting angle (Case1).

The jetting angle of fluidic oscillators is also strongly affected by the bending angle of the oscillator outlet nozzle. Figures 10 and 11 show that the jetting angle increases rapidly with the bending angle and reaches the maximum at $\gamma = 35°$ in both Case 1 and Case 3. In Figure 10, both curves for Case 1 and Case 3 show the similar variations, but there must be some contribution of pitch angle to the enhancement of jetting angle for $\gamma = -\beta > 15°$ in Case 3. The sudden decrease in the jetting angle at $\gamma = -\beta = 40°$ in Case 3 is affected strongly by the bending of the oscillators as shown in Case 1. At this angle, there must be a suppression of oscillation by the bending as reported by the previous work [27]. Therefore, the improved characteristics of the fluidic oscillators (F_{VR} and jetting angle) shown in

Figures 9 and 10, especially the jetting angle, must have increased C_L and thus C_L/C_D in Case 3 as shown in Figures 6 and 8, respectively.

Figure 12 shows the time-averaged velocity contours in the x-z plane on the flap for different γ and/or $-\beta$ at a flap angle of 40°. In Case 1, the separation area reduces as γ increases, but the pitch angle does not affect largely the separation region in Case 2. However, in Case 3, the separation region on the flap is greatly reduced as the angle increases from 15° to 35°. Therefore, it is found that the bending angle contributes to the improvement of aerodynamic performance in Case 3 by controlling the separation on the flap through changing the jetting angle and peak velocity ratio of the fluidic oscillators as shown in Figures 9 and 10.

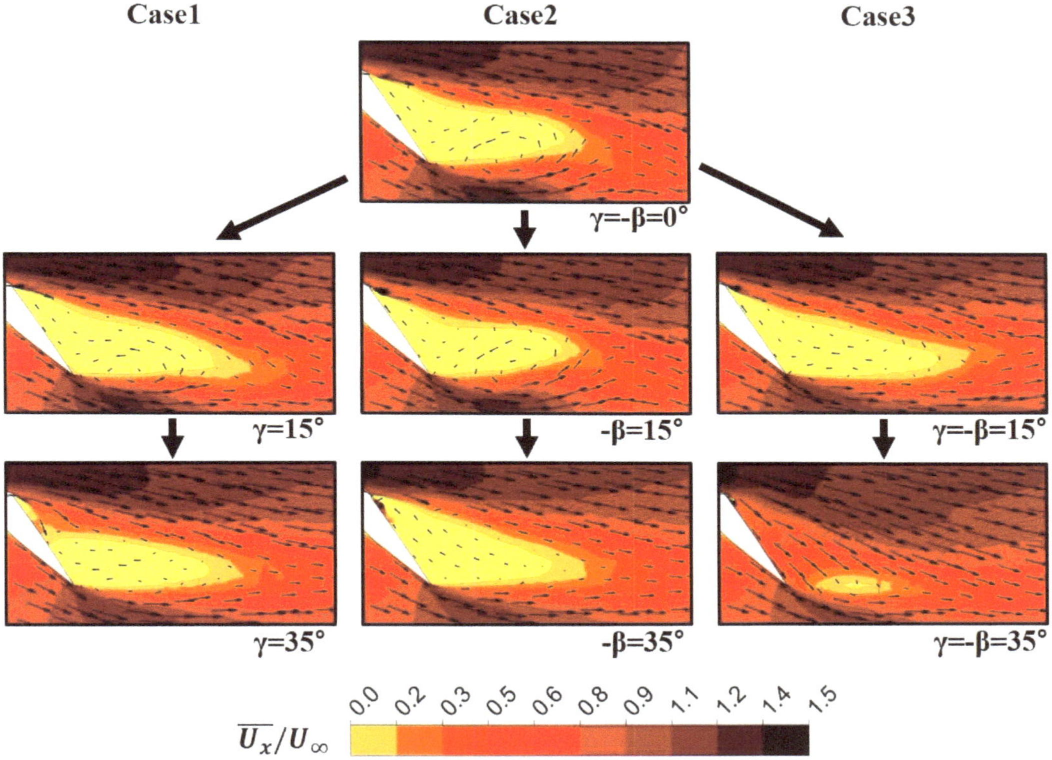

Figure 12. Time-averaged streamwise velocity contours and velocity vectors for different γ and/or $-\beta$ angles on x-z plane at the center of computational domain (α = 8°, δ_f = 40°).

In Figure 13, time-averaged span vorticity distributions around the airfoil are compared at various angles. It is known that positive vortices generally contribute to the thrust, and negative vortices affect the lift [32]. In Case 1, at γ = 35°, the area of negative vorticity is attached to the flap surface, but this kind of attachment is not found in Case 2 with the change of β. When γ and $-\beta$ are simultaneously changed in Case 3, the area of negative vorticity is attached to the flap surface at γ = 25° and 35°. This means that the coupling effect of γ and β promotes the attachment and thus enhances the aerodynamic performance.

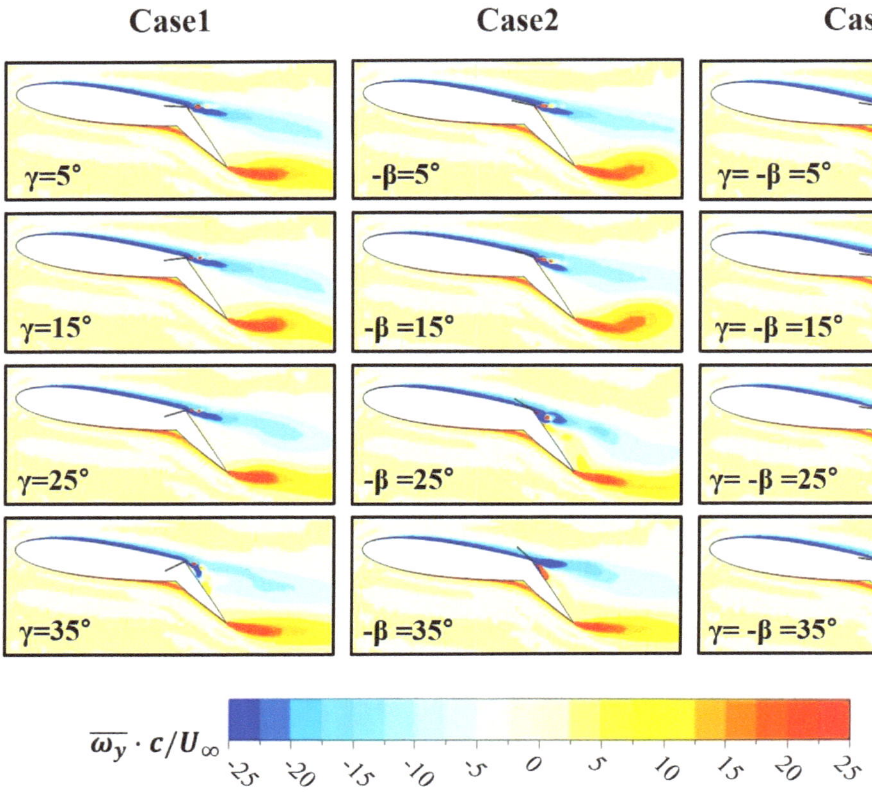

Figure 13. Time-averaged spanwise vorticity contours.

6. Conclusions

The present work investigated the effects of bending the outlet nozzle of fluidic oscillators on the flow control performance of an airfoil with a flap using URANS analysis. Calculations were performed for two cases: the first case changed the bending angle (γ) with zero pitch angle (β) of the oscillators (Case 1), and the second case changed the pitch angle without bending the outlet nozzles (Case 2). The results of these two cases were compared to the previously reported results for the case where both the bending and pitch angles changed with $-\beta = \gamma$ (Case 3).

It is noticeable that an increase in the bending angle enhanced the lift coefficient without changing the pitch angle in a range of $\gamma > 20°$ and the highest C_L was achieved at $\gamma = 35°$ (Case 1). Due to the coupling with this bending effect, Case 3 showed a much higher lift coefficient at $-\beta = 35°$ but a sudden drop to a lower value at $-\beta = 40°$ compared to the results of Case 2. Thus, the bending was not beneficial for the lift at $\gamma = 40°$, where suppression of oscillation is expected due to excessive bending. The drag coefficient did not show large variations with the angles in all the cases, except at large angles.

Characteristics of the fluidic oscillators were also studied to find their effects on the aerodynamic performance. The peak velocity ratio at the exit of the fluidic oscillators was far more affected by the bending angle than by the pitch angle. Case 3 showed higher peak velocity ratio than that of Case 1 for $-\beta > 15°$ due to the contribution of the pitch angle. The jetting angle of fluidic oscillators was found to be strongly affected by the bending angle. The jetting angle increased rapidly with the bending angle and reached the maximum at $\gamma = 35°$ in both Case 1 and Case 3. The sudden decrease in the jetting angle at $\gamma = -\beta = 40°$ in Case 3 was due to the excessive bending of the oscillators. Therefore, the improved characteristics of the fluidic oscillators, especially the jetting angle, increased the lift coefficient and the lift-to-drag ratio in Case 3, which showed the best overall performance among the tested cases.

Bending the outlet parts of the fluidic oscillators were introduced inevitably due to the geometric limitations in the installation in the previous work. However, it is interesting that the bending rather improved the aerodynamic performance of the airfoil through improving the characteristics of the fluidic oscillators, such as the jetting angle and peak velocity ratio.

Author Contributions: N.-H.K. and K.-Y.K. contributed to the overall composition and writing of the manuscript; N.-H.K. performed numerical analysis and analyzed the data; K.-Y.K. revised and finalized the manuscript. All authors have read and agreed to the published version of the manuscript.

Funding: This work was supported by the National Research Foundation of Korea (NRF) grant funded by the Korean government (MSIT) (No. 2019R1A2C1007657).

Conflicts of Interest: The authors declare there is no conflict of interest.

References

1. Tang, Y.; Liu, Y.; Lu, L.; Lu, H.; Wang, M. Passive separation control with blade-end slots in a highly loaded compressor cascade. *AIAA J.* **2020**, *58*, 85–97. [CrossRef]
2. Rosenblum, J.P.; Vrchota, P.; Prachar, A.; Peng, S.H.; Wallin, S.; Eliasson, P.; Iannelli, P.; Ciobaca, V.; Wild, J.; Hantrais-Gervois, J.L.; et al. Active flow separation control at the outer wing. *CEAS Aeronaut. J.* **2020**, *11*, 823–836. [CrossRef]
3. Ebrahimi, A.; Hajipour, M. Flow separation control over an airfoil using dual excitation of DBD plasma actuators. *Aerosp. Sci. Technol.* **2018**, *79*, 658–668. [CrossRef]
4. Akbarzadeh, A.M.; Borazjani, I. Controlling flow separation on a thick airfoil using backward traveling waves. *AIAA J.* **2020**, *58*, 3799–3807. [CrossRef]
5. Jones, G.; Santer, M.; Debiasi, M.; Papadakis, G. Control of flow separation around an airfoil at low Reynolds numbers using periodic surface morphing. *J. Fluids Struct.* **2018**, *76*, 536–557. [CrossRef]
6. James, P.J.; Nishi, M. Vortex generator jets-means for flow separation control. *AIAA J.* **1990**, *28*, 989–994.
7. Koklu, M.; Owens, L.R. Flow separation control over a ramp using sweeping jet actuators. In Proceedings of the 7th AIAA Flow Control Conference, Atlanta, GA, USA, 16–20 June 2014. [CrossRef]
8. Ganesh, R.; Surya, R. Cavity resonance suppression using miniature fluidic oscillators. *AIAA J.* **2004**, *42*. [CrossRef]
9. Hossain, M.A.; Prenter, R.; Lundgreen, R.K.; Ameri, A.; Gregory, J.W.; Bons, J.P. Experimental and numerical investigation of sweeping jet film cooling. *J. Turbomach.* **2017**, *140*. [CrossRef]
10. Wu, Y.; Yu, S.; Zuo, L. Large eddy simulation analysis of the heat transfer enhancement using self-oscillating fluidic oscillators. *Int. J. Heat Mass Transf.* **2019**, *131*, 463–471. [CrossRef]
11. Cerretelli, C.; Wuerz, W.; Gharaibah, E. unsteady separation control on wind turbine blades using fluidic oscillators. *AIAA J.* **2010**, *48*, 1302–1311. [CrossRef]
12. Feikema, D.; Culley, D. Computational fluid dynamic modeling of a fluidic actuator for flow control. In Proceedings of the 45th AIAA Aerospace Sciences Meeting and Exhibit, Reno, NV, USA, 7–10 January 2008; pp. 1–13.
13. Seifert, A.; Greenblatt, D.; Wygnanski, I.J. Active separation control: An overview of Reynolds and Mach numbers effects. *Aerosp. Sci. Technol.* **2004**, *8*, 569–582. [CrossRef]
14. Melton, L.G.P. Active flow separation control on a NACA 0015 wing using fluidic actuators. In Proceedings of the 7th AIAA Journal Flow Control Conference, Atlanta, GA, USA, 16–20 June 2014. [CrossRef]
15. Melton, L.P.; Koklu, M.; Andino, M.; Lin, J.C. Active flow control via discrete sweeping and steady jets on a simple-hinged flap. *AIAA J.* **2018**, *56*, 2961–2973. [CrossRef]
16. Seele, R.; Graff, E.; Lin, J.; Wygnanski, I. Performance enhancement of a vertical tail model with sweeping jet actuators. In Proceedings of the 51st AIAA Aerospace Sciences Meeting including the New Horizons Forum and Aerospace Exposition, Grapevine, TX, USA, 7–10 January 2013.
17. Meng, Q.; Du, X.; Chen, S.; Wang, S. Numerical study of dual sweeping jet actuators for corner separation control in compressor cascade. *J. Therm. Sci.* **2021**, *30*, 201–209. [CrossRef]
18. Kim, S.-H.; Kim, K.-Y. Effects of installation conditions of fluidic oscillators on control of flow separation. *AIAA J.* **2019**, *57*, 5208–5219. [CrossRef]
19. Kim, S.-H.; Kim, K.-Y. Effects of installation location of fluidic oscillators on aerodynamic performance of an airfoil. *Aerosp. Sci. Technol.* **2020**, *99*, 105735. [CrossRef]
20. Kim, N.-H.; Kim, K.-Y. Flow control using fluidic oscillators on an airfoil with a flap. *Eng. Appl. Comput. Fluid Mech.* **2021**, *15*, 377–390. [CrossRef]

21. Melton, L.G.P.; Koklu, M.; Andino, M.; Lin, J.C.; Edelman, L. Sweeping jet optimization studies. In Proceedings of the 8th AIAA Flow Control Conference, Washington, DC, USA, 13–17 June 2016. [CrossRef]
22. Ostermann, F.; Woszidlo, R.; Nayeri, C.; Paschereit, C.O. Experimental comparison between the flow field of two common fluidic oscillator designs. In Proceedings of the 53rd AIAA Aerospace Sciences Meeting, Kissimmee, FL, USA, 5–9 January 2015; Volume 10, p. 6. [CrossRef]
23. Jeong, H.-S.; Kim, K.-Y. Shape optimization of a feedback-channel fluidic oscillator. *Eng. Appl. Comput. Fluid Mech.* **2018**, *12*, 169–181. [CrossRef]
24. Pandey, R.J.; Kim, K.-Y. Numerical modeling of internal flow in a fluidic oscillator. *J. Mech. Sci. Technol.* **2018**, *32*, 1041–1048. [CrossRef]
25. Kim, N.-H.; Kim, K.-Y. Effects of bent outlet on characteristics of a fluidic oscillator with and without ex-ternal flow. *Energies* **2021**, *14*, 4342. [CrossRef]
26. ANSYS. *ANSYS CFX-Solver Theory Guide-Release 15.0*; ANSYS: Canonsburg, PA, USA, 2014.
27. Pandey, R.J.; Kim, K.-Y. Comparative analysis of flow in a fluidic oscillator using large eddy simulation and unsteady Reynolds-averaged Navier–Stokes analysis. *Fluid Dyn. Res.* **2018**, *50*, 065515. [CrossRef]
28. Menter, F.R. Two-equation eddy-viscosity turbulence models for engineering applications. *AIAA J.* **1994**, *32*, 1598–1605. [CrossRef]
29. Bardina, J.; Huang, P.; Coakley, T. Turbulence modeling validation. In Proceedings of the 28th Fluid Dynamics Conference, Snowmass Village, CO, USA, 29 June–2 July 1997; American Institute of Aeronautics and Astronautics: Reston, VA, USA, 1997.
30. Barth, T.; Jesperson, D. The design and application of upwind schemes on unstructured meshes. In Proceedings of the 27th Aerospace Sciences Meeting, Reno, NV, USA, 9–12 January 1989.
31. Celik, I.B.; Ghia, U.; Roache, P.J.; Freitas, C.J.; Coleman, H.; Raad, P.E. Procedure for estimation and reporting of uncertainty due to discretization in CFD applications. *J. Fluids Eng.* **2008**, *130*, 0780001.
32. Zha, G.; Yang, Y.; Ren, Y.; McBreen, B. Super-lift and thrusting airfoil of co-flow jet actuated by micro-compressors. In Proceedings of the 2018 Flow Control Conference, Atlanta, GA, USA, 25–29 June 2018.

MDPI AG
Grosspeteranlage 5
4052 Basel
Switzerland
Tel.: +41 61 683 77 34

Processes Editorial Office
E-mail: processes@mdpi.com
www.mdpi.com/journal/processes

www.ingramcontent.com/pod-product-compliance
Lightning Source LLC
Chambersburg PA
CBHW041431210926
51850CB00013B/3400